Introducing Pure Mathematics

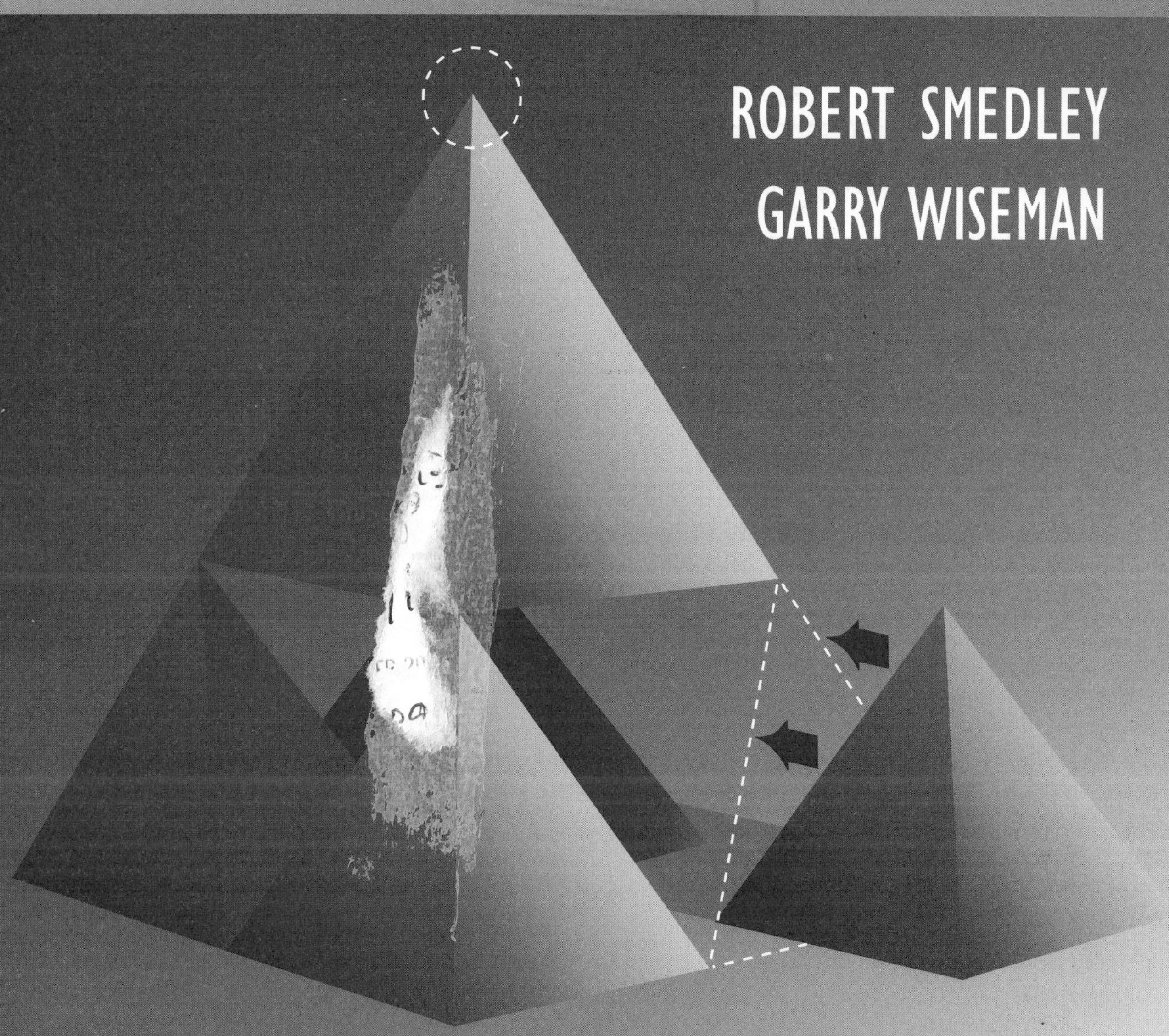

ROBERT SMEDLEY
GARRY WISEMAN

OXFORD

Oxford University Press, Great Clarendon Street,
Oxford OX2 6DP

Oxford New York

Athens Auckland Bangkok Bogota Buenos Aires
Calcutta Cape Town Chennai Dar es Salaam
Delhi Florence Hong Kong Istanbul Karachi
Kuala Lumpur Madrid Melbourne Mexico City
Mumbai Nairobi Paris São Paulo Singapore
Taipei Tokyo Toronto Warsaw

and associated companies in
Berlin Ibadan

Oxford is a trade mark of Oxford University Press

First published 1998

ISBN 0 19 914400 1 School and College edition
ISBN 0 19 914563 6 Bookshop edition

A CIP catalogue record for this book is available from the British Library.

Typeset and illustrated by Tech-Set Ltd, Gateshead, Tyne and Wear
Printed and bound in Great Britain

Contents

Preface

The aim has been to produce a single text which covers all the pure material of all the single-subject GCE mathematics syllabuses, linear and modular, as well as to progress into a number of further topics. Demands at this level are varied, and teachers and students may choose to omit certain sections depending upon the particular course they are following.

We have deliberately avoided tying ourselves to any one examination board, but have endeavoured to supply a comprehensive text which transcends the demands of the frequently changing syllabuses. Indeed, in some chapters, we have attempted to anticipate change by, for example, including a chapter on proof. The hope is that the present book will provide a useful teaching resource for many years to come.

A special feature is the two introductory chapters, the first on algebra and the second on geometry. These have been included as a result of comments from a number of teachers, with the intention of providing basic revision and practice in those important fundamental skills which are covered at GCSE, but which may not be solidly grounded. Some students may wish to work through these chapters before tackling the more advanced work; others may choose to dip into them as necessary.

The next three chapters cover functions, polynomials and straight-line geometry, preparing the way for the introduction to calculus in Chapter 6. The treatment of calculus is structured along clear lines. Differentiation and integration are in separate chapters, and the basic ideas are introduced through functions of the form x^n. Some teachers may find it useful to look at the first section of Chapter 18, which revises indices, before tackling this work. Differentiation moves through basic techniques into applications to tangents, normals, maxima and minima; and integration moves similarly through basic techniques into applications in areas and volumes. Progress is carefully structured, so those preparing for a single module can stop before the end of the chapter.

Further algebra – circles, sequences and series, binomial expansion and partial fractions – is then undertaken before we return to calculus again. This time methods of differentiation are introduced – function of a function, product and quotients – and these are then applied to revise the techniques introduced in Chapter 6. We then move on to self-contained units on parametric equations, implicit differentiation and rates of change before changing topics.

The groundwork in calculus having been laid, we take five chapters to cover two major areas – trigonometry and logarithms/exponentials. These blocks can be tackled in either order. Given the various demands of different teachers and examination boards, we have been careful to avoid combining the two ideas until the very end. Both begin with the traditional equation/identity approach before moving on to calculus. Again, in each area, all previous work on calculus is carefully revised through structured, clearly headed exercises. Thus if, for example, a teacher wishes to apply trigonometric techniques to parametric differentiation but not to maxima and minima, then there is a source of readily identifiable questions.

Chapter 20 offers thorough coverage of all relevant techniques of integration, culminating in work on differential equations with variables separable. And it is in this chapter, more than in any other, that we have probably been guilty of failing to stop when we might. It just didn't seem right for an A-level text to offer a 'toy-town' approach to such an important topic, and we make no apologies to those who think we may have got a little carried away. Each section is carefully structured, so those who weary may choose to 'cut and run' at any stage!

The final chapters cover numerical methods, vectors and probability. Vectors have been left to this late stage since they sadly seem to have disappeared from all single syllabuses. But having written Chapters 21 and 23, we felt the need to indulge ourselves in a little more 'proper' mathematics – and vectors seemed to be the answer. It is hoped that this chapter will be of use as an introduction to an applied course – or just as a worthwhile aside.

Although it is expected that most students will be using the present text with the help of a teacher, we have aimed to make it suitable for self-study. In each chapter, the theory sections are followed by a number of worked examples which are typical of, and lead to, the questions in the exercises. By reading the theory sections and following the worked examples, the student should be able to make considerable progress with the exercise that follows. Progress can then be assessed against the comprehensive selection of recent examination questions which appear at the end of each chapter. Most exercises also contain one or two starred questions. These aim to stretch the more able students beyond the routine and onto the more creative side of the subject.

We are grateful to AEB, EDEXCEL, MEI, NICCEA, NEAB, UCLES, UODLE, OCSEB and WJEC for permission to use their questions. The answers provided for these questions are the sole responsibility of the authors.

Finally, we wish to express our thanks to John Day for his work in editing the book, and to Ben Bramhall for working through the text and exercises and checking the answers. We also wish to thank the mathematics Department and pupils of Radley College for trialing the material over a number of years, and for providing innumerable valuable suggestions for improvements.

Garry Wiseman
Robert Smedley

February 1998

1 Algebra

Do you remember the famous toast, 'Here's to pure mathematics – may it never be of use to anybody'?
ARTHUR C. CLARKE

Linear equations

A linear equation is one which can be expressed in the form $ax + b = 0$. For example, each of the following equations can be rearranged into the form $ax + b = 0$:

$$x + 4 = 9 \qquad 3(x+2) - 7 = x + 1 \qquad \frac{x}{3} + 7 = 4(x-2)$$

Whatever is done to the left-hand side (LHS) of an equation must be done to the right-hand side (RHS). For example, the equation $x - 4 = 7$ can be manipulated so that x (the unknown) is the only term on the LHS. This is achieved in the following way:

$$x - 4 = 7$$

Add 4 to both sides to obtain

$$x - 4 + 4 = 7 + 4$$
$$\therefore \quad x + 0 = 11$$
$$\therefore \quad x = 11$$

The solution is $x = 11$.

Example 1 Solve the equation $3x - 7 = x + 3$.

SOLUTION

$$3x - 7 = x + 3$$

Add 7 to both sides to obtain

$$3x - 7 + 7 = x + 3 + 7$$
$$\therefore \quad 3x = x + 10$$

To obtain all the x terms on the LHS subtract x from both sides, which gives

$$3x - x = x + 10 - x$$
$$\therefore \quad 2x = 10$$

Dividing both sides by 2 gives

$$x = 5$$

The solution is $x = 5$.

Example 2 Solve the equation $4(x+1)-3(x-5)=17$.

SOLUTION

$$4(x+1)-3(x-5)=17$$

Expand the brackets and simplify to obtain

$$4x+4-3x+15=17$$

$$\therefore \quad x+19=17$$

Subtracting 19 from both sides gives

$$x+19-19=17-19$$

$$\therefore \quad x=-2$$

The solution is $x=-2$.

Example 3 Solve the equation $\dfrac{x+5}{2}=\dfrac{3x+11}{5}$.

SOLUTION

$$\frac{x+5}{2}=\frac{3x+11}{5}$$

Since the lowest common multiple of 2 and 5 is 10, multiply throughout by 10, which gives

$$10\left(\frac{x+5}{2}\right)=10\left(\frac{3x+11}{5}\right)$$

Simplifying the fractions gives

$$5(x+5)=2(3x+11)$$

Expanding the brackets gives

$$5x+25=6x+22$$

Rearrange to obtain all the x terms on the LHS and all the other terms on the RHS:

$$5x-6x=22-25$$

Simplifying gives

$$-x=-3$$

$$\therefore \quad x=3$$

The solution is $x=3$.

Alternatively, the technique of cross-multiplication can be used with the equation

$$\frac{x+5}{2}=\frac{3x+11}{5}$$

Cross-multiplying gives

$$5(x+5) = 2(3x+11)$$

Expanding and simplifying gives

$$5x + 25 = 6x + 22$$

$$\therefore \quad 5x - 6x = 22 - 25$$

$$\therefore \quad x = 3$$

The solution is $x = 3$, as before.

Example 4 Solve the equation $\dfrac{2}{x+1} = \dfrac{3}{5}$.

SOLUTION

$$\frac{2}{x+1} = \frac{3}{5}$$

Cross-multiply and expand the bracket to obtain

$$10 = 3(x+1)$$

$$\therefore \quad 10 = 3x + 3$$

Rearranging gives

$$3x = 7$$

$$\therefore \quad x = \frac{7}{3}$$

The solution is $x = \frac{7}{3}$.

Example 5 Solve the equation $2x + \dfrac{x+7}{3} = \dfrac{4x-19}{5}$.

SOLUTION

Cross-multiplication as used in Example 4 is not possible in this case because of the additional $2x$ term on the LHS.

Since the lowest common multiple of 3 and 5 is 15, multiply throughout by 15:

$$15(2x) + 15\left(\frac{x+7}{3}\right) = 15\left(\frac{4x-19}{5}\right)$$

Simplifying gives

$$30x + 5(x+7) = 3(4x-19)$$

$$\therefore \quad 30x + 5x + 35 = 12x - 57$$

$$\therefore \quad 23x = -92$$

$$\therefore \quad x = -4$$

The solution is $x = -4$.

Example 6 Solve the equation $\frac{1}{3}(5x - 4) + \frac{1}{7}(x + 2) = 13 - x$.

SOLUTION

Since the lowest common multiple of 3 and 7 is 21, multiply throughout by 21:

$$21 \times \frac{1}{3}(5x - 4) + 21 \times \frac{1}{7}(x + 2) = 21(13 - x)$$

Simplifying gives

$$\begin{aligned} 7(5x - 4) + 3(x + 2) &= 21(13 - x) \\ 35x - 28 + 3x + 6 &= 273 - 21x \\ \therefore \quad 59x &= 295 \\ \therefore \quad x &= 5 \end{aligned}$$

The solution is $x = 5$.

In the next example, terms in x^2 appear in the simplification of the equation. However, these terms cancel and the equation reduces to a linear equation.

Example 7 Solve the equation $\frac{6x + 1}{2x - 5} = \frac{3x - 2}{x + 1}$.

SOLUTION

Cross-multiply and expand the brackets:

$$\begin{aligned} (6x + 1)(x + 1) &= (3x - 2)(2x - 5) \\ \therefore \quad 6x^2 + 7x + 1 &= 6x^2 - 19x + 10 \end{aligned}$$

Rearranging gives

$$\begin{aligned} 26x &= 9 \\ \therefore \quad x &= \frac{9}{26} \end{aligned}$$

The solution is $x = \frac{9}{26}$.

Exercise 1A

In each of the following questions, solve the given equations for x.

1 **a)** $3x + 2 = 20$ **b)** $5x - 3 = 32$ **c)** $16 + 7x = 2$
d) $4 + 3x = 19$ **e)** $6 - x = 4$ **f)** $2x - 3 = 8$
g) $3x + 2 = x + 8$ **h)** $2x - 3 = 6x + 5$ **i)** $3x + 5 = 7x - 8$
j) $6x + 9 = 8 - 4x$ **k)** $2 - 5x = 8 - 3x$ **l)** $2x + 7 = 3 - 10x$

2 **a)** $2(x - 3) + 5(x - 1) = 3$ **b)** $3(5 - x) - 4(3x - 2) = 27$
c) $2(4x - 1) - 3(x - 2) = 14$ **d)** $3(x - 8) + 2(4x - 1) = 3$
e) $6(x + 4) + 5(2x - 1) = 7$ **f)** $3(2x + 5) - 4(x - 3) = 0$

g) $3(x-1)-4(x-2)-6(2x+3)=0$ **h)** $4(5-x)-2(x-3)-6(2x-1)=4$
i) $3(x+5)+2(x+1)-3x=22$ **j)** $3(2x-5)-4(x-2)=5(x-8)$
k) $7(x-4)+3(x-6)=6x-12$ **l)** $4(2x+1)+6(9x-2)=3(5x-4)+4$

3 a) $\frac{x+2}{3}=\frac{2x+1}{5}$ **b)** $\frac{5x-3}{4}=\frac{4x-3}{3}$ **c)** $\frac{3x+1}{4}=\frac{2-x}{3}$

d) $\frac{2x+3}{5}=\frac{4+3x}{3}$ **e)** $\frac{3}{x+1}=\frac{4}{x}$ **f)** $\frac{2}{3x-5}=\frac{5}{2x+3}$

g) $\frac{6}{x+8}=\frac{5}{3x+4}$ **h)** $\frac{7}{x-1}=\frac{3}{x+2}$ **i)** $\frac{2}{x+3}=\frac{4}{5}$

j) $\frac{5}{x-1}=\frac{2}{3}$ **k)** $\frac{6}{2x-3}=\frac{1}{3}$ **l)** $\frac{5}{2x-3}=\frac{3}{8}$

4 a) $3x+\frac{x+4}{3}=\frac{5x+6}{2}$ **b)** $5x-\frac{3x+1}{2}=\frac{7x+9}{3}$

c) $x+\frac{5x-1}{3}=\frac{11x-1}{4}$ **d)** $4x-\frac{2x+3}{2}=\frac{2x-1}{5}$

e) $6x+4+\frac{x-2}{4}=\frac{14-9x}{12}$ **f)** $x-4+\frac{3x-1}{2}=\frac{4x-5}{3}$

g) $2x+3+\frac{5x-1}{4}=\frac{3x-2}{8}$ **h)** $\frac{3x-2}{5}-\frac{5x-1}{2}=x+3$

i) $\frac{6x+5}{3}-\frac{x-1}{5}=2x-1$ **j)** $\frac{2x+1}{4}-x=\frac{6x+5}{3}$

k) $\frac{x-4}{6}-2x+1=\frac{3x-4}{2}$ **l)** $\frac{2x-3}{4}+\frac{6x-4}{3}=\frac{2x+5}{6}$

5 a) $\frac{1}{2}(2x-1)+\frac{1}{4}(x-2)=4$ **b)** $\frac{1}{3}(x-1)-\frac{1}{4}(2x-3)=1$
c) $\frac{1}{5}(2x-1)-\frac{1}{4}(3x-4)=0$ **d)** $\frac{2}{3}(x-1)-\frac{1}{5}(x-3)=x+1$
e) $\frac{2}{5}(2-x)-\frac{1}{4}(3-5x)=x-4$ **f)** $\frac{1}{3}(x-1)-\frac{1}{6}(3x-5)=2x+3$
g) $\frac{1}{2}(x-1)-\frac{2}{3}(x-4)=\frac{1}{6}(3x-1)$ **h)** $\frac{1}{5}(x-2)+\frac{1}{3}(5x-4)=\frac{1}{2}x$
i) $\frac{3}{4}(2x-1)-\frac{1}{2}(x-3)=\frac{1}{3}(4x-1)$ **j)** $\frac{1}{3}(x-4)-\frac{1}{18}(2x-3)=\frac{1}{9}(5-x)$
k) $\frac{2}{3}x+\frac{1}{2}(3-8x)=\frac{1}{6}-4x$ **l)** $\frac{1}{2}(x+1)+\frac{1}{3}(x+2)=\frac{1}{4}(x+3)+\frac{1}{5}(x+4)$

6 a) $\frac{x+3}{x-2}=\frac{x+4}{x-3}$ **b)** $\frac{x-6}{x+4}=\frac{x-2}{x+3}$ **c)** $\frac{2x+3}{x+2}=\frac{4x}{2x+5}$

d) $\frac{6x+5}{2x-1}=\frac{3x+4}{x-3}$ **e)** $\frac{x-3}{3x-2}=\frac{x-1}{3x-5}$ **f)** $\frac{x-4}{x+5}=\frac{x+3}{x-6}$

g) $\frac{2x+5}{x+4}=\frac{2x-1}{x-1}$ **h)** $\frac{4x+3}{x-2}=\frac{8x+1}{2x-3}$ **i)** $\frac{3x+1}{2x-3}=\frac{6x+1}{4x-5}$

j) $\frac{4-x}{2x+3}=\frac{x-1}{3-2x}$ **k)** $\frac{3-2x}{2-3x}=\frac{5-6x}{1-9x}$ **l)** $\frac{5-6x}{7+3x}=\frac{2(1-x)}{x}$

7 $3(x-2)-4(2x-3)-2(3x-1)=x+4$ **8** $3x-1+\frac{2x+3}{5}=\frac{x-4}{2}$

9 $5x + 7 = 2 - 3x$

10 $\dfrac{2x - 1}{3x + 2} = \dfrac{3 - 2x}{4 - 3x}$

11 $\dfrac{x + 5}{3} = \dfrac{2 - x}{4}$

12 $\frac{2}{5}(2x - 3) = \frac{1}{10}(x - 4) = 1$

13 $(x - 3)(x - 4) - 6 = x(x + 5)$

14 $\dfrac{3}{7 - 2x} = \dfrac{4}{x}$

15 $5(2 - x) = 3(6 - x) + 2$

16 $8x - 9 = 5 + 4x$

17 $\dfrac{x}{x + 1} = \dfrac{x + 4}{x}$

18 $\frac{1}{4}x + 3 = 1$

19 $3(5x - 4) + 2(3x - 1) = 8$

20 $\dfrac{2x - 3}{x - 4} = \dfrac{2x + 3}{x - 8}$

21 $\dfrac{x + 2}{4} = \dfrac{x - 1}{3}$

22 $2x + \dfrac{x - 4}{2} = \dfrac{3x + 5}{4}$

23 $(x - 2)^2 - 3x = (x + 1)(x + 4)$

24 $\dfrac{3}{x + 5} = \dfrac{2}{5}$

25 $\frac{1}{3}(2x - 1) - \frac{1}{6}(3x - 4) = 2$

26 $2(x - 1) - 3(3x - 4) = 5x + 2$

27 $\dfrac{5x - 2}{15x + 1} = \dfrac{x}{3x - 7}$

28 $\frac{1}{3}x + 5 = 9$

29 $x(x + 5) - x^2 = x - 3$

30 $\frac{1}{5}(2x - 3) - \frac{1}{4}(5x - 2) = x + 1$

Linear inequalities

An inequality states a relationship between two mathematical expressions as

less than	written as $<$
less than or equal to	written as $\leqslant$
greater than	written as $>$
greater than or equal to	written as $\geqslant$

For example,

$$3x + 7 > x - 8 \qquad a < 3b + 2 \qquad 5x - 7y \leqslant 4$$

Inequalities can be simplified using methods similar to those used for simplifying equations. However, there is one exception.

We know that the statement $7 > 5$ is true. If we multiply both sides of this statement by -1, it gives $-7 > -5$, which is obviously not true. The inequality sign must be reversed so that the new statement is true, i.e. $-7 < -5$.

In the previous section, we saw that the solution to an equation was a unique value for the unknown, whereas the 'solution' to an inequality is a range of possible values for the unknown.

Example 8 Simplify the inequality $3x + 7 \geqslant x + 2$.

SOLUTION

$$3x + 7 \geqslant x + 2$$

Subtracting 7 from both sides gives

$$3x + 7 - 7 \geqslant x + 2 - 7$$

$$\therefore \quad 3x \geqslant x - 5$$

Subtracting x from both sides gives

$$3x - x \geqslant x - 5 - x$$

$$\therefore \quad 2x \geqslant -5$$

$$\therefore \quad x \geqslant -\frac{5}{2}$$

This result tells us that provided $x \geqslant -\frac{5}{2}$, the original inequality will be satisfied.

Example 9 Simplify the inequality $4(3x + 1) - 3(x + 2) < 3x + 1$.

SOLUTION

$$4(3x + 1) - 3(x + 2) < 3x + 1$$

Expand the brackets, obtaining

$$12x + 4 - 3x - 6 < 3x + 1$$

Simplifying and rearranging give

$$9x - 2 < 3x + 1$$

$$\therefore \quad 6x < 3$$

$$\therefore \quad x < \frac{3}{6}$$

$$\therefore \quad x < \frac{1}{2}$$

This inequality may be written as $x < \frac{1}{2}$.

Example 10 Find the smallest positive integer which satisfies the inequality

$$3(x - 1) + 2(2x - 1) > 5x$$

SOLUTION

$$3(x - 1) + 2(2x - 1) > 5x$$

$$\therefore \quad 3x - 3 + 4x - 2 > 5x$$

$$\therefore \quad 7x - 5 > 5x$$

$$\therefore \quad 2x > 5$$

$$\therefore \quad x > \frac{5}{2}$$

The smallest positive integer which satisfies this inequality is 3.

Example 11 Simplify the inequality $3x + 7 \geqslant 5x - 3$.

SOLUTION

The x terms can be rearranged so that they are all on the RHS and therefore positive. Simplifying the inequality in this way avoids having to multiply throughout by -1.

$$3x + 7 \geqslant 5x - 3$$
$$\therefore \quad 7 + 3 \geqslant 5x - 3x$$
$$\therefore \quad 10 \geqslant 2x$$

Divide both sides by 2:

$$5 \geqslant x$$

Reading from right to left gives

$$x \leqslant 5$$

Alternatively, the x terms can be rearranged so that they are all on the LHS:

$$-2x \geqslant -10$$

Multiplying both sides by -1, but also remembering to reverse the inequality sign, gives

$$2x \leqslant 10$$

Dividing both sides by 2 gives

$$x \leqslant 5$$

Example 12 Find the set of integers which satisfy simultaneously both of the following inequalities:

$$\tfrac{1}{2}(5x - 3) > \tfrac{1}{5}(x + 1) \qquad [1]$$
$$\text{and} \quad 4(1 - x) \leqslant 7 - 5x \qquad [2]$$

SOLUTION

First, simplify [1].

Since the lowest common multiple of 2 and 5 is 10, multiply throughout by 10:

$$10 \times \frac{1}{2}(5x - 3) > 10 \times \frac{1}{5}(x + 1)$$
$$\therefore \quad 5(5x - 3) > 2(x + 1)$$
$$\therefore \quad 25x - 15 > 2x + 2$$
$$\therefore \quad 23x > 17$$
$$\therefore \quad x > \frac{17}{23}$$

Therefore, the integers which satisfy [1] are 1, 2, 3, 4, 5, ...

Next, simplify [2]:

$$4(1 - x) \leqslant 7 - 5x$$
$$\therefore \quad 4 - 4x \leqslant 7 - 5x$$
$$\therefore \quad x \leqslant 3$$

Therefore, the integers which satisfy [2] are 3, 2, 1, 0, −1, −2, . . .

Here are both results shown on the number line.

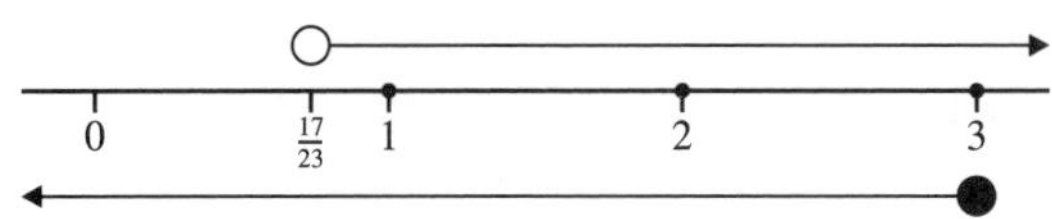

The set of integers which satisfy both inequalities is {1, 2, 3}.

Exercise 1B

In Questions **1** to **3** simplify each of the given inequalities.

1 **a)** $3x + 5 > x + 13$ **b)** $2x - 3 \leqslant 5x + 9$ **c)** $4x - 7 \geqslant 2x + 4$
d) $5x - 8 > x + 7$ **e)** $2x - 1 < x + 4$ **f)** $7x - 3 \geqslant 2x - 1$
g) $6 - 5x < 2 - 3x$ **h)** $3 - x \geqslant 9 + 6x$ **i)** $7x - 2 \geqslant 4x + 3$
j) $9 - 8x > 4$ **k)** $2 - 3x < 6x + 20$ **l)** $3x - 2 \geqslant 5x - 9$

2 **a)** $2(x + 3) - 3(x - 2) > 8$ **b)** $6(2x - 1) + 5(x + 1) < 33$
c) $5(x - 3) < 6(x - 4)$ **d)** $3(x + 4) \geqslant 6(x + 2)$
e) $3(x - 2) - 2(4 - 3x) > 5$ **f)** $7(1 - x) + 3(4 - 5x) \leqslant 41$
g) $3(2 - x) > 5(3 + 2x)$ **h)** $5(2 - x) - 2(3 - 6x) + 2(x - 1) > 0$
i) $2(3x - 1) - 2(x - 1) - x + 4 > 0$ **j)** $3(6x - 5) - 10(x - 4) \geqslant 3(x - 1)$
k) $5x - 2 + 3(2x - 7) < 2(5x - 3)$ **l)** $2(x - 3) - 3(5x - 2) \leqslant 6(3 - 2x)$

3 **a)** $\frac{1}{2}x + 2 < 7$ **b)** $\frac{1}{6}(x - 1) \geqslant \frac{1}{3}(x - 4)$
c) $\frac{1}{2}(x + 3) \leqslant \frac{1}{3}(x - 5)$ **d)** $\frac{1}{7}(2x + 5) > \frac{1}{8}(x + 3)$
e) $\dfrac{x - 2}{4} < \dfrac{2x - 3}{3}$ **f)** $\dfrac{4 - x}{2} \geqslant \dfrac{2 - x}{3}$
g) $\frac{1}{3}(6 - x) \leqslant \frac{1}{5}(2 - 3x)$ **h)** $\frac{1}{9}(2x - 1) > \frac{1}{3}(3 - x)$
i) $\frac{1}{3}(x - 2) + \frac{1}{2}(3x - 1) > 2$ **j)** $\frac{1}{2}(3x + 5) - \frac{1}{4}(2 - x) < 1$
k) $\frac{3}{4}(2 - x) + \frac{5}{6}(3 - 2x) \geqslant \frac{1}{2}$ **l)** $\frac{1}{2}(x - 1) + \frac{1}{3}(x - 2) \leqslant \frac{1}{4}(x - 3)$

4 Find the integers which simultaneously satisfy each of the following pairs of inequalities.

a) $4x + 3 \geqslant 2x + 5$ $\quad x + 4 \leqslant 7$ **b)** $5x + 3 > 3 - x$ $\quad 3x + 5 < 2x + 7$
c) $5 - 2x \leqslant 3 - x$ $\quad 1 - 2x \leqslant 11 - 4x$ **d)** $3x + 2 \geqslant 2x - 1$ $\quad 7x + 3 < 5x + 2$
e) $5x - 4 \geqslant 4x - 3$ $\quad \frac{1}{3}x < 1$ **f)** $\frac{1}{2}(x + 1) > 1$ $\quad 5x + 1 < 4(x + 2)$

5 Show that there are no real numbers which simultaneously satisfy the two inequalities $2x + 1 \geqslant x + 1$ and $\frac{1}{2}(x + 5) \leqslant 2$.

6 Prove that the pair of inequalities $\frac{1}{3}(4x + 1) > x + 2$ and $3 - x \leqslant 2(4 - x)$ cannot be solved simultaneously.

7 Show that there is just one number which simultaneously satisfies these three inequalities and find it.

$$\frac{1}{2}(x - 1) \geqslant 1 \qquad 2 - 3x \leqslant 7 - 4x \qquad \frac{1}{3}x \leqslant 1$$

Simultaneous linear equations

Consider the equation $x + y = 9$, where x and y are the unknowns.

There is an infinite number of pairs x and y which satisfy this equation. Some examples of pairs of solutions are: 0 and 9; 4 and 5; 0.25 and 8.75. The equation cannot be solved uniquely (i.e. does not have just one pair of solutions) until we have a second relationship between x and y. Suppose the second relationship is

$$2x + y = 13$$

Then we call the equations $x + y = 9$ and $2x + y = 13$ **simultaneous linear equations** in x and y. There are three methods of solving simultaneous equations of this type. We will now look at each method.

Method I: Elimination

This method involves eliminating either the x terms or the y terms by adding or subtracting the two equations.

In some cases, the coefficients of the x terms or the y terms may not be the same. In these cases, we multiply the equations by a suitable constant so that the coefficients of either the x terms or the y terms are the same. This process is called **balancing the coefficients**. The next example illustrates such a case.

Example 13 Solve the simultaneous equations

$$5x - 7y = 27 \qquad [1]$$

$$2x + 3y = 5 \qquad [2]$$

SOLUTION

To balance the coefficients of the y terms, multiply equation [1] by 3 and multiply equation [2] by 7. This gives

$$15x - 21y = 81 \qquad [3]$$

$$14x + 21y = 35 \qquad [4]$$

Adding [3] and [4] gives

$$29x = 116$$

$$\therefore \quad x = 4$$

Substituting $x = 4$ into [2] to find the corresponding y value gives

$$2(4) + 3y = 5$$

$$\therefore \quad 8 + 3y = 5$$

$$\therefore \quad 3y = -3$$

$$\therefore \quad y = \frac{-3}{3} = -1$$

The solution is $x = 4$, $y = -1$.

Alternatively, the coefficients of the x terms can be balanced by multiplying equation [1] by 2 and multiplying equation [2] by 5. This gives

$$10x - 14y = 54 \qquad [5]$$

$$10x + 15y = 25 \qquad [6]$$

Subtracting [6] from [5] gives

$$-29y = 29$$

$$\therefore \quad y = -1$$

Substituting $y = -1$ into [2] to find the corresponding x value gives

$$2x + 3(-1) = 5$$

$$\therefore \quad x = 4$$

The solution is $x = 4$, $y = -1$, as before.

Method II: Substitution

This method involves rearranging one of the equations so that we have either

i) y in terms of x, in which case we then substitute this expression for y into the second equation,

or

ii) x in terms of y, in which case we substitute this expression for x into the second equation.

The choice between (i) and (ii) usually depends on which gives the simpler expression after rearranging.

Example 14 Solve the simultaneous equations

$$3x + 4y - 27 = 0 \qquad [1]$$

$$5x + y - 11 = 0 \qquad [2]$$

SOLUTION

Rearranging [2] for y gives

$$y = 11 - 5x \qquad [3]$$

Substituting [3] into [1] gives

$$3x + 4(11 - 5x) - 27 = 0$$

$$\therefore \quad 3x + 44 - 20x - 27 = 0$$

$$\therefore \quad -17x = -17$$

$$\therefore \quad x = 1$$

Substituting $x = 1$ into [3] to find the corresponding y value gives

$$y = 11 - 5(1)$$

$$\therefore \quad y = 6$$

The solution is $x = 1$, $y = 6$.

Method III: Graphical

This method involves drawing graphs of the two linear equations and finding the coordinates of their point of intersection. The coordinates of the intersection point give the solution to the pair of equations.

Example 15 Solve the simultaneous equations

$$x + y = 9$$
$$2x + y = 13$$

SOLUTION

Derive a table of values for $x + y = 9$. Since $x + y = 9$ is a linear equation, we know that the graph will be a straight line. Therefore, we only need to have two points through which the line passes. Let $x = 0$ and 2, which give the following table of values for $x + y = 9$:

x	0	2
y	9	7

Deriving a table of values for $2x + y = 13$ gives

x	0	2
y	13	9

Plotting the graph of each equation gives the diagram on the right.

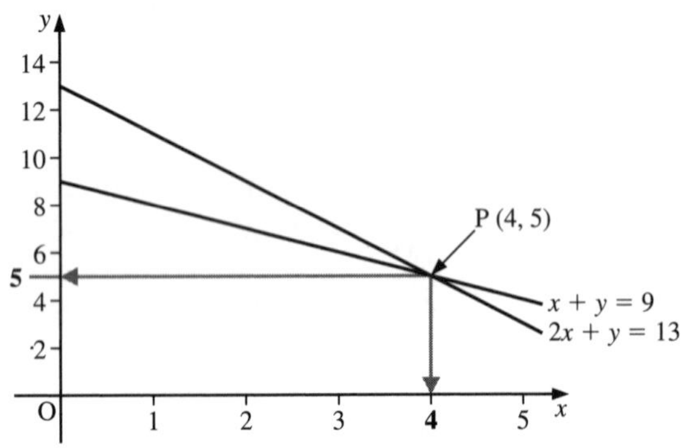

The coordinates of point P, where the two lines intersect, satisfy both equations and are therefore the solutions to both equations.

The point P has coordinates (4, 5), therefore the solution is $x = 4$, $y = 5$.

Exercise 1C

In Questions **1** and **2**, solve each of the given pairs of simultaneous equations for x and y.

1 **a)** $2x - 3y = 7$, $2x + 3y = 1$
b) $3x - y = 1$, $5x + y = 7$
c) $2x - 7y = 1$, $2x + 3y = 11$
d) $3x - 4y = 5$, $6x - 4y = 2$
e) $5x + 2y = 7$, $2x + y = 2$
f) $x - 2y = 5$, $3x + y = 8$
g) $2x + 3y = 1$, $3x + y = 5$
h) $3x - 2y = 5$, $2x + y = 8$
i) $3x + 5y = 1$, $2x + 3y = 0$
j) $5x + 4y = 1$, $7x + 5y = 2$
k) $6x - 5y = 12$, $5x - 4y = 11$
l) $3x + 2y = 3$, $x - 6y = 11$

2 **a)** $2x + y - 10 = 0$, $3x - 2y - 8 = 0$ **b)** $5x - 4y + 1 = 0$, $3x + y - 13 = 0$ **c)** $4x - 3y + 1 = 0$, $3x - 4y + 6 = 0$ **d)** $5x - 7y + 3 = 0$, $3x + 2y + 8 = 0$

e) $7x - 3y - 8 = 0$, $5x + 7y + 8 = 0$ **f)** $2x - 5y + 4 = 0$, $3x - 5y + 1 = 0$ **g)** $x - y - 1 = 0$, $5x - 2y - 3 = 0$ **h)** $3x + 2y - 2 = 0$, $x - 6y + 1 = 0$

i) $2x + 2y + 1 = 0$, $3x + 5y + 4 = 0$ **j)** $5x + 2y - 10 = 0$, $3x - 7y - 6 = 0$ **k)** $3x - 7y - 3 = 0$, $6x - 11y - 9 = 0$ **l)** $2x + 3y - 2 = 0$, $4x + 9y - 3 = 0$

3 Find the point of intersection of each of the following pairs of straight lines.

a) $y = x + 3$, $y = 2x + 1$ **b)** $y = x - 4$, $y = 3x - 16$ **c)** $y = 5 - 2x$, $y = x + 8$ **d)** $y = 4 - x$, $y = 2x + 10$

e) $y = x - 3$, $y = 6x + 2$ **f)** $y = 2x - 10$, $y = \frac{1}{3}x$ **g)** $2x - 5y = 4$, $3x + 2y = -13$ **h)** $7x + 3y = 6$, $5x + 4y = 8$

i) $6x + y + 8 = 0$, $5x - 4y + 26 = 0$ **j)** $2x + 3y - 3 = 0$, $3x - 2y + 15 = 0$ **k)** $5x + 6y + 39 = 0$, $3y = 4x$ **l)** $x - 6y + 23 = 0$, $y = 4x$

Quadratic equations

An expression of the form $ax^2 + bx + c$, where a, b and c are constants with $a \neq 0$, is called a **quadratic expression**.

An equation of the form $ax^2 + bx + c = 0$, where a, b and c are constants with $a \neq 0$, is called a **quadratic equation**. A quadratic equation will have at most two real solutions. The solutions are generally called the **roots** of the quadratic. We will look at the methods for solving the following three types of quadratic equation.

- Type I $\quad b = 0, a \neq 0, c \neq 0$ that is $ax^2 + c = 0$
- Type II $\quad c = 0, a \neq 0, b \neq 0$ that is $ax^2 + bx = 0$
- Type III $\quad a \neq 0, b \neq 0, c \neq 0$ that is $ax^2 + bx + c = 0$

Type I: $ax^2 + c = 0$

Example 16 Solve the equation $x^2 - 16 = 0$.

SOLUTION

Rearrange the equation to obtain

$$x^2 = 16$$

$$\therefore \quad x = \pm\sqrt{16}$$

$$\therefore \quad x = \pm 4$$

The negative solution must be considered since $(-4)^2 = 16$.

The solutions are $x = \pm 4$.

This equation can also be solved using factorisation techniques, namely 'the difference of two squares'. That is,

$$x^2 - y^2 = (x - y)(x + y) \qquad [1]$$

Since the equation $x^2 - 16 = 0$ can be written as the difference of two squares, that is

$$x^2 - 4^2 = 0$$

using [1] gives

$$x^2 - 4^2 = (x - 4)(x + 4) = 0$$

Using the result 'if $ab = 0$ then $a = 0$ or $b = 0$' gives

$$x - 4 = 0 \quad \text{or} \quad x + 4 = 0$$

$$\therefore \quad x = 4 \quad \text{or} \quad x = -4$$

The solutions are $x = \pm 4$, as before.

Example 17 Solve the equation $4x^2 - 24 = 0$.

SOLUTION

Rearranging gives

$$4x^2 = 24$$

$$\therefore \quad x^2 = 6$$

$$\therefore \quad x = \pm\sqrt{6}$$

The solutions are $x = \pm\sqrt{6}$.

Type II: $ax^2 + bx = 0$

Example 18 Solve the equation $x^2 - 7x = 0$.

SOLUTION

Factorising the LHS gives

$$x(x - 7) = 0$$

$$\therefore \quad x = 0 \quad \text{or} \quad x - 7 = 0$$

$$\therefore \quad x = 0 \quad \text{or} \quad x = 7$$

The solutions are $x = 0$ and $x = 7$.

Example 19 Solve the equation $3x^2 + 5x = 0$.

SOLUTION

Factorising the LHS gives

$$x(3x + 5) = 0$$

$$\therefore \quad x = 0 \quad \text{or} \quad 3x + 5 = 0$$

$$\therefore \quad x = 0 \quad \text{or} \quad x = -\frac{5}{3}$$

The solutions are $x = 0$ and $x = -\frac{5}{3}$.

Type III: $ax^2 + bx + c = 0$

First, we will look at solving quadratic equations of this form via factorisation methods. However, not all quadratics of this form will factorise. When a quadratic expression will not factorise, we say that the quadratic is **irreducible**. Other methods are required for solving irreducible quadratics, namely 'completing the square' and the formula method, both of which are met later in this chapter (see pages 20–6).

Example 20 Solve the equation $x^2 + 5x + 6 = 0$.

SOLUTION

Factorising the LHS gives

$$(x+2)(x+3) = 0$$

$$\therefore \quad x+2=0 \quad \text{or} \quad x+3=0$$

$$\therefore \quad x=-2 \quad \text{or} \quad x=-3$$

The solutions are $x = -2$ and $x = -3$.

Example 21 Solve the equation $2x^2 - 13x - 24 = 0$.

SOLUTION

Factorising the LHS gives

$$(2x+3)(x-8) = 0$$

$$\therefore \quad 2x+3=0 \quad \text{or} \quad x-8=0$$

$$\therefore \quad x=-\frac{3}{2} \quad \text{or} \quad x=8$$

The solutions are $x = -\frac{3}{2}$ and $x = 8$.

Example 22 Solve the equation $x^2 - x - 10 = x + 5$.

SOLUTION

Rearranging and factorising the LHS give

$$x^2 - 2x - 15 = 0$$

$$\therefore \quad (x+3)(x-5) = 0$$

$$\therefore \quad x+3=0 \quad \text{or} \quad x-5=0$$

$$\therefore \quad x=-3 \quad \text{or} \quad x=5$$

The solutions are $x = -3$ and $x = 5$.

Example 23 Solve the equation $(3x+1)(2x-1)-(x+2)^2=5$.

SOLUTION

Expanding the LHS and simplifying give

$$\begin{aligned} 6x^2-x-1-(x^2+4x+4)&=5\\ \therefore\quad 5x^2-5x-5&=5\\ \therefore\quad 5x^2-5x-10&=0\\ \therefore\quad 5(x^2-x-2)&=0\\ \therefore\quad 5(x+1)(x-2)&=0\\ \therefore\quad x+1=0 \quad &\text{or} \quad x-2=0\\ \therefore\quad x=-1 \quad &\text{or} \quad x=2 \end{aligned}$$

The solutions are $x=-1$ and $x=2$.

Example 24 Solve the equation $\dfrac{4}{x-1}+\dfrac{3}{x}=3$.

SOLUTION

Multiplying throughout by $x(x-1)$ gives

$$\begin{aligned} x(x-1)\left(\frac{4}{x-1}\right)+x(x-1)\left(\frac{3}{x}\right)&=3x(x-1)\\ \therefore\quad 4x+3(x-1)&=3x(x-1)\\ \therefore\quad 4x+3x-3&=3x^2-3x\\ \therefore\quad 3x^2-10x+3&=0\\ \therefore\quad (3x-1)(x-3)&=0\\ \therefore\quad 3x-1=0 \quad &\text{or} \quad x-3=0\\ \therefore\quad x=\frac{1}{3} \quad &\text{or} \quad x=3 \end{aligned}$$

The solutions are $x=\frac{1}{3}$ and $x=3$.

Example 25 Solve the equation $\dfrac{2x+1}{x+5}=\dfrac{3x-1}{x+7}$.

SOLUTION

Cross-multiply to obtain

$$(2x+1)(x+7)=(3x-1)(x+5)$$

Expanding and simplifying give

$$\begin{aligned} 2x^2+15x+7&=3x^2+14x-5\\ \therefore\quad x^2-x-12&=0\\ \therefore\quad (x+3)(x-4)&=0\\ \therefore\quad x+3=0 \quad &\text{or} \quad x-4=0\\ \therefore\quad x=-3 \quad &\text{or} \quad x=4 \end{aligned}$$

The solutions are $x=-3$ and $x=4$.

Example 26 A piece of wire of length 1 metre is cut into two parts and each part is bent to form a square. If the total area of the two squares formed is $325\,\text{cm}^2$, find the perimeter of each square.

SOLUTION

Let one of the pieces of wire be of length x cm. Then the other piece is of length $(100 - x)$ cm.

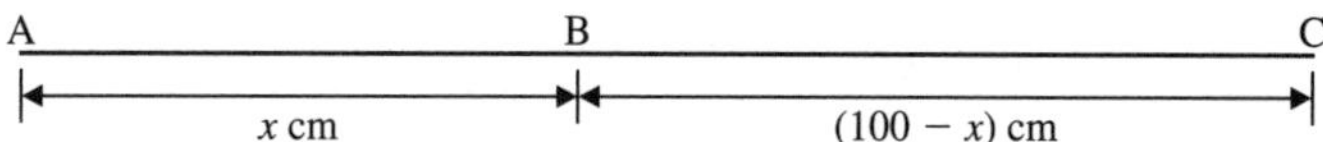

The square formed from the piece AB has sides of length $\frac{x}{4}$ cm.

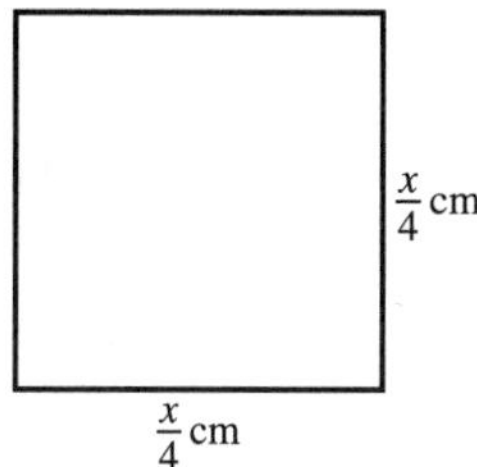

The area, A_1, of this square is given by

$$A_1 = \left(\frac{x}{4}\right)\left(\frac{x}{4}\right) = \frac{x^2}{16}\ \text{cm}^2$$

The square formed from the piece BC has sides of length $\left(\frac{100 - x}{4}\right)$ cm.

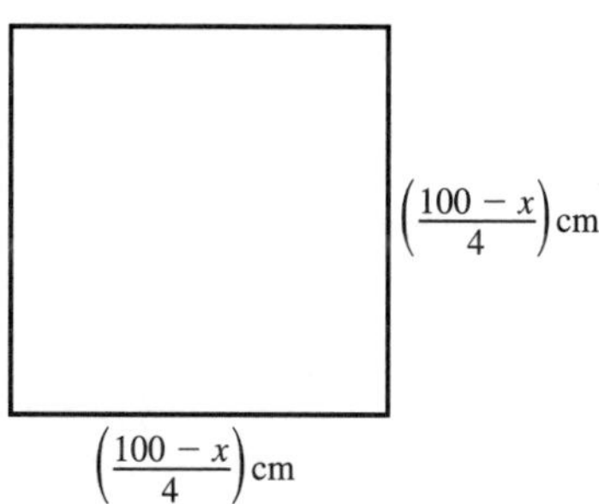

The area, A_2, of this square is given by

$$A_2 = \left(\frac{100 - x}{4}\right)\left(\frac{100 - x}{4}\right) = \frac{(100 - x)^2}{16}\ \text{cm}^2$$

Since the total area of the two squares is $325\,\text{cm}^2$,

$$A_1 + A_2 = 325$$

Therefore,

$$\frac{x^2}{16} + \frac{(100 - x)^2}{16} = 325$$

Multiplying throughout by 16 gives

$$x^2 + (100 - x)^2 = 5200$$

$$\therefore \quad x^2 + 10\,000 - 200x + x^2 = 5200$$

$$\therefore \quad 2x^2 - 200x + 4800 = 0$$

$$\therefore \quad 2(x^2 - 100x + 2400) = 0$$

$$\therefore \quad 2(x - 40)(x - 60) = 0$$

$$\therefore \quad x - 40 = 0 \quad \text{or} \quad x - 60 = 0$$

$$\therefore \quad x = 40 \quad \text{or} \quad x = 60$$

If $x = 40$ cm, the square formed from the piece of wire AB has perimeter 40 cm, and the square formed from the piece of wire BC has perimeter 60 cm.

If $x = 60$ cm, the square formed from the piece of wire AB has perimeter 60 cm, and the square formed from the piece of wire BC has perimeter 40 cm.

The perimeters of the squares are 40 cm and 60 cm.

Exercise 1D

In Questions **1** to **4**, solve each of the given quadratic equations for x.

1 a) $x^2 - 5x + 6 = 0$ **b)** $x^2 - 3x - 4 = 0$ **c)** $x^2 - 7x + 10 = 0$
d) $x^2 + 5x + 6 = 0$ **e)** $x^2 - 6x + 8 = 0$ **f)** $x^2 - 5x - 6 = 0$
g) $x^2 = 9$ **h)** $x^2 + 2x = 8$ **i)** $x^2 = x + 12$
j) $x^2 + 20 = 9x$ **k)** $x^2 = 4x$ **l)** $x^2 - 8 = 7x$

2 a) $2x^2 + 5x + 2 = 0$ **b)** $3x^2 - 7x + 2 = 0$ **c)** $2x^2 - 3x - 5 = 0$
d) $5x^2 + 14x - 3 = 0$ **e)** $4x^2 + 5x + 1 = 0$ **f)** $6x^2 - 5x + 1 = 0$
g) $3x^2 = 10x + 8$ **h)** $2x^2 + x = 15$ **i)** $16x^2 = 9$
j) $3x^2 - x = 10$ **k)** $5x^2 + 13x = 6$ **l)** $8x^2 + 3 = 14x$

3 a) $(x + 1)(x + 3) = 8$ **b)** $(x + 2)^2 = 2x + 12$
c) $(2x + 3)(x - 1) = 2(5x + 1)$ **d)** $(x - 3)(x - 4) + 7 = (2x + 5)(x - 1)$
e) $(3x + 5)(x - 1) = 2(x - 1)^2 - 12$ **f)** $(4x + 1)(x + 3) - (x - 3)(x - 1) = 6$
g) $(x - 2)(x + 1) + 3(2x - 1)(x - 3) = 4$ **h)** $(2x + 1)(x - 2) + 3x = x^2 - 1$
i) $x(4 - 3x) + (x + 1)(2x - 3) = 1 - x$ **j)** $(x + 3)(2x + 1) = (5x - 2)(x + 2) + 3x + 8$
k) $(x - 2)^2 = (5x + 3)(2x + 1) + 6x^2 + 1$ **l)** $x(x + 1) + (x + 1)(x + 2) + (x + 2)(x + 3) = 8$

4 a) $\dfrac{2}{x + 1} = \dfrac{x}{3 - 2x}$ **b)** $\dfrac{x + 2}{4} = \dfrac{x}{4 - x}$ **c)** $\dfrac{x^2}{x + 3} = \dfrac{2 - 3x}{2}$

d) $\dfrac{2x - 1}{8 - x} = \dfrac{5}{x + 2}$ **e)** $\dfrac{3x + 2}{2x - 1} = \dfrac{5x + 6}{x + 4}$ **f)** $\dfrac{1 - x}{x^2} = \dfrac{4}{2x + 1}$

g) $1 + \dfrac{4}{x} = \dfrac{3}{x - 1}$ **h)** $\dfrac{6}{x - 1} + \dfrac{1}{x - 4} = 2$ **i)** $\dfrac{2}{3x + 1} + \dfrac{3}{1 - x} = \dfrac{1}{2}$

j) $\dfrac{3}{x - 2} + \dfrac{4}{x + 1} = 4$ **k)** $\dfrac{4}{x} + \dfrac{3}{x - 1} = 5$ **l)** $\dfrac{5}{x + 3} + \dfrac{7}{x - 1} = 8$

5 The perimeter of a rectangle is 34 cm. Given that the diagonal is of length 13 cm, and that the width is x cm, derive the equation $x^2 - 17x + 60 = 0$. Hence find the dimensions of the rectangle.

6 A garden is in the shape of a rectangle, 20 metres by 8 metres. Around the outside is a border of uniform width, and in the middle is a square pond. The width of the border is the same as the width of the pond. The size of the area which is not occupied by either border or pond is $124\,\text{m}^2$. Letting the width of the border be x m, derive the equation $3x^2 - 56x + 36 = 0$. Solve this equation to find the value of x.

7 A metal sleeve of length 20 cm has rectangular cross-section 10 cm by 8 cm. The metal has uniform thickness, x cm, along the sleeve, and the total volume of metal in the sleeve is $495\,\text{cm}^3$.

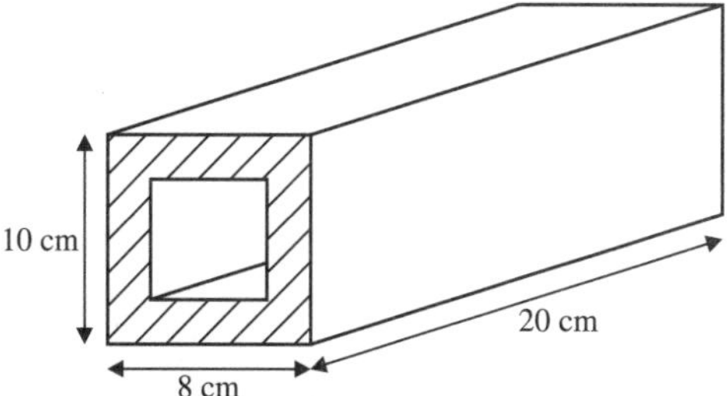

Derive the equation $16x^2 - 144x + 99 = 0$, and solve it to find the value of x.

8 A strand of wire of length 32 cm is cut into two pieces. One piece is bent to form a rectangle of width x cm and length $(x + 2)$ cm, and the other piece is bent to form a square.

a) Show that the square has sides of length $(7 - x)$ cm.

b) Given that the total of the areas enclosed by both the rectangle and the square is $31\,\text{cm}^2$, form an equation for x and solve it to find the value of x.

9 A train usually covers a journey of 240 km at a steady speed of $v\,\text{km h}^{-1}$. One day, due to adverse weather conditions, it reduces its speed by $40\,\text{km h}^{-1}$ and the journey takes one hour longer.

Derive the equation $v^2 - 40v - 9600 = 0$, and solve it to find the value of v.

10 As part of his training an athlete usually runs 80 km at a steady speed of $v\,\text{km h}^{-1}$. One day he decides to reduce his speed by $2.5\,\text{km h}^{-1}$ and his run takes him an extra 2 h 40 min.

Derive the equation $\dfrac{80}{v} + \dfrac{8}{3} = \dfrac{160}{2v - 5}$, and solve it to find the value of v.

Completing the square

Some examples of perfect squares are

$$25 = 5^2 \qquad x^4 = (x^2)^2 \qquad x^2 + 12x + 36 = (x+6)^2 \qquad (x+y)^6 = [(x+y)^3]^2$$

To make the quadratic expression $x^2 + 8x$ into a perfect square, we add half the coefficient of the x term squared. That is,

$$\begin{aligned} x^2 + 8x + \left(\frac{8}{2}\right)^2 &= x^2 + 8x + (4)^2 \\ &= x^2 + 8x + 16 \end{aligned}$$

Now

$$\begin{aligned} x^2 + 8x + 16 &= (x+4)(x+4) \\ &= (x+4)^2 \end{aligned}$$

which is a perfect square.

This process of adding half the coefficient of the x term squared is called **completing the square**.

In general, if we want to make $x^2 + bx$ into a perfect square we add $\left(\frac{b}{2}\right)^2$:

$$\begin{aligned} x^2 + bx + \left(\frac{b}{2}\right)^2 &= x^2 + bx + \frac{b^2}{4} \\ &= \left(x + \frac{b}{2}\right)\left(x + \frac{b}{2}\right) \\ &= \left(x + \frac{b}{2}\right)^2 \end{aligned}$$

which is a perfect square.

The process of completing the square is used to express a quadratic expression $ax^2 + bx + c$ in the form

$$a(x+p)^2 + q$$

where p and q are constants.

First, we will look at those quadratic expressions in which $a = 1$. In general,

$$\begin{aligned} x^2 + bx + c &= \left(x + \frac{b}{2}\right)^2 - \left(\frac{b}{2}\right)^2 + c \\ &= \left(x + \frac{b}{2}\right)^2 - \frac{b^2}{4} + c \end{aligned}$$

In the examples which follow, we will see why it is useful to express a quadratic expression in this form.

Example 27 Express $x^2 + 6x - 1$ in the form $a(x + p)^2 + q$. Hence solve the equation $x^2 + 6x - 1 = 0$.

SOLUTION

Completing the square gives

$$\begin{aligned} x^2 + 6x - 1 &= \left(x + \frac{6}{2}\right)^2 - \left(\frac{6}{2}\right)^2 - 1 \\ &= (x + 3)^2 - 9 - 1 \\ &= (x + 3)^2 - 10 \end{aligned}$$

Rewriting the equation $x^2 + 6x - 1 = 0$ gives

$$\begin{aligned} &(x + 3)^2 - 10 = 0 \\ \therefore\quad &(x + 3)^2 = 10 \\ \therefore\quad &x + 3 = \pm\sqrt{10} \\ \therefore\quad &x = -3 \pm \sqrt{10} \end{aligned}$$

The two solutions are $x = -3 + \sqrt{10}$ and $x = -3 - \sqrt{10}$.

Example 28 Use the method of completing the square to solve the equation $x^2 - 3x + 1 = 0$.

SOLUTION

Completing the square gives

$$\begin{aligned} x^2 - 3x + 1 &= \left(x - \frac{3}{2}\right)^2 - \left(-\frac{3}{2}\right)^2 + 1 \\ &= \left(x - \frac{3}{2}\right)^2 - \frac{9}{4} + 1 \\ &= \left(x - \frac{3}{2}\right)^2 - \frac{5}{4} \end{aligned}$$

Rewriting the equation $x^2 - 3x + 1 = 0$ gives

$$\begin{aligned} &\left(x - \frac{3}{2}\right)^2 - \frac{5}{4} = 0 \\ \therefore\quad &\left(x - \frac{3}{2}\right)^2 = \frac{5}{4} \\ \therefore\quad &x - \frac{3}{2} = \pm\sqrt{\frac{5}{4}} = \pm\frac{\sqrt{5}}{2} \\ \therefore\quad &x = \frac{3}{2} \pm \frac{\sqrt{5}}{2} \end{aligned}$$

The solutions are $x = \dfrac{3}{2} + \dfrac{\sqrt{5}}{2}$ and $x = \dfrac{3}{2} - \dfrac{\sqrt{5}}{2}$.

Example 29 Express $2x^2 + 8x + 5$ in the form $a(x + p)^2 + q$ and state the values of a, p and q.

SOLUTION

To use the techniques of the previous examples, we require the coefficient of the x^2 term to be 1. This can be achieved by taking a factor of 2 out of the expression. That is,

$$2x^2 + 8x + 5 = 2\left(x^2 + 4x + \frac{5}{2}\right)$$

We now proceed as before with the expression $x^2 + 4x + \frac{5}{2}$:

$$\begin{aligned} 2\left(x^2 + 4x + \frac{5}{2}\right) &= 2\left[(x + 2)^2 - (2)^2 + \frac{5}{2}\right] \\ &= 2\left[(x + 2)^2 - 4 + \frac{5}{2}\right] \\ &= 2\left[(x + 2)^2 - \frac{3}{2}\right] \\ &= 2(x + 2)^2 - 3 \end{aligned}$$

Therefore, $a = 2$, $p = 2$ and $q = -3$.

Example 30 Express $3x^2 + 15x + 20$ in the form $a(x + p)^2 + q$. Hence show that the equation $3x^2 + 15x + 20 = 0$ has no real roots.

SOLUTION

Taking out a factor of 3 gives

$$3x^2 + 15x + 20 = 3\left(x^2 + 5x + \frac{20}{3}\right)$$

Completing the square gives

$$\begin{aligned} 3\left(x^2 + 5x + \frac{20}{3}\right) &= 3\left[\left(x + \frac{5}{2}\right)^2 - \left(\frac{5}{2}\right)^2 + \frac{20}{3}\right] \\ &= 3\left[\left(x + \frac{5}{2}\right)^2 - \frac{25}{4} + \frac{20}{3}\right] \\ &= 3\left[\left(x + \frac{5}{2}\right)^2 + \frac{5}{12}\right] \\ &= 3\left(x + \frac{5}{2}\right)^2 + \frac{5}{4} \end{aligned}$$

Rewriting the equation $3x^2 + 15x + 20 = 0$ gives

$$3\left(x+\frac{5}{2}\right)^2 + \frac{5}{4} = 0$$

$$\therefore \quad 3\left(x+\frac{5}{2}\right)^2 = -\frac{5}{4}$$

$$\therefore \quad \left(x+\frac{5}{2}\right)^2 = -\frac{5}{12}$$

$$\therefore \quad x+\frac{5}{2} = \pm\sqrt{-\frac{5}{12}}$$

Since $\sqrt{-\frac{5}{12}}$ is not real, the equation $3x^2 + 15x + 20 = 0$ has no real roots.

You may be amused by the short song given below which outlines a basic method for solving quadratic equations by the method of completing the square.

SONG FOR QUADRATIC EQUATIONS

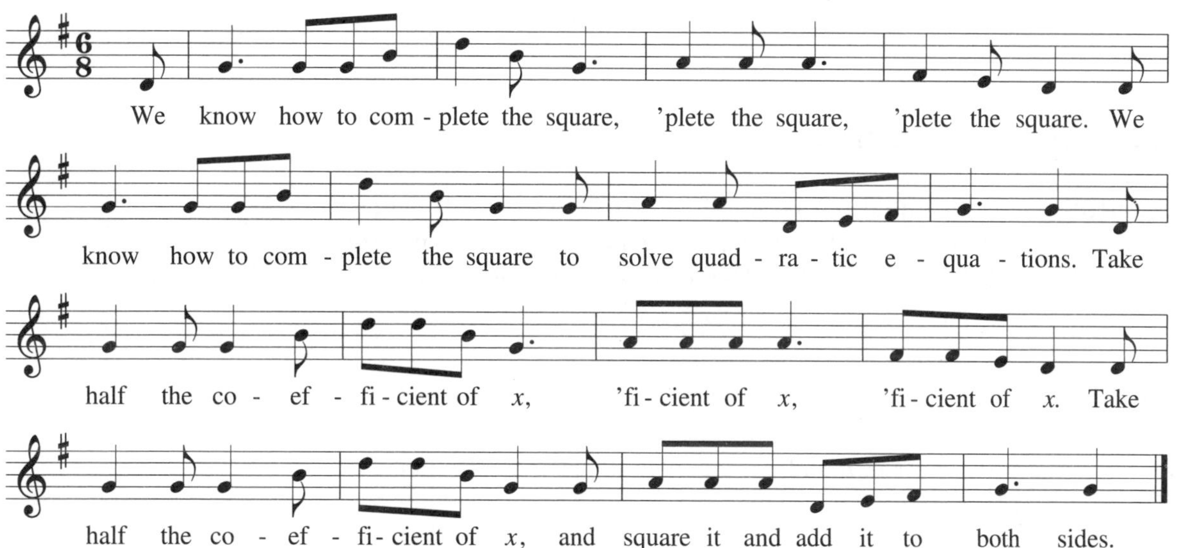
We know how to com - plete the square, 'plete the square, 'plete the square. We
know how to com - plete the square to solve quad - ra - tic e - qua - tions. Take
half the co - ef - fi - cient of x, 'fi - cient of x, 'fi - cient of x. Take
half the co - ef - fi - cient of x, and square it and add it to both sides.

Exercise 1E

1 Use the method of completing the square to express the solutions to each of the following quadratic equations in the form $a \pm b\sqrt{n}$, where a and b are rational, and n is an integer.

a) $x^2 - 4x - 1 = 0$ **b)** $x^2 + 6x + 2 = 0$ **c)** $x^2 - 2x - 1 = 0$

d) $x^2 - 8x - 3 = 0$ **e)** $x^2 + x - 1 = 0$ **f)** $x^2 + 3x + 1 = 0$

g) $x^2 - 5x - 2 = 0$ **h)** $x^2 - x - 3 = 0$ **i)** $x^2 + 5x + 1 = 0$

j) $x^2 + 12x + 5 = 0$ **k)** $x^2 - 9x + 10 = 0$ **l)** $x^2 - \frac{1}{2}x - \frac{1}{4} = 0$

2 Use the method of completing the square to solve each of the following quadratic equations, expressing your solutions in the form $a \pm b\sqrt{n}$, where a, b and n are rational.

a) $2x^2 - 3x - 3 = 0$ **b)** $3x^2 - 6x + 1 = 0$ **c)** $4x^2 + 4x - 5 = 0$
d) $3x^2 + 5x - 1 = 0$ **e)** $5x^2 + x - 3 = 0$ **f)** $2x^2 - 3x - 1 = 0$
g) $2x^2 - x - 2 = 0$ **h)** $4x^2 + 3x - 2 = 0$ **i)** $7x^2 - 14x + 5 = 0$
j) $6x^2 + 4x - 3 = 0$ **k)** $5x^2 - 20x + 17 = 0$ **l)** $2x^2 + 18x + 21 = 0$

3 Express $x^2 + 4x + 7$ in the form $(x + p)^2 + q$. Hence show that the equation $x^2 + 4x + 7 = 0$ has no real root.

4 Express $5x^2 - 30x + 47$ in the form $a(x + p)^2 + q$. Hence show that the equation $5x^2 - 30x + 47 = 0$ has no real root.

5 Show that the equation $\dfrac{x+2}{x-3} = \dfrac{x+4}{2x+3}$ has no real root.

***6** Given that the equation $x^2 + ax = b$, where a and b are real numbers, has a unique solution, prove that $a^2 + 4b = 0$.

Quadratic formula

The general quadratic equation has the form $ax^2 + bx + c = 0$, where a, b and c are constants with $a \neq 0$. Solving this general equation in terms of a, b and c will give a formula for the roots of any quadratic equation $ax^2 + bx + c = 0$.

Result

If $ax^2 + bx + c = 0$, where a, b and c are constants with $a \neq 0$, then

$$x = \frac{-b \pm \sqrt{b^2 - 4ac}}{2a}$$

Proof

Factorising out a in $ax^2 + bx + c = 0$ gives

$$a\left(x^2 + \frac{b}{a}x + \frac{c}{a}\right) = 0$$

Completing the square gives

$$a\left[\left(x + \frac{b}{2a}\right)^2 - \frac{b^2}{4a^2} + \frac{c}{a}\right] = 0$$

$$a\left[\left(x+\frac{b}{2a}\right)^2+\left(\frac{-b^2+4ac}{4a^2}\right)\right]=0$$

$$a\left(x+\frac{b}{2a}\right)^2+\frac{4ac-b^2}{4a}=0$$

$$\therefore\quad a\left(x+\frac{b}{2a}\right)^2=\frac{b^2-4ac}{4a}$$

$$\therefore\quad \left(x+\frac{b}{2a}\right)^2=\frac{b^2-4ac}{4a^2}$$

$$\therefore\quad x+\frac{b}{2a}=\pm\sqrt{\frac{b^2-4ac}{4a^2}}$$

$$\therefore\quad x=-\frac{b}{2a}\pm\frac{\sqrt{b^2-4ac}}{2a}$$

$$\therefore\quad x=\frac{-b\pm\sqrt{b^2-4ac}}{2a}$$

as required.

This is known as the **quadratic formula**.

Example 31 Solve the equation $x^2-8x+4=0$, giving your answers correct to two decimal places.

SOLUTION

Using the quadratic formula with $a=1$, $b=-8$ and $c=4$ gives

$$x=\frac{-(-8)\pm\sqrt{(-8)^2-4(1)(4)}}{2(1)}=\frac{8\pm\sqrt{48}}{2}$$

$$\therefore\quad x=7.46 \text{ or } 0.54\quad\text{(2 decimal places)}$$

Example 32 Using the quadratic formula, show that the equation $5x^2+4x+10=0$ has no real roots.

SOLUTION

Using the quadratic formula with $a=5$, $b=4$ and $c=10$ gives

$$x=\frac{-4\pm\sqrt{(4)^2-4(5)(10)}}{2(5)}$$

$$\therefore\quad x=\frac{-4\pm\sqrt{-184}}{10}$$

Since $\sqrt{-184}$ is not real, the equation $5x^2+4x+10=0$ has no real roots.

Example 33 Solve the equation $x^2 - 10x + 25 = 0$.

SOLUTION

Using the quadratic formula with $a = 1$, $b = -10$ and $c = 25$ gives

$$x = \frac{-(-10) \pm \sqrt{(-10)^2 - 4(1)(25)}}{2(1)}$$

$$= \frac{10 \pm \sqrt{0}}{2}$$

$$\therefore \quad x = 5 \quad \text{(called a \textbf{repeated root})}$$

Discriminant of a quadratic equation

The quantity

$$D = b^2 - 4ac$$

is called the **discriminant** of the quadratic equation $ax^2 + bx + c = 0$.

The type of root which arises from a quadratic equation depends on the value of the discriminant.

- In Example 31, $D = 48$ and the associated equation has two real roots.
- In Example 32, $D = -184$ and the associated equation has no real roots.
- In Example 33, $D = 0$ and the associated equation has a repeated root.

Consider the general quadratic equation

$$ax^2 + bx + c = 0$$

- When $b^2 - 4ac > 0$, the equation has two real roots.
- When $b^2 - 4ac < 0$, the equation has no real roots.
- When $b^2 - 4ac = 0$, the equation has one repeated root.

The discriminant of a quadratic indicates whether the graph of the quadratic expression cuts the x-axis at two different points, does not cut the x-axis at all or touches the x-axis at one point. Each case is illustrated below, for $a > 0$.

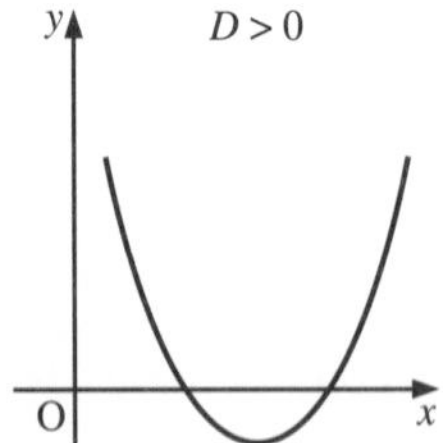

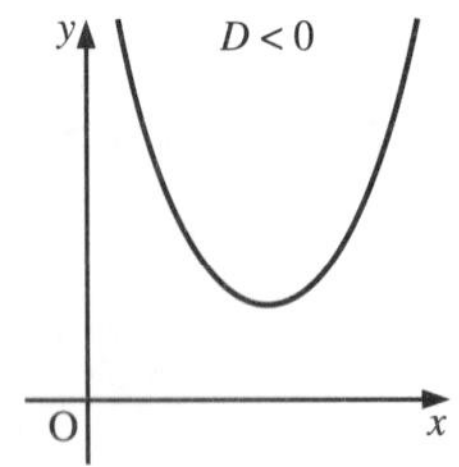

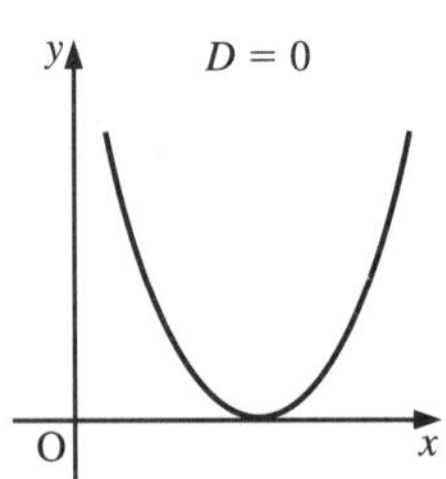

Example 34 Calculate the discriminant of the quadratic equation $2x^2 + 7x + 7 = 0$. Hence show that $2x^2 + 7x + 7$ is always positive.

SOLUTION

Calculating the discriminant with $a = 2$, $b = 7$ and $c = 7$ gives

$$\begin{aligned} D &= b^2 - 4ac \\ &= (7)^2 - 4(2)(7) \\ \therefore \quad D &= -7 \end{aligned}$$

The discriminant is -7.

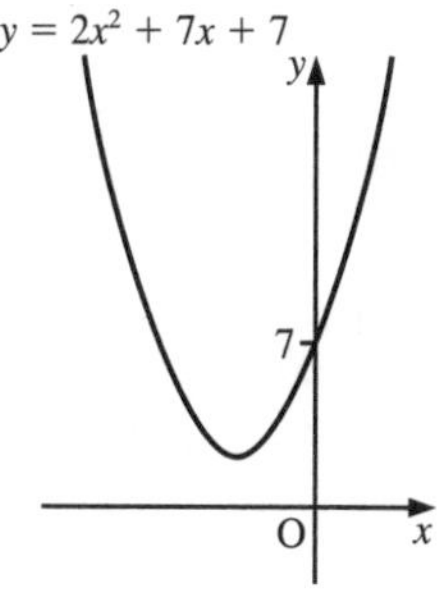

Since $D = -7 < 0$, the equation $2x^2 + 7x + 7 = 0$ has no real roots. Therefore, $y = 2x^2 + 7x + 7$ is never zero and the graph of $y = 2x^2 + 7x + 7$ does not cut the x-axis. Since the coefficient of the x^2 term is positive, we know that the curve is U-shaped. Therefore, the entire curve lies above the x-axis and is always positive.

Example 35 Find the values of the constant k given that the equation $(5k + 1)x^2 - 8kx + 3k = 0$ has a repeated root.

SOLUTION

The equation $(5k + 1)x^2 - 8kx + 3k = 0$ has a repeated root if the discriminant of the equation is zero.

Calculating the discriminant of the equation with $a = 5k + 1$, $b = -8k$ and $c = 3k$ gives

$$\begin{aligned} D &= b^2 - 4ac \\ &= (-8k)^2 - 4(5k + 1)(3k) \\ &= 64k^2 - 12k(5k + 1) \\ \therefore \quad D &= 4k^2 - 12k \end{aligned}$$

Putting $D = 0$ and factorising give

$$\begin{aligned} 4k^2 - 12k &= 0 \\ \therefore \quad 4k(k - 3) &= 0 \\ \therefore \quad k = 0 \quad &\text{or} \quad k - 3 = 0 \\ \therefore \quad k = 0 \quad &\text{or} \quad k = 3 \end{aligned}$$

The required values of k are 0 and 3.

Exercise 1F

In Questions **1** to **3**, use the quadratic formula to solve each of the given equations. Answers should be expressed correct to two decimal places.

1 **a)** $x^2 + 2x - 1 = 0$ **b)** $x^2 + 4x + 2 = 0$ **c)** $x^2 - 3x - 5 = 0$
d) $x^2 - 7x + 4 = 0$ **e)** $x^2 + 3x - 5 = 0$ **f)** $x^2 + 8x - 10 = 0$
g) $x^2 + x - 1 = 0$ **h)** $x^2 - 3x - 5 = 0$ **i)** $x^2 + 5x + 3 = 0$
j) $x^2 - 6x + 6 = 0$ **k)** $x^2 - 10x + 15 = 0$ **l)** $x^2 + 12x - 20 = 0$

2 **a)** $2x^2 + 3x - 4 = 0$ **b)** $3x^2 + x - 3 = 0$ **c)** $4x^2 + 5x - 7 = 0$
d) $2x^2 + 7x + 4 = 0$ **e)** $5x^2 + 2x - 1 = 0$ **f)** $6x^2 + 5x - 3 = 0$
g) $2x^2 + x - 8 = 0$ **h)** $6x^2 + 3x - 1 = 0$ **i)** $3x^2 + 7x + 3 = 0$
j) $2x^2 - 3x - 8 = 0$ **k)** $6x^2 + 9x + 2 = 0$ **l)** $5x^2 + 4x - 3 = 0$

3 **a)** $(x + 2)(x - 1) - 3x = 4$ **b)** $(x + 2)^2 + 5x = 6$
c) $(2x + 3)(x - 1) + 2(x - 2)(x + 3) = 4$ **d)** $(x - 5)(3x + 2) = 5 - (x - 3)(x + 2)$
e) $\dfrac{2x + 5}{x - 2} = \dfrac{x + 3}{x + 6}$ **f)** $\dfrac{2 - x}{3 - x} = \dfrac{4 - 3x}{5 + x}$
g) $\dfrac{x^2}{x + 3} = \dfrac{5x - 4}{2}$ **h)** $\dfrac{3}{x + 4} - \dfrac{2}{x} = 5$
i) $\dfrac{1}{2x - 3} + \dfrac{2}{x + 4} = 3$ **j)** $\dfrac{(x + 2)(x + 3)}{x} + 4 = 3x$
k) $\dfrac{(x - 1)(x - 5)}{x + 2} = 2x + 3$ **l)** $\dfrac{3}{(x - 2)^2} + \dfrac{4}{x - 2} = 5$

4 Calculate the discriminant of the quadratic $3x^2 + 5x + 8$. Hence show that $3x^2 + 5x + 8 > 0$, for all values of x.

5 Calculate the discriminant of the quadratic $5x^2 + 2x + 1$. Hence show that $5x^2 + 2x + 1 > 0$, for all values of x.

6 Show that $x^2 + 3x + 5 > 0$, for all values of x.

7 Show that $x^2 + 6x > 3x - 4$, for all values of x.

8 Show that $2x^2 + 6 > 4x + 1$, for all values of x.

9 Prove that the inequality $3x^2 + 13 < 12x$ has no real solution.

10 Find the possible values of the constant a given that the equation $ax^2 + (8 - a)x + 1 = 0$ has a repeated root.

11 Given that the equation $x^2 - 3bx + (4b + 1) = 0$ has a repeated root, find the possible values of the constant b.

12 Calculate the possible values of the constant k given that the equation

$$(k + 1)x^2 - 8kx + (3k + 5) = 0$$

has a repeated root.

13 Given that the equation $(4k + 1)x^2 - (k + 10)x + 2k = 0$ has a repeated root, calculate the possible values of the constant k.

14 Show that there is no real value of the constant c for which the equation

$$cx^2 + (4c + 1)x + (c + 2) = 0$$

has a repeated root.

***15** Given that the roots of the equation $x^2 + ax + (a + 2) = 0$ differ by 2, find the possible values of the constant a. Hence state the possible values of the roots of the equation.

Disguised quadratic equations

We have seen various types of quadratic equation but all have been quadratic equations in x. We now look at equations which don't appear to be quadratic, but in fact are.

Example 36 Solve the equation $x^4 + 5x^2 - 14 = 0$.

SOLUTION

This equation does not appear to be a quadratic, but writing it in the form

$$(x^2)^2 + 5(x^2) - 14 = 0$$

and letting $y = x^2$ gives

$$y^2 + 5y - 14 = 0$$

which is a quadratic equation in y. We say that the original equation is a 'quadratic equation in x^2'. To solve the equation in y, we factorise in the usual way giving

$$(y + 7)(y - 2) = 0$$

Solving gives $y = -7$ or $y = 2$. Now replacing y with x^2 gives $x^2 = -7$ or $x^2 = 2$.

We see that $x^2 = -7$ gives no real solutions.
However, $x^2 = 2$ gives $x = \pm\sqrt{2}$.

Example 37 Solve the equation $x - 9\sqrt{x} + 20 = 0$.

SOLUTION

Rewriting the equation gives

$$(\sqrt{x})^2 - 9(\sqrt{x}) + 20 = 0$$

We see that the original equation is a quadratic in $\sqrt{x}$. Letting $y = \sqrt{x}$ gives

$$y^2 - 9y + 20 = 0$$

Factorising and solving give

$$(y - 4)(y - 5) = 0$$

$$\therefore \quad y = 4 \quad \text{or} \quad y = 5$$

Replacing y with $\sqrt{x}$ gives $\sqrt{x} = 4$ or $\sqrt{x} = 5$.
Solving gives $x = 16$ or $x = 25$.

Exercise 1G

1 Solve each of these equations for x.

a) $x^4 - 13x^2 + 36 = 0$ **b)** $x^4 - 2x^2 - 3 = 0$ **c)** $x^6 - 28x^3 + 27 = 0$
d) $x^6 + 5x^3 - 24 = 0$ **e)** $x - 5\sqrt{x} + 6 = 0$ **f)** $x - 6\sqrt{x} + 5 = 0$
g) $x^4 + x^2 = 12$ **h)** $x = 4\sqrt{x} - 3$ **i)** $x^8 + 16 = 17x^4$
j) $x^6 = 8 + 2x^3$ **k)** $8\sqrt{x} = 15 + x$ **l)** $65x^4 = 16 + 4x^8$

2 Solve each of these equations for x.

a) $x^2 + 1 = \dfrac{6}{x^2}$ **b)** $x^3 + 7 = \dfrac{8}{x^3}$ **c)** $x = 12\sqrt{x} - 35$

d) $x^3 - 6x + \dfrac{8}{x} = 0$ **e)** $\sqrt{x} + \dfrac{10}{\sqrt{x}} = 7$ **f)** $x^2 + 3 = \dfrac{18}{x^2}$

g) $x^4(x^4 - 20) = 64$ **h)** $15 = \sqrt{x}(8 - \sqrt{x})$ **i)** $\dfrac{5}{x^2} = x^2 + \dfrac{4}{x^6}$

j) $2(x^4 + 6) = 11x^2$ **k)** $2 + \dfrac{10}{x} = \dfrac{9}{\sqrt{x}}$ **l)** $x = \dfrac{2(3x^3 + 8)}{x^5}$

3 Solve $(x + 3)^2 - 5(x + 3) + 4 = 0$.

4 Solve $(3x - 1)^2 + 6(3x - 1) - 7 = 0$.

5 **a)** Solve $y^2 - 7y + 10 = 0$.
b) Hence find the solutions to $(x^2 + 1)^2 - 7(x^2 + 1) + 10 = 0$.

6 **a)** Solve $y^2 - 5y - 14 = 0$.
b) Hence find the solutions to $(x^3 - 1)^2 - 5(x^3 - 1) - 14 = 0$.

7 Solve $x(x + 1) + \dfrac{24}{x(x + 1)} = 14$.

***8** **a)** By using the substitution $p = x + \dfrac{1}{x}$, show that the equation

$$2x^4 + x^3 - 6x^2 + x + 2 = 0$$

reduces to $2p^2 + p - 10 = 0$.

b) Hence solve $2x^4 + x^3 - 6x^2 + x + 2 = 0$.

Sketching the graph of a quadratic function

Expressing $y = ax^2 + bx + c$ in the form $a(x + p)^2 + q$ allows us to deduce details about the graph of $y = ax^2 + bx + c$.

Example 38 Express $y = x^2 - 2x - 8$ in the form $a(x + p)^2 + q$ and hence sketch the graph of $y = x^2 - 2x - 8$.

SOLUTION

Completing the square gives

$$x^2 - 2x - 8 = (x - 1)^2 - (-1)^2 - 8$$
$$= (x - 1)^2 - 9$$

Therefore, $y = (x - 1)^2 - 9$.

Since $(x - 1)^2 \geqslant 0$, with equality holding when $x = 1$, the minimum value of y is -9 and this occurs when

$$(x - 1)^2 = 0$$

That is, when $x = 1$.

Therefore, the point with coordinates $(1, -9)$ is the minimum point of the graph.

The graph cuts the x-axis when $y = 0$. That is, when

$$x^2 - 2x - 8 = 0$$
$$\therefore \quad (x + 2)(x - 4) = 0$$
$$\therefore \quad x = -2 \quad \text{or} \quad x = 4$$

Therefore, the graph cuts the x-axis at the points $(-2, 0)$ and $(4, 0)$.

The graph cuts the y-axis when $x = 0$. That is, when

$$y = (0)^2 - 2(0) - 8$$
$$\therefore \quad y = -8$$

Therefore, the graph cuts the y-axis at the point $(0, -8)$.

The graph of $y = x^2 - 2x - 8$ is sketched on the right.

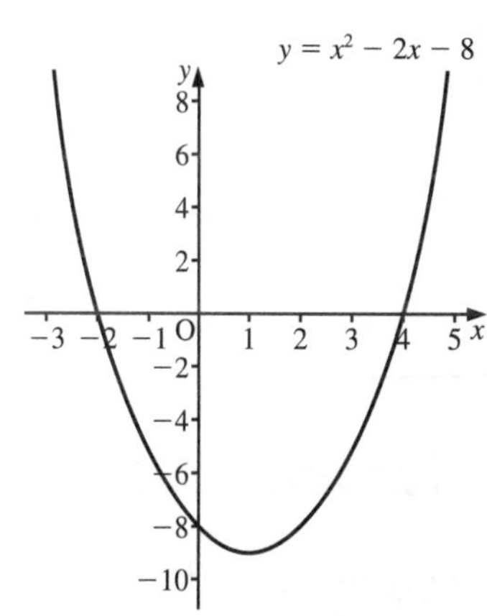

Example 39 Express $y = -x^2 + 10x - 21$ in the form $a(x+p)^2 + q$ and hence sketch the graph of $y = -x^2 + 10x - 21$.

SOLUTION

Completing the square gives

$$\begin{aligned} -x^2 + 10x - 21 &= -(x^2 - 10x + 21) \\ &= -[(x-5)^2 - (-5)^2 + 21] \\ &= -[(x-5)^2 - 4] \\ &= -(x-5)^2 + 4 \end{aligned}$$

Therefore, $y = -(x-5)^2 + 4$.

Since $-(x-5)^2 \leqslant 0$, with equality holding when $x = 5$, the maximum value of y is 4 and this occurs when $x = 5$. Therefore, the point with coordinates $(5, 4)$ is the maximum point of the graph.

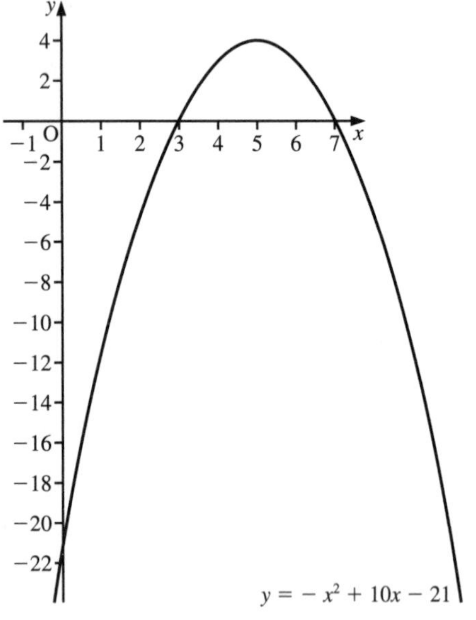

The graph cuts the x-axis when $y = 0$. That is, when

$$-x^2 + 10x - 21 = 0$$

$$\therefore \quad x^2 - 10x + 21 = 0$$

$$\therefore \quad (x-3)(x-7) = 0$$

$$\therefore \quad x = 3 \quad \text{or} \quad x = 7$$

The graph cuts the x-axis at the points $(3, 0)$ and $(7, 0)$.

The graph cuts the y-axis when $x = 0$. That is, when

$$y = -(0)^2 + 10(0) - 21$$

$$\therefore \quad y = -21$$

The graph cuts the y-axis at the point $(0, -21)$.

The graph of $y = -x^2 + 10x - 21$ is sketched on the right.

Maxima and minima

Many practical situations give rise to quadratic expressions. Sometimes these involve finding the maximum or minimum value of a quantity. The next two examples show how to do this by the method of completing the square.

Example 40 The net of an open box is given on the right. Show that the volume, V, of the box is given by

$$V = 75x - 10x^2$$

Hence find the value of x for which the volume of the box is a maximum. State the maximum volume of the box.

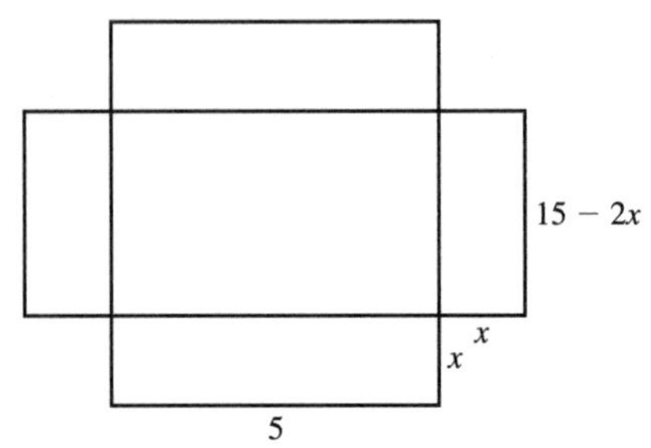

SOLUTION

When the net is folded, the box has the dimensions x, 5, and $15 - 2x$.

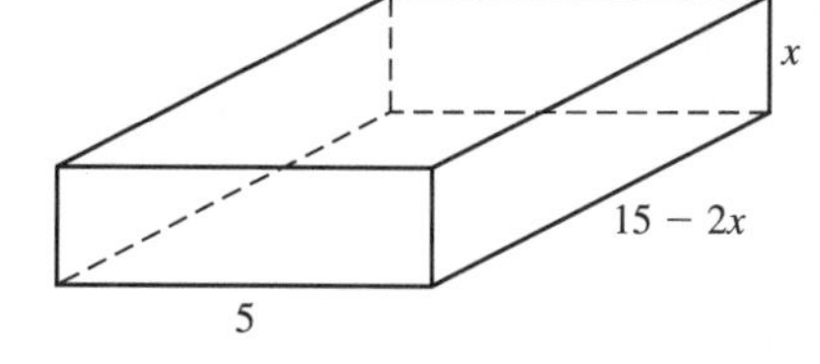

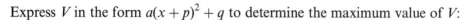

Therefore, the volume V is given by

$$V = 5x(15 - 2x)$$

$$\therefore \quad V = 75x - 10x^2$$

as required.

Express V in the form $a(x+p)^2 + q$ to determine the maximum value of V:

$$V = -10x^2 + 75x = -10\left(x^2 - \frac{75}{10}x\right)$$

$$= -10\left(x^2 - \frac{15}{2}x\right)$$

$$= -10\left[\left(x - \frac{15}{4}\right)^2 - \left(-\frac{15}{4}\right)^2\right]$$

$$= -10\left[\left(x - \frac{15}{4}\right)^2 - \frac{225}{16}\right]$$

Therefore, $$V = -10\left(x - \frac{15}{4}\right)^2 + \frac{1125}{8}$$

The maximum value of V occurs when $x = \frac{15}{4} = 3.75$ units. So the maximum value of V is $\frac{1125}{8} = 140.625$ units^3.

Example 41 Find the maximum area of a rectangle which has a perimeter of 28 units.

SOLUTION

Since the perimeter is 28 units, the sum of the length and the width is 14 units. Let the rectangle have length x. Then its width is $14 - x$.

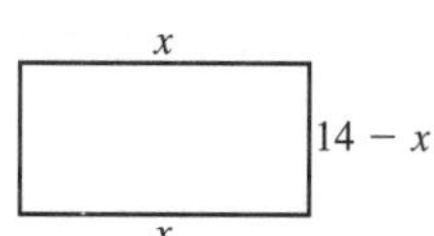

The area, A, of the rectangle is given by

$$A = x(14 - x)$$

$$= 14x - x^2$$

Express A in the form $a(x+p)^2 + q$ to determine the maximum value of A:

$$A = -x^2 + 14x = -(x^2 - 14x)$$

$$= -[(x - 7)^2 - 49]$$

$$= -(x - 7)^2 + 49$$

Therefore, $A = -(x - 7)^2 + 49$.

The maximum value of A is 49 units^2, which occurs when $x = 7$ units.

Notice that when $x = 7$, $y = 14 - 7 = 7$. In other words, the rectangle is a square when it attains its maximum area.

Exercise 1H

1 For each of the following find the minimum value of y and the value of x at which it occurs.

a) $y = x^2 + 4x + 6$ **b)** $y = x^2 - 6x + 13$ **c)** $y = x^2 - 10x + 40$
d) $y = x^2 + 2x - 5$ **e)** $y = x^2 + 3x + 8$ **f)** $y = x^2 - 7x + 15$
g) $y = 2x^2 + 10x - 5$ **h)** $y = 3x^2 + 6x + 14$ **i)** $y = 4x^2 + x - 7$
j) $y = 3x^2 - 2x + 9$ **k)** $y = 6x^2 + x + 5$ **l)** $y = 5x^2 - 2x + 8$

2 Find the maximum value for each of the following quadratics, and the value of x at which each occurs.

a) $y = 3 - 2x - x^2$ **b)** $y = 5 + 4x - x^2$ **c)** $y = 8 + 2x - x^2$
d) $y = 20 + 6x - x^2$ **e)** $y = 4 - 3x - x^2$ **f)** $y = -2 + 4x - x^2$
g) $y = 6 - x - x^2$ **h)** $y = 3 + 2x - 2x^2$ **i)** $y = 7 - 3x - 4x^2$
j) $y = 14 + x - 3x^2$ **k)** $y = 6 - 5x - 2x^2$ **l)** $y = 12 + 7x - 14x^2$

3 Use the method of completing the square to sketch the graphs of these quadratics.

a) $y = x^2 - 6x + 8$ **b)** $y = 2x^2 - 4x + 5$ **c)** $y = x^2 + 2x - 15$
d) $y = 8 + 2x - x^2$ **e)** $y = 3 - 4x - 4x^2$ **f)** $y = x^2 + 4x - 21$
g) $y = -x^2 + 2x - 3$ **h)** $y = x^2 - 5x$ **i)** $y = 3x^2 - 18x + 24$
j) $y = 15 - 4x - 4x^2$ **k)** $y = 10 + 3x - x^2$ **l)** $y = 4x^2 - 20x + 21$

4 A farmer has 40 m of fencing with which to enclose a rectangular pen. Given the pen is x m wide,

a) show that is area is $(20x - x^2)\,\text{m}^2$
b) deduce the maximum area that he can enclose.

5 Another farmer also has 40 m of fencing, and he also wishes to enclose a rectangular pen of maximum area, but one side of his pen will consist of part of a wall which is already in place.

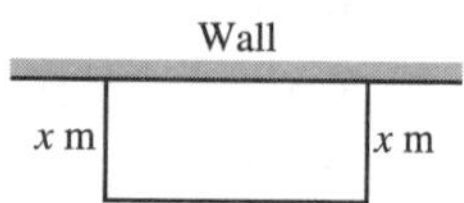

Given that the two sides of his pen touching the wall each have length x m, find an expression, in terms of x, for the area that he can enclose. Deduce that the maximum area is 200 m^2.

6 A third farmer also has 40 m of fencing but he decides to use a right-angled corner of a building, as in the diagram.

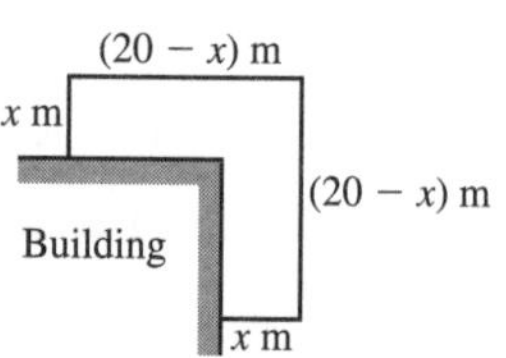

Show that the area which he can enclose is given by the expression $(40x - 3x^2)$ m^2, and deduce the maximum value of this area.

7 When a stone is projected vertically into the air with an initial speed of $30\,\text{m s}^{-1}$ its height, h metres, above the point of projection, at a time t seconds after the instant of projection, can be approximated by the formula $h = 30t - 5t^2$.

Find the maximum height reached by the stone, and the time at which this occurs.

8 A strip of wire of length 28 cm is cut into two pieces. One piece is bent to form a square of side x cm, and the other piece is bent to form a rectangle of width 3 cm.

a) Show that the lengths of the other two sides of the rectangle are given by $(11 - 2x)$ cm.

b) Deduce that the total combined area of the square and the rectangle is $(x^2 - 6x + 33)\,\text{cm}^2$.

c) Prove that the minimum total area which can be enclosed in this way is $24\,\text{cm}^2$.

9 A string of length 60 cm is cut into two pieces and each piece is formed into a rectangle. The first rectangle has width 6 cm, and the second rectangle is three times as long as it is wide. Given the width of the second rectangle is x cm,

a) deduce that the total combined area enclosed by the two rectangles may be expressed as $[3(x - 4)^2 + 96]\,\text{cm}^2$

b) show that the minimum area which can be enclosed in this way is $96\,\text{cm}^2$.

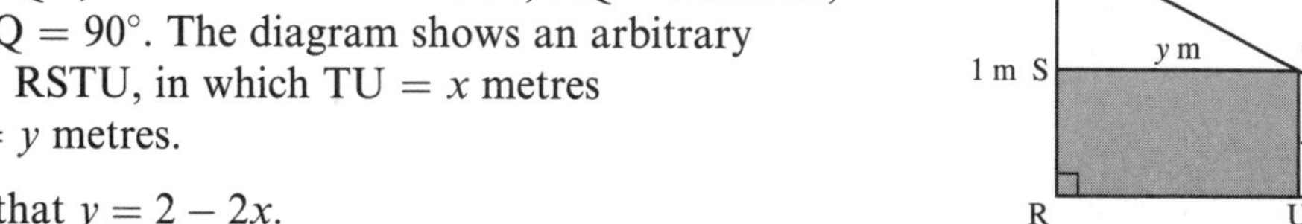

10 It is required to fit a rectangle of maximum area inside a triangle, PQR, in which PR = 1 metre, RQ = 2 metres, and $\angle$PRQ = 90°. The diagram shows an arbitrary rectangle, RSTU, in which TU = x metres and ST = y metres.

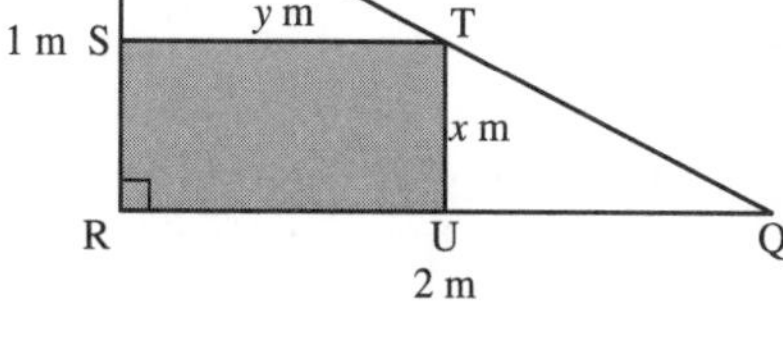

a) Show that $y = 2 - 2x$.

b) Find an expression, in terms of x, for the area of the rectangle, and deduce that the rectangle of maximum area which fits inside triangle PQR has area $\frac{1}{2}\,\text{m}^2$.

***11** Show that, in general, for any rectangle drawn inside any right-angled triangle, the area of the rectangle cannot exceed half the area of the triangle.

Quadratic inequalities

In order to solve a quadratic inequality, it is useful to have a sketch graph of the quadratic function involved and the points where the graph cuts the x-axis, if at all.

Example 42 Solve the inequality $x^2 + 3x - 4 < 0$.

SOLUTION

First, identify where the graph of $y = x^2 + 3x - 4$ cuts the x-axis. This occurs when $y = 0$. That is, when

$$x^2 + 3x - 4 = 0$$
$$(x + 4)(x - 1) = 0$$
$$\therefore \quad x = -4 \quad \text{or} \quad x = 1$$

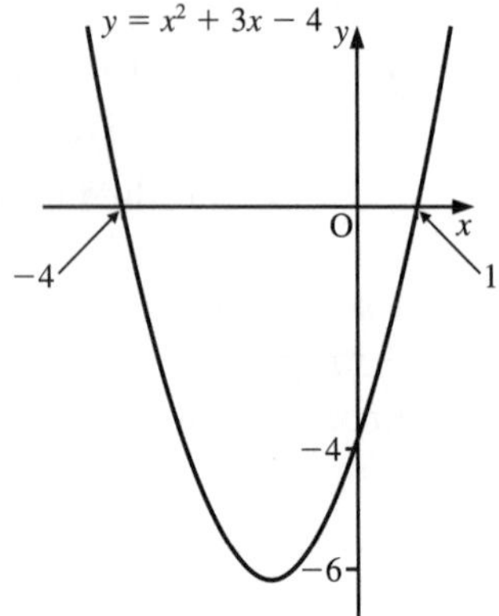

Since the coefficient of the x^2 term is positive, the graph of $y = x^2 + 3x - 4$ is U-shaped. Therefore, the sketch is as shown on the right.

By inspection of the graph, $x^2 + 3x - 4 < 0$ when $-4 < x < 1$.

Example 43 Solve the inequality $\dfrac{x+1}{7x-1} \leqslant \dfrac{2}{7}$.

SOLUTION

We multiply throughout by $(7x - 1)^2$ to ensure that the inequality is positive. This gives

$$(x + 1)(7x - 1) \leqslant \frac{2}{7}(7x - 1)^2$$
$$\therefore \quad 7x^2 + 6x - 1 \leqslant \frac{2}{7}(49x^2 - 14x + 1)$$
$$\therefore \quad 0 \leqslant 7x^2 - 10x + \frac{9}{7}$$
$$\therefore \quad 49x^2 - 70x + 9 \geqslant 0$$

The graph of $y = 49x^2 - 70x + 9$ cuts the x-axis when $y = 0$. That is, when

$$49x^2 - 70x + 9 = 0$$
$$(7x - 1)(7x - 9) = 0$$

Solving gives $x = \frac{1}{7}$ or $x = \frac{9}{7}$.

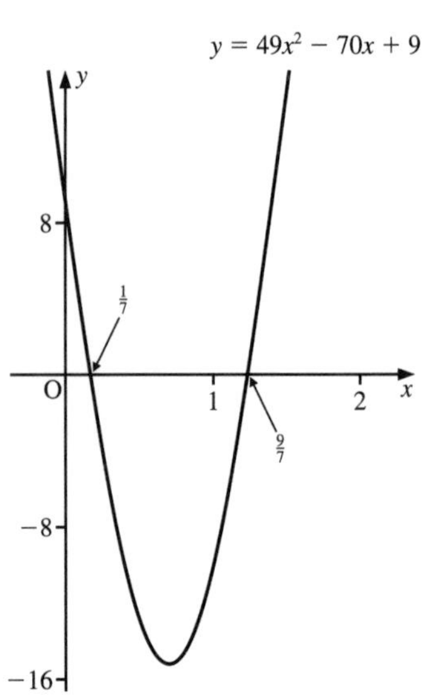

The sketch of $y = 49x^2 - 70x + 9$ is shown on the right.

By inspection of the graph, we see that $49x^2 - 70x + 9 \geqslant 0$ when $x \leqslant \frac{1}{7}$ or $x \geqslant \frac{9}{7}$.

Exercise 1I

Solve each of these inequalities for x.

1 a) $(x-3)(x-5)>0$ **b)** $(x+4)(x-3)<0$ **c)** $(x+3)(x+1)\leqslant 0$
d) $(2x+3)(x-1)<0$ **e)** $(3x+2)(2x-1)\geqslant 0$ **f)** $(3+x)(1-x)<0$
g) $(2+x)(3-x)>0$ **h)** $(4+3x)(5-2x)\leqslant 0$ **i)** $(3+x)(6-x)<0$
j) $(5+x)(4+5x)\geqslant 0$ **k)** $(x-3)(x+3)\leqslant 0$ **l)** $x(5-6x)>0$

2 a) $x^2-2x-8>0$ **b)** $x^2-2x-15\leqslant 0$ **c)** $x^2+7x+12<0$
d) $2x^2+3x-5\geqslant 0$ **e)** $5x^2+11x+6\leqslant 0$ **f)** $x^2-8x+16\leqslant 0$
g) $6-x-x^2>0$ **h)** $2+x-3x^2<0$ **i)** $10+23x-5x^2\geqslant 0$
j) $16-x^2<0$ **k)** $10x^2-3x-1\leqslant 0$ **l)** $4x^2-12x+9>0$

3 a) $x^2+3x\geqslant 10$ **b)** $x^2+21<10x$ **c)** $x^2+4>5x$
d) $54-11x\leqslant 30-x^2$ **e)** $5x>3-2x^2$ **f)** $x(10x-13)+4<0$
g) $2(x^2-6)\geqslant 5x$ **h)** $x^2<25$ **i)** $x+1>6x^2$
j) $2x^2+15x\leqslant 8$ **k)** $9x^2+1<6x$ **l)** $x^2\geqslant 5x$

4 a) $\dfrac{2}{x-3}\leqslant 1$ **b)** $\dfrac{5}{x+1}>2$ **c)** $\dfrac{1}{2x-5}\leqslant 3$
d) $\dfrac{2}{3x+1}<5$ **e)** $\dfrac{1}{x-2}-3\geqslant 0$ **f)** $3+\dfrac{2}{2x-1}>0$
g) $\dfrac{5x}{x-2}\leqslant 1$ **h)** $\dfrac{x-2}{x+3}>4$ **i)** $\dfrac{4-3x}{1-x}<2$
j) $4-\dfrac{x}{3-2x}\geqslant 0$ **k)** $\dfrac{x-1}{5x+3}-1<0$ **l)** $3+\dfrac{4-x}{x}>0$

***5 a)** $\dfrac{2+3x}{4-x}\leqslant 2x$ **b)** $\dfrac{x-1}{x-3}>\dfrac{x-4}{x-2}$

Further simultaneous equations

So far, we have looked at simultaneous equations where both equations are linear. We are now going to look at the techniques involved in solving simultaneous equations in which one or both of the equations is/are quadratic.

Example 44 Solve the simultaneous equations $y=x^2+3x+2$ and $y=2x+8$.

SOLUTION

Eliminate y, obtaining

$$x^2+3x+2=2x+8$$
$$\therefore\quad x^2+x-6=0$$
$$(x+3)(x-2)=0$$
$$\therefore\quad x=-3\quad\text{or}\quad x=2$$

To find the corresponding values of y, substitute each x value into $y = 2x + 8$.

When $x = -3$: $y = 2(-3) + 8 = 2$

When $x = 2$: $y = 2(2) + 8 = 12$

The solutions are $x = -3$, $y = 2$ and $x = 2$, $y = 12$.

Example 45 Solve the simultaneous equations $y = x^2 - 2x + 2$ and $y = 4x - 7$. Interpret your result geometrically.

SOLUTION

Eliminate y, obtaining

$$x^2 - 2x + 2 = 4x - 7$$

$$\therefore \quad x^2 - 6x + 9 = 0$$

$$\therefore \quad (x - 3)(x - 3) = 0$$

$$\therefore \quad x = 3 \quad \text{(Repeated root)}$$

Substituting $x = 3$ into $y = 4x - 7$ gives

$$y = 4(3) - 7 = 5$$

The solution is $x = 3$, $y = 5$.

The geometrical interpretation of this result is that the line $y = 4x - 7$ intersects the curve $y = x^2 - 2x + 2$ at only one point. Therefore, the line $y = 4x - 7$ is a tangent to the curve $y = x^2 - 2x + 2$ at the point (3, 5).

Sketching graphs of both $y = x^2 - 2x + 2$ and $y = 4x - 7$ on the same set of axes gives the diagram shown here.

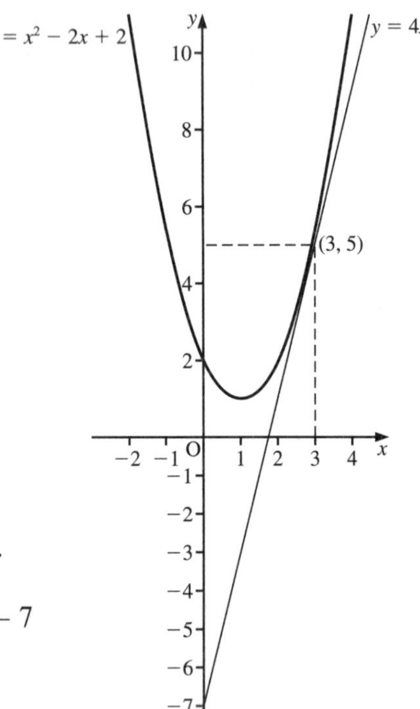

Example 46 Solve the simultaneous equations

$$x^2 + 2x - y = 14 \qquad [1]$$

$$2x^2 - 3y = 47 \qquad [2]$$

SOLUTION

Rearranging [1] for y gives

$$y = x^2 + 2x - 14 \qquad [3]$$

Substituting [3] into [2] gives

$$2x^2 - 3(x^2 + 2x - 14) = 47$$

$$\therefore \quad 2x^2 - 3x^2 - 6x + 42 = 47$$

$$\therefore \quad -x^2 - 6x - 5 = 0$$

$$\therefore \quad x^2 + 6x + 5 = 0$$

Factorising and solving give

$$(x+5)(x+1)=0$$

$$\therefore \quad x=-5 \quad \text{or} \quad x=-1$$

Substituting $x=-5$ into [3] gives

$$y=(-5)^2+2(-5)-14$$
$$=25-10-14$$
$$\therefore \quad y=1$$

Substituting $x=-1$ into [3] gives

$$y=(-1)^2+2(-1)-14$$
$$=1-2-14$$
$$\therefore \quad y=-15$$

The solutions are $x=-5$, $y=1$ and $x=-1$, $y=-15$.

Example 47 Solve the simultaneous equations $2x^2-xy+y^2=32$ and $y=-\dfrac{5}{x}$.

SOLUTION

Substituting $y=-\dfrac{5}{x}$ into $2x^2-xy+y^2=32$, gives

$$2x^2-x\left(-\frac{5}{x}\right)+\left(-\frac{5}{x}\right)^2=32$$

$$\therefore \quad 2x^2+5+\frac{25}{x^2}=32$$

$$\therefore \quad 2x^2+\frac{25}{x^2}-27=0$$

Multiplying throughout by x^2 gives

$$x^2(2x^2)+x^2\left(\frac{25}{x^2}\right)-27x^2=0$$

$$\therefore \quad 2x^4+25-27x^2=0$$

$$\therefore \quad (2x^2-25)(x^2-1)=0$$

$$\therefore \quad x^2=\frac{25}{2} \quad \text{or} \quad x^2=1$$

$$\therefore \quad x=\pm\sqrt{\frac{25}{2}} \quad \text{or} \quad x=\pm 1$$

$$x=\pm\frac{5}{\sqrt{2}} \quad \text{or} \quad x=\pm 1$$

Substituting each x value into $y = -\frac{5}{x}$ gives the corresponding y value.

When $x = \frac{5}{\sqrt{2}}$: $\quad y = -\frac{5}{\left(\frac{5}{\sqrt{2}}\right)} = -\sqrt{2}$

When $x = -\frac{5}{\sqrt{2}}$: $\quad y = -\frac{5}{\left(-\frac{5}{\sqrt{2}}\right)} = \sqrt{2}$

When $x = 1$: $\quad y = -\frac{5}{1} = -5$

When $x = -1$: $\quad y = -\frac{5}{(-1)} = 5$

The solutions are $x = \frac{5}{\sqrt{2}}$, $y = -\sqrt{2}$; $x = -\frac{5}{\sqrt{2}}$, $y = \sqrt{2}$; $x = 1$, $y = -5$; $x = -1$, $y = 5$.

Example 48 Shown on the right is the plan of a room ABCDEF.

The length of BC is half the length of AB, and the length of EF is half the length of AF. Given that the room has an area of $113\,\text{m}^2$ and a perimeter of 48 m, find the length of CD.

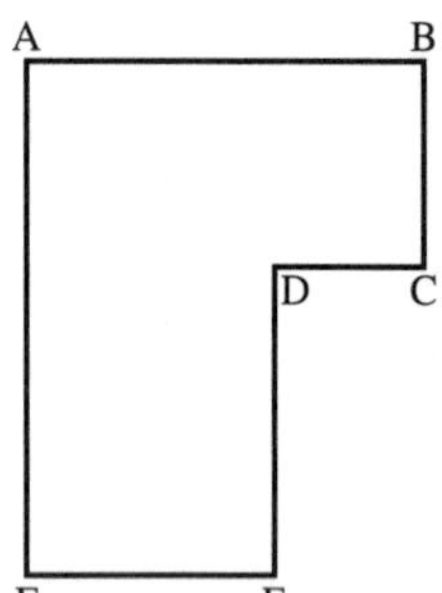

SOLUTION

Let AB $= 2x$ and AF $= 2y$ then BC $= x$ and EF $= y$. This gives DE $= (2y - x)$ and CD $= (2x - y)$.

The area A of the room is given by

$$A = 2x^2 + y(2y - x)$$

$$\therefore \quad A = 2x^2 + 2y^2 - xy$$

The perimeter P of the room is given by

$$P = x + 2x + 2y + y + (2y - x) + (2x - y)$$

$$\therefore \quad P = 4x + 4y$$

We know that $A = 113$ and $P = 48$. Therefore,

$$2x^2 + 2y^2 - xy = 113 \qquad [1]$$

and

$$4x + 4y = 48 \qquad [2]$$

From [2] we have $x = 12 - y$. Substituting this into [1] gives

$$2(12 - y)^2 + 2y^2 - y(12 - y) = 113$$
$$2(144 - 24y + y^2) + 2y^2 - 12y + y^2 = 113$$
$$\therefore \quad 5y^2 - 60y + 175 = 0$$
$$\therefore \quad y^2 - 12y + 35 = 0$$
$$\therefore \quad (y - 7)(y - 5) = 0$$
$$\therefore \quad y - 7 = 0 \quad \text{or} \quad y - 5 = 0$$
$$\therefore \quad y = 7 \quad \text{or} \quad y = 5$$

When $y = 7$, then, since $x = 12 - y$, we have $x = 12 - 7 = 5$.

When $y = 5$, then $x = 12 - 5 = 7$.

There are two possible answers for the length CD.

When $y = 7$ and $x = 5$, we have $\text{CD} = 2x - y$
$$= 2(5) - 7 = 3$$

When $y = 5$ and $x = 7$, we have $\text{CD} = 2x - y$
$$= 2(7) - 5 = 9$$

Therefore, the length of CD is 3 m or 9 m.

Exercise 1J

1 Solve each of the following pairs of simultaneous equations.

a) $y = 3x - 4 \quad y = x^2 - 4x + 6$ **b)** $y = 4x + 1 \quad y = 2x^2 - 3x + 4$

c) $y = 3x + 4 \quad y = 3x^2 - 8x$ **d)** $y = 4 - 11x \quad y = 2x^2 + 19$

e) $y = x - 3 \quad y^2 + xy + 4x = 7$ **f)** $y + 2x = 3 \quad y^2 + xy = 13 - 16x$

g) $x + y = 5 \quad xy = 6$ **h)** $xy = 2 \quad 3y - 2x = 11$

i) $x^2 + y^2 = 5 \quad y = \dfrac{2}{x}$ **j)** $xy = 8 \quad x^2 + y^2 - 3xy + 4 = 0$

k) $x^2 - 3x - y = 2 \quad x^2 + 3y + x = 6$

l) $x^2 - x - y - 2 = 0 \quad x^2 + 2x - 4y + 1 = 0$

2 In order to make a new type of beer, a brewer decides to mix x kg of malt with y kg of hops, in such a way that x and y satisfy the following equations:

$$x + y = 8 \qquad x - y = \frac{24}{x}$$

Find the pairs of values of x and y which satisfy these equations. Which of these answers can the brewer use in practice?

3 A rectangle, which is x cm long by y cm wide, has an area of 48 cm^2 and a perimeter of 32 cm. Derive the equations

$$xy = 48 \qquad [1]$$

$$x + y = 16 \qquad [2]$$

Hence deduce the dimensions of the rectangle.

4 A right-angled triangle has sides of length x cm, y cm and $(y - 2)$ cm, as shown in the diagram. Given that the perimeter of the triangle is 60 cm and its area is 120 cm^2, derive the equations

$$x + 2y = 62 \qquad [1]$$

$$xy - 2x = 240 \qquad [2]$$

Find the pairs of values of x and y which satisfy these equations. Which of the answers works in practice?

5 A box is in the shape of a cuboid, x cm wide by y cm long by 5 cm high. It has a volume of 40 cm^3 and an external surface area of 76 cm^2. Derive the equations

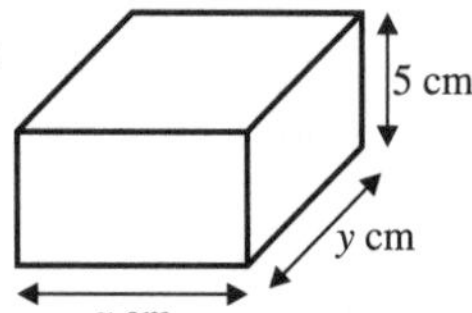

$$xy = 8 \qquad [1]$$

$$xy + 5x + 5y = 38 \qquad [2]$$

Find the pairs of values of x and y which satisfy these equations.

6 **a)** Find the maximum value of $y = 3 + 2x - x^2$, and the coordinates of the points where the curve $y = 3 + 2x - x^2$ cuts the coordinate axes.

b) Find the minimum value of $y = x^2 - 2x + 3$ and the coordinates of the point where $y = x^2 - 2x + 3$ cuts the y-axis.

c) Find the coordinates of the points of intersection of the curves $y = 3 + 2x - x^2$ and $y = x^2 - 2x + 3$.

d) Illustrate, on a sketch, each of the curves $y = 3 + 2x - x^2$ and $y = x^2 - 2x + 3$, labelling all points found in parts **a**, **b** and **c**.

7 **a)** Find the maximum value of $y = 5 + 4x - x^2$, and the coordinates at the point where the curve $y = 5 + 4x - x^2$ cuts the coordinate axes.

b) Find the minimum value of $y = x^2 - 2x + 5$ and the coordinates of the point where the curve $y = x^2 - 2x + 5$ cuts the y-axis.

c) Find the coordinates at the points of intersection of the curves $y = 5 + 4x - x^2$ and $y = x^2 - 2x + 5$.

d) Illustrate, on a sketch, each of the curves $y = 5 + 4x - x^2$ and $y = x^2 - 2x + 5$, labelling all the points found in parts **a**, **b** and **c**.

Variation

When a uniform elastic string is pulled by a force, the extension in the string is proportional to the force. In other words, the greater the force, the greater the extension. This is an example of **direct variation**. The extension in the string varies directly as the force. We use the symbol $\propto$ for 'varies as'. So, in this example, we could write $E \propto F$, where E is the extension and F the force. This means that the ratio $\frac{E}{F}$ is constant.

If $\frac{E}{F}$ is constant, we can write

$$\frac{E}{F} = k$$

where k is a constant. Therefore,

$$E = kF$$

Example 49 Two variables, A and B, are known to be directly proportional to each other. Given that $A = 8$ when $B = 12$, find A when $B = 21$.

SOLUTION

We know that $A \propto B$. That is, $A = kB$, where k is a constant.

We also know that $A = 8$ when $B = 12$. Therefore,

$$8 = k(12)$$

$$\therefore \quad k = \frac{8}{12} = \frac{2}{3}$$

The formula connecting A and B is $A = \frac{2}{3}B$. When $B = 21$, we have

$$A = \frac{2}{3}(21) = 14$$

Therefore, A is 14.

One quantity may vary directly as some power of another. For example, the area of a circle is given by $A = \pi r^2$. From this we see that $\frac{A}{r^2} = \pi$, a constant, and A is said to vary directly as the square of the radius.

Example 50 The resistance, R newtons, to the motion of a car varies directly as the square of the speed, $v\,\text{km h}^{-1}$. The resistance to motion of a car travelling at $40\,\text{km h}^{-1}$ is 3200 N. Find the resistance to motion of a car travelling at $55\,\text{km h}^{-1}$.

SOLUTION

We know that $R \propto v^2$. That is, $R = kv^2$, where k is a constant.

We also know that $R = 3200$ when $v = 40$. Therefore,

$$3200 = k(40)^2$$

$$\therefore \quad k = \frac{3200}{1600} = 2$$

The formula connecting R and v is $R = 2v^2$. When $v = 55$, we have

$$R = 2(55)^2 = 6050$$

Therefore, the resistance is 6050 N.

Example 51 Given that p varies directly as the square root of q and $p = 12$ when $q = 9$, find q when $p = 40$.

SOLUTION

We know that $p \propto \sqrt{q}$. That is, $p = k\sqrt{q}$, where k is a constant.

We also know that $p = 12$ when $q = 9$. Therefore,

$$12 = k\sqrt{9}$$

$$\therefore \quad k = \frac{12}{3} = 4$$

The formula connecting p and q is $p = 4\sqrt{q}$. When $p = 40$, we have

$$40 = 4\sqrt{q}$$

$$\therefore \quad \sqrt{q} = 10$$

$$\therefore \quad q = 10^2 = 100$$

Notice that when two variables are directly proportional to each other, then an increase in one of the variables results in an increase in the other. Also, of course, a decrease in one of the variables results in a decrease in the other. There are situations in which the increase in one variable results in a decrease in a connected variable: for example, the time to complete a job and the number of workers. In this situation, an increase in the number of workers will have the effect of reducing the time to complete the job. This is an example of **inverse variation**. We say that the time varies inversely as the number of workers.

When x varies inversely as y, then x varies directly as $\frac{1}{y}$, and we have $x = k \times \frac{1}{y}$, where k is a constant.

Example 52 If A varies inversely as B, and $A = 24$ when $B = \frac{1}{2}$, find A when $B = 6$.

SOLUTION

We know that $A \propto \frac{1}{B}$. That is, $A = k\left(\frac{1}{B}\right)$, where k is a constant.

We also know that $A = 24$ when $B = \frac{1}{2}$. Therefore,

$$24 = k\left(\frac{1}{\frac{1}{2}}\right)$$

$$\therefore \quad k = 24\left(\frac{1}{2}\right) = 12$$

The formula connecting A and B is $A = \frac{12}{B}$. When $B = 6$, we have

$$A = \frac{12}{6} = 2$$

One quantity may vary inversely as some power of another. The next example illustrates such a case.

Example 53 The electrical resistance of a wire varies inversely as the square of its radius. Given that the resistance is 0.6 ohm when the radius is 0.3 cm, find the resistance when the radius is 0.56 cm.

SOLUTION

We know that $R \propto \frac{1}{r^2}$; that is $R = k\left(\frac{1}{r^2}\right)$, where k is a constant.

We also know that $R = 0.6$ when $r = 0.3$. Therefore,

$$0.6 = k\left(\frac{1}{0.3^2}\right)$$

$$\therefore \quad k = 0.6 \times 0.09 = 0.054$$

The formula connecting R and r is $R = \frac{0.054}{r^2}$. When $r = 0.56$, we have

$$R = \frac{0.054}{0.56^2} \approx 0.17 \text{ ohm}$$

Exercise 1K

1 Given that y is proportional to x, and that $y = 12$ when $x = 3$, find the value of y when $x = 5$.

2 Given that p is proportional to q, and that $p = 2$ when $q = 4$, find the value of p when $q = 7$.

3 At low speeds, the air resistance experienced by a moving body is proportional to its speed. When moving at $2\,\text{m}\,\text{s}^{-1}$, a body experiences an air resistance of 5 N. Calculate the air resistance when the body is moving at $6\,\text{m}\,\text{s}^{-1}$.

4 The cost of photocopying a document is proportional to the number of pages copied. Given that it costs £4.50 to photocopy a document of 80 pages, calculate the cost of photocopying a document of 32 pages.

5 Two variables, A and B, are known to be proportional to each other. Given that $A = 35$ when $B = 5$, find a formula connecting A and B. Hence find the value of B when $A = 20$.

6 Hooke's Law states that the extension, x cm, of an elastic string is proportional to the force, F N, applied. Given that a string extends by 4 cm when a force of 3 N is applied, find a formula connecting F and x. Hence find the extension of the string when a force of $\frac{1}{2}$ N is applied.

7 Given that y is proportional to the square of x, and that $y = 20$ when $x = 2$, find the value of y when $x = 3$.

8 If w varies as the square of x, and $w = 15$ when $x = 5$, find the value of x when $w = 6$.

9 The air resistance, R newtons, experienced by a body falling under gravity is proportional to the square of the speed, $v\,\text{km}\,\text{h}^{-1}$. Given that $R = 50$ when $v = 6$, find the value of R when $v = 9$.

10 Given that A varies as the square root of B, and that $A = 10$ when $B = 25$, find a formula connecting A and B. Hence find the value of A when $B = 16$.

11 When a ball is dropped from a cliff the time, t seconds, it takes to reach the water below is proportional to the square root of the height, h metres, of the cliff.

a) Given it takes 2 s for a ball to drop from a cliff which is 16 m high, find a formula connecting h and t.

b) Hence find the height of a cliff from which it takes a ball 4.5 s to reach the water below.

12 Given that y is inversely proportional to x, and that $y = 3$ when $x = 6$, find the value of y when $x = 5$.

13 If C varies inversely as D, and $C = 10$ when $D = 2$, find the value of C when $D = 4$.

14 The volume, $V\,\text{m}^3$, of a given mass of gas varies inversely as the pressure, $P\,\text{N}\,\text{m}^{-2}$. Given that $P = 1000$ when $V = 3$, find the value of P when $V = 4$.

15 The magnetic force, F N, between two bodies is inversely proportional to the square of the distance, x cm between their centres. Given that $F = 30$ when $x = \frac{1}{2}$, find the value of F when $x = 2$.

16 P is inversely proportional to the square root of Q. Given that $P = 40$ when $Q = 25$, find a formula connecting P and Q. Hence find the value of Q when $P = 20$.

***17** Given that A is inversely proportional to B, that C is proportional to the square root of B, and that D is inversely proportional to the square of C, prove that A is directly proportional to D.

Exercise 1L: Examination questions

1 **i)** Write $x^2 + 6x + 16$ in the form $(x + a)^2 + b$, where a and b are integers to be found.

ii) Find the minimum value of $x^2 + 6x + 16$ and state the value of x for which this minimum occurs.

iii) Write down the maximum value of the function $\dfrac{1}{x^2 + 6x + 16}$. (MEI)

2 Show that $2x^2 - 36x + 175$ may be written in the form $a(x - b)^2 + c$, where the values of a, b and c are to be found.

State, with a reason, the least value of $2x^2 - 36x + 175$. (WJEC)

3 An enclosure PQRS is to be made as shown in the figure. PQ and QR are fences of total length 300 m. The other two sides are hedges. The angles at Q and R are right angles and the angle at S is 135°. The length of QR is x metres.

P Q x 135° S R

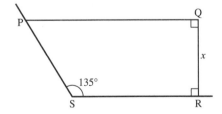

a) Show that the area, $A\,\text{m}^2$, of the enclosure is given by

$$A = 300x - \frac{3x^2}{2}$$

b) Show that A can be written as $-\frac{3}{2}[(x - a)^2 - b]$, where a and b are constants whose values you should determine. Hence show that A cannot exceed 15 000. (AEB Spec)

4 Given that, for all values of x,

$$3x^2 + 12x + 5 \equiv p(x+q)^2 + r$$

a) Find the values of p, q and r.
b) Hence, or otherwise, find the minimum value of $3x^2 + 12x + 5$.
c) Solve the equation $3x^2 + 12x + 5 = 0$, giving your answers to one decimal place. (EDEXCEL)

5 Find the set of values of x for which $2(x^2 - 5) < x^2 + 6$. (EDEXCEL)

6 Find the set of values of x for which $x^2 - x - 12 > 0$. (NEAB)

7 Find the set of values of x for which $2x(x+3) > (x+2)(x-3)$. (EDEXCEL)

8 Find the set of values of x for which $\dfrac{x}{x+4} > 2$. (EDEXCEL)

9 Find all the values of x for which $\dfrac{3x^2 - 1}{x^2 + 1} > 1$. (UODLE)

10 A rectangular tile has length x cm and breadth $(6 - x)$ cm. Given that the area of the tile must be at least $5\,\text{cm}^2$, form a quadratic inequality in x and hence find the set of possible values of x. (EDEXCEL)

11 A landscape gardener is given the following instructions about laying a rectangular lawn. The length x m is to be 2 m longer than the width. The width must be greater than 6.4 m and the area is to be less than $63\,\text{m}^2$.

By forming an inequality in x, find the set of possible values of x. (EDEXCEL)

12 Find the range of values of k for which the quadratic equation

$$(3+k)x^2 + 4x + k = 0$$

has real distinct roots. (WJEC)

13 Find the range of values of k for which the quadratic equation

$$x^2 + (k-4)x + (k-1) = 0$$

has real distinct roots. (WJEC)

14 The quadratic equation $x^2 + 6x + 1 = k(x^2 + 1)$ has equal roots. Find the possible values of the constant k. (AEB 94)

15 Solve the simultaneous equations $y = x - 2$, $y^2 = x$. (UCLES)

16 Show that the elimination of x from the simultaneous equations

$$x - 2y = 1$$

$$3xy - y^2 = 8$$

produces the equation

$$5y^2 + 3y - 8 = 0$$

Solve this quadratic equation and hence find the pairs (x, y) for which the simultaneous equations are satisfied. (EDEXCEL)

17 Solve the simultaneous equations $x + y = 2$, $x^2 + 2y^2 = 11$. (UCLES)

2 Geometry

Expression and shape are almost more to me than knowledge itself. My work has always tried to unite the true with the beautiful, and when I have had to choose one or the other, I usually chose the beautiful.
H. WEYL

Pythagoras' theorem

In any right-angled triangle ABC, the square of the hypotenuse is equal to the sum of the squares of each of the other two sides.

$$AB^2 = BC^2 + AC^2$$

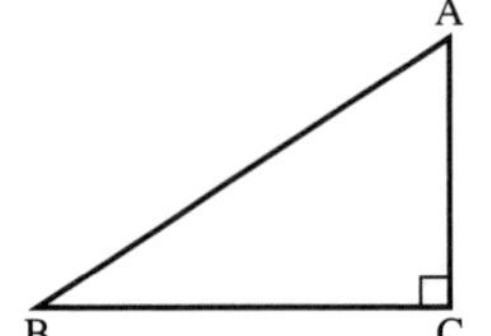

Trigonometric ratios

In the right-angled triangle ABC:

- side AB is called the **hypotenuse**
- side BC is called the **adjacent**, since it is adjacent to (next to) the angle ABC (θ)
- side AC is called the **opposite** (relative to the angle ABC), since it is opposite the angle ABC.

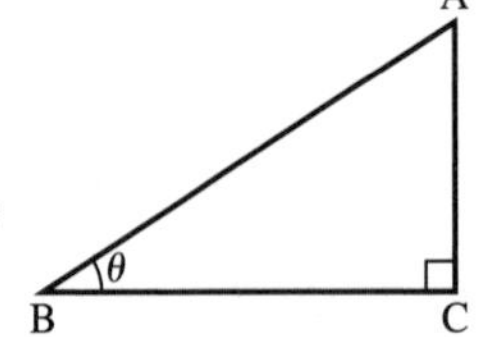

The trigonometric ratios are

$$\sin\theta = \frac{\text{Opposite}}{\text{Hypotenuse}} \qquad \cos\theta = \frac{\text{Adjacent}}{\text{Hypotenuse}} \qquad \tan\theta = \frac{\text{Opposite}}{\text{Adjacent}}$$

Example 1 In the triangle ABC, BC = 6 cm, angle ABC = 60°, and angle ACB = 90°.
Calculate **a)** the length AB, **b)** the length AC (to one decimal place).

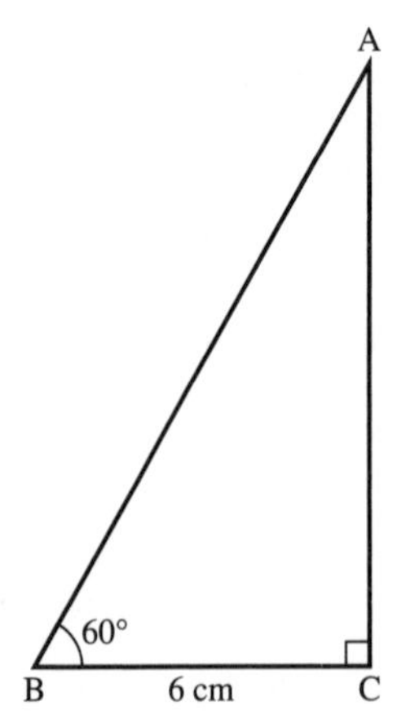

SOLUTION

a) Since the side AB is the hypotenuse and BC is adjacent to the 60° angle, we use the cosine ratio. That is,

$$\cos\theta = \frac{\text{Adjacent}}{\text{Hypotenuse}}$$

Therefore, $\cos 60° = \dfrac{6}{AB}$

$$\therefore \quad 0.5 = \frac{6}{AB}$$

$$\therefore \quad AB = \frac{6}{0.5} = 12$$

The length AB is 12 cm.

b) To find the length AC, we could use trigonometry again but using Pythagoras' theorem is easier.

By Pythagoras, we have

$$AB^2 = BC^2 + AC^2$$
$$\therefore \quad 12^2 = 6^2 + AC^2$$
$$\therefore \quad AC^2 = 144 - 36$$
$$\therefore \quad AC = \sqrt{108} = 10.4$$

The length AC is 10.4 cm.

Trigonometric ratios of 45°

Consider the right-angled, isosceles triangle ABC.

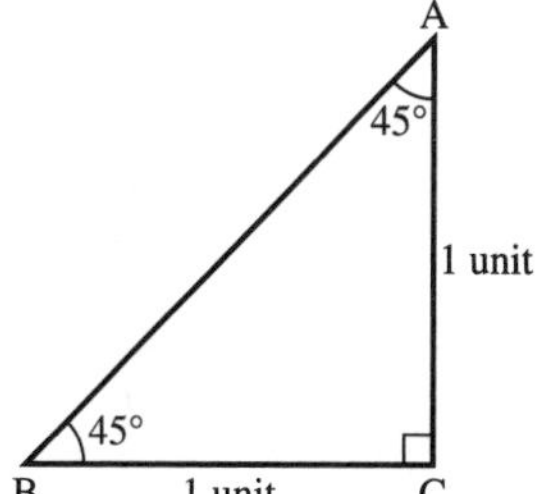

Using Pythagoras' theorem gives

$$AB^2 = BC^2 + AC^2$$
$$= 1^2 + 1^2$$
$$\therefore \quad AB = \sqrt{2}$$

Using the sine ratio gives

$$\sin 45° = \frac{AC}{AB} = \frac{1}{\sqrt{2}}$$

Therefore, $\sin 45° = \frac{1}{\sqrt{2}}$.

Using the cosine ratio gives

$$\cos 45° = \frac{BC}{AB} = \frac{1}{\sqrt{2}}$$

Therefore, $\cos 45° = \frac{1}{\sqrt{2}}$.

Using the tangent ratio gives

$$\tan 45° = \frac{AC}{BC} = 1$$

Therefore, $\tan 45° = 1$.

Summary:

$$\sin 45° = \frac{1}{\sqrt{2}} \qquad \cos 45° = \frac{1}{\sqrt{2}} \qquad \tan 45° = 1$$

Trigonometric ratios of 30° and 60°

Consider the equilateral triangle ABC with sides of length 2 units.

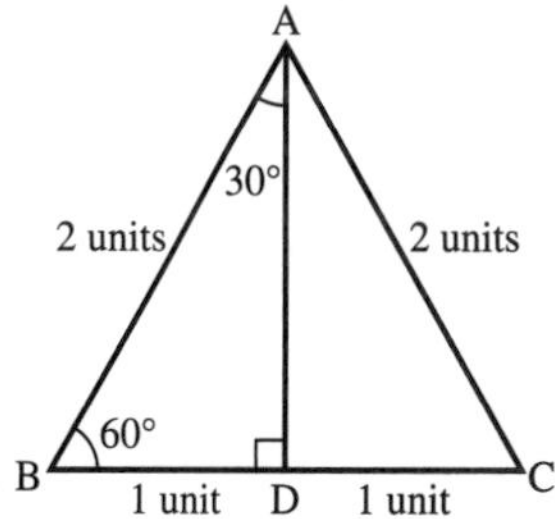

Let D be the point where the perpendicular from A meets the base BC.

From Pythagoras' theorem, we obtain the length AD:

$$AD^2 = AB^2 - BD^2$$
$$= 2^2 - 1^2$$
$$\therefore \quad AD = \sqrt{3}$$

Using the sine ratio gives

$$\sin 30^\circ = \frac{BD}{AB} = \frac{1}{2} \quad \text{and} \quad \sin 60^\circ = \frac{AD}{AB} = \frac{\sqrt{3}}{2}$$

Therefore, $\sin 30^\circ = \frac{1}{2}$ and $\sin 60^\circ = \frac{\sqrt{3}}{2}$.

Using the cosine ratio gives

$$\cos 30^\circ = \frac{AD}{AB} = \frac{\sqrt{3}}{2} \quad \text{and} \quad \cos 60^\circ = \frac{BD}{AB} = \frac{1}{2}$$

Therefore, $\cos 30^\circ = \frac{\sqrt{3}}{2}$ and $\cos 60^\circ = \frac{1}{2}$.

Using the tangent ratio gives

$$\tan 30^\circ = \frac{BD}{AD} = \frac{1}{\sqrt{3}} \quad \text{and} \quad \tan 60^\circ = \frac{AD}{BD} = \sqrt{3}$$

Therefore, $\tan 30^\circ = \frac{1}{\sqrt{3}}$ and $\tan 60^\circ = \sqrt{3}$.

The trigonometric ratios of 30°, 45° and 60° are summarised below. These ratios should be memorised, as many questions involve these angles.

Ratio	$\boldsymbol{\theta = 30^\circ}$	$\boldsymbol{\theta = 45^\circ}$	$\boldsymbol{\theta = 60^\circ}$
$\sin\theta$	$\frac{1}{2}$	$\frac{1}{\sqrt{2}}$	$\frac{\sqrt{3}}{2}$
$\cos\theta$	$\frac{\sqrt{3}}{2}$	$\frac{1}{\sqrt{2}}$	$\frac{1}{2}$
$\tan\theta$	$\frac{1}{\sqrt{3}}$	1	$\sqrt{3}$

Example 2 Find the perimeter of triangle ABC shown on the right, expressing your answer in the form $a + b\sqrt{c}$, where a, b and c are integers.

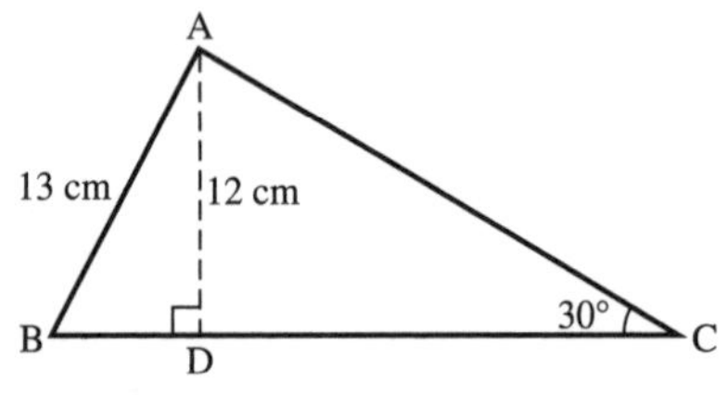

SOLUTION

The length BD can be found by applying Pythagoras' theorem to the right-angled triangle ABD:

$$AB^2 = AD^2 + BD^2$$
$$\therefore \quad 13^2 = 12^2 + BD^2$$
$$\therefore \quad BD^2 = 25$$
$$\therefore \quad BD = 5$$

The length CD is found by applying the tangent ratio to triangle ACD:

$$\tan A\hat{C}D = \frac{AD}{CD}$$

$$\therefore \quad \tan 30° = \frac{12}{CD}$$

$$\therefore \quad CD = \frac{12}{\left(\frac{1}{\sqrt{3}}\right)} = 12\sqrt{3}$$

Therefore, BC $= (5 + 12\sqrt{3})$ cm.

The length AC is found by applying the sine ratio to triangle ACD:

$$\sin A\hat{C}D = \frac{AD}{AC}$$

$$\therefore \quad \sin 30° = \frac{12}{AC}$$

$$\therefore \quad AC = \frac{12}{\left(\frac{1}{2}\right)} = 24$$

The length AC is 24 cm.

The perimeter, P, of triangle ABC is given by

$$P = AB + BC + CA$$
$$= 13 + (5 + 12\sqrt{3}) + 24$$
$$= (42 + 12\sqrt{3})$$

The perimeter of the triangle is $(42 + 12\sqrt{3})$ cm.

Three-dimensional applications

The angle between a line and a plane

Consider a line l meeting a plane at a point P, as shown in the diagram.

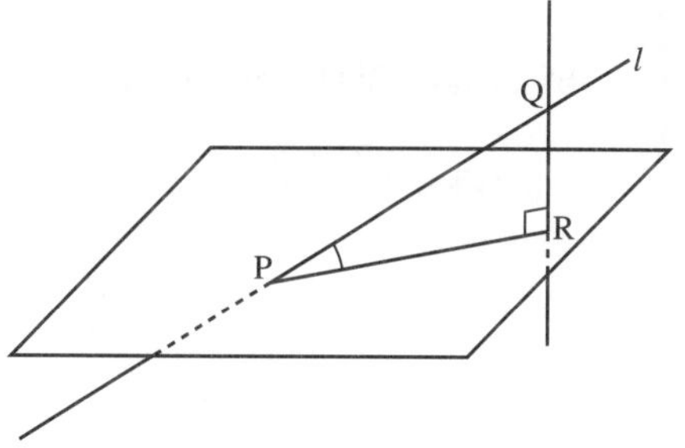

Choose another point Q on l. Let the line through Q which is perpendicular to the plane meet the plane at R.

The line PR is the projection of the line PQ onto the plane.

Angle QPR is the angle which PQ makes with the plane.

The angle between two planes

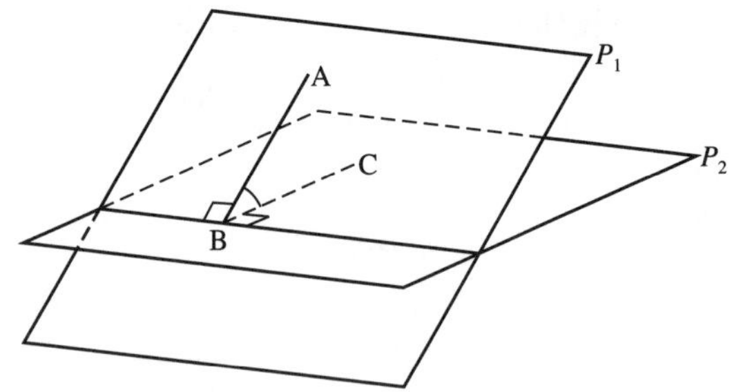

To identify the angle between two planes, choose a line AB, in the plane P_1, which is perpendicular to the line of intersection of the two planes P_1 and P_2. In the plane P_2, choose the line BC which is also perpendicular to the line of intersection of the two planes P_1 and P_2. The angle ABC is the angle between the two planes P_1 and P_2.

Example 3 Shown is a cuboid in which AB = 6 cm, BF = 5 cm and AD = 8 cm. Calculate **a)** the length of the diagonal BH, **b)** the angle that BH makes with the base.

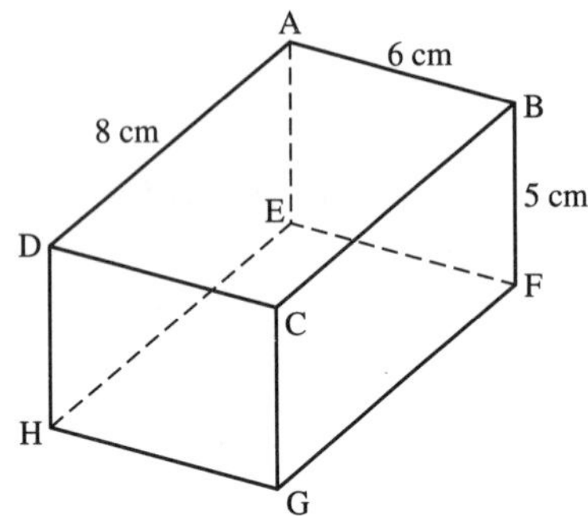

SOLUTION

a) BH is the hypotenuse of the right-angled triangle BFH. BH can be calculated once FH is found. FH is the hypotenuse of the right-angled triangle FGH.

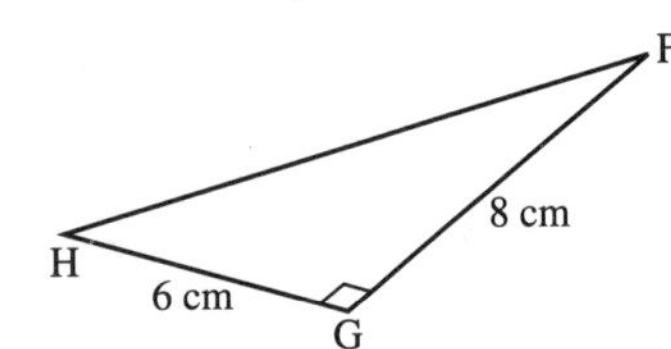

Using Pythagoras on triangle FGH gives

$$FH^2 = FG^2 + GH^2$$
$$= 8^2 + 6^2$$
$$\therefore \quad FH = \sqrt{100} = 10$$

Applying Pythagoras' theorem to triangle BFH gives

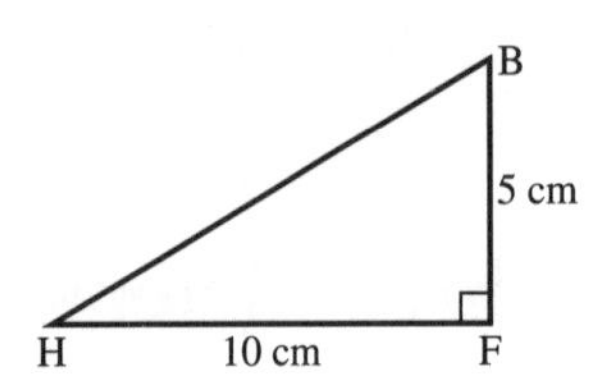

$$BH^2 = BF^2 + FH^2$$
$$= 5^2 + 10^2$$
$$\therefore \quad BH = \sqrt{125} = 11.2$$

The length of the diagonal BH is 11.2 cm.

b) Angle BHF represents the angle between BH and the base.

Applying the tangent ratio to triangle BHF gives

$$\tan B\hat{H}F = \frac{5}{10} = 0.5$$

$$\therefore \quad B\hat{H}F = 26.6^\circ$$

The angle between BH and the base is 26.6°.

Example 4 The pyramid VABCD has a square base ABCD of side 10 cm and vertex V. The vertical height of the pyramid is 15 cm. Find

a) the angle between AV and the base
b) the angle between the plane VAB and the base
c) the angle between the planes VAD and VCD.

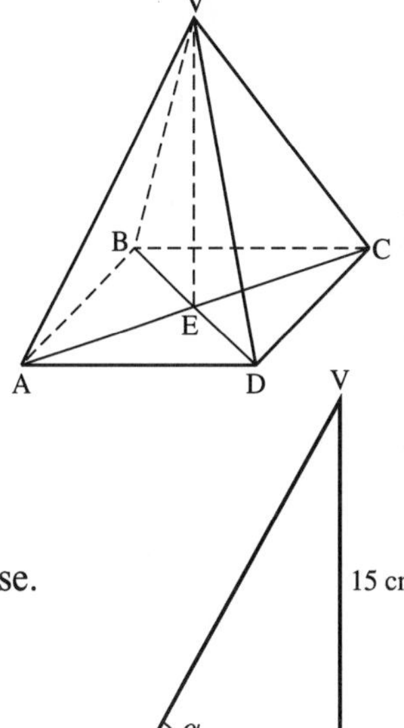

SOLUTION

a) Let the diagonals of the base AC and BD meet at E. The point E is vertically below V and VE = 15 cm.

By symmetry, all the sloping edges make equal angles with the base.

Let $\alpha = V\hat{A}E$, the required angle in triangle VAE

Since $AE = \frac{1}{2}AC$, we need to find length AC from triangle ACD. Using Pythagoras, we have

$$AC^2 = AD^2 + CD^2$$
$$= 10^2 + 10^2$$
$$\therefore \quad AC = \sqrt{200}$$
$$= \sqrt{100 \times 2} = 10\sqrt{2}$$

Therefore, $AE = \frac{1}{2}(10\sqrt{2}) = 5\sqrt{2}$.

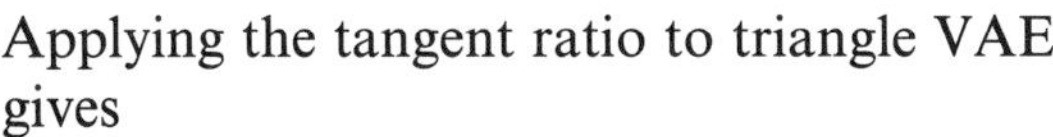
Applying the tangent ratio to triangle VAE gives

$$\tan\alpha = \frac{VE}{AE} = \frac{15}{5\sqrt{2}}$$
$$\therefore \quad \alpha = 64.8°$$

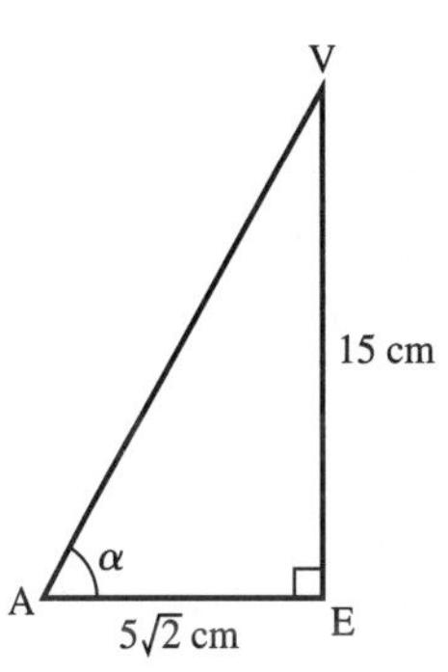

The angle between a sloping edge and the base is 64.8°.

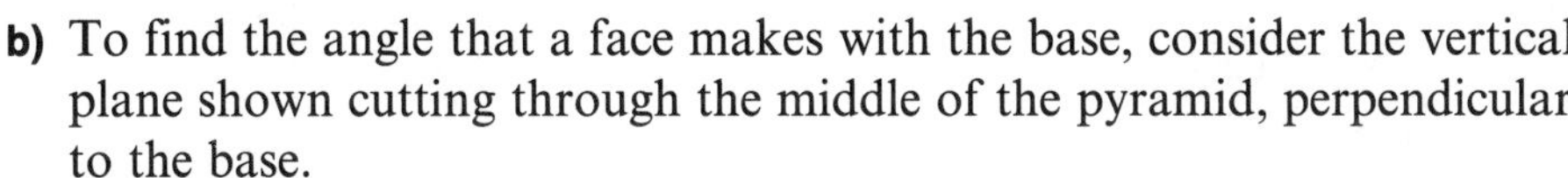
b) To find the angle that a face makes with the base, consider the vertical plane shown cutting through the middle of the pyramid, perpendicular to the base.

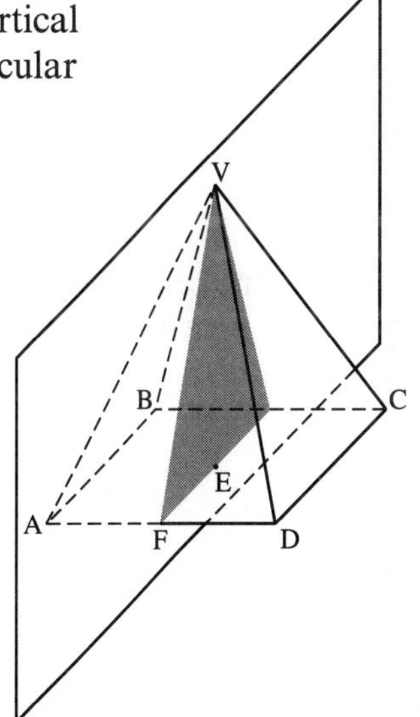

Let $\beta = \mathrm{V\widehat{F}E}$, the required angle in triangle VEF.

Since $\mathrm{FE} = \frac{1}{2}\mathrm{DC}$, $\mathrm{FE} = 5\,\mathrm{cm}$.

Using the tangent ratio gives

$$\tan\beta = \frac{\mathrm{VE}}{\mathrm{FE}} = \frac{15}{5}$$

$$\therefore \quad \beta = 71.6^\circ$$

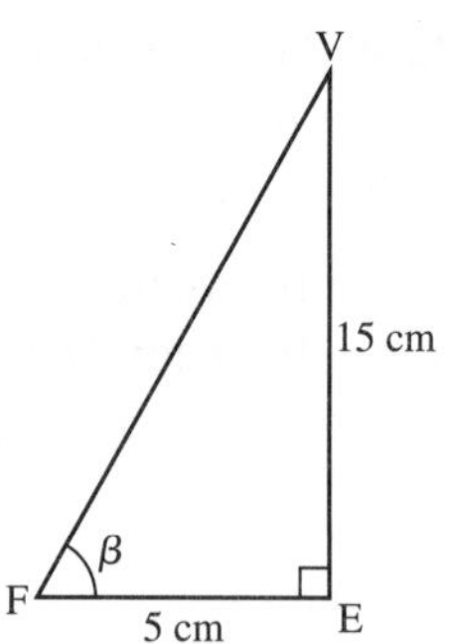

The angle between a face and the base is 71.6°.

c) The angle between two faces is represented by the angle FPG in triangle PFG.

Let $\gamma = \mathrm{F\widehat{P}G}$, the required angle.

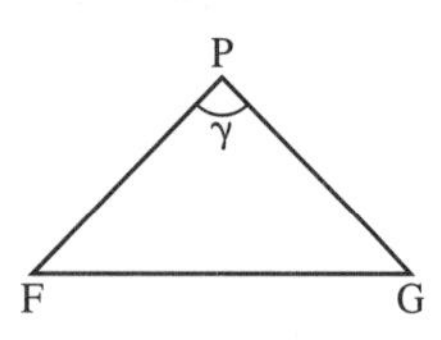

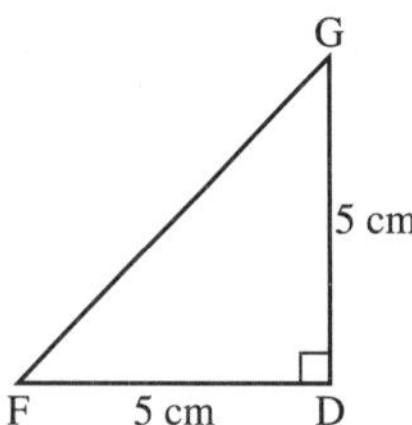

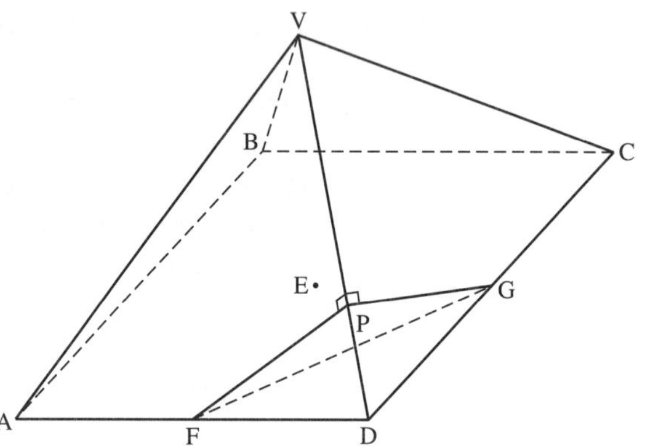

The length FG can be found from the right-angled triangle FDG. Using Pythagoras gives

$$\mathrm{FG}^2 = \mathrm{FD}^2 + \mathrm{DG}^2$$

$$= 5^2 + 5^2$$

$$\therefore \quad \mathrm{FG} = \sqrt{50} = 7.1$$

The length PF(=PG) is a perpendicular height of triangle VFD.

The length VF can be found from triangle VEF. Using Pythagoras gives

$$\mathrm{VF}^2 = \mathrm{VE}^2 + \mathrm{EF}^2$$

$$= 15^2 + 5^2$$

$$\therefore \quad \mathrm{VF} = \sqrt{250} = 15.8$$

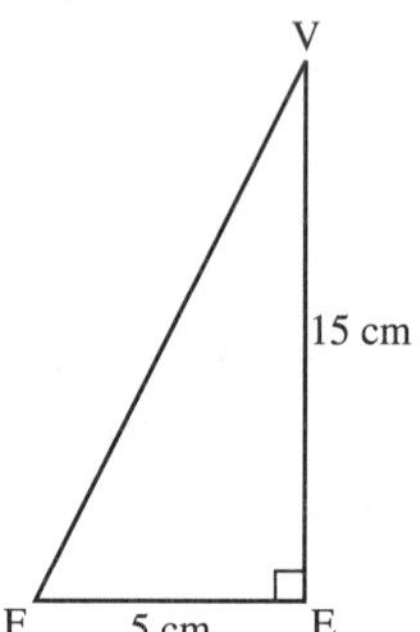

The length VD can be found by applying Pythagoras' theorem to triangle VFD.

$$\mathrm{VD}^2 = \mathrm{VF}^2 + \mathrm{FD}^2$$

$$= 250 + 5^2$$

$$\therefore \quad \mathrm{VD} = \sqrt{275} = 16.6$$

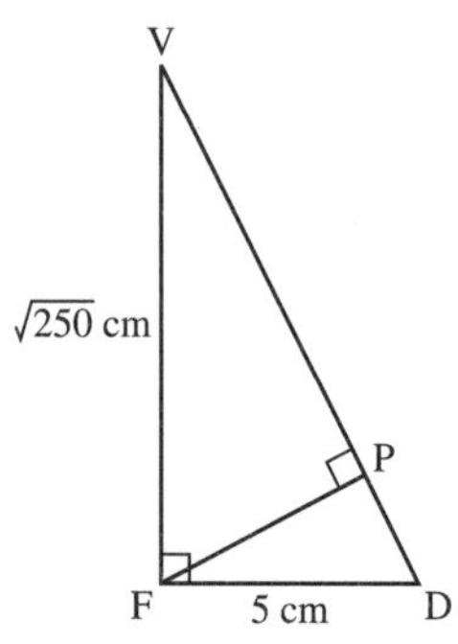

Triangles VFD and FPD are similar. Therefore,

$$\frac{\text{FP}}{\sqrt{250}} = \frac{5}{\sqrt{275}}$$

$$\therefore \quad \text{FP} = \frac{5 \times \sqrt{250}}{\sqrt{275}} = 4.767$$

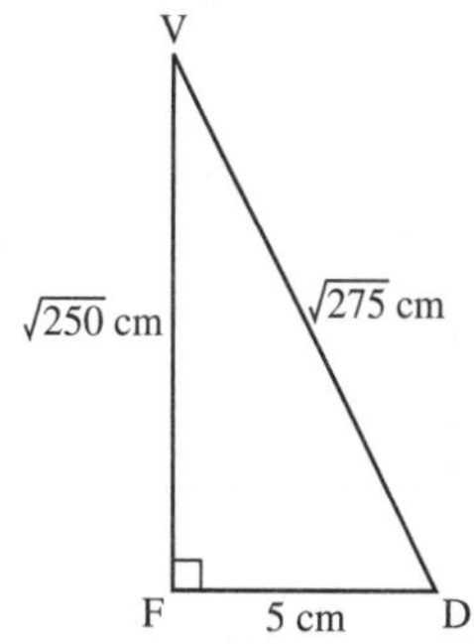

Returning to triangle PFG, let the perpendicular from P meet FG at S. Applying the sine ratio to triangle PFS gives

$$\sin\frac{\gamma}{2} = \frac{\text{FS}}{\text{FP}} = \frac{\frac{1}{2}\sqrt{50}}{4.767}$$

$$\therefore \quad \frac{\gamma}{2} = 47.87°$$

$$\therefore \quad \gamma = 95.7°$$

The angle between two faces of the pyramid is 95.7°.

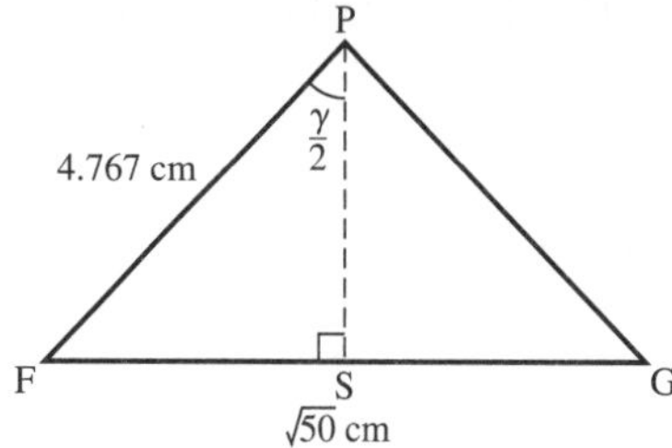

Exercise 2A

Answers should be stated **correct to three significant figures** throughout this exercise.

1 Find the perimeter of triangle ABC, expressing your answer in the form $a + b\sqrt{2} + c\sqrt{3}$, where a, b and c are integers.

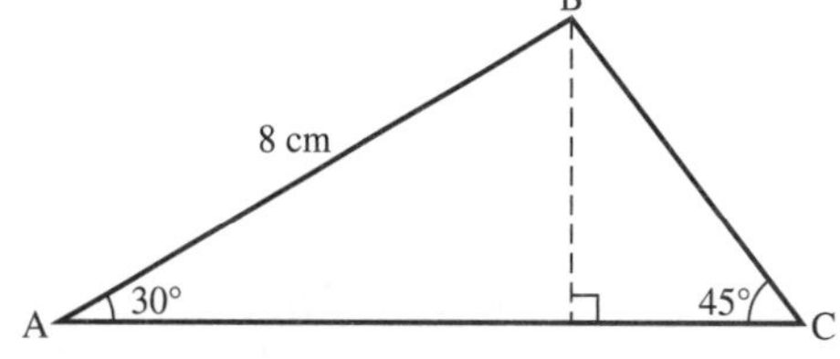

2 Find the perimeter of the quadrilateral ABCD, expressing your answer in the form $a + b\sqrt{2} + c\sqrt{6}$, where a, b and c are integers.

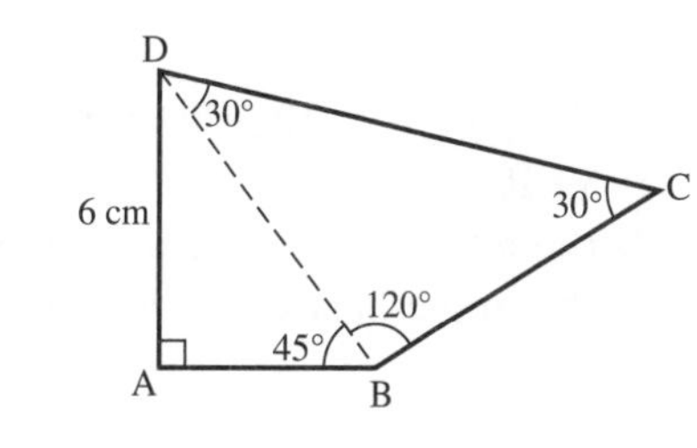

3 In the cube ABCDEFGH calculate **a)** HF, **b)** HB, **c)** $\text{B}\widehat{\text{H}}\text{F}$, **d)** $\text{H}\widehat{\text{F}}\text{A}$.

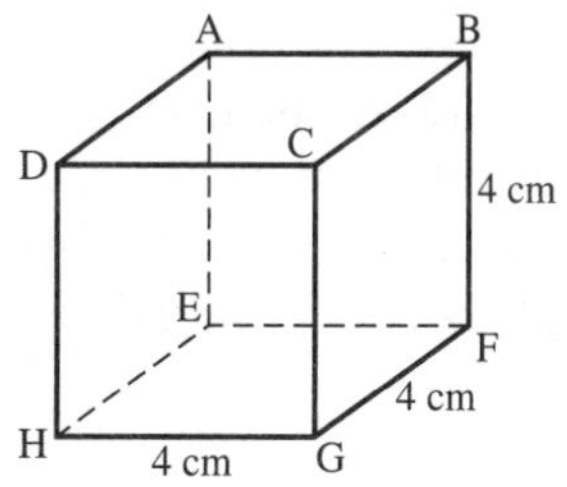

4

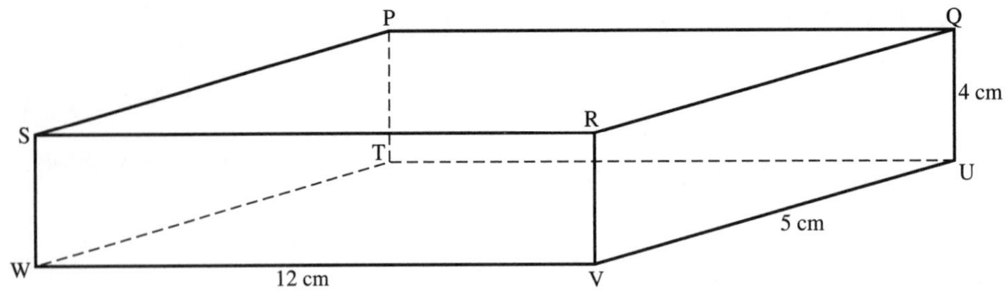

In the cuboid PQRSTUVW calculate **a)** TV, **b)** PV, **c)** $V\hat{P}T$, **d)** $P\hat{V}W$, **e)** $V\hat{Q}W$.

5 In the cuboid ABCDEFGH calculate **a)** BD, **b)** DF, **c)** $B\hat{D}F$, **d)** BG, **e)** $B\hat{G}D$.

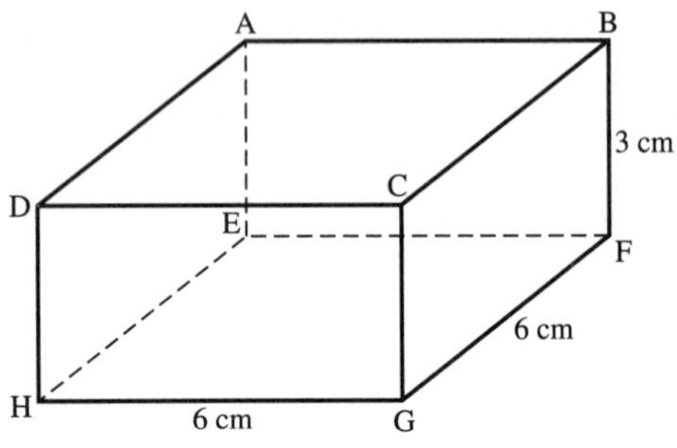

6 In the pyramid VABCD, the vertex V is at a height of 7 cm above the centre of the square base ABCD. Calculate **a)** AC, **b)** VA, **c)** $V\hat{A}C$, **d)** $V\hat{D}A$, **e)** the angle between the planes VAD and VCD.

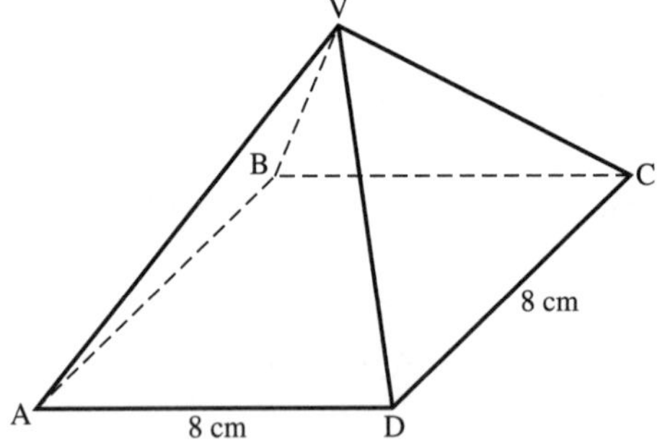

7 In the pyramid VPQRS, the vertex V is directly above the centre of the square base PQRS. Calculate

a) PR
b) the height of the pyramid
c) $V\hat{P}R$
d) the angle between the planes VPS and VRQ
e) the angle between the planes VPQ and VRQ.

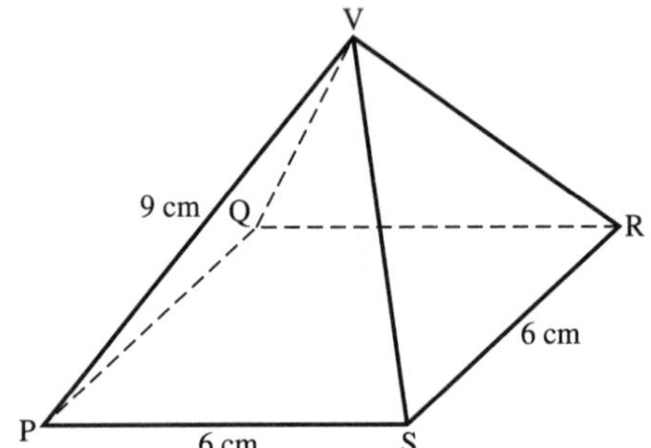

8 In the pyramid VABCD, the vertex V is at a height of 12 cm above the centre of the rectangular base ABCD. Calculate **a)** AC, **b)** VA, **c)** $V\hat{A}C$, **d)** $V\hat{D}A$, **e)** the angle between the planes VAB and VCD.

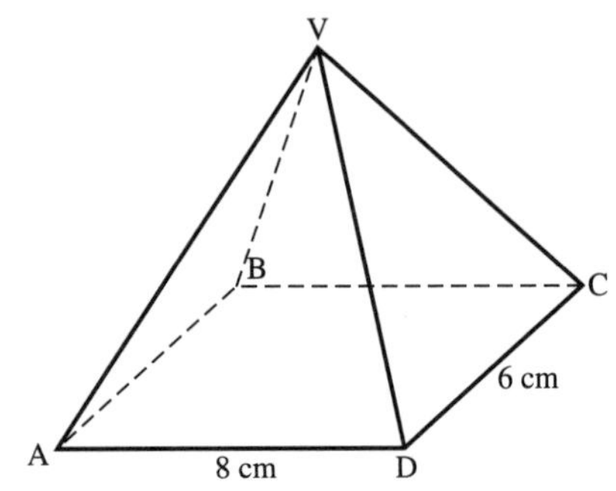

9 ABCD represents part of a uniform ski-slope which makes an angle of 26° with the horizontal rectangular plane CDPQ.

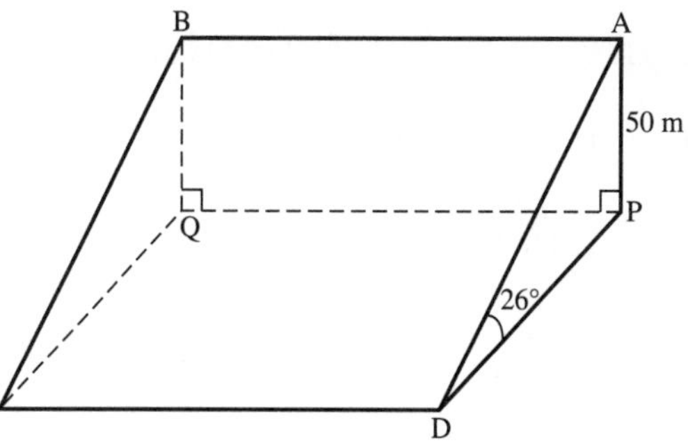

a) Calculate the distance AD.

A beginner decides that the slope AD is too steep and is only prepared to risk an angle of descent of 10°. He achieves this by skiing directly from A to C.

b) Calculate the distance AC,
c) Calculate the distance CD.

10 The diagram shows a tent with rectangular base ABCD, resting on horizontal ground. The vertex, V, of the tent is vertically above the centre of the base of the rectangle. Calculate

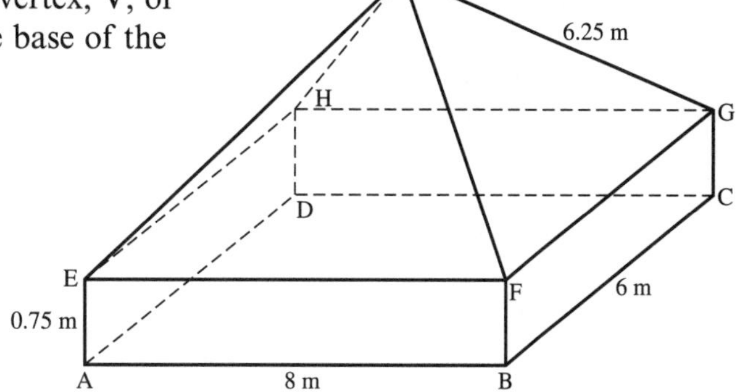

a) the length EG
b) the height of V above the ground
c) the angle $V\hat{A}C$
d) the angle between the planes VEH and VFG.

11 A hot-air balloon, B, is observed simultaneously from two points, P and Q, on horizontal ground. From P the bearing of B is 060° at an angle of elevation of 45°. From Q the bearing of B is 330° at an angle of elevation of 60°. The distance BQ is 800 m.

a) Draw a sketch showing the positions of P, Q and B.
b) Calculate the height of the balloon above the ground.
c) Calculate the bearing of Q from P.

12 A surveyor is attempting to calculate the height of a point, P, on a building by taking measurements on horizontal, level ground. From a point A he records the angle of elevation of P as 30°. He then advances 20 m to a point B, from which he records the angle of elevation of P as 45°. Calculate the height of P above the ground.

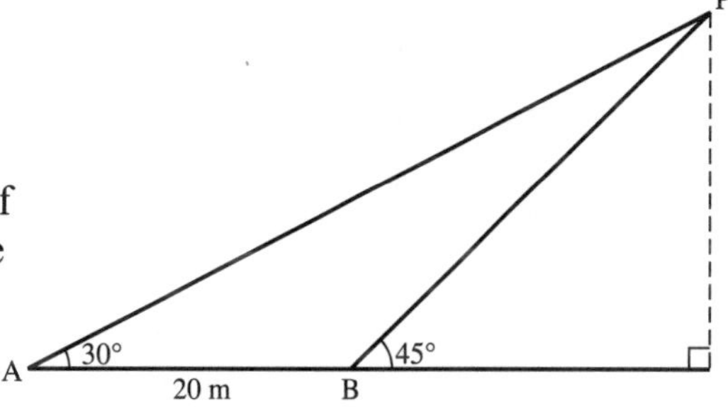

13 A man is attempting to calculate the height of a kite, K, which is flying above horizontal ground. From a point A he records the angle of elevation of K as 23°. He then advances 80 m to a point B from which he records the angle of elevation of K as 34°.
Calculate the height of the kite above the ground.

Sine and cosine rules

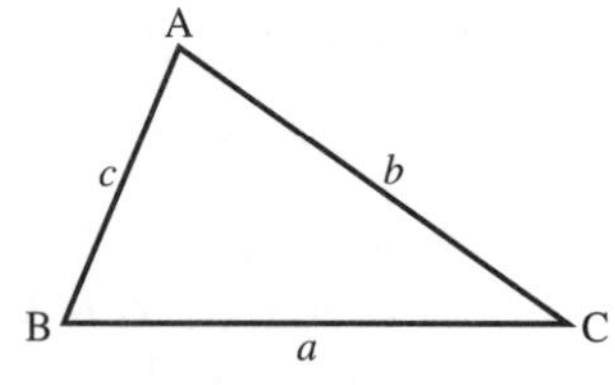

Result I: The sine rule

In any triangle ABC,

$$\frac{a}{\sin A} = \frac{b}{\sin B} = \frac{c}{\sin C}$$

Proof

Two cases have to be considered:

i) when triangle ABC is an acute-angled triangle
ii) when triangle ABC is not an acute-angled triangle.

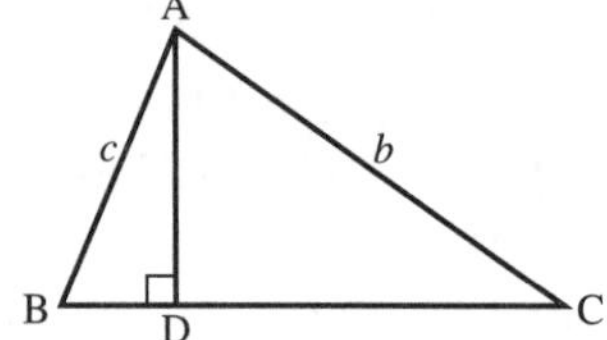

Case (i) Triangle ABC is acute-angled.
Let the perpendicular from A meet BC at D. Applying the sine ratio to triangle ABD gives

$$\sin B = \frac{\text{AD}}{c}$$

$$\therefore \quad \text{AD} = c \sin B \qquad [1]$$

Applying the sine ratio to triangle ADC gives

$$\sin C = \frac{\text{AD}}{b}$$

$$\therefore \quad \text{AD} = b \sin C \qquad [2]$$

Eliminating AD from [1] and [2] gives

$$c \sin B = b \sin C$$

$$\therefore \quad \frac{b}{\sin B} = \frac{c}{\sin C} \qquad [3]$$

Similarly, dropping a perpendicular from B to meet AC gives

$$\frac{c}{\sin C} = \frac{a}{\sin A} \qquad [4]$$

Therefore, from [3] and [4] we have

$$\frac{a}{\sin A} = \frac{b}{\sin B} = \frac{c}{\sin C}$$

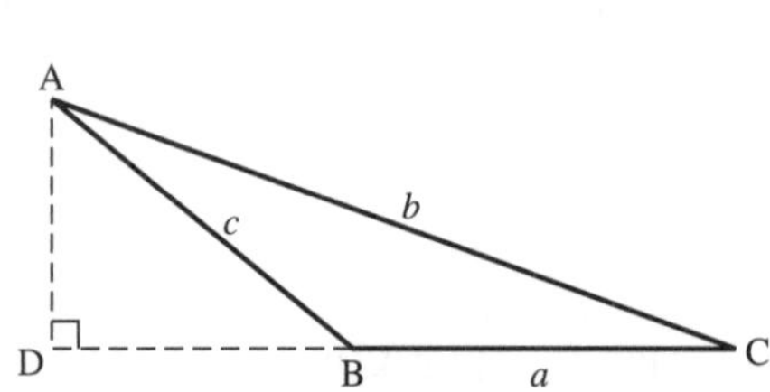

Case (ii) Triangle ABC is not acute-angled.
Let the perpendicular from A meet CB extended at D.
Applying the sine ratio to triangle ABD gives

$$\sin \text{A}\hat{\text{B}}\text{D} = \frac{\text{AD}}{c}$$

$$\therefore \quad \text{AD} = c \sin \text{A}\hat{\text{B}}\text{D}$$

Now $\text{A}\hat{\text{B}}\text{D} = (180° - B)$ and since $\sin(180° - B) = \sin B$, we have

$$\text{AD} = c \sin B \qquad [5]$$

Applying the sine ratio to triangle ACD gives

$$\sin C = \frac{\text{AD}}{b}$$

$$\therefore \quad \text{AD} = b \sin C \qquad [6]$$

Eliminating AD from [5] and [6] gives

$$c \sin B = b \sin C$$

$$\therefore \quad \frac{c}{\sin C} = \frac{b}{\sin B} \qquad [7]$$

By dropping a perpendicular from B to meet AC at E and proceeding as in case (i), we obtain

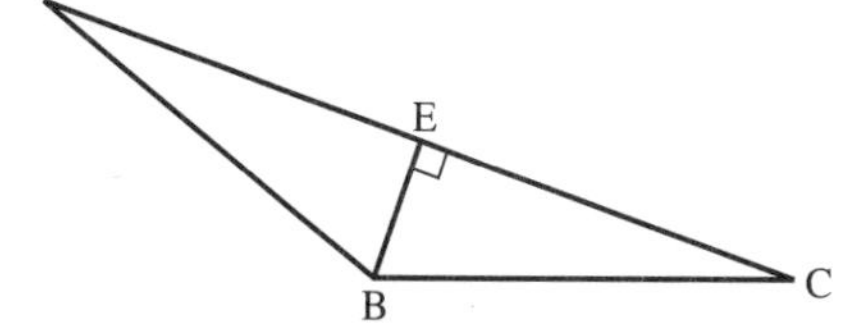

$$\frac{a}{\sin A} = \frac{c}{\sin C} \qquad [8]$$

Therefore, from [7] and [8] we have

$$\frac{a}{\sin A} = \frac{b}{\sin B} = \frac{c}{\sin C}$$

as required.

Example 5 In triangle ABC, $A = 40°$, $B = 75°$ and AB = 6 cm. Calculate **a)** the length AC, **b)** the length BC.

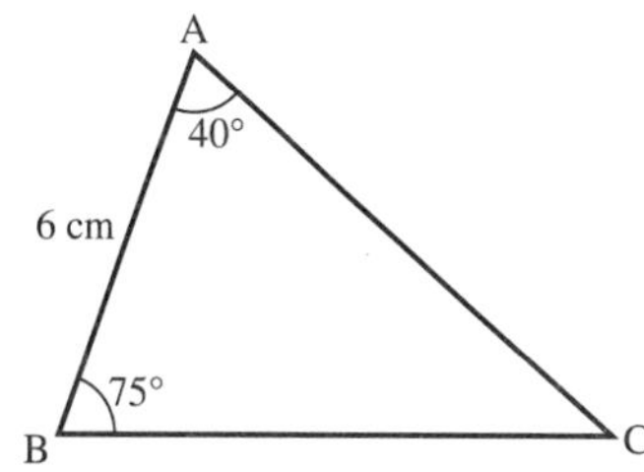

SOLUTION

a) Now $C = 180° - (40° + 75°) = 65°$.
Therefore, applying the sine rule gives

$$\frac{\text{AC}}{\sin 75°} = \frac{6}{\sin 65°}$$

$$\therefore \quad \text{AC} = \frac{6 \sin 75°}{\sin 65°} = 6.4$$

The length of AC is 6.4 cm.

b) Applying the sine rule again gives

$$\frac{\text{BC}}{\sin 40°} = \frac{6}{\sin 65°}$$

$$\therefore \quad \text{BC} = \frac{6 \sin 40°}{\sin 65°} = 4.3$$

The length of BC is 4.3 cm.

The ambiguous case

Suppose we are given triangle ABC such that AC = 7 cm, BC = 12 cm and B = 30°. Constructing this triangle with ruler and compasses gives the diagram on the right.

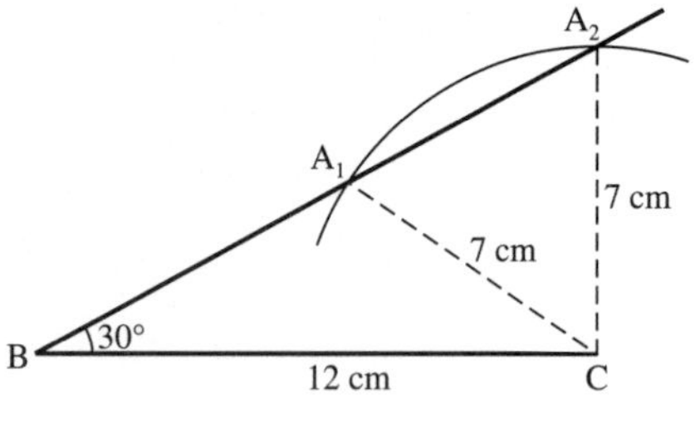

Since triangle A_1CA_2 is isosceles, $A_1\widehat{A_2}C = A_2\widehat{A_1}C$. Angles BA_1C and A_2A_1C are supplementary (i.e. their sum is 180°). In other words, there are two possible positions for vertex A, namely A_1 and A_2.

This is known as the **ambiguous case**. This situation arises when you are given two sides and a non-included angle (i.e. not the angle between the two sides).

Example 6 In triangle ABC, AB = 8 cm, BC = 10 cm and angle ACB = 42°. Calculate the length of AC.

SOLUTION

The sketch on the right shows the two possible triangles which can be drawn from this information.

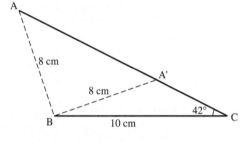

Applying the sine rule gives

$$\frac{10}{\sin A} = \frac{8}{\sin 42°}$$

$$\therefore \quad \sin A = \frac{10 \sin 42°}{8}$$

$$\therefore \quad A = 56.8°$$

Therefore, $B\widehat{A'}C = 180° - 56.8° = 123.2°$.

The two possible cases are shown below.

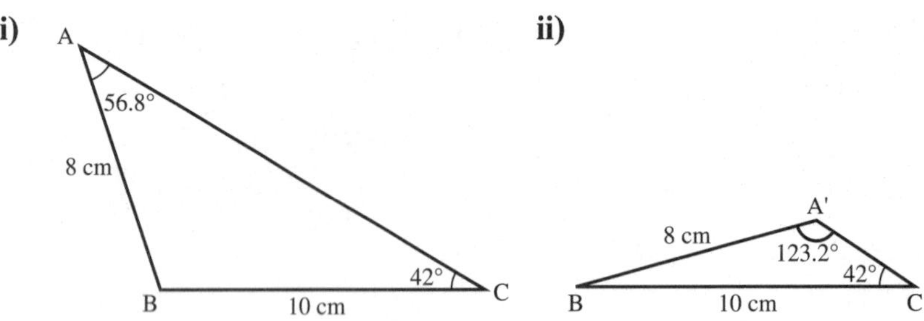

In case (i), $A\widehat{B}C = 180° - (42° + 56.8°) = 81.2°$.

Applying the sine rule to triangle ABC gives

$$\frac{\text{AC}}{\sin 81.2^\circ} = \frac{8}{\sin 42^\circ}$$

$$\therefore \quad \text{AC} = \frac{8 \sin 81.2^\circ}{\sin 42^\circ} = 11.8$$

In case (ii), $\text{A}'\hat{\text{B}}\text{C} = 180^\circ - (42^\circ + 123.2^\circ) = 14.8^\circ$.

Applying the sine rule to triangle A′BC gives

$$\frac{\text{A}'\text{C}}{\sin 14.8^\circ} = \frac{8}{\sin 42^\circ}$$

$$\therefore \quad \text{A}'\text{C} = \frac{8 \sin 14.8^\circ}{\sin 42^\circ} = 3.1$$

The two possible lengths of AC are 11.8 cm and 3.1 cm.

Result II: The cosine rule

In any triangle ABC,

$$c^2 = a^2 + b^2 - 2ab \cos C$$

and similarly we have

$$a^2 = b^2 + c^2 - 2bc \cos A$$

and $$b^2 = a^2 + c^2 - 2ac \cos B$$

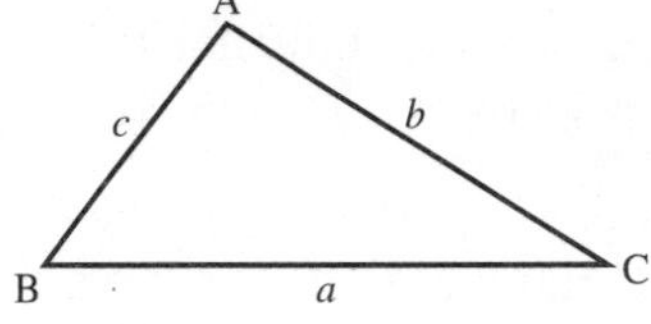

Proof

Two cases have to be considered:

i) when triangle ABC is an acute-angled triangle
ii) when triangle ABC is not an acute-angled triangle.

Case (i) Triangle ABC is acute-angled.
Let the perpendicular from A meet BC at D.
Let DC = x, then BD = $a - x$.

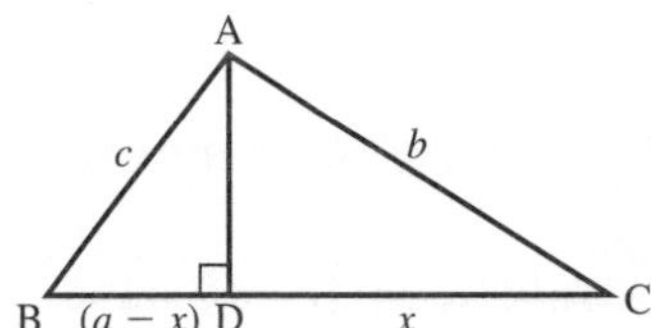

In triangle ABD, we have

$$\begin{aligned} \text{AD}^2 &= c^2 - (a - x)^2 \\ &= c^2 - a^2 + 2ax - x^2 \qquad [1] \end{aligned}$$

In triangle ADC, we have

$$\text{AD}^2 = b^2 - x^2 \qquad [2]$$

Eliminating AD^2 from [1] and [2] gives

$$b^2 - x^2 = c^2 - a^2 + 2ax - x^2$$

$$\therefore \quad b^2 = c^2 - a^2 + 2ax$$

$$\therefore \quad c^2 = a^2 + b^2 - 2ax \qquad [3]$$

In triangle ADC, we have

$$\cos C = \frac{x}{b}$$

$$\therefore \quad x = b\cos C$$

Substituting $x = b\cos C$ into [3] gives

$$c^2 = a^2 + b^2 - 2a(b\cos C)$$

$$\therefore \quad c^2 = a^2 + b^2 - 2ab\cos C$$

as required.

Case (ii) Triangle ABC is not acute-angled.
Let the perpendicular from A meet CB extended at D.
Let DB = x.

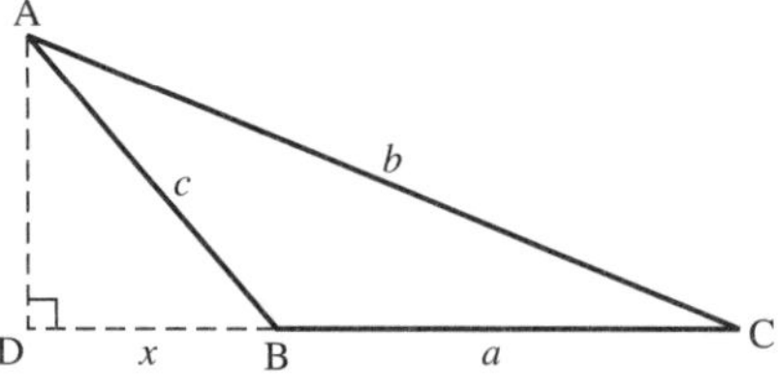

In triangle ADB, we have

$$AD^2 = c^2 - x^2 \qquad [4]$$

In triangle ADC, we have

$$AD^2 = b^2 - (a + x)^2$$

$$\therefore \quad AD^2 = b^2 - a^2 - 2ax - x^2 \qquad [5]$$

Eliminating AD^2 from [4] and [5] gives

$$c^2 - x^2 = b^2 - a^2 - 2ax - x^2$$

$$\therefore \quad c^2 = b^2 - a^2 - 2ax$$

$$\therefore \quad b^2 = a^2 + c^2 + 2ax \qquad [6]$$

In triangle ADB, we have

$$\cos A\hat{B}D = \frac{x}{c}$$

$$\therefore \quad x = c\cos A\hat{B}D$$

Now $A\hat{B}D = 180° - B$ and since $\cos(180° - B) = -\cos B$, we have $x = -c\cos B$.

Substituting $x = -c\cos B$ into [6] gives

$$b^2 = a^2 + c^2 + 2a(-c\cos B)$$

$$\therefore \quad b^2 = a^2 + c^2 - 2ac\cos B$$

as required.

Example 7 In triangle ABC, AC = 20 cm, BC = 11 cm and angle ACB = 20°. Calculate **a)** the length AB, **b)** angle ABC.

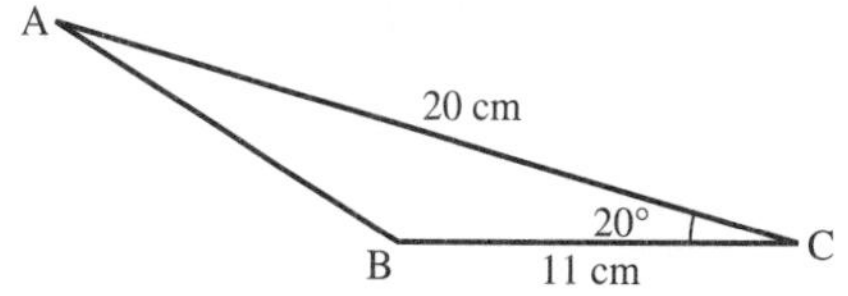

SOLUTION

a) Applying the cosine rule to triangle ABC gives

$$AB^2 = 11^2 + 20^2 - 2(11)(20)\cos 20^\circ$$
$$= 121 + 400 - 413.46$$
$$\therefore \quad AB = \sqrt{107.54} = 10.37$$

The length AB is 10.37 cm.

b) To find angle ABC, we must rearrange the cosine formula for $\cos B$. Now

$$b^2 = a^2 + c^2 - 2ac\cos B$$

Rearranging gives

$$\cos B = \frac{a^2 + c^2 - b^2}{2ac}$$

Applying this formula to triangle ABC gives

$$\cos B = \frac{11^2 + 10.37^2 - 20^2}{2(11)(10.37)} = -0.7516$$
$$\therefore \quad B = 138.7^\circ$$

Three-dimensional applications

Example 8 Sarah is standing on a cliff 14 m above sea level when she sees a boat anchored in the sea. The angle of depression of the boat from Sarah is 45° and the bearing of the boat from Sarah is 060°. Sarah then walks in a northerly direction along the edge of the cliff for 6 m to a tea shop. The tea shop is 11 m above sea level. Find

a) the angle of depression of the boat from the tea shop
b) the bearing of the boat from the tea shop.

SOLUTION

a) Let the boat be at a point B and the tea shop at a point T.

The angle of depression of the boat from T is given by $T\hat{B}C$. To find $T\hat{B}C$, we require the distance BC.

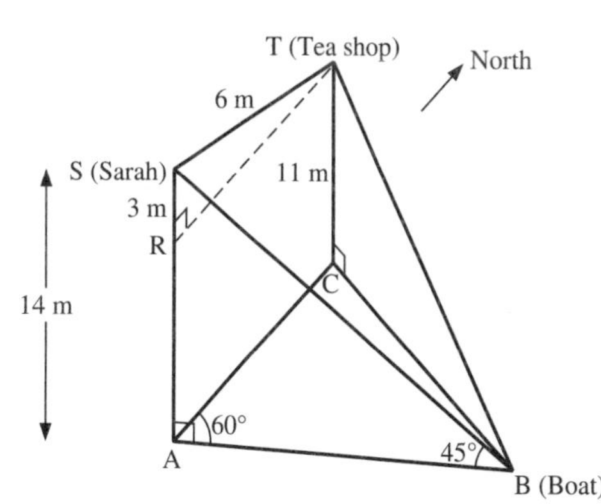

Applying Pythagoras to triangle SRT gives

$$RT^2 = 6^2 - 3^2 = 27$$

$$\therefore \quad RT = 3\sqrt{3}$$

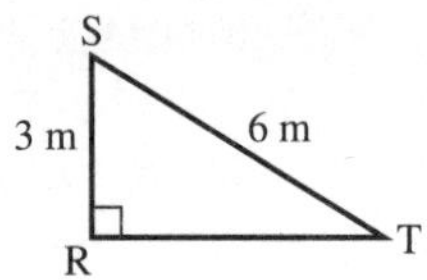

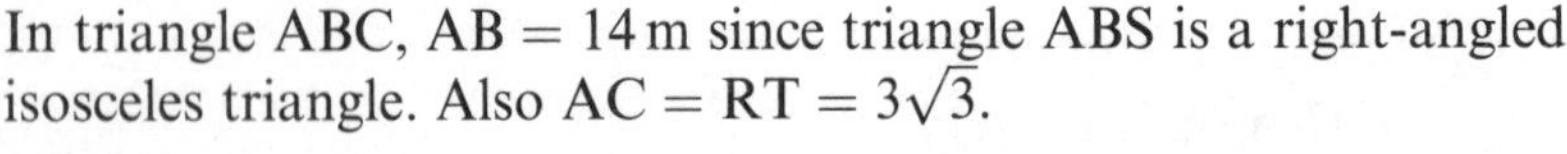

In triangle ABC, AB = 14 m since triangle ABS is a right-angled isosceles triangle. Also $AC = RT = 3\sqrt{3}$.

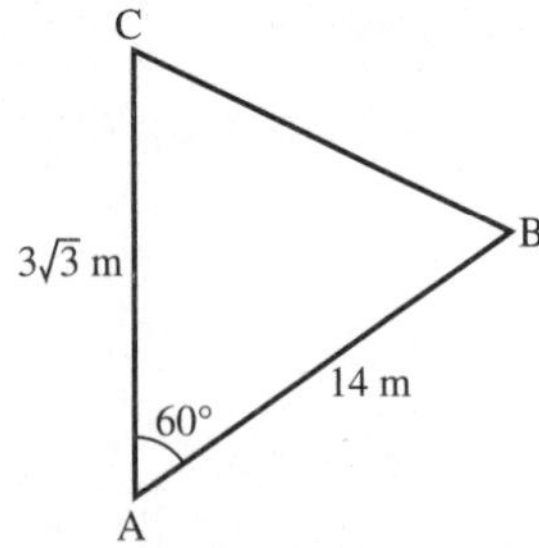

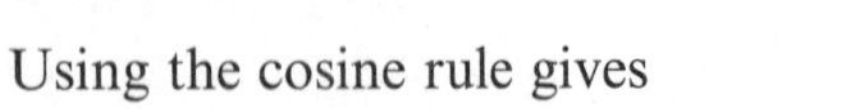

Using the cosine rule gives

$$BC^2 = 14^2 + (3\sqrt{3})^2 - 2(14)(3\sqrt{3})\cos 60^\circ$$

$$\therefore \quad BC = 12.26$$

Applying the tangent ratio to triangle CBT gives

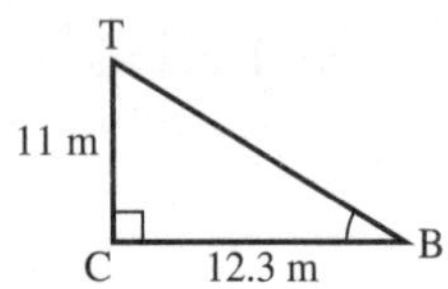

$$\tan C\widehat{B}T = \frac{11}{12.26}$$

$$\therefore \quad C\widehat{B}T = 41.9^\circ$$

The angle of depression of the boat from the tea shop is 41.9°.

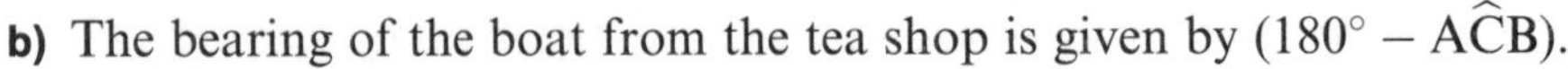

b) The bearing of the boat from the tea shop is given by $(180^\circ - A\widehat{C}B)$.

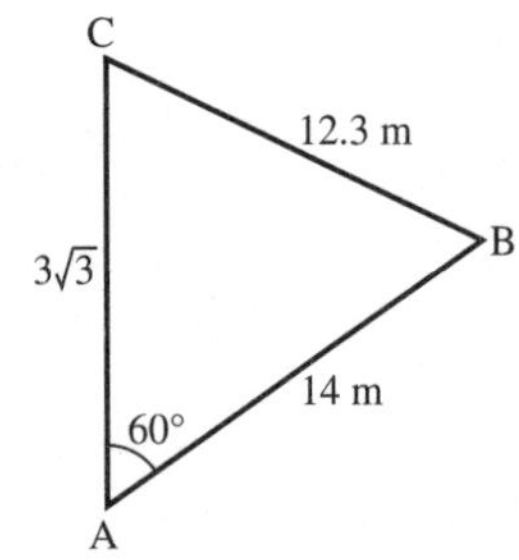

Applying the sine rule to triangle ABC gives

$$\frac{14}{\sin A\widehat{C}B} = \frac{12.26}{\sin 60^\circ}$$

$$\therefore \quad \sin A\widehat{C}B = \frac{14\sin 60^\circ}{12.26}$$

$$\therefore \quad A\widehat{C}B = 81.5^\circ$$

The required bearing is 180° − 81.5° = 098.5°.

Exercise 2B

Answers should be stated **correct to three significant figures** throughout this exercise.

1 Use the sine rule to find each of the unknown labelled sides or angles. In any ambiguous cases, give both alternatives.

a)

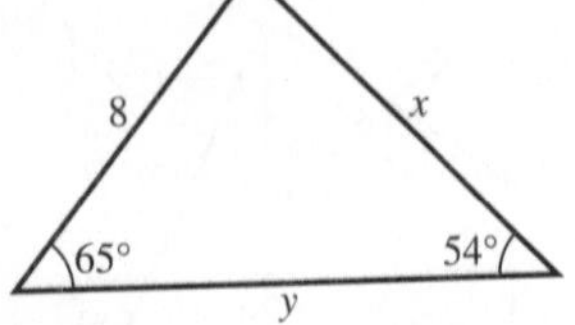

b)

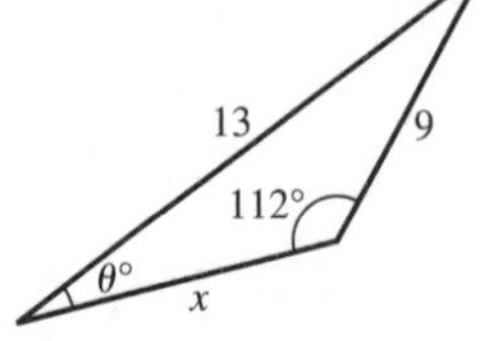

c)

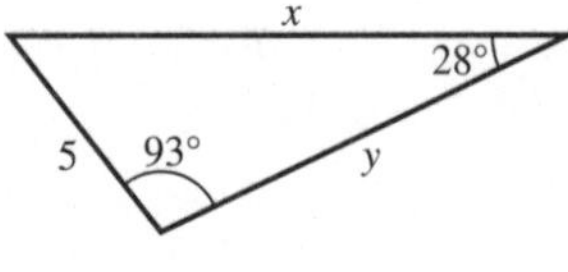

d)

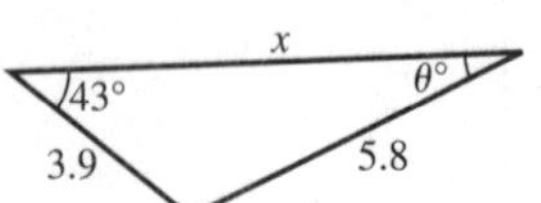

e)

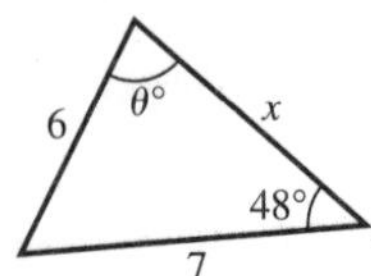

f)

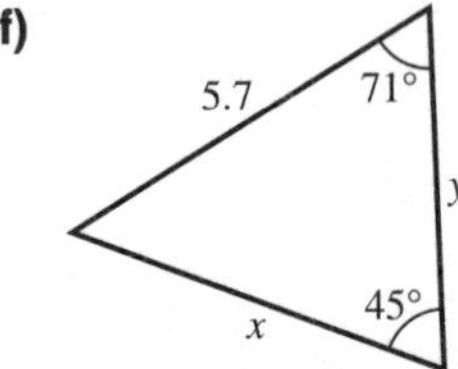

g) **h)** **i)**

j)

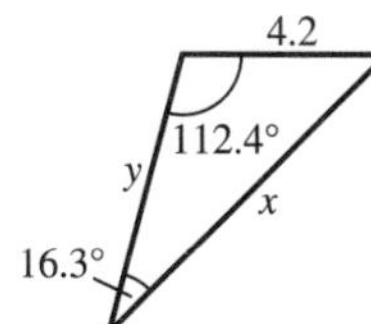

k)

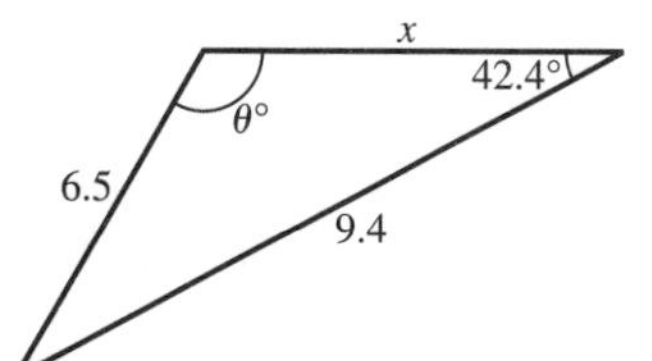

l)

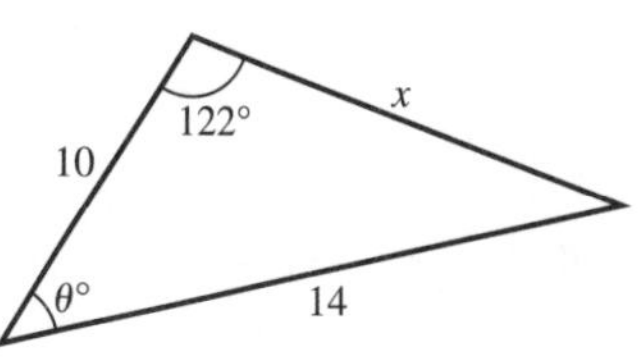

2 Use the cosine rule to find each of the unknown labelled sides or angles.

a)

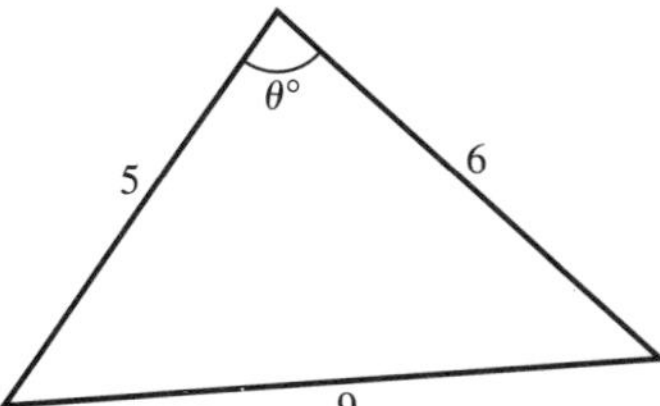

b)

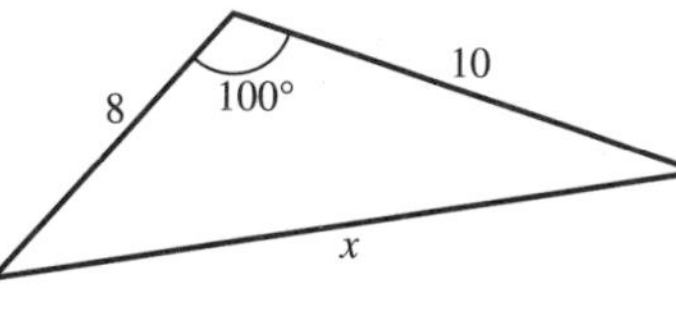

c)

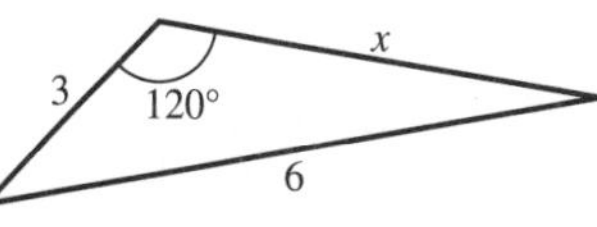

d)

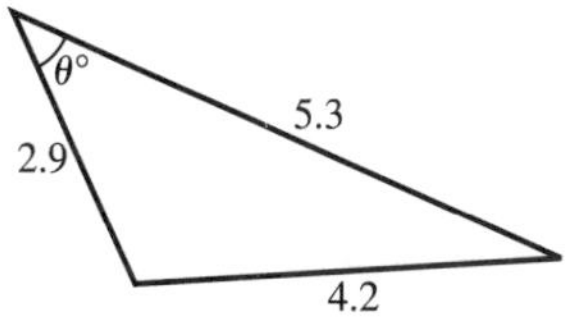

e)

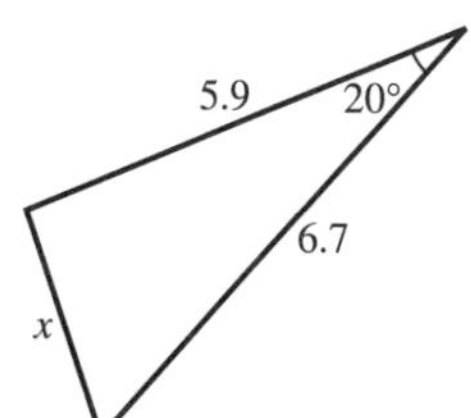

f)

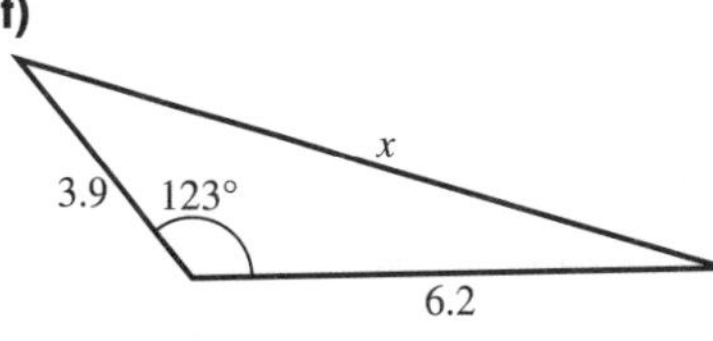

g)

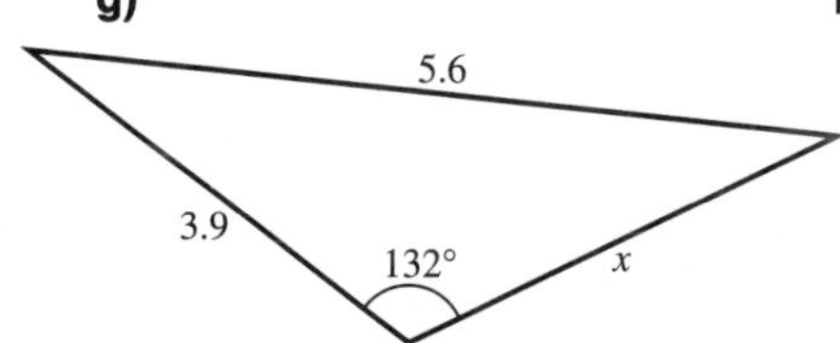

h)

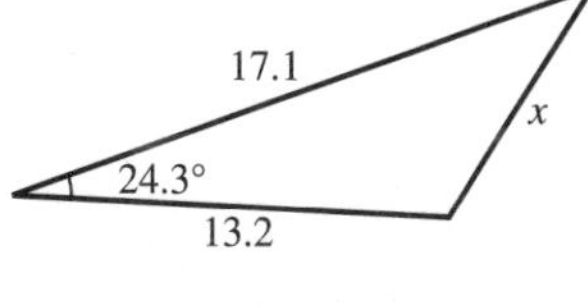

i)

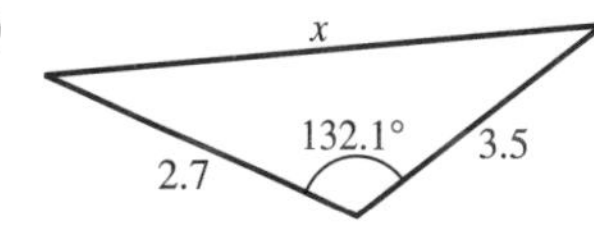

j)

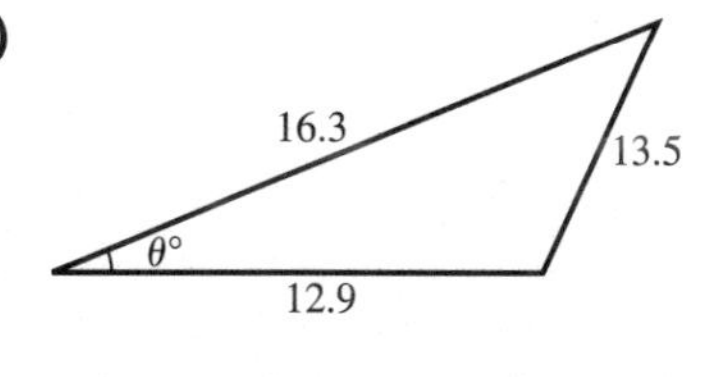

k)

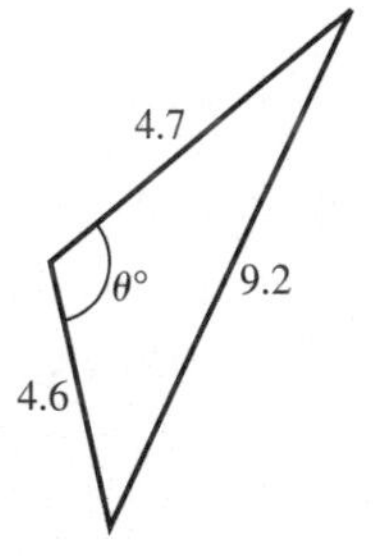

l)

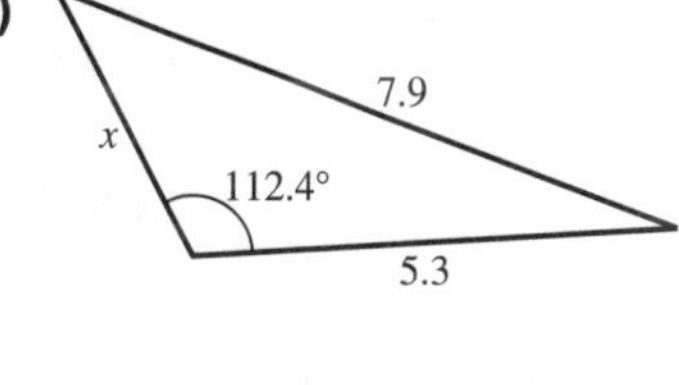

3 In triangle ABC, AB = 6 cm, BC = 8 cm and angle B = 50°.

a) Find the length of AC.

b) Find the size of angle A.

4 In triangle PQR, PQ = 12 cm, QR = 14 cm and PR = 11 cm.

a) Find the size of angle P.
b) Find the size of angle Q.

5 In triangle LMN, LM = 5 cm, MN = 8 cm and angle N = 20°. Given that angle L is obtuse,

a) calculate the size of angle L
b) calculate the length of LN.

6 Given that PQR is a triangle in which PQ = 5 cm, QR = 8 cm and angle R = 30°, calculate the two possible values of the length of PR.

7 A ship leaves a harbour, H, and sails for 32 km on a bearing of 025° to a point X. At X it changes course and then sails for 45 km on a bearing of 280° to a port P.

a) Sketch a diagram showing H, X and P.
b) Calculate the direct distance from H to P.
c) Calculate the bearing of P from H.

8 A bird leaves a nest, N, and flies 800 m on a bearing of 132° to a tree T. It then leaves T and flies 650 m on a bearing of 209° to a pylon P. Assuming that N, T and P are at the same height above the ground, calculate the distance and bearing on which the bird must fly in order to return directly from P to N.

9 An army cadet is involved in a compass exercise. He leaves a point O and walks 50 m due west to a point A. He then walks 80 m due north to a point B, and finally 60 m, on a bearing of 320°, to a point C.

a) Illustrate this information on a sketch.
b) Calculate the distance and bearing of B from O.
c) Calculate the distance and bearing on which he must walk in order to return from C to O.

10 A ship travelling south-west with constant speed observes the flash of a lighthouse on a bearing of 240°. 8 km further on the ship observes the flash of the same lighthouse, due west.

a) How far is the ship from the lighthouse at this time?
b) How close to the lighthouse will it pass?

11 In the cuboid ABCDEFGH calculate **a)** BD, **b)** BG, **c)** DG, **d)** $B\hat{G}D$.

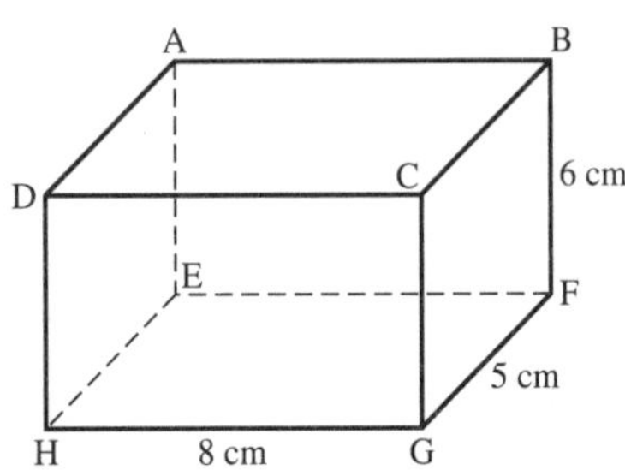

12 In the pyramid VPQRS, V is directly above the centre of the rectangular base, PQRS. Calculate **a)** $P\hat{S}V$, **b)** $R\hat{S}V$, **c)** PR, **d)** the angle between the planes VPS and VRS.

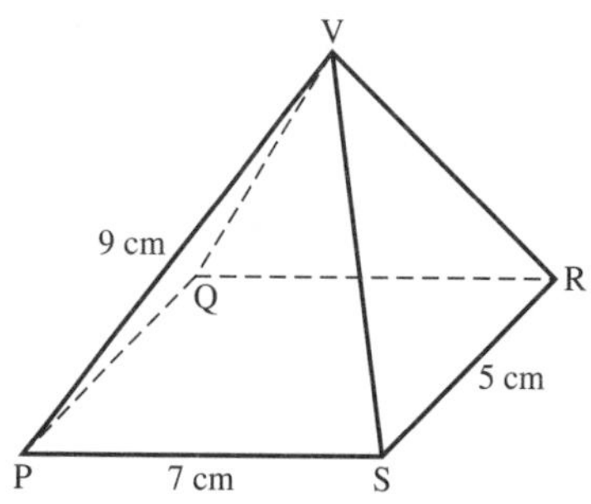

13 In the pyramid VABCD, the vertex, V, is at a height of 11 cm above the centre of the rectangular base ABCD. Calculate **a)** AC, **b)** VA, **c)** $A\hat{D}V$, **d)** $C\hat{D}V$, **e)** the angle between the planes VAD and VCD.

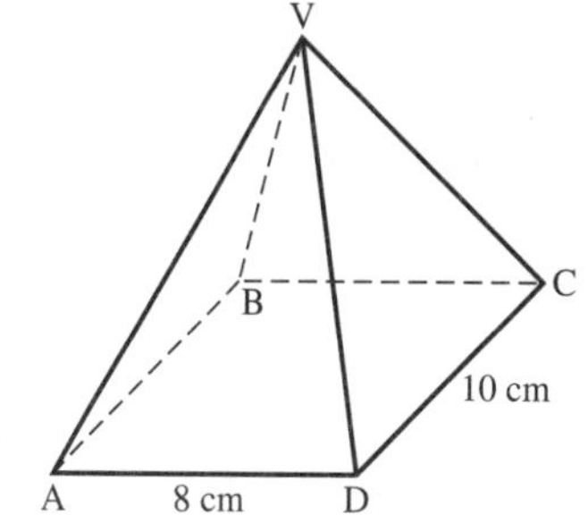

14 In the pyramid VJKLM, the vertex, V, is directly above the centre of the rectangular base, JKLM and VJ = 10 cm. Calculate **a)** $J\hat{M}V$, **b)** $L\hat{M}V$, **c)** JL, **d)** the angle between the planes VJM and VLM.

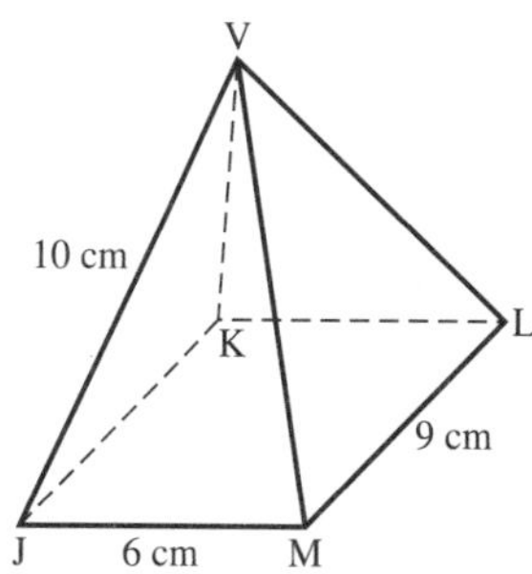

***15** ABC is the triangular base of the pyramid VABC in which the vertex, V, is 12 cm directly above the point A. Given that $B\hat{A}C = 90°$, calculate the angle between plane VBC and plane VAB.

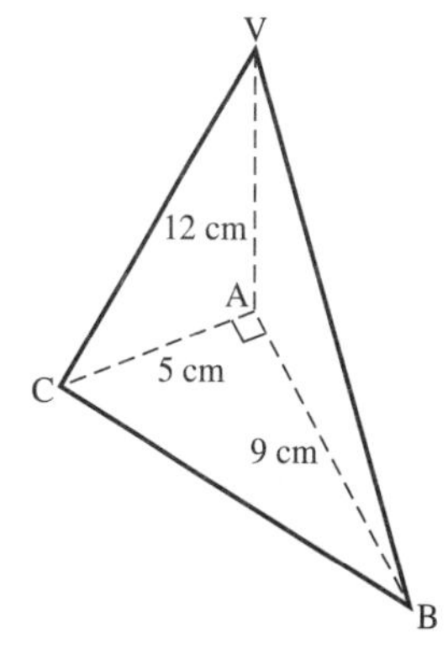

Plane shapes

A plane shape bounded by straight lines is called a **polygon**.

When the shape is bounded by three straight lines, the polygon is called a **triangle**.

When the shape is bounded by four straight lines, the polygon is called a **quadrilateral**.

Area of a triangle

The area of triangle ABC is given by

$$A = \tfrac{1}{2} \times a \times h$$

where a is the length of the base and h is the perpendicular height. That is,

$$A = \tfrac{1}{2}ah$$

in the usual notation.

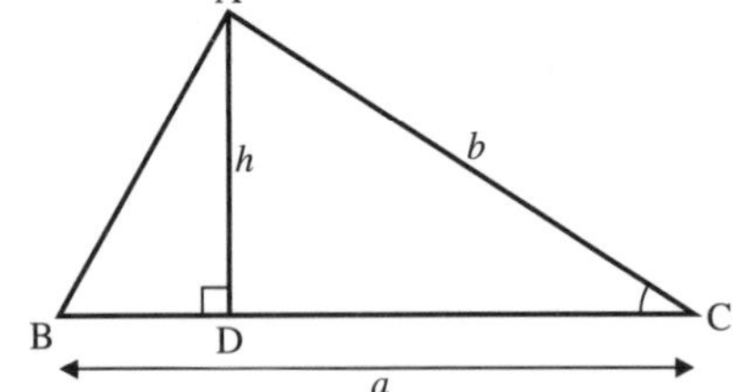

In triangle ADC,

$$\sin C = \frac{\text{AD}}{\text{AC}}$$

$$\therefore \quad \sin C = \frac{h}{b}$$

$$\therefore \quad h = b \sin C$$

Substituting $h = b \sin C$ into $A = \tfrac{1}{2}ah$ gives

$$A = \tfrac{1}{2}ab \sin C$$

Hence, the area of triangle ABC is also given by

$$A = \tfrac{1}{2}ab \sin C$$

where C is the included angle between sides a and b.

Quadrilaterals

Parallelogram

A parallelogram is a quadrilateral with both pairs of opposite sides parallel.

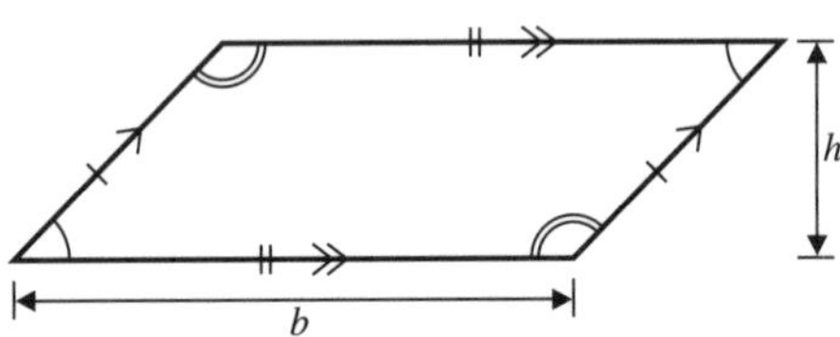

The area of a parallelogram is given by

$$A = bh$$

Rhombus

A rhombus is a quadrilateral in which all the sides are equal in length.

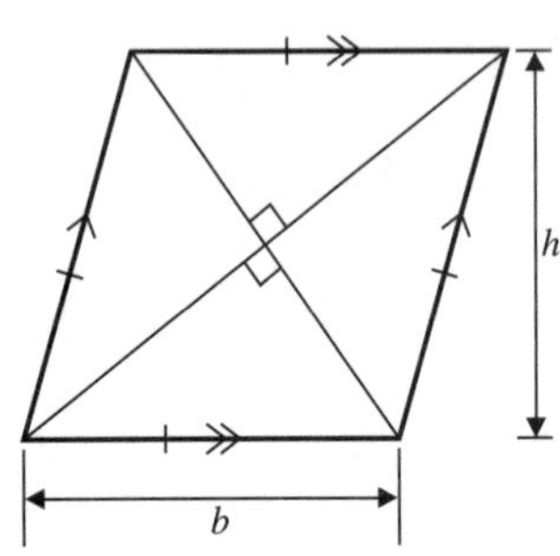

The area of a rhombus is given by

$$A = bh$$

Trapezium

A trapezium is a quadrilateral which has one pair of parallel sides.

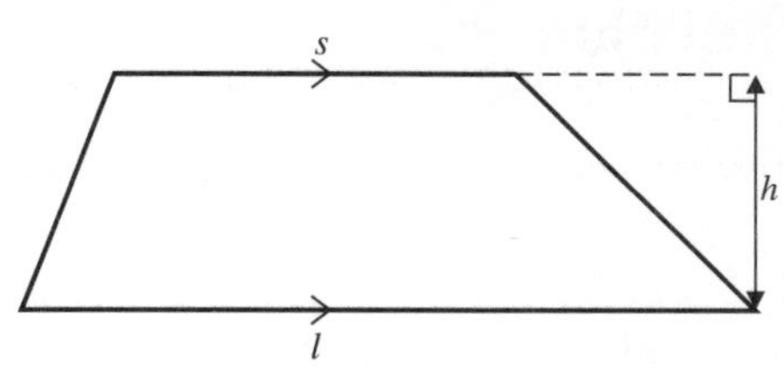

The area of the trapezium is given by

$$A = \frac{h}{2}(l + s)$$

Example 9 Shown on the right is the quadrilateral ACDE comprising a triangle ABE together with a trapezium BCDE. Given that AB = 6 cm, BC = 7 cm, DE = 12 cm and the area of triangle ABE is 12 cm^2, find the area of the quadrilateral ACDE.

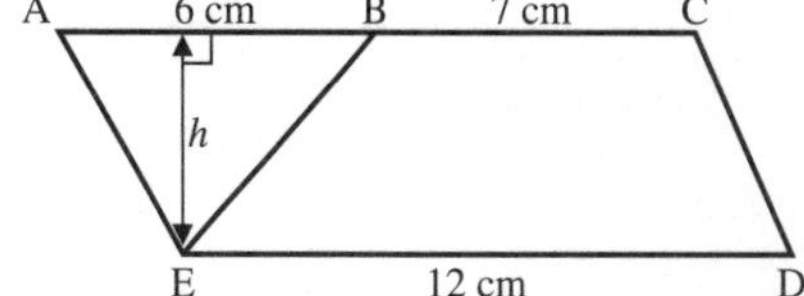

SOLUTION

The area of triangle ABE is given by $\frac{1}{2}bh$, where b is the base of the triangle and h is the perpendicular height. Therefore,

$$\frac{1}{2}bh = 12$$

$$\therefore \quad \frac{1}{2}(6)h = 12$$

$$\therefore \quad h = 4$$

The perpendicular height of triangle ABE is 4 cm.

Therefore, the perpendicular distance between the sides BC and DE of the trapezium BCDE is 4 cm.

The area of trapezium BCDE is given by

$$A = \frac{h}{2}(l + s)$$

$$\therefore \quad A = \frac{4}{2}(12 + 7) = 38$$

Therefore, the area of the quadrilateral ACDE is 12 cm^2 + 38 cm^2 = 50 cm^2.

The circle and circular measure

Consider a circle of radius r, diameter d and circumference C.

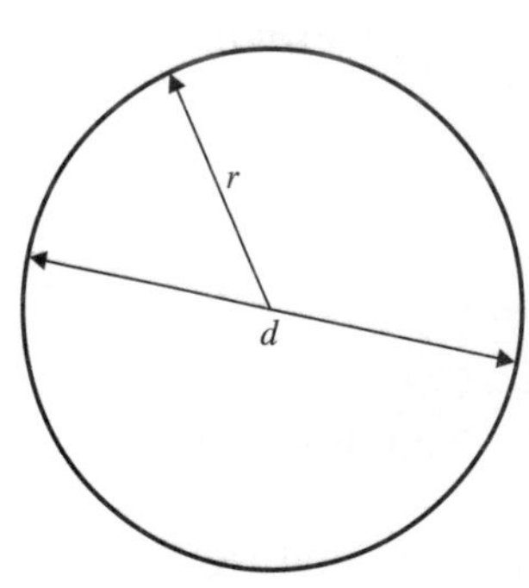

The circumference of the circle is given by

$$C = 2\pi r$$

Since $r = \frac{d}{2}$, we also have $C = \pi d$.

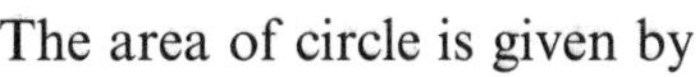

The area of circle is given by

$$A = \pi r^2$$

Exercise 2C

Answers should be stated **correct to three significant figures** throughout this exercise.

1 A pennant in the shape of the letter C is formed by cutting a rhombus from a semicircular plate, as in the diagram. Calculate the area of the pennant.

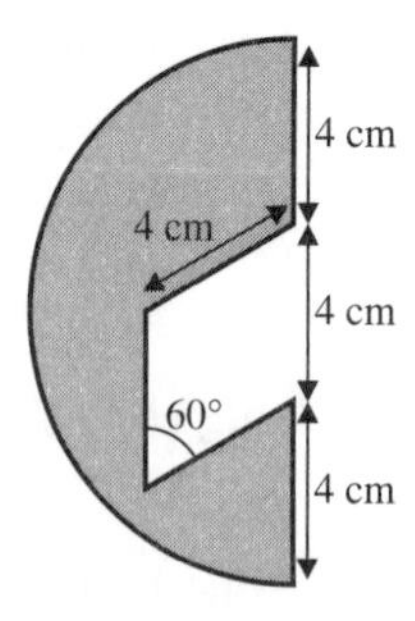

2 The diagram shows the component for a model, which is to be made from a square of metal of side b units, by removing a circular hole of radius a units. Given that the area of the circle removed is to be half the area of the square. Show that

$$\frac{b}{a} = \sqrt{2\pi}$$

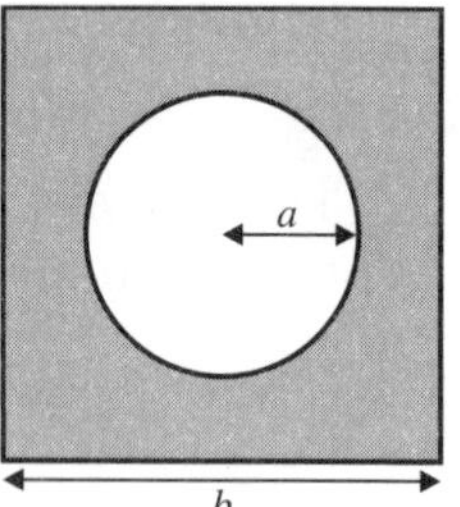

3 The diagram shows a regular pentagon ABCDE with centre O and sides of length 6 cm.

a) Explain why $\text{A}\hat{\text{O}}\text{B} = 72°$.
b) Calculate the area of the triangle AOB.
c) Hence find the area of the pentagon ABCDE.

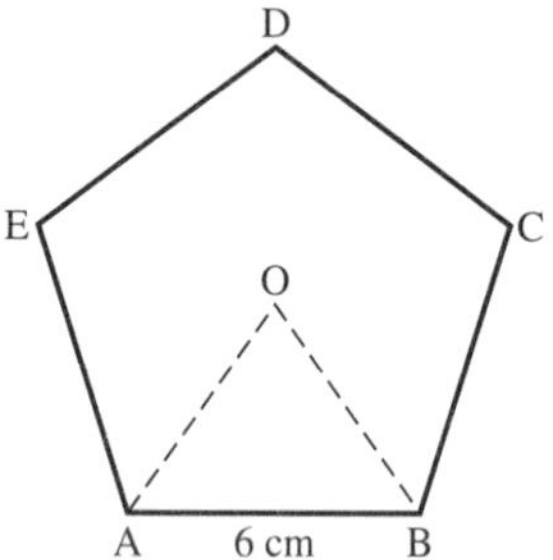

4 Calculate the area of a regular hexagon with sides of length 8 cm.

5 Calculate the area of a regular octagon with sides of length 9 cm.

6 In the triangle ABC, AB = 4.8 cm, BC = 3.8 cm and $\text{A}\hat{\text{B}}\text{C} = 42°$. P is the foot of the perpendicular from A to BC.

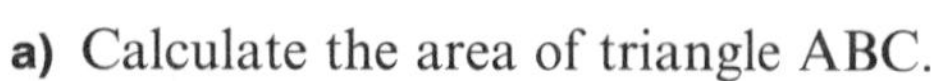

a) Calculate the area of triangle ABC.
b) Deduce the length of AP.

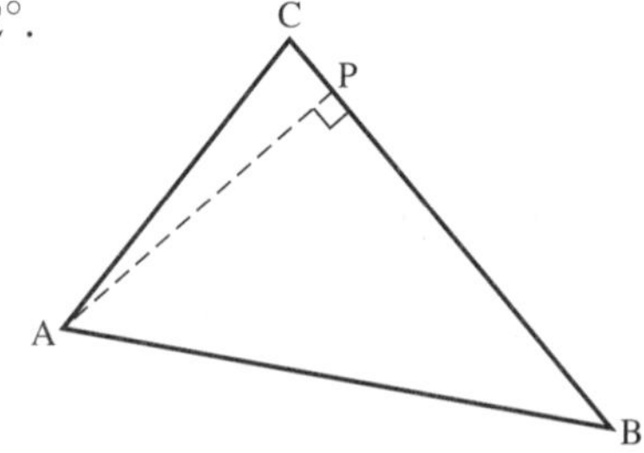

7 PQR is a triangle in which PQ = 5.9 cm, QR = 6.4 cm and $\text{P}\hat{\text{Q}}\text{R} = 63°$. Calculate the area of triangle PQR and the length of the perpendicular from P to QR.

8 In triangle KLM, KL = 9.3 cm, LM = 7.2 cm and $\text{K}\hat{\text{L}}\text{M} = 82°$. Calculate the area of triangle KLM and length of the perpendicular from K to ML.

9 The diagram shows a tent which has a rectangular base with vertical sides of height 2 m, surmounted by a pyramid. The vertex, V, of the pyramid is 3 m above the centre of the base of the tent.

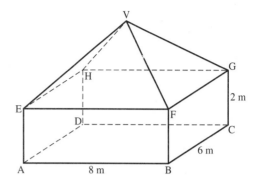

a) Calculate the area of the triangular face VEF.
b) Calculate the area of the triangular face VFG.

The canvas on the sides and the roof is to be waterproofed at a cost of £1.25 per square metre.

c) Calculate the cost of the waterproofing.

10 An item of jewellery consists of two identical pyramids, PABCD and QABCD, joined at their common rectangular base. AB = 8 mm, BC = 12 mm and PQ = 30 mm. The line PQ passes through the centre of the rectangle ABCD.

a) Calculate the area of triangle ABP.
b) Calculate the area of triangle ADP.

The item is to be dipped in silver at a cost of 0.4p per mm^2.

c) Calculate the cost of the dipping.

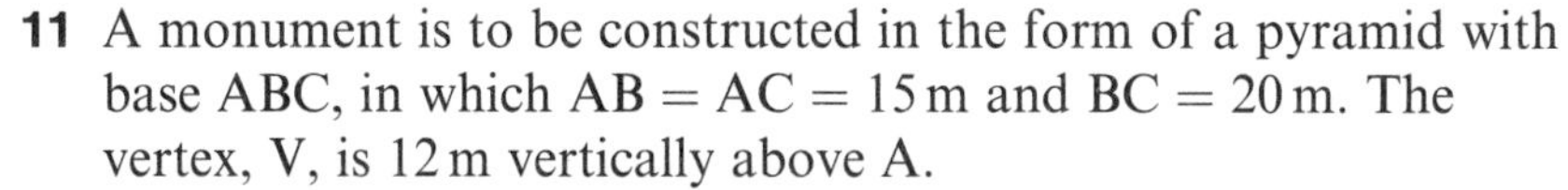

11 A monument is to be constructed in the form of a pyramid with base ABC, in which AB = AC = 15 m and BC = 20 m. The vertex, V, is 12 m vertically above A.

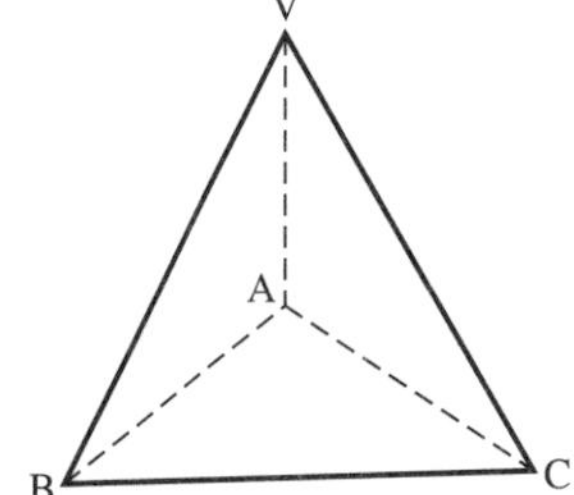

a) Calculate the area of triangle VAB.
b) Calculate the area of triangle VBC.

Special bricks for the outer faces are required and it is known that 14 bricks will cover $1\,\text{m}^2$.

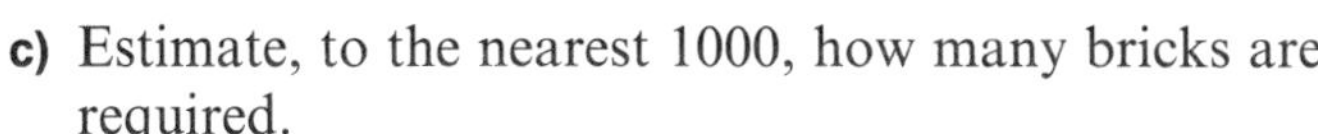

c) Estimate, to the nearest 1000, how many bricks are required.

***12** The triangle ABC has sides of length a, b and c. Show that p, the perimeter of triangle ABC, satisfies

$$p^2 = 2bc(1 + \cos A) + 2ac(1 + \cos B) + 2ab(1 + \cos C)$$

***13** The triangle ABC has sides of length a, b and c and $A = 45°$, $B = 60°$.

a) Show that the lengths a and b must satisfy $3a^2 - 2b^2 = 0$.
b) Show further that the lengths a and c must satisfy $a^2 - 2c^2 + 2ac = 0$.

The perpendicular from B meets AC at D. Given further that the perpendicular from C meets AB at E and BD at O,

c) show that the area of triangle BDC is $(\sqrt{3} - 1)$ units^2.

***14** Prove that a triangle with sides of length a, b and c has area

$$\sqrt{s(s-a)(s-b)(s-c)}$$

where $s = \frac{1}{2}(a + b + c)$. [This is known as Hero's formula.]

Radian measure

Consider an arc of length 1 unit of a circle of radius 1 unit.

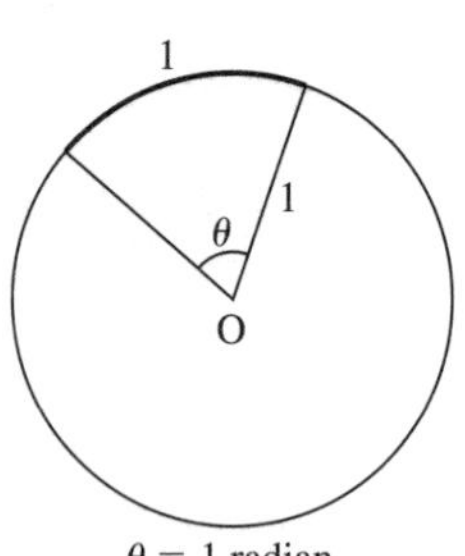

$\theta = 1$ radian

The angle θ subtended at the centre of the circle by the arc of length 1 unit is called 1 radian, written as 1 rad.

The circumference of the circle is given by

$$C = 2\pi r$$

Substituting $r = 1$ gives

$$C = 2\pi(1)$$

$$\therefore \quad C = 2\pi$$

Therefore, there are 2π radians at the centre O of the circle. In other words, 2π radians are equivalent to 360°.

Since 2π radians = 360°, we have

$$\pi \text{ radians} = 180^\circ$$

$$\frac{\pi}{2} \text{ radians} = 90^\circ$$

From these results we see that 1 radian $\approx 57.3^\circ$.

- To convert degrees into radians, we multiply by $\dfrac{\pi}{180}$.
- To convert radians into degrees, we multiply by $\dfrac{180}{\pi}$.

Example 10 Express each of the following angles in radians:
a) 45°, **b)** 60°, **c)** 270°.

SOLUTION

a) $45^\circ = 45 \times \dfrac{\pi}{180} = \dfrac{\pi}{4}$ rad **b)** $60^\circ = 60 \times \dfrac{\pi}{180} = \dfrac{\pi}{3}$ rad

c) $270^\circ = 270 \times \dfrac{\pi}{180} = \dfrac{3\pi}{2}$ rad

Example 11 Express each of the following angles in degrees:
a) $\dfrac{\pi}{6}$ radians, **b)** $\dfrac{5\pi}{6}$ radians, **c)** $\dfrac{4\pi}{3}$ radians.

SOLUTION

a) $\dfrac{\pi}{6}$ rad $= \dfrac{\pi}{6} \times \dfrac{180}{\pi} = 30^\circ$ **b)** $\dfrac{5\pi}{6}$ rad $= \dfrac{5\pi}{6} \times \dfrac{180}{\pi} = 150^\circ$

c) $\dfrac{4\pi}{3}$ rad $= \dfrac{4\pi}{3} \times \dfrac{180}{\pi} = 240^\circ$

Sectors and segments

Consider the sector of a circle, of radius r, which subtends an angle of θ at the centre.

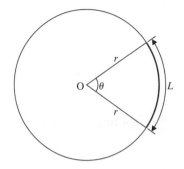

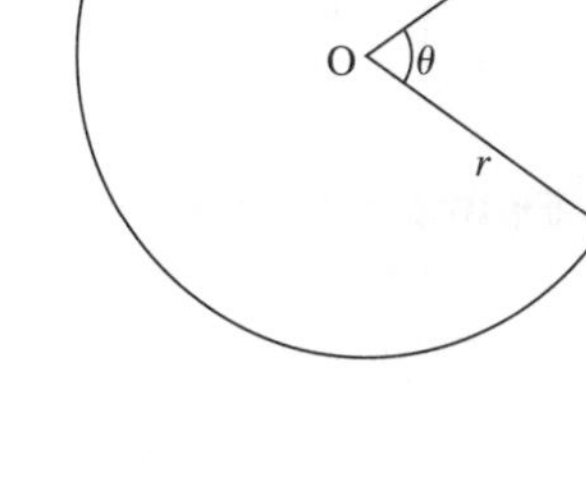

The length, L, of the arc is given by

$$L = \frac{\theta}{360} \times 2\pi r$$
$$= \frac{\theta \pi r}{180}$$

where θ is measured in degrees.

L is also given by

$$L = \frac{\theta}{2\pi} \times 2\pi r$$
$$= \theta r$$

where θ is measured in radians.

The area, A, of the sector is given by

$$A = \frac{\theta}{360} \times \pi r^2$$
$$= \frac{\theta \pi r^2}{360}$$

where θ is measured in degrees.

A is also given by

$$A = \frac{\theta}{2\pi} \times \pi r^2$$
$$= \frac{\theta r^2}{2}$$

where θ is measured in radians.

Example 12 The sector of a circle of radius 3 cm subtends an angle of $\frac{5\pi}{18}$ radians at the centre. Find

a) the length of the arc of the sector
b) the area of the sector of the circle.

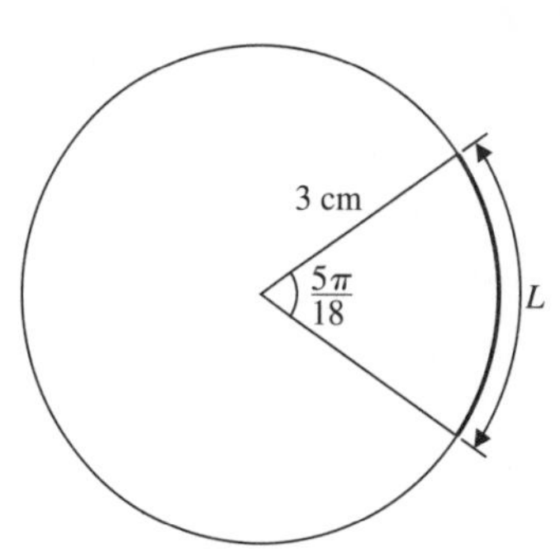

SOLUTION

a) The length, L, of the arc is given by

$$L = \theta r$$
$$= \left(\frac{5\pi}{18}\right)(3) = \frac{5\pi}{6} = 2.6$$

The length of the arc is 2.6 cm.

b) The area, A, of the sector is given by

$$A = \frac{\theta r^2}{2}$$

$$= \frac{\left(\frac{5\pi}{18}\right)(3)^2}{2} = \frac{5\pi}{4} = 3.9$$

The area of the sector is $3.9\,\text{cm}^2$.

Example 13 Shown tinted on the right is a segment of a circle of radius r. Show that the area of the segment is given by

$$\frac{r^2(\pi - 3)}{12}$$

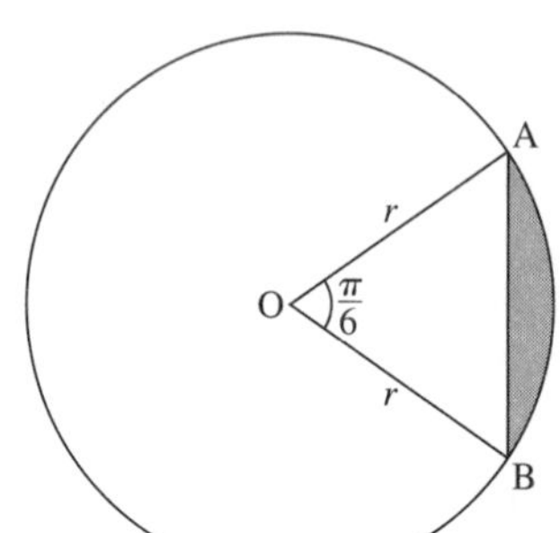

SOLUTION

The area, A_s, of the sector of the circle, is given by

$$A_s = \frac{\theta r^2}{2} = \frac{\left(\frac{\pi}{6}\right)r^2}{2}$$

$$\therefore \quad A_s = \frac{\pi r^2}{12}$$

The area of triangle OAB, A_t, is given by

$$A_t = \frac{1}{2}ab\sin C = \frac{1}{2}(r)(r)\sin\left(\frac{\pi}{6}\right)$$

$$\therefore \quad A_t = \frac{r^2}{4}$$

The area, A, of the shaded segment, is given by

$$A = A_s - A_t = \frac{\pi r^2}{12} - \frac{3r^2}{12}$$

$$\therefore \quad A = \frac{r^2(\pi - 3)}{12}$$

as required.

Exercise 2D

1 Express each of the following angles in radians, giving your answers in terms of π.

a) 30°	**b)** 90°	**c)** 120°	**d)** 10°
e) 80°	**f)** 300°	**g)** 36°	**h)** 240°
i) 72°	**j)** 360°	**k)** 342°	**l)** 1°

2 Express each of these angles in degrees.

a) π rad **b)** $\frac{\pi}{4}$ rad **c)** 3π rad **d)** $\frac{\pi}{6}$ rad
e) $\frac{4\pi}{5}$ rad **f)** $\frac{\pi}{12}$ rad **g)** $\frac{5\pi}{3}$ rad **h)** π rad
i) $\frac{5\pi}{12}$ rad **j)** $\frac{\pi}{90}$ rad **k)** $\frac{3\pi}{2}$ rad **l)** $\frac{7\pi}{6}$ rad

3 Express each of the following angles in degrees correct to 1 decimal place.

a) 4 rad **b)** 0.2 rad **c)** 4.3 rad **d)** 0.5 rad
e) 0.7 rad **f)** 3 rad **g)** 5.2 rad **h)** 2.1 rad
i) 5 rad **j)** 0.04 rad **k)** 16 rad **l)** 1 rad

4 A sector of a circle of radius 5 cm subtends an angle of $\frac{3\pi}{10}$ rad at the centre. Calculate

a) the length of the arc of the sector
b) the area of the sector.

5 A circle of radius 9 cm is divided into three equal sectors. Calculate

a) the length of the arc of each sector
b) the area of each sector.

6 A sector of angle $\frac{5\pi}{12}$ rad is cut from a circle of radius 6 cm. Calculate

a) the perimeter of the sector
b) the area of the sector.

7 OAB is a sector of a circle, centre O, and is such that OA = OB = 7 cm and $A\hat{O}B = \frac{5\pi}{14}$ rad. Calculate

a) the perimeter of the sector OAB
b) the area of the sector OAB.

8 The sector of a circle of radius 8 cm subtends an angle of 30° at the centre. Calculate

a) the length of the arc of the sector
b) the area of the sector.

9 OPQ is a sector of a circle, centre O, and is such that OP = OQ = 12 cm and $P\hat{O}Q = 45°$. Calculate

a) the perimeter of the sector POQ
b) the area of the sector POQ.

10 Calculate the area of a segment of angle $\frac{\pi}{2}$ rad cut from a circle of radius 5 cm.

11 Calculate the area of a segment of angle $\frac{\pi}{3}$ rad cut from a circle of radius 10 cm.

12 OMN is a sector of a circle, centre O, and is such that OM = ON = 12 cm, and $M\hat{O}N = \frac{\pi}{2}$ rad. S is the segment bounded by the chord MN and the arc MN. Calculate **a)** the area of S, **b)** the perimeter of S.

13 OAB is a sector of a circle, centre O, and is such that OA = OB = 8 cm, and $A\hat{O}B = \frac{\pi}{5}$ rad. S is the segment bounded by the chord AB and the arc AB. Calculate **a)** the area of S, **b)** the perimeter of S.

14 The shaded region in the diagram shows a component for a machine, which is to be cut from a square piece of metal, PQRS, of centre O and side 8 cm. The outer edge of the component is the regular octagon ABCDEFGH, and the inner edge is the circle centre O and radius 3 cm. Calculate

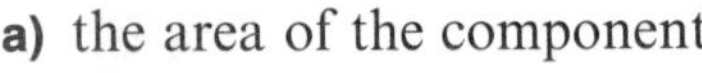

a) the area of the component
b) the outer perimeter of the component.

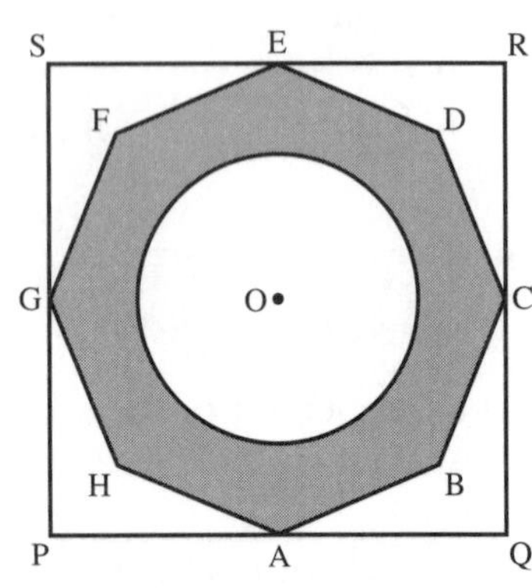

15 The diagram shows a pennant ABC, which has a triangular hole in the middle. The hole is an equilateral triangle ABC of side 8 cm. AB, BC and CA are circular arcs with centres at C, A and B respectively. Calculate

a) the area of the triangle ABC
b) the area of the sector ABC
c) the area of the shaded region.

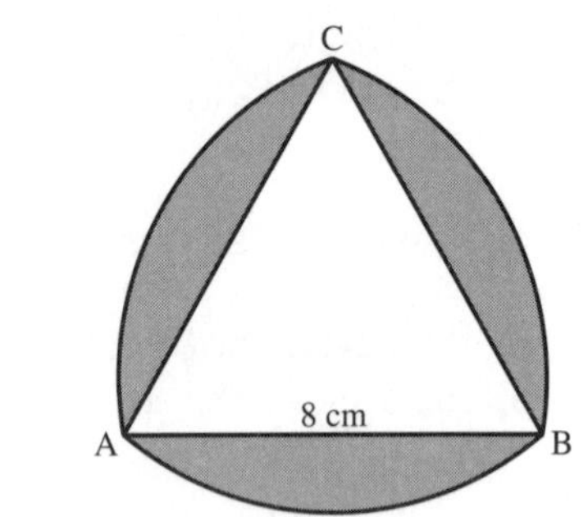

16 An ornamental mirror, CPQ, is to be cut from a triangle ABC in which AB = 6 m and $A\hat{B}C = C\hat{A}B = \frac{\pi}{4}$ rad. The circular arcs CP and CQ have centres B and A respectively. Calculate

a) the area of the sector BCP
b) the area of the triangle ABC
c) the area of the mirror CPQ
d) the length of the arc CP
e) the length of PQ
f) the perimeter of the mirror.

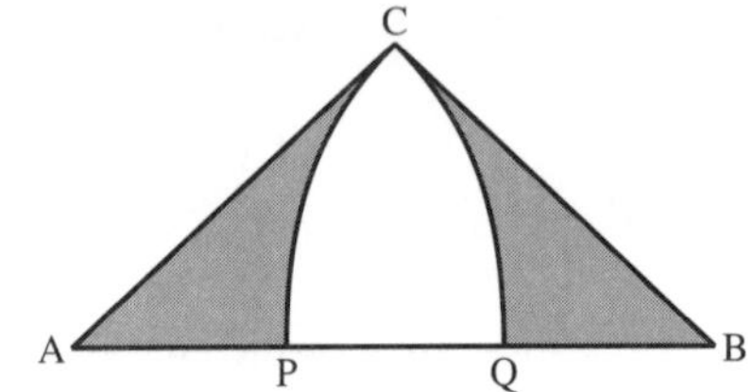

***17** The diagram shows two circles, each of radius r and such that the centre of one circle is on the circumference of the other circle. Prove that the shaded area enclosed by the two circles is given by the formula $\frac{r^2}{6}(4\pi - 3\sqrt{3})$.

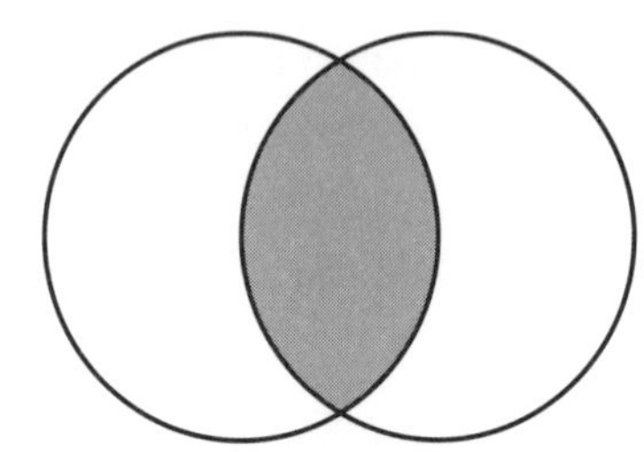

Volume

Prisms

If, when a solid is cut by a plane perpendicular to its length, the area of its cross-section is always the same, the solid is said to be a **prism**.

The volume of a prism is given by

$$\text{Volume} = \text{Area of cross-section} \times \text{Length}$$

- **Cuboid**

$$\text{Volume} = \text{Area of cross-section} \times \text{Length}$$
$$= (bh)l = lbh$$

The volume of a cuboid is given by

$$V = lbh$$

- **Cylinder**

Consider a cylinder of base radius r and height (length) h.

$$\text{Volume} = \text{Area of cross-section} \times \text{Length}$$
$$= (\pi r^2)h = \pi r^2 h$$

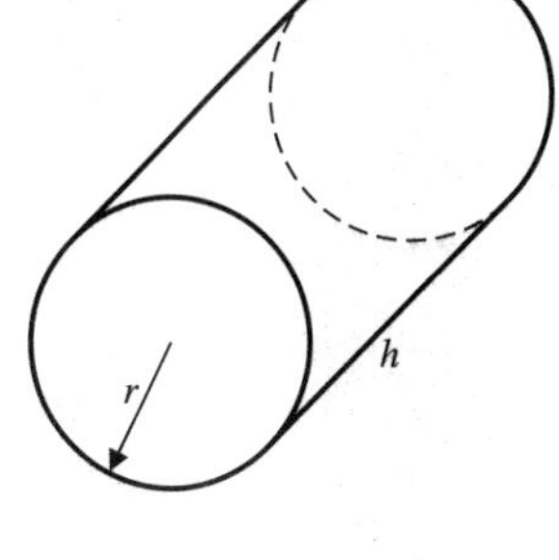

The volume of a cylinder is given by

$$V = \pi r^2 h$$

Shapes with non-uniform cross-section

- **Cone**

$$V = \tfrac{1}{3}\pi r^2 h$$

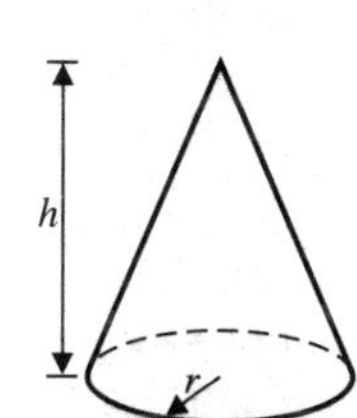

- **Pyramid**

$$V = \tfrac{1}{3} \times \text{Base area} \times \text{Height}$$

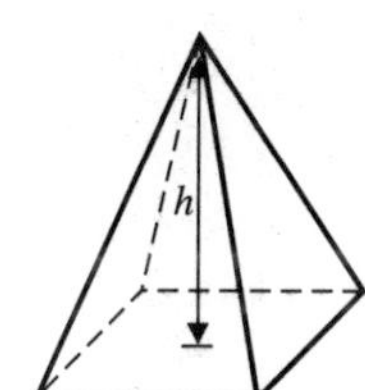

- **Sphere**

$$V = \tfrac{4}{3}\pi r^3$$

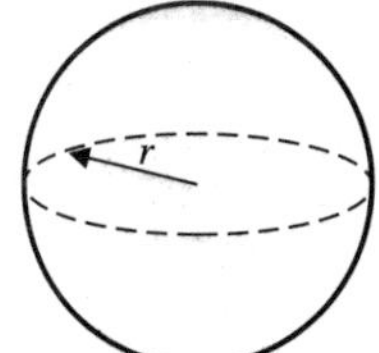

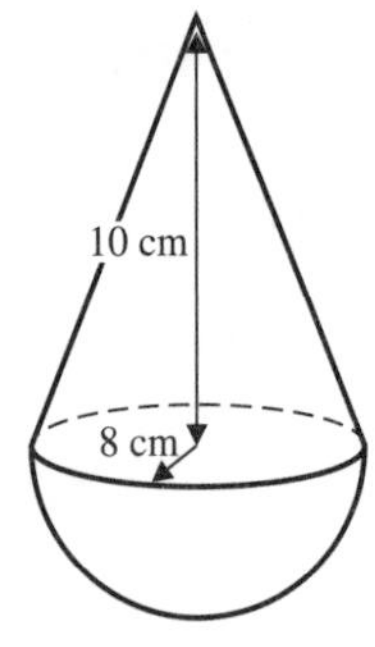

Example 14 The child's toy shown on the right comprises a hemisphere of radius 8 cm and a cone of height 10 cm. Calculate the volume of the toy.

SOLUTION

The volume, V_h, of the hemisphere, is given by

$$V_h = \frac{1}{2}\left(\frac{4}{3}\pi r^3\right)$$

$$= \frac{2}{3}\pi(8)^3 = 1072.3$$

The volume, V_c, of the cone, is given by

$$V_c = \frac{1}{3}\pi r^2 h$$

$$= \frac{1}{3}\pi(8)^2 10 = 670.2$$

The volume, V, of the toy, is given by

$$V = V_h + V_c$$

$$= 1072.3 + 670.2 = 1742.5$$

The volume of the toy is 1742.5 cm^3.

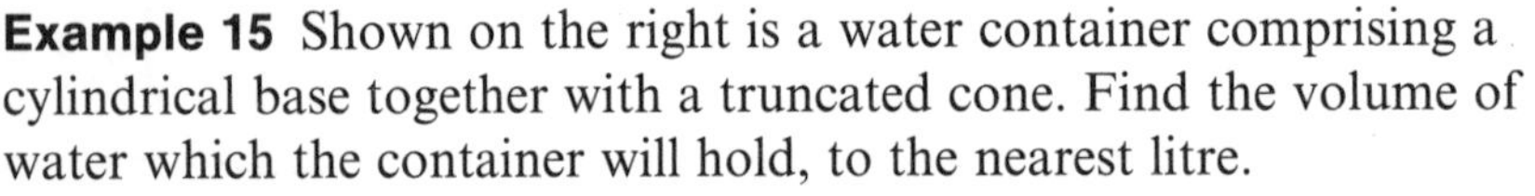

Example 15 Shown on the right is a water container comprising a cylindrical base together with a truncated cone. Find the volume of water which the container will hold, to the nearest litre.

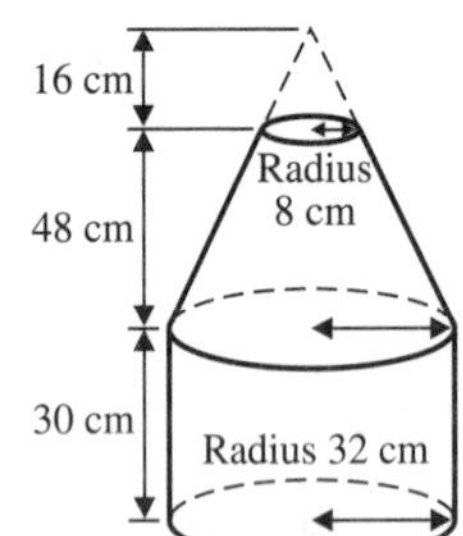

SOLUTION

The volume, V_b, of the cylindrical base is given by

$$V_b = \pi r^2 h$$

$$= \pi(32)^2 30 = 96\,509.7$$

The volume, V_t, of the truncated cone is given by

$$V_t = \frac{1}{3}\pi(32)^2 64 - \frac{1}{3}\pi(8)^2 16$$

$$= 68\,629.1 - 1072.3 = 67\,556.8$$

The capacity, V, of the container is given by

$$V = V_b + V_t$$

$$= 96\,509.7 + 67\,556.8 = 164\,066.5$$

The capacity of the container is 164 litres.

Exercise 2E

1 A pencil case comprises a cylinder of length 16 cm and radius 3 cm and a cone of height 5 cm, as shown. Calculate the volume of the pencil case.

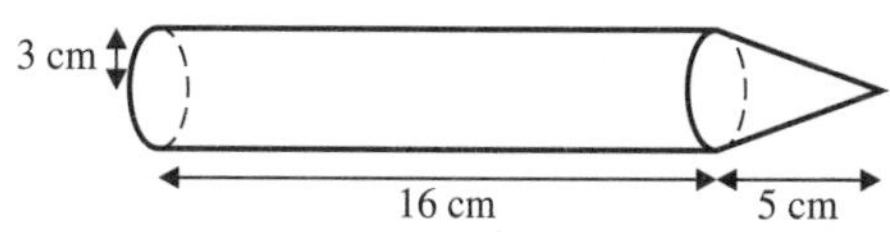

2 The diagram shows a swimming pool with rectangular surface ABCD in which AB = 10 m and BC = 20 m. The pool has uniform cross-section BFGC such that BF = 1 m and CG = 4 m. Calculate the volume of the pool.

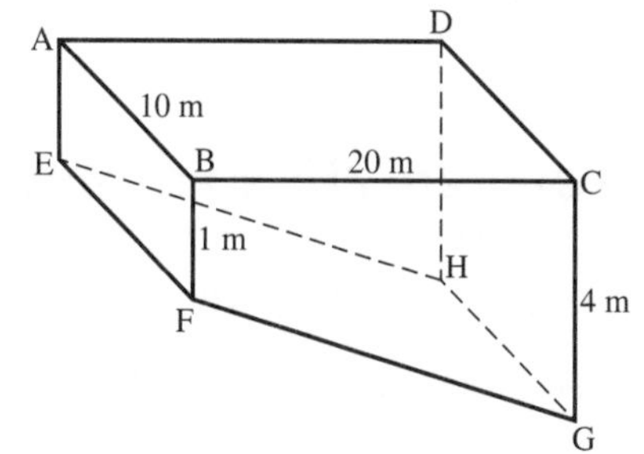

3 The base for a model windmill is to be cut from a wooden cone, of height 20 cm and base radius 8 cm, by removing a cone of height 5 cm from the top. Calculate

a) the volume of the original cone
b) the volume of the removed top
c) the volume of the model.

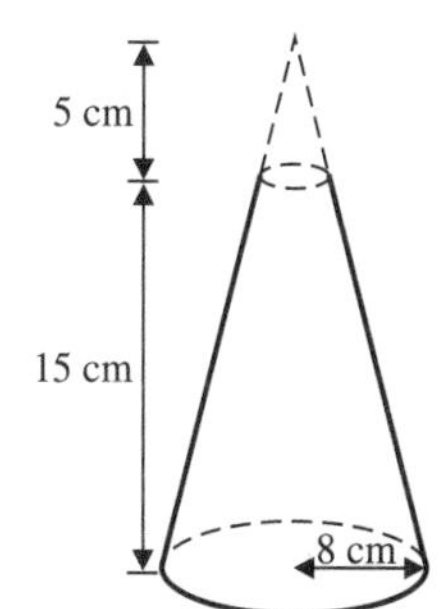

4 The diagram shows the uniform cross-section of a tunnel, ABCD, drawn inside a circle of centre O and radius 6 m. The width of the tunnel AB = 6 m. Calculate

a) the length AD
b) the angle $C\hat{O}D$
c) the area of the cross-section ABCD.

Given also that the tunnel is 120 m long,

d) calculate the volume of the tunnel.

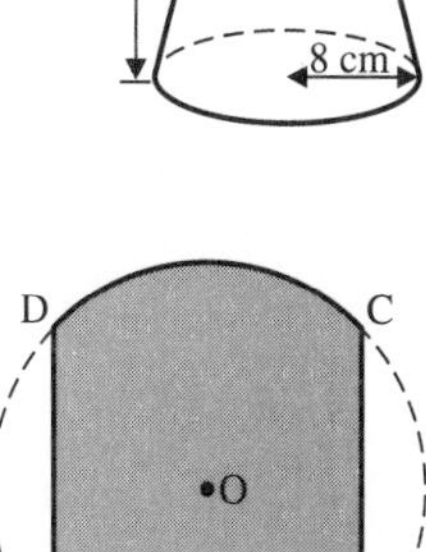

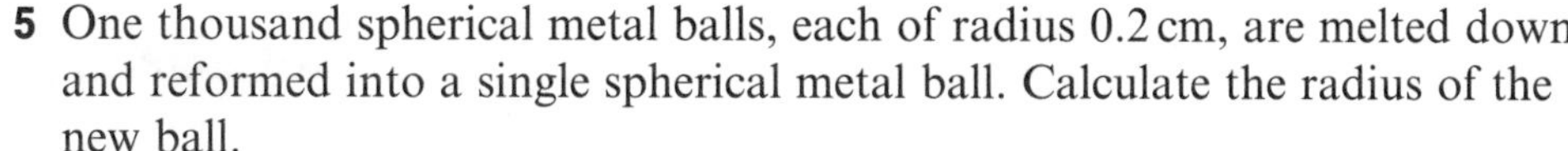

5 One thousand spherical metal balls, each of radius 0.2 cm, are melted down and reformed into a single spherical metal ball. Calculate the radius of the new ball.

6 The diagram shows a metal sphere of radius 3 cm, completely submerged in water, and resting at the base of a cylinder of radius 4 cm. Calculate the amount by which the depth of the water in the cylinder will fall when the sphere is removed.

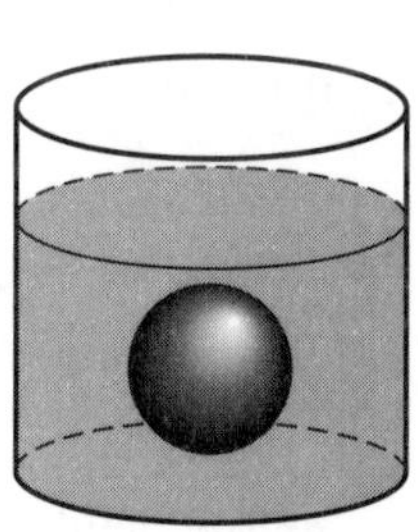

*7 The diagram shows a cone of height h sitting inside a sphere of radius r. Prove that the volume of the cone is given by the formula $\frac{1}{3}\pi h^2(2r - h)$.

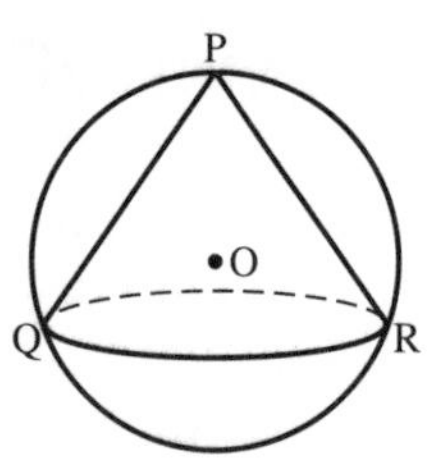

Exercise 2F: Examination questions

1 In triangle ABC, AB = 8 cm, BC = 6 cm and angle B = 30°.

a) Find the length of AC.
b) Find the size of angle A. (UODLE)

2 In the gales last year, a tree started to lean and needed to be supported by struts which were wedged as shown below. There is also a simplified diagram of the struts with approximate dimensions.

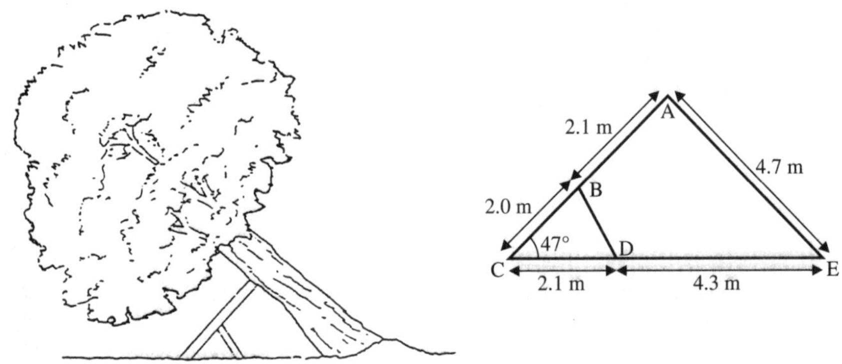

Give all your answers correct to two significant figures.

i) Use the cosine rule to calculate the length BD.
ii) Calculate the angle CAE. (MEI)

3 An inn sign is to be supported by iron struts as shown in the diagram.

QR is 156 cm, PQ is 85 cm, QS is 100 cm and angle SPQ = 36°. QR is perpendicular to PS.

i) Show that angle PQR is 126°.
ii) Calculate PR, using the cosine rule in triangle PQR.
iii) Calculate angle PSQ. (MEI)

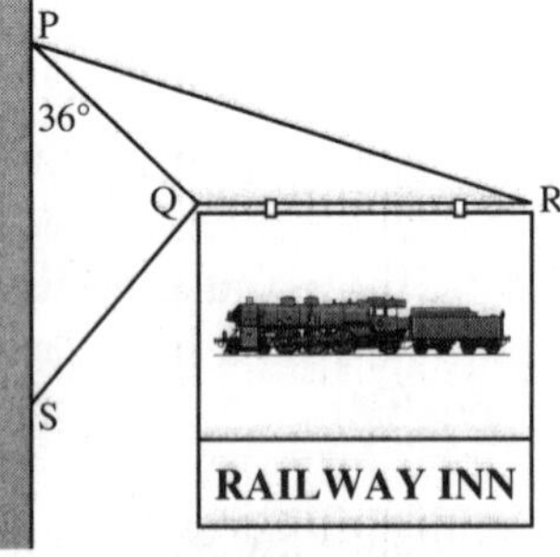

4 A man in a boat measures the angle of elevation of the top of a cliff as 12°. He rows the boat 800 m directly towards the cliff and then measures the angle of elevation of the top of the cliff as 65°. Calculate

a) the distance of the **top** of the cliff from the **second position** of the boat
b) the height of the cliff. (WJEC)

5 The points A, B and C lie on horizontal ground and are such that AB = 19 m, BC = 16 m and CA = 21 m.

a) Calculate the size of angle ACB.

A vertical mast AH of height 11 m is placed at A.

b) Calculate the size of the acute angle between the planes HBC and ABC, giving your answer to the nearest 0.1°. (EDEXCEL)

6 A tent is erected, as shown in the figure (right). The base ABCD is rectangular and horizontal and the top edge EF is also horizontal.

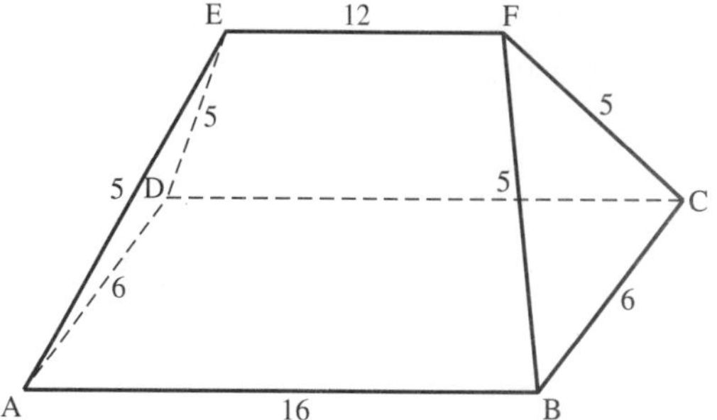

The length, in metres, of the edges are

$$AE = BF = CF = DE = 5$$

$$AB = CD = 16 \qquad AD = BC = 6$$

$$EF = 12$$

a) Calculate the size of angle ADE, giving your answer to the nearest degree.

b) Show that the vertical height of EF above the base ABCD is $2\sqrt{3}$ m.

Calculate, to the nearest degree, the size of the acute angle between

c) the face ADE and the horizontal

d) the edge AE and the horizontal. (EDEXCEL)

7 A pyramid has a horizontal square base ABCD of side 10 cm and the vertex V is 6 cm vertically above the centre of the base.

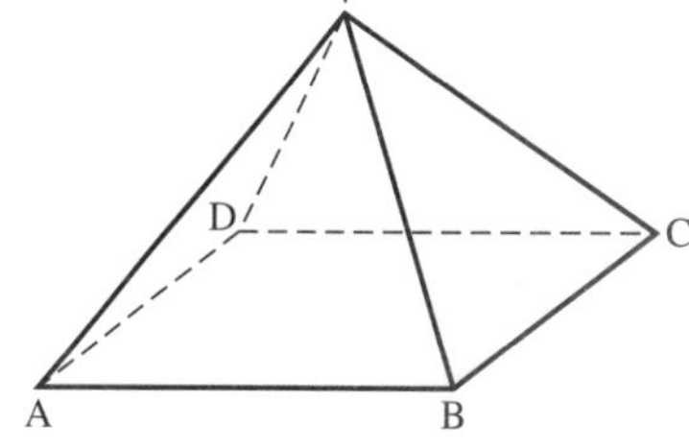

a) Calculate the angle between VA and the horizontal, giving your answer to the nearest 0.1°.

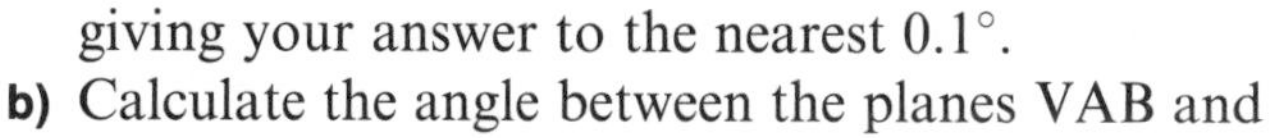

b) Calculate the angle between the planes VAB and VCD, giving your answer to the nearest 0.1°.

c) Calculate the perpendicular distance from C to the edge AV (extended if necessary), giving your answer to the nearest 0.01 cm. (UODLE)

8 A pyramid has a horizontal square base of side 8 cm. Its vertex V is 10 cm vertically above the centre of the base.

a) Calculate the angle between the faces VPQ and VRS.

b) Calculate the length VQ.

c) Calculate the size of the angle PQV, and hence find the perpendicular distance from P to the line VQ.

d) Deduce that the angle between the faces VPQ and VQR is approximately 97°.

9 The diagram shows the cross-section of a tunnel. The cross-section has the shape of a major segment of a circle, and the point O is the centre of the circle. The radius of the circle is 4 m, and the size of angle AOB is 1.5 radians. Calculate the perimeter of the cross-section. (UCLES)

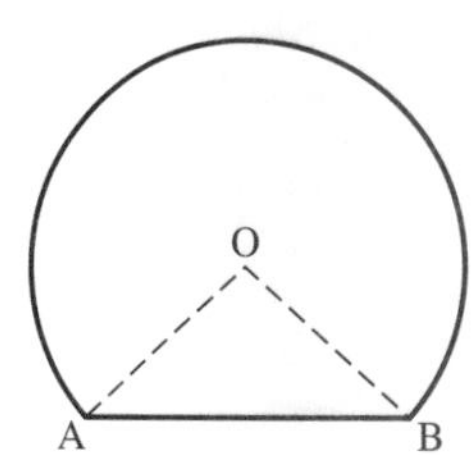

10 The diagram shows part of a circle, centre O and radius 4 cm. Given that the length of the arc ABC is 5 cm, calculate

i) the size of $A\hat{O}C$ in radians,
ii) the area of the shaded region. (WJEC)

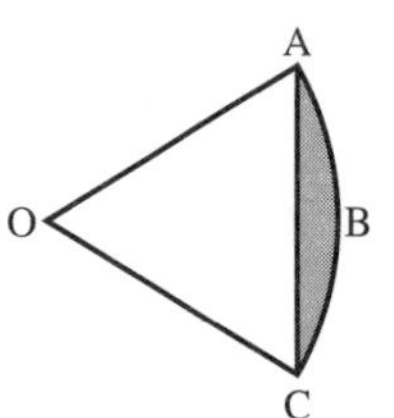

11 A circle, centre O, has an arc AB of length 13.44 cm and $A\hat{O}B = 1.6$ radians.

a) Calculate the radius of the circle.
b) Find the area of the region enclosed by the arc AB and the chord AB. (WJEC)

12 A partly filled cylindrical oil drum of diameter 80 cm lies with its curved surface in contact with a horizontal floor. The diagram shows a circular cross-section of the drum with centre O and the chord AB denotes the horizontal surface of the oil.

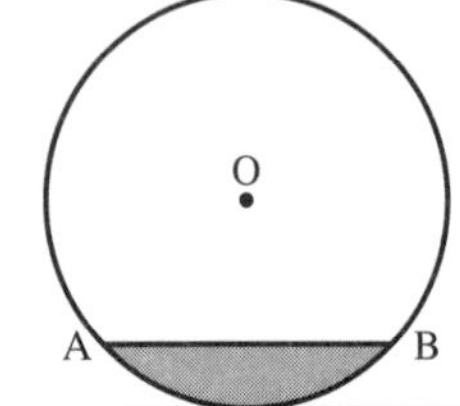

a) Given that the oil surface is 20 cm above the floor level, find the value of angle AOB in radians.
b) The length of the drum is 173 cm. Calculate the volume of oil in the drum, giving your answer in cm^3 to three significant figures. (AEB 94)

13 The diagram shows two circles, with centres A, B and radii 4 m, 6 m respectively, intersecting at C and D. The angles CAD and CBD are 1.4 radians and 0.7 radians respectively.

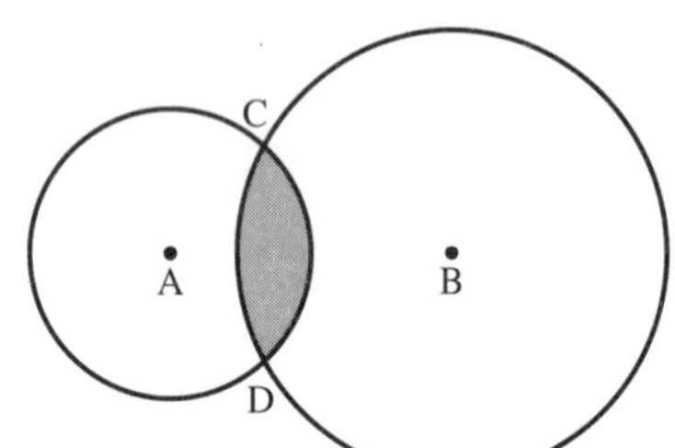

a) Calculate the perimeter of the shaded region.
b) Determine the area of the shaded region. (WJEC)

14 The figure shows a circle, centre O, inscribed in the triangle ABC, where P, Q, R are the points of contact so that AR = AQ = 2 cm, CP = CQ = 3 cm and BR = BP = 4 cm.

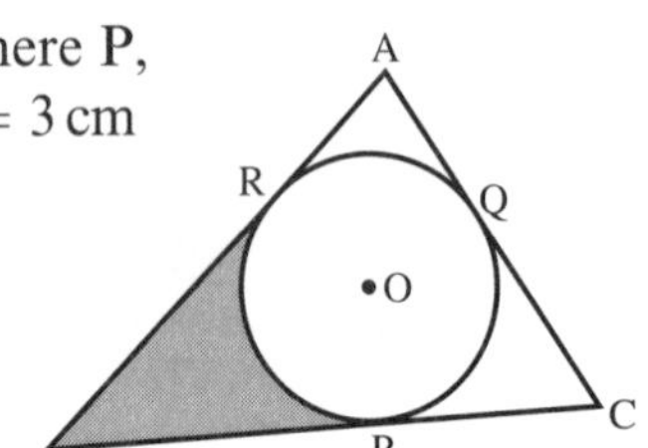

a) Show that the cosine of angle ABC is $\frac{5}{7}$ and hence find the value of angle OBR, to the nearest 0.1°.
b) Calculate the radius of the circle, giving your answer in cm to three decimal places.
c) Find the area of the shaded region bounded by BR, BP and the minor arc RP, giving your answer in cm^2 to three significant figures. (UODLE)

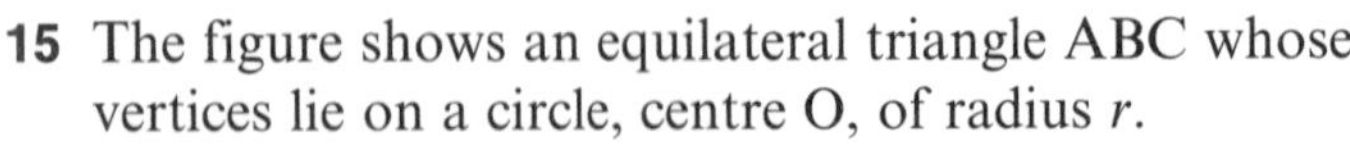

15 The figure shows an equilateral triangle ABC whose vertices lie on a circle, centre O, of radius r.

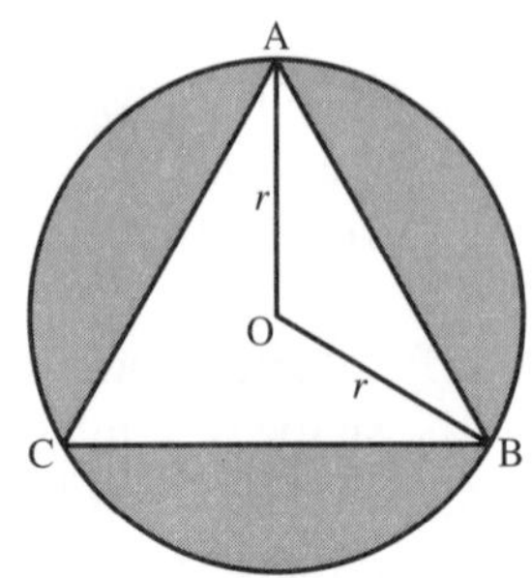

a) Show that the length of a side of this triangle is $r\sqrt{3}$.
b) Show that the ratio of the area of the shaded region to the area of the triangle is

$$[(4\pi\sqrt{3}) - 9] : 9 \quad \text{(EDEXCEL)}$$

16 The left edge of the shaded crescent-shaped region, shown in the figure, consists of an arc of a circle of radius r cm with centre O. The angle AOB $= \frac{2}{3}\pi$ radians. The right edge of the shaded region is a circular arc with centre X, where OX $= r$ cm.

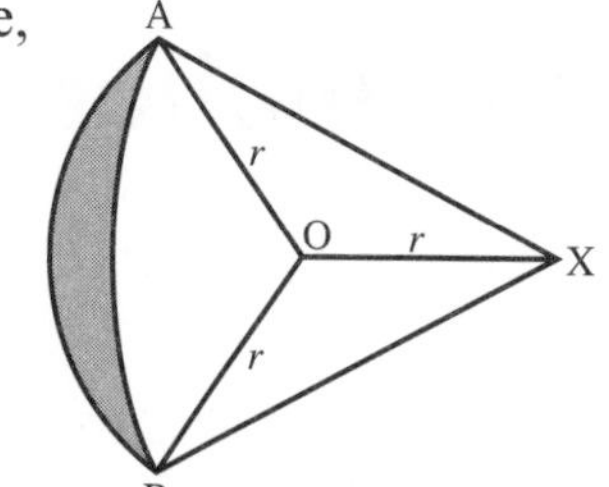

a) Show that angle AXB $= \frac{1}{3}\pi$ radians.
b) Show that AX $= r\sqrt{3}$ cm.
c) Calculate, in terms of r, π and $\sqrt{3}$, the area of the shaded region. (UODLE)

17 The figure shows a large circle of radius 2.5 cm and a smaller concentric circle of radius r cm. The radii OAB and OCD are inclined at $\frac{1}{3}\pi$ radians.

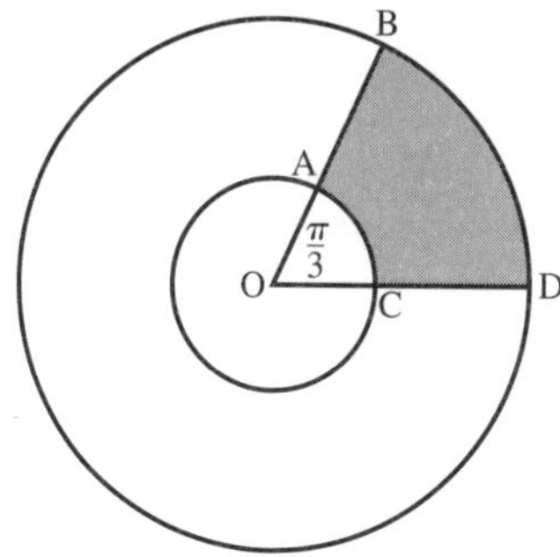

a) Find, in terms of r, expressions for:
i) the perimeter of the shaded region ABDC
ii) the area of the shaded region.
b) The area of the shaded region is equal to the area of the small circle. Find r, giving your answer correct to two decimal places. (UODLE)

3 Functions

I have yet to see any problem, however complicated, which, when you looked at it in the right way, did not become still more complicated.
PAUL ANDERSON

We have so far written the equation of a curve in the form

$y =$ 'Some expression in x'

Another way of writing this is to use **functional notation**. That is,

$$y = \underbrace{f(x)}_{\text{Meaning a function of } x}$$

For example, $y = x^2$ could be written as $f(x) = x^2$. To evaluate the function when $x = 3$ say, we would write

$$f(3) = 3^2$$
$$\therefore \quad f(3) = 9$$

We say that 9 is the image of 3 under the function f. This is identical to saying when $x = 3$, $y = 9$.

In general:

- $f(x)$ is called the **image** of x.
- The set of permitted x values is called the **domain** of the function.
- The set of all images is called the **range** of the function.

When a function is defined for all real values, we write the domain of f as

$$\left\{\text{The set of } x\text{'s such that } x \underbrace{\in}_{\text{Meaning belongs to}} \mathbb{R}\right\} = \{x : x \in \mathbb{R}\}$$

or simply $x \in \mathbb{R}$, where $\mathbb{R}$ is the set of all real numbers.

If a function f is defined for all real values except one particular value, say c, then we write the domain of f as $x \in \mathbb{R},\ x \neq c$.

Transformation of the graph of a function

Here are four functions the graphs of which you need to know:

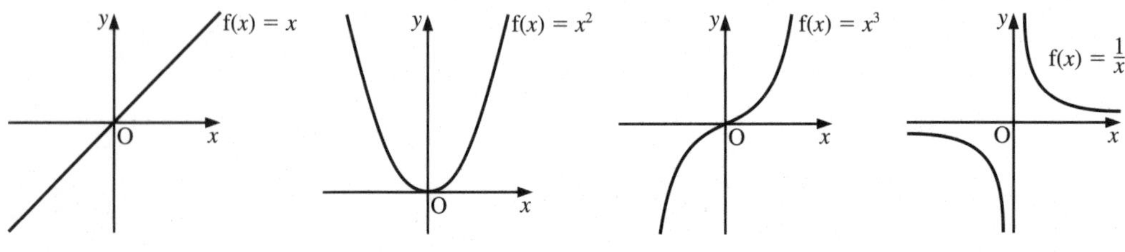

Translations parallel to the y-axis

Consider the function f defined by $f(x) = x^2$, $x \in \mathbb{R}$. Plotting the graph of f gives the curve shown on the first of the two diagrams below.

If we now simplify expressions for (i) $f(x) + 1$ and (ii) $f(x) - 1$, we have

$$\text{i) } f(x) + 1 = x^2 + 1 \quad \text{and} \quad \text{ii) } f(x) - 1 = x^2 - 1$$

Plotting graphs of both (i) and (ii) on the same set of axes gives the curves shown on the second of the two diagrams below.

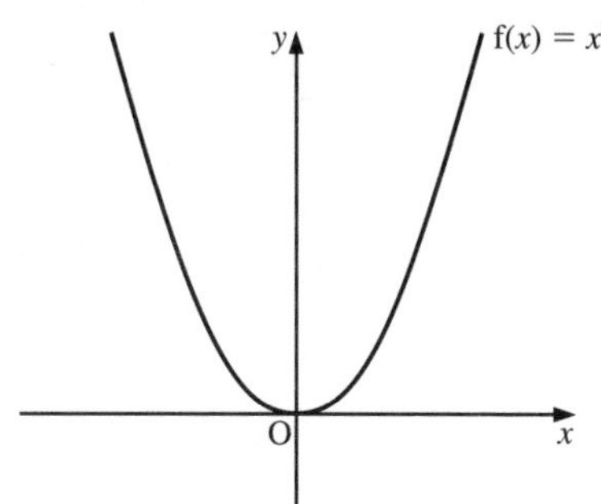

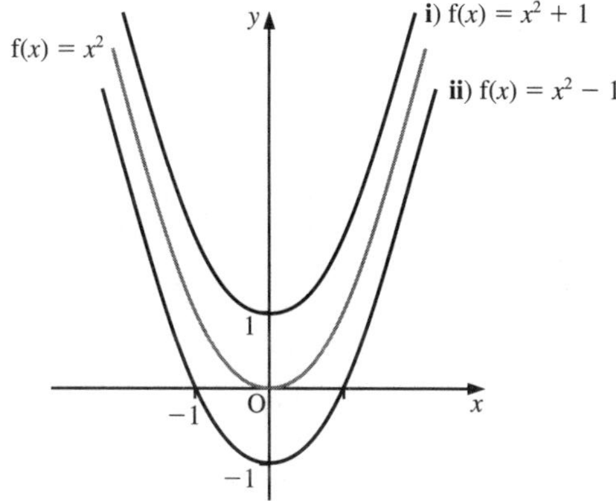

In case (i), the graph of f has been translated 1 unit parallel to the y-axis.

In case (ii), the graph of f has been translated -1 unit parallel to the y-axis.

In general:

- The algebraic transformation $f(x) + a$, where a is a constant, causes a geometric transformation of the graph of f, namely a **translation of a units parallel to the y-axis**.
- The algebraic transformation $f(x) - a$, where a is a constant, causes a geometric transformation of the graph of f, namely a **translation of $-a$ units parallel to the y-axis**.

Example 1 The function f is defined by $f(x) = x^3$, $x \in \mathbb{R}$. Sketch the graph of f. Hence sketch the graph of $g(x) = x^3 - 4$.

SOLUTION

The graph of f(x) is shown below left.

Since $g(x) = f(x) - 4$, we obtain the graph of g by translating the graph of f by -4 units parallel to the y-axis. This is shown below right.

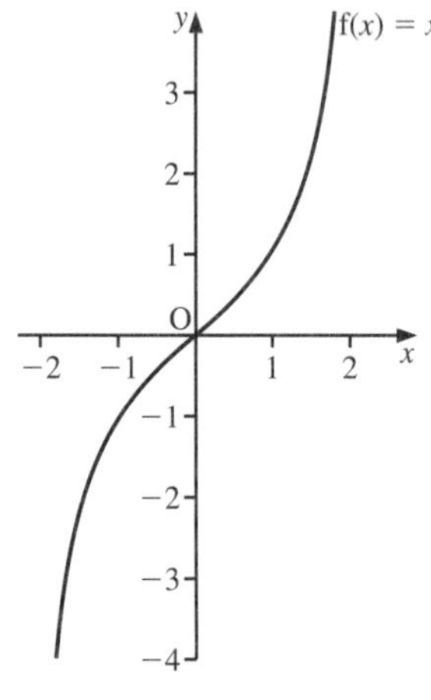

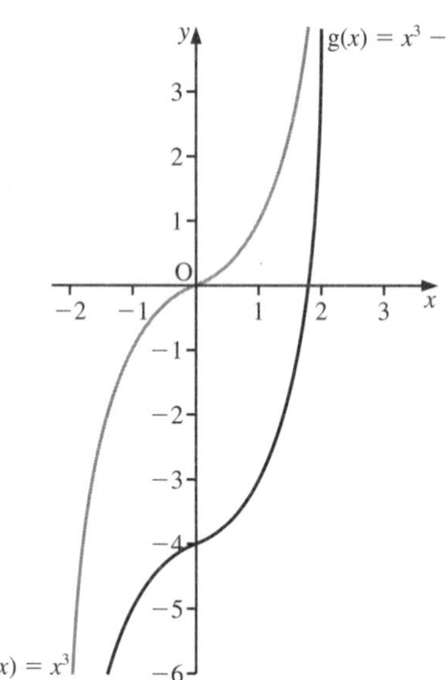

Translations parallel to the x-axis

Consider the function f defined by $f(x) = 4x - 1$, $x \in \mathbb{R}$.

Plotting the graph of f gives the line shown on the left in the two diagrams below.

If we now simplify expressions for (i) $f(x + 2)$ and (ii) $f(x - 2)$, we have

i) $f(x+2) = 4(x+2) - 1 = 4x + 7$ and ii) $f(x-2) = 4(x-2) - 1 = 4x - 9$

Plotting graphs of both (i) and (ii) on the same set of axes gives the parallel lines shown below right.

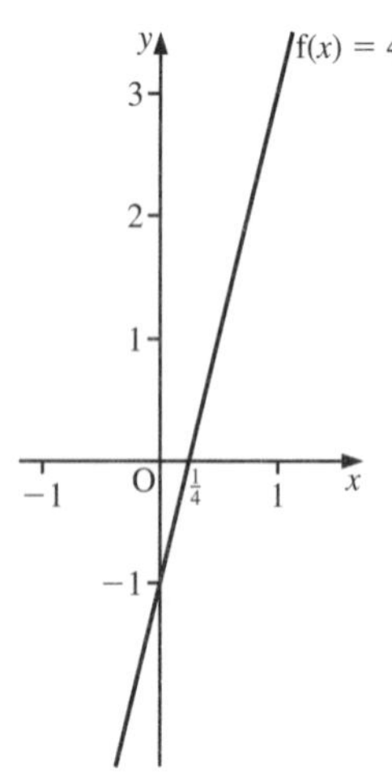

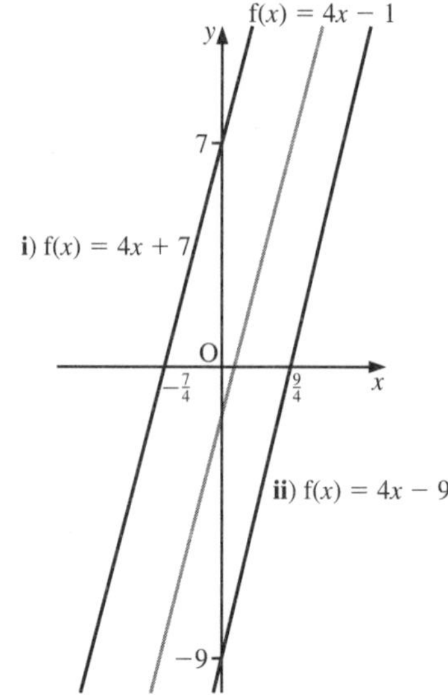

In case (i), the graph of f has been translated -2 units parallel to the x-axis.

In case (ii), the graph of f has been translated 2 units parallel to the x-axis.

In general:

- The algebraic transformation $f(x + a)$, where a is a constant, causes a geometric transformation of the graph of f, namely a **translation of $-a$ units parallel to the x-axis**.
- The algebraic transformation $f(x - a)$, where a is a constant, causes a geometric transformation of the graph of f, namely a **translation of a units parallel to the x-axis**.

Example 2 The function f is defined by $f(x) = x^2 + 1, \ x \in \mathbb{R}$.

a) Sketch the graph of f.

b) The function g is defined by $g(x) = f(x + 3)$. Find $g(x)$ in its simplest form and hence sketch the graph of g.

SOLUTION

a) Sketching the graph of $f(x) = x^2 + 1$ gives the curve shown on the right.

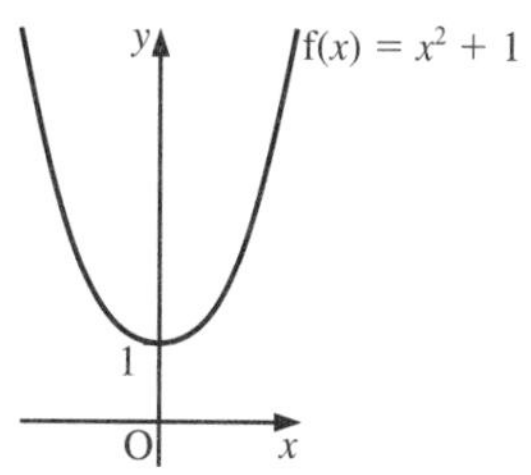

b) The function g is given by

$$\begin{aligned} g(x) &= f(x + 3) \\ &= (x + 3)^2 + 1 \\ &= x^2 + 6x + 9 + 1 \\ \therefore \quad g(x) &= x^2 + 6x + 10 \end{aligned}$$

Since $g(x) = f(x + 3)$, the graph of g can be obtained from the graph of f by a translation of -3 units parallel to the x-axis. Therefore, the graph of g is as sketched on the right.

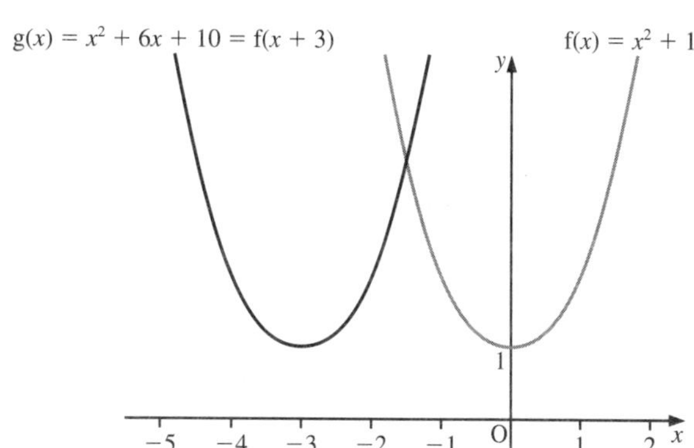

Example 3 The function f is defined by $f(x) = \dfrac{1}{x+3}$, $x \in \mathbb{R}$, $x \neq -3$.
Sketch the graph of f.

SOLUTION

We know that the graph of $\dfrac{1}{x}$ is as shown below left.

The graph of f is obtained by translating the graph of $\dfrac{1}{x}$ by -3 units parallel to the x-axis. Therefore, the graph of f is given as shown below right.

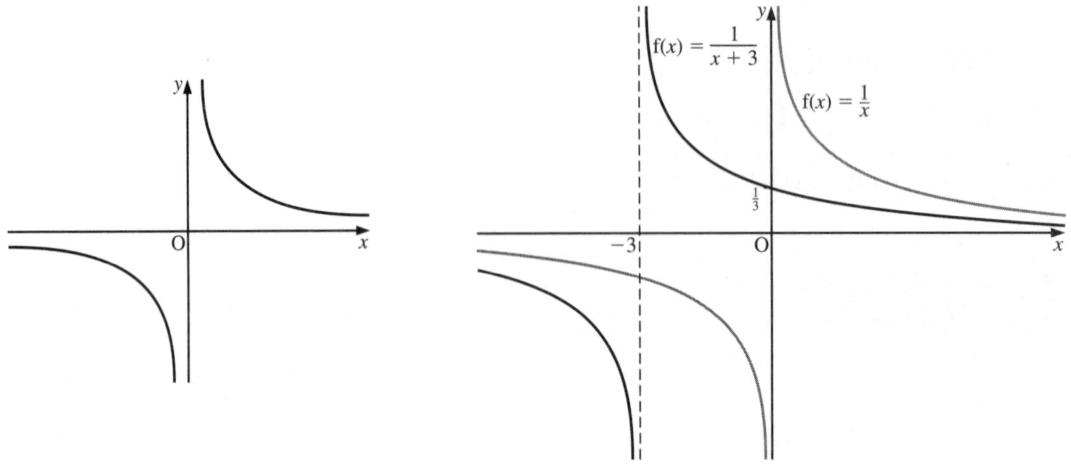

Reflection transformations

Consider the function f defined by $f(x) = x + 1$, $x \in \mathbb{R}$. Plotting the graph of f gives the line shown below left.

Now consider the functions

i) $-f(x) = -(x + 1) = -x - 1$
ii) $f(-x) = (-x) + 1 = -x + 1$

Plotting graphs of both (i) and (ii) on the same set of axes gives the lines shown below right.

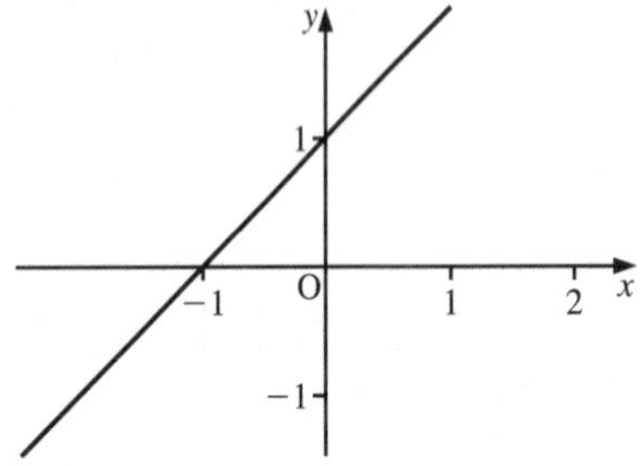

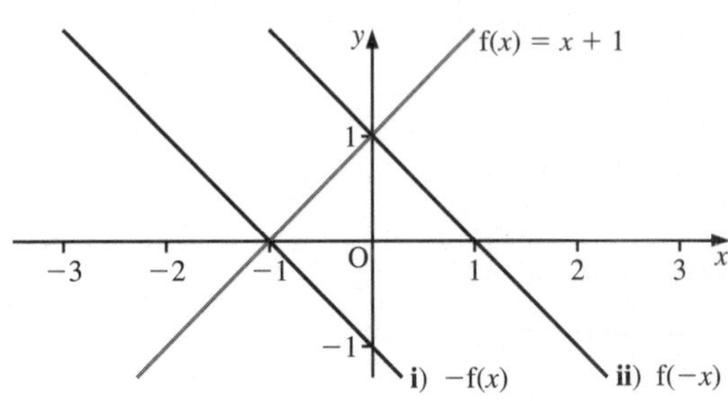

In case (i), the graph of f has been reflected in the x-axis.

In case (ii), the graph of f has been reflected in the y-axis.

In general:

- The algebraic transformation $-f(x)$ causes a geometric transformation of the graph of f, namely a **reflection in the x-axis**.
- The algebraic transformation $f(-x)$ causes a geometric transformation of the graph of f, namely a **reflection in the y-axis**.

Stretch transformations

Consider the function f defined by $f(x) = x + 1,\ x \in \mathbb{R}$. Plotting the graph of f gives the line shown on the left in the two diagrams below.

Now consider the functions (i) $f(2x)$ and (ii) $2f(x)$.

We have

$$\text{i) } f(2x) = (2x) + 1 = 2x + 1 \quad \text{and} \quad \text{ii) } 2f(x) = 2(x + 1) = 2x + 2$$

Plotting the graphs of both (i) and (ii) on the same set of axes gives the lines shown below right.

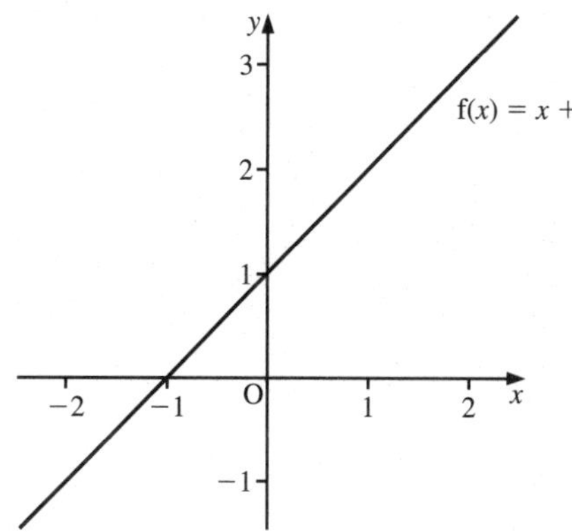

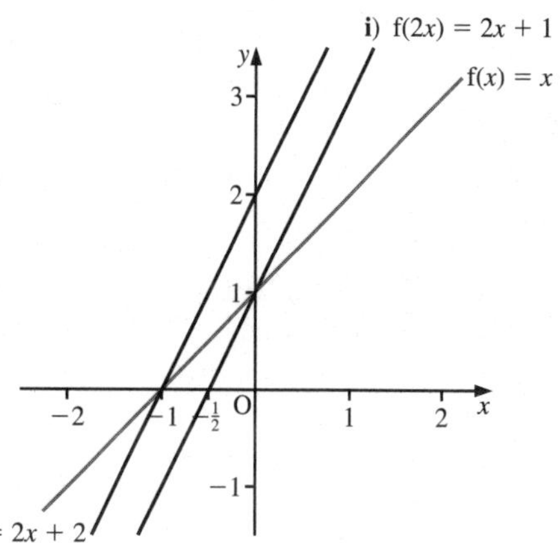

In case (i), the graph of f has been stretched parallel to the x-axis by a scale factor of $\frac{1}{2}$.

In case (ii), the graph of f has been stretched parallel to the y-axis by a scale factor of 2.

In general:

- The algebraic transformation $f(ax)$, where a is a constant, causes a geometric transformation of the graph of f, namely a **stretch parallel to the x-axis by a scale factor of $\frac{1}{a}$.**
- The algebraic transformation $a\,f(x)$, where a is a constant, causes a geometric transformation of the graph of f, namely a **stretch parallel to the y-axis by a scale factor of a.**

Example 4 The function f is defined by $f(x) = 2x + 5,\ x \in \mathbb{R}$. Sketch the graph of f. Describe a sequence of geometric transformations which, when applied to the graph of f, will give the graph of

a) $g(x) = 6x + 5$ **b)** $h(x) = 6x + 16$

SOLUTION

Sketching the graph of $f(x) = 2x + 5$ gives the line shown below left.

a) Since

$$\begin{aligned} g(x) &= 6x + 5 \\ &= 2(3x) + 5 \\ \therefore \quad g(x) &= f(3x) \end{aligned}$$

Therefore, the graph of g can be obtained by stretching the graph of f parallel to the x-axis by a scale factor of $\frac{1}{3}$. This gives the graph of g as shown below right.

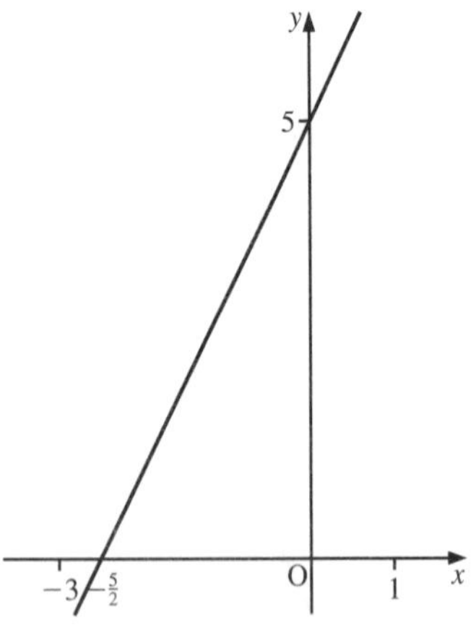

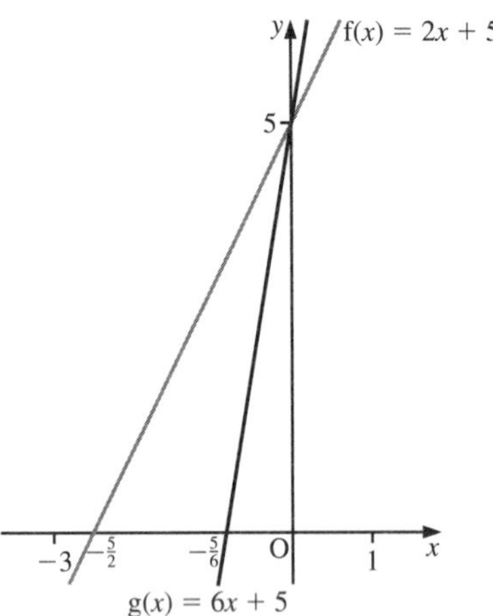

b) Since

$$\begin{aligned} h(x) &= 6x + 16 \\ &= 3(2x + 5) + 1 \\ \therefore \quad h(x) &= 3f(x) + 1 \end{aligned}$$

Therefore, the graph of h can be obtained by stretching the graph of f parallel to the y-axis by a scale factor of 3 followed by a translation of 1 unit parallel to the y-axis. This gives the graph of h as shown on the right.

Alternatively,

$$\begin{aligned} h(x) &= 6x + 16 \\ &= (2(3x) + 5) + 11 \\ \therefore \quad h(x) &= f(3x) + 11 \end{aligned}$$

Therefore, the graph of h can also be obtained by stretching the graph of f parallel to the x-axis by a scale factor of $\frac{1}{3}$ followed by a translation of 11 units parallel to the y-axis.

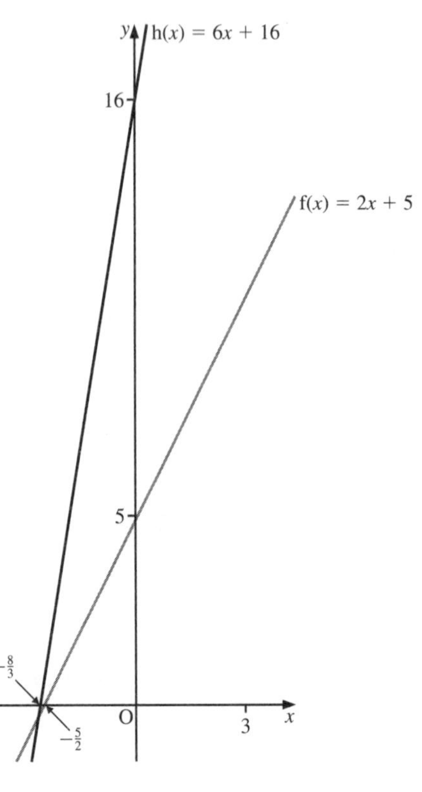

Example 5 The function f is defined by $f(x) = 2x^2 + 4x + 7,\ x \in \mathbb{R}$. Express $f(x)$ in the form $a(x+p)^2 + q$, where a, p and q are constants. Hence describe a sequence of geometrical transformations which when applied to the graph of $g(x) = x^2$ will give the graph of f.

SOLUTION

Completing the square gives

$$\begin{aligned} f(x) &= 2x^2 + 4x + 7 \\ &= 2\left(x^2 + 2x + \frac{7}{2}\right) \\ &= 2\left[(x+1)^2 - 1 + \frac{7}{2}\right] \\ \therefore\quad f(x) &= 2(x+1)^2 + 5 \end{aligned}$$

Therefore,

$$f(x) = 2g(x+1) + 5$$

The graph of f is obtained from the graph of $g(x) = x^2$ by applying the following geometric transformations in the stated order:

i) a translation of -1 unit parallel to the x-axis
ii) a stretch parallel to the y-axis by a scale factor of 2
iii) a translation of 5 units parallel to the y-axis.

Exercise 3A

In each part of Questions **1** to **4**, sketch the curves of the given functions on the same set of axes.

1 a) $y = x^2;\quad y = x^2 + 3;\quad y = x^2 - 2$
b) $y = x^2;\quad y = 2x^2;\quad y = 2x^2 + 5$
c) $y = x^2;\quad y = -x^2;\quad y = 4 - x^2$
d) $y = x^2;\quad y = (x-6)^2;\quad y = (x+3)^2$
e) $y = x^2;\quad y = (3x+1)^2;\quad y = (2x-5)^2$
f) $y = x^2;\quad y = (x-4)^2;\quad y = (x-4)^2 + 7$

2 a) $y = x^3;\quad y = x^3 + 2;\quad y = x^3 - 5$
b) $y = x^3;\quad y = \frac{1}{2}x^3;\quad y = \frac{1}{2}x^3 + 4$
c) $y = x^3;\quad y = -3x^3;\quad y = -3x^3 - 5$
d) $y = x^3;\quad y = (x+1)^3;\quad y = (x-2)^3$
e) $y = x^3;\quad y = (2x-3)^3;\quad y = (3x+5)^3$
f) $y = x^3;\quad y = -(x-1)^3;\quad y = 3 - (x-1)^3$

3 a) $y = \dfrac{1}{x},\ x \neq 0;\quad y = \dfrac{1}{x} + 2,\ x \neq 0$
b) $y = \dfrac{1}{x},\ x \neq 0;\quad y = -\dfrac{1}{x},\ x \neq 0$
c) $y = \dfrac{1}{x},\ x \neq 0;\quad y = \dfrac{2}{x},\ x \neq 0$
d) $y = \dfrac{1}{x},\ x \neq 0;\quad y = \dfrac{1}{x-2},\ x \neq 2$
e) $y = \dfrac{1}{x},\ x \neq 0;\quad y = 1 + \dfrac{1}{x+2},\ x \neq -2$
f) $y = \dfrac{1}{x},\ x \neq 0;\quad y = \dfrac{1}{1-x},\ x \neq 1$

4 a) $y = \sqrt{x},\ x \geqslant 0;\quad y = \sqrt{x} - 2,\ x \geqslant 0;\quad y = \sqrt{x} + 5,\ x \geqslant 0$
b) $y = \sqrt{x},\ x \geqslant 0;\quad y = -\sqrt{x},\ x \geqslant 0;\quad y = 2 - \sqrt{x},\ x \geqslant 0$
c) $y = \sqrt{x},\ x \geqslant 0;\quad y = \frac{1}{2}\sqrt{x},\ x \geqslant 0;\quad y = \frac{1}{2}\sqrt{x} + 3,\ x \geqslant 0$
d) $y = \sqrt{x},\ x \geqslant 0;\quad y = \sqrt{x+3},\ x \geqslant -3;\quad y = \sqrt{x-1},\ x \geqslant 1$

e) $y = \sqrt{x},\ x \geqslant 0;\quad y = \sqrt{2x-5},\ x \geqslant \frac{5}{2};\quad y = \sqrt{4x+3},\ x \geqslant -\frac{3}{4}$

f) $y = \sqrt{x},\ x \geqslant 0;\quad y = \sqrt{5-x},\ x \leqslant 5;\quad y = 2 + \sqrt{5-x},\ x \leqslant 5$

5 Given $f(x) = x^2,\ x \in \mathbb{R}$, sketch the graph of each of the following functions on the same set of axes.

a) $y = f(x)$ **b)** $y = -f(x)$ **c)** $y = f(x-2)$ **d)** $y = 3 + f(x-2)$

6 The function f is defined by $f\colon x \rightarrow \dfrac{1}{x},\ x \in \mathbb{R},\ x > 0$.

a) Sketch the graph of $y = f(x)$.

b) On the same set of axes sketch the graphs of

i) $y = f(x+1)$ **ii)** $y = f(x+1) - 2$

7 A function g is given by $g\colon x \rightarrow x^3,\ x \in \mathbb{R}$. On a single diagram sketch each of these graphs.

a) $y = g(x)$ **b)** $y = g\left(\dfrac{x}{2}\right)$ **c)** $y = g\left(\dfrac{x}{2}\right) - 8$

In each of Questions **8** to **13**, a diagram is given for the graph of a function f(x), where $f(x) = 0$ for $x < 0$ or $x > 4$. On separate axes, sketch the graphs of the functions listed below each diagram.

8

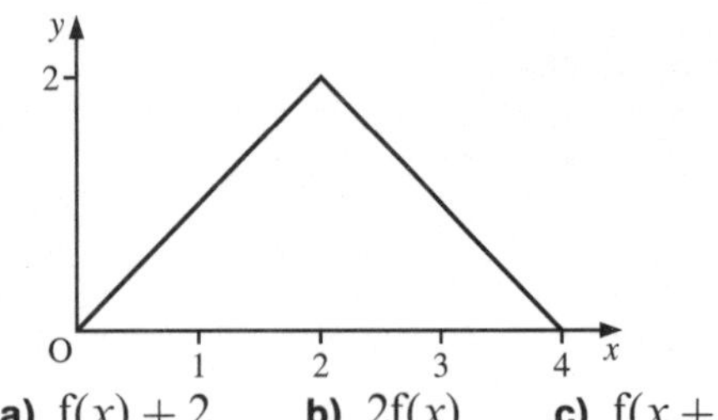

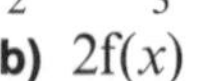

a) $f(x) + 2$ **b)** $2f(x)$ **c)** $f(x+2)$

9

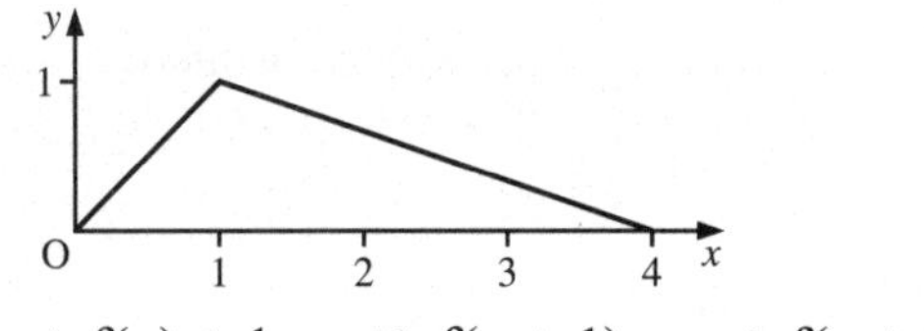

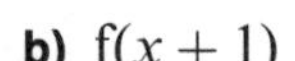

a) $f(x) + 1$ **b)** $f(x+1)$ **c)** $f(x+1) + 1$

10

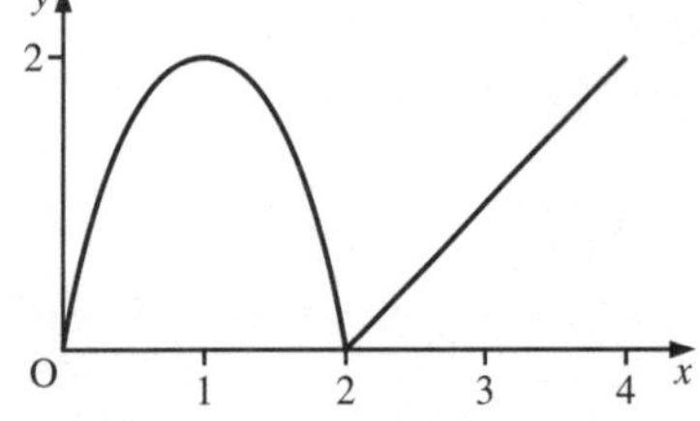

a) $f(x) - 2$ **b)** $f(4-x)$ **c)** $f(2x)$

11

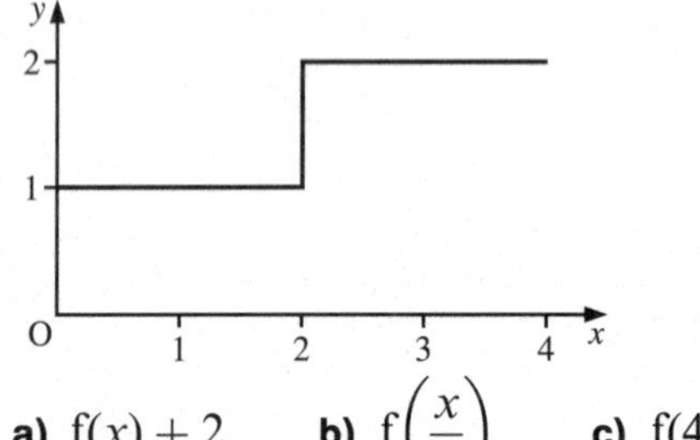

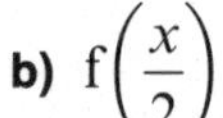

a) $f(x) + 2$ **b)** $f\left(\dfrac{x}{2}\right)$ **c)** $f(4x)$

12

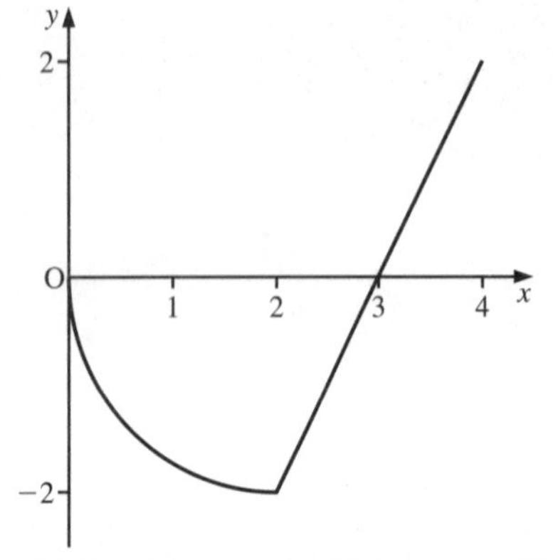

a) $f(x) - 2$ **b)** $-f(x)$ **c)** $f(-x)$

13

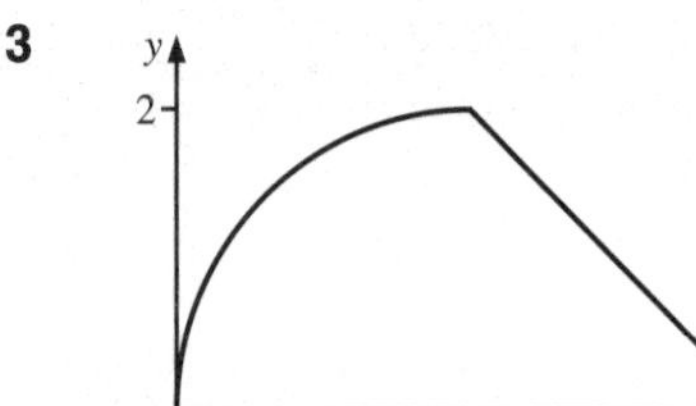

a) $f(x-2)$ **b)** $f(4-x)$ **c)** $f(2-x)$

14 The function f is given by $f : x \longrightarrow 3x - 2,\ x \in \mathbb{R}$. Sketch the graph of f. Find a combination of geometrical transformations which, when applied to the graph of f, will give the graph of $g(x) = 6x + 1$.

15 The functions f and g are defined for all real numbers by $f(x) = -x^2$ and $g(x) = x^2 + 2x + 8$.

a) Express $g(x)$ in the form $(x + a)^2 + b$, where a and b are constants.
b) Describe two transformations in detail, and the order in which they should be applied, whereby the graph of g may be obtained from the graph of f.

16 The function f is defined by $f(x) = x^2,\ x \in \mathbb{R}$. The graph of $g(x)$ is obtained by reflecting the graph of $f(x)$ in the x-axis, and the graph of $h(x)$ is obtained by translating the graph of $g(x)$ by $+2$ units parallel to the y-axis.

a) Sketch the graphs of $f(x)$, $g(x)$ and $h(x)$ on the same set of axes.
b) Find the equations of $g(x)$ and $h(x)$.

17 The function f is defined by $f(x) = \frac{3}{x},\ x \in \mathbb{R},\ x > 0$. The graph of $g(x)$ is obtained by translating the graph of $f(x)$ by -4 units parallel to the x-axis, and the graph of $h(x)$ is obtained by reflecting the graph of $g(x)$ in the x-axis.

a) Sketch the graphs of $f(x)$, $g(x)$ and $h(x)$ on the same set of axes.
b) Find the equations of $g(x)$ and $h(x)$.

***18** Find an expression for the image of the function $f(x)$ under a translation $\begin{pmatrix} p \\ q \end{pmatrix}$.

Mappings

Consider two non-empty sets A and B. A mapping from A to B is a rule which associates with each element of A an element of B.

A mapping can be represented by a mapping diagram. Consider the following mappings all from the set $A = \{-2, -1, 0, 1, 2\}$ to the set $B = \{0, 1, 2, 3, 4, 5, 6\}$.

Case (i)

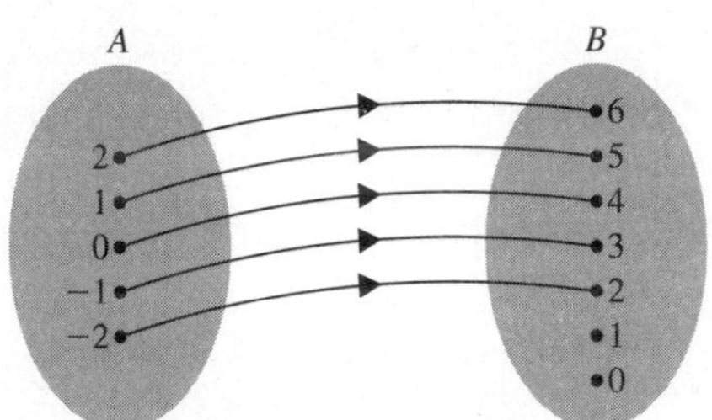

In case (i) we notice that each element of A maps to one and only one element of B. This is called a **one-to-one mapping**. It doesn't matter that no element of A maps to either of the elements 0 or 1 in B.

Case (ii)

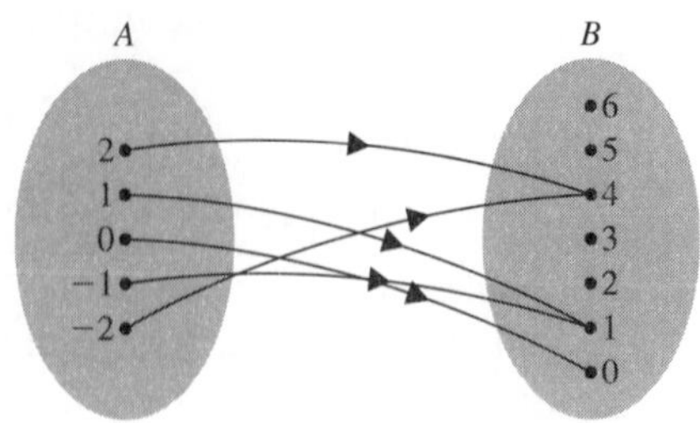

In case (ii) we notice that two elements of A map to one element of B. This is called a **two-to-one mapping**, or a **many-to-one mapping**.

A one-to-one mapping or a many-to-one mapping is called a **function**. We usually denote the rule which associates each element of A to an element B by f.

For example, in case (i) the rule is 'add 4'. Using 'functional notation', we would write

$$f(x) = x + 4 \quad \text{or} \quad f: x \to x + 4$$

In case (ii) we would write the rule as

$$f(x) = x^2 \quad \text{or} \quad f: x \to x^2$$

Example 6 For each of the following mappings f, determine whether f is one-to-one.

a) $f(x) = x^2, \ x \in \mathbb{R}$

b) $f(x) = \dfrac{x}{2} + 1, \ x \in \mathbb{R}$

SOLUTION

a) Consider $f(x) = x^2$. Since $f(-1) = (-1)^2 = 1$ and $f(1) = (1)^2 = 1$, the mapping $f(x) = x^2$ is not one-to-one.

This can also be seen from the graph of the mapping.

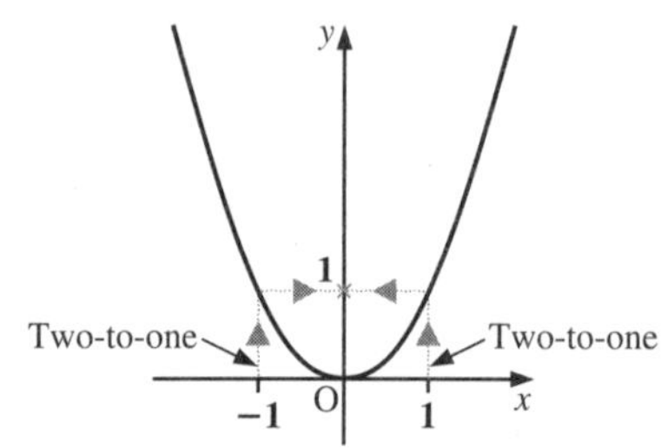

b) The graph of $f(x) = \dfrac{x}{2} + 1$ is a straight line.

It is clear from the graph that this mapping is one-to-one.

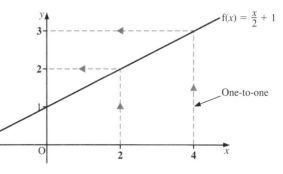

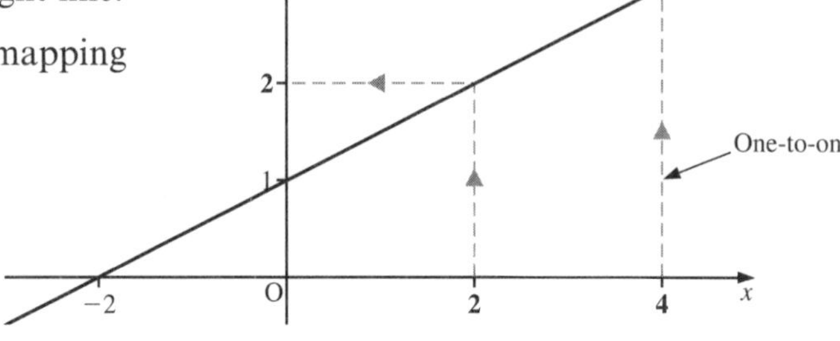

Range of a function

To identify the range of a function, it is very useful to have the graph of the function. For example, if the function $f(x) = 2x$ is defined for all real values of x then the graph of f is as shown on the right, and the range of the function is the set of all images of the function. (In other words, 'that part of the y-axis which is used up by the function'.) Therefore, the range of the function is the set of all real values. We write this as

$$\{f(x) : f(x) \in \mathbb{R}\}$$

or simply $f(x) \in \mathbb{R}$.

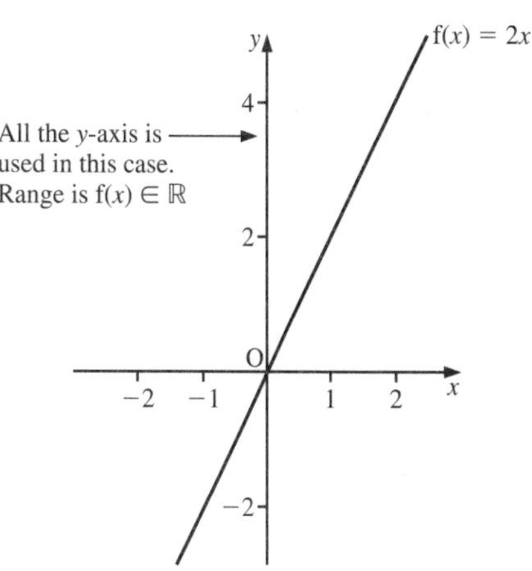

f(x) = 2x
All the y-axis is used in this case.
Range is f(x) ∈ ℝ

A function may be defined on a restricted domain. For example, consider the function f defined by

$$f(x) = 2x, \quad -1 < x < 4$$

Now the graph of f is as shown on the right, and the range of the function is the set of real values from -2 to 8, excluding -2 and 8, since -1 and 4 are excluded in the domain. We write this as $-2 < f(x) < 8$.

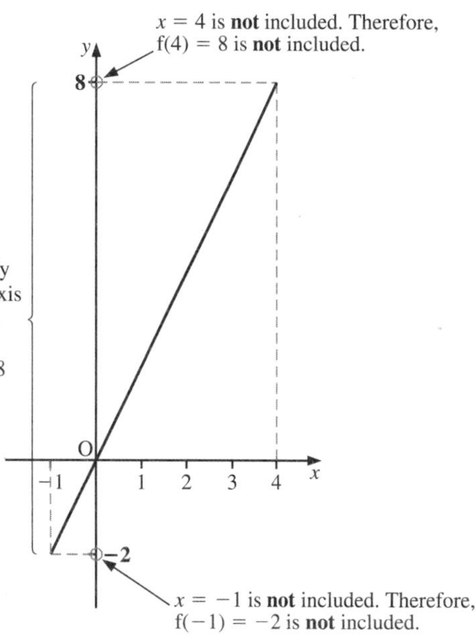

x = 4 is not included. Therefore, f(4) = 8 is not included.
This is the only part of the y-axis which is used. Range is −2 < f(x) < 8
x = −1 is not included. Therefore, f(−1) = −2 is not included.

Example 7 Find the range of each of the following functions.

a) $f(x) = 2x - 1$, for $x \geqslant 0$

b) $f(x) = \dfrac{x}{4}$, for $x < 1$

c) $f(x) = x^2$, for $1 \leqslant x < 3$

SOLUTION

a) The graph of $f(x) = 2x - 1$, for $x \geqslant 0$, is shown on the right.

From the graph it can be seen that when $x \geqslant 0$, $f(x) \geqslant -1$. The range of the function is $f(x) \geqslant -1$.

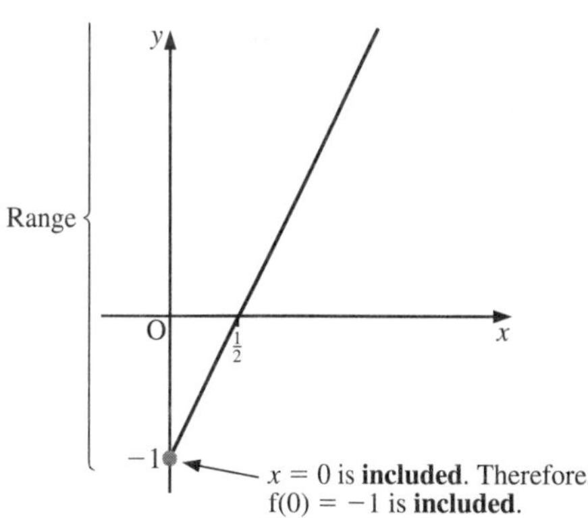

Range
x = 0 is included. Therefore, f(0) = −1 is included.

b) The graph of $f(x) = \dfrac{x}{4}$, for $x < 1$, is shown on the right.

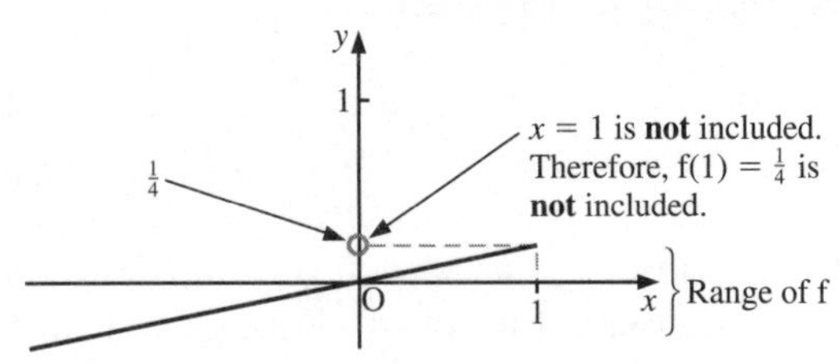

y
1
$\frac{1}{4}$
$x = 1$ is **not** included. Therefore, $f(1) = \frac{1}{4}$ is **not** included.
O
1
x
Range of f

From the graph it can be seen that if $x < 1$ then $f(x) < \frac{1}{4}$. (Notice that since $x = 1$ is **not** included in the domain, the value $f(1) = \frac{1}{4}$ is **not** included in the range.)

The range of the function is $f(x) < \frac{1}{4}$.

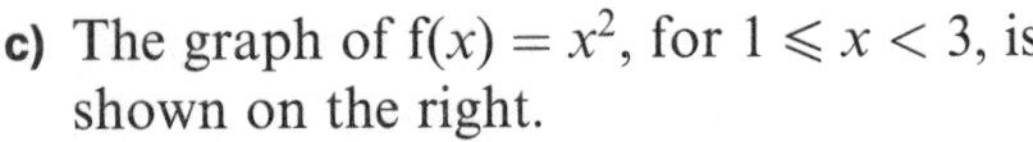

c) The graph of $f(x) = x^2$, for $1 \leqslant x < 3$, is shown on the right.

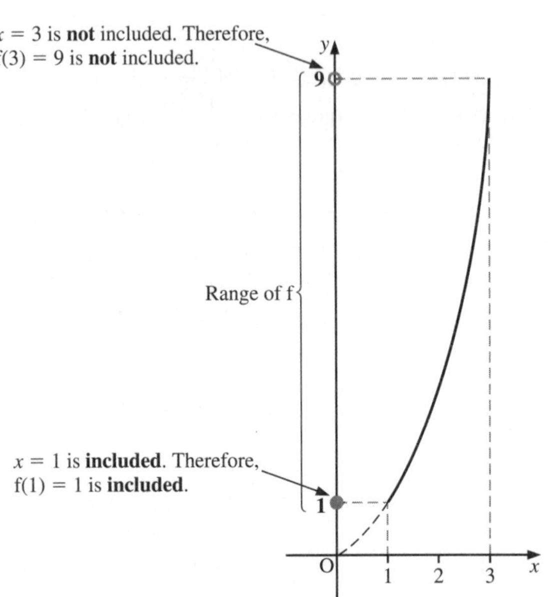

$x = 3$ is **not** included. Therefore, $f(3) = 9$ is **not** included.
y
9
Range of f
$x = 1$ is **included**. Therefore, $f(1) = 1$ is **included**.
1
O
1
2
3
x

From the graph it can be seen that if $1 \leqslant x < 3$ then $1 \leqslant f(x) < 9$.

The range of the function is $1 \leqslant f(x) < 9$.

Example 8 The function f is defined as

$$f(x) = \begin{cases} x+2 & \text{for} \quad 0 \leqslant x \leqslant 2 \\ x^2 & \text{for} \quad 2 \leqslant x \leqslant 4 \end{cases}$$

Draw a sketch graph of the function and state the range of f.

SOLUTION

The sketch graph of f is shown on the right. From the graph it can be seen that the range is $2 \leqslant f(x) \leqslant 16$.

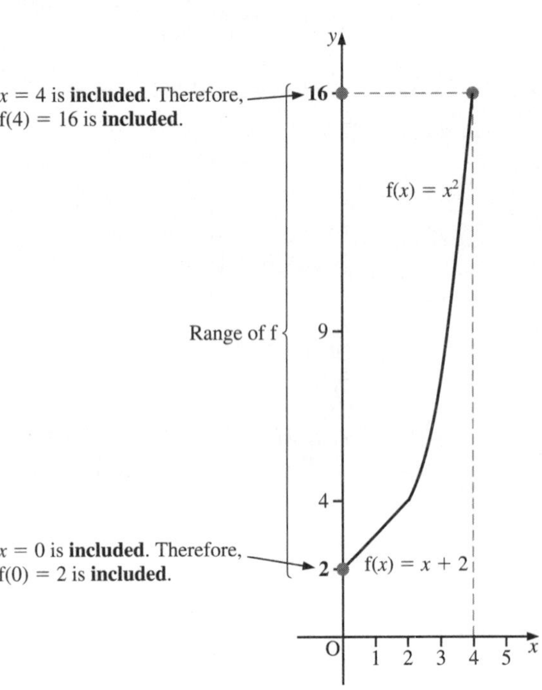

y
$x = 4$ is **included**. Therefore, $f(4) = 16$ is **included**.
16
$f(x) = x^2$
Range of f
9
4
$x = 0$ is **included**. Therefore, $f(0) = 2$ is **included**.
2
$f(x) = x + 2$
O
1
2
3
4
5
x

Example 9 The function f is defined by $f(x) = x + \frac{3}{x}$, for $x \geqslant 2$.

a) Evaluate f(2).
b) Find the value of x for which $f(x) = 4$.

SOLUTION

a) $f(2) = 2 + \frac{3}{2} = \frac{7}{2}$

b) If $f(x) = 4$, then

$$x + \frac{3}{x} = 4$$

$$\therefore \quad x^2 - 4x + 3 = 0$$

$$\therefore \quad (x - 1)(x - 3) = 0$$

Solving gives $x = 1$ and $x = 3$.

Since the domain of f is $\{x : x \geqslant 2\}$ the only value of x that is required is $x = 3$.

Exercise 3B

1 Determine which of the following functions are one-to-one and which are two-to-one.

a) $f: x \rightarrow x + 3,\ x \in \mathbb{R}$
b) $f: x \rightarrow x^2 + 3,\ x \in \mathbb{R}$
c) $f: x \rightarrow \frac{1}{x},\ x \in \mathbb{R},\ x \neq 0$
d) $f: x \rightarrow (x - 4)^2,\ x \in \mathbb{R},\ 2 \leqslant x \leqslant 6$
e) $f: x \rightarrow x^2 - 4x,\ x \in \mathbb{R},\ 0 < x < 4$
f) $f: x \rightarrow x^2 - 4x,\ x \in \mathbb{R},\ 0 < x < 2$
g) $f: x \rightarrow x^4 - 3,\ x \in \mathbb{R},\ 3 \leqslant x \leqslant 6$
h) $f: x \rightarrow \frac{2}{x - 3},\ x \in \mathbb{R},\ -1 < x < 2$
i) $f: x \rightarrow x^3 - x^2,\ x \in \mathbb{R},\ 0 \leqslant x \leqslant 1$
j) $f: x \rightarrow x^6,\ x \in \mathbb{R},\ -2 < x < 0$
k) $f: x \rightarrow x^6,\ x \in \mathbb{R},\ -2 < x < 2$
l) $f: x \rightarrow (x^4 + 1)^2 - 3,\ x \in \mathbb{R}$

2 Determine the range of each of the following functions.

a) $f: x \rightarrow x + 4,\ x \in \mathbb{R},\ 0 < x < 5$
b) $f: x \rightarrow x^2 + 7,\ x \in \mathbb{R}$
c) $f: x \rightarrow 2x - 3,\ x \in \mathbb{R},\ 2 < x \leqslant 6$
d) $f: x \rightarrow \frac{1}{x^2 + 2},\ x \in \mathbb{R},\ 1 \leqslant x \leqslant 4$
e) $f: x \rightarrow (x^2 + 3)^2,\ x \in \mathbb{R}$
f) $f: x \rightarrow 5x^3 - 1,\ x \in \mathbb{R},\ 1 < x < 3$
g) $f: x \rightarrow x^2 - 6x,\ x \in \mathbb{R},\ 0 \leqslant x \leqslant 6$
h) $f: x \rightarrow \frac{1}{x + 1},\ x \in \mathbb{R},\ 1 \leqslant x < 9$
i) $f: x \rightarrow 3\sqrt{x} - 4,\ x \in \mathbb{R},\ 0 < x < \infty$
j) $f: x \rightarrow \sqrt{3x - 2},\ x \in \mathbb{R},\ 2 \leqslant x \leqslant 9$
k) $f: x \rightarrow x^4 + x^2,\ x \in \mathbb{R},\ 0 < x \leqslant 2$
l) $f: x \rightarrow \frac{1}{3 + x^4},\ x \in \mathbb{R}$

3 Sketch the graph of each of the following functions and state its range.

a) $f(x) = \begin{cases} 3x + 4 & \text{for } 0 \leqslant x \leqslant 4 \\ x^2 & \text{for } 4 \leqslant x \leqslant 6 \end{cases}$
b) $g(x) = \begin{cases} x^2 & \text{for } 0 \leqslant x \leqslant 3 \\ 12 - x & \text{for } 3 \leqslant x \leqslant 12 \end{cases}$

c) $h(x) = \begin{cases} x + 3 & \text{for} \quad -3 \leqslant x \leqslant 0 \\ x^2 + 3 & \text{for} \quad 0 \leqslant x \leqslant 2 \end{cases}$

d) $f(x) = \begin{cases} -3(x + 2) & \text{for} \quad -3 \leqslant x \leqslant -2 \\ 4 - x^2 & \text{for} \quad -2 \leqslant x \leqslant 2 \\ 3(x - 2) & \text{for} \quad 2 \leqslant x \leqslant 3 \end{cases}$

e) $g(x) = \begin{cases} (x + 2)^2 & \text{for} \quad -1 \leqslant x \leqslant 0 \\ 4 & \text{for} \quad 0 \leqslant x \leqslant 3 \\ 7 - x & \text{for} \quad 3 \leqslant x \leqslant 6 \end{cases}$

f) $h(x) = \begin{cases} x^3 & \text{for} \quad 0 \leqslant x \leqslant 2 \\ 2x + 4 & \text{for} \quad 2 \leqslant x \leqslant 6 \\ 16 - (x - 6)^2 & \text{for} \quad 6 \leqslant x \leqslant 10 \end{cases}$

***4** Use the method of completing the square to find the range of each of these functions.

a) $f: x \longrightarrow x^2 + 8x + 23, \ x \in \mathbb{R}$
b) $f: x \longrightarrow x^4 - 6x^2 + 13, \ x \in \mathbb{R}$
c) $f: x \longrightarrow \dfrac{1}{x^2 + 4x + 7}, \ x \in \mathbb{R}$

Modulus function

The modulus of x, written $|x|$, is defined as

$$|x| = \begin{cases} x & \text{for} \quad x \geqslant 0 \\ -x & \text{for} \quad x < 0 \end{cases}$$

In other words, $|x|$ means the magnitude of x. For example,

$$|-2| = 2 \qquad |2| = 2 \quad \text{and} \quad |-\tfrac{1}{2}| = \tfrac{1}{2}$$

The modulus function is sometimes called the **absolute value function**.

The graph of $f(x) = |x|$ is shown on the right.

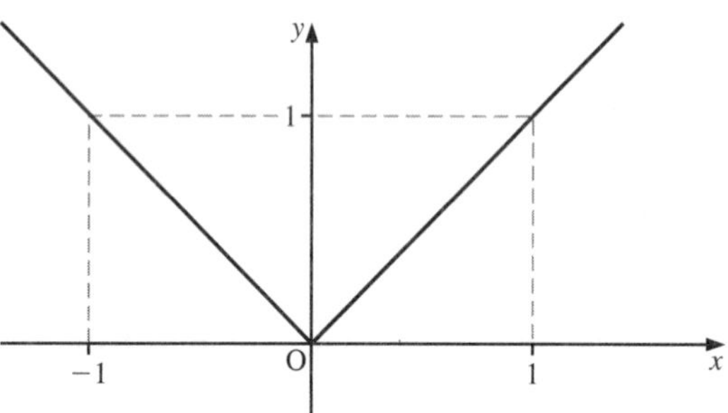

From the graph it can be seen that $|-1| = |1| = 1$. In other words, the mapping $x \longrightarrow |x|$ is not a one-to-one mapping.

The graph of $f(x) = |x|$ is obtained from the graph of $f(x) = x$ by reflecting in the x-axis that part of the graph for which $f(x) < 0$, as shown lower right.

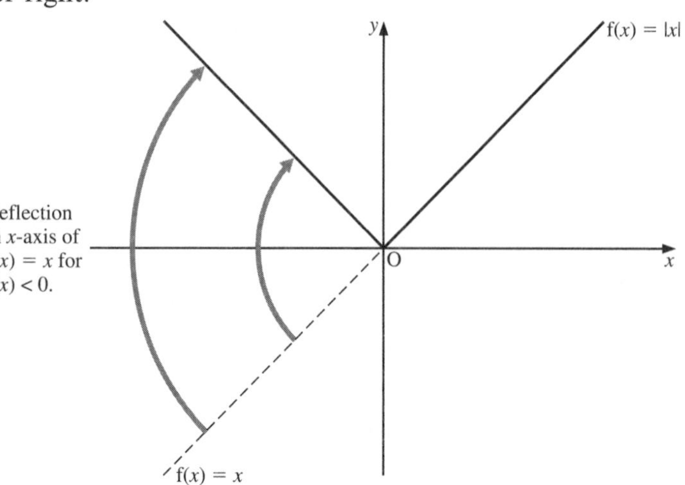

Example 10 Solve the equation $|x - 1| = 4$.

SOLUTION

Squaring both sides will ensure that the LHS is positive:

$$|x - 1|^2 = 4^2$$
$$\therefore \quad (x - 1)^2 = 4^2$$
$$\therefore \quad x^2 - 2x + 1 = 16$$
$$\therefore \quad x^2 - 2x - 15 = 0$$
$$\therefore \quad (x - 5)(x + 3) = 0$$

Solving gives $x = 5$ and $x = -3$.

Example 11 Sketch the graph of $f(x) = |2x - 3|$ and hence solve the equation $|2x - 3| = 2$.

SOLUTION

The graph of $f(x) = 2x - 3$ is shown below left.

Reflecting in the x-axis that part of the graph for which $f(x) < 0$ gives the graph of $f(x) = |2x - 3|$, as shown below right.

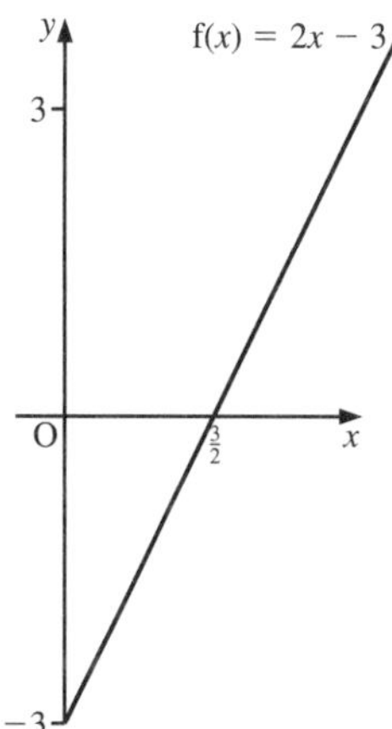

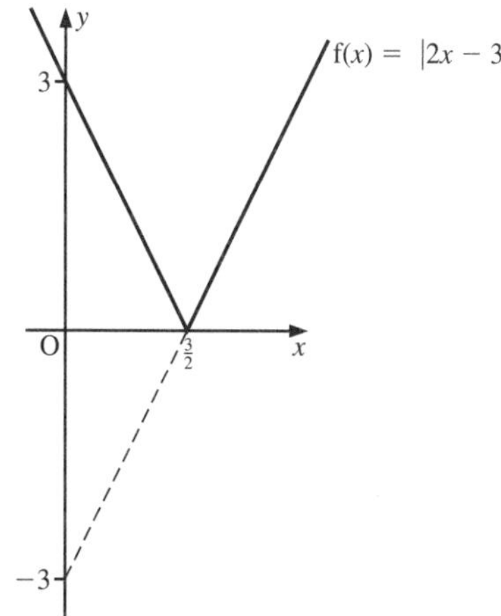

To solve the equation $|2x - 3| = 2$, we draw the line $f(x) = 2$ on the graph of $|2x - 3|$, giving the diagram on the right.

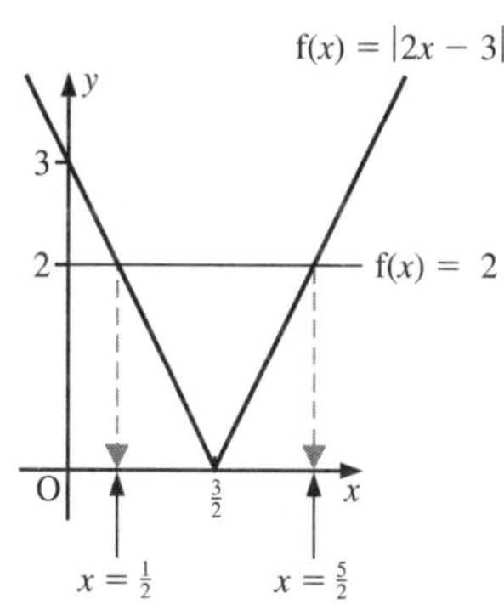

The x-coordinates of the intersection points of $f(x) = 2$ and $f(x) = |2x - 3|$ give the solutions of the equation $|2x - 3| = 2$. The x-coordinates of the intersection points are $x = \frac{1}{2}$ and $x = \frac{5}{2}$. Therefore, the solutions are $x = \frac{1}{2}$ and $x = \frac{5}{2}$.

Example 12 Sketch the graph of $f(x) = |3x + 1|$ and hence solve the inequality $|3x + 1| \leqslant 2$.

SOLUTION

The graph of $f(x) = |3x + 1|$ is shown below left.

Drawing the line $f(x) = 2$ on the same set of axes gives the graph shown below right.

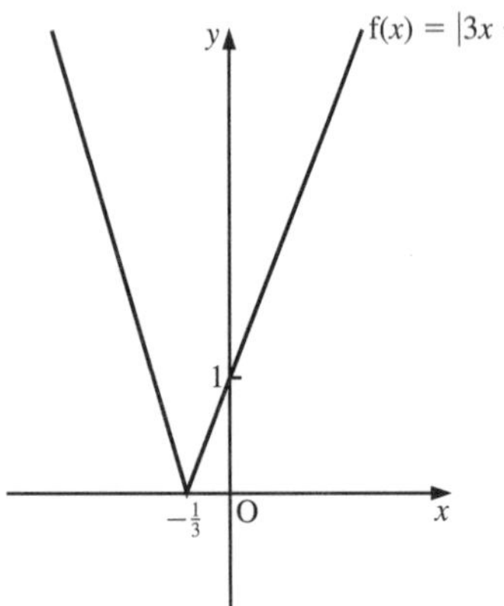

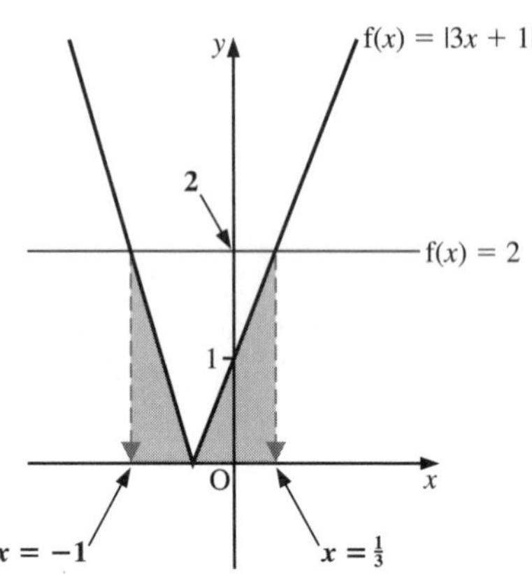

The region for which $|3x + 1| \leqslant 2$ is shaded on the graph above right. Therefore, the inequality $|3x + 1| \leqslant 2$ is satisfied provided that $-1 \leqslant x \leqslant \frac{1}{3}$.

Example 13 Solve the inequality $|2x + 1| \geqslant |x + 3|$.

SOLUTION

Squaring both sides of the inequality will ensure that both the LHS and the RHS are positive. That is,

$$|2x + 1|^2 \geqslant |x + 3|^2$$

$$\therefore \quad (2x + 1)^2 \geqslant (x + 3)^2$$

$$4x^2 + 4x + 1 \geqslant x^2 + 6x + 9$$

$$\therefore \quad 3x^2 - 2x - 8 \geqslant 0$$

$$\therefore \quad (3x + 4)(x - 2) \geqslant 0$$

On the number line we have the situation shown on the right.

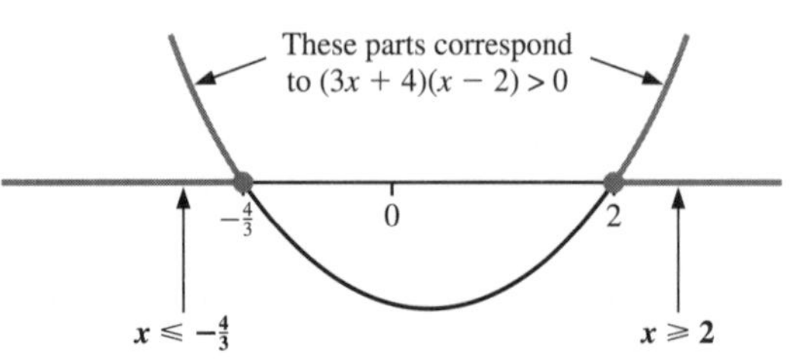

The solution sets are $x \leqslant -\frac{4}{3}$ and $x \geqslant 2$.

Exercise 3C

1 Solve each of the following equations.

a) $|x-2|=4$ **b)** $|x+4|=5$ **c)** $|3-x|=6$
d) $|4-x|=2$ **e)** $|3x+1|=4$ **f)** $|5x-3|=7$
g) $|x+1|=|x-3|$ **h)** $|x-4|=|6-x|$ **i)** $|2x-1|=|x|$
j) $|3x+1|=|x+4|$ **k)** $|2x-5|=|2-3x|$ **l)** $|2x-1|=|4x+3|$

2 Sketch the graph of $f(x)=|3x-2|$, and hence solve the equation $|3x-2|=5$.

3 Sketch the graph of $f(x)=|2x-1|$, and hence solve the equation $|2x-1|=3$.

4 By first sketching the graph of $y=|\frac{1}{2}x-3|$, find the solutions to the equation $|\frac{1}{2}x-3|=2$.

5 On the same set of axes sketch the graphs of the functions $f\colon x \longrightarrow |x-2|$, and $g\colon x \longrightarrow |x-6|$. Hence solve the equation $|x-2|=|x-6|$.

6 Sketch the graph of $y=|2x+5|$, and hence solve the inequality $|2x+5|<7$.

7 Use a graph to solve the inequality $|8x-3|>9$.

8 By first sketching the graph of the function $f(x)=|\frac{1}{4}x+3|$, find the solution to the inequality $|\frac{1}{4}x+3| \geqslant 3$.

9 On one set of axes sketch the graphs of the functions $f(x)=|x|-4$, and $g(x)=\frac{1}{2}x$. Hence solve the inequality $|x|-4 \leqslant \frac{1}{2}x$.

10 The diagram shows the graph of $y=|f(x)|$. Sketch, on two separate diagrams, two possibilities for the graph of $y=f(x)$.

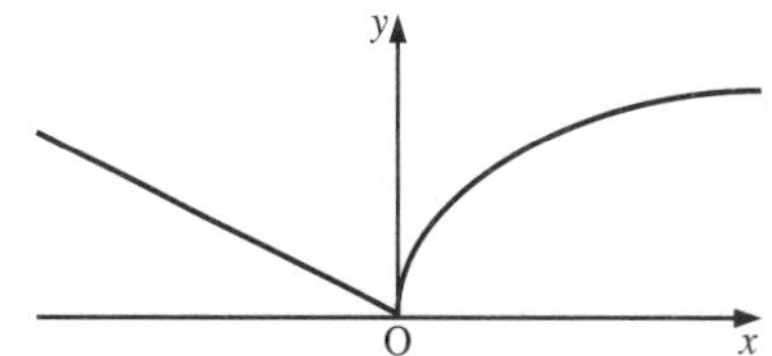

11 Solve each of the following inequalities.

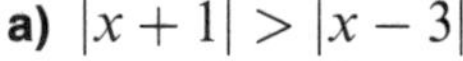
a) $|x+1|>|x-3|$ **b)** $|2x+3| \leqslant |2x-1|$ 
c) $|x+4| \geqslant |2x-3|$
d) $|x-5| \leqslant |3x+2|$ **e)** $|x-1|>|2x+5|$ **f)** $|3-x|<|x-4|$
g) $|2x-3| \geqslant |1-4x|$ **h)** $|2x-7|>|x|$ **i)** $|4x-1|>|3x+1|$
j) $|x-4| \geqslant |2x+5|$ **k)** $|x+3| \leqslant |3x-7|$ **l)** $|2x+5|>|3-4x|$

***12** The diagram shows the graph of $y=f(x)$. On separate diagrams sketch the graphs of the following:

i) $y=|f(x)|$ **ii)** $y=f(|x|)$ **iii)** $y=|f(-|x|)|$

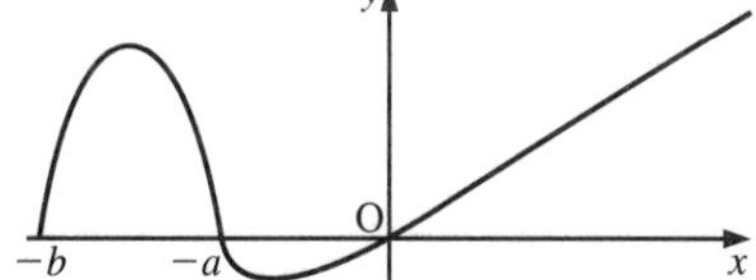

***13** Solve these inequalities.

a) $|x|+3 \geqslant |4x-1|$ **b)** $|2-|x||<|1+2x|$

Even and odd functions

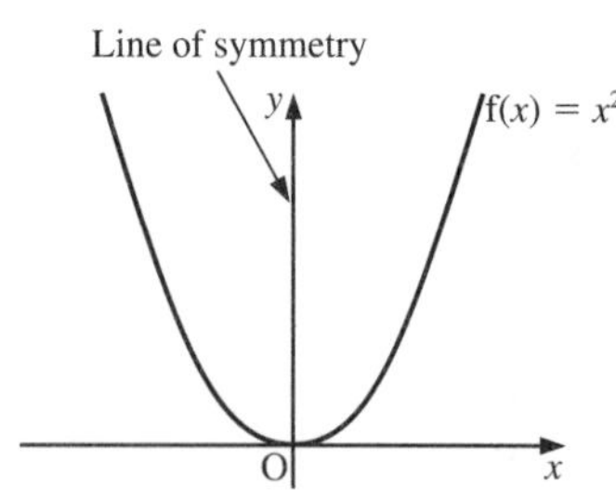

A function f is called **even** if $f(-x) = f(x)$ for all x belonging to the domain of f.

The graph of an even function is symmetrical about the y-axis. For example, the function $f(x) = x^2$ is an even function since

$$f(-x) = (-x)^2 = x^2 = f(x)$$

The graph of $f(x) = x^2$ is symmetrical about the y-axis, as shown.

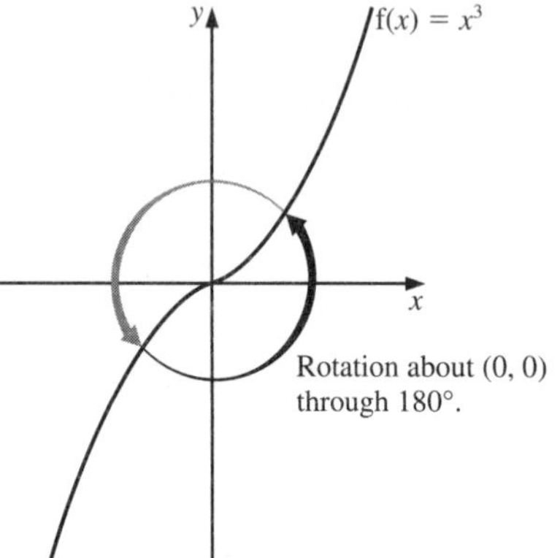

A function f is called **odd** if $f(-x) = -f(x)$ for all x belonging to the domain of f. For example, the function $f(x) = x^3$ is an odd function since

$$f(-x) = (-x)^3 = -x^3 = -f(x)$$

The graph of an odd function is unchanged under a 180° rotation about the origin, as shown.

Example 14 Show that the following functions are both even functions.

a) $f(x) = 2x^4 + x^2 - 1, \; x \in \mathbb{R}$
b) $f(x) = 3x^2 - |x|, \; x \in \mathbb{R}$

SOLUTION

a) We must check that $f(-x) = f(x)$. Now

$$f(-x) = 2(-x)^4 + (-x)^2 - 1$$
$$= 2x^4 + x^2 - 1$$
$$\therefore \quad f(-x) = f(x) \quad \text{for all } x$$

Therefore, the function f is even.

b) Again we must check that $f(-x) = f(x)$. Now

$$f(-x) = 3(-x)^2 - |-x|$$
$$= 3x^2 - |x|$$
$$\therefore \quad f(-x) = f(x) \quad \text{for all } x$$

Therefore, the function f is even.

Example 15 Show that the following functions are both odd functions.

a) $f(x) = 4x^3 - x, \ x \in \mathbb{R}$

b) $f(x) = \dfrac{1}{x} + x, \ x \in \mathbb{R}, \ x \neq 0$

SOLUTION

a) We must check that $f(-x) = -f(x)$. Now

$$\begin{aligned} f(-x) &= 4(-x)^3 - (-x) \\ &= -4x^3 + x = -(4x^3 - x) \end{aligned}$$

$$\therefore \quad f(-x) = -f(x)$$

Therefore, the function f is odd.

b) Again we must check that $f(-x) = -f(x)$. Now

$$\begin{aligned} f(-x) &= \frac{1}{(-x)} + (-x) \\ &= -\frac{1}{x} - x = -\left(\frac{1}{x} + x\right) \end{aligned}$$

$$\therefore \quad f(-x) = -f(x)$$

Therefore, the function f is odd.

Periodic functions

A function whose graph repeats itself at regular intervals is called **periodic**. For example, the graph below is of a periodic function.

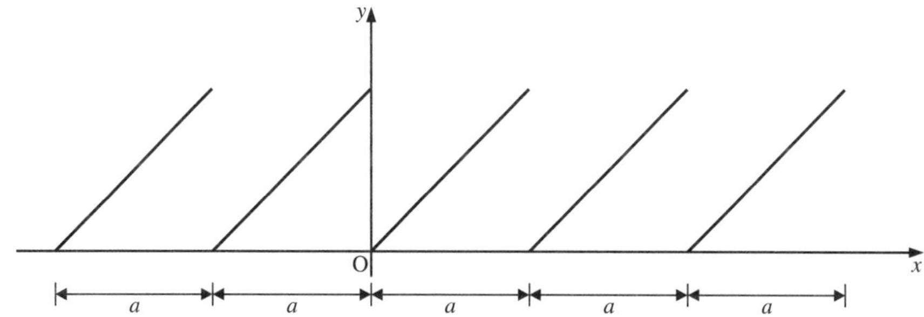

Notice that the graph repeats itself after a distance of a. We say that the **period of the function** is a.

Example 16 The function f is defined by $f(x) = 1 - x$, for $0 < x \leqslant 2$. Given that f is periodic with a period of 2, sketch the graph of f for $-4 < x \leqslant 6$.

SOLUTION

We can plot the graph of $f(x) = 1 - x$ between $x = 0$ and $x = 2$, as shown below left.

Since the function is periodic of period 2, the graph repeats itself every 2 units , as shown below right.

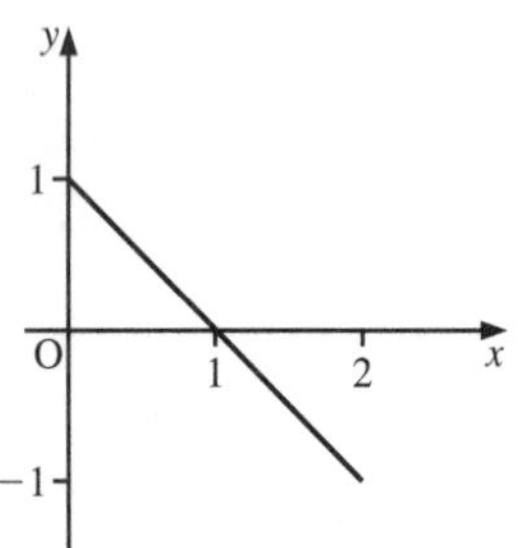

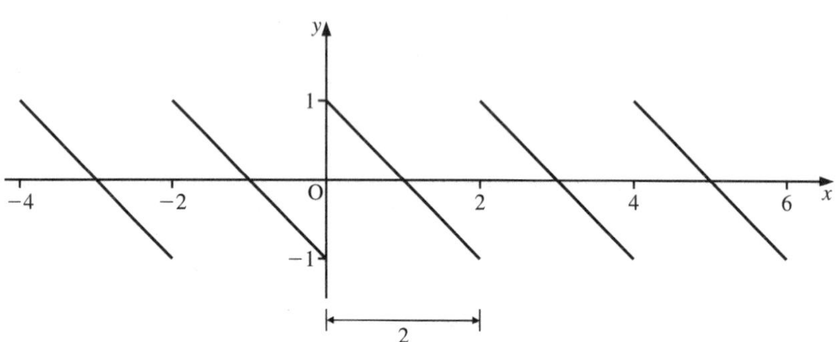

Exercise 3D

1 Determine which of the following functions are odd, which are even, and which are neither odd nor even:

a) $f(x) = 2x^2, \ x \in \mathbb{R}$

b) $f(x) = x^3, \ x \in \mathbb{R}$

c) $f(x) = x^2 - x, \ x \in \mathbb{R}$

d) $f(x) = 1 + |x|, \ x \in \mathbb{R}$

e) $f(x) = x^5 - 1, \ x \in \mathbb{R}$

f) $f(x) = \dfrac{1}{1 + x^2}, \ x \in \mathbb{R}$

g) $f(x) = 1 + \dfrac{1}{x}, \ x \in \mathbb{R}, \ x \neq 0$

h) $f(x) = (x^3 - 5)^2, \ x \in \mathbb{R}$

i) $f(x) = (x^2 - 5)^3, \ x \in \mathbb{R}$

j) $f(x) = \dfrac{1}{x + 3}, \ x \in \mathbb{R}, \ x \neq -3$

k) $f(x) = x^3(1 + x^2), \ x \in \mathbb{R}$

l) $f(x) = \left(\dfrac{x}{x^4 + 3}\right)^3, \ x \in \mathbb{R}$

2 Determine which of the following functions are odd, which are even, and which are neither odd nor even.

a)

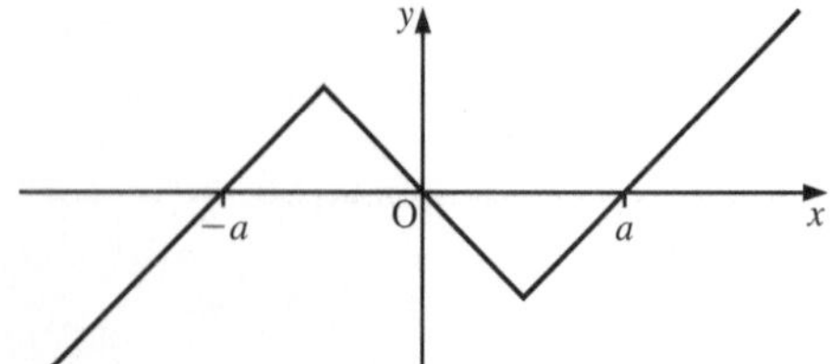

b)

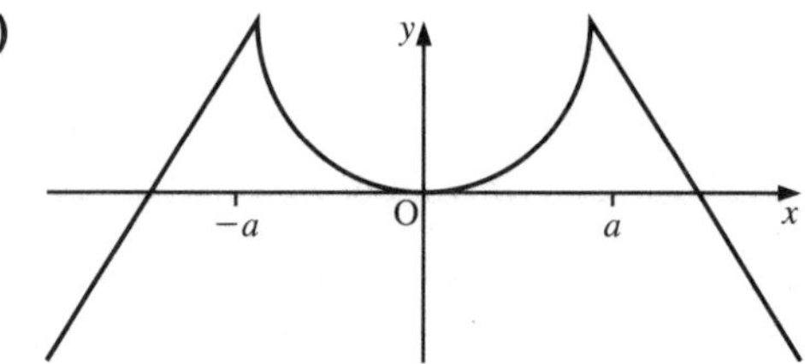

c)

d)

e)

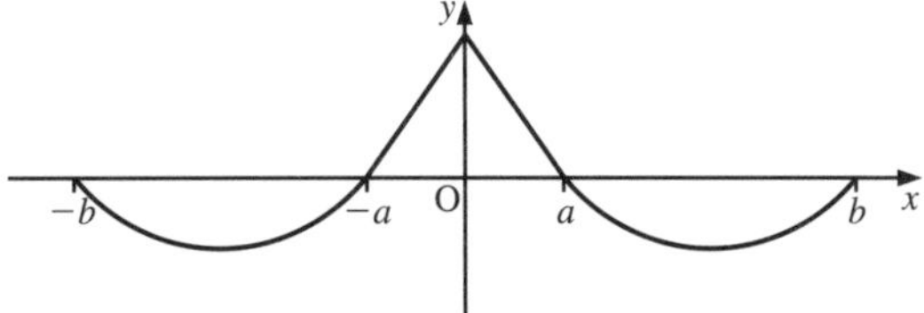

f)

g)

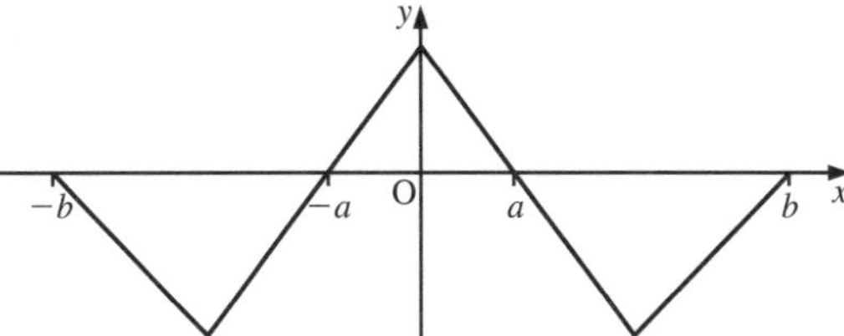

h)

i)

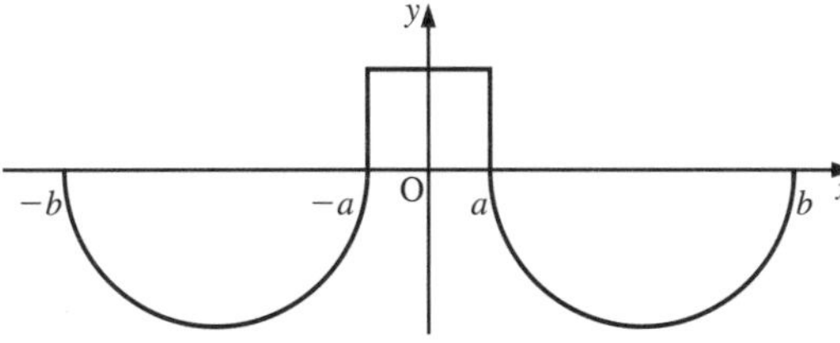

j)

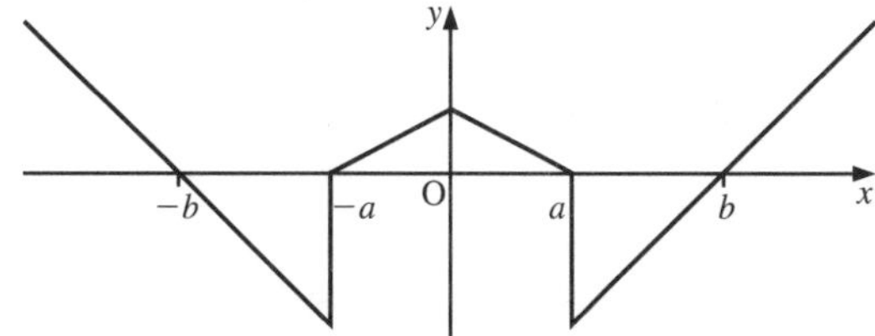

k)

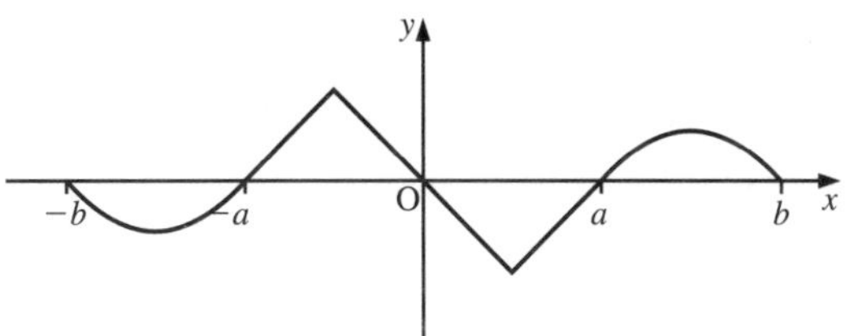

l)

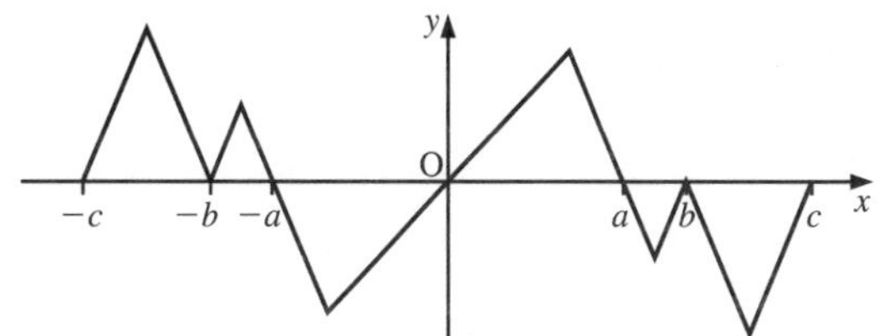

3 The diagram shows part of the graph of a function f(x) for $0 \leqslant x \leqslant 3a$. On separate diagrams draw sketches of f(x) for $-3a \leqslant x \leqslant 3a$ in each of these cases:

a) f is odd **b)** f is even **c)** f is periodic of period $3a$

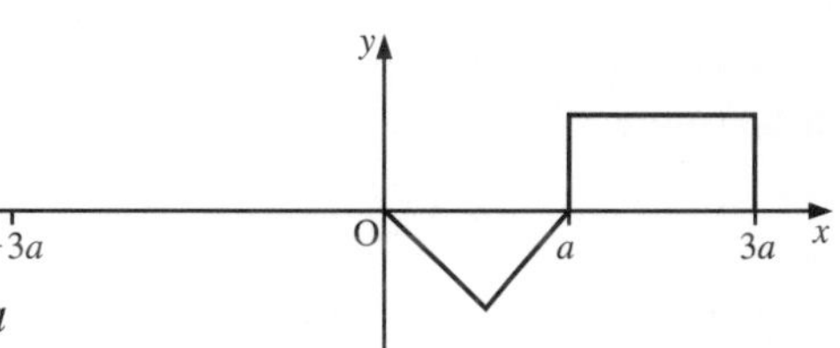

4 The diagram shows part of the graph of a function f(x). On separate diagrams complete the graph of f(x) for $-4 \leqslant x \leqslant 4$ in each of these cases:

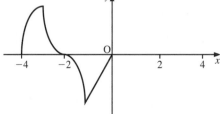

a) f is odd **b)** f is even **c)** f is periodic of period 4

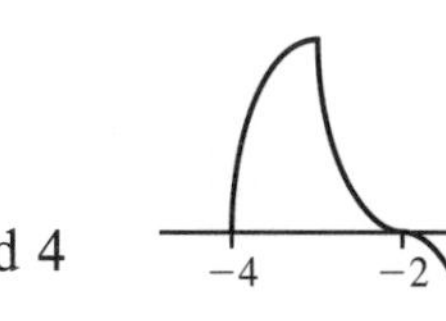

5 The function $g(x)$ is defined for all real numbers. The diagram on the right shows that part of $g(x)$ for which $-a \leqslant x \leqslant 2a$. On separate diagrams complete the graph of $g(x)$ for $-2a \leqslant x \leqslant 2a$ in each of these cases:

a) g is odd **b)** g is periodic of period $3a$

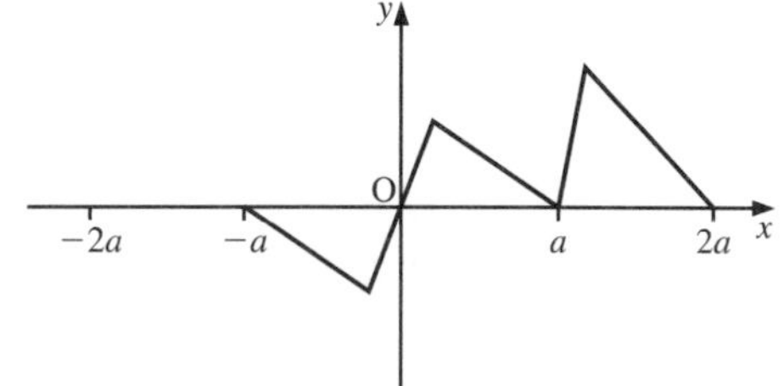

Composite functions

Consider the two functions $f(x) = 2x + 5$ and $g(x) = x - 3$, where the domain of f is $\{1, 2, 3, 4\}$ and the domain of g is the range of f. This can be illustrated using a mapping diagram, as shown below.

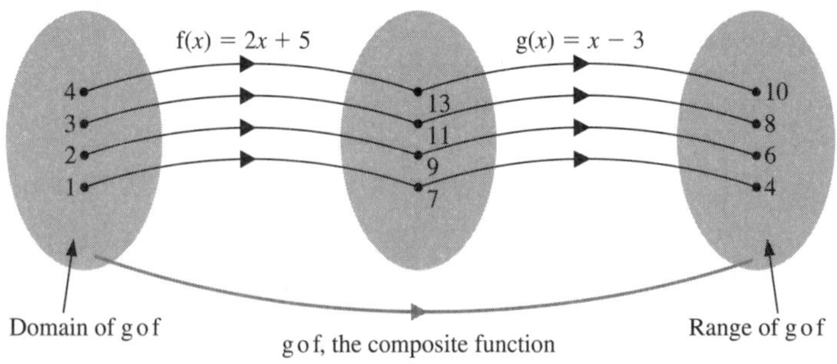

The function indicated on the diagram with domain $\{1, 2, 3, 4\}$ and range $\{4, 6, 8, 10\}$ is called the **composite function**. This function is denoted by gf or g o f.

A single 'rule' for the composite function $gf(x)$ can be obtained in terms of x.

Notice that f is nearest to the variable x since f is the first function to operate on the set $\{1, 2, 3, 4\}$. The rule for the composite function is given by

$$\begin{aligned} gf(x) &= g(2x + 5) \\ &= (2x + 5) - 3 \\ \therefore \quad gf(x) &= 2x + 2 \end{aligned}$$

The composite function gf is defined by $gf(x) = 2x + 2$ with domain $\{1, 2, 3, 4\}$ and range $\{4, 6, 8, 10\}$.

If the function g were to operate first on the set $\{1, 2, 3, 4\}$ and then f were to operate on the range of g, the mapping diagram would be as shown below.

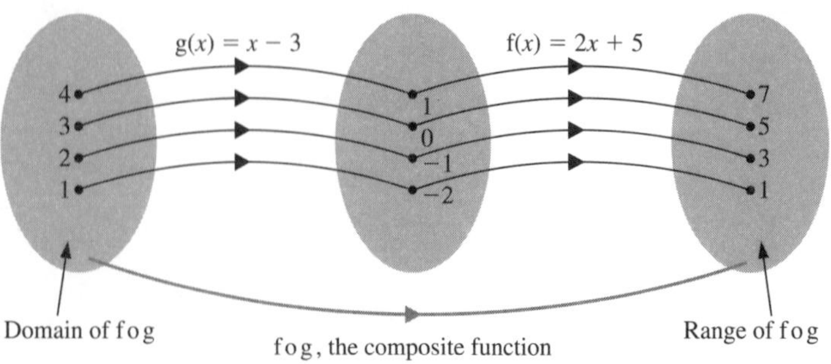

In this case, the composite function which has domain $\{1, 2, 3, 4\}$ and range $\{1, 3, 5, 7\}$ is written as fg or f∘g.

The rule for fg(x) is given by

$$\begin{aligned} \text{fg}(x) &= \text{f}(x-3) \\ &= 2(x-3)+5 \\ &= 2x-6+5 \\ \therefore \quad \text{fg}(x) &= 2x-1 \end{aligned}$$

The composite function fg is defined by $\text{fg}(x) = 2x - 1$ with domain $\{1, 2, 3, 4\}$ and range $\{1, 3, 5, 7\}$.

Example 17 The functions f and g are defined by $\text{f}(x) = 3x - 5$, $x \in \mathbb{R}$ and $\text{g}(x) = 3 - 2x$, $x \in \mathbb{R}$.

a) Evaluate **i)** f(2), **ii)** fg(3).
b) The composite function h is defined by h = gf. Find h(x).

SOLUTION

a) **i)** Since $\text{f}(x) = 3x - 5$, we have

$$\begin{aligned} \text{f}(2) &= 3(2) - 5 \\ \therefore \quad \text{f}(2) &= 1 \end{aligned}$$

ii) To find fg(3), we first evaluate g(3). Since $\text{g}(x) = 3 - 2x$, we have

$$\begin{aligned} \text{g}(3) &= 3 - 2(3) \\ \therefore \quad \text{g}(3) &= -3 \end{aligned}$$

Therefore,

$$\begin{aligned} \text{fg}(3) &= \text{f}(-3) \\ &= 3(-3) - 5 \\ \therefore \quad \text{fg}(3) &= -14 \end{aligned}$$

b) We are given that h = gf. Therefore,

$$\begin{aligned} \text{h}(x) &= \text{gf}(x) \\ &= \text{g}(3x-5) \\ &= 3 - 2(3x-5) \\ &= 3 - 6x + 10 \\ \therefore \quad \text{h}(x) &= 13 - 6x \end{aligned}$$

Example 18 The functions f and g are defined by

$$\text{f}(x) = x^2 \quad 0 \leqslant x \leqslant 4$$

and $\quad \text{g}(x) = x + 3 \quad x \in \mathbb{R}$

Find the composite function gf(x) and state the range of this function.

SOLUTION

Now

$$\text{gf}(x) = \text{g}(x^2)$$
$$= x^2 + 3$$
$$\therefore \quad \text{gf}(x) = x^2 + 3$$

Since f is the first function to operate in the composite function gf, we need the range of f, as this will be the domain of g.

Sketching the graph of $\text{f}(x) = x^2$, for $0 \leqslant x \leqslant 4$, gives the diagram shown on the right.

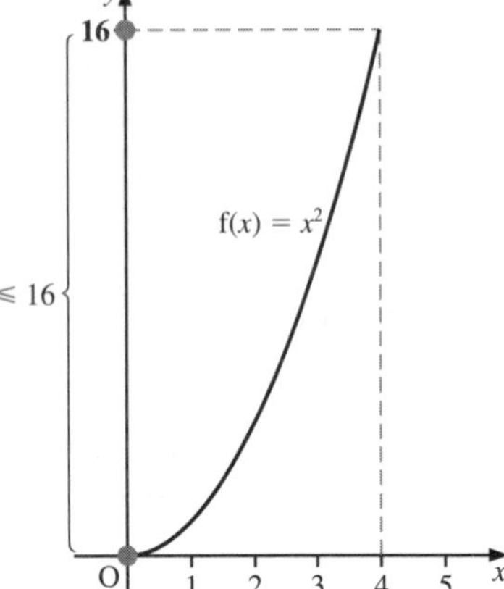

From the graph of f it can be seen that the range of f is $0 \leqslant \text{f}(x) \leqslant 16$. Therefore, the domain of g is $0 \leqslant x \leqslant 16$, giving the graph of g as shown on the left in the two diagrams below.

From the graph of g it can be seen that the range of g (when its domain is $0 \leqslant x \leqslant 16$) is $3 \leqslant x \leqslant 19$. Therefore, the composite function gf has range $3 \leqslant \text{gf}(x) \leqslant 19$.

Alternatively, since we know that $\text{gf}(x) = x^2 + 3$ and it has domain $0 \leqslant x \leqslant 4$, we can sketch the graph of $\text{gf}(x)$, as shown below right.

From the graph it can be seen that the range is $3 \leqslant \text{gf}(x) \leqslant 19$, as before.

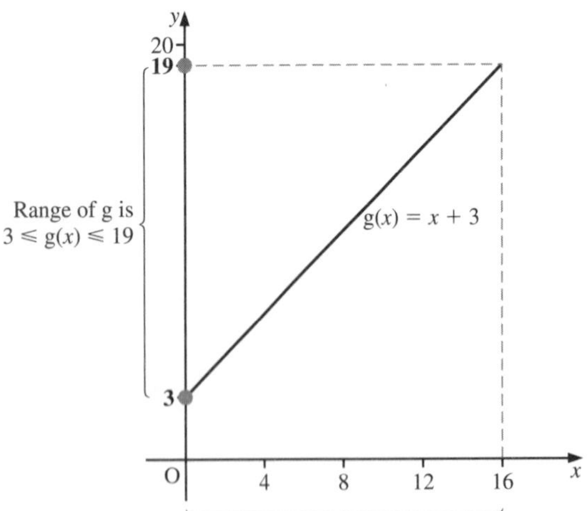

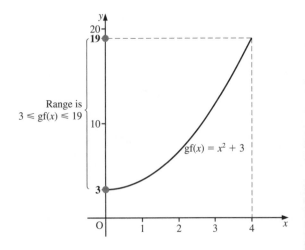

Note that sometimes the composite function is more complicated and therefore the alternative method shown above is not quite as straightforward.

Exercise 3E

Throughout this exercise, the domain of each function is the set of real numbers unless specifically stated otherwise.

1 Given $f(x) = 2x + 1$, $g(x) = x^2$ and $h(x) = \frac{1}{x}$ evaluate each of the following.

a) f(3) **b)** g(2) **c)** hg(2) **d)** fg(−3) **e)** gf(1) **f)** gh(−2)
g) hf(4) **h)** ff(5) **i)** gg(−3) **j)** hh(12) **k)** fgh(2) **l)** hfg(4)

2 Given $f: x \to 3x - 1$, $g: x \to x^2$ and $h: x \to \frac{2}{x}$, write down and simplify expressions for each of the following.

a) $fg(x)$ **b)** $gf(x)$ **c)** $fh(x)$ **d)** $hg(x)$ **e)** $gg(x)$ **f)** $ff(x)$

3 Functions f and g are defined by

$$f: x \to x^2 + 3 \qquad g: x \to x + 5$$

a) Write down and simplify expressions for **i)** $fg(x)$, **ii)** $gf(x)$.
b) Hence solve the equation $fg(x) = gf(x)$.

4 Functions h and k are defined by

$$h: x \to \frac{3}{x} \qquad k: x \to x + 5$$

a) Write down an expression for $hk(x)$, and hence solve the equation $hk(x) = 1$.
b) Write down an expression for $kh(x)$, and hence solve the equation $kh(x) = 6$.

5 Given $f(x) = x^2$ and $g(x) = 2x + 5$, solve the following equations.

a) $fg(x) = 9$ **b)** $gg(x) = 21$

6 Given

$$f(x) = x^2,\ x \in \mathbb{R},\ 1 \leqslant x \leqslant 5 \quad \text{and} \quad g(x) = 2x + 5,\ x \in \mathbb{R}$$

find an expression for the composite function $gf(x)$. State the domain and range of $gf(x)$.

7 Functions p and q are defined by

$$p: x \to 3x^2 + 1,\ x \in \mathbb{R},\ 0 \leqslant x \leqslant 2 \qquad \text{and} \qquad q: x \to x^2 - 2,\ x \in \mathbb{R}$$

Find the composite function $qp(x)$ and state its range.

8 Given

$$f: x \to x^2 + 4,\ x \in \mathbb{R} \quad \text{and} \quad g: x \to \frac{1}{x - 3},\ x \in \mathbb{R},\ x \geqslant 4$$

find an expression for the composite function $gf(x)$ and state its range.

9 Functions g and h are defined by

$$\text{g}: x \longrightarrow x^2 + 3, \ x \in \mathbb{R} \qquad \text{and} \qquad \text{h}: x \longrightarrow |x| - 5, \ x \in \mathbb{R}$$

a) Write down an expression for hg(x) and state its range.
b) Write down an expression for gg(x) and state its range.

10 Given

$$\text{f}(x) = \sqrt{x+1}, \ x \in \mathbb{R}, \ x > 0 \quad \text{and} \quad \text{g}(x) = x^2, \ x \in \mathbb{R}$$

a) find an expression for fg(x) and state its range
b) find an expression for gf(x) and state its range.

11 Functions h and k are defined by h(x) $= 3x + 5$, and k(x) $= 2 - x$.

a) Write down and simplify expressions for hh(x) and kk(x).
b) Hence solve the equation hh(x) = kk(x).

12 Functions f and g are defined by

$$\text{f}: x \longrightarrow x + 1, \ x \in \mathbb{R} \qquad \text{and} \qquad \text{g}: x \longrightarrow x^2 - 3, \ x \in \mathbb{R}$$

a) Show that fg(x) + gf(x) $= 2x^2 + 2x - 4$.
b) Hence solve the equation fg(x) + gf(x) = 0.

13 Given f(x) $= x^2 + 3$, g(x) $= 2x + a$ and fg(x) $= 4x^2 - 8x + 7$, calculate the value of the constant a.

14 Functions p, q and r are defined by

$$\text{p}: x \longrightarrow \frac{3}{x+1} \qquad \text{and} \qquad \text{q}: x \longrightarrow \frac{b}{x^2} \qquad \text{and} \qquad \text{r}: x \longrightarrow \frac{3x^2}{2+x^2}$$

Given pq(x) = r(x), find the value of the constant b.

***15** Functions f and g are defined for all real numbers and are such that g(x) $= x^2 + 7$, and gf(x) $= 9x^2 + 6x + 8$. Find possible expressions for f(x).

***16** **a)** Given f(x) $= ax + b$, and $\text{f}^{(3)}(x) = 64x + 21$, find the values of the constants a and b.
b) Suggest a rule for $\text{f}^{(n)}(x)$.
[Note: for $\text{f}^{(3)}(x)$ read fff(x).]

Inverse functions

Consider the function f defined by f(x) $= x + 3$ with domain $\{1, 2, 3\}$. The range of f is $\{4, 5, 6\}$. We now want a function f^{-1}, called the **inverse function**, which has domain $\{4, 5, 6\}$ and range $\{1, 2, 3\}$ such that

$$\text{f}^{-1}(4) = 1 \qquad \text{f}^{-1}(5) = 2 \quad \text{and} \quad \text{f}^{-1}(6) = 3$$

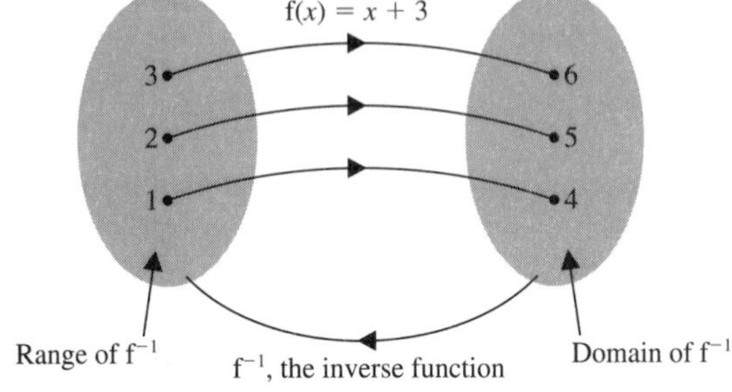

In this case, it is easy to see that the inverse function, f^{-1}, is given by

$$f^{-1}(x) = x - 3$$

However, in some examples it is not quite so easy to identify a formula for f^{-1}. It is for this reason that we require a technique for finding such a formula. Consider the function $y = f(x)$ whose mapping diagram is shown below.

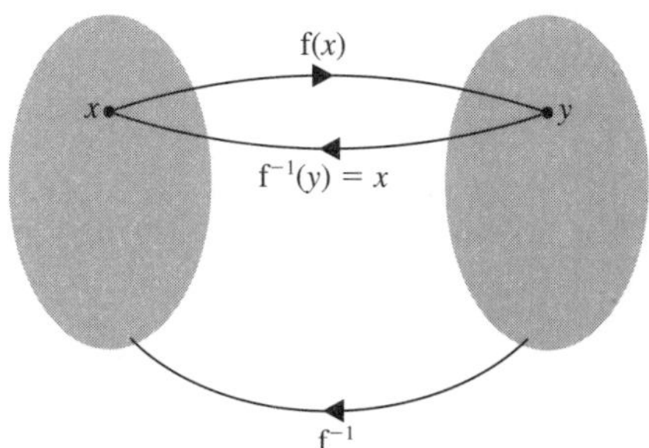

We require the function f^{-1} such that $f^{-1}(y) = x$. In other words, we require x to be expressed as a function of y. Therefore, a useful technique for finding the formula for an inverse function is to let $y = f(x)$ and rearrange for x.

In this example, let $y = x + 3$, then rearranging for x gives $x = y - 3$. (In other words, if we are given y, the corresponding x value can be found using $x = y - 3$.) Therefore, the inverse function is given by $f^{-1}(x) = x - 3$.

The function $f(x) = x - 3$ is an example of a one-to-one function. Therefore, the inverse function is also a one-to-one function. However, if we attempted to find the inverse of a many-to-one function, we would have a one-to-many mapping as the inverse. But a one-to-many mapping is not a function. It is for this reason that only one-to-one functions can have inverses.

If a function f has an inverse f^{-1}, then the composite function ff^{-1} is given by $ff^{-1}(x) = x$, and similarly the composite function $f^{-1}f$ is given by $f^{-1}f(x) = x$.

Example 19 The function f is defined by $f(x) = 5x + 4$, $x \in \mathbb{R}$. Find $f^{-1}(x)$ and verify that $ff^{-1}(x) = x$.

SOLUTION

To find $f^{-1}(x)$, let $y = 5x + 4$. Then rearranging for x gives $x = \dfrac{y-4}{5}$.

Therefore, the inverse function is given by $f^{-1}(x) = \dfrac{x-4}{5}$.

The composite function $ff^{-1}(x)$ is given by

$$\begin{aligned} ff^{-1}(x) &= f\left(\frac{x-4}{5}\right) \\ &= 5\left(\frac{x-4}{5}\right) + 4 = x \end{aligned}$$

Therefore, $ff^{-1}(x) = x$, as required.

Example 20 Two functions f and g are defined by

$$f(x) = 7x + 1, \quad x \in \mathbb{R}$$

and $$g(x) = \frac{x}{3} - 1, \quad x \in \mathbb{R}$$

Find the inverse functions f^{-1} and g^{-1} and verify that $(fg)^{-1} = g^{-1}f^{-1}$.

SOLUTION

To find the inverse function of f, let $y = 7x + 1$. Then rearranging for x gives $x = \frac{y-1}{7}$.

Therefore, $f^{-1}(x) = \frac{x-1}{7}$.

To find the inverse function of g, let $y = \frac{x}{3} - 1$. Then rearranging for x gives $x = 3y + 3$.

Therefore, $g^{-1}(x) = 3x + 3$.

To show that $(fg)^{-1} = g^{-1}f^{-1}$, first look at the LHS. We need the composite function $fg(x)$, which is given by

$$\begin{aligned} fg(x) &= f\left(\frac{x}{3} - 1\right) \\ &= 7\left(\frac{x}{3} - 1\right) + 1 \\ &= \frac{7x}{3} - 7 + 1 \\ \therefore \quad fg(x) &= \frac{7x}{3} - 6 \end{aligned}$$

To find the inverse of $fg(x) = \frac{7x}{3} - 6$, let $y = \frac{7x}{3} - 6$. Then rearranging for x gives

$$x = \frac{3y + 18}{7}$$

Therefore,

$$(fg)^{-1}(x) = \frac{3x + 18}{7}$$

Next, look at the RHS. We need the composite function $g^{-1}(f^{-1}(x))$, which is given by

$$g^{-1}(f^{-1}(x)) = g^{-1}\left(\frac{x-1}{7}\right)$$
$$= 3\left(\frac{x-1}{7}\right) + 3$$
$$= \frac{3x-3}{7} + 3$$
$$= \frac{3x-3+21}{7}$$

Therefore $\quad g^{-1}(f^{-1}(x)) = \dfrac{3x+18}{7} = (fg)^{-1}(x)$

as required.

We have said that only one-to-one functions have inverses. However, many-to-one functions can have inverses by restricting the domain of the function so that it is one-to-one. For example, the function $f(x) = x^2$ defined for all real x is a two-to-one function and has the graph shown below left.

If we restrict the domain to $x \geqslant 0$, the graph of the function becomes that shown below right. We now have a one-to-one function which will have an inverse. The inverse is $f^{-1}(x) = +\sqrt{x}$.

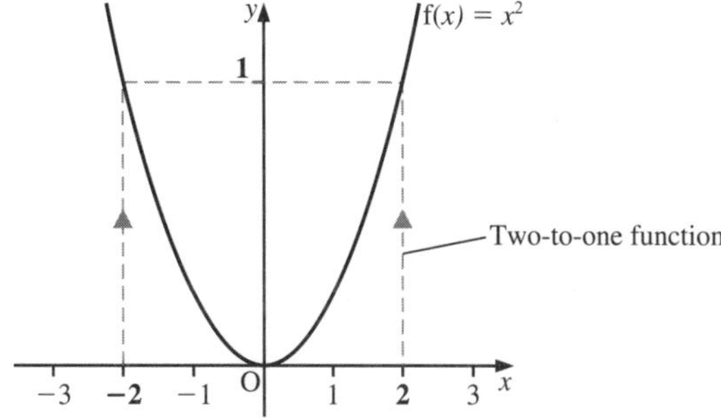

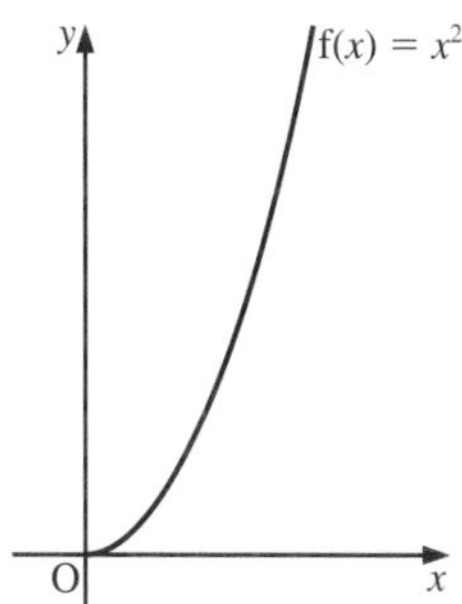

Example 21 The function f is defined by $f(x) = x^2 - 2x$, for $x \geqslant 1$. Explain why f^{-1} exists and find $f^{-1}(x)$. State the range of the function f^{-1}.

SOLUTION

The graph of $f(x) = x^2 - 2x$ for $x \geqslant 1$ is sketched on the right.

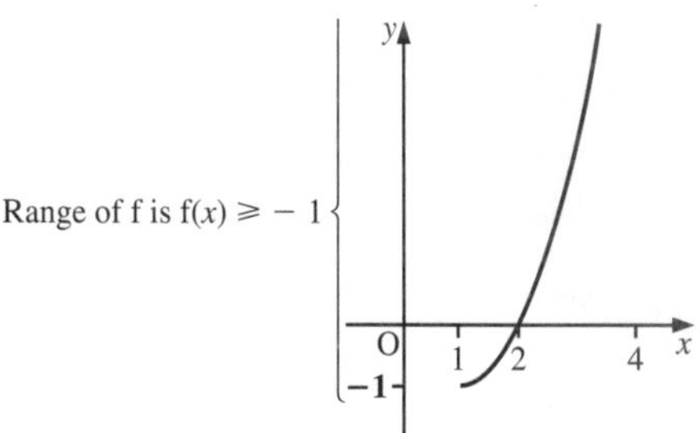

From the graph it can be seen that the range of f is $f(x) \geqslant -1$. It can also be seen that f is a one-to-one function and therefore f^{-1} exists.

To find $f^{-1}(x)$, let $y = x^2 - 2x$. Then rearranging for x gives a quadratic in x. That is,

$$x^2 - 2x - y = 0$$

Using the quadratic formula, we have

$$x = \frac{-(-2) \pm \sqrt{(-2)^2 - 4(1)(-y)}}{2(1)}$$

$$= \frac{2 \pm \sqrt{4+4y}}{2}$$

$$= \frac{2 \pm 2\sqrt{1+y}}{2}$$

$$\therefore \quad x = 1 \pm \sqrt{1+y}$$

We want the positive square root. Therefore,

$$f^{-1}(x) = 1 + \sqrt{1+x}$$

The range of f^{-1} is the domain of f. Therefore, the range of f^{-1} is the set $\{x : x \geqslant 1\}$.

Alternatively, the method of completing the square could be used. Starting with $y = x^2 - 2x$, we have

$$y = (x-1)^2 - 1$$

$$y + 1 = (x-1)^2$$

$$\therefore \quad x - 1 = \pm\sqrt{y+1}$$

$$\therefore \quad x = 1 \pm \sqrt{y+1}$$

as before.

Graph of an inverse function

Consider the function $f(x) = x + 3$ with inverse $f^{-1}(x) = x - 3$. Plotting graphs of both functions on the same set of axes gives the lines shown on the right.

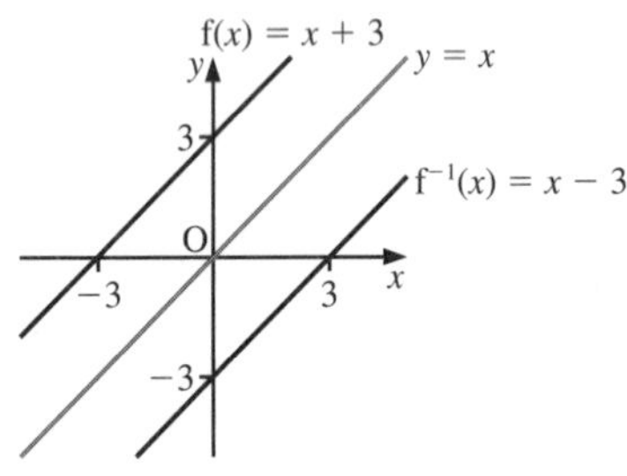

The graph of the inverse function is a reflection of the graph of f in the line $y = x$. This is because for every point (x, y) on the graph of the function f there is a point (y, x) on the graph of the function f^{-1}.

Example 22 The function f is defined by $f(x) = 3x - 6$ for all real values of x. Find the inverse function f^{-1}. Sketch the graphs of f and f^{-1} on the same set of axes and hence find the coordinates of the point of intersection of the graphs of f and f^{-1}.

SOLUTION

To find f^{-1}, let $y = 3x - 6$. Then rearranging for x gives $x = \dfrac{y+6}{3}$.

Therefore, $f^{-1}(x) = \dfrac{x+6}{3}$.

Sketching graphs of f and f^{-1} on the same set of axes gives the lines shown on the right.

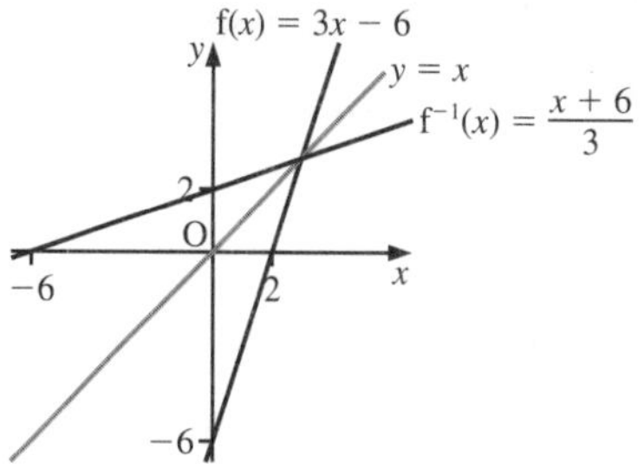

The point of intersection of the graphs of f and f^{-1} is also the point of intersection of the line $y = x$ with each of the graphs of f and f^{-1}. Therefore, to find the coordinates of this intersection point, we need to solve simultaneously the equations

$$y = x \quad \text{and} \quad y = 3x - 6$$

Eliminating y gives

$$x = 3x - 6$$

$$\therefore \quad 2x = 6$$

$$\therefore \quad x = 3$$

Substituting $x = 3$ into $y = x$ gives $y = 3$.

The coordinates of the point of intersection of the graphs of f and f^{-1} are $(3, 3)$.

Example 23 The function f is defined by

$$f(x) = \frac{2x+1}{x+2}, \quad \text{for } x > -2$$

Find the inverse function f^{-1} and find the coordinates of the points of intersection of the graphs of f and f^{-1}.

SOLUTION

To find f^{-1}, let $y = \dfrac{2x+1}{x+2}$. Then rearranging for x gives

$$y(x+2) = 2x + 1$$

$$\therefore \quad yx + 2y = 2x + 1$$

$$\therefore \quad yx - 2x = 1 - 2y$$

$$\therefore \quad x(y-2) = 1 - 2y$$

$$\therefore \quad x = \frac{1-2y}{y-2}$$

Therefore, $f^{-1}(x) = \dfrac{1-2x}{x-2}$.

The graphs of f and f^{-1} intersect at the points where the graphs of $y = x$ and $y = f(x)$ intersect. To find the x-coordinates of the points of intersection, we solve simultaneously the equations

$$y = x \quad \text{and} \quad y = \frac{2x+1}{x+2}$$

Eliminating y gives

$$x = \frac{2x+1}{x+2}$$

$$\therefore \quad x(x+2) = 2x+1$$

$$\therefore \quad x^2 - 1 = 0$$

$$\therefore \quad (x-1)(x+1) = 0$$

Solving gives $x = 1$ or $x = -1$.

When $x = 1$, $y = 1$ and when $x = -1$, $y = -1$. Therefore, the coordinates of the points of intersection of the graphs of f and f^{-1} are $(1, 1)$ and $(-1, -1)$.

Two useful techniques for sketching the graph of an inverse function are as follows.

i) Reflect the graph of the function f in the line $y = x$.
ii) Sketch the graph of $y = f(x)$, turn the page over and then turn it through 90° clockwise. What you see through the page is the graph of the inverse function. (Note that (i) a reflection in the y-axis followed by a rotation through 90° clockwise is equivalent to a reflection in the line $y = x$, and (ii) a reflection in the x-axis followed by a rotation through 90° anticlockwise is equivalent to a reflection in the line $y = x$.)

Exercise 3F

Throughout this exercise, the domain of each function is the set of real numbers unless specifically stated otherwise.

1 Find the inverse of each of the following functions.

a) $f: x \longrightarrow 3x + 2$ **b)** $f: x \longrightarrow 5x - 1$ **c)** $f: x \longrightarrow 4 - 3x$

d) $f: x \longrightarrow \frac{2}{x}, \ x \neq 0$ **e)** $f: x \longrightarrow \frac{3}{x-1}, \ x \neq 1$ **f)** $f: x \longrightarrow \frac{5}{2-3x}, \ x \neq \frac{3}{2}$

g) $f: x \longrightarrow \frac{x}{2+x}, \ x \neq -2$ **h)** $f: x \longrightarrow \frac{2x}{5-x}, \ x \neq 5$ **i)** $f: x \longrightarrow \frac{3x}{2x+1}, \ x \neq -\frac{1}{2}$

j) $f: x \longrightarrow 1 + \frac{1}{x}, \ x \neq 0$ **k)** $f: x \longrightarrow 3 + \frac{x}{1+x}, \ x \neq -1$ **l)** $f: x \longrightarrow 2 - \frac{3}{4+x}, \ x \neq -4$

2 Find the inverse of each of the following functions, and state the domain on which each inverse is defined.

a) $f(x) = x^2, \ x \in \mathbb{R}, \ x > 2$ **b)** $f(x) = \frac{1}{2+x}, \ x \in \mathbb{R}, \ x > 0$

c) $f(x) = \sqrt{x-2}, \ x \in \mathbb{R}, \ x > 3$ **d)** $f(x) = 3x^2 - 1, \ x \in \mathbb{R}, \ 1 < x < 4$

e) $f(x) = \sqrt{2x+3}, \ x \in \mathbb{R}, \ x \geqslant 11$ **f)** $f(x) = \frac{1}{x} - 3, \ x \in \mathbb{R}, \ 2 < x < 5$

g) $f(x) = (x+2)^2 + 3, \ x \in \mathbb{R}, \ x \geqslant -2$ **h)** $f(x) = x^3 + 1, \ x \in \mathbb{R}$

i) $f(x) = \dfrac{1}{x-2},\ x \in \mathbb{R},\ x > 3$

j) $f(x) = \sqrt{3-x},\ x \in \mathbb{R},\ x \leqslant 2$

k) $f(x) = (x-3)^2 + 5,\ x \in \mathbb{R},\ 4 \leqslant x \leqslant 6$

l) $f(x) = 5 - \sqrt{x+3},\ x \in \mathbb{R},\ x \geqslant -3$

3 Given $f: x \longrightarrow 3x - 4,\ x \in \mathbb{R}$,

a) find an expression for the inverse function $f^{-1}(x)$
b) sketch the graphs of $f(x)$ and $f^{-1}(x)$ on the same set of axes
c) solve the equation $f(x) = f^{-1}(x)$.

4 **a)** Sketch the graph of the function defined by

$$f(x) = 10 - 2x,\ x \in \mathbb{R},\ x \geqslant 0$$

b) Find an expression for the inverse function $f^{-1}(x)$, and sketch the graph of $f^{-1}(x)$ on the same set of axes.
c) Calculate the value of x for which $f(x) = f^{-1}(x)$.

5 A function is defined by $f(x) = x^2 - 6,\ x \in \mathbb{R},\ x > 0$.

a) Find an expression for the inverse function $f^{-1}(x)$.
b) Sketch the graphs of $f(x)$ and $f^{-1}(x)$ on the same set of axes.
c) Calculate the value of x for which $f(x) = f^{-1}(x)$.

6 **a)** Sketch the graph of the function defined by

$$f: x \longrightarrow (x-2)^2\ \ x \in \mathbb{R},\ x \geqslant 2$$

b) Find an expression for the inverse function $f^{-1}(x)$, and sketch the graph of $f^{-1}(x)$ on the same set of axes.
c) Calculate the value of x for which $f(x) = f^{-1}(x)$.

7 The functions f and g are defined by

$$f: x \longrightarrow 2x - 5,\ x \in \mathbb{R} \quad \text{and} \quad g: x \longrightarrow 7 - 4x,\ x \in \mathbb{R}$$

a) Solve the equation $f(x) = g(x)$.
b) Write down expressions for $f^{-1}(x)$ and $g^{-1}(x)$.
c) Solve the equation $f^{-1}(x) = g^{-1}(x)$, and comment on your answer.

8 The function h with domain $\{x: x \geqslant 0\}$ is defined by $h(x) = \dfrac{4}{x+3}$.

a) Sketch the graph of h and state its range.
b) Find an expression for $h^{-1}(x)$.
c) Calculate the value of x for which $h(x) = h^{-1}(x)$.

9 Functions f and g are defined by

$$f: x \longrightarrow 3x + 1,\ x \in \mathbb{R} \quad \text{and} \quad g: x \longrightarrow x - 2,\ x \in \mathbb{R}$$

a) Write down and simplify an expression for the composite function $fg(x)$.
b) Find expressions for each of these inverse functions.
i) $f^{-1}(x)$ **ii)** $g^{-1}(x)$ **iii)** $(fg)^{-1}(x)$
c) Verify that $(fg)^{-1}(x) = g^{-1}f^{-1}(x)$.

10 The function g is defined by $g(x) = 2x^2 - 3,\ x \in \mathbb{R},\ x \geqslant 0$.

a) State the range of g and sketch its graph.
b) Explain why the inverse function g^{-1} exists and sketch its graph.
c) Given also that h is defined by $h(x) = \sqrt{5x+2},\ x \in \mathbb{R},\ x \geqslant -\frac{2}{5}$, solve the inequality $gh(x) \geqslant x$.

11 Functions f and g are defined by

$$f: x \longrightarrow 2x + 3,\ x \in \mathbb{R} \qquad \text{and} \qquad g: x \longrightarrow \frac{1}{x-1},\ x \in \mathbb{R},\ x \neq 1$$

a) Find an expression for the inverse function $f^{-1}(x)$.
b) Find an expression for the composite function $gf(x)$.
c) Solve the equation $f^{-1}(x) = gf(x) - 1$.

12 Given $f(x) = \dfrac{1}{1-x},\ x \in \mathbb{R},\ x \neq 0,\ x \neq 1$

a) find expressions for **i)** $ff(x)$ **ii)** $fff(x)$ **iii)** $f^{(4)}(x)$
[Note: $f^{(4)}(x) = ffff(x)$]
b) Hence write down expressions for **i)** $f^{-1}(x)$ **ii)** $f^{(13)}(x)$ **iii)** $f^{(360)}(x)$

13 Functions g and h are defined by

$$g(x) = \frac{5}{x-3},\ x \in \mathbb{R},\ x \neq 3 \qquad \text{and} \qquad h(x) = x^2 + 4,\ x \in \mathbb{R},\ x > 0$$

Find

a) an expression for the inverse function $g^{-1}(x)$
b) an expression for the composite function $gh(x)$
c) the solutions to the equation $3g^{-1}(x) = 10gh(x) + 9$.

14 Functions f and g are defined by

$$f: x \longrightarrow \frac{1}{x+3},\ x \in \mathbb{R},\ x \neq -3 \qquad \text{and} \qquad g: x \longrightarrow \frac{2}{x-4},\ x \in \mathbb{R},\ x \neq 4$$

a) Show that $fg: x \longrightarrow \dfrac{x-4}{3x-10},\ x \in \mathbb{R},\ x \neq \dfrac{10}{3}$.
b) Find an expression for $(fg)^{-1}(x)$.

***15** Given $f: x \longrightarrow \dfrac{a}{x} + b,\ x \in \mathbb{R},\ x \neq 0,\ x \neq b,\ x \neq -\dfrac{a}{b}$, and $ff(x) = f^{-1}(x)$, prove that the constants a and b satisfy the equation $a + b^2 = 0$.

***16** The function f is defined by $f(x) = \dfrac{ax+b}{cx+d},\ x \in \mathbb{R},\ x \neq -\dfrac{d}{c},\ b \neq 0,\ c \neq 0$.

a) Prove that if $a + d = 0$, then $f(x) = f^{-1}(x)$.
b) Prove that if $a + d \neq 0$ and $(a-d)^2 + 4bc = 0$, then the graph of $y = f(x)$ intersects the graph of $y = f^{-1}(x)$ in exactly one point.

Exercise 3G: Examination questions

1 i) Sketch the curve with equation $y = x^2$.

Given $f(x) = (x - 2)^2 + 1$, sketch the curves with the following equations on separate diagrams. Label each curve and give the coordinates of the vertex and the equation of the axis of symmetry for each curve.

ii) $y = f(x)$ **iii)** $y = -f(x)$ **iv)** $y = f(x + 1) + 2$ (MEI)

2 The diagram shows a sketch of the curve with equation $y = f(x)$, where $f(x) = 0$ for $x \leqslant 1$ and $x \geqslant 4$. Sketch, on separate axes, the graph of the curves with equations

a) $y = f(x - 1)$ **b)** $y = 2f(\frac{1}{2}x)$. (AEB Spec)

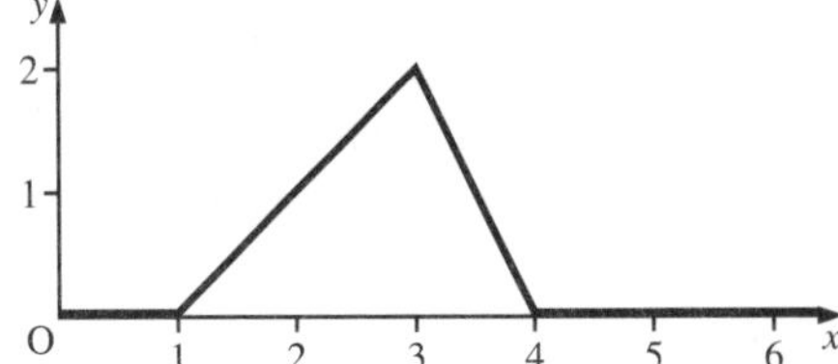

3 The figure shows the graph of $y = f(x)$, with $f(x) = 0$ for $|x| \geqslant 2$. On separate diagrams, sketch the graphs of

a) $y = f(x + 2)$ **b)** $y = f(2x)$. (UODLE)

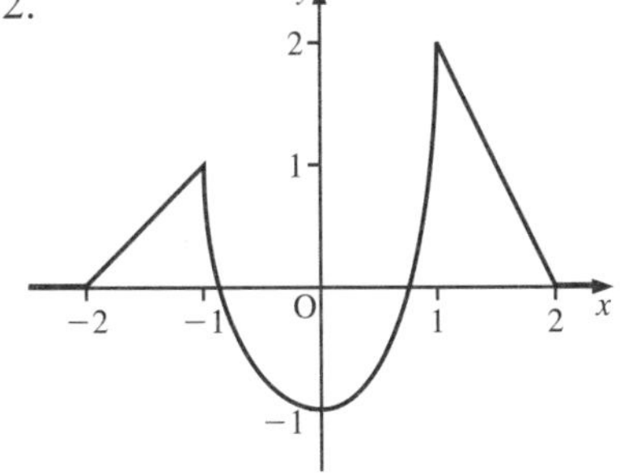

4 The function $f(x)$ is defined for all values of x except $x = 0$ and is an odd function, i.e. $f(-x) = -f(x)$.

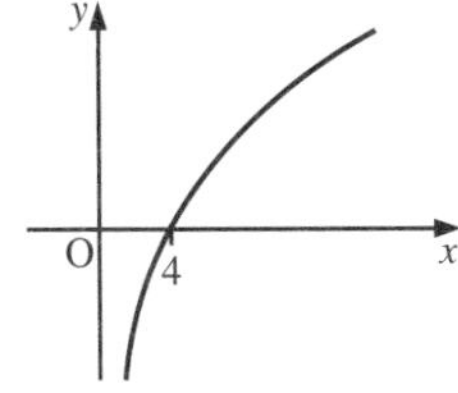

a) Part of the graph of $y = f(x)$ is given on the right. Copy and complete the sketch.

b) Draw a separate sketch to illustrate the graph of

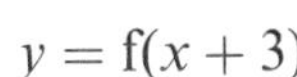

$$y = f(x + 3)$$

showing clearly where the graph will intersect the x-axis. (NEAB)

5 The figure shows a sketch of the curve with equation $y = f(x)$. In separate diagrams show, for $-3 \leqslant x \leqslant 3$, sketches of the curves with equation

a) $y = f(-x)$ **b)** $y = -f(x)$ **c)** $y = f(|x|)$

Mark on each sketch the x-coordinate of any point, or points, where a curve touches or crosses the x-axis. (EDEXCEL)

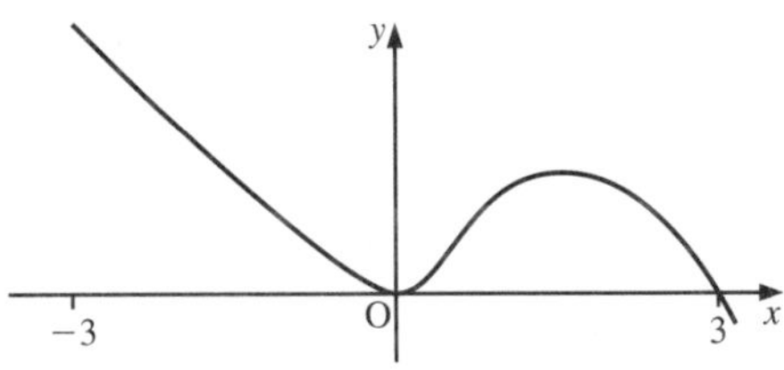

6 Solve the inequality $|x + 1| < |x - 2|$. (UCLES)

7 The function f is defined by $f(x) = |x - 3|,\ x \in \mathbb{R}$. Sketch the graph of f. Solve the inequality $|x - 3| < \frac{1}{2}x$. (AEB 92)

8 On the same diagram, sketch the graphs of

$$y = x \quad \text{and} \quad y = |2x - 1|$$

a) Find the coordinates of the points of intersection of the two graphs.
b) Hence, or otherwise, find the set of values of x for which $|2x - 1| > x$.
(EDEXCEL)

9 The functions f and g are defined by

$$\mathrm{f}: x \rightarrow x^2 - 10, \ x \in \mathbb{R} \qquad \text{and} \qquad \mathrm{g}: x \rightarrow |x - 2|, \ x \in \mathbb{R}$$

a) Show that $\mathrm{f} \circ \mathrm{f}: x \rightarrow x^4 - 20x^2 + 90, \ x \in \mathbb{R}$. Find all the values of x for which $\mathrm{f} \circ \mathrm{f}(x) = 26$.
b) Show that $\mathrm{g} \circ \mathrm{f}(x) = |x^2 - 12|$. Sketch the graph of $\mathrm{g} \circ \mathrm{f}$. Hence, or otherwise, solve the equation $\mathrm{g} \circ \mathrm{f}(x) = x$. (AEB Spec)

10 i) Show that $x^2 + 4x + 7 = (x + 2)^2 + a$, where a is to be determined.
ii) Sketch the graph of $y = x^2 + 4x + 7$, giving the equation of the axis of symmetry and the coordinates of its vertex.

The function f is defined by $\mathrm{f}: x \rightarrow x^2 + 4x + 7$ and has as its domain the set of all real numbers.

iii) Find the range of f.
iv) Explain, with reference to your sketch, why f has no inverse with its given domain. Suggest a domain for f for which it has an inverse. (MEI)

11 Functions f and g are defined by

$$\mathrm{f}: x \rightarrow 4 - x, \ x \in \mathbb{R} \qquad \text{and} \qquad \mathrm{g}: x \rightarrow 3x^2, \ x \in \mathbb{R}$$

a) Find the range of g.
b) Solve $\mathrm{gf}(x) = 48$.
c) Sketch the graph of $y = |\mathrm{f}(x)|$ and hence find the values of x for which $|\mathrm{f}(x)| = 2$. (EDEXCEL)

12 The functions f and g are defined by

$$\mathrm{f}: x \rightarrow x^2 + 3, \ x \in \mathbb{R} \qquad \text{and} \qquad \mathrm{g}: x \rightarrow 2x + 1, \ x \in \mathbb{R}$$

a) Find, in a similar form, the function fg.
b) Find the range of the function fg.
c) Solve the equation $\mathrm{f}(x) = 12\mathrm{g}^{-1}(x)$. (EDEXCEL)

13 The function f is given by

$$\mathrm{f}: x \rightarrow x^2 - 8x, \ x \in \mathbb{R}, \ x \leqslant 4$$

a) Determine the range of f.
b) Find the value of x for which $\mathrm{f}(x) = 20$.
c) Find $\mathrm{f}^{-1}(x)$ in terms of x. (EDEXCEL)

14 The functions f and g are defined by

$$\mathrm{f}: x \rightarrow 3x - 1, \ x \in \mathbb{R} \qquad \text{and} \qquad \mathrm{g}: x \rightarrow x^2 + 1, \ x \in \mathbb{R}$$

a) Find the range of g.
b) Determine the values of x for which $\mathrm{gf}(x) = \mathrm{fg}(x)$.

c) Determine the values of x for which $|f(x)| = 8$.

The function $h: x \longrightarrow x^2 + 3x, \ x \in \mathbb{R}, \ x \geqslant q$, is one-one.

d) Find the least value of q and sketch the graph of this function. (EDEXCEL)

15 The function f with domain $\{x: x \geqslant 0\}$ is defined by $f(x) = \dfrac{8}{x+2}$.

a) Sketch the graph of f and state the range of f.
b) Find $f^{-1}(x)$, where f^{-1} denotes the inverse of f.
c) Calculate the value of x for which $f(x) = f^{-1}(x)$. (AEB 93)

16 The functions f and g are defined with their respective domains by

$$f: x \longrightarrow \frac{3}{2x-1}, \quad x \in \mathbb{R}, \ x \neq \tfrac{1}{2} \qquad \text{and} \qquad g: x \longrightarrow x^2 + 1, \quad x \in \mathbb{R}$$

a) Find the values of x for which $f(x) = x$.
b) Find the range of g.
c) The domain of the composite function $f \circ g$ is $\mathbb{R}$. Find $f \circ g(x)$ and state the range of $f \circ g$. (AEB 95)

17 The function f is defined by

$$f(x) = \frac{x}{x-1} \qquad (x \neq 1)$$

i) Find and simplify an expression for $f \circ f(x)$.
ii) Hence, or otherwise, find an expression for $f^{-1}(x)$, where f^{-1} is the inverse function of f.
iii) State the range of f. (WJEC)

18 **a)** On the same diagram, sketch the graphs of

$$y = \frac{1}{x-a} \quad \text{and} \quad y = 4|x-a|$$

where a is a positive constant. Show clearly the coordinates of any points of intersection with the coordinate axes.

b) Hence, or otherwise, find the set of values of x for which

$$\frac{1}{x-a} < 4|x-a| \qquad \text{(EDEXCEL)}$$

4 Polynomials

An interesting theorem of mathematics differs from interesting results in other fields because over and above the surprise and beauty of what it says, it has an 'aspect of eternity'; it is always part of an infinite chain of results.
LEO ZIPPIN

An expression of the form

$$a_n x^n + a_{n-1}x^{n-1} + a_{n-2}x^{n-2} + \ldots + a_0$$

where $a_n, a_{n-1}, \ldots, a_0$ are real numbers with $a_n \neq 0$ and n is a positive integer, is called a **polynomial of degree *n***.

When $n = 2$, the polynomial is called a **quadratic**.

When $n = 3$, the polynomial is called a **cubic**.

When $n = 4$, the polynomial is called a **quartic**.

Example 1 Find the degree of each of these polynomials.

a) $4x^6 + 3x^5 + x^3 - x^2 + 5x$ **b)** $x^4 - 3x^3 + 2x^9 - 7$

SOLUTION

a) The highest power of x which occurs is 6. Therefore, the degree of the polynomial is 6.

b) Rearranging the terms in descending order gives

$$2x^9 + x^4 - 3x^3 - 7$$

The highest power of x which occurs is 9. Therefore, the degree of the polynomial is 9.

From now on we will write polynomials in descending powers of x.

Polynomials can be evaluated for particular values of x. We will use functional notation and write $f(x)$ for the polynomial.

Example 2 Given the polynomial $f(x) = 3x^4 + 2x^2 - x + 7$, evaluate

a) $f(0)$ **b)** $f(-2)$

SOLUTION

a) $f(0) = 3(0)^4 + 2(0)^2 - (0) + 7 = 7$

b) $f(-2) = 3(-2)^4 + 2(-2)^2 - (-2) + 7$

$= 48 + 8 + 2 + 7 = 65$

Addition and subtraction of polynomials

Polynomials can be added or subtracted by collecting together terms of the same degree. In general, the result of adding and subtracting any number of polynomials is also a polynomial.

Example 3 Given the two polynomials $f(x) = 3x^3 + 2x^2 - x + 4$ and $g(x) = x^3 - x^2 + 7$, find

a) $f(x) + g(x)$ **b)** $f(x) - g(x)$

SOLUTION

a)
$$\begin{aligned} f(x) + g(x) &= (3x^3 + 2x^2 - x + 4) + (x^3 - x^2 + 7) \\ &= (3x^3 + x^3) + (2x^2 - x^2) + (-x) + (4 + 7) \end{aligned}$$

Therefore,

$$f(x) + g(x) = 4x^3 + x^2 - x + 11$$

b)
$$\begin{aligned} f(x) - g(x) &= (3x^3 + 2x^2 - x + 4) - (x^3 - x^2 + 7) \\ &= (3x^3 - x^3) + (2x^2 + x^2) + (-x) + (4 - 7) \end{aligned}$$

Therefore,

$$f(x) - g(x) = 2x^3 + 3x^2 - x - 3$$

Multiplication of polynomials

When two polynomials are multiplied together each term of one polynomial is multiplied by each term of the other polynomial. In general, the product of any number of polynomials is also a polynomial.

Example 4 Given the two polynomials $f(x) = x^4 + 4x - 1$ and $g(x) = 2x^4 + x^3 - 4x$, find in simplest form each of the polynomials

a) $f(x)\,g(x)$ **b)** $3f(x) + 4x\,g(x)$

SOLUTION

a)
$$\begin{aligned} f(x)\,g(x) &= (x^4 + 4x - 1)(2x^4 + x^3 - 4x) \\ &= x^4(2x^4 + x^3 - 4x) + 4x(2x^4 + x^3 - 4x) - 1(2x^4 + x^3 - 4x) \\ &= 2x^8 + x^7 - 4x^5 + 8x^5 + 4x^4 - 16x^2 - 2x^4 - x^3 + 4x \end{aligned}$$

Therefore,

$$f(x)\,g(x) = 2x^8 + x^7 + 4x^5 + 2x^4 - x^3 - 16x^2 + 4x$$

b)
$$\begin{aligned} 3f(x) + 4x\,g(x) &= 3(x^4 + 4x - 1) + 4x(2x^4 + x^3 - 4x) \\ &= 3x^4 + 12x - 3 + 8x^5 + 4x^4 - 16x^2 \end{aligned}$$

Therefore,

$$3f(x) + 4x\,g(x) = 8x^5 + 7x^4 - 16x^2 + 12x - 3$$

It is sometimes useful to be able to identify the coefficient of a particular term in a polynomial expansion, without expanding the whole expression.

Example 5 Find the coefficient of the x^3 term in the expansion of

a) $(x^3 + 4x^2 - 7x + 1)(x + 2)$

b) $(2x + 1)(x^4 + x^3 + 3x^2 - 2) + x(x^2 + 3x - 4)$

SOLUTION

a) The terms contributing to the x^3 term are indicated below.

$$(x^3 + 4x^2 - 7x + 1)(x + 2)$$

The x^3 term is $2x^3 + 4x^3 = 6x^3$. Therefore, the coefficient of the x^3 term is 6.

b) The terms contributing to the x^3 term are indicated below.

$$(2x + 1)(x^4 + x^3 + 3x^2 - 2) + x(x^2 + 3x - 4)$$

The x^3 term is $6x^3 + x^3 + x^3 = 8x^3$. Therefore, the coefficient of the x^3 term is 8.

Exercise 4A

1 Find the degree of each of the following polynomials.

a) $2x^3 + 3x^2 - 2x + 4$
b) $x^2 + 3x - 2$
c) $5x + 7$
d) $x^{12} - 4$
e) $4 + x^2$
f) $4x - 3x^4$

2 Given

a) $f(x) = x^2 + 3x + 4$, evaluate $f(2)$.
b) $f(x) = x^3 - 2x^2 + 5x + 1$, evaluate $f(3)$.
c) $f(x) = 2x^2 + 5x - 1$, evaluate $f(-1)$.
d) $f(x) = 3x^2 - 5x + 2$, evaluate $f(-4)$.
e) $f(x) = 5x^4 + 2x - 1$, evaluate $f(0)$.
f) $f(x) = 6x^2 - 3x + 2$, evaluate $f(3)$.

3 Given

a) $f(x) = x^3 + 2x^2 - 3x + 2$ and $g(x) = 2x^3 - x^2 + 5x - 4$, find $f(x) + g(x)$.
b) $f(x) = 5x^3 - 4x^2 + 3x + 2$ and $g(x) = x^3 - 2x^2 + 4x + 7$, find $f(x) - g(x)$.
c) $f(x) = 2x^3 - 5x^2 + 6x$ and $g(x) = x^3 - 6x^2 + 5x + 1$, find $f(x) - g(x)$.
d) $f(x) = 2x^3 + 3x^2 + 7x - 5$ and $g(x) = x^2 + 3x - 5$, find $2\,f(x) + g(x)$.
e) $f(x) = 3x^4 + 2x^2 + 6x - 8$ and $g(x) = 2x^3 + 7x^2 + 5x - 4$, find $3\,f(x) - g(x)$.
f) $f(x) = 3x^5 + 7x^2 - 2$ and $g(x) = x^4 + x^2 - 7$, find $3\,f(x) + 2\,g(x)$.

4 Expand and simplify each of the following products.

a) $(x^3 + 2x - 1)(x^2 - 3x + 2)$
b) $(x^3 + 5x^2 - 2)(x^2 + 3x - 5)$
c) $(x^4 + 3x^2 - 2)(x^2 + 5x - 3)$
d) $(2x^2 + 7x - 1)(3x^3 - x - 2)$
e) $(2x^3 + 7x - 3)(x^4 + x^2 + x)$
f) $(2x^3 + 5x^2 + 7x - 2)(3x^3 - 2x^2 + 4x + 3)$

5 Given

a) $f(x) = x^2 - 2x + 5$ and $g(x) = x^3 + 6x - 4$, expand and simplify $x\,f(x) + 3\,g(x)$.
b) $f(x) = x^3 + x^2 - 3$ and $g(x) = x^2 - 2x + 5$, expand and simplify $x\,f(x) - x^2g(x)$.
c) $f(x) = x^5 + 3x^2 - 6x + 4$ and $g(x) = x^2 + 4x + 3$, expand and simplify $2\,f(x) - x^3g(x)$.
d) $f(x) = 3x^2 - 2x + 3$ and $g(x) = x^2 + 6x - 2$, expand and simplify $(x + 1)\,f(x) - 2\,g(x)$.
e) $f(x) = x^2 + 3$ and $g(x) = x^3 - 5x^2 + 3x + 1$, expand and simplify $[f(x)]^2 + 2xg(x)$.
f) $f(x) = x^2 + 3x + 2$ and $g(x) = x^2 + 2x + 1$, expand and simplify $[f(x)]^2 - [g(x)]^2$.

6 Find the coefficients of the terms indicated in square brackets in the expansions of the expressions below.

a) $(x^3 + 5x^2 + 6x - 1)(x + 3)$ $[x^2]$
b) $(x^3 - 3x^2 + 2x - 5)(x - 2)$ $[x^3]$
c) $(x^2 + 5x - 4)(x^2 - 3x + 6)$ $[x^2]$
d) $(2x^3 - 4x^2 + 7)(x^2 - 5x - 2)$ $[x^4]$
e) $(x^2 + 6x - 5)^2$ $[x^2]$
f) $(3x - 1)(x^3 - 4x^2 + 6x + 8)$ $[x^3]$

Dividing polynomials

Before we look at the techniques of dividing polynomials, it will be useful to recall a technique used in the division of numbers.

One way of writing ‘19 divided by 5’ is

$$\frac{19}{5} = 3 \text{ remainder } 4$$

or $$19 = 5 \times \underbrace{3}_{\text{Called the quotient}} + \underbrace{4}_{\text{Called the remainder}}$$

The same method can also be applied to polynomial division.

Example 6 Find the quotient and the remainder when the polynomial $x^2 + 4x - 5$ is divided by $x + 3$.

SOLUTION

Writing $x^2 + 4x - 5$ in terms of a quotient and a remainder gives an expression of the form

$$x^2 + 4x - 5 \equiv (x + 3)\underbrace{(\quad)}_{\text{Quotient}} + \underbrace{(\quad)}_{\text{Remainder}}$$

We can see that the quotient is of the form $ax + b$. Therefore,

$$x^2 + 4x - 5 \equiv (x + 3)(ax + b) + r$$

Expanding and collecting like terms give

$$x^2 + 4x - 5 \equiv ax^2 + bx + 3ax + 3b + r$$
$$\equiv ax^2 + (3a + b)x + 3b + r$$

Comparing the coefficients of the x^2 terms gives

$$a = 1 \quad [1]$$

Comparing the coefficients of the x terms gives

$$3a + b = 4 \quad [2]$$

Comparing the constant terms gives

$$3b + r = -5 \quad [3]$$

Substituting $a = 1$ into [2] gives $b = 1$. Substituting $b = 1$ into [3] gives $r = -8$.

Therefore, the quotient is $x + 1$ and the remainder is -8.

Note the use in Examples 6 and 7 of the symbol $\equiv$, which means 'is identically equal to'. Strictly speaking, $\equiv$ should always be used instead of $=$ when dealing with identities.

Example 7

a) Find the remainder when the polynomial $x^3 + x^2 - 14x - 24$ is divided by **i)** $x + 1$, **ii)** $x + 3$.

b) Hence factorise $x^3 + x^2 - 14x - 24$.

SOLUTION

a) i) $x^3 + x^2 - 14x - 24 \equiv (x + 1)\underbrace{(\quad)}_{\text{Quotient}} + \underbrace{(\quad)}_{\text{Remainder}}$

We can see that the quotient is of the form $ax^2 + bx + c$. Therefore,

$$x^3 + x^2 - 14x - 24 \equiv (x + 1)(ax^2 + bx + c) + r$$

Expanding and collecting like terms gives

$$\begin{aligned} x^3 + x^2 - 14x - 24 &\equiv ax^3 + bx^2 + cx + ax^2 + bx + c + r \\ &\equiv ax^3 + (a + b)x^2 + (b + c)x + c + r \end{aligned}$$

Comparing the coefficients of the x^3 terms gives

$$a = 1 \quad [1]$$

Comparing the coefficients of the x^2 terms gives

$$a + b = 1 \quad [2]$$

Comparing the coefficients of the x terms gives

$$b + c = -14 \quad [3]$$

Comparing the constant terms gives

$$c + r = -24 \quad [4]$$

Substituting $a = 1$ into [2] gives $b = 0$. Substituting $b = 0$ into [3] gives $c = -14$ and substituting $c = -14$ into [4] gives $r = -10$.

When $x^3 + x^2 - 14x - 24$ is divided by $x + 1$, the remainder is -10.

ii) When $x^3 + x^2 - 14x - 24$ is divided by $x + 3$, the quotient is also of the form $ax^2 + bx + c$. Therefore,

$$x^3 + x^2 - 14x - 24 \equiv (x + 3)(ax^2 + bx + c) + r$$
$$\equiv ax^3 + (3a + b)x^2 + (3b + c)x + 3c + r$$

Comparing coefficients and solving gives $a = 1$, $b = -2$, $c = -8$ and $r = 0$. Therefore,

$$x^3 + x^2 - 14x - 24 \equiv (x + 3)(x^2 - 2x - 8) + 0$$

When $x^3 + x^2 - 14x - 24$ is divided by $x + 3$, the remainder is 0.

b) Since the remainder is 0 when $x^3 + x^2 - 14x - 24$ is divided by $x + 3$, $x + 3$ is a factor of $x^3 + x^2 - 14x - 24$.

To factorise $x^3 + x^2 - 14x - 24$, it only remains to see whether $x^2 - 2x - 8$ will factorise into two linear factors. Since

$$x^2 - 2x - 8 \equiv (x + 2)(x - 4)$$

we have

$$x^3 + x^2 - 14x - 24 \equiv (x + 3)(x + 2)(x - 4)$$

Result: The remainder theorem

When the polynomial $f(x)$ is divided by $(ax - b)$, the remainder is $f\left(\dfrac{b}{a}\right)$.

Proof

This can be seen by writing

$$f(x) \equiv (ax - b)(\text{Quotient}) + (\text{Remainder})$$

Thus, when $x = \dfrac{b}{a}$,

$$f\left(\frac{b}{a}\right) = \left[a\left(\frac{b}{a}\right) - b\right](\text{Quotient}) + (\text{Remainder})$$
$$= (0)(\text{Quotient}) + (\text{Remainder})$$
$$= \text{Remainder}$$

as required.

Example 8 Find each of the remainders when the polynomial $x^3 + 5x^2 - 17x - 21$ is divided by

a) $x + 1$ **b)** $x - 4$ **c)** $2x + 1$

SOLUTION

Let $f(x) = x^3 + 5x^2 - 17x - 21$.

a) By the remainder theorem, the remainder when $f(x)$ is divided by $x + 1$ is $f(-1)$. Now

$$\begin{aligned} f(-1) &= (-1)^3 + 5(-1)^2 - 17(-1) - 21 \\ &= -1 + 5 + 17 - 21 \\ &= 0 \end{aligned}$$

Therefore, the remainder is 0.

b) By the remainder theorem, the remainder when $f(x)$ is divided by $x - 4$ is $f(4)$. Now

$$\begin{aligned} f(4) &= (4)^3 + 5(4)^2 - 17(4) - 21 \\ &= 64 + 80 - 68 - 21 \\ &= 55 \end{aligned}$$

Therefore, the remainder is 55.

c) By the remainder theorem, the remainder when $f(x)$ is divided by $2x + 1$ is $f(-\tfrac{1}{2})$. Now

$$\begin{aligned} f(-\tfrac{1}{2}) &= (-\tfrac{1}{2})^3 + 5(-\tfrac{1}{2})^2 - 17(-\tfrac{1}{2}) - 21 \\ &= -\frac{1}{8} + \frac{5}{4} + \frac{17}{2} - 21 \\ &= -\frac{91}{8} \end{aligned}$$

Therefore, the remainder is $-\dfrac{91}{8}$.

Example 9 Find the remainder when the polynomial

$$f(x) = 2x^3 + 5x^2 - 39x + 18$$

is divided by $x + 6$. Hence solve the equation $2x^3 + 5x^2 - 39x + 18 = 0$.

SOLUTION

By the remainder theorem, the remainder when $f(x)$ is divided by $x + 6$ is $f(-6)$. Now

$$\begin{aligned} f(-6) &= 2(-6)^3 + 5(-6)^2 - 39(-6) + 18 \\ &= -432 + 180 + 234 + 18 \\ &= 0 \end{aligned}$$

Since the remainder is 0, $x + 6$ is a factor of $f(x)$ and

$$2x^3 + 5x^2 - 39x + 18 \equiv (x + 6)(ax^2 + bx + c).$$

Expanding and comparing coefficients (or by inspection) give $a = 2$, $b = -7$ and $c = 3$. Therefore,

$$2x^3 + 5x^2 - 39x + 18 \equiv (x+6)(2x^2 - 7x + 3)$$
$$\equiv (x+6)(2x-1)(x-3).$$

The equation $2x^3 + 5x^2 - 39x + 18 = 0$ can be written as

$$(x+6)(2x-1)(x-3) = 0$$

Solving gives $x = -6$, $x = \frac{1}{2}$ or $x = 3$.

Example 10 Given that when the polynomial $f(x) = x^3 + ax^2 + bx + 2$ is divided by $x - 1$ the remainder is 4 and when it is divided by $x + 2$ the remainder is also 4, find the values of the constants a and b.

SOLUTION

By the remainder theorem, $f(1) = 4$. That is

$$(1)^3 + a(1)^2 + b(1) + 2 = 4$$
$$\therefore \quad a + b = 1 \qquad [1]$$

Also by the remainder theorem, $f(-2) = 4$. That is,

$$(-2)^3 + a(-2)^2 + b(-2) + 2 = 4$$
$$\therefore \quad 4a - 2b = 10 \qquad [2]$$

Dividing [2] by 2 gives

$$2a - b = 5 \qquad [3]$$

Solving [1] and [3] simultaneously gives $a = 2$ and $b = -1$.

Exercise 4B

1 Find the quotient and the remainder when

a) $x^2 + 6x + 5$ is divided by $x + 2$.
b) $x^2 - 4x + 3$ is divided by $x + 1$.
c) $2x^2 + 5x - 4$ is divided by $x + 2$.
d) $2x^2 - 5x + 8$ is divided by x.
e) $6x^2 - x + 2$ is divided by $2x + 1$.
f) $6x^2 - 7x + 5$ is divided by $2x - 3$.

2 Find the quotient and the remainder when

a) $x^3 + 3x^2 - 2x + 1$ is divided by $x - 2$.
b) $x^3 + 5x^2 - 6x + 3$ is divided by $x + 3$.
c) $2x^3 - 3x^2 - 4x + 1$ is divided by $x - 4$.
d) $2x^3 + x^2 - 3x - 14$ is divided by $x - 2$.
e) $2x^3 + x^2 + 5x - 4$ is divided by $2x - 1$.
f) $4x^4 - 3x^2 + x + 2$ is divided by $2x + 3$.

3 Use the remainder theorem to find the remainder when

a) $6x^2 + 5x - 1$ is divided by $x - 1$.
b) $3x^3 + 2x - 4$ is divided by $x - 2$.
c) $3x^2 + 6x - 8$ is divided by $x + 3$.
d) $2x^3 + 4x^2 - 6x + 5$ is divided by $x - 1$.
e) $6x^3 - 2x^2 + 5x - 4$ is divided by x.
f) $8x^3 + 4x + 3$ is divided by $2x - 1$.

4 The expression $2x^3 - 3x^2 + ax - 5$ gives a remainder of 7 when divided by $x - 2$. Find the value of the constant a.

5 The cubic $3x^3 + bx^2 - 7x + 5$ gives a remainder of 17 when divided by $x + 3$. Find the value of the constant b.

6 The remainder when $x^3 - 2x^2 + ax + 5$ is divided by $x - 3$ is twice the remainder when the same expression is divided by $x + 1$. Find the value of the constant a.

7 The remainder when $cx^3 + 2x^2 - 5x + 7$ is divided by $x - 2$ is equal to the remainder when the same expression is divided by $x + 1$. Find the value of the constant c.

8 Given that $x - 2$ is a factor of $x^3 + 4x^2 - 2x + k$, find the value of the constant k.

9 The expression $x^3 - 5x^2 + ax + 9$, where a is a constant, gives a remainder of 6 when divided by $x - 3$. Find the remainder when the same expression is divided by $x - 4$.

10 Given that $x - 4$ is a factor of $2x^3 - 3x^2 - 7x + b$, where b is a constant, find the remainder when the same expression is divided by $2x - 1$.

11 The expression $2x^3 + 3x^2 + ax + b$ leaves a remainder of 7 when divided by $x - 2$ and a remainder of -3 when divided by $x - 1$. Find the values of the constants a and b.

12 The cubic $cx^3 + dx^2 + 3x + 8$ leaves a remainder of -6 when divided by $x - 2$, and a remainder of -34 when divided by $x + 2$. Find the values of the constants c and d.

13 Given that $x + 3$ and $x + 4$ are both factors of the expression $ax^3 + bx^2 + 5x - 12$, find the values of the constants a and b.

14 The expression $x^3 + px^2 + qx - 10$ is divisible by $x - 2$, and leaves a remainder of 5 when divided by $x + 3$. Find the values of the constants p and q.

15 The expression $x^3 - x^2 + ax + b$ has a factor of $x + 3$, and leaves a remainder of 6 when divided by $x - 3$. Find the values of the constants a and b, and hence factorise the expression.

16 Given that $x^2 - 4$ is a factor of the cubic $x^3 + cx^2 + dx - 12$, find the values of the constants c and d, and hence factorise the cubic.

17 The remainder when the expression $x^3 - 2x^2 + ax + b$ is divided by $x - 2$ is five times the remainder when the same expression is divided by $x - 1$, and 12 less than the remainder when the same expression is divided by $x - 3$. Find the values of the constants a and b.

Factorising polynomials

Example 11 Factorise the polynomial $f(x) = 2x^3 - 5x^2 - 19x + 42$. Hence solve the inequality $f(x) \leqslant 0$.

SOLUTION

We know that if $f(x)$ has a linear factor $ax + b$, the constant b will be a factor of 42. In other words, one of ± 1, ± 2, ± 3, ± 6, ± 7, ± 14, ± 21 or ± 42.

To see whether $x \pm 1$ are factors, we evaluate $f(\mp 1)$. Since $f(-1) = 54$ and $f(1) = 20$, $x \pm 1$ are not factors.

To see whether $x \pm 2$ are factors, we evaluate $f(\mp 2)$. Since $f(-2) = 44$, $x + 2$ is not a factor. But $f(2) = 0$, therefore $x - 2$ is a factor. So, we have

$$f(x) = (x - 2)(ax^2 + bx + c)$$

Expanding and comparing coefficients (or by inspection) give $a = 2$, $b = -1$ and $c = -21$. Therefore,

$$\begin{aligned} 2x^3 - 5x^2 - 19x + 42 &\equiv (x - 2)(2x^2 - x - 21) \\ &\equiv (x - 2)(2x - 7)(x + 3) \end{aligned}$$

To solve the inequality $f(x) \leqslant 0$, a sketch graph of $f(x)$ would be useful.

Since $f(x) = (x + 3)(x - 2)(2x - 7)$, solving $f(x) = 0$ gives $x = -3$, $x = 2$ or $x = \frac{7}{2}$. Therefore, the curve cuts the x-axis at $x = -3$, $x = 2$ and $x = \frac{7}{2}$. Also $f(0) = 42$. Therefore, the curve cuts the y-axis at 42. The sketch is shown on the right.

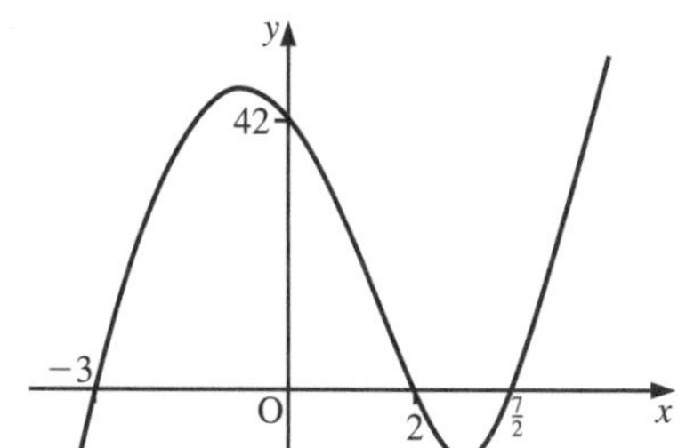

By inspection of the graph, it can be seen that $f(x) \leqslant 0$ when $x \leqslant -3$ and when $2 \leqslant x \leqslant \frac{7}{2}$.

Exercise 4C

1 Factorise each of these expressions.

a) $x^3 - x^2 - 9x + 9$ **b)** $x^3 + 6x^2 + 11x + 6$ **c)** $x^3 - 2x^2 + 2x - 4$
d) $x^3 - 4x^2 - x + 4$ **e)** $x^3 - 2x^2 - 5x + 6$ **f)** $x^3 - 4x^2 - 21x$
g) $x^3 - 5x^2 + 8x - 4$ **h)** $x^3 - 9x^2 + 27x - 27$ **i)** $x^3 + 6x^2 - x - 30$
j) $x^3 + x^2 + 4x + 4$ **k)** $x^3 - x^2 - 5x - 3$ **l)** $x^3 + 7x^2 - x - 7$

2 Factorise each of these expressions.

a) $2x^3 - x^2 - 2x + 1$ **b)** $2x^3 - x^2 - 5x - 2$ **c)** $3x^3 + 2x^2 - 7x + 2$
d) $2x^3 + 7x^2 + 2x - 3$ **e)** $3x^3 - x^2 + 3x - 1$ **f)** $5x^3 + 14x^2 + 7x - 2$
g) $x^4 + x^3 - 3x^2 - x + 2$ **h)** $x^4 + 2x^3 - 7x^2 - 8x + 12$ **i)** $x^4 + 3x^3 + x^2 - 3x - 2$
j) $x^4 - 7x^3 + 15x^2 - 13x + 4$ **k)** $4x^4 + 11x^3 - 7x^2 - 11x + 3$ **l)** $x^4 - 16$

3 Find the real solutions to each of the following equations.

a) $x^3 + x^2 - 10x + 8 = 0$ **b)** $x^3 - 6x^2 + 12x - 8 = 0$ **c)** $x^3 - 7x - 6 = 0$
d) $x^3 - 3x = 2$ **e)** $x^3 + 7x^2 = 15 - 7x$ **f)** $2x^3 - 3x^2 - 5x + 6 = 0$
g) $3x^3 - 20x^2 + 29x + 12 = 0$ **h)** $2x^3 + x^2 = 16x + 15$ **i)** $5x^3 + 23x = 34x^2 - 6$
j) $x^3 + 8x^2 = 2 - 11x$ **k)** $x^3 + 30x = 10x^2 + 27$ **l)** $x^3 + 5 = 8x - 2x^2$

4 Given $\mathrm{p}(x) \equiv x(x^2 - 13) + 12$, express $\mathrm{p}(x)$ as a product of linear factors. Hence solve the equation $\mathrm{p}(x) = 0$.

5 Express the function $(x - 1)(x^2 - 2x - 11) - 16$ as the product of linear factors. Hence solve the equation $(x - 1)(x^2 - 2x - 11) = 16$.

6 Show that the equation $x^3 - 5x^2 + 2x - 10 = 0$ has only one real solution, and state its value.

7 Given

$$(x + a)(x + 3)(x - 2) \equiv x^3 + bx^2 + cx - 30$$

find the values of the constants a, b and c. Hence, with these values of b and c, solve the equation $x^3 + bx^2 + cx - 30 = 0$.

8 Given

$$(x + c)(x + d)(x - 1) \equiv x^3 + 2x^2 - 13x + e$$

find the possible values of the constants c, d and e. With this value of e, solve the equation $x^3 + 2x^2 - 13x + e = 0$.

9 Find the values of the constants a, b and c for which

$$(x - 4)(x - 2)(x + a) \equiv x^3 - 7x^2 + bx + c$$

Taking these values for b and c, solve the equation $x^3 - 7x^2 + bx + c = 0$.

10 Given

$$(x^2 + a)(x - 4) \equiv x^3 + bx^2 + cx - 20$$

find the values of the constants a, b and c. Hence show that, with these values of b and c, the equation $x^3 + bx^2 + cx - 20$ has only one real solution.

11 Find the values of the constants b and c for which

$$(x^2 + b)(x + c) \equiv x^3 - 3x^2 + bx - 15$$

With this value of b, solve the equation $x^3 - 3x^2 + bx - 15 = 0$.

12 Solve each of the following inequalities for x.

a) $x^3 + 2x^2 - 5x - 6 \geqslant 0$ **b)** $x^3 - 5x^2 + 2x + 8 > 0$ **c)** $x^3 - 6x^2 + 11x - 6 > 0$
d) $x^3 - 6x^2 + 3x + 10 \leqslant 0$ **e)** $x^3 + 3x^2 - 9x + 5 < 0$ **f)** $2x^3 + x^2 - 13x + 6 > 0$
g) $3x^3 + 23x \geqslant 16x^2 + 6$ **h)** $2x^3 - 7x < 7x^2 - 12$ **i)** $x^3 - 4x^2 + 2x - 8 > 0$
j) $2x^3 + 12x^2 + 3x + 18 \leqslant 0$ **k)** $x^3 - 6x^2 + 12x < 8$ **l)** $6(x^3 + 1) < 17x^2 + 5x$

13 Express $(2x - 3)(x^2 - 5x - 1) + 7$ as the product of linear factors. Hence solve the inequality $(2x - 3)(x^2 - 5x - 1) \geqslant -7$.

14 Given $p(x) \equiv (x^2 - 12)(x + 1) + 4x$, express $p(x)$ as a product of linear factors. Hence solve the inequality $p(x) \geqslant 0$.

15 Find the values of the constants a, b and c for which

$$(x + 5)(x - 4)(x - a) \equiv x^3 + bx^2 - 23x + c$$

Taking these values for b and c, solve the inequality

$$x^3 + bx^2 - 23x + c < 0$$

16 Given

$$(2x + c)(x + c)(x - 4) \equiv 2x^3 + dx^2 + ex - 36$$

where c, d and e are constants and $c > 0$, find the values of c, d and e. With these values of d and e, solve the inequality

$$2x^3 + dx^2 + ex > 36.$$

17 Given that $a > b$ and

$$(3x + 1)(ax - 1)(bx - 1) \equiv 24x^3 - 10x^2 + cx + 1$$

find the values of the constants a, b and c. With this value of c, solve the inequality

$$24x^3 - 10x^2 + cx + 1 \leqslant 0$$

Exercise 4D: Examination questions

1 Use the remainder theorem to find one of the factors of the cubic

$$f(x) \equiv 2x^3 - 9x^2 + 7x + 6$$

Hence factorise $f(x)$ into its linear factors. (WJEC)

2 a) When the cubic expression

$$x^3 + ax^2 - (2a^2 + 12)x + (7a + 10)$$

is divided by $(x - 1)$ the remainder is 7. Find a.

b) Find the three linear factors of $x^3 + 2x^2 - 20x + 24$. (WJEC)

3 i) When the expression $6x^3 + ax^2 + bx + 4$ is divided by $(x + 1)$ the remainder is -15, and when the expression is divided by $(x - 3)$ the remainder is 49. Use the remainder theorem to show that $a = -13$ and $b = 0$.

ii) Find the three linear factors of $6x^3 - 13x^2 + 4$. (WJEC)

4 Given that the expression $ax^3 + 8x^2 + bx + 6$ is exactly divisible by $x^2 - 2x - 3$, find the values of a and b. (UODLE)

5 $$f(x) \equiv 2x^3 + x^2 - 8x - 4$$

a) Show that $(2x + 1)$ is a factor of $f(x)$.

b) Factorise $f(x)$ completely.

c) Hence find the values of x for which $f(x) = 0$. (EDEXCEL)

6 $$f(x) \equiv x^3 + 2x^2 - 11x - 12$$

a) Show that $(x + 1)$ is a factor of $f(x)$.
b) Solve the equation $f(x) = 0$. (EDEXCEL)

7 The cubic polynomial $x^3 + Ax - 12$ is exactly divisible by $x + 3$. Find the constant A, and solve the equation

$$x^3 + Ax - 12 = 0$$

for this value of A. (MEI)

8 A function is defined as $f(x) = x^3 - 3x^2 + x + 1$.

i) Find $f(1)$ and $f(-2)$ and hence state a factor of $f(x)$.
ii) Express $f(x)$ in the form $(x + p)(x^2 + qx + r)$, where p, q and r are numbers to be determined.
iii) Solve the equation $f(x) = 0$. (MEI)

9 You are given that $f(x) = x^3 - 19x + 30$.

i) Calculate $f(0)$ and $f(3)$. Hence write down a factor of $f(x)$.
ii) Find p and q such that $f(x) \equiv (x - 2)(x^2 + px + q)$.
iii) Solve the equation $x^3 - 19x + 30 = 0$.
iv) Without further calculation draw a sketch of $y = f(x)$. (MEI)

10 **a)** Show that $(x + 2)$ is a factor of the polynomial $f(x)$ given by

$$f(x) = 2x^3 - 3x^2 - 11x + 6$$

b) Express $f(x)$ as the product of three linear factors.
c) By considering the graph of $y = f(x)$, or otherwise, solve the inequality $f(x) \leqslant 0$. (NEAB)

11 Prove that $x + 1$ is a factor of $x^3 - 6x^2 + 3x + 10$. Solve the inequality $x^3 - 6x^2 + 3x + 10 > 0$.

12 Show that $(x - 2)$ is a factor of $x^3 - 9x^2 + 26x - 24$. Find the set of values of x for which

$$x^3 - 9x^2 + 26x - 24 < 0 \qquad \text{(AEB Spec)}$$

5 Coordinate geometry

He is unworthy of the name of man who is ignorant of the fact that the diagonal of a square is incommensurable with its sides.

PLATO

Distance between two points

Given two points $A(x_1, y_1)$ and $B(x_2, y_2)$ in the xy-plane, we require a general formula for the distance AB. The following example illustrates the technique for finding the distance between two points in the xy-plane.

Example 1 Find the distance between the points $A(1, 3)$ and $B(6, 15)$.

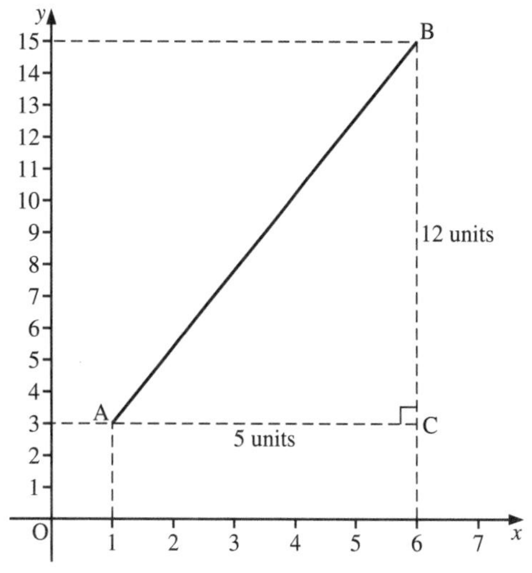

SOLUTION

Construct a right-angled triangle ABC, with AB as the hypotenuse:

$$\text{Length AC} = 6 - 1 = 5 \text{ units}$$

$$\text{Length BC} = 15 - 3 = 12 \text{ units}$$

Using Pythagoras gives

$$AB^2 = 5^2 + 12^2$$

$$= 169$$

$$\therefore \quad AB = \sqrt{169} = 13 \text{ units}$$

The distance between the points A and B is 13 units.

Result I: General formula for the distance between two points

The distance between points $A(x_1, y_1)$ and $B(x_2, y_2)$ is given by

$$\sqrt{(x_2 - x_1)^2 + (y_2 - y_1)^2}$$

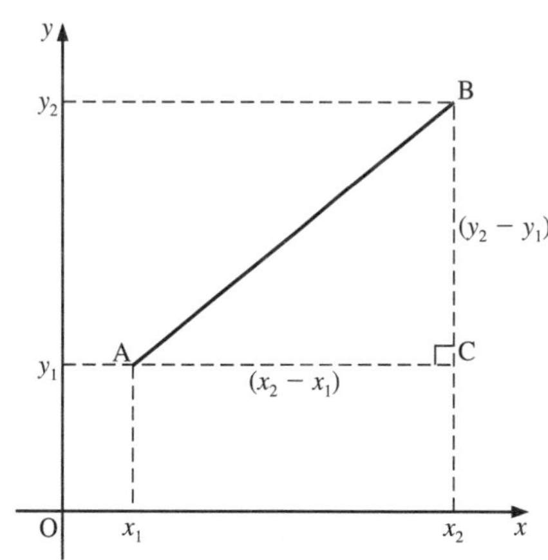

Proof

The lengths AC and BC are given by

$$AC = x_2 - x_1$$

$$BC = y_2 - y_1$$

Using Pythagoras gives

$$AB^2 = (x_2 - x_1)^2 + (y_2 - y_1)^2$$

$$\therefore \quad AB = \sqrt{(x_2 - x_1)^2 + (y_2 - y_1)^2}$$

Therefore, the distance between the two points $A(x_1, y_1)$ and $B(x_2, y_2)$ is given by

$$\sqrt{(x_2 - x_1)^2 + (y_2 - y_1)^2}$$

Example 2 Find the distance between the points A(−1, 4) and B(4, −6).

SOLUTION

Using the result $AB^2 = (x_2 - x_1)^2 + (y_2 - y_1)^2$ gives

$$AB^2 = (4 - (-1))^2 + (-6 - 4)^2$$

$$= 125$$

$$\therefore \quad AB = \sqrt{125} = 5\sqrt{5} \text{ units}$$

Therefore, the distance between the points A and B is $5\sqrt{5}$ units.

Example 3 Prove that the points A(−3, 4), B(1, 1) and C(7, 9) are the vertices of a right-angled triangle.

SOLUTION

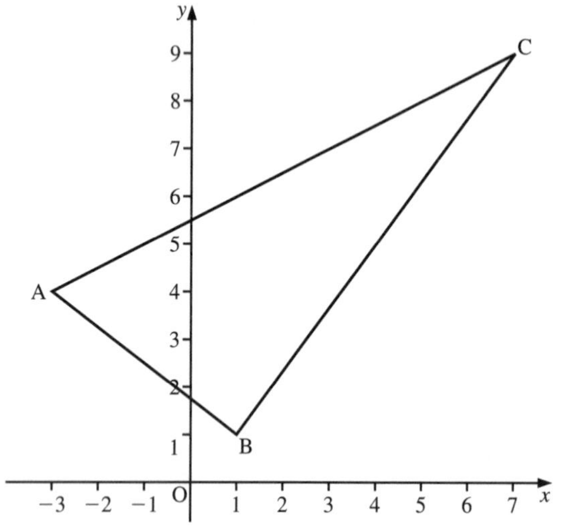

The distance AB is given by

$$AB^2 = (-3 - 1)^2 + (4 - 1)^2$$

$$= 25$$

$$\therefore \quad AB = 5 \text{ units}$$

The distance BC is given by

$$BC^2 = (1 - 7)^2 + (1 - 9)^2$$

$$= 100$$

$$\therefore \quad BC = 10 \text{ units}$$

The distance AC is given by

$$AC^2 = (-3 - 7)^2 + (4 - 9)^2$$

$$= 125$$

$$\therefore \quad AC = \sqrt{125} \text{ units}$$

When the lengths of the sides of triangle ABC satisfy Pythagoras' theorem, then triangle ABC is right-angled. Since AC is the longest side, we need to check that $AC^2 = AB^2 + BC^2$.

Now

$$AC^2 = 125 \quad \text{and} \quad AB^2 + BC^2 = 25 + 100 = 125$$

$$\therefore \quad AC^2 = AB^2 + BC^2$$

Therefore, the triangle ABC is right-angled.

Example 4 The points A, B and C have coordinates $(-3, 2)$, $(-1, -2)$ and $(0, k)$ respectively, where k is a constant. Given that $AC = 5BC$, find the possible values of k.

SOLUTION

If $AC = 5BC$, then squaring both sides gives

$$AC^2 = 25BC^2 \qquad [1]$$

The lengths AC^2 and BC^2 are found in the usual way, giving

$$AC^2 = (0 - (-3))^2 + (k - 2)^2$$

$$\therefore \quad AC^2 = k^2 - 4k + 13 \qquad [2]$$

Similarly,

$$BC^2 = (0 - (-1))^2 + (k - (-2))^2$$

$$\therefore \quad BC^2 = k^2 + 4k + 5 \qquad [3]$$

Substituting [2] and [3] into [1] gives

$$k^2 - 4k + 13 = 25(k^2 + 4k + 5)$$

$$k^2 - 4k + 13 = 25k^2 + 100k + 125$$

$$\therefore \quad 24k^2 + 104k + 112 = 0$$

Dividing through by 8 gives

$$3k^2 + 13k + 14 = 0$$

Factorising and solving give

$$(3k + 7)(k + 2) = 0$$

$$\therefore \quad k = -\tfrac{7}{3} \quad \text{or} \quad k = -2$$

Therefore, the required values of the constant are $-\frac{7}{3}$ and -2.

Exercise 5A

1 Find the distance between each of the following pairs of points.

a) $(2, 1)$ and $(5, 5)$ **b)** $(3, 6)$ and $(8, 18)$ **c)** $(-3, 2)$ and $(5, 8)$
d) $(0, -2)$ and $(8, 13)$ **e)** $(-3, -4)$ and $(-15, 12)$ **f)** $(2, 5)$ and $(6, 1)$
g) $(-7, 3)$ and $(-2, 5)$ **h)** $(6, 0)$ and $(-4, 0)$ **i)** $(2, -3)$ and $(7, 7)$
j) $(-7, 4)$ and $(-1, 1)$ **k)** $(4, -1)$ and $(-2, 1)$ **l)** $(5, 8)$ and $(8, 5)$

2 The three points A, B and C have coordinates $(-1, 3)$, $(6, 4)$ and $(1, -1)$ respectively. Show that the distance AB is equal to the distance BC.

3 $P(-2, -6)$, $Q(6, 9)$ and $R(1, -3)$ are the vertices of a triangle. Find the number of units by which the length of PQ exceeds the length of QR.

4 Prove that the points $A(2, 3)$, $B(5, 6)$ and $C(8, 3)$ are the vertices of a right-angled triangle.

5 A triangle has vertices at $(1, 2)$, $(13, 7)$ and $(6, 14)$. Prove that the triangle is isosceles.

6 Prove that the triangle with vertices P(2, 1), Q(5, –1) and R(9, 5) is right-angled.

7 Point C has coordinates (1, 3), point D has coordinates (5, –1), and point E has coordinates (–1, –3). Prove that the triangle CDE is isosceles.

8 By first showing that the triangle with vertices A(–3, –1), B(1, –4) and C(7, 4) is right-angled, deduce that the area of the triangle ABC is 25 units^2.

9 Calculate the area of the right-angled triangle whose vertices are P(2, 5), Q(5, 4) and R(8, 13).

10 Prove that the points A(–2, 0), B($0, 2\sqrt{3}$) and C(2, 0) are the vertices of an equilateral triangle.

11 The four points A(5, 4), B(6, 2), C(12, 5) and D(11, 7) are the vertices of a quadrilateral. Prove that the quadrilateral is a rectangle, and calculate its area.

12 The points L, M and N have coordinates (3, 1), (2, 6) and (x, 5) respectively. Given that the distance LM is equal to the distance MN, calculate the possible values of x.

13 Given that the distance between P(p, 4) and Q(2, 3) is equal to the distance between R(3, –1) and S(–2, 4), calculate the possible values of p.

14 A triangle has vertices A(6, 2), B(b, 6) and C(–2, 6). Given that the triangle is isosceles with AB = BC, calculate the value of b.

15 F(5, 1), G(x, 7) and H(8, 2) are the vertices of a triangle. Given that the length of the side FG is twice the length of the side FH, calculate the possible values of x.

16 Given that the distance from A(13, 10) to B(1, b) is three times the distance from B to C(–3, –2), calculate the possible values of b.

Mid-point of a straight line

Example 5 Find the coordinates of the mid-point M of the straight line joining A(1, 3) and B(5, 11).

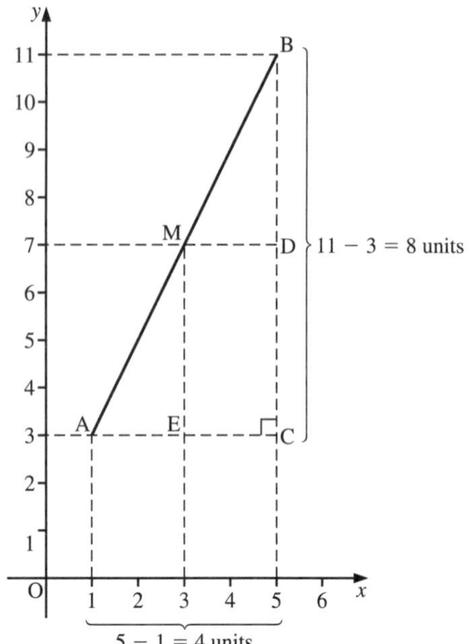

SOLUTION

Let M(X, Y) be the mid-point of the straight line joining A(1, 3) and B(5, 11). Since M is the mid-point of AB, E is the mid-point of AC. Therefore,

$$AE = \frac{1}{2}(5 - 1) = 2$$

$$\therefore \quad X = 1 + 2 = 3$$

Similarly, D is the mid-point of BC. Therefore

$$CD = \frac{1}{2}(11 - 3) = 4$$

$$\therefore \quad Y = 3 + 4 = 7$$

The coordinates of M are therefore (3, 7)

Result II: Coordinates of the mid-point of a line

The coordinates of the mid-point of the straight line joining $A(x_1, y_1)$ and $B(x_2, y_2)$ are

$$\left(\tfrac{1}{2}(x_1 + x_2), \tfrac{1}{2}(y_1 + y_2)\right)$$

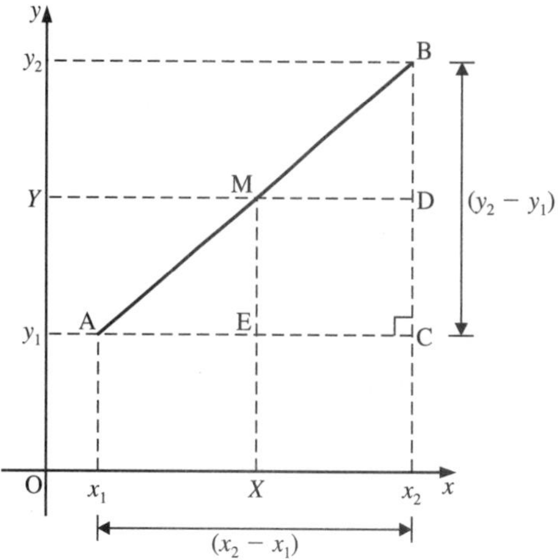

Proof

Let the mid-point M of AB have coordinates (X, Y). Since M is the mid-point of AB, E is the mid-point of AC. Therefore,

$$AE = \frac{1}{2}(x_2 - x_1)$$

$$\therefore \quad X = x_1 + \frac{1}{2}(x_2 - x_1)$$

$$= \frac{1}{2}(x_1 + x_2)$$

Similarly, D is the mid-point of BC. Therefore,

$$DC = \frac{1}{2}(y_2 - y_1)$$

$$\therefore \quad Y = y_1 + \frac{1}{2}(y_2 - y_1)$$

$$= \frac{1}{2}(y_1 + y_2)$$

The coordinates of M are therefore $\left(\tfrac{1}{2}(x_1 + x_2), \tfrac{1}{2}(y_1 + y_2)\right)$, as required.

Example 6 Find the coordinates of the mid-point M of the straight line joining A(4, −2) and B(−3, −1). Verify that the distance AM is equal to the distance BM.

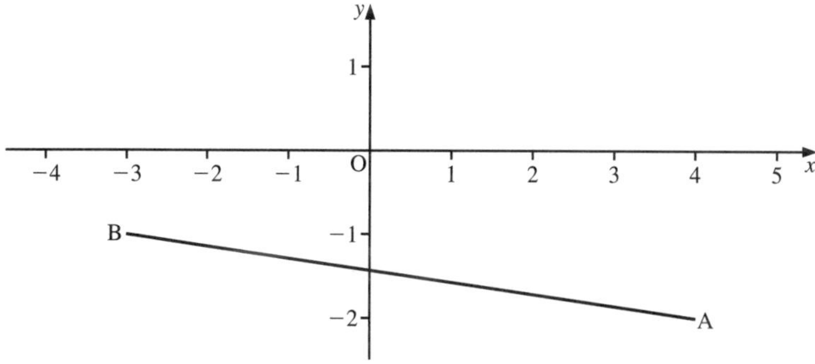

SOLUTION

The coordinates of the mid-point M are

$$\left(\frac{1}{2}(x_1 + x_2), \frac{1}{2}(y_1 + y_2)\right) = \left(\frac{1}{2}(4 + (-3)), \frac{1}{2}(-2 + (-1))\right)$$

$$= \left(\tfrac{1}{2}, -\tfrac{3}{2}\right)$$

The distances AM and BM are found using $\sqrt{(x_2 - x_1)^2 + (y_2 - y_1)^2}$. This gives

$$AM^2 = \left[4 - \frac{1}{2}\right]^2 + \left[-2 - \left(-\frac{3}{2}\right)\right]^2$$
$$= \frac{50}{4}$$
$$\therefore \quad AM = \sqrt{\frac{50}{4}} = \frac{5\sqrt{2}}{2}$$

Similarly,

$$BM^2 = \left[-3 - \frac{1}{2}\right]^2 + \left[-1 - \left(-\frac{3}{2}\right)\right]^2$$
$$= \frac{50}{4}$$
$$\therefore \quad BM = \sqrt{\frac{50}{4}} = \frac{5\sqrt{2}}{2}$$

Therefore, the distance AM is equal to the distance BM.

Exercise 5B

1 Find the coordinates of the mid-point of the straight line joining each of the following pairs of points.

a) $(3, 2)$ and $(7, 4)$ **b)** $(3, 5)$ and $(7, 7)$ **c)** $(-2, 3)$ and $(4, 1)$
d) $(6, 8)$ and $(2, -4)$ **e)** $(7, -5)$ and $(-2, -3)$ **f)** $(-6, -7)$ and $(-4, -3)$
g) $(7, 0)$ and $(3, 0)$ **h)** $(5, -2)$ and $(6, 3)$ **i)** $(-7, 2)$ and $(7, -2)$
j) $(3, -6)$ and $(5, -4)$ **k)** $(3, -5)$ and $(4, 9)$ **l)** $(-2, 5)$ and $(9, -4)$

2 M(6, 5) is the mid-point of the straight line joining the point A(2, 3) to the point B. Find the coordinates of B.

3 P is the mid-point of the straight line joining the point C(−5, 3) to the point D. Given that P has coordinates (2, 1), find the coordinates of D.

4 Find the coordinates of the point S given that M(3, −2) is the mid-point of the straight line joining S to T(9, −2).

5 Prove that the points A(1, −1), B(2, −5) and C(−2, −4) are the vertices of an isosceles triangle. Given that M is the mid-point of the longest side of the triangle ABC, calculate the coordinates of M.

6 A triangle has vertices A(4, 4), B(7, 6) and C(5, 3). Prove that the triangle ABC is isosceles. Given that P is the mid-point of AB, Q is the mid-point of BC and R is the mid-point of CA, prove that the triangle PQR is also isosceles.

7 A quadrilateral has vertices A(2, 5), B(6, 9), C(10, 3) and D(4, –5). The points E, F, G and H are the mid-points of AB, BC, CD and DA, respectively. Prove that

a) the length EF is equal to the length GH
b) the length EH is equal to the length FG.

8 C(2, 1), D(6, 5), E(10, 3) and F(8, 3) are the vertices of the quadrilateral CDEF. The points P, Q, R and S are the mid-points of CD, DE, EF and FC respectively.

a) Calculate the coordinates of P, Q, R and S.
b) Prove that the length of PQ is equal to the length of RS.
c) Prove that the length of PS is equal to the length of QR.

***9** PQR is an isosceles triangle in the positive quadrant of the xy-plane, with PQ equal to PR. Given that the coordinates of Q and R are (a, b) and (c, b) respectively, where $c > a$, and that the perpendicular distance of P from QR is 2 units, find, giving your answers in terms of a, b and c as appropriate,

a) the coordinates of M, the mid-point of the line QR
b) the coordinates of the vertex P
c) an expression for A, the area of the triangle PQR.

Deduce that

d) $A^2 = 4\,\text{PQ}^2 - 16$.

Gradient of a straight line

The gradient of a straight line is given by

$$\frac{\text{Change in } y}{\text{Change in } x}$$

in moving from one point on the line to another point on the line. The gradient is denoted by m.

The gradient m can be negative or positive, as shown in the diagrams.

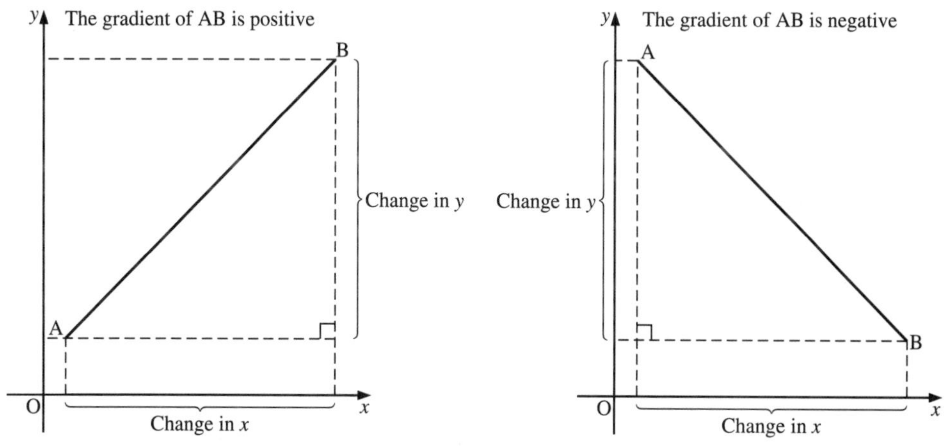

Example 7 Find the gradient of the straight line joining A(−4, 1) and B(6, 6).

SOLUTION

In moving from A to B,

the change in y is 5

the change in x is 10

$$\therefore \quad \text{Gradient} = \frac{5}{10} = \frac{1}{2}$$

The gradient of the line AB is $\frac{1}{2}$.

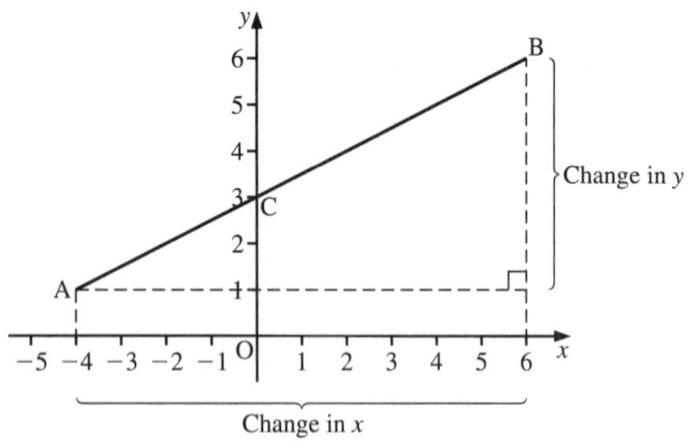

The gradient is independent of the choice of points on the line. To see this, consider point C(0, 3), which lies on the line AB. In moving from A to C,

the change in y is 2

the change in x is 4

$$\therefore \quad \text{Gradient} = \frac{2}{4} = \frac{1}{2}$$

The gradient of the line AC is also $\frac{1}{2}$.

Result III: General formula for the gradient of a straight line

The gradient of the straight line joining $A(x_1, y_1)$ and $B(x_2, y_2)$ is given by

$$\frac{y_2 - y_1}{x_2 - x_1}$$

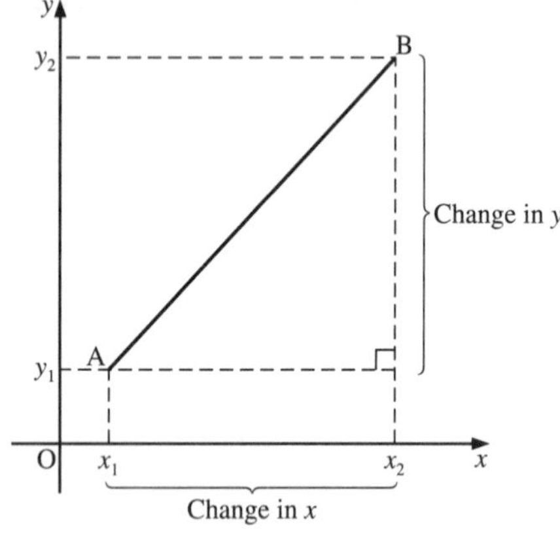

Proof

In moving from A to B,

the change in y is $(y_2 - y_1)$

the change in x is $(x_2 - x_1)$

$$\therefore \quad \text{Gradient} = \frac{y_2 - y_1}{x_2 - x_1}$$

The gradient of the straight line AB is

$$\frac{y_2 - y_1}{x_2 - x_1}$$

as required.

Angle of a straight line to the horizontal

In the right-angled triangle ABC, θ is the angle which the line AB makes with the horizontal.

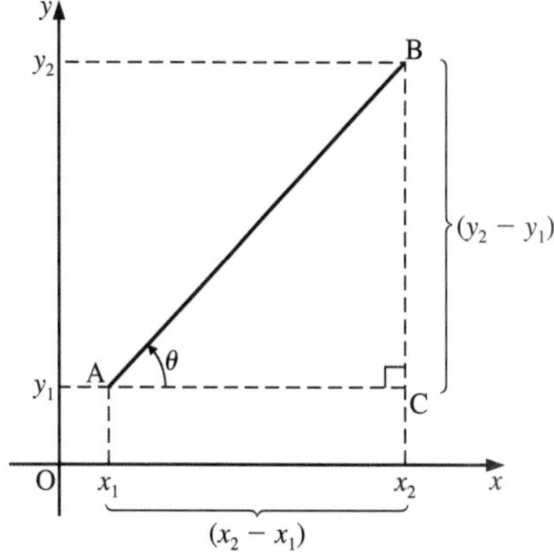

Using trigonometry, we obtain $\tan\theta = \dfrac{BC}{AC}$.

Now $\dfrac{BC}{AC} = \dfrac{y_2 - y_1}{x_2 - x_1}$ is the gradient of the line AB.

Therefore,

$$\tan\theta = \frac{y_2 - y_1}{x_2 - x_1}$$

$$\therefore \quad \theta = \tan^{-1}\left(\frac{y_2 - y_1}{x_2 - x_1}\right)$$

- If m is the gradient of the straight line AB, then the angle AB makes with the horizontal is given by $\tan^{-1}m$.
- When $\tan^{-1}m$ is negative, this indicates that the angle θ is measured from the horizontal to the straight line in a clockwise direction.

Example 8 Find the gradient of the straight line joining A(−7, −5) and B(5, −3). Find also the angle which AB makes with the horizontal.

SOLUTION

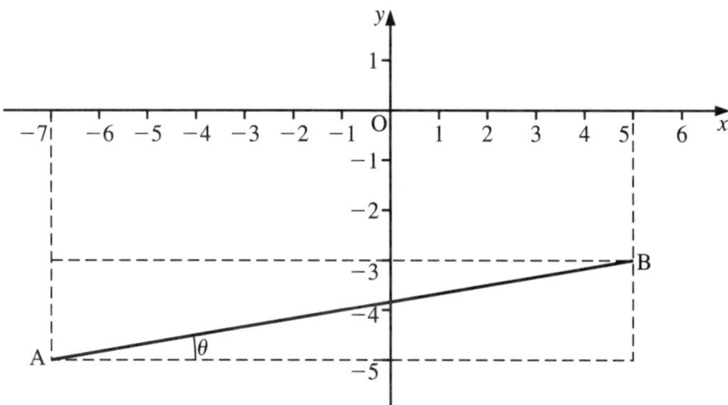

The gradient is found using $\dfrac{y_2 - y_1}{x_2 - x_1}$. Therefore,

$$\text{Gradient of AB} = m = \frac{-3 - (-5)}{5 - (-7)}$$

$$= \frac{-3 + 5}{5 + 7}$$

$$= \frac{2}{12} = \frac{1}{6}$$

The gradient of the line AB is $\frac{1}{6}$.

If θ is the angle that AB makes with the horizontal, then

$$\tan\theta = \text{gradient of AB}$$

$$\therefore \quad \tan\theta = \tfrac{1}{6}$$

$$\therefore \quad \theta = 9.46°$$

Parallel and perpendicular lines

Two lines AB and CD which are parallel make equal angles with the horizontal. Therefore, parallel lines have the **same gradient**.

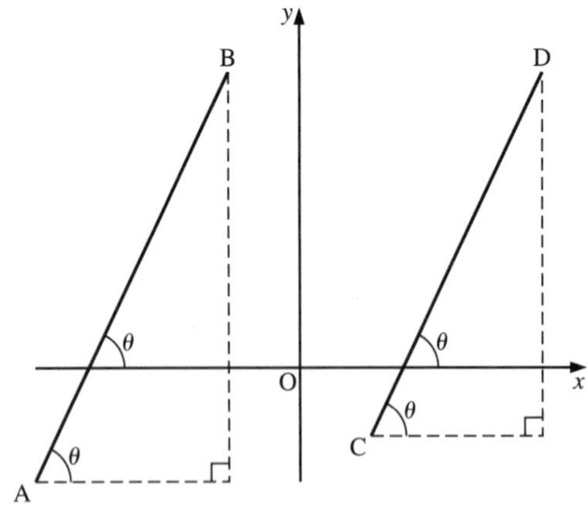

Result IV: Product of the gradients of two perpendicular lines

When two straight lines are perpendicular, the product of their gradients is -1.

Proof

Let PQ and PS be two straight lines which are perpendicular, as shown.

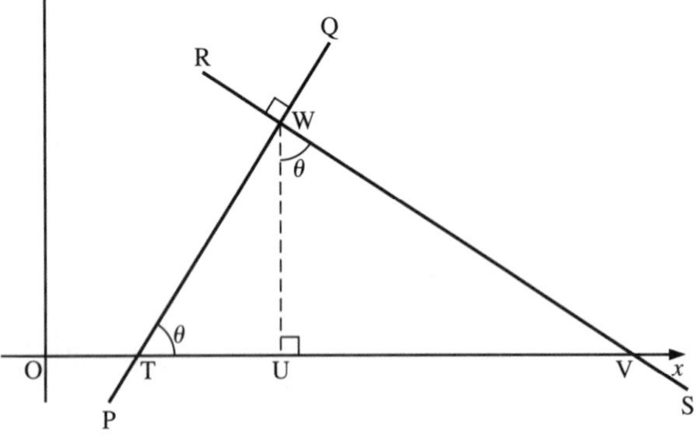

By inspection of the diagram, angle UWV $= \theta$.

Let the gradient of PQ be m. The gradient of RS, m_{RS}, is given by

$$m_{\mathrm{RS}} = -\frac{\mathrm{WU}}{\mathrm{UV}} \qquad [1]$$

From triangle WUV,

$$\tan\theta = \frac{\mathrm{UV}}{\mathrm{WU}}$$

$$\therefore \quad \mathrm{UV} = \mathrm{WU}\tan\theta \qquad [2]$$

Substituting [2] into [1] gives

$$m_{\mathrm{RS}} = -\frac{\mathrm{WU}}{\mathrm{WU}\tan\theta} = -\frac{1}{\tan\theta} = -\frac{1}{m}$$

Now

$$\text{Gradient of PQ} \times \text{Gradient of RS} = m \times \left(-\frac{1}{m}\right) = -1$$

Therefore, when two straight lines are perpendicular, the product of their gradients is -1.

Example 9 The four points A(0, 5), B(−2, 4), C(−2, −1) and D(0, 0) are the vertices of a quadrilateral. Show that AB is perpendicular to BD and that AB is parallel to CD. Find the area of the quadrilateral ABCD.

SOLUTION

Calculating the gradients of AB and BD gives

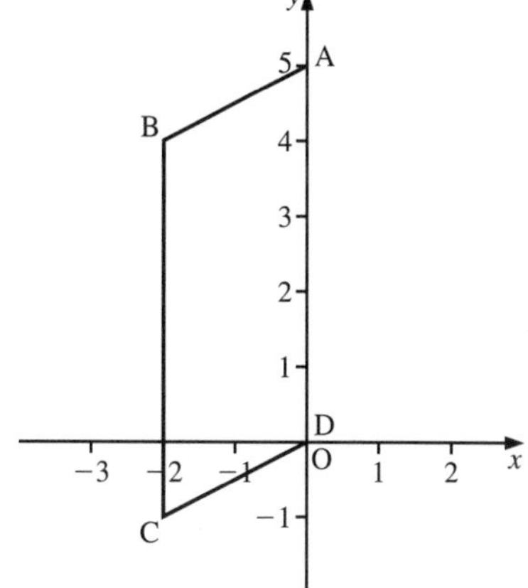

$$m_{AB} = \frac{4-5}{-2-0} = \frac{1}{2}$$

and

$$m_{BD} = \frac{0-4}{0-(-2)} = -2$$

Now

$$m_{AB} \times m_{BD} = \frac{1}{2} \times (-2)$$
$$= -1$$

Therefore, the lines AB and BD are perpendicular.

The lines AB and CD will be parallel if their gradients are equal. Calculating the gradients of each gives

$$m_{AB} = \frac{1}{2}$$

and

$$m_{CD} = \frac{0-(-1)}{0-(-2)} = \frac{1}{2}$$

Since $m_{AB} = m_{CD}$, AB and CD are parallel.

From the diagram, it can be seen that ABCD is a parallelogram. The area of ABCD is calculated using

$$\text{Area} = \text{Base} \times \text{Perpendicular height}$$

This gives

$$\text{Area} = \text{AD} \times 2$$
$$= 5 \times 2 = 10$$

Therefore, the area of the quadrilateral ABCD is 10 units^2.

Exercise 5C

1 Find the gradient of the straight line joining each of the following pairs of points.

a) (2, 3) and (4, 7) **b)** (−1, 2) and (1, 8) **c)** (5, 4) and (3, 3)

d) (7, 4) and (−1, −2) **e)** (3, 2) and (−5, 4) **f)** (−2, −1) and (5, 3)

g) (7, 4) and (−3, 2) **h)** (3, 8) and (5, 8) **i)** (−3, −2) and (−4, −5)

j) (−2, 5) and (5, −3) **k)** (3, 7) and (7, −4) **l)** (6, 3) and (6, 4)

2 Find the angle which the straight line joining each of the following pairs of points makes with the horizontal.

a) $(5, 2)$ and $(7, 10)$
b) $(-3, -2)$ and $(4, 8)$
c) $(3, 7)$ and $(2, 5)$
d) $(5, -4)$ and $(-6, 7)$
e) $(7, -2)$ and $(3, -5)$
f) $(6, 7)$ and $(12, 7)$
g) $(3, -3)$ and $(5, -4)$
h) $(7, -2)$ and $(-3, 5)$
i) $(-6, 3)$ and $(-5, 5)$
j) $(2, 5)$ and $(-3, -2)$
k) $(3, 7)$ and $(-6, 11)$
l) $(5, -3)$ and $(5, 2)$

3 Given the points A(2, 3), B(5, 5), C(7, 2) and D(4, 0),

a) prove that AB is parallel to DC
b) prove that AC is perpendicular to BD.

4 Show that the line joining the point A(2, 5) to the point B(5, 12) is parallel to the line joining the point C(−2, −6) to the point D(1, 1).

5 A triangle has vertices A(3, −2), B(2, −14) and C(−2, −4). Find the gradients of the straight lines AB, BC and CA. Hence prove that the triangle is right-angled.

6 The straight line joining the point A(a, 3) to the point B(5, 7) is parallel to the straight line joining the point B to the point C(−3, −1). Calculate the value of a.

7 The straight line joining the point P(5, 6) to the point Q(q, 2) is perpendicular to the straight line joining the point Q to the point R(9, −1). Calculate the possible values of q.

8 A triangle has vertices D(x, −1), E(1, 1) and F(2, −7). Given that the triangle is right-angled at D, calculate the possible values of x.

9 Prove that the quadrilateral PQRS with vertices P(−1, 3), Q(2, 4), R(4, −2) and S(1, −3) is a rectangle, and calculate its area.

10 Prove that the points A(2, 3), B(4, 8), C(8, 9) and D(4, −1) form a trapezium.

11 Prove that the points W(1, 3), X(3, 4), Y(5, 0) and Z(3, −1) form a parallelogram.

12 The quadrilateral ABCD has vertices A(−2, −3), B(1, −1), C(7, −10) and D(2, −9).

a) Prove that AD is parallel to BC.
b) Prove that AB is perpendicular to BC.
c) Prove that the area of the quadrilateral ABCD is $32\frac{1}{2}$ units^2.

13 A quadrilateral has vertices A(−2, 1), B(0, 4), C(3, 2) and D(1, −1).

a) Prove that each side of the quadrilateral is of the same length.
b) Prove that AB is parallel to DC.
c) Prove that AD is parallel to BC.
d) What is the name given to the quadrilateral ABCD?

14 S(1, 1), T(4, 5), U(12, −1) and V(1, −1) are the vertices of a quadrilateral STUV.

a) Prove that ST is perpendicular to TU, and that SV is perpendicular to UV.
b) Calculate the lengths of each of the sides, ST, TU, UV and VS.
c) Prove that the area of the quadrilateral STUV is 36 units^2.

15 The quadrilateral CDEF has vertices C(4, 0), D(8, 4), E(2, −8) and F(0, 2). The points P, Q, R and S are the mid-points of the sides CD, DE, EF and FC, respectively. Prove that the quadrilateral PQRS is a rhombus, and show that its area is 15 units^2.

General equation of a straight line

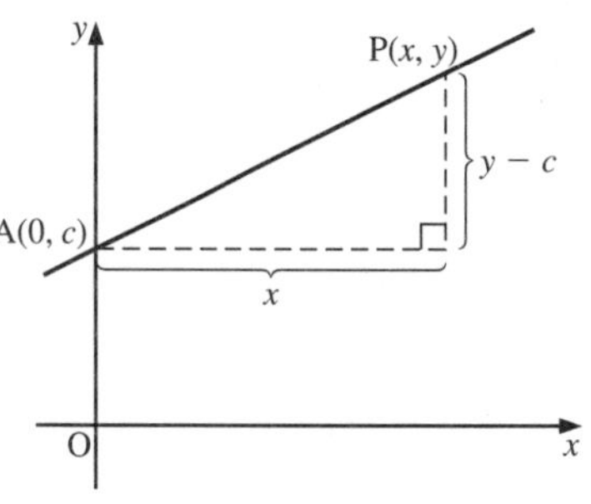

Consider the straight line with gradient m, passing through the point A$(0, c)$ on the y-axis. Let point P(x, y) be a general point on the line.

Since m is the gradient of the line,

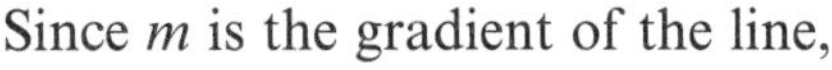

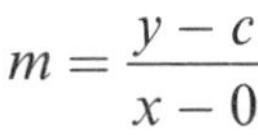

$$m = \frac{y - c}{x - 0}$$

$$\therefore \quad y = mx + c$$

This is the general cartesian form of the equation of a straight line, where m is the gradient of the line and c is the y-intercept.

Example 10 Find the gradient of each of the following straight lines.

a) $y = 5x - 6$ **b)** $2y + 4x = 3$ **c)** $5y - 3x + 4 = 0$

SOLUTION

The general equation is $y = mx + c$. Therefore, each equation needs to be expressed in this form so that m can be identified.

a)

$$y = 5x - 6$$

$$\therefore \quad m = 5$$

The gradient of the line $y = 5x - 6$ is 5.

b)

$$2y + 4x = 3$$

$$\therefore \quad 2y = -4x + 3$$

$$\therefore \quad y = -2x + \tfrac{3}{2}$$

$$\therefore \quad m = -2$$

The gradient of the line $2y + 4x = 3$ is -2.

c)

$$5y - 3x + 4 = 0$$

$$\therefore \quad 5y = 3x - 4$$

$$\therefore \quad y = \tfrac{3}{5}x - \tfrac{4}{5}$$

$$\therefore \quad m = \tfrac{3}{5}$$

The gradient of the line $5y - 3x + 4 = 0$ is $\tfrac{3}{5}$.

Example 11 Find the equation of the straight line with gradient 3 which passes through the point P$(4, 2)$.

SOLUTION

The general equation of the line is $y = mx + c$. We are given that the gradient is 3 (i.e. $m = 3$). Therefore, the equation of the line is

$$y = 3x + c \qquad [1]$$

Since the line passes through P(4, 2), $x = 4$, $y = 2$ satisfies [1]. Therefore,

$$2 = 3(4) + c$$

$$\therefore \quad c = -10$$

The equation of the straight line is $y = 3x - 10$.

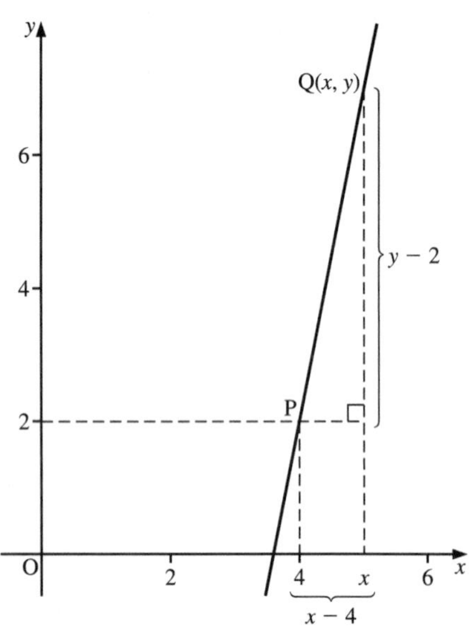

An alternative method is to let Q(x, y) be a general point on the straight line.

Since the gradient of the line is 3, the gradient of PQ must also be 3. Therefore,

$$\frac{y-2}{x-4} = 3$$

$$\therefore \quad y - 2 = 3(x - 4)$$

$$\therefore \quad y = 3x - 10$$

as before.

Example 12 Find the equation of the straight line joining A(1, 4) and B(7, 10).

SOLUTION

The general equation of the straight line is given by

$$y = mx + c \qquad [1]$$

The gradient of the line AB is

$$m = \frac{10-4}{7-1} = 1$$

Since the line AB passes through A(1, 4), $x = 1$, $y = 4$ satisfies [1]. Therefore,

$$4 = 1(1) + c$$

$$\therefore \quad c = 3$$

The equation of the straight line AB is $y = x + 3$.

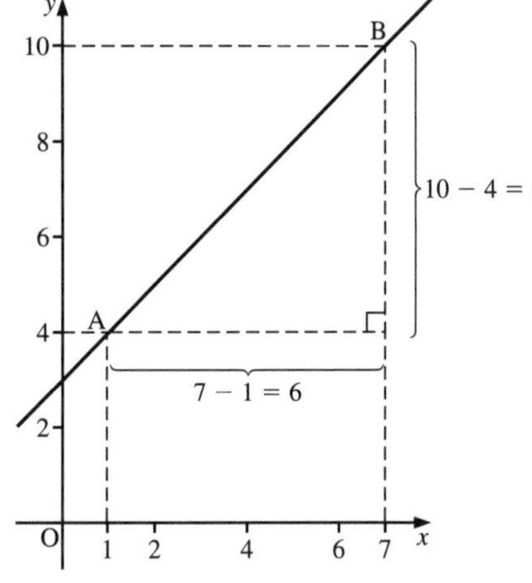

Example 13 Find the equation of the perpendicular bisector of the straight line joining A(−4, 1) and B(2, 6).

SOLUTION

The perpendicular bisector of a line AB is the line which is both perpendicular to AB and which passes through the mid-point of AB.

Let the perpendicular bisector of line AB be line l. To find the equation of line l, we need:

- the gradient of l
- a point through which line l passes, which in this case is the mid-point, M, of AB.

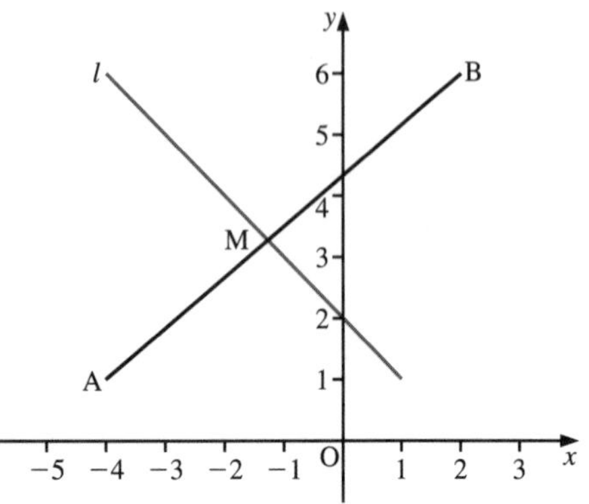

The gradient of AB, m_{AB}, is given by

$$m_{\text{AB}} = \frac{6-1}{2-(-4)} = \frac{5}{6}$$

Since line l is perpendicular to AB, the gradient, m_l, is given by

$$m_l = -\frac{1}{m_{\text{AB}}}$$

$$\therefore \quad m_l = -\frac{1}{\left(\frac{5}{6}\right)} = -\frac{6}{5}$$

Line l passes through the mid-point M of AB. The coordinates of M are

$$\left(\frac{1}{2}(-4+2), \frac{1}{2}(1+6)\right) = \left(-1, \frac{7}{2}\right)$$

Using the result $y = mx + c$ gives

$$y = -\frac{6}{5}x + c \qquad [1]$$

Since line l passes through M, the coordinates of M must satisfy [1]. Substituting $x = -1$, $y = \frac{7}{2}$ gives

$$\frac{7}{2} = -\frac{6}{5}(-1) + c$$

$$\therefore \quad c = \frac{23}{10}$$

Therefore, the equation of line l is

$$y = -\frac{6}{5}x + \frac{23}{10} \quad \text{or} \quad 10y + 12x - 23 = 0$$

Example 14 A straight line l passes through the point $(-3, 5)$ and makes an angle of 45° with the horizontal. Find the equation of line l.

Given that line l intersects the x-axis at A and the y-axis at B, find the distance AB.

SOLUTION

The gradient of line l is $\tan 45° = 1$.

Line l passes through the point $(-3, 5)$ and has a gradient of 1. Therefore,

$$5 = 1(-3) + c$$

$$\therefore \quad c = 8$$

Therefore, the equation of line l is

$$y = x + 8 \qquad [1]$$

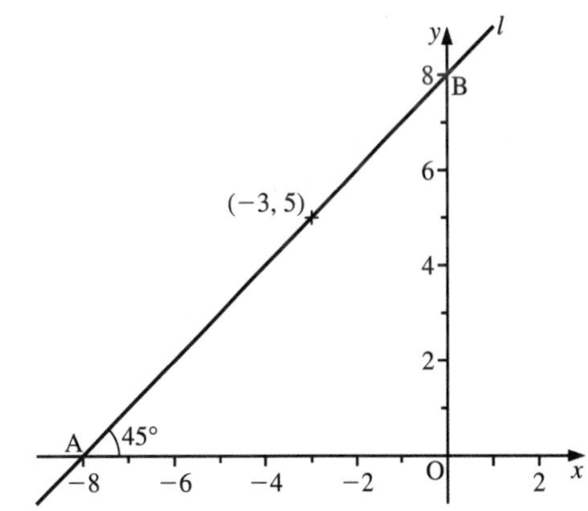

At point A (the intersection with the x-axis), $y = 0$. Substituting into [1] gives

$$0 = x + 8$$
$$\therefore \quad x = -8$$

Therefore, the coordinates of point A are $(-8, 0)$.

At point B (the intersection with the y-axis), $x = 0$. Substituting into [1] gives

$$y = 0 + 8$$
$$\therefore \quad y = 8$$

Therefore, the coordinates of point B are $(0, 8)$.

The distance between the points $A(-8, 0)$ and $B(0, 8)$ is

$$AB^2 = (0 - (-8))^2 + (8 - 0)^2$$
$$= 128$$
$$\therefore \quad AB = \sqrt{128} = \sqrt{64 \times 2} = 8\sqrt{2}$$

The distance AB is $8\sqrt{2}$ units.

Exercise 5D

1 Find the gradient of each of the following straight lines.

a) $y = 5x - 2$ **b)** $y = 3x + 5$ **c)** $y = 8 - 4x$
d) $y = 3 + 7x$ **e)** $y = 5(x - 2)$ **f)** $4y = x - 2$
g) $2y = 5x$ **h)** $7y + 2x = 4$ **i)** $2y + 5x + 6 = 0$
j) $4y + 5x - 3 = 0$ **k)** $5 + 3x + 2y = 0$ **l)** $\frac{y}{3} + \frac{x}{5} = 4$

2 Find the equation of the straight line that has the following properties.

a) Gradient 2 and passes through $(5, 3)$.
b) Gradient 5 and passes through $(2, 4)$.
c) Gradient -2 and passes through $(6, -3)$.
d) Gradient $\frac{1}{2}$ and passes through $(6, -2)$.
e) Gradient $\frac{1}{4}$ and passes through $(2, 5)$.
f) Gradient $-\frac{2}{3}$ and passes through $(-4, 2)$.

3 Find the equation of the straight line that has the following properties.

a) Passes through $(4, 3)$ and is parallel to $y = 3x + 5$.
b) Passes through $(4, -1)$ and is parallel to $y = 3 - 2x$.
c) Passes through $(6, -2)$ and is perpendicular to $y = -3x + 4$.
d) Passes through $(-6, 3)$ and is perpendicular to $4y + 3x = 8$.
e) Passes through $(3, -5)$ and is parallel to $2y + 5x + 7 = 0$.
f) Passes through $(-\frac{1}{3}, -\frac{2}{5})$ and is perpendicular to $3y + 10x = 8$.

4 Find the equations of the straight lines joining each of the following pairs of points.

a) $(2, 3)$ and $(4, 7)$ **b)** $(6, 2)$ and $(2, 0)$ **c)** $(2, -3)$ and $(1, 6)$
d) $(-2, -4)$ and $(-3, -8)$ **e)** $(3, 7)$ and $(4, -5)$ **f)** $(3, 9)$ and $(5, 9)$

5 Find the equation of the perpendicular bisector of the straight line joining each of the following pairs of points.

a) $(2, 3)$ and $(6, 5)$ **b)** $(2, 0)$ and $(6, 4)$ **c)** $(2, -5)$ and $(4, -1)$
d) $(5, 4)$ and $(2, -2)$ **e)** $(-1, 4)$ and $(3, 3)$ **f)** $(3, 2)$ and $(-4, 1)$

6 A straight line l, of positive gradient, passes through the point $(2, 5)$ and makes an angle of $45°$ with the horizontal. Find the equation of l.

7 A straight line l, of negative gradient, passes through the point $(3, -1)$ and makes an angle of $45°$ with the horizontal. Find the equation of l.

8 Find the equation of the straight line, p, which is the perpendicular bisector of the straight line joining the points $(1, 2)$ and $(5, 4)$. The line p meets the x-axis at A and the y-axis at B. Calculate the area of the triangle OAB.

9 Find the equation of the straight line, p_1, which is the perpendicular bisector of the points A$(-2, 3)$ and B$(1, -5)$, and the equation of the straight line, p_2, which is the perpendicular bisector of the points B$(1, -5)$ and C$(17, 1)$. Show that p_1 is perpendicular to p_2.

10 The perpendicular bisector of the straight line joining the points $(3, 2)$ and $(5, 6)$ meets the x-axis at A and the y-axis at B. Prove that the distance AB is equal to $6\sqrt{5}$.

11 A is the point $(1, 2)$ and B is the point $(7, 4)$. The straight line l_1 passes through B and is perpendicular to AB; the straight line l_2 passes through A and is also perpendicular to AB. The line l_1 meets the x-axis at P and the y-axis at Q; the line l_2 meets the x-axis at R and the y-axis at S.

a) Find the equations of each of the lines l_1 and l_2.
b) Calculate the area of the triangle OPQ.
c) Calculate the area of the triangle ORS.
d) Deduce that the area of the trapezium PQSR is 100.

12 P is the point with coordinates $(2, 1)$, and l is the straight line which is perpendicular to OP and which passes through the point P.

a) Find the equation of l.

Given further that the line l meets the x-axis at A and the y-axis at B,

b) calculate the area of the triangle OAP
c) calculate the area of the triangle OBP
d) deduce that the ratio (Area of triangle OAP) : (Area of triangle OBP) is $1:4$.

Intersection of two straight lines

Example 15

a) The straight lines l_1 and l_2 have equations $y = 3x + 4$, $y = -x - 8$ respectively. Find the coordinates of the point of intersection of l_1 and l_2.

b) The straight line l_3 has equation $y = 2x + 1$. Show that l_1, l_2 and l_3 are concurrent.

a) The coordinates of the point of intersection of l_1 and l_2 are found by solving the equations $y = 3x + 4$ and $y = -x - 8$ simultaneously.

Eliminating y gives

$$-x - 8 = 3x + 4$$

$$\therefore \quad x = -3$$

Substituting $x = -3$ into either of the equations gives $y = -5$.

Therefore, the coordinates of the point of intersection are $(-3, -5)$.

b) The three lines are concurrent if they all pass through the same point.

To check that l_3 also passes through the point of intersection, we must check that $x = -3$ and $y = -5$ satisfy the equation of l_3. On substituting these values for x and y, it is clear that the line l_3 does pass through the point of intersection $(-3, -5)$. Therefore, the three lines are concurrent.

Angle between two straight lines

Example 16 Find the angle between the two straight lines $y = 2x - 4$ and $y = x - 1$.

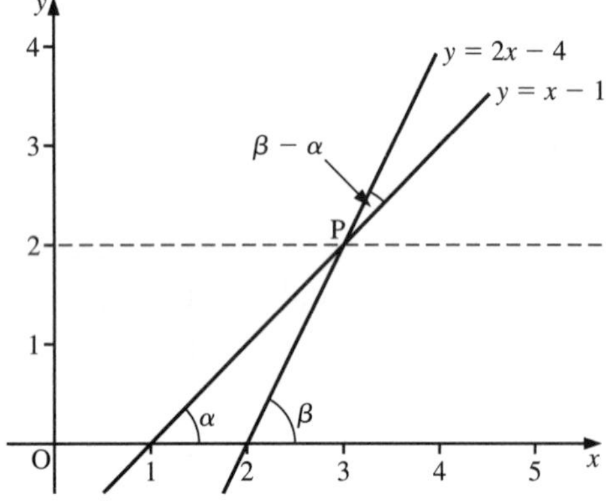

SOLUTION

Plotting both lines on the same set of axes gives the graph on the right.

Let α be the angle that the straight line $y = x - 1$ makes with the horizontal. The gradient of the line $y = x - 1$ is 1. Therefore,

$$\tan \alpha = 1$$

$$\therefore \quad \alpha = 45°$$

Let β be the angle that the straight line $y = 2x - 4$ makes with the horizontal. The gradient of the line $y = 2x - 4$ is 2. Therefore,

$$\tan \beta = 2$$

$$\therefore \quad \beta = 63.4°$$

Drawing a horizontal line through point P, the point of intersection of the two straight lines, shows that the acute angle between the lines $y = x - 1$ and $y = 2x - 4$ is given by

$$\beta - \alpha = 63.4° - 45°$$

$$= 18.4°$$

The obtuse angle between the two straight lines is $180° - 18.4° = 161.6°$.

Exercise 5E

1 Find the coordinates of the point of intersection of each of the following pairs of straight lines.

a) $y = 2x + 3$ and $y = 4x + 1$
b) $y = 3x + 2$ and $y = 5x - 2$
c) $y = x + 3$ and $y = 4x + 6$
d) $y = 5 - x$ and $y = 6x - 2$
e) $y = 3 - 2x$ and $y = 3x + 8$
f) $y = x$ and $y = 2 - 3x$
g) $2y + 3x = 1$ and $3y + x = 5$
h) $5y - 3x = 1$ and $2y + x = 7$
i) $2x - 3y = 7$ and $3x - 7y = 13$
j) $5x + 2y - 9 = 0$ and $3x + 4y - 4 = 0$
k) $x + 3y - 2 = 0$ and $3x + 5y - 8 = 0$
l) $2x + 5y + 6 = 0$ and $3x + 4y - 2 = 0$

2 Find the acute angle between each of the following pairs of straight lines.

a) $y = x$ and $y = 3x - 4$
b) $y = 2x - 5$ and $y = 5x + 6$
c) $y = 3 + x$ and $y = 6 + 2x$
d) $y = 6x - 7$ and $y = 2 - x$
e) $y = 2x + 3$ and $y = 5 - 2x$
f) $y = 4 - 2x$ and $y = 9 - 3x$
g) $y = 5 - 6x$ and $y = 2 + 3x$
h) $y = 4$ and $x = 3$
i) $3y + 2x = 4$ and $5y + 3x = 7$
j) $4y - 3x = 6$ and $2y + x = 3$
k) $3y + 7x - 2 = 0$ and $5y - 6x + 4 = 0$
l) $y = 4$ and $3y + 2x - 6 = 0$

3 **a)** Find the coordinates of the point of intersection, P, of the lines $2x + y = 7$ and $3x - 4y = 5$.
b) Show that the line $x + 2y = 5$ also passes through the point P.

4 Show that the three lines $2x + 3y = 4$, $x - 2y = 9$ and $3x + 7y = 1$ are concurrent.

5 Find the equation of the line which is parallel to the line $y - 3x + 5 = 0$, and which passes through the point of intersection of the lines $2y + 3x - 5 = 0$ and $3y - 2x - 14 = 0$.

6 **a)** Find the equation of the straight line, l, which passes through the point $(2, 4)$ and which is perpendicular to the line $5y + x = 7$.

Given that the line l meets the line $y = x + 6$ at the point S,

b) find the coordinates of the point S.

7 The line $y = 2x$ meets the line $x + 3y = 14$ at the point A.

a) Find the coordinates of A.

The line $x + 3y = 14$ meets the x-axis at B.

b) Find the coordinates of B.
c) Calculate the area of triangle OAB when O is the origin.

8 The line $12x + 9y = 108$ meets the x-axis at A and the y-axis at B.

a) Find the coordinates of A and B.

The line $12x + 9y = 108$ intersects the lines $3y = 2x$ and $3y = 8x$ at the points P and Q respectively.

b) Find the coordinates of P and Q.
c) Show that AP = PQ = QB.

9 Find the coordinates of the vertices of the triangle which has sides given by the equations $x + y = 3$, $x - 2y = 1$ and $3x = 2y$.

10 Calculate the area of the triangle which has sides given by the equations $2y - x = 1$, $y + 2x = 8$ and $4y + 3x = 7$.

11 Prove that the triangle which has sides given by the equations $3y + x = 8$, $y + 3x = 24$ and $y = x$ is isosceles.

12 Three points have coordinates L(2, 5), M(−2, 3) and N(4, 9). Find

a) the equation of the perpendicular bisector, p, of LM
b) the coordinates of the point where p meets MN.

13 Calculate the size of the smallest angle of the triangle which has sides given by the equations $x + y = 6$, $2y = x$ and $2x - 5y = 2$.

14 P, Q and R are the points (3, 8), (−3, 4) and (5, 6) respectively. Find the equations of the perpendicular bisectors of PQ, QR and RS and show that all three are concurrent.

15 Points P, Q and R have coordinates (1, −2), (7, 6) and (9, 2) respectively.

a) Find the coordinates of the point S, where the perpendicular bisector of the line PQ meets the perpendicular bisector of the line QR.
b) Explain why the points P, Q and R lie on a circle with centre S, and calculate the radius of the circle.

***16** Points C, D and E have coordinates $(6, p)$, $(2, p + 2)$ and $(-1, p + 1)$ respectively, and lie on a circle with centre (p, q).

a) Find the equation of the perpendicular bisector of CD and deduce that $3p - q = 7$.
b) Find the equation of the perpendicular bisector of DE and deduce that $2p + q = 3$.
c) Hence find the values of p and q.

Exercise 5F: Examination questions

1 The points O, P and Q have coordinates (0, 0), (4, 3) and (a, b) respectively. Given that OQ is perpendicular to PQ,

a) show that $a^2 + b^2 = 4a + 3b$.

Given also that $a = 1$,

b) find, to 2 decimal places, the possible values of b. (EDEXCEL)

2 The coordinates of the points A, B and C are (8, −2), (−8, 10), (14, 6) respectively.

a) Find the gradients of the lines AB, BC, CA.
b) Show that one of the angles of the triangle ABC is a right-angle.
c) Find the coordinates of the centre of the circle which passes through A, B, C. (WJEC)

3 The points A, B, C, D have coordinates (3, 3), (8, 0), (−1, 1), (−6, 4) respectively.

a) Find the gradients of the lines AB and CD.
b) Show that ABCD is a parallelogram.
c) Find the coordinates of the point of intersection of the diagonals AC and BD. (WJEC)

4 Find the equation of the straight line passing through the origin and perpendicular to the line $x + 2y = 3$. (UCLES)

5 **a)** Find the gradient of the straight line with equation $2x + 3y = 5$.
b) Find the equation of the straight line which passes through the point (4, 5) and is perpendicular to $2x + 3y = 5$. (UODLE)

6 The points P, Q and R have coordinates (2, 4), (7, −2) and (6, 2) respectively.
Find the equation of the straight line l which is perpendicular to the line PQ and which passes through the mid-point of PR. (AEB Spec)

7 Find the equation of the straight line that passes through the points (3, −1) and (−2, 2), giving your answer in the form $ax + by + c = 0$. Hence find the coordinates of the point of intersection of the line and the x-axis. (UCLES)

8 **a)** **i)** Find the gradient of the straight line $2x + 3y = -11$.
ii) Find the equation of the line through (9, −1) perpendicular to $2x + 3y = -11$.
b) Calculate the coordinates of the point where these two straight lines meet. (UODLE)

9 The line l has equation $2x - y - 1 = 0$. The line m passes through the point A(0, 4) and is perpendicular to the line l.

a) Find an equation of m and show that the lines l and m intersect at the point P(2, 3).

The line n passes through the point B(3, 0) and is parallel to the line m.

b) Find an equation of n and hence find the coordinates of the point Q where the lines l and n intersect.
c) Prove that AP = BQ = PQ. (EDEXCEL)

10 The line L_1 has gradient $\frac{1}{7}$ and passes through the point A(2, 2). The line L_2 has gradient −1 and passes through the point B(4, 8). The lines L_1 and L_2 intersect at the point C.

a) Find an equation for L_1 and an equation for L_2.
b) Determine the coordinates of C.
c) Verify, by calculation, that AC and BC are equal in length. (EDEXCEL)

11 The points A and B have coordinates (8, 7) and (−2, 2) respectively. A straight line l passes through A and B and meets the coordinate axes at the points C and D.

a) Find, in the form $y = mx + c$, the equation of l.
b) Find the length CD, giving your answer in the form $p\sqrt{q}$, where p and q are integers and q is prime. (EDEXCEL)

12 Points A, B and C have coordinates (1, 1), (2, 6) and (4, 2) respectively.

a) Find an equation of the line through A and the mid-point M of BC.
b) Find an equation of the line through C which is perpendicular to BC.
c) Solve these two equations to find the coordinates of the point P where these two lines meet.
d) Find the ratio PA : AM, giving your answer in the form $l : k$, where k is an integer. (EDEXCEL)

13 The points A and B have coordinates $(-4, 6)$ and $(2, 8)$ respectively. A line p is drawn through B perpendicular to AB to meet the y-axis at the point C.

a) Find an equation of the line p.
b) Determine the coordinates of C.
c) Show, by calculation, that $AB = BC$.

Given that ABCD is a square whose diagonals intersect at the point M, calculate the coordinates of

d) the point M
e) the point D. (EDEXCEL)

14 The point A has coordinates $(2, -5)$. The straight line $3x + 4y - 36 = 0$ cuts the x-axis at B and the y-axis at C. Find

a) the equation of the line through A which is perpendicular to the line BC
b) the perpendicular distance from A to the line BC
c) the area of the triangle ABC. (UODLE)

15 You are given the coordinates of the four points A(6, 2), B(2, 4), C(−6, −2) and D(−2, −4).

i) Calculate the gradients of the lines AB, CB, DC and DA. Hence describe the shape of the figure ABCD.
ii) Show that the equation of the line DA is $4y - 3x = -10$, and find the length DA.
iii) Calculate the gradient of a line which is perpendicular to DA, and hence find the equation of the line l through B which is perpendicular to DA.
iv) Calculate the coordinates of the point P where l meets DA.
v) Calculate the area of the figure ABCD. (MEI)

6 Differentiation I

A proof tells us where to concentrate our doubts. Logic is the art of going wrong with confidence. An elegantly executed proof is a poem in all but the form in which it is written.
MORRIS KLINE

Some of the examples in this chapter use the properties of surds and indices. The reader may find it helpful first to work through these topics in Chapter 18.

Gradients of curves

We already know that the gradient of a straight line which passes through the points A(x_1, y_2) and B(x_2, y_2) is given by

$$m_{AB} = \frac{y_2 - y_1}{x_2 - x_1}$$

We have also seen that the gradient of a straight line is the same at all points on the line. However, this is not true on a curve.

Consider the curve shown below.

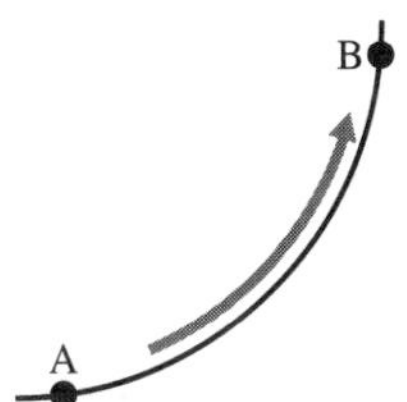

It is clear from the sketch that, as you move along the curve from point A to point B, the gradient of the curve changes, in fact the curve becomes steeper.

We define the gradient of a curve at a point P to be the gradient of the tangent line to the curve at point P.

For example, to find the gradient of the curve $y = x^2$ at the point P(3, 9), we need the gradient of the tangent line to the curve at point P.

It is obvious that drawing a tangent line by eye is not a very accurate method for finding the gradient of a curve. We therefore need an algebraic method for finding the gradient of any curve.

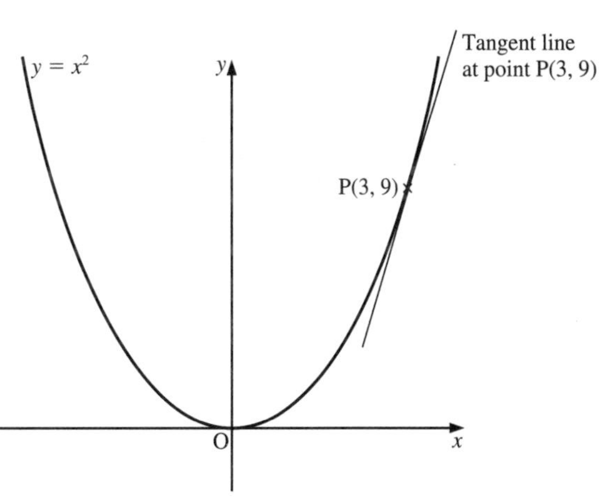

Differentiation from first principles

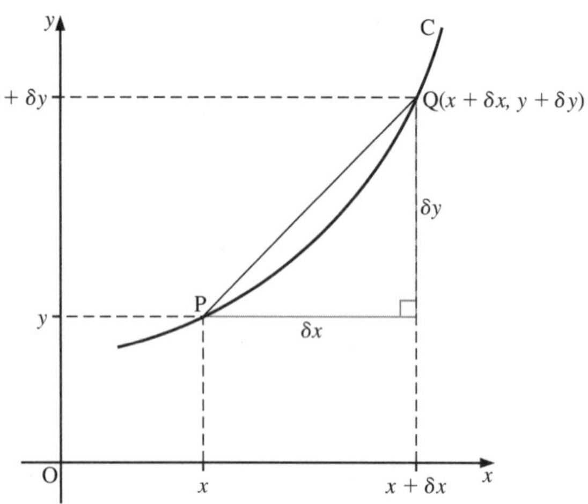

Consider the point P on the curve C, shown on the right. We want the gradient of the curve at point P.

Let Q be a different point on the curve such that it has coordinates $(x + \delta x,\ y + \delta y)$, where δx and δy are small. The straight line PQ is called a **chord** of the curve C.

As the distance δx becomes smaller, point Q moves closer to P and the chord PQ approaches the position of the tangent at P.

The gradient of PQ is given by

$$m_{\mathrm{PQ}} = \frac{(y + \delta x) - y}{(x + \delta x) - x}$$

$$\therefore \quad m_{\mathrm{PQ}} = \frac{\delta y}{\delta x}$$

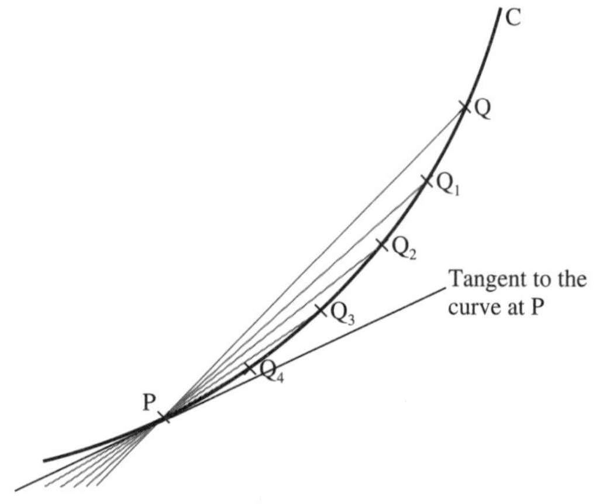

As δx tends to zero ($\delta x \to 0$), $\dfrac{\delta y}{\delta x}$ approaches the value of the gradient of the tangent line at P. This value is called the limiting value of $\dfrac{\delta y}{\delta x}$ and is written as

$$\lim_{\delta x \to 0} \frac{\delta y}{\delta x}$$

The limiting value of $\dfrac{\delta y}{\delta x}$ is called the **differential coefficient** or **first derivative of *y* with respect to *x*** and is denoted by $\dfrac{\mathrm{d}y}{\mathrm{d}x}$.

The process of finding this limiting value is called **differentiation**.

Note It is important to understand that $\dfrac{\mathrm{d}y}{\mathrm{d}x}$ does not mean dy divided by dx. It means the derivative of y with respect to x. The $\dfrac{\mathrm{d}}{\mathrm{d}x}$ is an operator, operating on the function y. To see this more clearly, we can write $\dfrac{\mathrm{d}y}{\mathrm{d}x}$ as $\dfrac{\mathrm{d}(y)}{\mathrm{d}x}$.

Example 1 Given that $y = x^2$, find $\frac{dy}{dx}$.

SOLUTION

We want

$$\frac{dy}{dx} = \lim_{\delta x \to 0} \frac{\delta y}{\delta x}$$

Consider points $P(x, y)$ and $Q(x + \delta x, y + \delta y)$ on the curve $y = x^2$.

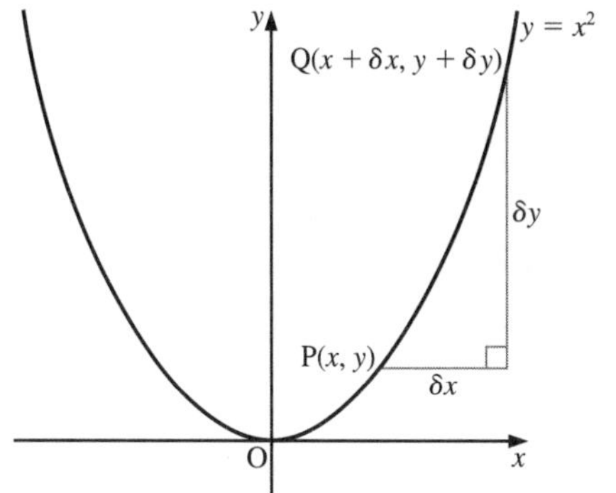

Since the coordinates of points P and Q satisfy the equation $y = x^2$, their coordinates can be written as $P(x, x^2)$ and $Q(x + \delta x, (x + \delta x)^2)$. Therefore,

$$\delta y = (x + \delta x)^2 - x^2$$

$$\therefore \quad \delta y = x^2 + 2x\delta x + (\delta x)^2 - x^2$$

$$\therefore \quad \delta y = 2x\delta x + (\delta x)^2$$

We want $\frac{\delta y}{\delta x}$, so divide throughout by δx, giving

$$\frac{\delta y}{\delta x} = 2x + \delta x$$

We have

$$\frac{dy}{dx} = \lim_{\delta x \to 0} \frac{\delta y}{\delta x}$$

$$= \lim_{\delta x \to 0} (2x + \delta x)$$

As $\delta x \to 0$, the expression $2x + \delta x \to 2x$. Therefore,

$$\frac{dy}{dx} = 2x$$

The expression $\frac{dy}{dx} = 2x$ is called the **derived expression** or the **gradient function** or the **first derivative** of y with respect to x.

It tells us that the gradient of any point on the curve $y = x^2$ is twice the x coordinate of the point. For example, the gradient of the curve $y = x^2$ at the point P(3, 9) is $2 \times 3 = 6$.

We write

$$\underbrace{\left.\frac{dy}{dx}\right|_{x=3}}_{\text{Meaning } \frac{dy}{dx} \text{ evaluated when } x=3} = 2(3) = 6$$

Alternative notation

If the equation of a curve is denoted by $y = f(x)$, it is sometimes more convenient to denote $\frac{dy}{dx}$ by f'.

Result: Derivative of $y = ax^n$

If $y = ax^n$, then

$$\frac{dy}{dx} = anx^{n-1}$$

for all rational n.

Notice that if $y = a$, a constant, then this can be written as $y = ax^0$ and

$$\frac{dy}{dx} = (a \times 0)x^{-1} = 0$$

In other words, the derivative of a constant is **always zero**. Thinking about this result geometrically, $y = a$ is a horizontal line which has a gradient of zero.

Example 2 Find $\frac{dy}{dx}$ for each of the following.

a) $y = x^3$ **b)** $y = 6x^4$ **c)** $y = \frac{1}{x^5}$

d) $y = \frac{3}{4x^2}$ **e)** $y = x^{\frac{1}{3}}$ **f)** $y = \frac{1}{\sqrt{x}}$

SOLUTION

a) When $y = x^3$,

$$\frac{dy}{dx} = 3x^{3-1}$$

$$\therefore \quad \frac{dy}{dx} = 3x^2$$

b) When $y = 6x^4$,

$$\frac{dy}{dx} = 6 \times (4x^{4-1})$$

$$\therefore \quad \frac{dy}{dx} = 24x^3$$

c) When $y = \frac{1}{x^5} = x^{-5}$,

$$\frac{dy}{dx} = -5x^{-5-1}$$

$$= -5x^{-6}$$

$$\therefore \quad \frac{dy}{dx} = -\frac{5}{x^6}$$

d) When $y = \frac{3}{4x^2} = \frac{3}{4}x^{-2}$,

$$\frac{dy}{dx} = \frac{3}{4} \times -2x^{-2-1}$$

$$= -\frac{3}{2}x^{-3}$$

$$\therefore \quad \frac{dy}{dx} = -\frac{3}{2x^3}$$

e) When $y = x^{\frac{1}{3}}$,

$$\frac{dy}{dx} = \frac{1}{3}x^{\frac{1}{3}-1}$$

$$\therefore \quad \frac{dy}{dx} = \frac{1}{3}x^{-\frac{2}{3}}$$

$$= \frac{1}{3x^{\frac{2}{3}}}$$

f) When $y = \frac{1}{\sqrt{x}} = \frac{1}{x^{\frac{1}{2}}} = x^{-\frac{1}{2}}$,

$$\frac{dy}{dx} = -\frac{1}{2}x^{-\frac{1}{2}-1}$$

$$\therefore \quad \frac{dy}{dx} = -\frac{1}{2x^{\frac{3}{2}}} \quad \left(= -\frac{1}{2\sqrt{x^3}}\right)$$

Sum or difference of two functions

When y comprises more than one function, to find the first derivative we differentiate each function in turn. In other words, when

$$y = \mathrm{f}(x) \pm \mathrm{g}(x) \quad \text{then} \quad \frac{\mathrm{d}y}{\mathrm{d}x} = \mathrm{f}'(x) \pm \mathrm{g}'(x)$$

This applies to the sum or difference of any number of functions.

Example 3 Find $\mathrm{f}'(x)$ for each of the following.

a) $\mathrm{f}(x) = 4x^2 + 1$ **b)** $\mathrm{f}(x) = 2x^3 + \sqrt{x}$ **c)** $\mathrm{f}(x) = x + \dfrac{1}{x}$

d) $\mathrm{f}(x) = x^2 + 6x^{\frac{1}{3}} - 3$ **e)** $\mathrm{f}(x) = \dfrac{2}{\sqrt{x}} + \dfrac{3}{x^2} - 1$

SOLUTION

a) When $\mathrm{f}(x) = 4x^2 + 1$,

$$\mathrm{f}'(x) = 8x$$

b) When $\mathrm{f}(x) = 2x^3 + \sqrt{x} = 2x^3 + x^{\frac{1}{2}}$,

$$\mathrm{f}'(x) = 6x^2 + \frac{1}{2}x^{-\frac{1}{2}}$$

$$= 6x^2 + \frac{1}{2x^{\frac{1}{2}}}$$

$$\therefore \quad \mathrm{f}'(x) = 6x^2 + \frac{1}{2\sqrt{x}}$$

c) When $\mathrm{f}(x) = x + \dfrac{1}{x} = x + x^{-1}$,

$$\mathrm{f}'(x) = 1 - x^{-2}$$

$$\therefore \quad \mathrm{f}'(x) = 1 - \frac{1}{x^2}$$

d) When $\mathrm{f}(x) = x^2 + 6x^{\frac{1}{3}}$,

$$\mathrm{f}'(x) = 2x + 2x^{-\frac{2}{3}}$$

e) When $\mathrm{f}(x) = \dfrac{2}{\sqrt{x}} + \dfrac{3}{x^2} - 1 = 2x^{-\frac{1}{2}} + 3x^{-2} - 1$,

$$\mathrm{f}'(x) = -x^{-\frac{3}{2}} - 6x^{-3}$$

$$= -\frac{1}{x^{\frac{3}{2}}} - \frac{6}{x^3}$$

$$\therefore \quad \mathrm{f}'(x) = -\frac{1}{\sqrt{x^3}} - \frac{6}{x^3}$$

A function may not be given in the form ax^n. In this case, it becomes necessary to manipulate the expression for y and write it as a sum of functions, each in the form ax^n.

Example 4 Find $\dfrac{dy}{dx}$ for each of the following.

a) $y = (x+3)^2$ **b)** $y = \sqrt{x}(x^2 - 1)$ **c)** $y = \dfrac{x^3 + 6}{x}$

SOLUTION

a) To differentiate $y = (x+3)^2$, first expand the bracket:

$$y = (x+3)(x+3)$$
$$= x^2 + 6x + 9$$
$$\therefore \quad \frac{dy}{dx} = 2x + 6$$

b) When $y = \sqrt{x}(x^2 - 1) = x^{\frac{1}{2}}(x^2 - 1)$, then expanding the bracket will give y in the required form:

$$y = x^{\frac{5}{2}} - x^{\frac{1}{2}}$$
$$\therefore \quad \frac{dy}{dx} = \frac{5}{2}x^{\frac{3}{2}} - \frac{1}{2}x^{-\frac{1}{2}}$$

This expression for $\dfrac{dy}{dx}$ can be factorised by taking out $\frac{1}{2}x^{-\frac{1}{2}}$. That is, the lowest power of x to which it occurs. This gives

$$\frac{dy}{dx} = \frac{1}{2}x^{-\frac{1}{2}}(5x^2 - 1)$$
$$= \frac{1}{2x^{\frac{1}{2}}}(5x^2 - 1)$$
$$\therefore \quad \frac{dy}{dx} = \frac{1}{2\sqrt{x}}(5x^2 - 1)$$

c) When $y = \dfrac{x^3 + 6}{x} = \dfrac{x^3}{x} + \dfrac{6}{x} = x^2 + \dfrac{6}{x} = x^2 + 6x^{-1}$,

$$\frac{dy}{dx} = 2x - 6x^{-2}$$
$$= 2x^{-2}(x^3 - 3)$$
$$\therefore \quad \frac{dy}{dx} = \frac{2(x^3 - 3)}{x^2}$$

Examples **b** and **c** illustrate the technique of manipulating the derivative to obtain a more mathematically tidy result. It should be noted that any form of the correct derivative is acceptable. It is useful to understand such manipulation since many examination questions ask for the derivative in a particular form.

Second derivative

The derivative of $\frac{dy}{dx}$, that is $\frac{d}{dx}\left(\frac{dy}{dx}\right)$, is denoted by $\frac{d^2y}{dx^2}$ and is called the **second derivative of *y* with respect to *x***. In this section, we will look at how to find the second derivative in preparation for its use in determining maximum and minimum turning points on a graph (see pages 174–80).

The derivative of $f'(x)$ is denoted by $f''(x)$ and is called the **second derivative of f(*x*) with respect to *x***.

Example 5 Given that $f(x) = x + \frac{1}{x}$, find $f'(x)$ and $f''(x)$.

SOLUTION

Now $f(x) = x + \frac{1}{x} = x + x^{-1}$. Therefore,

$$f'(x) = 1 - x^{-2}$$
$$= 1 - \frac{1}{x^2}$$

Given that $f'(x) = 1 - x^{-2}$, then

$$f''(x) = 2x^{-3}$$
$$= \frac{2}{x^3}$$

Example 6 If $y = 4x^3$, find $\frac{dy}{dx}$ and $\frac{d^2y}{dx^2}$. Hence show that y satisfies

$$3y\frac{d^2y}{dx^2} - 2\left(\frac{dy}{dx}\right)^2 \equiv 0$$

SOLUTION

When $y = 4x^3$,

$$\frac{dy}{dx} = 12x^2 \quad \text{and} \quad \frac{d^2y}{dx^2} = 24x$$

Substituting into the LHS of

$$3y\frac{d^2y}{dx^2} - 2\left(\frac{dy}{dx}\right)^2 = 0$$

gives

$$3y\frac{d^2y}{dx^2} - 2\left(\frac{dy}{dx}\right)^2 \equiv 3(4x^3)(24x) - 2(12x^2)^2$$
$$\equiv 288x^4 - 288x^4 \equiv 0$$

as required.

Exercise 6A

1 Differentiate each of the following from first principles.

a) $y = x^3$ **b)** $y = 2x^2 + 3$ **c)** $y = 1 - x^2$

d) $y = x^3 - 6x$ **e)** $y = \dfrac{1}{x}$ ***f)** $y = \sqrt{x}$

2 Find $\dfrac{dy}{dx}$ for each of the following.

a) $y = x^4$ **b)** $y = x^6$ **c)** $y = 6x^2$ **d)** $y = -5x^3$ **e)** $y = 3x$ **f)** $y = 2x^6$

g) $y = -7x^2$ **h)** $y = 2$ **i)** $y = \frac{1}{2}x^4$ **j)** $y = \frac{2}{3}x^6$ **k)** $y = -\frac{1}{4}x^3$ **l)** $y = \frac{2}{5}x$

3 Differentiate each of the following with respect to x.

a) x^{-2} **b)** x^{-4} **c)** $2x^{-3}$ **d)** $4x^{-1}$ **e)** $\dfrac{1}{x^3}$ **f)** $-\dfrac{1}{x^2}$

g) $\dfrac{3}{x^3}$ **h)** $-\dfrac{2}{x}$ **i)** $\dfrac{3}{2x^2}$ **j)** $\dfrac{9}{2x^3}$ **k)** $-\dfrac{3}{4x^4}$ **l)** $\dfrac{2}{5x}$

4 Find $f'(x)$ for each of the following.

a) $f(x) = x^{\frac{1}{2}}$ **b)** $f(x) = 6x^{\frac{1}{3}}$ **c)** $f(x) = x^{-\frac{2}{3}}$

d) $f(x) = -10x^{-\frac{1}{5}}$ **e)** $f(x) = 7\sqrt{x}$ **f)** $f'(x) = \sqrt[3]{x}$

g) $f(x) = \dfrac{4}{5\sqrt{x}}$ **h)** $f(x) = -\dfrac{6}{\sqrt[3]{x}}$ **i)** $f(x) = \dfrac{5}{2\sqrt{x}}$

j) $f(x) = -\dfrac{15}{\sqrt[5]{x}}$ **k)** $f(x) = \sqrt{x^5}$ **l)** $f(x) = (5x)^2$

5 Find $\dfrac{dy}{dx}$ for each of the following.

a) $y = x^2 + 2x$ **b)** $y = 3x^2 - 5x$ **c)** $y = x^2 + 1$

d) $y = 5 - 4x^3$ **e)** $y = x^2 + 2x + 3$ **f)** $y = x^7 + 3x^4$

g) $y = x^4 - 3x^2 + 2$ **h)** $y = x + \dfrac{1}{x}$ **i)** $y = 5x^2 - \dfrac{2}{x^3}$

j) $y = x^5 - \dfrac{3}{x^3}$ **k)** $y = x^2 - 2x^4$ **l)** $y = \dfrac{3}{x} - 1 + 4x^3$

6 Differentiate each of the following with respect to x.

a) $3x - 5x^3$ **b)** $2 - \dfrac{3}{x}$ **c)** $\dfrac{4}{x} - 2x^3$

d) $2\sqrt{x} + 1$ **e)** $\sqrt{x} + \dfrac{1}{\sqrt{x}}$ **f)** $4x^{-2} - 3x$

g) $3x^{\frac{1}{3}} - 4x^{-\frac{1}{3}}$ **h)** $4x^{\frac{1}{2}} + 2x - 1$ **i)** $6x^{\frac{2}{3}} - 4x^{\frac{5}{2}}$

j) $\dfrac{9}{\sqrt[3]{x}} - \dfrac{8}{\sqrt[4]{x}}$ **k)** $\dfrac{6}{\sqrt{x}} - 4\sqrt{x}$ **l)** $\sqrt{x} + 1 + \dfrac{1}{\sqrt{x}}$

7 Find $f'(x)$ for each of the following.

a) $f(x) = 4x - 7$ **b)** $f(x) = 2\sqrt{x} + \dfrac{5}{2x}$ **c)** $f(x) = 4x^2 + 7x - 3$

d) $f(x) = (\sqrt[6]{x})^5$ **e)** $f(x) = 5x^{\frac{2}{5}} - 2x^{\frac{5}{2}}$ **f)** $f(x) = 6\sqrt{x} - \dfrac{3}{2x^2}$

g) $f(x) = 2x^{-7} - 5x^{-3} + x$ **h)** $f(x) = \dfrac{3}{\sqrt[6]{x}} - \dfrac{2}{\sqrt[4]{x}}$ **i)** $f(x) = \dfrac{5}{x^2} - \dfrac{2}{x} + 3$

j) $f(x) = 9x^{\frac{4}{3}} + 3$ **k)** $f(x) = 2x^{-4} - 4x^{-2}$ **l)** $f(x) = \dfrac{3}{\sqrt{x}} + \dfrac{5}{\sqrt{x^3}}$

8 Find $\dfrac{dy}{dx}$ for each of the following.

a) $y = x^2(x + 3)$ **b)** $y = x(2 - x)$ **c)** $y = x^3(4 - x^2)$

d) $y = \sqrt{x}(5 + x)$ **e)** $y = 6\sqrt{x}(x^3 - 2x + 1)$ **f)** $y = x^{\frac{1}{3}}(2x - 5)$

g) $y = 2x^{\frac{1}{4}}(x^2 - 2)$ **h)** $y = (x + 3)(x - 4)$ **i)** $y = (x + 4)^2$

j) $y = (x + 5)(2x - 1)$ **k)** $y = 2(x - 3)^2$ **l)** $y = (x + 8)(x - 2)$

9 Differentiate each of the following with respect to x.

a) $3x(x - 4)$ **b)** $x^3(3x^2 - 1)$ **c)** $\sqrt{x}(x^2 + 3)$

d) $x^{-\frac{1}{2}}(x + 1)$ **e)** $\dfrac{x^2 + 7}{x}$ **f)** $\dfrac{x + 5}{x^2}$

g) $\dfrac{3x^2 + 2}{x}$ **h)** $\dfrac{6x^3 - 7}{x^2}$ **i)** $\dfrac{2x + 3}{5x}$

j) $\dfrac{6x^2 - 7x^3}{3\sqrt{x}}$ **k)** $x^2(2 - x)^2$ **l)** $\dfrac{(x + 5)^2}{x}$

10 Find $f'(x)$ for each of the following.

a) $f(x) = x^3(3x - 1)$ **b)** $f(x) = 2x^2(x - 1)^2$ **c)** $f(x) = \dfrac{\sqrt{x} + 1}{\sqrt{x}}$

d) $f(x) = \dfrac{3x^3 + 5}{x^2}$ **e)** $f(x) = \dfrac{(x + 3)(x - 4)}{x}$ **f)** $f(x) = \dfrac{(2x - 5)(x - 4)}{x^3}$

g) $f(x) = \dfrac{(3x - 1)^2}{2x}$ **h)** $f(x) = \dfrac{5x + 3}{2\sqrt{x}}$ **i)** $f(x) = \dfrac{(x + 1)^2}{\sqrt{x}}$

j) $f(x) = \dfrac{x^2 + 5}{3\sqrt{x}}$ **k)** $f(x) = \dfrac{3\sqrt{x} - 7}{2\sqrt{x}}$ **l)** $f(x) = \dfrac{2x^2 - 5}{(3\sqrt{x})^2}$

11 Find $\dfrac{d^2y}{dx^2}$ for each of the following.

a) $y = 3x^3 + 5x$ **b)** $y = x^2 - 4x^6$ **c)** $y = \dfrac{1}{x}$

d) $y = \dfrac{2}{x^2} - x$ **e)** $y = \dfrac{1}{\sqrt{x}} - \sqrt{x}$ **f)** $y = \sqrt[3]{x} - x^3$

g) $y = x^3(x^2 + 5)$ **h)** $y = (x^2 - 1)(2x + 3)$ **i)** $y = \dfrac{x^2 - 1}{x}$

j) $y = \dfrac{\sqrt{x} + 5}{\sqrt{x}}$ **k)** $y = \dfrac{6x - 5}{x^2}$ **l)** $y = \dfrac{5x - 4}{3\sqrt{x}}$

12 Given that a, b, c and d are constants, find the values of $\frac{dy}{dx}$ and $\frac{d^2y}{dx^2}$ for each of the following.

a) $y = ax^2 + bx + c$ **b)** $y = \frac{a}{x} + \frac{b}{x^2}$ **c)** $y = a\sqrt{x} + \frac{b}{\sqrt{x}}$

d) $y = (ax + b)(cx + d)$ **e)** $y = \frac{ax^3 + bx^2}{cx}$

13 Given that $y = \frac{1}{\sqrt{x}}$, show that

$$2x\left(\frac{d^2y}{dx^2}\right) + 3\frac{dy}{dx} \equiv 0$$

14 Given that $y = x^4$, show that

$$\frac{4y}{3}\left(\frac{d^2y}{dx^2}\right) - \left(\frac{dy}{dx}\right)^2 \equiv 0$$

15 Given that $y = \frac{1}{x^2}$, show that

$$y\left(\frac{d^2y}{dx^2}\right) + \left(\frac{dy}{dx}\right)^2 - 10y^3 \equiv 0$$

16 Given that $y = \frac{x+1}{x^2}$, show that

$$\frac{d^2y}{dx^2} + \frac{4}{x}\left(\frac{dy}{dx}\right) + \frac{2}{x^2}y \equiv 0$$

***17** Given that $y = ax^2 + bx$ and

$$\frac{d^2y}{dx^2} \equiv 4\left(\frac{dy}{dx}\right)^2 - 32y$$

find the possible values of the constants a and b.

***18** Let $f(x) = 2x^3 - 3x^2 - 32x - 15$.

a) Find an expression for $f'(x)$.
b) Given that the roots of the equation $f(x) = 0$ are p, q and r, find p, q and r.
c) Show that

$$f'(x) \equiv \frac{f(x)}{x-p} + \frac{f(x)}{x-q} + \frac{f(x)}{x-r}$$

Tangents and normals to a curve

As we showed on page 158, the gradient of a curve at a point P is the gradient of the tangent to the curve at the point P.

Example 7 Find the gradient of the curve $f(x) = x^2 + \frac{1}{x}$ at the point P(1, 2).

SOLUTION

To find the gradient of the curve, we must first find $f'(x)$:

$$f(x) = x^2 + \frac{1}{x}$$
$$= x^2 + x^{-1}$$
$$\therefore \quad f'(x) = 2x - x^{-2}$$
$$= 2x - \frac{1}{x^2}$$

The gradient of the curve at the point P(1, 2) is given by f′(1):

$$f'(1) = 2(1) - \frac{1}{(1)^2} = 1$$

The gradient of the curve at point P is 1.

Example 8 The gradient of the curve $y = 3x^2 + x - 3$ at the point P is 13. Find the coordinates of point P.

SOLUTION

When $y = 3x^2 + x - 3$, then $\frac{dy}{dx} = 6x + 1$.

We are given that $\frac{dy}{dx} = 13$ at point P. Therefore,

$$6x + 1 = 13$$

$$\therefore \quad x = 2$$

To find the y coordinate of point P substitute $x = 2$ into $y = 3x^2 + x - 3$, which gives

$$y = 3(2)^2 + 2 - 3$$

$$\therefore \quad y = 11$$

The coordinates of point P are (2, 11).

Example 9 The curve C is given by $y = ax^2 + b\sqrt{x}$, where a and b are constants. Given that the gradient of C at the point (1, 1) is 5, find a and b.

SOLUTION

Rewriting $y = ax^2 + b\sqrt{x}$ as $y = ax^2 + bx^{\frac{1}{2}}$ and then differentiating gives

$$\frac{dy}{dx} = 2ax + \frac{b}{2}x^{-\frac{1}{2}}$$

$$= 2ax + \frac{b}{2x^{\frac{1}{2}}}$$

Now the point (1, 1) is on the curve C. Therefore,

$$1 = a(1)^2 + b(1)^{\frac{1}{2}}$$

$$\therefore \quad 1 = a + b \qquad [1]$$

The gradient of the curve when $x = 1$ is 5. Therefore,

$$2a(1) + \frac{b}{2(1)^{\frac{1}{2}}} = 5$$

$$\therefore \quad 4a + b = 10 \qquad [2]$$

Solving [1] and [2] simultaneously gives $a = 3$ and $b = -2$.

Equation of a tangent

We now look at how to find the equation of the tangent line to a curve at a particular point.

Example 10 Find the equation of the tangent to the curve $y = x^2$ at the point P(3, 9).

SOLUTION

The gradient of the tangent to the curve at point P is given by $\frac{dy}{dx}$ evaluated when $x = 3$.

When $y = x^2$, then $\frac{dy}{dx} = 2x$. At the point P(3, 9),

$$\left.\frac{dy}{dx}\right|_{x=3} = 2(3) = 6$$

The tangent is a straight line, so its equation is given by $y = mx + c$. Since we know that the gradient of the tangent at point P is 6, the equation of the line is given by

$$y = 6x + c$$

The tangent passes through P(3, 9). Therefore,

$$9 = 6(3) + c$$

$$\therefore \quad c = -9$$

The equation of the tangent is therefore $y = 6x - 9$.

Example 11 Find the equation of the tangent to the curve $f(x) = \frac{1}{x^2}$ at the point P(−1, 1). Find the coordinates of the point where this tangent meets the curve again.

SOLUTION

Now $f(x) = \frac{1}{x^2} = x^{-2}$. Therefore,

$$f'(x) = -2x^{-3} = -\frac{2}{x^3}$$

The gradient of the tangent to the curve at the point P(−1, 1) is

$$f'(-1) = -\frac{2}{(-1)^3} = 2$$

Therefore, the equation of the tangent is $y = 2x + c$.

The tangent passes through P(−1, 1). Therefore,

$$1 = 2(-1) + c$$

$$\therefore \quad c = 3$$

Therefore, the equation of the tangent to the curve at the point P is $y = 2x + 3$.

The tangent meets the curve again at the points whose x coordinates satisfy

$$2x + 3 = \frac{1}{x^2}$$

$$\therefore \quad 2x^3 + 3x^2 - 1 = 0$$

$$\therefore \quad (2x - 1)(x^2 + 2x + 1) = 0$$

Solving this equation gives

$$2x - 1 = 0 \quad \text{or} \quad x^2 + 2x + 1 = 0$$

$$\therefore \quad x = \tfrac{1}{2} \quad \text{or} \quad (x + 1)(x + 1) = 0$$

$$\therefore \quad x = -1$$

When $x = \frac{1}{2}$, $f(\frac{1}{2}) = 4$.

When $x = -1$, $f(-1) = 1$. This is, in fact, just point P.

The tangent meets the curve again at the point with coordinates $(\frac{1}{2}, 4)$.

Equation of a normal

The **normal** to a curve at a point P is the straight line through P which is perpendicular to the tangent at P.

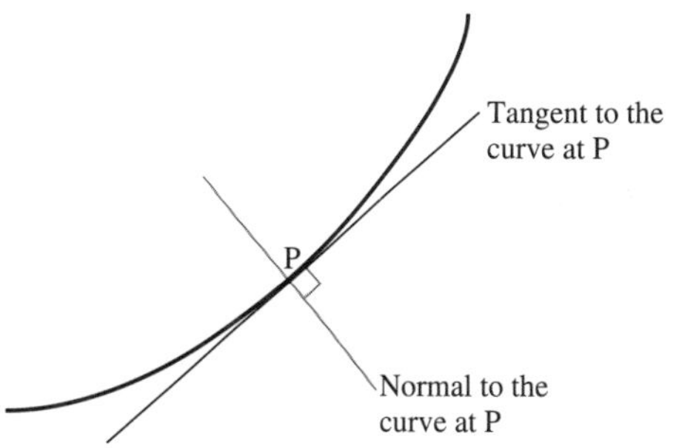

Since the tangent and normal are perpendicular to each other, if the gradient of the tangent is m, then the gradient of the normal is $-\dfrac{1}{m}$.

Example 12 Find the equation of the normal to the curve $y = 3x^2 + 7x - 2$ at the point P where $x = -1$.

SOLUTION

When $x = -1$,

$$y = 3(-1)^2 + 7(-1) - 2 = -6$$

Therefore, P has coordinates $(-1, -6)$.

When $y = 3x^2 + 7x - 2$, then $\dfrac{dy}{dx} = 6x + 7$. At the point $P(-1, -6)$,

$$\left.\frac{dy}{dx}\right|_{x=-1} = 6(-1) + 7 = 1$$

The gradient of the tangent line at P is 1. Therefore, the gradient of the normal at P is $-\dfrac{1}{(1)} = -1$.

The equation of the normal at P is

$$y = -x + c$$

The normal passes through P(−1, −6). Therefore,

$$-6 = -(-1) + c$$

$$\therefore \quad c = -7$$

Therefore, the equation of the normal to the curve at the point P(−1, −6) is $y = -x - 7$.

Example 13 Find the equations of the normals to the curve $y = x^3 - 3x^2 + 4$ which are perpendicular to the line $y - 24x = 1$.

SOLUTION

The equation $y - 24x = 1$ can be written as $y = 24x + 1$, giving the gradient of the line as 24.

If the required normals are perpendicular to the line $y = 24x + 1$, they must have a gradient of $-\frac{1}{24}$. Therefore, the gradient of the corresponding tangents to the curve is 24.

When $y = x^3 - 3x^2 + 4$, then $\dfrac{dy}{dx} = 3x^2 - 6x$.

Now $\dfrac{dy}{dx} = 24$. Therefore,

$$3x^2 - 6x = 24$$

$$\therefore \quad 3x^2 - 6x - 24 = 0$$

$$\therefore \quad x^2 - 2x - 8 = 0$$

$$\therefore \quad (x + 2)(x - 4) = 0$$

Solving gives $x = -2$ or $x = 4$.

When $x = -2$: $\quad y = (-2)^3 - 3(-2)^2 + 4 = -16$

When $x = 4$: $\quad y = (4)^3 - 3(4)^2 + 4 = 20$

We want the normals to the curve at the points P(−2, −16) and Q(4, 20).

Both of the normals have equations of the form $y = -\frac{1}{24}x + c$. Since one of the normals passes through P(−2, −16), it follows that

$$-16 = -\frac{1}{24}(-2) + c$$

$$\therefore \quad c = -\frac{193}{12}$$

Therefore, the equation of the normal through P(−2, −16) is

$$y = -\frac{1}{24}x - \frac{193}{12} \quad \text{or} \quad 24y + x + 386 = 0$$

The second normal passes through Q(4, 20). Therefore,

$$20 = -\frac{1}{24}(4) + c$$

$$\therefore \quad c = \frac{121}{6}$$

Therefore, the equation of the normal through Q(4, 20) is

$$y = -\frac{1}{24}x + \frac{121}{6} \quad \text{or} \quad 24y + x - 484 = 0$$

Exercise 6B

1 Find the gradient of each of the following curves at the point given.

a) $y = x^2$, at (3, 9) **b)** $y = 2x^3 - 4$, at (2, 12) **c)** $y = \sqrt{x} + 2$, at (9, 5)

d) $y = \dfrac{1}{x}$, at $(3, \frac{1}{3})$ **e)** $y = 5 - x^2$, at $(-2, 1)$ **f)** $y = 3 - \dfrac{2}{x}$, at $(4, \frac{5}{2})$

g) $y = x + \dfrac{3}{x}$, at (3, 4) **h)** $y = 2 - \dfrac{4}{x^2}$, at $(-2, 1)$ **i)** $y = \dfrac{x+5}{x}$, at $(-1, -4)$

j) $y = 3x + 7$, at $(-3, -2)$ **k)** $y = 6\sqrt{x} + \dfrac{1}{2\sqrt{x}}$, at $(\frac{1}{9}, \frac{7}{2})$ **l)** $y = \dfrac{4 - x^3}{x^2}$, at $(-2, 3)$

2 Find the coordinates of any points on each of the following curves where the gradient is as stated.

a) $y = x^3$, grad 12 **b)** $y = 3x^2$, grad -6 **c)** $y = x^4 + 1$, grad 32

d) $y = \dfrac{4}{x}$, grad -16 **e)** $y = \dfrac{16}{x^2}$, grad 4 **f)** $y = x^3 + 2x - 1$, grad 29

g) $y = x^3 - x^2 + 3$, grad 0 **h)** $y = 2x^3 - 4x^2 + 3x + 2$, grad 1 **i)** $y = \sqrt{x} + 5$, grad 1

j) $4\sqrt{x} - x$, grad 5 **k)** $y = \dfrac{4 - x}{x}$, grad -1 **l)** $y = \dfrac{x^2 + 3}{2x^2}$, grad 3

3 The curve C is defined by $y = ax^2 + b$, where a and b are constants. Given that the gradient of the curve at the point $(2, -2)$ is 3, find the values of a and b.

4 Given that the curve with equation $y = Ax^2 + Bx$ has gradient 7 at the point (6, 8), find the values of the constants A and B.

5 A curve whose equation is $y = \dfrac{a}{x} + c$, passes through the point (3, 9) with gradient 5. Find the values of the constants a and c.

6 Given that the curve with equation $y = a\sqrt{x} + b$ has gradient 3 at the point (4, 6), find the values of the constants a and b.

7 A curve with equation $y = A\sqrt{x} + \dfrac{B}{\sqrt{x}}$, for constants A and B, passes through the point (1, 6) with gradient -1. Find A and B.

8 Find the equation of the tangent to each of the following curves at the point indicated by the given value of x.

a) $y = x^2 + 3$, where $x = 2$

b) $y = 2x^3 - 1$, where $x = 1$

c) $y = \dfrac{9}{x}$, where $x = -3$

d) $y = 6x - x^2$, where $x = 4$

e) $y = 5 - \dfrac{8}{x^2}$, where $x = -2$

f) $y = 6\sqrt{x}$, where $x = 4$

g) $y = x^3 - x^2 + 2$, where $x = 1$

h) $y = x^2 - 10x + 30$, where $x = 5$

i) $y = \dfrac{x+4}{x}$, where $x = -2$

j) $y = x(x^2 - 3)$, where $x = 2$

k) $y = \dfrac{x+5}{\sqrt{x}}$, where $x = 25$

l) $y = \dfrac{1}{\sqrt[3]{x}}$, where $x = \frac{1}{8}$

9 Find the equation of the normal to each of the following curves at the point indicated by the given value of x.

a) $y = x^2 - 3x$, where $x = 2$

b) $y = x^3 + 4$, where $x = -1$

c) $y = \dfrac{6}{x}$, where $x = 3$

d) $y = 2\sqrt{x}$, where $x = 9$

e) $y = 6 - \dfrac{1}{x^2}$, where $x = 1$

f) $y = x^3 + 2x^2 - 3$, where $x = -2$

g) $y = x^4 - 9x^2$, where $x = 3$

h) $y = \dfrac{x-3}{x^2}$, where $x = -1$

i) $y = \sqrt{x}(x^2 - 2)$, where $x = 1$

j) $y = 3 - \dfrac{36}{x^2}$, where $x = -3$

k) $y = \dfrac{4+x^2}{x}$, where $x = -2$

l) $y = 3\sqrt{x} + \dfrac{1}{\sqrt{x}}$, where $x = \frac{1}{4}$

10 Find the equation of the tangent to the curve $y = x^2 - 3x + 1$ at the point where the curve cuts the y-axis.

11 Find the equation of the normal to the curve $y = x^3 - 8$ at the point where the curve cuts the x-axis.

12 a) Find the equations of the two tangents to the curve $y = x^2 - 5x + 4$ at the points where the curve cuts the x-axis.

b) Find also the coordinates of the point of intersection of the two tangents.

13 The two tangents to the curve $y = x^2$ at the points where $y = 9$, intersect at the point P. Find the coordinates of P.

14 The normal to the curve $y = x^2 + 5x - 2$ at the point where $x = -3$, and the tangent to the same curve at the point where $x = 1$, meet at the point Q. Find the coordinates of Q.

15 **a)** Find the values of x at which the gradient of the curve $y = x^3 - 6x^2 + 9x + 2$ is zero.
b) Hence find the equations of the tangents to the curve which are parallel to the x-axis.

16 **a)** Find the values of x for which the gradient of the curve $y = x^3 - 3x^2 - 9x + 10$ is 15.
b) Hence find the equations of the tangents to the curve which have gradient 15.

17 Find the equations of the two normals to the curve $y = 8 + 15x + 3x^2 - x^3$ which have gradient $\frac{1}{9}$.

18 Show that the curve $y = 3x^4 - 4x^3 + 6x^2 - 18x + 10$ has just one normal of gradient $\frac{1}{6}$ and find its equation.

19 Show that there is no point on the curve $y = x^3 + 6x - 1$ at which the gradient of the tangent is 3.

20 **a)** Find the equation of the tangent, t, to the curve $y = x^2 + 5x + 2$, which is perpendicular to the line, l, with equation $3y + x = 5$.
b) Find also the coordinates of the point where t meets l.

21 Find the coordinates of the point of intersection of the two normals to the curve $y = x^2 + 3x + 5$ which make an angle of 45° with the x-axis.

22 **a)** Find the equation of the tangent at the point (1, 2) on the curve $y = x^3 + 3x - 2$.
b) Find also the coordinates of the point where this tangent meets the curve again.

23 Find the coordinates of the point, P, where the tangent to the curve $y = x^3 + x - 3$ at (2, 7) meets the curve again.

24 **a)** Find the equation of the normal at the point (2, 3) on the curve $y = 2x^3 - 12x^2 + 23x - 11$.
b) Find also the coordinates of the points where the normal meets the curve again.

25 Find the coordinates of the two points, where the normal to the curve $y = \frac{1}{2}x^3 - x + 3$ at the point (0, 3) meets the curve again.

26 The tangent to the curve $y = ax^2 + 1$ at the point $(1, b)$ has gradient 6. Find the values of the constants a and b.

27 The normal to the curve $y = x^3 + cx$ at the point $(2, d)$ has gradient $\frac{1}{2}$. Find the values of the constants c and d.

28 The tangent to the curve $y = a\sqrt{x} - 5$ at the point $(4, b)$ is parallel to the line $y = 2x + 1$. Find the values of the constants a and b.

***29** A curve has equation $y = Ax^3 + Bx^2 + Cx + D$, where A, B, C and D are constants. Given that the curve has gradient -4 at the point (1, 2), and gradient 8 at the point $(-1, 6)$, find A, B, C and D.

***30** **a)** Show that the tangent at the point (a, a^3) on the curve $y = x^3$ has equation $y = a^2(3x - 2a)$.
b) Prove that the tangents to the curve $y = x^3$ at the points (a, a^3) and $(-a, -a^3)$ are parallel, and show that the distance between these two tangents is $\dfrac{6a^3}{\sqrt{1 + 9a^4}}$.
c) Hence find the equations of two parallel tangents to the curve $y = x^3$ which are a distance of $\sqrt{\dfrac{2}{3}}$ units apart.

Maximum, minimum and point of inflexion

A point on a curve at which the gradient is zero, i.e. where $\frac{dy}{dx} = 0$, is called a **stationary point**. At a stationary point, the tangent to the curve is horizontal and the curve is 'flat'.

There are three types of stationary point and you must know how to distinguish one from another.

Minimum point

In this case, the gradient of the curve is negative to the left of point P. To the right of point P, the gradient of the curve is positive.

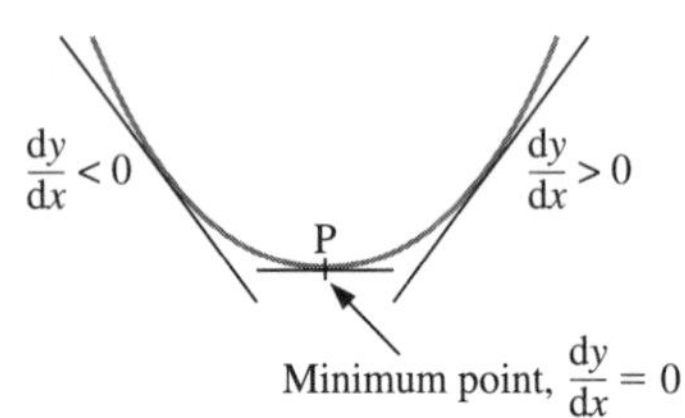

To the left of P	At point P	To the right of P
$\frac{dy}{dx} < 0$	$\frac{dy}{dx} = 0$	$\frac{dy}{dx} > 0$

Maximum point

In this case, the gradient of the curve is positive to the left of point P. To the right of point P, the gradient of the curve is negative.

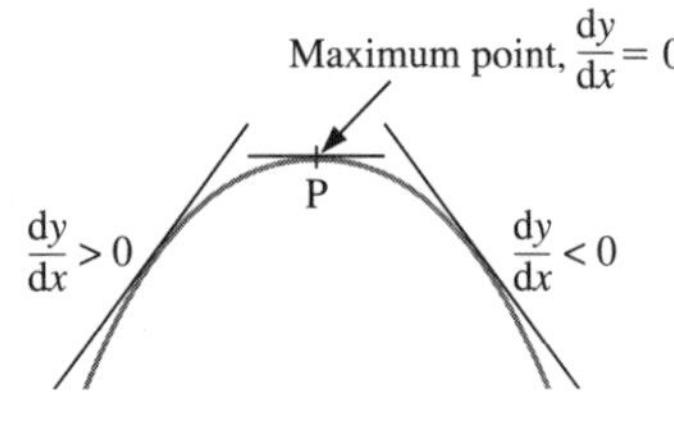

To the left of P	At point P	To the right of P
$\frac{dy}{dx} > 0$	$\frac{dy}{dx} = 0$	$\frac{dy}{dx} < 0$

Point of inflexion

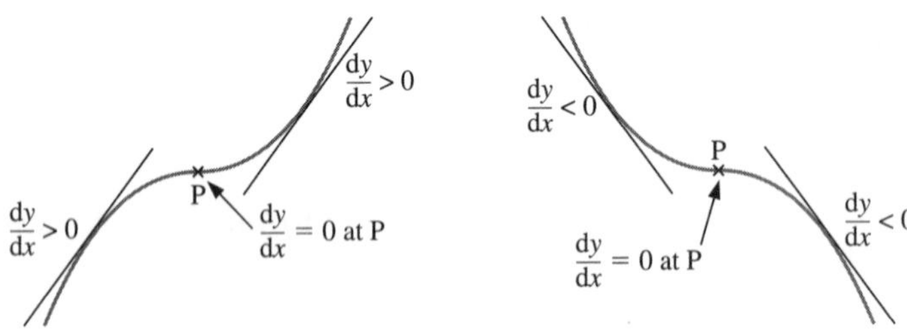

In this case, the gradient has the **same sign** each side of the stationary point. A point of inflexion which has zero gradient (such as this one) is called a **horizontal point of inflexion** or **saddle point**. This distinguishes it from a point of inflexion which has non-zero gradient. However, you should be aware of the fact that, generally, no distinction is made between these two types and all are just called points of inflexion.

Example 14 Find the coordinates of the stationary points on the curve $y = x^3 + 3x^2 + 1$ and determine their nature. Sketch the curve.

SOLUTION

When $y = x^3 + 3x^2 + 1$, then $\frac{dy}{dx} = 3x^2 + 6x$. At a stationary point, $\frac{dy}{dx} = 0$. Therefore,

$$3x^2 + 6x = 0$$
$$\therefore \quad 3x(x+2) = 0$$

Solving gives $x = 0$ or $x = -2$.

When $x = 0$: $\quad y = (0)^3 + 3(0)^2 + 1 = 1$

When $x = -2$: $\quad y = (-2)^3 + 3(-2)^2 + 1 = 5$

The coordinates of the stationary points are (0, 1) and (−2, 5).

To determine the nature of each stationary point, we must examine the gradient each side of each point.

For the point (0, 1),

x	-1	0	1
$\frac{dy}{dx}$	-3 Negative ╲	0 —	9 Positive ╱

Note When choosing x values each side of $x = 0$, the chosen interval must **not** include any other stationary points.

The stationary point (0, 1) is a minimum.

For the point (−2, 5),

x	-3	-2	-1
$\frac{dy}{dx}$	9 Positive ╱	0 —	-3 Negative ╲

The stationary point (−2, 5) is a maximum.

There are two stationary points on the curve: (0, 1) a minimum and (−2, 5) a maximum.

To sketch the curve, plot the stationary points and notice that the curve $y = x^3 + 3x^2 + 1$ cuts the y-axis at the point (0, 1).

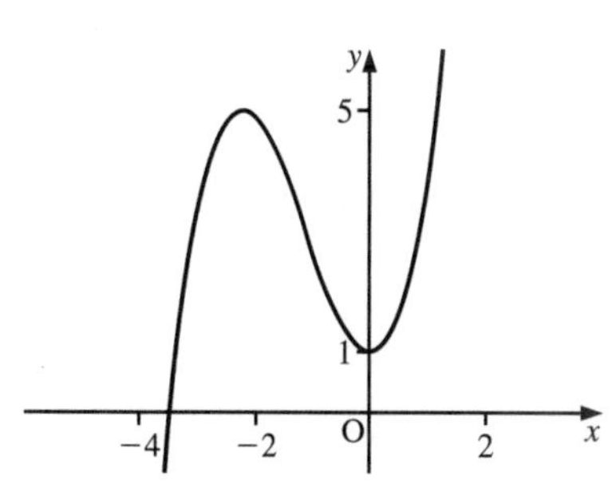

Second derivative and stationary points

Consider the graph of some function $y = \mathrm{f}(x)$ which possesses a maximum, a minimum and a point of inflexion. Consider also the graph of the derived function as shown below right.

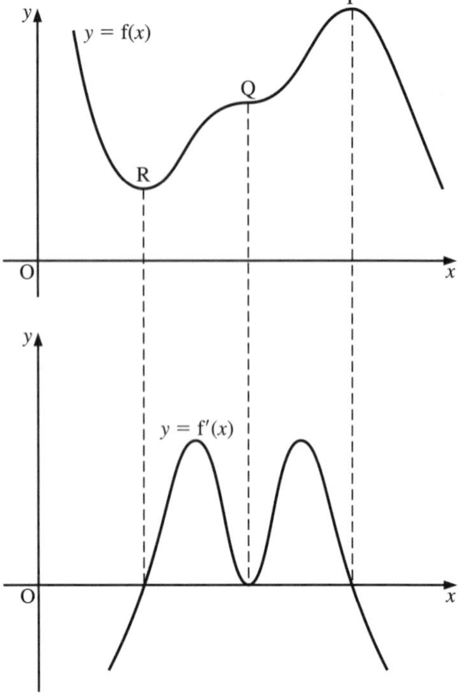

- At the maximum point P, the gradient of the derived function is negative. That is,

$$\frac{\mathrm{d}^2y}{\mathrm{d}x^2} < 0 \quad \text{at a maximum}$$

- At the point of inflexion Q, the gradient of the derived function is zero. That is,

$$\frac{\mathrm{d}^2y}{\mathrm{d}x^2} = 0$$

However, the second derivative can also be zero at a maximum or a minimum. For this reason, we must examine the sign of $\frac{\mathrm{d}y}{\mathrm{d}x}$ each side of the point.

- At the minimum point R, the gradient of the derived function is positive. That is,

$$\frac{\mathrm{d}^2y}{\mathrm{d}x^2} > 0 \quad \text{at a minimum}$$

This provides an easier method for classifying stationary points, as Example 15 illustrates.

Example 15 Find the coordinates of the stationary points on the curve $\mathrm{f}(x) = x^3 - 6x^2 - 15x + 1$, and using the second derivative determine their nature.

SOLUTION

When $\mathrm{f}(x) = x^3 - 6x^2 - 15x + 1$, then

$$\mathrm{f}'(x) = 3x^2 - 12x - 15$$

At stationary points, $\mathrm{f}'(x) = 0$. That is,

$$3x^2 - 12x - 15 = 0$$
$$\therefore \quad 3(x^2 - 4x - 5) = 0$$
$$\therefore \quad 3(x - 5)(x + 1) = 0$$

Solving gives $x = 5$ or $x = -1$.

When $x = 5$: $\quad \mathrm{f}(5) = (5)^3 - 6(5)^2 - 15(5) + 1 = -99$
When $x = -1$: $\quad \mathrm{f}(-1) = (-1)^3 - 6(-1)^2 - 15(-1) + 1 = 9$

The stationary points are $(5, -99)$ and $(-1, 9)$.

The second derivative is given by $\mathrm{f}''(x) = 6x - 12$. At the point $(5, -99)$,

$$\mathrm{f}''(5) = 6(5) - 12 = 18 > 0$$

Therefore, the stationary point $(5, -99)$ is a minimum.

At the point $(-1, 9)$,

$$f''(-1) = 6(-1) - 12 = -18 < 0$$

Therefore, the stationary point $(-1, 9)$ is a maximum.

Example 16 Find the coordinates of the stationary points on the curve $y = x^4 - 4x^3$ and determine their nature. Sketch the curve.

SOLUTION

When $y = x^4 - 4x^3$, then $\frac{dy}{dx} = 4x^3 - 12x^2$. At a stationary point, $\frac{dy}{dx} = 0$. That is,

$$4x^3 - 12x^2 = 0$$

$$\therefore \quad 4x^2(x - 3) = 0$$

Solving gives $x = 0$ or $x = 3$.

When $x = 0$: $\quad y = (0)^4 - 4(0)^3 = 0$

When $x = 3$: $\quad y = (3)^4 - 4(3)^3 = -27$

The coordinates of the stationary points are $(0, 0)$ and $(3, -27)$. To determine the nature of each stationary point, we check the sign of $\frac{d^2y}{dx^2}$ at each point. Now

$$\frac{dy}{dx} = 4x^3 - 12x^2$$

$$\therefore \quad \frac{d^2y}{dx^2} = 12x^2 - 24x$$

For the point $(0, 0)$,

$$\left.\frac{d^2y}{dx^2}\right|_{x=0} = 12(0)^2 - 24(0) = 0$$

This is not sufficient to conclude that the point $(0, 0)$ is a point of inflexion. We must examine the sign of $\frac{dy}{dx}$ each side of the point $(0, 0)$:

x	-1	0	1
$\frac{dy}{dx}$	-16	0	-8
	Negative		Negative
	$\searrow$	—	$\searrow$

The stationary point $(0, 0)$ is a point of inflexion, because the gradient of the curve has the same sign each side of $x = 0$.

For the point $(3, -27)$,

$$\left.\frac{d^2y}{dx^2}\right|_{x=3} = 12(3)^2 - 24(3) = 36 > 0$$

The stationary point $(3, -27)$ is a minimum.

There are two stationary points on the curve: (0, 0) a point of inflexion and (3, −27) a minimum.

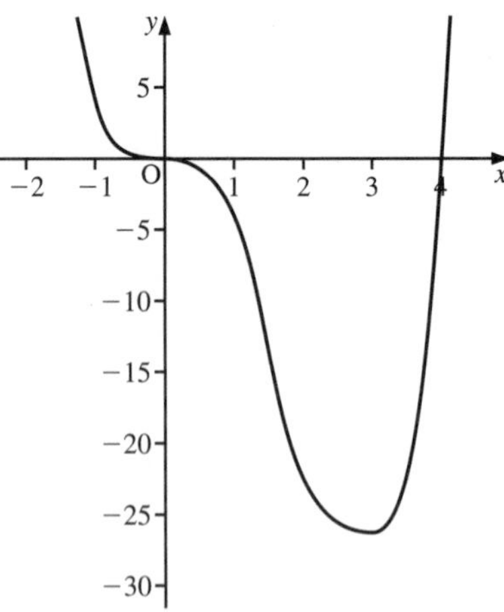

To sketch the curve, plot the stationary points and notice that the curve crosses the x-axis when $y = 0$. That is,

$$x^4 - 4x^3 = 0$$

$$\therefore \quad x^3(x-4) = 0$$

Solving gives $x = 0$ or $x = 4$. The curve crosses the x-axis at (0, 0) and (4, 0).

Example 17 In the right-angled triangle ABC shown opposite, the lengths AB and BC vary such that their sum is always 6 cm.

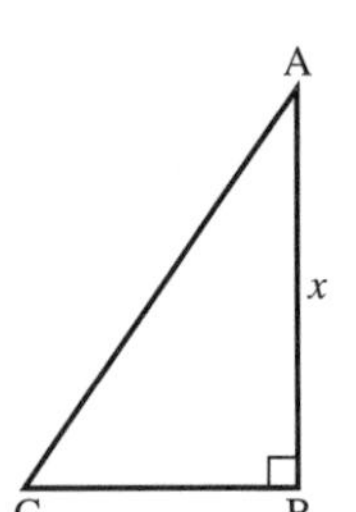

a) If the length of AB is x cm, write down, in terms of x, the length BC.
b) Find the maximum area of triangle ABC.

SOLUTION

a) We know that

$$AB + BC = 6$$

$$\therefore \quad x + BC = 6$$

$$\therefore \quad BC = 6 - x$$

b) The area, A, of triangle ABC is given by

$$A = \tfrac{1}{2} \times BC \times AB$$

$$= \tfrac{1}{2}(6-x)x$$

$$\therefore \quad A = 3x - \frac{x^2}{2} \qquad [1]$$

We can see that A attains a maximum value since the coefficient of the x^2 term is negative. (That is, the graph of x against A is ∩ shaped).

The maximum value occurs when $\frac{dA}{dx} = 0$. Now

$$\frac{dA}{dx} = 3 - x$$

When $\frac{dA}{dx} = 0$: $\quad 3 - x = 0$

$$\therefore \quad x = 3$$

The area of the triangle is a maximum when $x = 3$. Therefore, the maximum area of triangle ABC is found by substituting $x = 3$ into [1], which gives

$$A_{max} = 3(3) - \frac{(3)^2}{2} = \frac{9}{2} = 4.5$$

The maximum area of triangle ABC is $4.5\,\text{cm}^2$.

Example 18 A closed, right circular cylinder of base radius r cm and height h cm has a volume of $54\pi\,\text{cm}^3$. Show that S, the total surface area of the cylinder, is given by

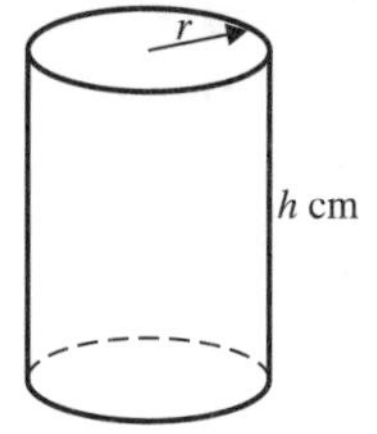

$$S = \frac{108\pi}{r} + 2\pi r^2$$

Hence find the radius and height which make the surface area a minimum.

SOLUTION

The volume, V, is given by

$$V = \pi r^2 h \qquad [1]$$

The total surface area, S, is given by

$$S = 2\pi rh + 2\pi r^2 \qquad [2]$$

We know that $V = 54\pi$. Therefore, from [1]

$$\pi r^2 h = 54\pi$$

$$\therefore \quad h = \frac{54}{r^2}$$

Substituting $h = \frac{54}{r^2}$ into [2] gives

$$S = 2\pi r\left(\frac{54}{r^2}\right) + 2\pi r^2$$

$$\therefore \quad S = \frac{108\pi}{r} + 2\pi r^2$$

as required.

Maximum/minimum surface area occurs when $\frac{\mathrm{d}S}{\mathrm{d}r} = 0$. Now

$$S = 108\pi r^{-1} + 2\pi r^2$$

$$\therefore \quad \frac{\mathrm{d}S}{\mathrm{d}r} = -108\pi r^{-2} + 4\pi r = -\frac{108\pi}{r^2} + 4\pi r$$

When $\frac{\mathrm{d}S}{\mathrm{d}r} = 0$: $\quad -\frac{108\pi}{r^2} + 4\pi r = 0$

$$\therefore \quad 4\pi r^3 = 108\pi$$

$$\therefore \quad r^3 = 27$$

Therefore, the radius, r, is 3 cm.

When $r = 3$: $\quad h = \frac{54}{(3)^2} = 6$

In other words, the height is 6 cm.

We must check that this value of r corresponds to a minimum. We will do this by checking the second derivative:

$$\frac{d^2S}{dr^2} = 216\pi r^{-3} + 4\pi$$

When $r = 3$,

$$\frac{d^2S}{dr^2} = \frac{216\pi}{(3)^3} + 4\pi = 12\pi$$

which is positive. That is, $\frac{d^2S}{dr^2} > 0$ when $r = 3$, which implies that $r = 3$ corresponds to a minimum S.

So, the surface area is a minimum when the radius is 3 cm and the height is 6 cm.

Exercise 6C

1 Find the coordinates of the points on each of the following curves at which the gradient is zero.

a) $y = x^2 - 4x + 3$ **b)** $y = x^2 + 6x + 5$ **c)** $y = 6 - x^2$
d) $y = 3 - 5x + x^2$ **e)** $y = 2x^2 - 3x + 1$ **f)** $y = x^3 - 3x + 2$
g) $y = x^3 - 6x^2 - 36x$ **h)** $y = 6 + 9x - 3x^2 - x^3$ **i)** $y = 5 + 3x^2 - x^3$
j) $y = x^4 - 2x^2 + 3$ **k)** $y = x^4 - 32x + 3$ **l)** $y = 1 - 6x + 6x^2 + 2x^3 - 3x^4$

2 Find the coordinates of the stationary points on each of the following curves, and determine their nature.

a) $y = x^2 - 2x + 5$ **b)** $y = x^2 + 4x + 2$ **c)** $y = 3 + x - x^2$
d) $y = (x - 4)(x - 2)$ **e)** $y = 3(x + 3)(2x - 1)$ **f)** $y = (x - 5)^2$
g) $y = x^3 + 6x^2 - 36x$ **h)** $y = x^3 - 5x^2 + 3x + 1$ **i)** $y = 3 + 15x - 6x^2 - x^3$
j) $y = x^4 - 8x^2 + 3$ **k)** $y = x^4 + 4x^3 + 1$ **l)** $y = x^4 - 14x^2 + 24x - 10$

3 Find the coordinates of the stationary points on the following curves. In each case, determine their nature and sketch the curve.

a) $y = x^3 - 3x + 3$ **b)** $y = -x^3 + 6x^2 - 9x$ **c)** $y = x^3 - 9x^2 + 27x - 19$
d) $y = x^4 - 8x^2 - 9$ **e)** $y = 8x^3 - x^4$ **f)** $y = x^4 + x^3 - 3x^2 - 5x - 2$

4 Find the coordinates of the stationary points on each of the following curves, and determine their nature.

a) $y = x + \frac{1}{x}$ **b)** $y = x^2 + \frac{16}{x}$ **c)** $y = \frac{1}{x} - \frac{3}{x^2}$
d) $y = \frac{2}{x^3} - \frac{1}{x^2}$ **e)** $y = \frac{12x^2 - 1}{x^3}$ **f)** $y = \frac{2 - x^3}{x^4}$

5 By investigating the stationary points of $f(x) = x^3 + 3x^2 + 6x - 30$ and sketching the curve $y = f(x)$, show that the equation $f(x) = 0$ has only one real solution.

6 Show that the equation $x^4 - 4x^3 - 2x^2 + 12x + 12 = 0$ has no real solution.

7 Show that the equation $3x^4 + 4x^3 - 36x^2 + 64 = 0$ has precisely three real solutions.

8 The profit, £y, generated from the sale of x items of a certain luxury product is given by the formula $y = 600x + 15x^2 - x^3$. Calculate the value of x which gives a maximum profit, and determine that maximum profit.

9 The profit, y hundred pounds, generated from the sale of x thousand items of a certain product is given by the formula $y = 72x + 3x^2 - 2x^3$. Calculate how many items should be sold in order to maximise the profit, and determine that maximum profit.

10 At a speed of x mph a certain car will travel y miles on each gallon of petrol, where

$$y = 15 + x - \frac{x^2}{110}$$

Calculate the speed at which the car should aim to travel in order to maximise the distance it can cover on a single tank of petrol.

11 At a speed of x mph, a transporter can cover y miles on 1 gallon of diesel fuel, where

$$y = 5 + \frac{x}{2} + \frac{x^2}{60} - \frac{x^3}{1800}$$

Calculate the maximum distance which the transporter can travel on 30 gallons of diesel fuel.

12 A ball is thrown vertically upwards. At time t seconds after the instant of projection, its height, y metres above the point of projection, is given by the formula $y = 15t - 5t^2$. Calculate the time at which the ball is at its maximum height, and find the value of y at that time.

13 An unpowered missile is launched vertically from the ground. At a time t seconds after its launch its height, y metres, above the ground is given by the formula $y = 80t - 5t^2$. Calculate the maximum height reached by the missile.

14 A piece of string which is 40 cm long is cut into two lengths. Each length is laid out to form a square. Given that the length of the sides of one of the squares is x cm, find expressions in terms of x for

a) the length of the sides of the other square
b) the total area enclosed by the two squares.

Given also that the sum of the two areas is a minimum,

c) calculate the value of x.

15 A stick of length 24 cm is cut into three pieces, two of which are of equal length. The two pieces of equal length are then each cut into four equal lengths and constructed into squares of side x cm. The remaining piece is cut and constructed into a rectangle of width 3 cm. Find expressions, in terms of x, for

a) the length of the rectangle
b) the total area enclosed by the three shapes.

Given also that the sum of the three areas is a minimum,

c) calculate the value of x.

16 A strip of wire of length 150 cm is cut into two pieces. One piece is bent to form a square of side x cm, and the other piece is bent to form a rectangle which is twice as long as it is wide. Find expressions, in terms of x, for

a) the width of the rectangle
b) the length of the rectangle
c) the area of the rectangle.

Given also that the sum of the two areas enclosed is a minimum,

d) calculate the value of x.

17 A rectangular enclosure is formed from 1000 m of fencing. Given that each of the two opposite sides of the rectangle has length x metres,

a) find, in terms of x, the length of the other two sides.

Given also that the area enclosed is a maximum,

b) find the value of x, and hence calculate the area enclosed.

18 A rectangular pen is formed from 40 m of fencing with a long wall forming one side of the pen, as shown in the diagram on the right.

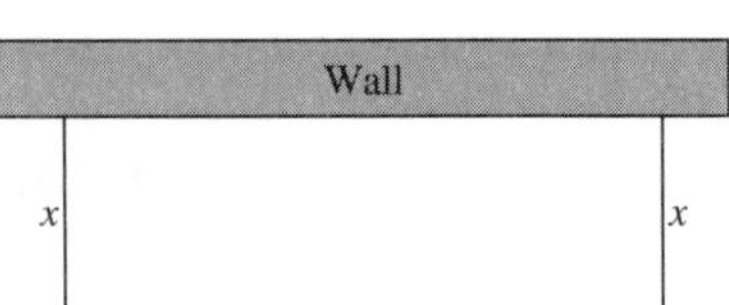

Given that the two opposite sides of the pen which touch the wall each have length x metres,

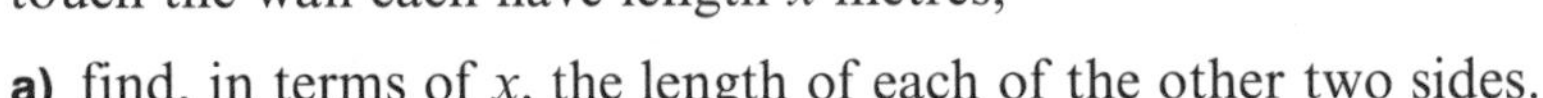

a) find, in terms of x, the length of each of the other two sides.

Given also that the area enclosed is a maximum,

b) find the value of x, and hence calculate the area enclosed.

19 A closed cuboidal box of square base has volume 8 m^3. Given that the square base has sides of length x metres, find expressions, in terms of x, for

a) the height of the box
b) the surface area of the box.

Given also that the surface area of the box is a minimum,

c) find the value of x.

20 An *open* metal tank of square base has volume 108 m^3. Given that the square base has sides of length x metres, find expressions, in terms of x, for

a) the height of the tank
b) the surface area of the tank.

Given also that the surface area is a minimum,

c) find the value of x.

21 A silver bar of volume 576 cm^3 is cuboidal in shape, and has a length which is twice its breadth. Given that the breadth of the bar is x cm, find expressions, in terms of x, for

a) the length of the bar
b) the height of the bar
c) the surface area of the bar.

Given also that the surface area is a minimum,

d) find the value of x.

22 An *open* cuboidal tank of rectangular base is to be made with an external surface area of $36\,\text{m}^2$. The base is to be such that its length is three times its breadth. Find the length of the base of the tank for the volume of the tank to be a maximum, and find this maximum volume.

23 A closed cuboidal plastic box is to be made with an external surface area of $216\,\text{cm}^2$. The base is to be such that its length is four times its breadth. Find the length of the base of the box if the volume of the box is to be a maximum, and find this maximum volume.

24 A closed cuboidal box of square base and volume $36\,\text{cm}^3$ is to be constructed and silver plated on the outside. Silver plating for the top and the base costs 40 p per cm^2, and silver plating for the sides costs 30 p per cm^2. Given that the length of the sides of the base is to be x cm, find expressions, in terms of x, for

a) the height of the box
b) the cost of plating the top
c) the cost of plating a side
d) the total cost of plating the box.

Given also that this cost is to be a minimum,

e) find the value of x
f) calculate the cost of plating the box.

25 An *open* cuboidal fish tank of rectangular base and volume $2.5\,\text{m}^3$ is to be made in such a way that its length is three times its breadth. Glass for the sides costs £4 per m^2, and glass for the base costs £15 per m^2. Given that the base has breadth x m, find expressions, in terms of x, for

a) the height of the tank
b) the cost of all of the glass for the sides
c) the cost of glass for the base.

Given also that the cost is to be a minimum,

d) find the value of x
e) calculate the cost of the glass for the tank.

26 *Open* cuboidal metal boxes of square base are to be made such that each box has a volume of $750\,\text{cm}^3$. The metal sheeting used for the sides costs 2 p per cm^2, and the metal sheeting used for the base costs 3 p per cm^2. Calculate the dimensions of the boxes which should be made if the cost of the metal sheeting is to be a minimum.

27 An *open* cardboard box is to be made by cutting small squares of side x cm from each of the four corners of a larger square of card of side 10 cm, and folding along the dashed lines, as shown in the diagram on the right. Find the value of x such that the box has a maximum volume, and state the value of that maximum volume.

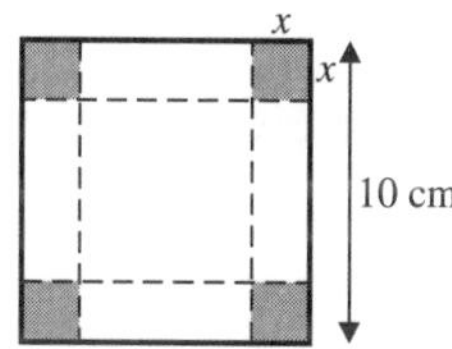

28 An *open* metal tray is to be made by cutting squares of side x cm from each of the four corners of a rectangular piece of metal measuring 8 cm by 5 cm, and folding the resulting shape as in Question **27**. Find the value of x that will give the box a maximum volume.

***29** A cylinder is to be made of circular cross-section with a specified volume. Prove that if the surface area is to be a minimum, then the height of the cylinder will be equal to the diameter of the cross-section of the cylinder.

Exercise 6D: Examination questions

1 Find an equation of the tangent to the curve with equation $y = x^2 - 9x^{-1}$ at the point (3, 6). (EDEXCEL)

2 Given that $y = x^3 - 4x^2 + 5x - 2$, find $\frac{dy}{dx}$.
P is the point on the curve where $x = 3$.

i) Calculate the y coordinate of P.
ii) Calculate the gradient at P.
iii) Find the equation of the tangent at P.
iv) Find the equation of the normal at P.

Find the values of x for which the curve has a gradient of 5. (MEI)

3 The curve C has equation $y = x^3 - 2x^2$.

a) Show that N, the normal to C at the point $(1, -1)$, has equation $y = x - 2$.
b) Show that the x coordinates of the points of intersection of C and N satisfy

$$x^3 - 2x^2 - x + 2 = 0$$

Solve this equation and hence find the coordinates of the points where N meets C again. (WJEC)

4 The equation of the curve C is $y = \frac{4}{x}$.

i) Show that the tangent to C at the point P(1, 4) has equation $4x + y = 8$.
ii) Show that the normal to C at P has equation $-x + 4y = 15$.
iii) The tangent to C at P intersects the x-axis at S and the y-axis at T. Find the coordinates of S and T.
iv) The normal to C at P intersects the line $y = x$ at Q, and the line $y = -x$ at R. Find the coordinates of Q and R.
v) Show that QSRT is a square. (WJEC)

5 Use differentiation to find the coordinates of the stationary point on the curve $y = x^2 + \frac{16}{x}$. (UCLES)

6 a) Given $y = 5x^3 - 2x^2 + 1$, find $\frac{dy}{dx}$.

b) Hence find the exact values of x at which the graph of

$$y = 5x^3 - 2x^2 + 1$$

has stationary points. (NEAB)

7 A curve has equation $y = 2x^3 - 9x^2 + 12x - 5$.

a) Calculate the coordinates of the turning points of the curve, showing that one of these points lies on the x-axis.
b) Determine the values of x for which $y = 0$. (UODLE)

8 Given the function $y = 3x^4 + 4x^3$,

i) find $\dfrac{dy}{dx}$

ii) show that the graph of the function y has stationary points at $x = 0$ and $x = -1$ and find their coordinates

iii) determine whether each of the stationary points is a maximum, minimum or point of inflexion, giving reasons for your answer

iv) sketch the graph of the function y, giving the coordinates of the stationary points and the points where the curve cuts the axes. (MEI)

9 Use differentiation to find the coordinates of the stationary points on the curve

$$y = x + \frac{4}{x}$$

and determine whether each stationary point is a maximum point or a minimum point. Find the set of values of x for which y increases as x increases.

10 You are given that $y = x^3 - 12x + 5$.

i) Find $\dfrac{dy}{dx}$.

The curve with equation $y = x^3 - 12x + 5$ has two stationary points.

ii) Find the coordinates of these two stationary points.

The coordinates of the stationary points give the impression that the left-hand point is a maximum and the right-hand point is a minimum.

iii) State how you would decide whether this is the case.

iv) Draw a sketch of the curve with equation $y = x^3 - 12x + 5$.

v) Find the equation of the tangent to this curve at the point where $x = 0$. (MEI)

11 A curve has the equation $y = 2x^3 - 3x^2 - 36x + 120$.

i) Calculate the values of y when x is 3 and when x is -2.

ii) Find $\dfrac{dy}{dx}$.

iii) Use your expression for $\dfrac{dy}{dx}$ to find the coordinates of the two stationary points on the curve.

iv) By considering the values of $\dfrac{dy}{dx}$ near the stationary points, decide which type of stationary point each is.

v) Sketch the curve. Deduce the range of values of k for which the equation

$$2x^3 - 3x^2 - 36x + 120 = k$$

has three real roots. (MEI)

12 A straight line having a positive gradient m passes through the point $(-1, 2)$ and cuts the coordinate axes at X and Y. Find, in terms of m, the coordinates of X and Y.

i) Show that the area A of the triangle OXY, where O is the origin, is given by

$$A = \frac{(m+2)^2}{2m}$$

ii) Show that A has a minimum value of 4. (WJEC)

13 The figure shows the flat surface of a tray consisting of a rectangular region WXYZ and a semicircular region at each end with WX and YZ as diameters. The rectangle WXYZ has area $200\,\text{cm}^2$.

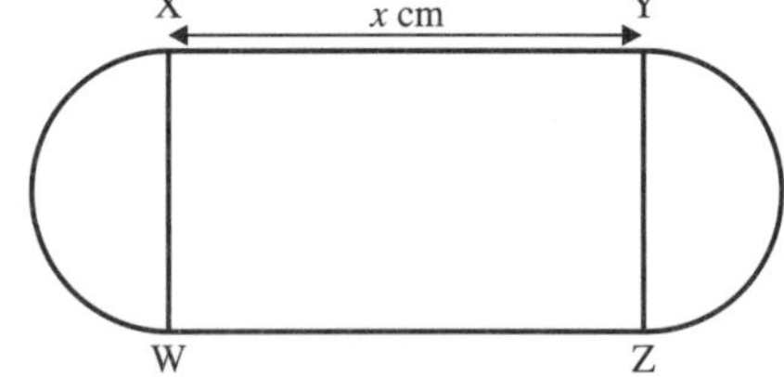

a) Given that $XY = x\,\text{cm}$, show that the perimeter $P\,\text{cm}$ of the tray is given by the formula

$$P = 2x + \frac{200\pi}{x}$$

b) Find the minimum value of P as x varies. (EDEXCEL)

14 The figure shows a sector POQ of a circle of radius r metres, the angle POQ being equal to θ radians. The perimeter of the sector is of length 3 metres and the area of the sector is A square metres.

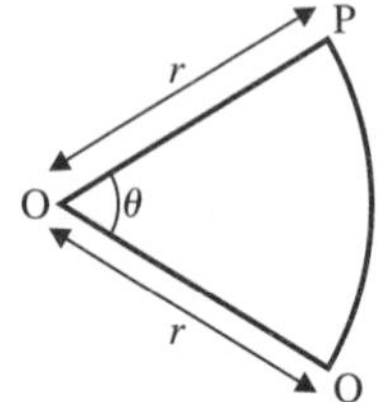

a) Show that $A = \frac{1}{2}r(3 - 2r)$.

b) Show that, as r varies, the maximum value of A is $\frac{9}{16}$.

c) Find the corresponding value of θ. (WJEC)

15 The figure shows a minor sector OMN of a circle centre O and radius $r\,\text{cm}$. The perimeter of the sector is $100\,\text{cm}$ and the area of the sector is $A\,\text{cm}^2$.

a) Show that $A = 50r - r^2$.

Given that r varies, find

b) the value of r for which A is a maximum and show that A is a maximum

c) the value of $\angle$MON for this maximum area

d) the maximum area of the sector OMN. (EDEXCEL)

16 The figure shows a rectangular cake-box, with no top, which is made from thin card. The volume of the box is $500\,\text{cm}^3$. The base of the box is a square with sides of length $x\,\text{cm}$.

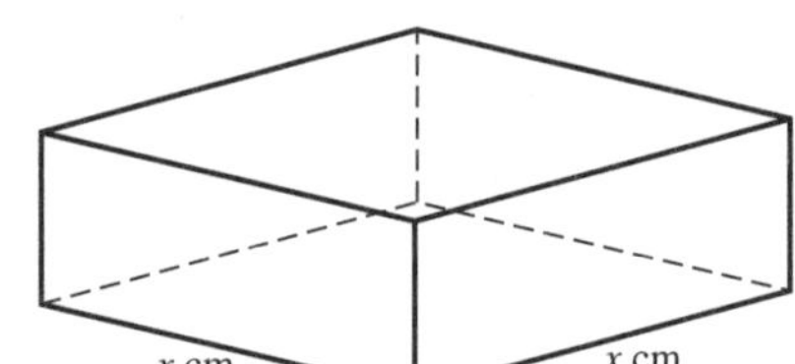

a) Show that the area, $A\,\text{cm}^2$, of card used to make such an open box is given by

$$A = x^2 + \frac{2000}{x}$$

b) Given that x varies, find the value of x for which $\frac{\mathrm{d}A}{\mathrm{d}x} = 0$.

c) Find the height of the box when x has this value.

d) Show that when x has this value, the area of card used is least. (EDEXCEL)

17 A large tank in the shape of a cuboid is to be made from $54\,\text{m}^2$ of sheet metal. The tank has a horizontal rectangular base and no top. The height of the tank is x metres. Two of the opposite vertical faces are squares.

a) Show that the volume, $V\,\text{m}^3$, of the tank is given by

$$V = 18x - \tfrac{2}{3}x^3$$

b) Given that x can vary, use differentiation to find the maximum value of V.

c) Justify that the value of V you have found is a maximum. (EDEXCEL)

18 The figure shows a brick in the shape of a cuboid with base x cm by $2x$ cm and height h cm. The total surface area of the brick is 300 cm^2.

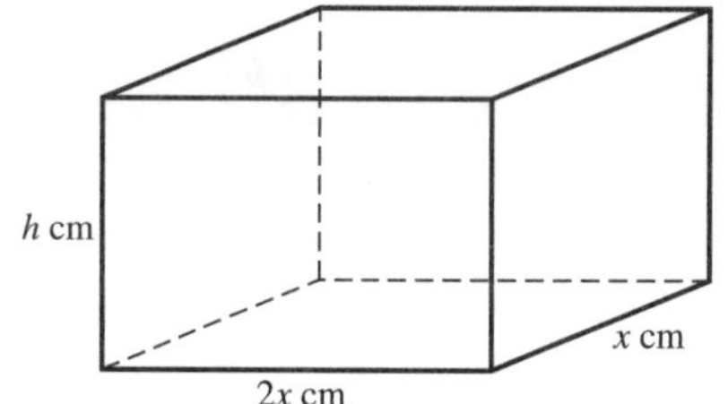

a) Show that $h = \dfrac{50}{x} - \dfrac{2x}{3}$.

The volume of the brick is V cm^3.

b) Express V in terms of x only.
c) Given that x can vary, find the maximum value of V.
d) Explain why the value of V you have found is the maximum. (EDEXCEL)

19 A cylindrical can, with no lid, has a circular base of radius r cm. The total surface area of the can is (300π) cm^2.

a) Show that the volume V cm^3 of the can is given by

$$V = \frac{\pi r}{2}(300 - r^2)$$

b) Given that r may vary, find the positive value of r for which $\dfrac{dV}{dr} = 0$.
c) Show that this value of r gives a maximum value of V.
d) Find the maximum value of V. (EDEXCEL)

20 A cylindrical tank, open at the top and of height h m and radius r m, has a capacity of one cubic metre. Show that $h = 1/(\pi r^2)$.

Its total internal surface area is S m^2. Show that $S = \dfrac{2}{r} + \pi r^2$.

Determine the value of r which makes the surface area S as small as possible. (MEI)

21 The figure shows a right circular cone of base radius 2 cm and height 6 cm standing on a horizontal table. A cylinder of radius x cm stands inside the cone with its axis coincident with the axis of symmetry of the cone and such that the cylinder touches the curved surface of the cone as shown. The volume of the cylinder is V cm^3.

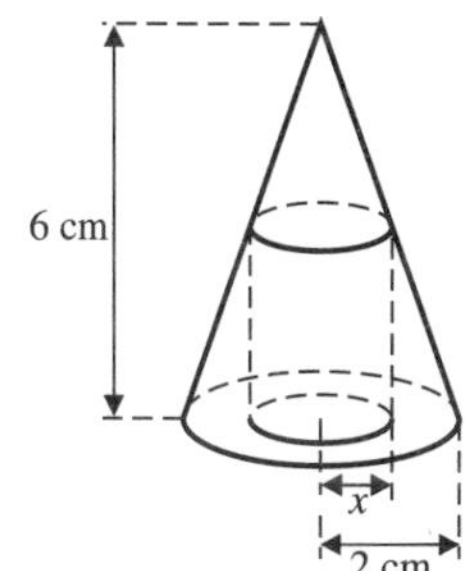

a) Show that $V = 6\pi x^2 - 3\pi x^3$.
b) Given that x can vary, obtain the maximum value of V. (EDEXCEL)

7 Integration I

God does not care about our mathematical difficulties – he integrates empirically
ALBERT EINSTEIN

We know that if $y = x^2$, then $\frac{dy}{dx} = 2x$. Now suppose that we are given $\frac{dy}{dx} = 2x$ and asked to find y in terms of x. This process is the reverse of differentiation and is called **integration**.

In this particular case, we know that $y = x^2$ will satisfy $\frac{dy}{dx} = 2x$, but so will $y = x^2 + 1$ and $y = x^2 + 2$. In fact, $y = x^2 + c$, where c is a constant, will also satisfy $\frac{dy}{dx} = 2x$.

In other words, we do not know whether the original function contained a constant term or not. For this reason, we write $y = x^2 + c$, where c is called the **constant of integration**.

Notation $y = x^2 + c$ is called the **integral** of $2x$ with respect to x. This is written as

$$\int 2x \quad dx \quad = x^2 + c$$

Integral sign — $\int$

Indicating that the integration is with respect to the variable x — dx

To find $\int x^4\,dx$, we notice that x^5 differentiates to give $5x^4$, i.e. the required power of x. However, the constant 5 is not required, therefore we multiply x^5 by $\frac{1}{5}$. Now we have

$\frac{1}{5}x^5$ differentiates to give x^4

which is the required result. Therefore, we have

$$\int x^4\,dx = \tfrac{1}{5}x^5 + c$$

Result

If $\frac{dy}{dx} = ax^n$, then

$$y = \frac{ax^{n+1}}{n+1} + c \qquad (n \neq -1)$$

That is

$$\int ax^n\,dx = \frac{ax^{n+1}}{n+1} + c \qquad (n \neq -1)$$

One way of remembering this is 'add one to the power and divide by the new power'.

Proof

This can be shown by differentiating y with respect to x:

$$y = \frac{ax^{n+1}}{n+1} + c$$

Therefore,

$$\frac{\mathrm{d}y}{\mathrm{d}x} = \frac{a(n+1)x^n}{(n+1)} = ax^n$$

as required.

Example 1 Find

a) $\int 3x^2\,\mathrm{d}x$ **b)** $\int \frac{1}{x^2}\,\mathrm{d}x$ **c)** $\int 6\sqrt{x}\,\mathrm{d}x$

SOLUTION

a) To find $\int 3x^2\,\mathrm{d}x$, we notice that x^3 differentiates to give $3x^2$. Therefore,

$$\int 3x^2\,\mathrm{d}x = x^3 + c$$

Alternatively, using Result I, we have

$$\int 3x^2\,\mathrm{d}x = \frac{3x^{2+1}}{(2+1)} + c$$

$$\therefore \quad \int 3x^2\,\mathrm{d}x = x^3 + c$$

b) Using Result I, we have

$$\int \frac{1}{x^2}\,\mathrm{d}x = \frac{x^{-2+1}}{(-2+1)} + c = \frac{x^{-1}}{-1} + c$$

$$\therefore \quad \int \frac{1}{x^2}\,\mathrm{d}x = -\frac{1}{x} + c$$

c) Using Result I, we have

$$\int 6\sqrt{x}\,\mathrm{d}x = \frac{6x^{\frac{1}{2}+1}}{\left(\frac{1}{2}+1\right)} + c = \frac{6x^{\frac{3}{2}}}{\left(\frac{3}{2}\right)} + c$$

$$\therefore \quad \int 6\sqrt{x}\,\mathrm{d}x = 4x^{\frac{3}{2}} + c = 4\sqrt{x^3} + c$$

On page 161, we saw that

$$\frac{d}{dx}[f(x)+g(x)] = f'(x) + g'(x)$$

The integral behaves in exactly the same way and we have the results

- $\int a\,f(x)\,dx = a\int f(x)\,dx$
- $\int [f(x) \pm g(x)]\,dx = \int f(x)\,dx \pm \int g(x)\,dx$

The second result applies generally to any number of functions.

Example 2 Find

a) $\int (x^2 + 6x - 3)\,dx$ **b)** $\int \left(1 - \frac{1}{x^2}\right) dx$

SOLUTION

a) From the second result, we have

$$\int (x^2 + 6x - 3)\,dx = \frac{x^3}{3} + \frac{6x^2}{2} - 3x + c$$

$$\therefore \quad \int (x^2 + 6x - 3)\,dx = \frac{x^3}{3} + 3x^2 - 3x + c$$

b) From the second result, we have

$$\int \left(1 - \frac{1}{x^2}\right) dx = \int 1\,dx - \int x^{-2}\,dx$$

$$= x + x^{-1} + c$$

$$\therefore \quad \int \left(1 - \frac{1}{x^2}\right) dx = x + \frac{1}{x} + c$$

Some expressions are not in the form ax^n or written as a sum of functions of this form. In these cases, it is necessary to manipulate the integrand into this form. Example 3 illustrates such a case.

Example 3 Find

a) $\int (x-4)^2\,dx$ **b)** $\int \left(\frac{x^3 - 3x}{x}\right) dx$ **c)** $\int \sqrt{x}(x+1)\,dx$

SOLUTION

a)

$$\int (x-4)^2\,dx = \int (x^2 - 8x + 16)\,dx$$

$$= \frac{x^3}{3} - 4x^2 + 16x + c$$

b)

$$\int\left(\frac{x^3-3x}{x}\right)dx = \int\left(\frac{x^3}{x}-\frac{3x}{x}\right)dx$$

$$= \int(x^2-3)\,dx$$

$$\therefore \quad \int\left(\frac{x^3-3x}{x}\right)dx = \frac{x^3}{3}-3x+c$$

c)

$$\int\sqrt{x}(x+1)\,dx = \int x^{\frac{1}{2}}(x+1)\,dx$$

$$= \int\left(x^{\frac{3}{2}}+x^{\frac{1}{2}}\right)dx$$

$$= \frac{x^{\frac{5}{2}}}{\left(\frac{5}{2}\right)}+\frac{x^{\frac{3}{2}}}{\left(\frac{3}{2}\right)}+c = 2x^{\frac{3}{2}}\left(\frac{x}{5}+\frac{1}{3}\right)+c$$

$$\therefore \quad \int\sqrt{x}(x+1)\,dx = 2\sqrt{x^3}\left(\frac{x}{5}+\frac{1}{3}\right)+c \quad \text{or} \quad \frac{2}{15}\sqrt{x^3}(3x+5)+c$$

Example 4 The gradient of a curve at the point (x, y) is $12x^3-\frac{1}{x^2}$ and the curve passes through the point $(1, 2)$. Find the equation of the curve.

SOLUTION

We know that

$$\frac{dy}{dx} = 12x^3-\frac{1}{x^2} = 12x^3-x^{-2}$$

$$\therefore \quad y = \int(12x^3-x^{-2})\,dx$$

$$= \frac{12x^4}{4}-\frac{x^{-1}}{(-1)}+c$$

$$\therefore \quad y = 3x^4+\frac{1}{x}+c \qquad [1]$$

Since the curve passes through the point $(1, 2)$ we know that $x = 1$, $y = 2$ satisfies [1]. Therefore,

$$2 = 3(1)^4+\frac{1}{1}+c$$

$$\therefore \quad c = -2$$

The equation of the curve is $y = 3x^4+\frac{1}{x}-2$.

Example 5 Given that

$$f''(x) = 2 - \frac{2}{\sqrt{x^3}} \quad \text{and} \quad f'(1) = 0$$

find $f'(x)$. Given further that $f(1) = 8$, find $f(x)$.

SOLUTION

When $f''(x) = 2 - \frac{2}{\sqrt{x^3}} = 2 - 2x^{-\frac{3}{2}}$,

$$f'(x) = \int (2 - 2x^{-\frac{3}{2}})\,dx$$

$$= 2x - \frac{2x^{-\frac{1}{2}}}{\left(-\frac{1}{2}\right)} + c_1$$

$$\therefore \quad f'(x) = 2x + 4x^{-\frac{1}{2}} + c_1 \qquad [1]$$

Since $f'(1) = 0$, substituting into [1] gives $c_1 = -6$ and so

$$f'(x) = 2x + \frac{4}{\sqrt{x}} - 6 = 2x + 4x^{-\frac{1}{2}} - 6$$

Therefore,

$$f(x) = \int (2x + 4x^{-\frac{1}{2}} - 6)\,dx = x^2 + 8x^{\frac{1}{2}} - 6x + c_2$$

Since $f(1) = 8$, we have

$$8 = 1 + 8 - 6 + c_2$$

$$\therefore \quad c_2 = 5$$

Therefore, $f(x) = x^2 + 8\sqrt{x} - 6x + 5$.

Exercise 7A

1 Integrate each of the following with respect to x.

a) x^3 **b)** x^4 **c)** $3x^2$ **d)** $12x^5$ **e)** $-4x$ **f)** $15x^4$
g) $2x^3$ **h)** 3 **i)** $\frac{1}{2}x^5$ **j)** $\frac{2}{3}x^3$ **k)** $-\frac{1}{3}x^2$ **l)** $\frac{2}{3}$

2 Find each of these integrals.

a) $\int x^{-2}\,dx$ **b)** $\int x^{-4}\,dx$ **c)** $\int 2x^{-3}\,dx$ **d)** $\int -6x^{-4}\,dx$

e) $\int \frac{1}{x^3}\,dx$ **f)** $\int -\frac{1}{x^5}\,dx$ **g)** $\int \frac{3}{x^2}\,dx$ **h)** $\int -\frac{2}{x^3}\,dx$

i) $\int \frac{4}{x^7}\,dx$ **j)** $\int \frac{3}{2x^4}\,dx$ **k)** $\int -\frac{5}{3x^2}\,dx$ **l)** $\int \frac{2}{3x^4}\,dx$

3 Integrate each of the following functions with respect to x.

a) $f(x) = x^{\frac{1}{3}}$ **b)** $f(x) = 3x^{\frac{1}{2}}$ **c)** $f(x) = x^{-\frac{2}{3}}$ **d)** $f(x) = -4x^{-\frac{1}{5}}$

e) $f(x) = -3\sqrt{x}$ **f)** $f(x) = \sqrt[4]{x}$ **g)** $f(x) = \dfrac{4}{\sqrt[3]{x}}$ **h)** $f(x) = -\dfrac{2}{\sqrt[5]{x}}$

i) $f(x) = \dfrac{3}{7\sqrt{x}}$ **j)** $f(x) = \dfrac{6}{5\sqrt[3]{x}}$ **k)** $f(x) = \sqrt{x^3}$ **l)** $f(x) = \sqrt{9x}$

4 Integrate each of the following with respect to x.

a) $x^3 + 2x$ **b)** $3x^2 - 4x$ **c)** $x^3 - 1$ **d)** $6 + 3x^5$

e) $x^2 - 5x + 3$ **f)** $x^8 + 2x^5$ **g)** $x^4 - 3x + 2$ **h)** $x^2 - \dfrac{1}{x^2}$

i) $5x^4 - \dfrac{2}{x^3}$ **j)** $2x^6 + \dfrac{8}{x^5}$ **k)** $x^2 - \dfrac{3}{x^2}$ **l)** $\dfrac{5}{x^2} - 2 - 2x^3$

5 Find each of these integrals.

a) $\int (3\sqrt{x} - 4)\,dx$ **b)** $\int \left(\sqrt{x} + \dfrac{1}{\sqrt{x}}\right) dx$ **c)** $\int (3x^{\frac{1}{3}} - 2x^{\frac{1}{4}})\,dx$

d) $\int (5x^{-\frac{1}{2}} + 2x^{-\frac{1}{3}})\,dx$ **e)** $\int \left(4\sqrt{x} - \dfrac{2}{3x^2}\right) dx$ **f)** $\int \left(2\sqrt[3]{x} - \dfrac{6}{\sqrt{x}}\right) dx$

g) $\int \left(\dfrac{4}{\sqrt[3]{x}} - 8\sqrt{x}\right) dx$ **h)** $\int \left(\sqrt[4]{x} - \dfrac{1}{\sqrt[4]{x}}\right) dx$ **i)** $\int \left(\dfrac{4}{3\sqrt[3]{x}} - 2\sqrt[7]{x}\right) dx$

j) $\int \left(\dfrac{2}{\sqrt[6]{x}} - \dfrac{8}{\sqrt[5]{x}}\right) dx$ **k)** $\int (2x^{\frac{1}{3}} - 3x^{\frac{4}{3}} - 5x^{\frac{7}{3}})\,dx$ **l)** $\int \left(\dfrac{5}{(\sqrt{x})^3} - \dfrac{2}{(\sqrt[3]{x})^2}\right) dx$

6 Find $\int y\,dx$ for each of these.

a) $y = x(3 - x)$ **b)** $y = x^2(x + 5)$ **c)** $y = x^3(2 - x^2)$

d) $y = \sqrt{x}(x + 3)$ **e)** $y = 3\sqrt{x}(x^2 - x + 1)$ **f)** $y = x^{\frac{1}{3}}(2x + 3)$

g) $y = 3x^{\frac{1}{4}}(x - 2)$ **h)** $y = (x + 3)(x + 5)$ **i)** $y = (x - 2)^2$

j) $y = 2(x + 5)^2$ **k)** $y = x(x - 1)^2$ **l)** $y = (\sqrt{x} - 3)(\sqrt{x} + 5)$

7 Integrate each of the following functions with respect to x.

a) $f(x) = 5x(x - 2)$ **b)** $f(x) = x^3(6x^2 - 1)$ **c)** $f(x) = \sqrt{x}(x^2 + 1)$

d) $f(x) = x^{-\frac{1}{3}}(2x + 3)$ **e)** $f(x) = \dfrac{x^2 + 5}{x^2}$ **f)** $f(x) = \dfrac{x - 4}{x^3}$

g) $f(x) = \dfrac{3x^2 + 5}{x^2}$ **h)** $f(x) = \dfrac{4x^3 - 3x^2}{2x}$ **i)** $f(x) = \dfrac{5x^2 - 4}{\sqrt{x}}$

j) $f(x) = \dfrac{6x - 3}{2\sqrt{x}}$ **k)** $f(x) = x^3(5 - \sqrt{x})^2$ **l)** $f(x) = \dfrac{(x - 1)^2}{\sqrt{x}}$

8 Given that a, b, c and d are constants write down expressions for each of the following.

a) $\int (ax + b)\,dx$ **b)** $\int \left(a\sqrt{x} + \dfrac{b}{\sqrt{x}}\right) dx$ **c)** $\int (ax + b)(cx + d)\,dx$

9 Show that

$$\int x^{n-1} + x^{2n-1}\,\mathrm{d}x = \frac{x^n(2+x^n)}{2n} + c$$

for any non-zero constant, n.

10 In each of the following parts use the information given to find an expression for y in terms of x.

a) $\frac{\mathrm{d}y}{\mathrm{d}x} = 3x^2 + 1,\ y = 12$ when $x = 2$

b) $\frac{\mathrm{d}y}{\mathrm{d}x} = 4x - 3,\ y = 6$ when $x = -1$

c) $\frac{\mathrm{d}y}{\mathrm{d}x} = 6x^2 - 4x,\ y = 24$ when $x = 3$

d) $\frac{\mathrm{d}y}{\mathrm{d}x} = 4 - 6x,\ y = -4$ when $x = -2$

e) $\frac{\mathrm{d}y}{\mathrm{d}x} = \frac{2}{x^2} - 1,\ x \neq 0;\ y = 5$ when $x = 1$

f) $\frac{\mathrm{d}y}{\mathrm{d}x} = -\frac{10}{x^3},\ x \neq 0;\ y = 13$ when $x = \frac{1}{2}$

g) $\frac{\mathrm{d}y}{\mathrm{d}x} = \sqrt{x} - 5,\ x > 0;\ y = -18$ when $x = 9$

h) $\frac{\mathrm{d}y}{\mathrm{d}x} = x - \frac{1}{2}\sqrt{x},\ x > 0;\ y = \frac{2}{3}$ when $x = 4$

i) $\frac{\mathrm{d}y}{\mathrm{d}x} = x^2(3 - x),\ y = 2\frac{3}{4}$ when $x = -1$

j) $\frac{\mathrm{d}y}{\mathrm{d}x} = (2x - 3)^2,\ y = 13$ when $x = 3$

k) $\frac{\mathrm{d}y}{\mathrm{d}x} = \frac{\sqrt{x} - 1}{\sqrt{x}},\ x > 0;\ y = 5\frac{1}{4}$ when $x = \frac{1}{4}$

l) $\frac{\mathrm{d}y}{\mathrm{d}x} = \frac{x - 2}{x^3},\ x \neq 0;\ y = 3\frac{1}{2}$ when $x = -2$

11 The gradient of a curve at the point (x, y) on the curve is given by $\frac{\mathrm{d}y}{\mathrm{d}x} = 3x^2 + 4$. Given that the point $(1, 7)$ lies on the curve, determine the equation of the curve.

12 A curve passes through the point $(-2, 8)$ and its gradient function is $4x^3 - 6x$. Find the equation of the curve.

13 The gradient of a curve at the point (x, y) is $16x^3 + 2x + 1$. Given that the curve passes through the point $(\frac{1}{2}, 3)$ find the equation of the curve.

14 Find y as a function of x given that $\frac{\mathrm{d}y}{\mathrm{d}x} = \frac{5}{x^2} - 4$, $(x \neq 0)$, and that $y = -12$ when $x = 5$.

15 A function $\mathrm{f}(x)$ is such that $\mathrm{f}'(x) = 3\sqrt{x} - 5,\ x \in \mathbb{R},\ x \geqslant 0$. Given that $\mathrm{f}(4) = 3$, find an expression for $\mathrm{f}(x)$.

16 A curve has an equation which satisfies $\frac{\mathrm{d}^2y}{\mathrm{d}x^2} = 6x - 4$. The point $\mathrm{P}(2, 11)$ lies on the curve, and the gradient of the curve at the point P is 9. Determine the equation of the curve.

17 Find y as a function of x given that $\frac{\mathrm{d}^2y}{\mathrm{d}x^2} = 12x^2 - 6$, and that when $x = 1$, $\frac{\mathrm{d}y}{\mathrm{d}x} = -2$ and $y = 1$.

18 Given that $\frac{\mathrm{d}^2y}{\mathrm{d}x^2} = 6x + \frac{4}{x^3}$, $(x \neq 0)$, and that $y = 1$ when $x = 1$, and that $y = 5$ when $x = 2$, find an expression for y in terms of x.

***19** The curve with equation $y = ax^2 + bx + c$ passes through the points $\mathrm{P}(2, 6)$ and $\mathrm{Q}(3, 16)$, and has a gradient of 7 at the point P. Find the values of the constants a, b and c.

Area under a curve

Consider the area A under a curve $f(x)$ as shown in the diagram.

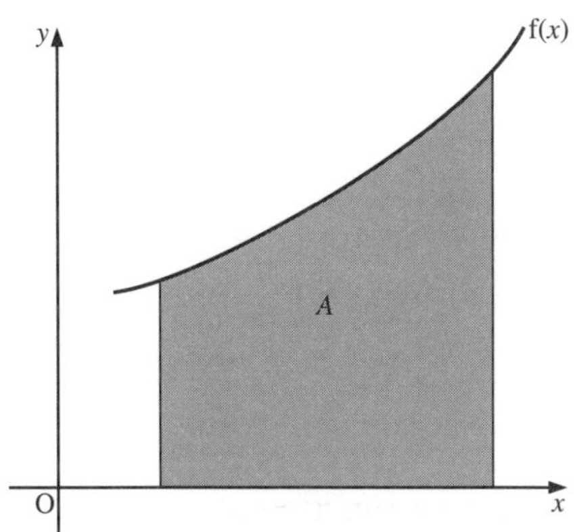

The area A can be approximated by splitting the shaded region into rectangles, and summing the areas of these rectangles. There are two cases to consider:

i)

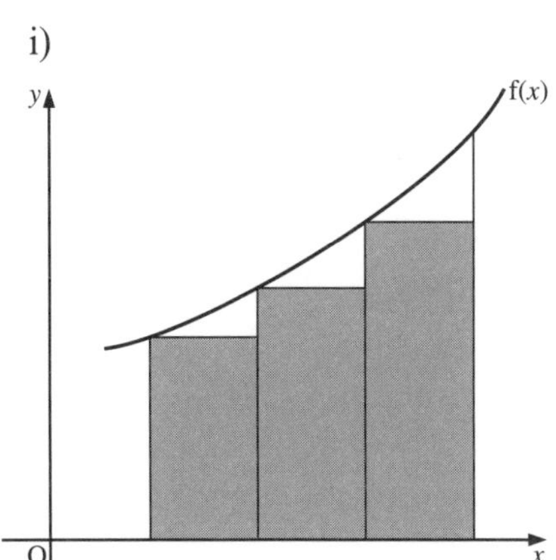

ii)

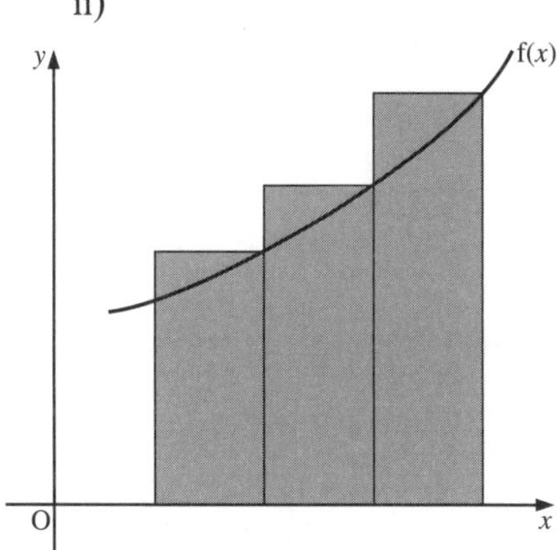

In case (i) the approximation is less than A, whereas in case (ii) the approximation is greater than A. In both cases, as each rectangle is made narrower (i.e. more rectangles are used) the approximation approaches a limiting value, namely A.

Consider one such rectangle of width δx. Let δA be the shaded area.

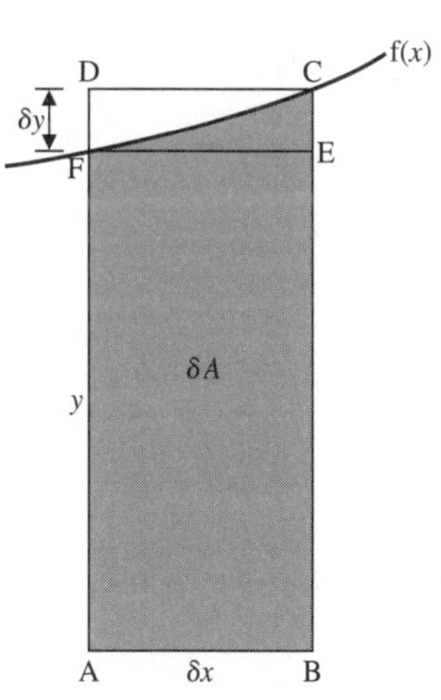

The area of rectangle ABEF is $y\delta x$.

The area of rectangle ABCD is $(y + \delta y)\delta x$.

Therefore, we have

$$y\delta x < \delta A < (y + \delta y)\delta x$$

Since $\delta x > 0$, we can divide throughout by δx, giving

$$y < \frac{\delta A}{\delta x} < y + \delta y$$

Letting $\delta x \to 0$ (i.e. increasing the number of rectangles) gives $\dfrac{\delta A}{\delta x} \to \dfrac{dA}{dx}$ and $\delta y \to 0$. Therefore,

$$\frac{dA}{dx} = y$$

Integrating each side with respect to x gives

$$\int \frac{dA}{dx}\,dx = \int y\,dx$$

$$\therefore \quad A = \int y\,dx$$

This expression for the area will not give a definite value but give a function of x. This is because $\int y\,dx$ gives the area measured from an arbitrary origin to the point x. Therefore,

$$A = (\text{Area up to the ordinate } x = b) - (\text{Area up to the ordinate } x = a)$$

$$= A(b) - A(a)$$

We write this as

$$A = \int_a^b y\,dx \quad \text{or} \quad A = \int_a^b f(x)\,dx$$

We call $\displaystyle\int_a^b f(x)\,dx$ a **definite integral** since it gives a definite answer.

- The dx indicates that the limits a and b are x limits.
- The constant a is called the **lower limit** of the integral.
- The constant b is called the **upper limit** of the integral.

For example, to evaluate the definite integral $\displaystyle\int_0^1 2x\,dx$, we first integrate to obtain

$$\int_0^1 2x\,dx = \underbrace{\left[x^2 + c\right]_0^1}_{\text{We use square brackets here}}$$

Substituting the values $x = 1$ and $x = 0$ gives

$$\int_0^1 2x\,dx = [(1)^2 + c] - [(0)^2 + c] = 1 - 0$$

$$\therefore \quad \int_0^1 2x\,dx = 1$$

Notice that the constants of integration cancel and it is for this reason that we exclude the constants of integration when working with definite integrals.

Example 6 Evaluate the following definite integrals.

a) $\displaystyle\int_0^2 4x^3\,dx$ **b)** $\displaystyle\int_{-1}^1 (3x^2-5)\,dx$

c) $\displaystyle\int_{-3}^{-2} \frac{1}{x^2}\,dx$ **d)** $\displaystyle\int_2^8 \left(x-\frac{3}{\sqrt{x}}\right)dx$

SOLUTION

a)

$$\int_0^2 4x^3\,dx = \left[x^4\right]_0^2$$
$$= (2)^4 - (0)^4$$
$$\therefore \quad \int_0^2 4x^3\,dx = 16$$

b)

$$\int_{-1}^1 (3x^2-5)\,dx = \left[x^3-5x\right]_{-1}^1$$
$$= (1-5) - (-1+5)$$
$$= -4-4$$
$$\therefore \quad \int_{-1}^1 (3x^2-5)\,dx = -8$$

c)

$$\int_{-3}^{-2} \frac{1}{x^2}\,dx = \int_{-3}^{-2} x^{-2}\,dx$$
$$= \left[-x^{-1}\right]_{-3}^{-2}$$
$$= \frac{1}{2} - \frac{1}{3}$$
$$\therefore \quad \int_{-3}^{-2} \frac{1}{x^2}\,dx = \frac{1}{6}$$

d)

$$\int_2^8 \left(x-\frac{3}{\sqrt{x}}\right)dx = \int_2^8 (x-3x^{-\frac{1}{2}})\,dx$$
$$= \left[\frac{x^2}{2} - 6x^{\frac{1}{2}}\right]_2^8$$
$$= (32-6\sqrt{8}) - (2-6\sqrt{2})$$
$$= (32-12\sqrt{2}) - (2-6\sqrt{2})$$
$$\therefore \quad \int_2^8 \left(x-\frac{3}{\sqrt{x}}\right)dx = 30-6\sqrt{2}$$

Example 7 Find the area under the curve $y = x^2$ between $x = 1$ and $x = 3$.

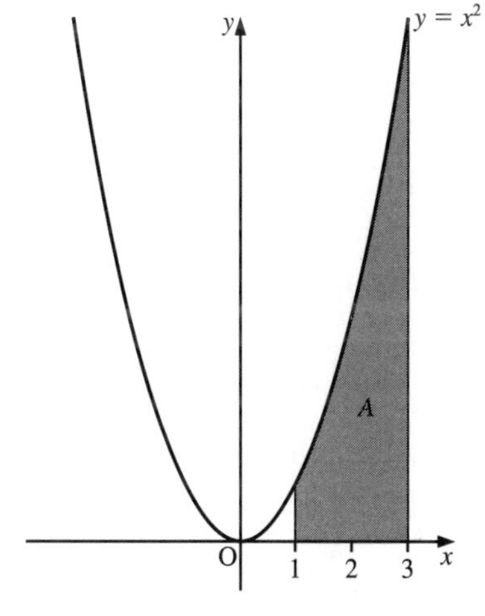

SOLUTION

Let A be the required area, then

$$A = \int_1^3 x^2\,dx$$
$$= \left[\frac{x^3}{3}\right]_1^3 = 9 - \frac{1}{3}$$
$$\therefore \quad A = \frac{26}{3}$$

The required area is $\frac{26}{3}$ or $8\frac{2}{3}$.

Example 8 Find the area between the curve $y = x^2 + 4x$ and the x-axis from

a) $x = -2$ to $x = 0$ **b)** $x = -2$ to $x = 2$

SOLUTION

a) The required area is shown shaded on the right. Notice that it lies below the x-axis.

Let A be the required area, then

$$A = \int_{-2}^{0} (x^2 + 4x)\,dx$$

$$= \left[\frac{x^3}{3} + 2x^2\right]_{-2}^{0}$$

$$= 0 - \frac{16}{3}$$

$$\therefore \quad A = -\left(-\frac{8}{3} + 8\right)$$

The minus sign tells us that the area is below the x-axis. Therefore, the required area is $\frac{16}{3}$.

b) In this case, we see that the required area comprises two parts, one part below the x-axis and one part above the x-axis.

Evaluating $\int_{-2}^{2} (x^2 + 4x)\,dx$ gives

$$\int_{-2}^{2} (x^2 + 4x)\,dx = \left[\frac{x^3}{3} + 2x^2\right]_{-2}^{2}$$

$$= \left(\frac{8}{3} + 8\right) - \left(-\frac{8}{3} + 8\right)$$

$$= \frac{32}{3} - \frac{16}{3} = \frac{16}{3}$$

It is obvious that this cannot be the total shaded area, since we know from part **a** that the area between the curve and the x-axis from $x = -2$ to $x = 0$ is $\frac{16}{3}$.

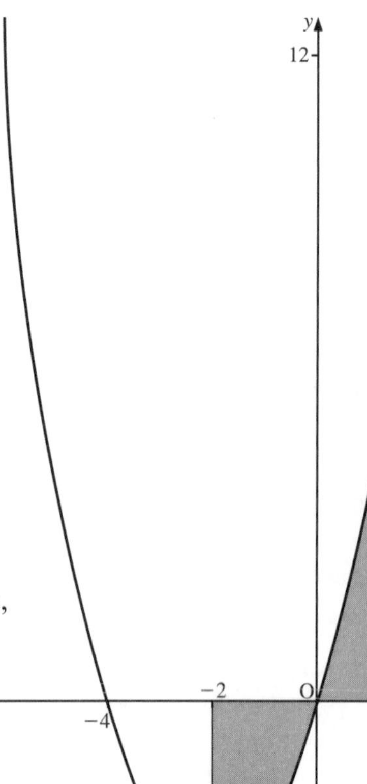

Calculating each of the integrals $\int_{-2}^{0} (x^2 + 4x)\,dx$ and $\int_{0}^{2} (x^2 + 4x)\,dx$ will explain the mystery.

We know that

$$\int_{-2}^{0} (x^2 + 4x)\,dx = -\frac{16}{3}$$

and

$$\int_{0}^{2} (x^2 + 4x)\,dx = \left[\frac{x^3}{3} + 2x^2\right]_0^2 = \left(\frac{8}{3} + 8\right) - (0) = \frac{32}{3}$$

Therefore, the required area, A, is given by

$$A = \frac{16}{3} + \frac{32}{3} = 16$$

Notice that

$$\int_{-2}^{0} (x^2 + 4x)\,dx + \int_{0}^{2} (x^2 + 4x)\,dx = -\frac{16}{3} + \frac{32}{3} = \frac{16}{3}$$

$$\therefore \quad \int_{-2}^{0} (x^2 + 4x)\,dx + \int_{0}^{2} (x^2 + 4x)\,dx = \int_{-2}^{2} (x^2 + 4x)\,dx$$

This example illustrates the importance of a sketch in order to identify whether part of the required area lies below the x-axis or not.

Example 9 Shown is a sketch of the curve given by $y = x^3 - 4x^2 + 3x$. Find the area between the curve and the x-axis from $x = 0$ to $x = 3$.

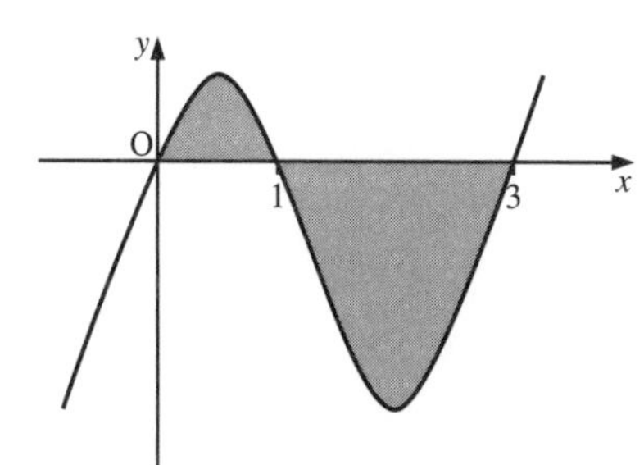

SOLUTION

The required area comprises two parts, A_1 and A_2, as shown on the right.

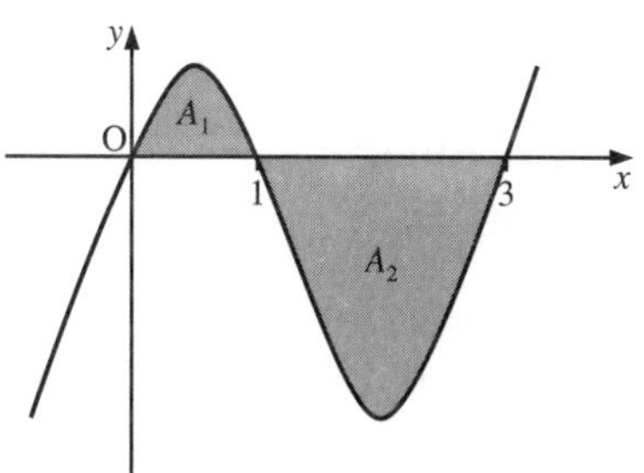

Calculating A_1 gives

$$\int_{0}^{1} (x^3 - 4x^2 + 3x)\,dx = \left[\frac{x^4}{4} - \frac{4x^3}{3} + \frac{3x^2}{2}\right]_0^1$$

$$= \left(\frac{1}{4} - \frac{4}{3} + \frac{3}{2}\right) - (0)$$

$$\therefore \quad A_1 = \frac{5}{12}$$

Calculating A_2 gives

$$\int_{1}^{3} (x^3 - 4x^2 + 3x)\,dx = \left[\frac{x^4}{4} - \frac{4x^3}{3} + \frac{3x^2}{2}\right]_1^3 = \left(\frac{81}{4} - 36 + \frac{27}{2}\right) - \left(\frac{1}{4} - \frac{4}{3} + \frac{3}{2}\right)$$

$$\therefore \quad \int_{1}^{3} (x^3 - 4x^2 + 3x)\,dx = -\frac{9}{4} - \frac{5}{12} = -\frac{8}{3}$$

$$\therefore \quad A_2 = \frac{8}{3}$$

The required area A is given by

$$A = A_1 + A_2 = \frac{5}{12} + \frac{8}{3} = \frac{37}{12}$$

Area between a curve and the y-axis

Consider the area shown below, which is bounded by the curve $y = \mathrm{f}(x)$ and the y-axis between $y_1 = \mathrm{f}(a)$ and $y_2 = \mathrm{f}(b)$.

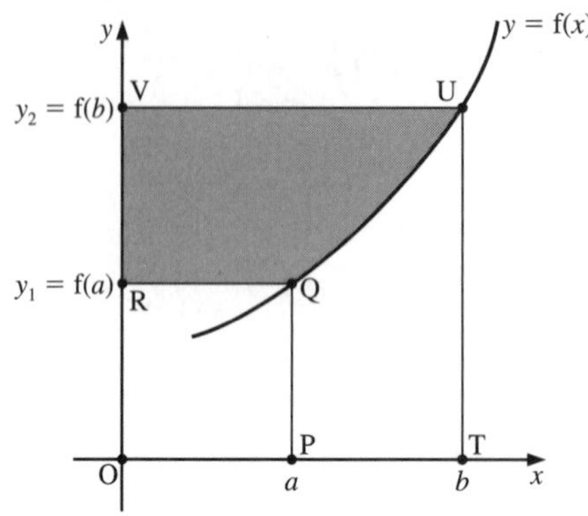

This area can be found in two ways. In the first, using rectangles, the required area, A, is given by

$$A = (\text{Area of OTUV}) - (\text{Area of OPQR}) - (\text{Area under } \mathrm{f}(x) \text{ between } x = a \text{ and } x = b)$$

Therefore,

$$A = b\,\mathrm{f}(b) - a\,\mathrm{f}(a) - \int_a^b \mathrm{f}(x)\,\mathrm{d}x$$

In the second method, the required area, A, is given by

$$\int_{\mathrm{f}(a)}^{\mathrm{f}(b)} x\,\mathrm{d}y$$

In this case, x must be expressed as a function of y before we can integrate.

Example 10 Find the area between the curve $y = x^2$ and the y-axis between $y = 1$ and $y = 4$.

SOLUTION

Using the first method, we see that the required area, A, is given by

$$A = (4 \times 2) - (1 \times 1) - \int_1^2 x^2\,\mathrm{d}x$$

Note The limits 1 and 2 come from the fact that when $x = 1$, $y = 1$ and when $x = 2$, $y = 4$.

Now

$$\int_1^2 x^2\,\mathrm{d}x = \left[\frac{x^3}{3}\right]_1^2 = \left(\frac{8}{3}\right) - \left(\frac{1}{3}\right)$$

$$\therefore \quad \int_1^2 x^2\,\mathrm{d}x = \frac{7}{3}$$

Therefore, we have

$$A = 8 - 1 - \frac{7}{3} = \frac{14}{3}$$

The required area is $\frac{14}{3}$.

Using the second method gives

$$A = \int_1^4 x\,dy$$

$$= \int_1^4 y^{\frac{1}{2}}\,dy$$

$$= \left[\frac{2y^{\frac{3}{2}}}{3}\right]_1^4 = \frac{2(4)^{\frac{3}{2}}}{3} - \frac{2}{3}$$

$$\therefore \quad A = \frac{16}{3} - \frac{2}{3} = \frac{14}{3}$$

Area between two curves

Consider two intersecting curves f(x) and g(x) as shown on the right.

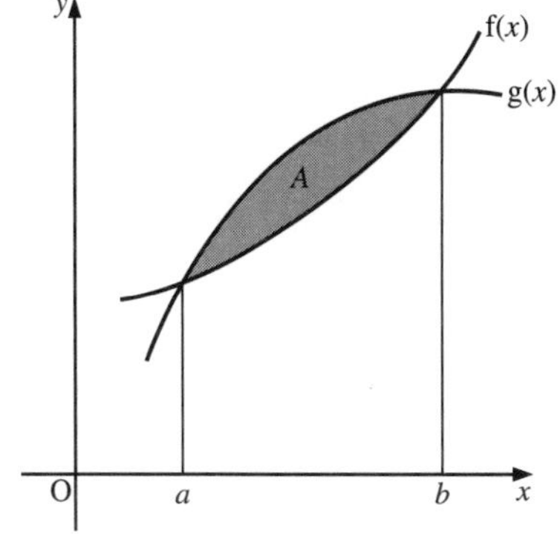

The shaded area, A, between the two curves is given by

$$A = \int_a^b \mathrm{g}(x)\,dx - \int_a^b \mathrm{f}(x)\,dx$$

$$\therefore \quad A = \int_a^b (\mathrm{g}(x) - \mathrm{f}(x))\,dx$$

Example 11 Find the area enclosed between the curves $y = x^2 + 2x + 2$ and $y = -x^2 + 2x + 10$.

SOLUTION

We must first find the points of intersection of the two curves. The x coordinates of the points of intersection satisfy the equation

$$x^2 + 2x + 2 = -x^2 + 2x + 10$$

Simplifying gives

$$2x^2 - 8 = 0$$

$$\therefore \quad x^2 = 4$$

$$\therefore \quad x = \pm 2$$

The sketch on the right shows the two curves.

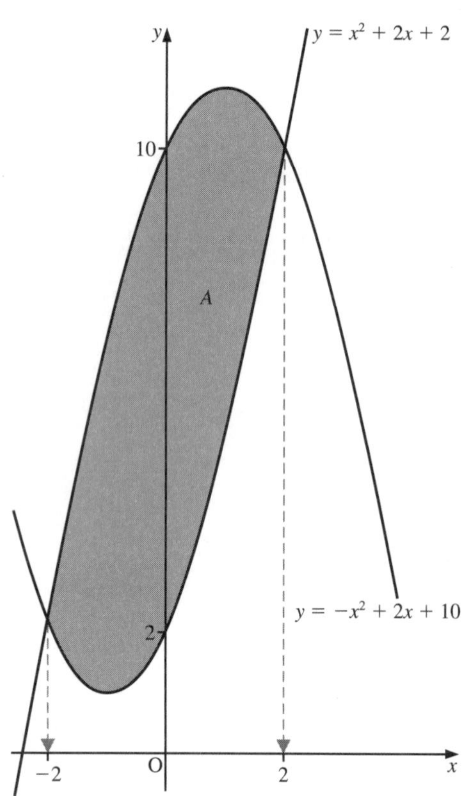

The shaded area A is given by

$$A = \int_{-2}^{2} (-x^2 + 2x + 10)\,dx - \int_{-2}^{2} (x^2 + 2x + 2)\,dx$$

$$= \int_{-2}^{2} [(-x^2 + 2x + 10) - (x^2 + 2x + 2)]\,dx$$

$$= \int_{-2}^{2} (-2x^2 + 8)\,dx$$

$$= \left[-\frac{2x^3}{3} + 8x\right]_{-2}^{2} = \left(-\frac{16}{3} + 16\right) - \left(\frac{16}{3} - 16\right)$$

$$\therefore \quad A = \frac{32}{3} + \frac{32}{3} = \frac{64}{3}$$

The area enclosed between the two curves is $\frac{64}{3}$.

Example 12 Find the area enclosed between the curve $y = x^2 - 2x - 3$ and the line $y = x + 1$.

SOLUTION

We must first find the coordinates of the points of intersection. The x coordinates of the points of intersection satisfy the equation

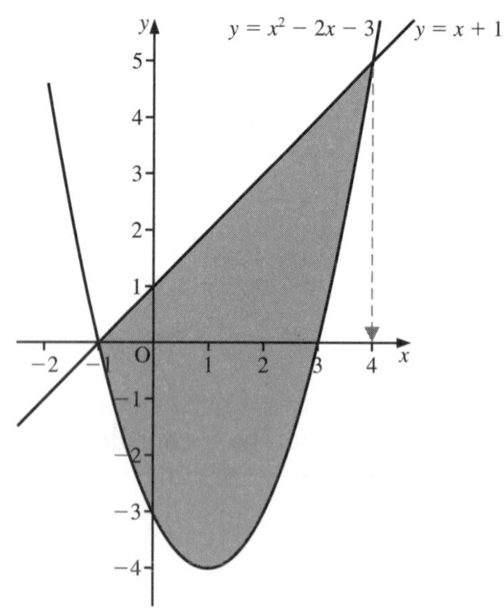

$$x^2 - 2x - 3 = x + 1$$

Simplifying and factorising give

$$x^2 - 3x - 4 = 0$$

$$\therefore \quad (x - 4)(x + 1) = 0$$

Solving gives $x = 4$ or $x = -1$. Sketching both on the same set of axes gives the diagram on the right.

Notice that part of the required area lies below the x-axis. There are two ways of finding this required area. First, we can consider the two parts A_1 and A_2 together with triangle PQR, as shown at the top of page 203.

The area A_1 is given by

$$-A_1 = \int_{-1}^{3} (x^2 - 2x - 3)\,dx = \left[\frac{x^3}{3} - x^2 - 3x\right]_{-1}^{3} = (9 - 9 - 9) - \left(\frac{1}{3} - 1 + 3\right)$$

$$= -9 - \frac{5}{3}$$

Therefore,

$$A_1 = \frac{32}{3}$$

Calculating A_2 gives

$$A_2 = \int_3^4 (x^2 - 2x - 3)\,dx = \left[\frac{x^3}{3} - x^2 - 3x\right]_3^4$$

$$= -\frac{20}{3} - (-9)$$

$$\therefore \quad A_2 = \frac{7}{3}$$

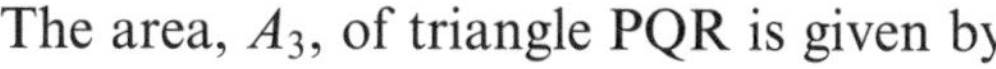

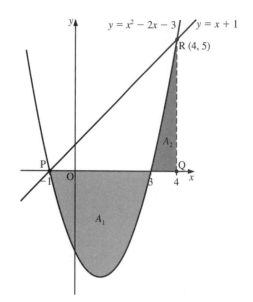

The area, A_3, of triangle PQR is given by

$$A_3 = \frac{1}{2}(5)(5) = \frac{25}{2}$$

The required area A is given by

$$A = A_1 + (A_3 - A_2)$$

$$= \frac{32}{3} + \left(\frac{25}{2} - \frac{7}{3}\right) = \frac{125}{6}$$

The area between the curve and the line is $\frac{125}{6} = 20\frac{5}{6}$.

The second method for finding the area between the two curves uses the result

$$A = \int_a^b [(g(x) - f(x)]\,dx$$

This gives

$$A = \int_{-1}^4 (x + 1)\,dx - \int_{-1}^4 (x^2 - 2x - 3)\,dx$$

$$= \int_{-1}^4 [(x + 1) - (x^2 - 2x - 3)]\,dx$$

$$= \int_{-1}^4 (-x^2 + 3x + 4)\,dx$$

$$= \left[-\frac{x^3}{3} + \frac{3x^2}{2} + 4x\right]_{-1}^4 = \left(-\frac{64}{3} + 24 + 16\right) - \left(\frac{1}{3} + \frac{3}{2} - 4\right)$$

$$= \frac{56}{3} - \left(-\frac{13}{6}\right)$$

$$\therefore \quad A = \frac{125}{6}$$

as before.

Exercise 7B

1 Work out each of these definite integrals.

a) $\int_0^2 x^2\,dx$ **b)** $\int_0^3 4x^3\,dx$ **c)** $\int_1^4 6x\,dx$ **d)** $\int_2^3 (6x^2 - 1)\,dx$

e) $\int_4^5 (4x + 3)\,dx$ **f)** $\int_2^3 (4 - 3x^2)\,dx$ **g)** $\int_2^8 \frac{1}{x^2}\,dx$ **h)** $\int_1^2 \frac{4}{x^3}\,dx$

i) $\int_4^9 \sqrt{x}\,dx$ **j)** $\int_1^4 \left(3 - \frac{1}{\sqrt{x}}\right)dx$ **k)** $\int_{\frac{1}{2}}^1 1 + \frac{1}{x^2}\,dx$ **l)** $\int_1^8 \sqrt[3]{x}\,dx$

2 Evaluate each of the following.

a) $\int_{-1}^3 4x\,dx$ **b)** $\int_{-2}^3 6x^2\,dx$ **c)** $\int_{-3}^{-1} 2x^3\,dx$ **d)** $\int_{-2}^{-1} \frac{2}{x^3}\,dx$

e) $\int_{-4}^5 (x - 1)\,dx$ **f)** $\int_{-1}^2 (2x - x^5)\,dx$ **g)** $\int_{-3}^3 (x^3 + x)\,dx$ **h)** $\int_{-4}^{-1} \frac{16}{x^5}\,dx$

i) $\int_1^5 \left(\frac{x^3 - 1}{x^2}\right)dx$ **j)** $\int_1^{16} \left(\frac{\sqrt{x} - 4}{\sqrt{x}}\right)dx$ **k)** $\int_{-1}^1 x^3(2x - 1)\,dx$ **l)** $\int_{-2}^4 (x - 2)^2\,dx$

3 Work out the shaded area on each of these diagrams.

a)

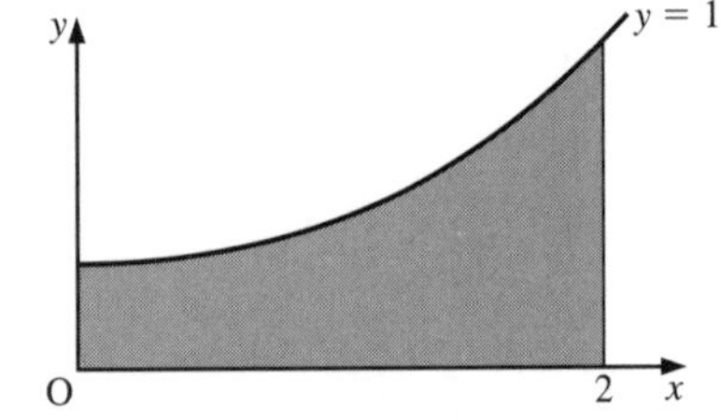

b)

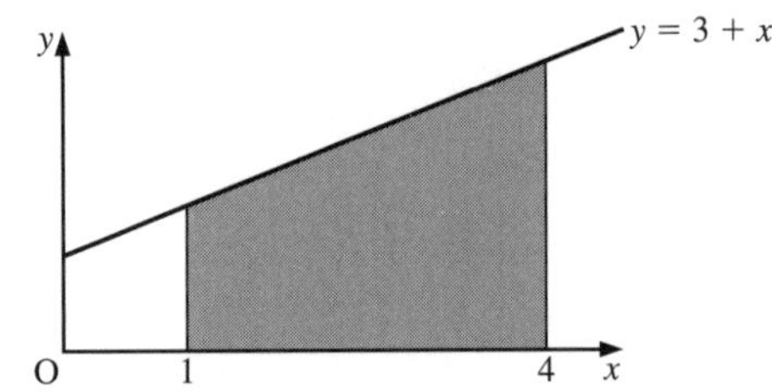

c)

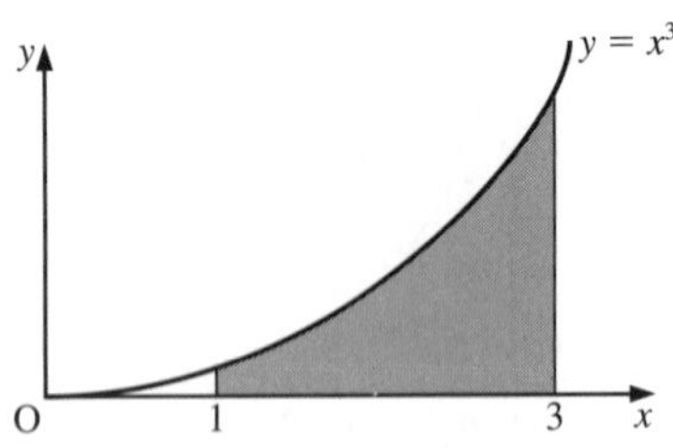

d)

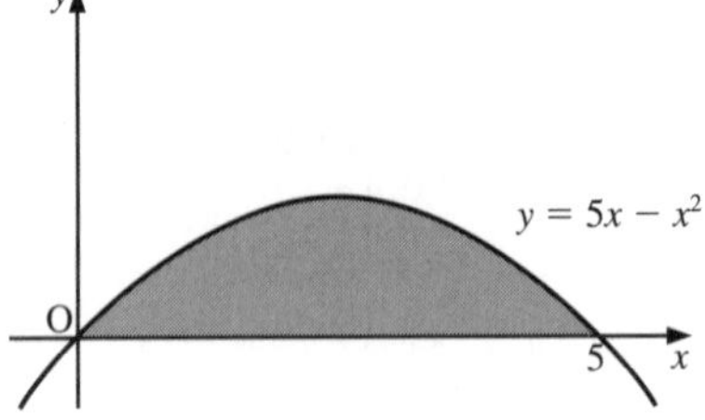

e)

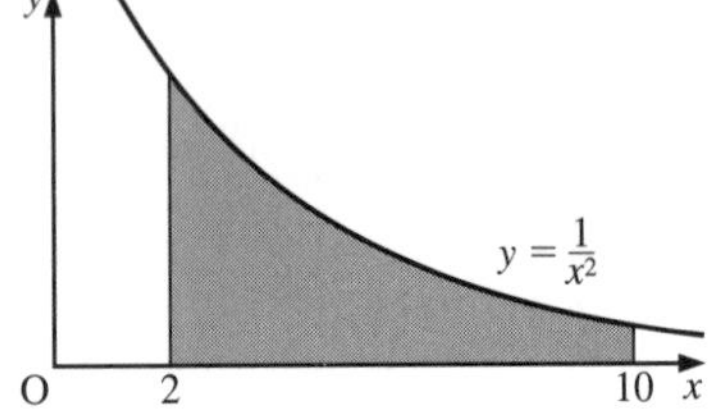

f)

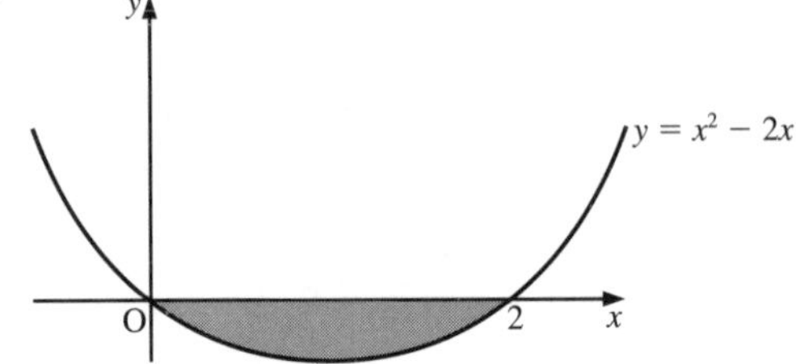

g)

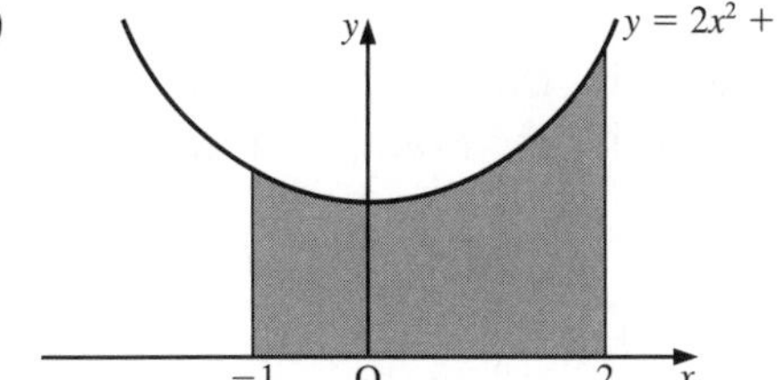

h)

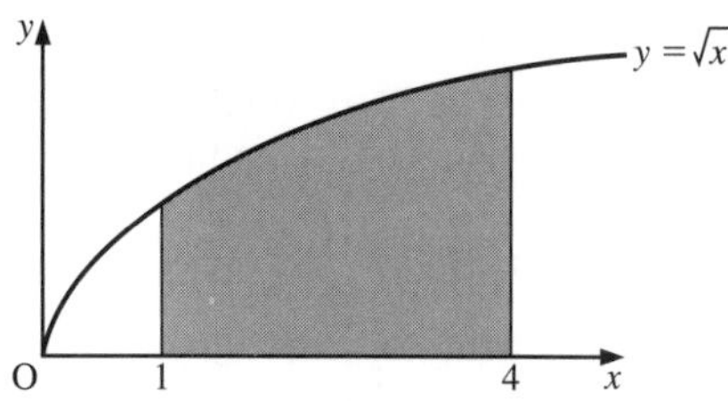

i)

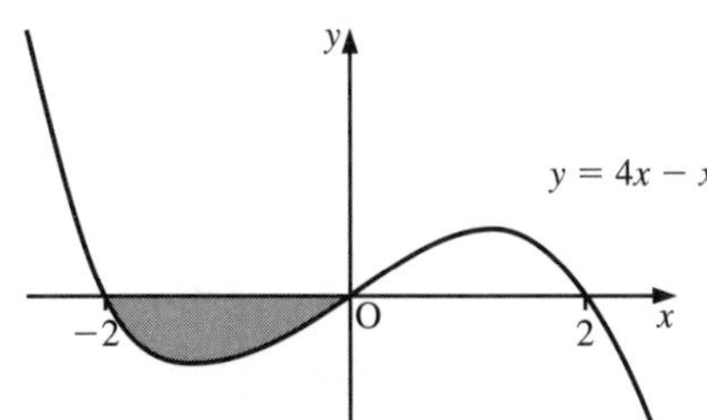

j)

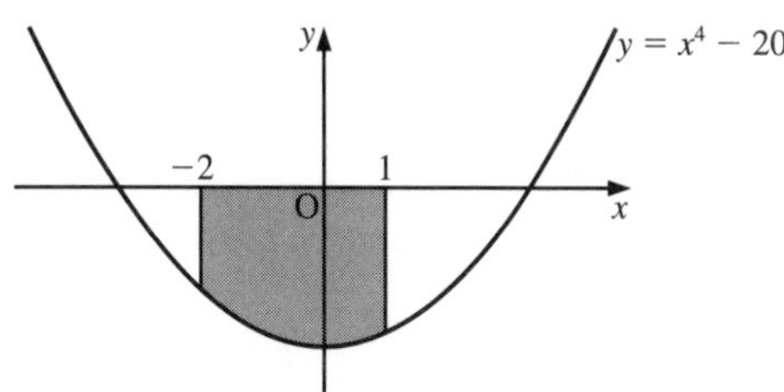

k)

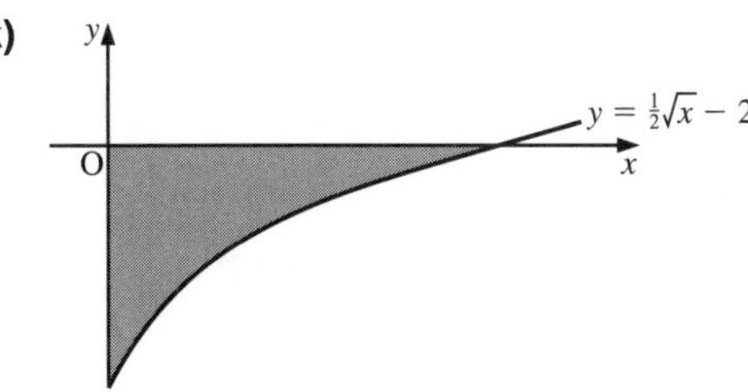

l)

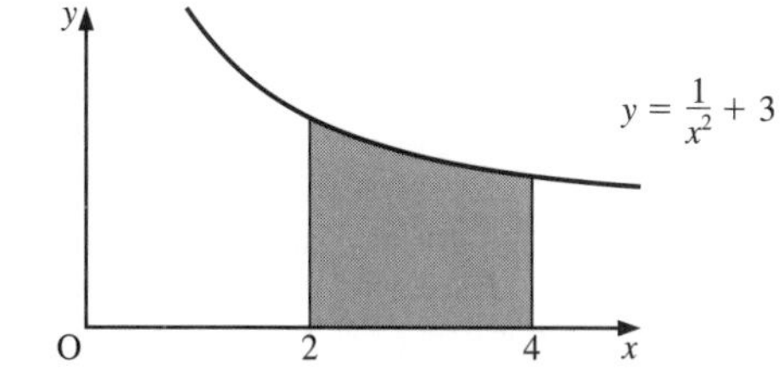

4 Find the areas enclosed by the x-axis and the following curves and straight lines.

a) $y = x^2 + 3x,\ x = 2,\ x = 5$
b) $y = \frac{1}{8}x^3 + 2x,\ x = 2,\ x = 4$
c) $y = 2 - x^3,\ x = -3,\ x = -2$
d) $x = \dfrac{4}{x^3},\ x = \frac{1}{4},\ x = \frac{1}{2}$
e) $y = 6 - \dfrac{1}{\sqrt{x}},\ x = 16,\ x = 25$
f) $y = (3x - 4)^2,\ x = 1,\ x = 3$

5 Sketch the graph of the region bounded by the curve $y = x^3 - 5$, the lines $x = 2$ and $x = 4$, and the x-axis. Find the area of the region.

6 Find the area enclosed above the x-axis and below the curve $y = 16 - x^2$.

7 Sketch the curve $y = (x - 2)(x - 3)$, showing where it crosses the x-axis. Hence find the area enclosed below the x-axis and above the curve.

8 Sketch the curve with equation $y = (x - 2)^2$. Calculate the area of the region bounded by the curve and the x- and y-axes.

9 Sketch the curve $y = 3x^2 - x^3$. Hence find the area of the region bounded by the curve and the x-axis.

10 Sketch the graph of the function $f(x) = \sqrt{x} - 3$ for $x > 0$. Calculate the area of the region bounded by the curve and the x- and y-axes.

11 Calculate the area of the region bounded by the x-axis and the function $f: x \rightarrow 5 + \dfrac{1}{x^2},\ x \in \mathbb{R}$, $2 \leqslant x \leqslant 8$.

12 a) Sketch the curve $y = x(x+1)(x-3)$, showing where it cuts the x-axis.
b) Calculate the area of the region, above the x-axis, bounded by the x-axis and the curve.
c) Calculate the area of the region, below the x-axis, bounded by the x-axis and the curve.

13 a) Sketch the curve $y = x^2(x-1)(x+2)$.
b) Calculate the area of the region bounded by the positive x-axis and the curve.
c) Calculate the area of the region bounded by the negative x-axis and the curve.

14 Use $\int x\,dy$ to work out the shaded area in each of these diagrams.

a)

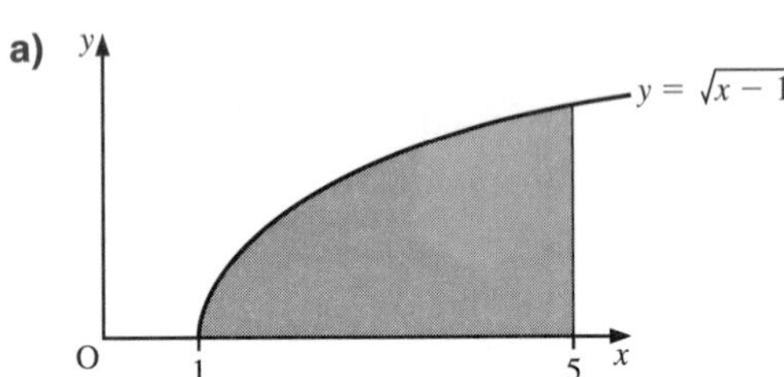

b)

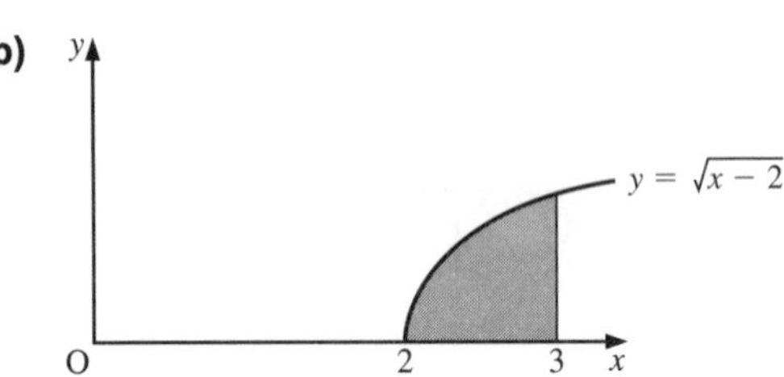

c)

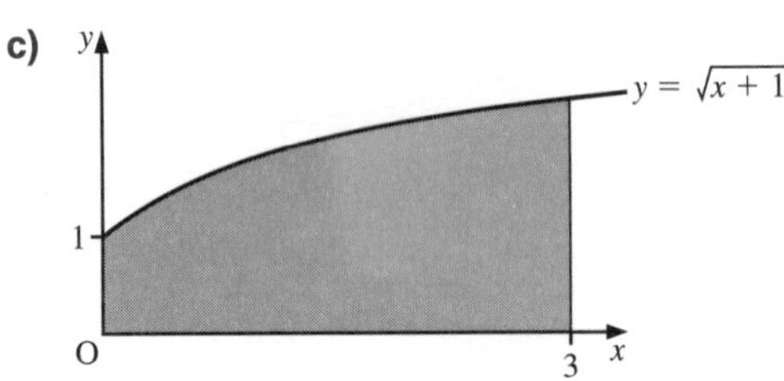

d)

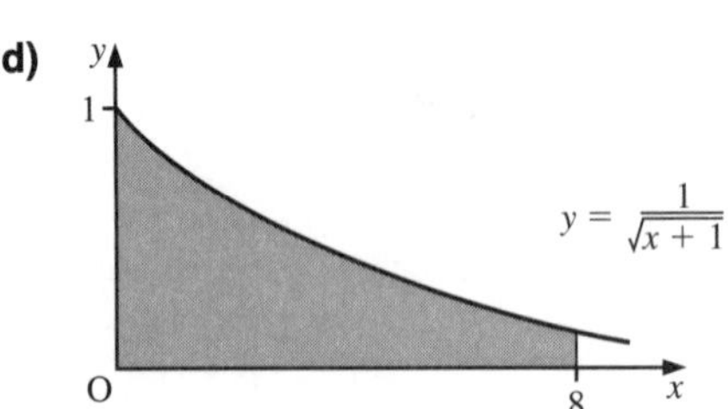

e)

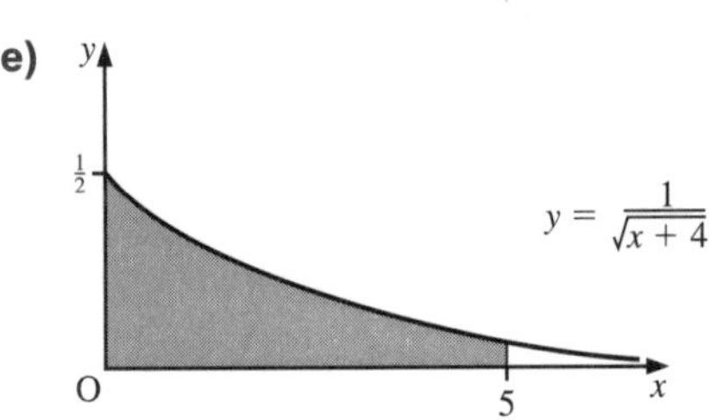

f)

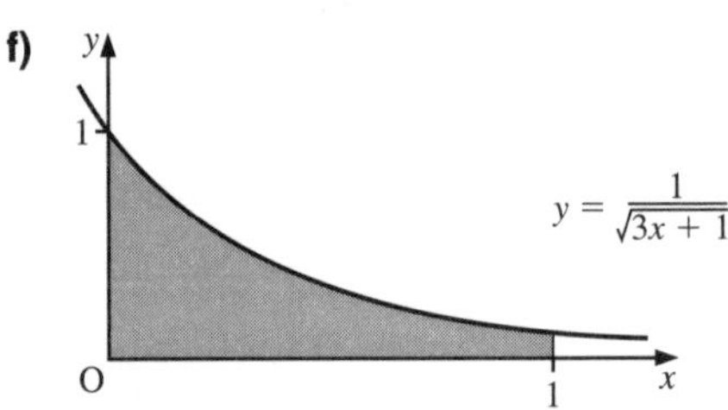

15 The line $y = 3x + 1$ meets the curve $y = x^2 + 3$ at the points P and Q.

a) Calculate the coordinates of P and Q.
b) Sketch the line and the curve on the same set of axes.
c) Calculate the area of the finite region bounded by the line and the curve.

16 The curve $y = x^2 - 2x + 3$ meets the line $y = 9 - x$ at the points A and B.

a) Find the coordinates of A and B.
b) Sketch the line and the curve on the same set of axes.
c) Calculate the area of the finite region bounded by the line and the curve.

17 The curve $y = x^2 + 16$ meets the curve $y = x(12 - x)$ at the points C and D.

a) Find the coordinates of C and D.
b) Sketch the two curves on the same set of axes.
c) Calculate the area bounded by the two curves.

18 a) Sketch, on the same diagram, the curves $y = x^2 - 5x$ and $y = 3 - x^2$, and find their points of intersection.
b) Find the area of the region bounded by the two curves.

19 **a)** On the same diagram sketch the graphs of the line $y = \frac{1}{3}x$ and the curve $y = \sqrt{x}$ for positive values of x, and find the coordinates of their points of intersection.
b) Find the area of the region bounded by the line and the curve.

20 Find the area enclosed between the curves $y = 2x^2 - 7$ and $y = 5 - x^2$.

21 Find the area enclosed between the curves $y = (x-1)^2$ and $y = 8 - (x-1)^2$.

22 Calculate the area bounded by the y-axis, the line $y = 8$ and the curve $y = x^3$.

23 Calculate the area bounded by the curve $y = \dfrac{16}{x^2}$, the line $x = 4$ and the line $y = 16$.

24 The curve $y = x^2 - 2x$ cuts the x-axis at the points O and P, and meets the line $y = 2x$ at the point Q, as in the diagram on the right.

a) Calculate the coordinates of P and Q.
b) Find the area of the shaded region.

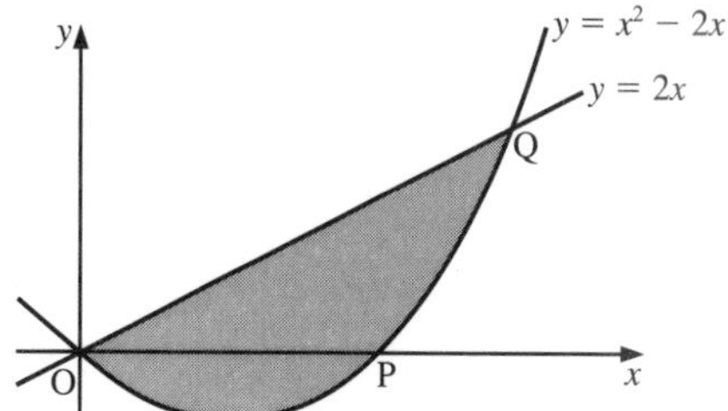

25 The curve $y = 3x - x^2$ cuts the x-axis at the points O and A, and meets the line $y = -3x$ at the point B, as in the diagram on the right.

a) Calculate the coordinate of A and B.
b) Find the area of the shaded region.

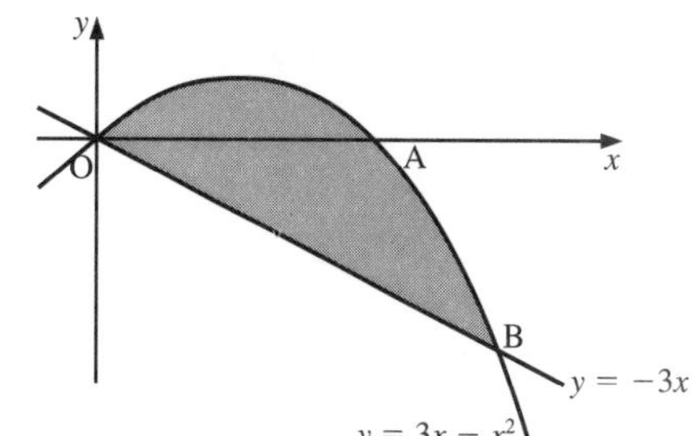

26 The curve $y = x^2 - 1$ cuts the x-axis at the points P and Q, and meets the line $y = x + 1$ at the points P and R, as in the diagram on the right.

a) Calculate the coordinates of P, Q and R.
b) Find the area of the shaded region.

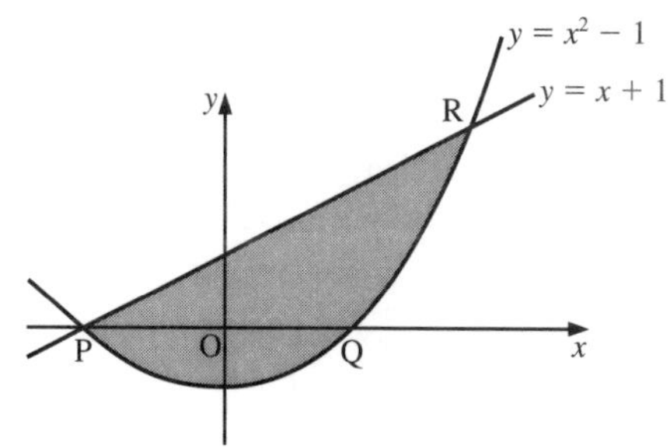

***27** **a)** Given f is an odd function, g is an even function and a is a constant, explain the following results.

$$\int_{-a}^{a} f(x)\,dx = 0 \qquad \int_{-a}^{a} g(x)\,dx = 2\int_{0}^{a} g(x)\,dx$$

b) Deduce the value of each of the following.

i) $\displaystyle\int_{-2}^{2} x^3(1 + x^2)\,dx$ **ii)** $\displaystyle\int_{-2}^{2} x^3(1 + x)^2\,dx$ **iii)** $\displaystyle\int_{-2}^{2} x^2(1 + x^3)\,dx$ **iv)** $\displaystyle\int_{-2}^{2} x^2(1 + x)^3\,dx$

***28 a)** Sketch the curve of $y = f(x)$, where $f(x) = x(x+1)(x-2)$.

b) Hence evaluate the following.

i) $\int_0^2 f(x)\,dx$ **ii)** $\int_{-1}^0 f(x)\,dx$ **iii)** $\int_{-1}^2 f(x)\,dx$ **iv)** $\int_{-1}^2 |f(x)|\,dx$ **v)** $\int_{-2}^2 f(|x|)\,dx$

Volume of revolution about the x-axis

Consider the area under the curve $y = x^2$ between $x = 1$ and $x = 2$, as shown on the right.

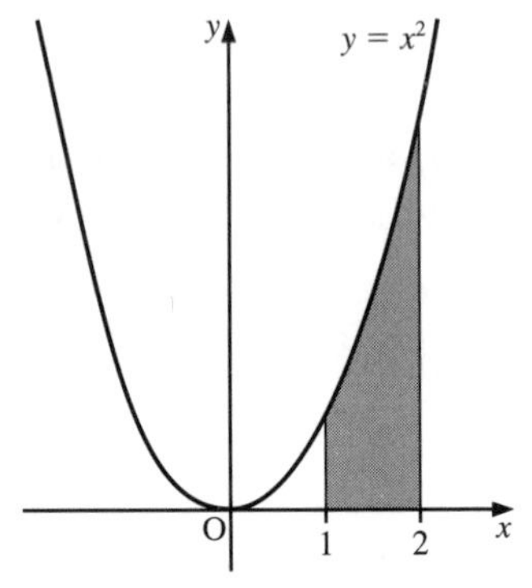

Now consider the solid formed when this area is rotated through 2π radians about the x-axis (centre right). The volume of this solid can be calculated using calculus methods.

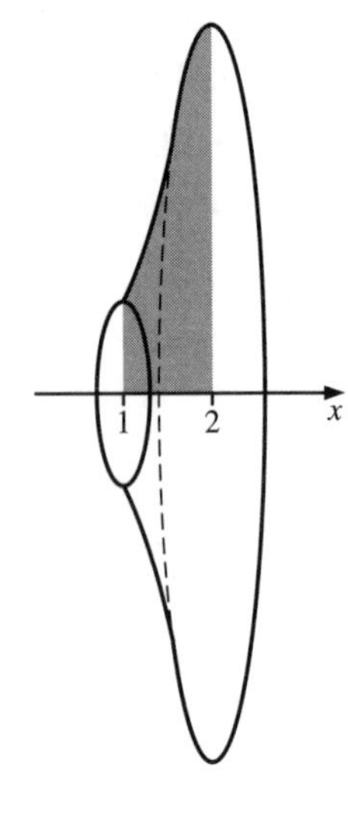

Consider a small strip of width δx under the curve f(x), as shown bottom right. When this small area is rotated through 2π radians about the x-axis, a disc is formed of radius y and thickness δx. The volume, δV, of the disc is given by

$$\delta V = \pi y^2 \delta x$$

To find the volume, V, of the total solid, we must find the sum of all such discs from $x = a$ to $x = b$. Therefore,

$$V = \sum_{x=a}^{b} \pi y^2 \,\delta x$$

As $\delta x \to 0$, this summation approaches a limiting value, namely V. Therefore,

$$V = \lim_{\delta x \to 0} \pi \sum_{x=a}^{b} y^2 \,\delta x$$

which gives

$$V = \pi \int_a^b y^2 \, dx$$

The volume, V, of the solid of revolution is given by

$$\pi \int_a^b y^2 \, dx$$

where $y = f(x)$.

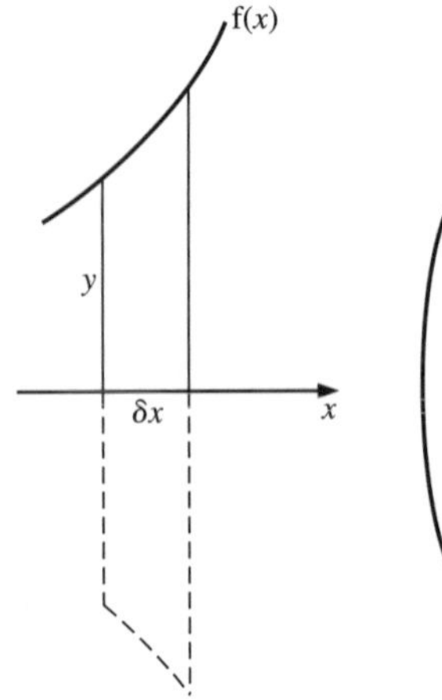

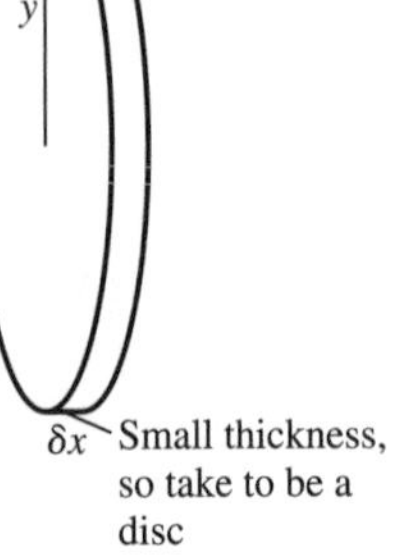

Example 13 Find the volume of the solid formed when the area between the curve $y = x^2 + 2$ and the x-axis from $x = 1$ to $x = 3$ is rotated through 2π radians about the x-axis.

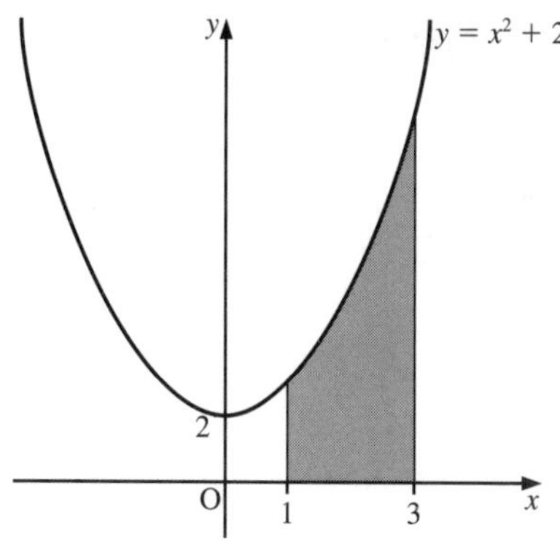

SOLUTION

The volume V is given by

$$V = \pi \int_1^3 y^2 \, dx$$

Now $y^2 = (x^2 + 2)^2 = x^4 + 4x^2 + 4$. Therefore,

$$V = \pi \int_1^3 (x^4 + 4x^2 + 4) \, dx$$

$$= \pi \left[\frac{x^5}{5} + \frac{4x^3}{3} + 4x \right]_1^3$$

$$= \pi \left(\frac{483}{5} - \frac{83}{15} \right)$$

$$\therefore \quad V = \frac{1366\pi}{15}$$

The volume of the solid formed is $\dfrac{1366\pi}{15}$.

Example 14 The area enclosed between the curve $y = 4 - x^2$ and the line $y = 4 - 2x$ is rotated through 2π radians about the x-axis. Find the volume of the solid generated.

SOLUTION

The sketch of both the curve and the line on the same set of axes shows the area to be rotated.

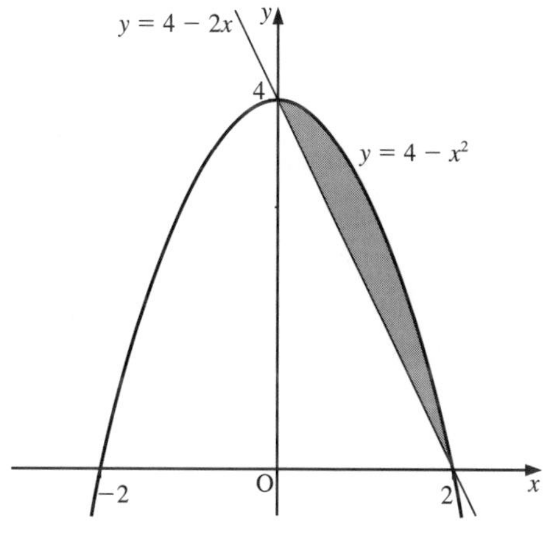

The required volume V is given by

$$V = \pi \int_0^2 (4 - x^2)^2 \, dx - \pi \int_0^2 (4 - 2x)^2 \, dx$$

$$= \pi \int_0^2 [(4 - x^2)^2 - (4 - 2x)^2] \, dx$$

$$= \pi \int_0^2 (x^4 - 12x^2 + 16x) \, dx$$

$$= \pi \left[\frac{x^5}{5} - 4x^3 + 8x^2 \right]_0^2$$

$$\therefore \quad V = \frac{32\pi}{5}$$

The volume of the solid of revolution is $\dfrac{32\pi}{5}$.

Volumes of revolution about other axes

The volume of the solid of revolution formed by rotating an area through 2π radians about the y-axis can be found in a way similar to that about the x-axis. The volume of such a solid of revolution is given by

$$V = \pi \int_a^b x^2 \, dy$$

Remember that dy implies that the limits a and b are y limits.

Example 15 Find the volume of the solid formed when the area between the curve $y = x^3$ and the y-axis from $y = 1$ to $y = 8$ is rotated through 2π radians about the y-axis.

SOLUTION

The required volume V is given by

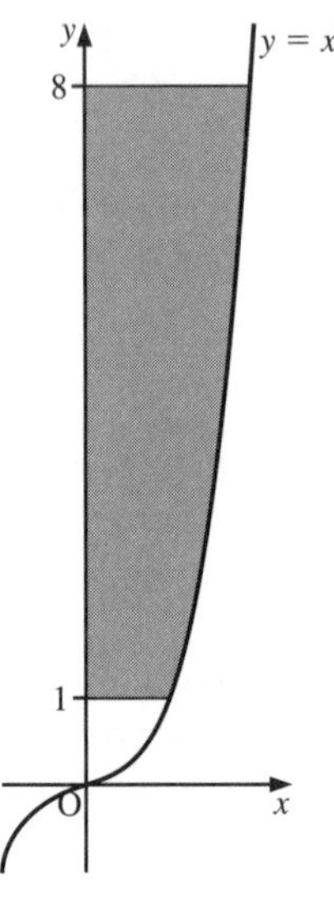

$$V = \pi \int_1^8 x^2 \, dy$$

Now $y = x^3$. Therefore,

$$y^{\frac{2}{3}} = (x^3)^{\frac{2}{3}}$$

$$\therefore \quad y^{\frac{2}{3}} = x^2$$

So

$$V = \pi \int_1^8 y^{\frac{2}{3}} \, dy$$

$$= \pi \left[\frac{3}{5} y^{\frac{5}{3}} \right]_1^8$$

$$= \pi \left[\frac{3(8)^{\frac{5}{3}}}{5} - \frac{3(1)^{\frac{5}{3}}}{5} \right]$$

$$\therefore \quad V = \frac{93\pi}{5}$$

The volume of the solid formed is $\dfrac{93\pi}{5}$.

Example 16 Find the volume generated when the region bounded by the curve $y = x^2$, the x-axis and the line $x = 2$, is rotated through 180° about the line $x = 3$.

SOLUTION

The sketches show the regions to be rotated.

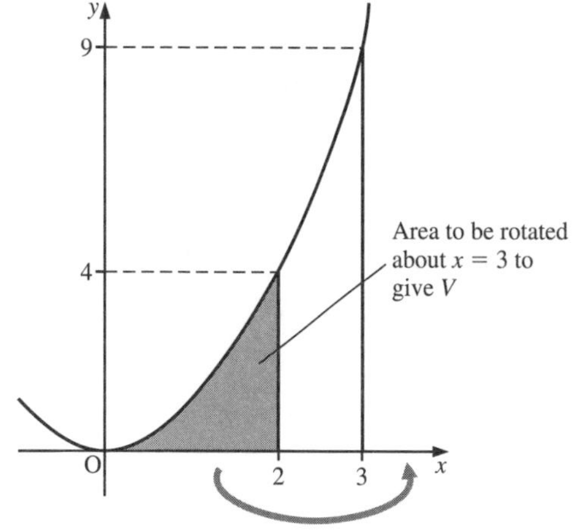

The required volume, V, is found by first calculating the volume, V_1, of the solid generated when the area between the curve $y = x^2$, the x-axis and the line $x = 3$ is rotated through 180° about the line $x = 3$, and then subtracting the volume, V_2, of the half-cylinder with radius $(3 - 2) = 1$ and height 4.

Volume V_1 is given by

$$V_1 = \frac{\pi}{2}\int_0^4 (3 - x)^2 \, dy$$

$$= \frac{\pi}{2}\int_0^4 (3 - \sqrt{y})^2 \, dy$$

$$= \frac{\pi}{2}\int_0^4 (9 - 6\sqrt{y} + y) \, dy$$

$$= \frac{\pi}{2}\left[9y - 4y^{\frac{3}{2}} + \frac{y^2}{2}\right]_0^4$$

$$= \frac{\pi}{2}((36 - 32 + 8) - (0))$$

$$\therefore \quad V_1 = 6\pi$$

Volume V_1

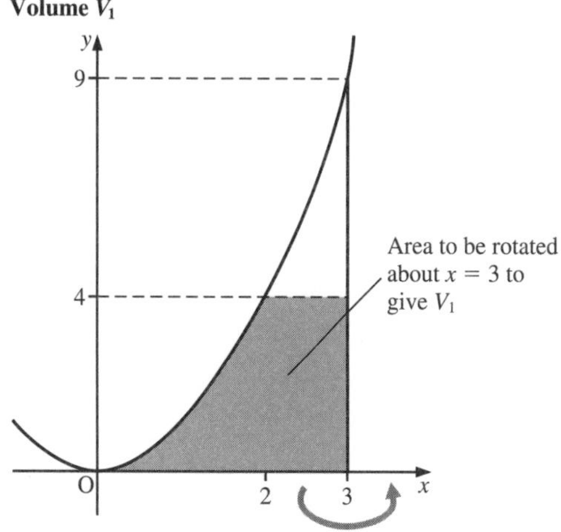

The volume, V_2, of the half-cylinder is given by

$$V_2 = \frac{\pi}{2}(1)^2 4 = 2\pi$$

Volume V_2

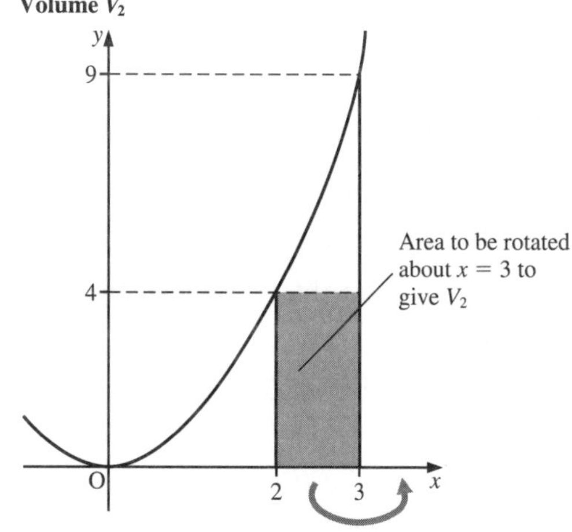

Therefore, the required volume, V, is given by

$$V = V_1 - V_2$$

$$= 6\pi - 2\pi$$

$$= 4\pi$$

The volume of the solid generated is 4π.

Exercise 7C

Throughout this exercise leave your answers as **multiples of** π.

1 Find the volume generated when each of the areas, bounded by the following curves and the x-axis, is rotated through 360° about the x-axis between the given lines.

a) $y = x$; $x = 0$ and $x = 6$

b) $y = x^2$; $x = 0$ and $x = 5$

c) $y = \sqrt{x}$; $x = 0$ and $x = 4$

d) $y = \dfrac{1}{x^2}$; $x = 1$ and $x = 2$

e) $y = 3\sqrt{x}$; $x = 2$ and $x = 4$

f) $y = 2x + 1$; $x = 1$ and $x = 3$

g) $y = 5 - x$; $x = 2$ and $x = 5$

h) $y = x^2 + 1$; $x = 0$ and $x = 3$

i) $y = \sqrt{x^2 + 3x}$; $x = 2$ and $x = 6$

j) $y = \sqrt{3x^2 + 8}$; $x = 1$ and $x = 3$

k) $y = \dfrac{x^2 - 2}{x^2}$; $x = \frac{1}{4}$ and $x = \frac{1}{2}$

l) $y = \sqrt{x} - 3$; $x = 4$ and $x = 9$

2 Find the volume generated when each of the areas in the positive quadrant, bounded by the following curves and lines, is rotated through 360° about the y-axis.

a) $y = \frac{1}{2}x$; $y = 0$ and $y = 6$

b) $y = x^2$; $y = 0$ and $y = 9$

c) $y = x^3$; $y = 0$ and $y = 8$

d) $y = \sqrt{x}$; $y = 0$ and $y = 3$

e) $y = 3x$; $y = 3$ and $y = 6$

f) $y = x^4$; $y = 1$ and $y = 4$

g) $y = \dfrac{1}{x}$; $y = 2$ and $y = 4$

h) $y = x - 1$; $y = 2$ and $y = 5$

i) $y = \frac{1}{2}x + 3$; $y = 4$ and $y = 6$

j) $y = x^2 + 2$; $y = 2$ and $y = 6$

k) $y = \sqrt{x^2 + 1}$; $y = 1$ and $y = 3$

l) $y = \sqrt{2x^2 - 1}$; $y = 1$ and $y = 6$

3 The curve $y = x^2$ meets the line $y = 4$ at the points P and Q.

a) Find the coordinates of P and Q.

b) Calculate the volume generated when the region bounded by the curve and the line is rotated through 360° about the x-axis.

4 The curve $y = x^2 + 1$ meets the line $y = 2$ at the points A and B.

a) Find the coordinates of A and B.

The region bounded by the curve and the line is rotated through 360° about the x-axis.

b) Calculate the volume of the solid generated.

5 The region bounded by the lines $y = x + 1$, $y = 3$ and the y-axis is rotated through 360° about the x-axis. Calculate the volume of the solid generated.

6 Calculate the volume generated when the region bounded by the curve $y = \dfrac{4}{x}$ and the lines $x = 1$ and $y = 1$ is rotated through 360° about the x-axis.

7 The region bounded by the curve $y = \sqrt{x}$, the x-axis and the line $x = 4$ is rotated 360° about the y-axis. Calculate the volume of the solid generated.

8 The region R is bounded by the curve $y = x^2 + 2$, the line $x = 1$, and the x- and y-axes. Calculate the volume of the solid generated when R is rotated through 360° about the y-axis.

9 Calculate the volume generated when the region bounded by the curve $y = 9 - x^2$, the line $x = 2$ and the x-axis is rotated through 360° about the y-axis.

10 The line $y = 3x$ meets the curve $y = x^2$ at the points O and P.

a) Calculate the coordinates of P.
b) Find the volume of the solid generated when the area enclosed by the line and the curve is rotated through 360° about **i)** the x-axis, **ii)** the y-axis.

11 **a)** On one set of axes sketch the graphs of the curves $y = x(1 - x)$ and $y = 2x(1 - x)$.
b) Calculate the volume generated when the finite region bounded by the two curves is rotated through 360° about the x-axis.

12 Find the volume generated when the region bounded by the curve $y = \sqrt{x}$ and the line $y = \frac{1}{5}x$ is rotated through 360° about the y-axis.

13 The curve $y = x^2$ meets the curve $y = 8 - x^2$ at the points P and Q.

a) Find the coordinates of P and Q.
b) Calculate the volume generated when the region bounded by the two curves is rotated through 180° about **i)** the x-axis, **ii)** the y-axis.

14 Find the volume generated when the region bounded by the curves $y = 2x^2$ and $y = 3 - x^2$ is rotated through 180° about **a)** the x-axis, **b)** the y-axis.

15 The curve $y = (x - 2)(x - 4)$ meets the line $y = 8$ at the points A and B.

a) Find the cordinates of A and B.

The region bounded by the curve and the line $y = 8$ is rotated through 360° about the line $y = 8$.

b) Calculate the volume of the solid generated.

16 The curve $y = 2 + (x - 2)^2$ meets the line $y = 6$ at the points C and D.

a) Find the cordinates of C and D.

The region bounded by the curve and the line $y = 6$ is rotated through 180° about the line $y = 6$.

b) Calculate the volume of the solid generated.

17 Calculate the volume of the solid generated when the region bounded by the curve $y = x^2$, the line $x = 3$ and the x-axis is rotated through 360° about the line $x = 3$.

18 The region bounded by the lines $y = x - 3$, $x = 6$ and the x-axis is rotated through 360° about the line $x = 6$. Calculate the volume of the solid generated.

19 The region R is bounded by the curve $y = \sqrt{x}$, the y-axis and the line $y = 2$. Calculate the volume of the solid formed by rotating R through 360° about the line $y = 2$.

20 The area in the positive quadrant, bounded by the curve $y = x^2$, the x-axis and the line $x = 3$, is rotated through 360° about the line $x = 4$. Calculate the volume of the solid generated.

21 Calculate the volume of the solid generated when the region bounded by the curve $y = \sqrt{x}$, and the lines $x = 4$ and $y = 0$, is rotated through $180°$ about the line $x = -2$.

22 Find the volume generated when the region bounded by the curve $y = x^3$, the x-axis and the line $x = 2$, is rotated through $360°$ about the line $x = 3$.

23 The area bounded by the curve $y = 1 - x^2$ and the x-axis is rotated through $360°$ about the line $y = -2$. Calculate the volume of the solid generated.

***24** When the region bounded by the curve $y = x^2$, the line $x = a$ and the x-axis, is rotated through $360°$ about the line $x = a$, the volume of the solid generated is 216π. Calculate the possible values of the constant a.

***25** The region R in the positive quadrant is bounded by the curve $y = f(x)\,(f(x) > 0)$, the x-axis, and the lines $x = a$ and $x = b$. A denotes the area of R, V denotes the volume generated when R is rotated through $360°$ about the x-axis and $\tilde{V}$ denotes the volume generated when R is rotated through $360°$ about the line $y = -c\,(c > 0)$.

Show that $\tilde{V} = V + 2c\pi A$.

Exercise 7D: Examination questions

1
$$f(x) \equiv \frac{(3x^2 + 1)^2}{x^2}.$$

a) Express f(x) in the form $Ax^2 + B + \frac{C}{x^2}$, where the constants A, B and C are to be found.

b) Find $\int f(x)\,dx$.

c) Hence evaluate $\int_1^2 f(x)\,dx$. (EDEXCEL)

2 A curve passes through the point (2, 3). The gradient of the curve is given by

$$\frac{dy}{dx} = 3x^2 - 2x - 1$$

i) Find y in terms of x.
ii) Find the coordinates of any stationary points of the graph of y.
iii) Sketch the graph of y against x, marking the coordinates of any stationary points and the point where the curve cuts the y-axis. (MEI)

3 A certain curve has an equation which satisfies $\frac{d^2y}{dx^2} = 24x^2 - 2$ for all values of x.

The point P(1, 4) lies on the curve and the gradient of the curve at P is 5. Determine the equation of the curve. (AEB 93)

4 A curve has an equation which satisfies $\frac{dy}{dx} = kx(x-1)$, where k is a constant. Given that the gradient of the curve at the point (2, 1) is 12,

a) find the value of k
b) find the equation of the curve.

5 The sketch shows the graph of $y = x(3-x)$.

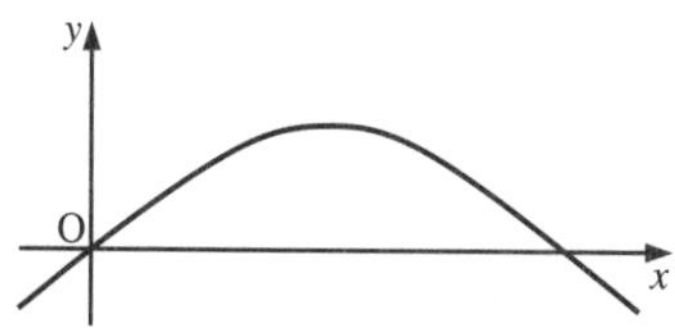

a) Find $\int x(3-x)\,dx$.
b) Hence calculate the area enclosed between the curve and the x-axis. (NEAB)

6 The curve C has equation $y = x^2 - 6x + 8$.

a) Find the coordinates of the turning point of C and determine its nature.
b) Sketch the curve C.
c) Find the area of the region bounded by C and the x-axis. (WJEC)

7 The equation of the curve C is $y = 2x^3 - 9x^2 + 12x$.

i) Find the coordinates of the maximum and minimum points of C, carefully distinguishing between them.
ii) Sketch C.
iii) Find the area of the region enclosed by C, the x-axis and the line $x = 2$. (WJEC)

8 The curve with equation $y = \sqrt{x} + \frac{3}{\sqrt{x}}$ is sketched for $x > 0$.

The region R, shaded in the diagram, is bounded by the curve, the x-axis and the lines $x = 1$ and $x = 4$. Use integration to determine R. (UODLE)

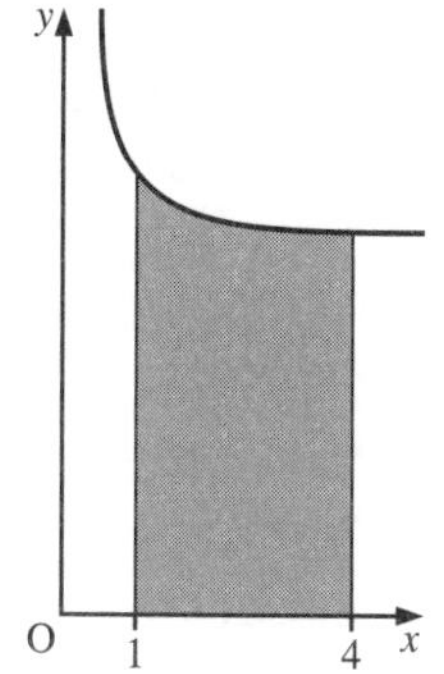

9 The figure shows the curve C with equation $y = x(8-x)$ and the line with equation $y = 12$ which meet at the points L and M.

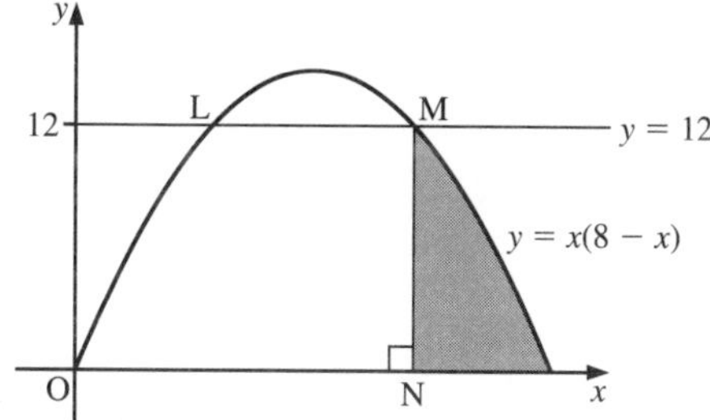

a) Determine the coordinates of the point M.

Given that N is the foot of the perpendicular from M onto the x-axis,

b) calculate the area of the shaded region which is bounded by NM, the curve C and the x-axis. (EDEXCEL)

10

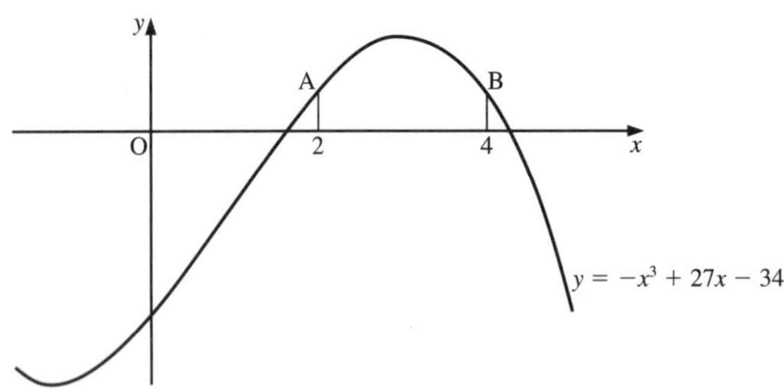

The figure shows a sketch of part of the curve with equation $y = f(x)$, where

$$f(x) = -x^3 + 27x - 34$$

a) Find $\int f(x)\,dx$.

The lines $x = 2$ and $x = 4$ meet the curve at points A and B, as shown.

b) Find the area of the finite region bounded by the curve and the lines $x = 2$, $x = 4$ and $y = 0$.
c) Find the area of the finite region bounded by the curve and the straight line AB.

(EDEXCEL)

11 i) Differentiate $y = -x^2 + 2x + 3$.
ii) Find the maximum value of y.

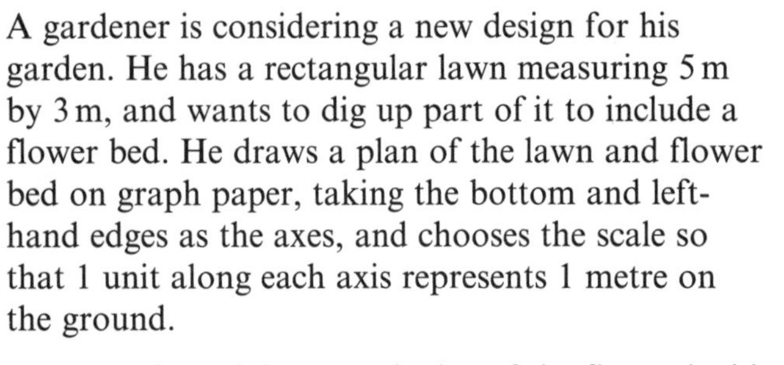

A gardener is considering a new design for his garden. He has a rectangular lawn measuring 5 m by 3 m, and wants to dig up part of it to include a flower bed. He draws a plan of the lawn and flower bed on graph paper, taking the bottom and left-hand edges as the axes, and chooses the scale so that 1 unit along each axis represents 1 metre on the ground.

iii) The equation of the curved edge of the flower bed is

$$y = -x^2 + 2x + 3$$

Calculate the area of the flower bed. (MEI)

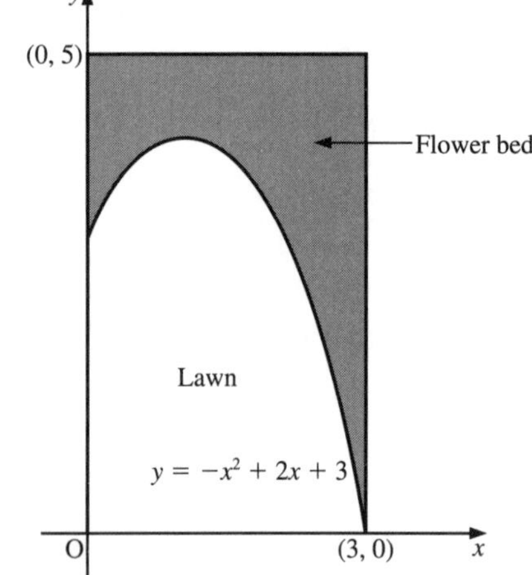

12 a) Draw a sketch to show the line $y = x$ and the curve $y = x^2$.
b) Express the finite area between the curves as an integral.
c) Calculate the value of this area by evaluating the integral in part **b**, showing all your working. (UODLE)

13 i) Find the coordinates of the points of intersection of the line $y = 2x$ with the curve $y = x^2 - 3x$.
ii) Sketch the line and the curve for the domain $0 \leqslant x \leqslant 6$.
iii) Shade in the region enclosed between the line and the curve and calculate the area of this region. (MEI)

14 The figure shows the curve with equation $y = x^2 - 6x + 5$ which meets the y-axis at the point A and the x-axis at the points B and C. The line L, with equation $y = x - 1$, meets the curve at the points B and D. The line M is perpendicular to L and passes through C. The lines L and M meet at the point E.

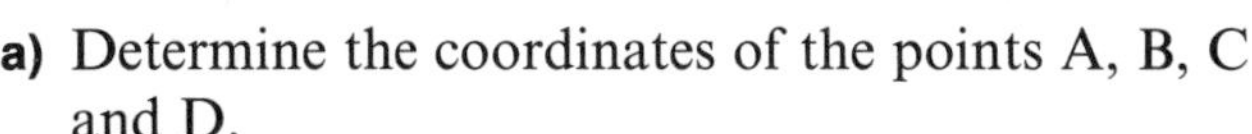

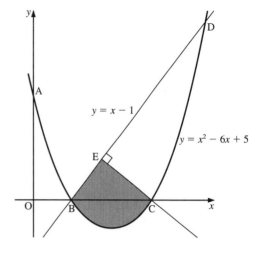

a) Determine the coordinates of the points A, B, C and D.

b) Find an equation of the line M.

c) Show, by calculation, that the coordinates of E are $(3, 2)$.

The shaded region is bounded by the line segments BE and CE and the arc of the curve between B and C.

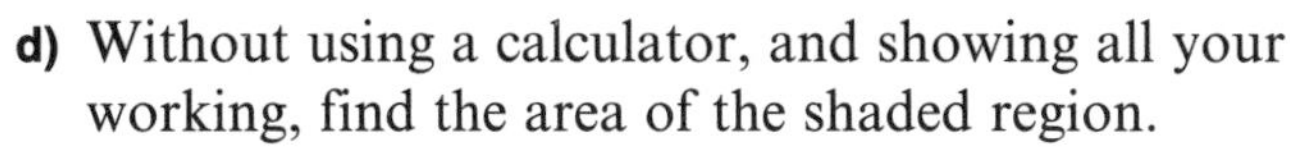

d) Without using a calculator, and showing all your working, find the area of the shaded region.

(EDEXCEL)

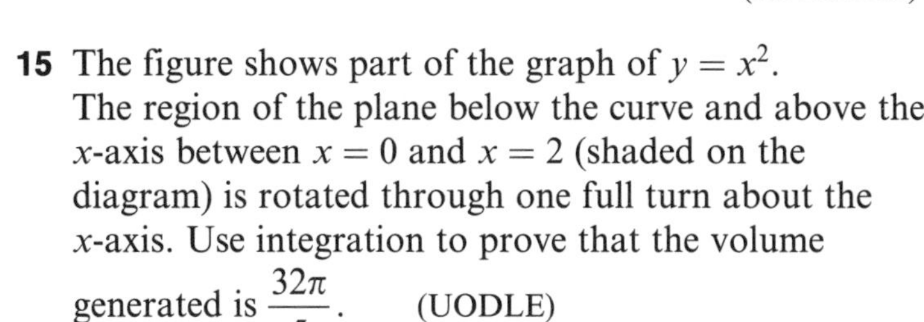

15 The figure shows part of the graph of $y = x^2$. The region of the plane below the curve and above the x-axis between $x = 0$ and $x = 2$ (shaded on the diagram) is rotated through one full turn about the x-axis. Use integration to prove that the volume generated is $\dfrac{32\pi}{5}$. (UODLE)

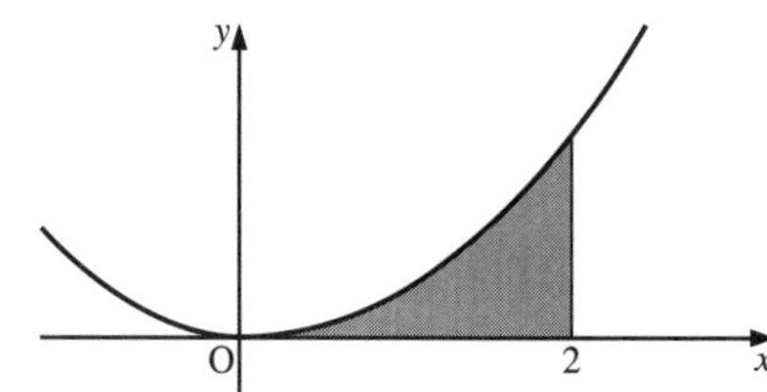

16 The functions f and g are defined on the domain $-1 \leqslant x \leqslant 1$ by

$$f(x) = x^2 \qquad \text{and} \qquad g(x) = 2 - x^2$$

The graphs of $y = f(x)$ and $y = g(x)$ are shown in the diagram.

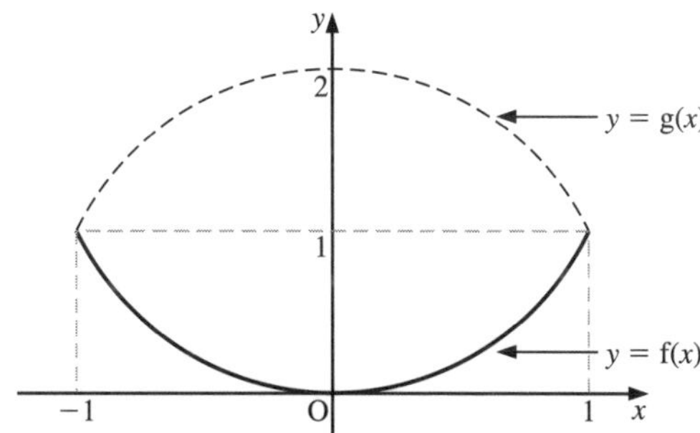

i) Describe, either as a single transformation or as two separate transformations, how the graph of $y = g(x)$ can be obtained from the graph of $y = f(x)$.

ii) Find the area of the region enclosed by the two graphs.

iii) Show that the volume generated when this region is rotated through 360° about the x-axis is $\dfrac{16\pi}{3}$. (NEAB)

17 i) Sketch, on the same diagram, the curves

$$y = 2 + x^2 \quad \text{and} \quad y = 6 - x^2$$

and find their points of intersection

ii) Find the area of the region bounded by these two curves.

iii) Find also the volume generated when this region is rotated through π radians about the y-axis. (NICCEA)

18 The region R is bounded by the x-axis, the y-axis, the line $y = 12$ and the part of the curve whose equation is $y = x^2 - 4$ which lies between $x = 2$ and $x = 4$.

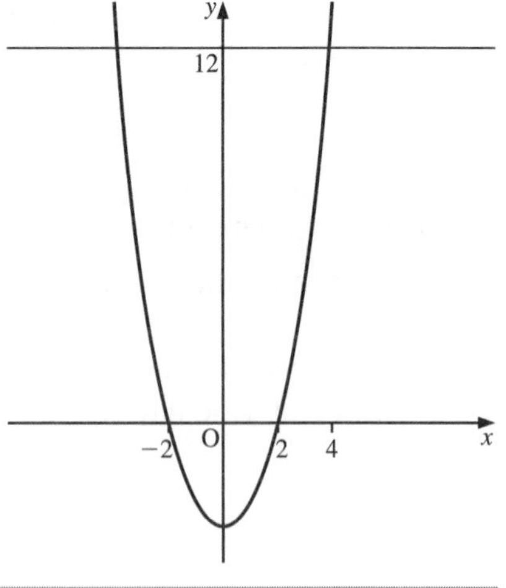

i) Copy the sketch graph and shade the region R.
The inside of a vase is formed by rotating the region R through 360° about the y-axis.
Each unit of x and y represents 5 cm

ii) Write down an expression for the volume of revolution of the region R about the y-axis.

iii) Find the capacity of the vase in litres.

iv) Show that when the vase is filled to $\frac{5}{6}$ of its internal height it is three-quarters full. (MEI)

8 The circle

Beauty is the first test; there is no permanent place in the world for ugly mathematics.
G. H. HARDY

Let C (a, b) be the centre of a circle radius r, and let P(x, y) be any point on the circumference of the circle.

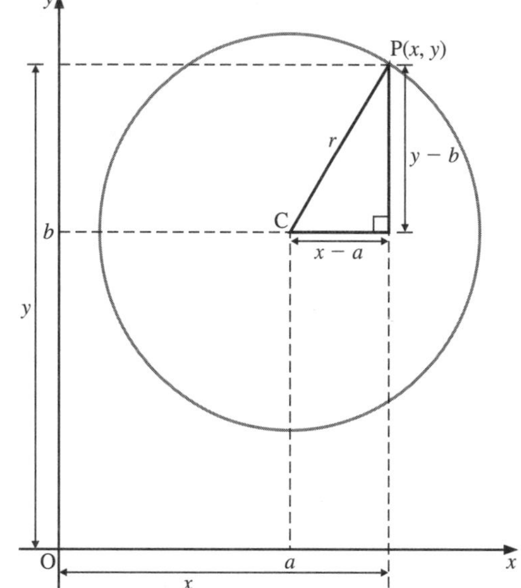

Using the formula (see page 135)

$$d = \sqrt{(x_2 - x_1)^2 + (y_2 - y_1)^2}$$

for the distance between two points, we have

$$r = \sqrt{(x - a)^2 + (y - b)^2}$$

Therefore,

$$r^2 = (x - a)^2 + (y - b)^2$$

This is one standard form of the equation of a circle. Expanding and simplifying give

$$x^2 - 2ax + a^2 + y^2 - 2by + b^2 - r^2 = 0$$

$$\therefore \quad x^2 + y^2 - 2ax - 2by + a^2 + b^2 - r^2 = 0$$

The general equation of a circle centred on (a, b) with radius r is given by

$$x^2 + y^2 - 2ax - 2by + c = 0$$

where $c = a^2 + b^2 - r^2$.

This is an alternative standard form for the equation of a circle.

Example 1 Find the equation of the circle with centre $(3, -1)$ and radius 4.

SOLUTION

The equation is

$$(x - 3)^2 + (y - (-1))^2 = 4^2$$

$$\therefore \quad (x - 3)^2 + (y + 1)^2 = 16$$

which is one way of writing the equation of the circle. Expanding and simplifying give the alternative form

$$x^2 - 6x + 9 + y^2 + 2x + 1 = 16$$

$$\therefore \quad x^2 + y^2 - 6x + 2y - 6 = 0$$

Example 2 Find the centre and radius of each of these circles.

a) $x^2 + y^2 + 2x - 4y - 4 = 0$ **b)** $9x^2 + 9y^2 - 12x + 18y + 4 = 0$

SOLUTION

a) $$x^2 + 2x + y^2 - 4y - 4 = 0$$

Completing the square gives

$$(x+1)^2 - 1 + (y-2)^2 - 4 - 4 = 0$$

$$\therefore \quad (x+1)^2 + (y-2)^2 = 9$$

The centre of the circle is $(-1, 2)$ and its radius is $\sqrt{9} = 3$.

b) We must express

$$9x^2 + 9y^2 - 12x + 18y + 4 = 0$$

in the form

$$x^2 + y^2 - 2ax - 2by + c = 0$$

Dividing throughout by 9 and rearranging give

$$x^2 - \frac{4}{3}x + y^2 + 2y + \frac{4}{9} = 0$$

Completing the square gives

$$\left(x - \frac{2}{3}\right)^2 - \frac{4}{9} + (y+1)^2 - 1 + \frac{4}{9} = 0$$

$$\therefore \quad \left(x - \frac{2}{3}\right)^2 + (y+1)^2 = 1$$

The centre of the circle is $(\frac{2}{3}, -1)$ and its radius is 1.

Example 3 Find the equation of the circle which passes through the points A(1, 2), B(2, 5) and C(−3, 4).

SOLUTION

The circle has equation $(x-a)^2 + (y-b)^2 = r^2$, where (a, b) is the centre of the circle and r is the radius. Since the circle passes through (1, 2),

$$(1-a)^2 + (2-b)^2 = r^2$$

$$\therefore \quad a^2 + b^2 - 2a - 4b + 5 = r^2 \qquad [1]$$

Since the circle also passes through (2, 5),

$$(2-a)^2 + (5-b)^2 = r^2$$

$$\therefore \quad a^2 + b^2 - 4a - 10b + 29 = r^2 \qquad [2]$$

And since the circle also passes through (−3, 4),

$$(-3-a)^2 + (4-b)^2 = r^2$$

$$\therefore \quad a^2 + b^2 + 6a - 8b + 25 = r^2 \qquad [3]$$

Solving [1], [2] and [3] simultaneously will give the values of a, b and c.
Subtracting [2] from [1] gives

$$2a + 6b - 24 = 0$$

$$\therefore \quad a + 3b - 12 = 0 \qquad [4]$$

Subtracting [3] from [2] gives

$$-10a - 2b + 4 = 0$$

$$\therefore \quad -5a - b + 2 = 0 \qquad [5]$$

Solving [4] and [5] gives $a = -\frac{3}{7}$ and $b = \frac{29}{7}$. Substituting into [1] gives

$$r^2 = \left(-\frac{3}{7}\right)^2 + \left(\frac{29}{7}\right)^2 - 2\left(-\frac{3}{7}\right) - 4\left(\frac{29}{7}\right) + 5$$

$$\therefore \quad r^2 = \frac{325}{49}$$

$$\therefore \quad r = \frac{5\sqrt{13}}{7}$$

Therefore, the equation of the circle is

$$\left(x + \frac{3}{7}\right)^2 + \left(y - \frac{29}{7}\right)^2 = \frac{325}{49}$$

$$\therefore \quad x^2 + y^2 + \frac{6}{7}x - \frac{58}{7}y + \frac{75}{7} = 0$$

$$\therefore \quad 7x^2 + 7y^2 + 6x - 58y + 75 = 0$$

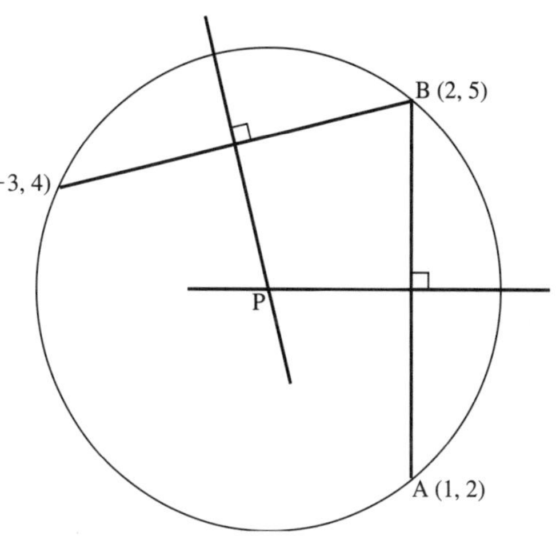

An alternative method is to use the fact that the perpendicular bisectors of two chords intersect at the centre of the circle.

We can find the perpendicular bisectors of AB and BC and then find their point of intersection P, which will be the centre of the circle. The radius is then given by PA (= PB = PB).

To find the perpendicular bisectors of AB and AC, we need their mid-points and gradients:

Coordinates of the mid-point of AB are given by

$$\left(\frac{1+2}{2}, \frac{2+5}{2}\right) = \left(\frac{3}{2}, \frac{7}{2}\right)$$

Coordinates of the mid-point of BC are given by

$$\left(\frac{2-3}{2}, \frac{5+4}{2}\right) = \left(-\frac{1}{2}, \frac{9}{2}\right)$$

Gradient of AB is given by $m_{\mathrm{AB}} = \dfrac{5-2}{2-1} = 3$

Gradient of BC is given by $m_{\mathrm{BC}} = \dfrac{4-5}{-3-2} = \dfrac{-1}{-5} = \dfrac{1}{5}$

The perpendicular bisector of AB has gradient $-\frac{1}{3}$ and passes through the point $\left(\frac{3}{2}, \frac{7}{2}\right)$. Its equation is given by $y - y_1 = m(x - x_1)$. That is,

$$y - \frac{7}{2} = -\frac{1}{3}\left(x - \frac{3}{2}\right)$$

$$\therefore \quad y = -\frac{1}{3}x + \frac{1}{2} + \frac{7}{2}$$

$$\therefore \quad y = -\frac{1}{3}x + 4 \qquad [6]$$

The perpendicular bisector of BC has gradient -5 and passes through the point $\left(-\frac{1}{2}, \frac{9}{2}\right)$. its equation is given by $y - y_1 = m(x - x_1)$. That is,

$$y - \frac{9}{2} = -5\left(x + \frac{1}{2}\right)$$

$$\therefore \quad y = -5x - \frac{5}{2} + \frac{9}{2}$$

$$\therefore \quad y = -5x + 2 \qquad [7]$$

The centre of the circle is the point, P at which the two perpendicular bisectors intersect. We must solve simultaneously equations [6] and [7]. That is,

$$-\frac{1}{3}x + 4 = -5x + 2$$

$$\therefore \quad \frac{14}{3}x = -2$$

$$\therefore \quad x = -\frac{6}{14} = -\frac{3}{7}$$

$$\text{and} \quad y = -5\left(-\frac{3}{7}\right) + 2 = \frac{15}{7} + 2 = \frac{29}{7}$$

The coordinates of the centre are $\left(-\frac{3}{7}, \frac{29}{7}\right)$, as before.

The radius of the circle is the distance from the centre to any one of the points A, B or C.

Using $\left(-\frac{3}{7}, \frac{29}{7}\right)$ and A(1, 2) gives the radius r as

$$r = \sqrt{\left(1 + \frac{3}{7}\right)^2 + \left(2 - \frac{29}{7}\right)^2}$$

$$= \sqrt{\frac{100}{49} + \frac{225}{49}} = \sqrt{\frac{325}{49}}$$

$$\therefore \quad r = \frac{5\sqrt{13}}{7}$$

as before.

Therefore, the equation of the circle is

$$\left(x+\frac{3}{7}\right)^2+\left(y-\frac{29}{7}\right)^2=\frac{325}{49}$$

which simplifies to

$$7x^2+7y^2+6x-58y+75=0$$

as before.

Exercise 8A

1 Find the equations of the circles with the following centres and radii.

a) Centre (1, 2), radius 3
b) Centre (3, 1), radius 4
c) Centre (−2, 3), radius 1
d) Centre (1, −3), radius 5
e) Centre (−4, 0), radius 4
f) Centre (2, −4), radius 7
g) Centre (−3, 5), radius 6
h) Centre (4, −1), radius 3
i) Centre (0, −2), radius 1
j) Centre (−5, −3), radius 7
k) Centre (−8, 7), radius 10
l) Centre $(\frac{1}{2}, \frac{3}{2})$, radius 2

2 Find the centre and radius of each of the following circles.

a) $x^2+y^2-4x-2y+1=0$
b) $x^2+y^2-2x-8y+8=0$
c) $x^2+y^2+6x-4y+12=0$
d) $x^2+y^2-4x=0$
e) $x^2+y^2+6y-16=0$
f) $x^2+y^2-6x+8y-11=0$
g) $x^2+y^2+14x-10y-7=0$
h) $x^2+y^2-12x-12y+8=0$
i) $x^2+y^2+16x+12y=0$
j) $x^2+y^2-2x+2y-2=0$
k) $x^2+y^2-14x+16y-31=0$
l) $x^2+y^2-5y+4=0$

3 Find the equation of the circle whose centre is at the point (5, 4) and which passes through the point (9, 7).

4 Find the equation of the circle whose centre is at the point (1, −7) and which passes through the point (−4, 5).

5 Find the equation of the circle whose centre is at the point (5, 7) and which touches the x-axis.

6 Find the equation of the circle whose centre is at the point (−2, −3) and which touches the y-axis.

7 Find the equation of the circle which has the points A(2, 5) and B(10, 11) as the ends of a diameter.

8 Find the equation of the circle which has the points P(−2, 3) and Q(4, 5) as the ends of a diameter.

9 Find the equations of the circles of radius 5, which touch the x-axis, and pass through the point (3, 1).

10 **a)** Find the equation of the perpendicular bisector, p_1, to the points (3, 9) and (5, 9).
b) Find the equation of the perpendicular bisector, p_2, to the points (5, 9) and (5, 3).
c) Find the coordinates of the point of intersection of the lines p_1 and p_2.
d) Hence find the equation of the circle which passes through the points (3, 9), (5, 9) and (5, 3).

11 **a)** Find the equation of the perpendicular bisector, p_1, to the points (1, 11) and (9, 7).
b) Find the equation of the perpendicular bisector, p_2, to the points (9, 7) and (5, −1).
c) Find the coordinates of the point of intersection of the lines p_1 and p_2.
d) Hence find the equation of the circle which passes through the points (1, 11), (9, 7) and (5, −1).

12 **a)** Find the equation of the perpendicular bisector, p_1, to the points (−1, 6) and (1, 2).
b) Find the equation of the perpendicular bisector, p_2, to the points (1, 2) and (−5, 4).
c) Find the coordinates of the point of intersection of the lines p_1 and p_2.
d) Hence find the equation of the circle which passes through the points (−1, 6), (1, 2) and (−5, 4).

13 Find the equation of the circle which passes through the points (2, 3), (8, 3) and (8, −1).

***14** **a)** Find the centre and radius of each of the circles $x^2 + y^2 - 4x - 2y - 20 = 0$, and $x^2 + y^2 - 16x - 18y + 120 = 0$.
b) Deduce that the two circles touch externally.
c) Find the coordinates of the centres of the two circles, of radius 8, which touch each of the original circles externally.

Tangents to a circle

Example 4 Find the equation of the tangent to the circle $x^2 + y^2 + 2x - 4y - 20 = 0$ at the point P(2, 6).

SOLUTION

We can see that the point P(2, 6) lies on the circle since

$$(2)^2 + (6)^2 + 2(2) - 4(6) - 20 = 0$$

Now

$$x^2 + 2x + y^2 - 4y - 20 = 0$$

Completing the square gives

$$(x + 1)^2 - 1 + (y - 2)^2 - 4 - 20 = 0$$

$$\therefore \quad (x + 1)^2 + (y - 2)^2 = 25$$

The circle has centre (−1, 2) and radius $r = 5$.

The gradient, m, of the radius through P(2, 6) is given by

$$m = \frac{6 - 2}{2 - -1} = \frac{4}{3}$$

Therefore, the gradient of the tangent to the circle at P(2, 6) is $-\frac{3}{4}$. The equation of the tangent through P is of the form $y = -\frac{3}{4}x + c$. Using P(2, 6), we have

$$6 = -\frac{3}{4}(2) + c \quad \therefore \quad c = \frac{15}{2}$$

The equation of the tangent at the point P is $y = -\frac{3}{4}x + \frac{15}{2}$.

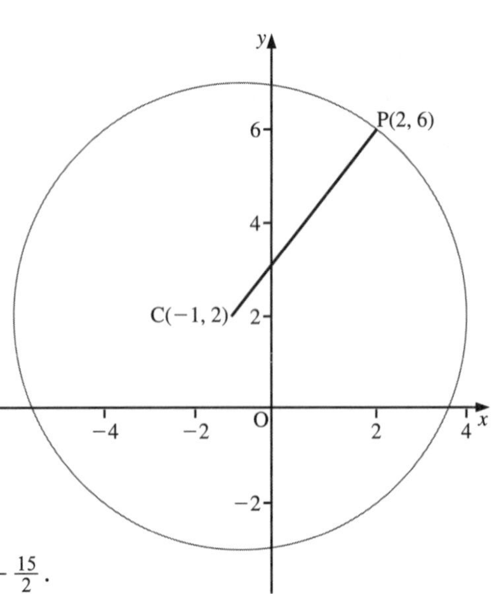

Example 5 Find the length of the tangents from the point (8, 4) to the circle with centre (3, 0) and radius 2.

SOLUTION

The situation is shown on the right, where x is the length to be found. The distance d is given by

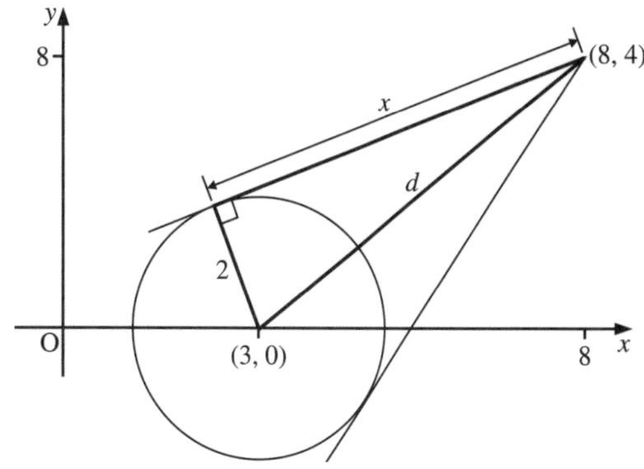

$$d = \sqrt{(8-3)^2 + (4-0)^2}$$

$$\therefore \quad d = \sqrt{41}$$

By Pythagoras' theorem, we have

$$d^2 = x^2 + 2^2$$

$$\therefore \quad 41 = x^2 + 4$$

$$\therefore \quad x = \sqrt{37}$$

The length of the tangent is $\sqrt{37}$.

Intersection of two circles

Example 6 Find the coordinates of the points of intersection of the two circles with equations

$$x^2 + y^2 - 3x + 13y - 48 = 0 \qquad [1]$$

and $$x^2 + y^2 + x - 3y = 0 \qquad [2]$$

SOLUTION

The coordinates of the points of intersection, P and Q, of the two circles satisfy both equations. Subtracting [2] from [1] gives

$$-4x + 16y - 48 = 0$$

$$\therefore \quad -x + 4y - 12 = 0 \qquad [3]$$

The coordinates of P and Q also satisfy this linear equation. Since it is the equation of a straight line and is satisfied by both P and Q, it must be the equation of the common chord.

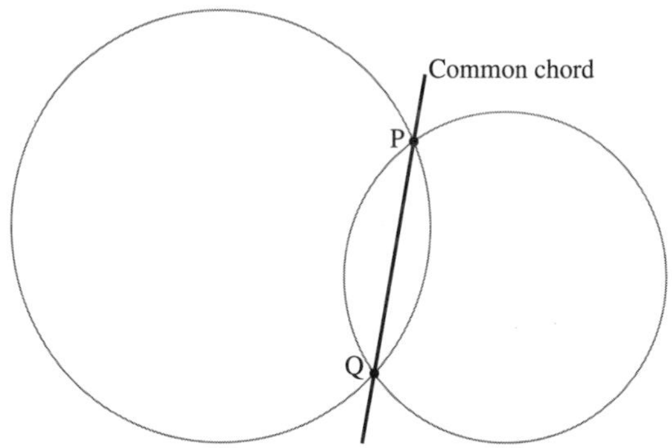

To find the coordinates of P and Q, we solve simultaneously [3] and one of the circle equations.

From [3] we have $x = 4y - 12$. Substituting this into [2] gives

$$(4y - 12)^2 + y^2 + (4y - 12) - 3y = 0$$

$$\therefore \quad 17y^2 - 95y + 132 = 0$$

$$\therefore \quad (17y - 44)(y - 3) = 0$$

Solving gives $y = \frac{44}{17}$ and $y = 3$.

When $y = \frac{44}{17}$, $x = -\frac{28}{17}$, and when $y = 3$, $x = 0$.

The points of intersection of the two circles are $(-\frac{28}{17}, \frac{44}{17})$ and $(0, 3)$.

Suppose we have two circles which do not intersect, such as

$C_1: x^2 + y^2 - 6x - 6y + 14 = 0$ and $C_2: x^2 + y^2 + 6x + 4y + 12 = 0$

Now C_1 has centre (3, 3) and radius 2, and C_2 has centre (−3, −2) and radius 1 (see diagram).

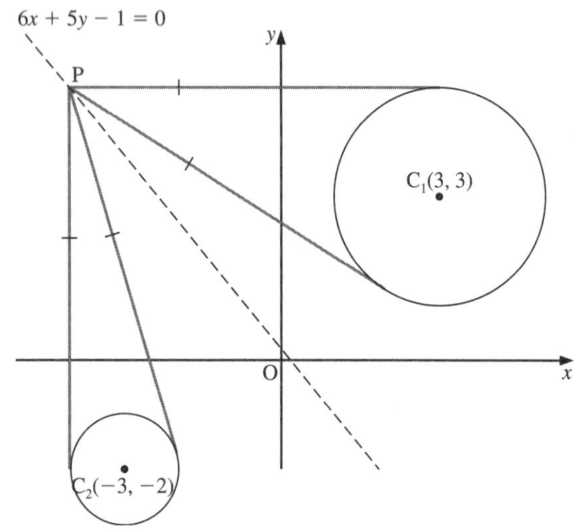

Subtracting the equations for C_1 and C_2 gives

$$-12x - 10y + 2 = 0$$

$$\therefore \quad 6x + 5y - 1 = 0$$

This line is shown dashed on the diagram. The question now asked is 'What does this line represent if the two circles do not intersect?'

If we choose any point P on $6x + 5y - 1 = 0$, then the lengths of the tangents from P to each of the circles are equal.

Exercise 8B

1 Find the equations of the tangents to the following circles at the points given.

a) $x^2 + y^2 - 2x - 6y + 8 = 0$, at (2, 2)

b) $x^2 + y^2 + 4x + 6y - 21 = 0$, at (1, 2)

c) $x^2 + y^2 + 6x - 4y + 8 = 0$, at (−1, 1)

d) $x^2 + y^2 - 8x + 2y + 15 = 0$, at (3, −2)

e) $x^2 + y^2 + 10x + 8y + 39 = 0$, at (−4, −3)

f) $x^2 + y^2 + 8x - 8y - 17 = 0$, at (3, 4)

g) $x^2 + y^2 + 10y + 20 = 0$, at (2, −4)

h) $x^2 + y^2 + 4x - 14 = 0$, at (−5, 3)

i) $x^2 + y^2 - 14x + 8y + 57 = 0$, at (9, −2)

j) $x^2 + y^2 + 10x + 12 = 0$, at (−7, 3)

k) $x^2 + y^2 - 12x - 16y = 0$, at (0, 0)

l) $x^2 + y^2 - x - 5y + 4 = 0$, at (2, 3)

2 Find the equations of the tangents to the circle

$$x^2 + y^2 - 10x - 8y + 21 = 0$$

at the points where the circle cuts the x-axis.

3 Find the equations of the tangents to the circle

$$x^2 + y^2 - 6x - 8y + 15 = 0$$

at the points where the circle cuts the y-axis.

4 Find the equations of the tangents to the circle

$$x^2 + y^2 + 4x - 2y - 24 = 0$$

at the points where the circle cuts the line $y = x$.

5 The tangent to the circle $x^2 + y^2 - 4x + 6y - 7 = 0$ at the point (4, 1) meets the x-axis at A and the y-axis at B. Find the area of the triangle AOB, where O is the origin.

6 Find the length of the tangents from the point (2, 5) to the circle $x^2 + y^2 - 14x - 2y + 34 = 0$.

7 Find the length of the tangents from the point (3, −1) to the circle $x^2 + y^2 + 4x + 8y + 9 = 0$.

8 **a)** Find the length of the tangents from the point (2, −3) to the circle

$$x^2 + y^2 - 10x - 4y + 12 = 0.$$

b) Deduce that the shape formed by these two tangents, and the two radii through the points of contact of the circle and the tangents, is a square.

9 Find the coordinates of the points of intersection of each of the following circles with the corresponding straight lines.

a) $x^2 + y^2 - 6x - 4y + 9 = 0$ and $y = 7 - x$
b) $x^2 + y^2 + 8x + 2y - 8 = 0$ and $y = 7x + 2$
c) $x^2 + y^2 - 8y - 9 = 0$ and $y = 11 - x$
d) $x^2 + y^2 + 10x - 6y - 31 = 0$ and $y = 5x + 15$

10 Find the coordinates of the points of intersection of each of the following pairs of circles.

a) $x^2 + y^2 - 3x + 5y - 4 = 0$ and $x^2 + y^2 - x + 4y - 7 = 0$
b) $x^2 + y^2 - 5x + 3y - 4 = 0$ and $x^2 + y^2 - 4x + 6y - 12 = 0$
c) $x^2 + y^2 - 4x + 3y + 5 = 0$ and $x^2 + y^2 - 6x + 5y + 9 = 0$
d) $x^2 + y^2 + 3x - 2y - 7 = 0$ and $x^2 + y^2 + x - y - 8 = 0$

11 Show that the circles

$$x^2 + y^2 - 10x - 8y + 18 = 0 \quad \text{and} \quad x^2 + y^2 - 8x - 4y + 14 = 0$$

do not intersect.

***12** Given that the line $y = mx + c$ is a tangent to the circle

$$(x - a)^2 + (y - b)^2 = r^2$$

show that

$$(1 + m^2)r^2 = (c - b + ma)^2$$

Exercise 8C: Examination questions

1 Two circles C_1, C_2 have equations

$$(x+1)^2 + y^2 = 9$$
$$x^2 + y^2 - 16x - 80 = 0$$

respectively.

a) Find the radius of C_1 and the radius of C_2.
b) Calculate the distance between the centres of C_1 and C_2.
c) Show that C_1 and C_2 touch each other and find the coordinates of T, their point of contact. (WJEC)

2 The equation of the circle C is $x^2 + y^2 - 2x + 6y = 27$. Find the equation of the tangent to C at the point (2, 3). (WJEC)

3 Determine the coordinates of the centre and the radius of the circle with equation

$$x^2 + y^2 + 2x - 6y - 26 = 0$$

Find the distance from the point P(7, −9) to the centre of the circle. Hence find the length of the tangents from P to the circle. (AEB 92)

4 Determine the coordinates of the centre C and the radius of the circle with equation

$$x^2 + y^2 + 4x - 6y = 12$$

The circle cuts the x-axis at the points A and B. Calculate the area of the triangle ABC.

Calculate the area of the minor segment of the circle cut off by the chord AB, giving your answer to three significant figures. (AEB 95)

5 A circle has equation $x^2 + y^2 + 2x - 8y = 152$.

a) Find the radius and the coordinates of the centre.
b) The point P lies on the circle and has coordinates $(k, 16)$, where $k > 0$.
  i) Determine the value of k.
  ii) Find the coordinates of the point Q at the opposite end of the diameter from P. (AEB 95)

6 The circle C has equation $x^2 + y^2 - 18x - 6y + 45 = 0$.

i) Find the radius of C and the coordinates of its centre P.
ii) Show that the line $y = 2x$ is a tangent to C and find the coordinates of its point of contact Q.
iii) Show that the tangent to C at the point R(6, 9) has equation $x - 2y + 12 = 0$.
iv) Given that the tangents to C at Q and R meet at S show that $QS = \sqrt{5}$. (WJEC)

7 The points P, Q and R have coordinates (2, 4), (8, −2) and (6, 2) respectively.

a) Find the equation of the straight line l which is perpendicular to the line PQ and which passes through the mid-point of PR.
b) The line l cuts PQ at S. Find the ratio PS : SQ.
c) The circle passing through P, Q and R has centre C. Find the coordinates of C and the radius of the circle.
d) Given that angle PCQ = θ radians, show that $\tan\theta = 24/7$.
Prove that the smaller segment of the circle cut off by the chord PQ has area $25\theta - 24$.
(AEB 91)

9 Sequences and series

Mathematicians are like lovers ... grant a mathematician the least principle, and he will draw from it a consequence which you must also grant him, and from this consequence another.
B. LE BOVIER FONTERELLE

A **sequence** is a set of numbers in a particular order with each number in the sequence being derived from a particular rule. For example, consider the sequence

$$3, 6, 9, 12, \ldots$$

The first term is 3, the second term is 6, the nth term is given by $3n$.

We write the nth term as u_n. In this example, we would write

$$u_n = 3n \quad \text{for } n \geqslant 1$$

Example 1 Write down the first three terms of the sequence whose nth term is given by

$$u_n = n^2 + 6n \quad \text{for } n \geqslant 1$$

SOLUTION

Let $n = 1$, then $\quad u_1 = (1)^2 + 6(1) = 7$

Let $n = 2$, then $\quad u_2 = (2)^2 + 6(2) = 16$

Let $n = 3$, then $\quad u_3 = (3)^2 + 6(3) = 27$

The first three terms of the sequence are 7, 16 and 27.

Example 2 Write down an expression, in terms of n, for the nth term of the sequence 5, 9, 13, 17, ...

SOLUTION

Finding the difference between each of the consecutive terms gives 4. Consider the sequence defined by $u_n = 4n$:

$$4, 8, 12, 16, \ldots$$

It is clear that we require the sequence defined by $u_n = 4n + 1$.

A sequence can be defined by giving one general term of the sequence as an expression using other terms of the sequence. This relationship between the terms occurs throughout the sequence and is therefore called a **recurrence relation**. The next two examples illustrate how a recurrence relation can be used to define a sequence.

Example 3 A sequence is defined by $u_n = 2u_{n-1} + 1$, where the first term is $u_1 = 3$. Write down the first four terms of the sequence.

SOLUTION

Let $n = 2$, then $u_2 = 2u_1 + 1$. Since $u_1 = 3$, we have

$$u_2 = 2(3) + 1$$
$$\therefore \quad u_2 = 7$$

Let $n = 3$, then $u_3 = 2u_2 + 1$. Since $u_2 = 7$, we have

$$u_3 = 2(7) + 1$$
$$\therefore \quad u_3 = 15$$

Let $n = 4$, then $u_4 = 2u_3 + 1$. Since $u_3 = 15$, we have

$$u_4 = 2(15) + 1$$
$$\therefore \quad u_4 = 31$$

The first four terms of the sequence are 3, 7, 15 and 31.

Example 4 Write down a recurrence relation between the terms of the sequence 5, 14, 41, 122, ...

SOLUTION

We see that $u_1 = 5$ and that

$$14 = 3(5) - 1$$
$$41 = 3(14) - 1$$
$$122 = 3(41) - 1$$
$$\therefore \quad u_n = 3u_{n-1} - 1$$

The recurrence relation is $u_n = 3u_{n-1} - 1$, where the first term is $u_1 = 5$.

Convergent and divergent sequences

Consider the sequence

$$3, 2 + \frac{1}{5}, 2 + \frac{1}{25}, \ldots, 2 + \frac{1}{5^{n-1}}, \ldots$$

The terms in the sequence are getting closer and closer to the value 2. This can be seen on a graph of u_n against n.

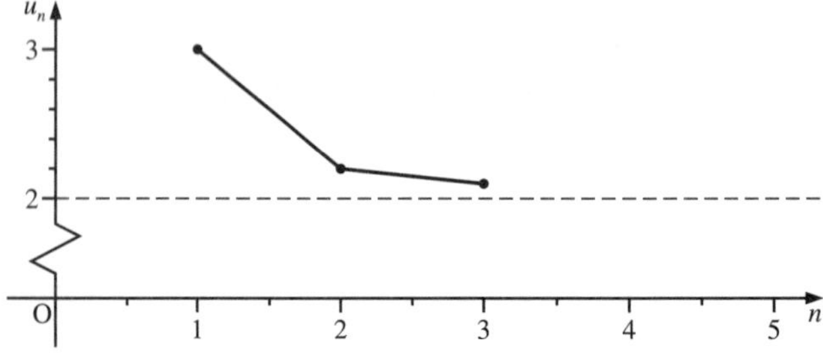

The sequence is said to be a **convergent sequence** since, as the number of terms increases, the values of the terms tend to a definite finite limit. The value 2 is said to be the **limit of the sequence**.

Not all sequences are convergent. Consider the sequence

$$5, 9, 13, 17, \ldots, 4n + 1, \ldots$$

In this sequence, as the number of terms increases, the values of the terms increase and tend to infinity. The graph of u_n against n shows this clearly.

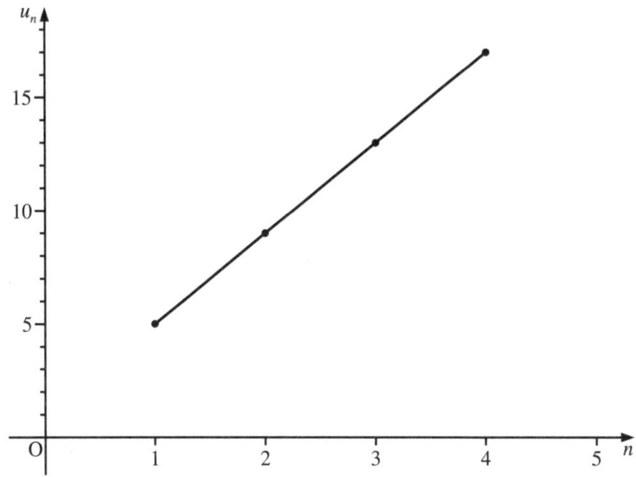

This sequence does not converge to a limit and is called a **divergent sequence**.

Oscillating sequences

Consider the sequence

$$1 + \frac{1}{2}, 1 - \frac{1}{4}, 1 + \frac{1}{8}, \ldots, 1 + (-1)^{n+1}\left(\frac{1}{2}\right)^n, \ldots$$

As the number of terms increases, the sequence is oscillating about the value 1, but at the same time getting closer and closer to the value 1. The graph of u_n against n shows this clearly.

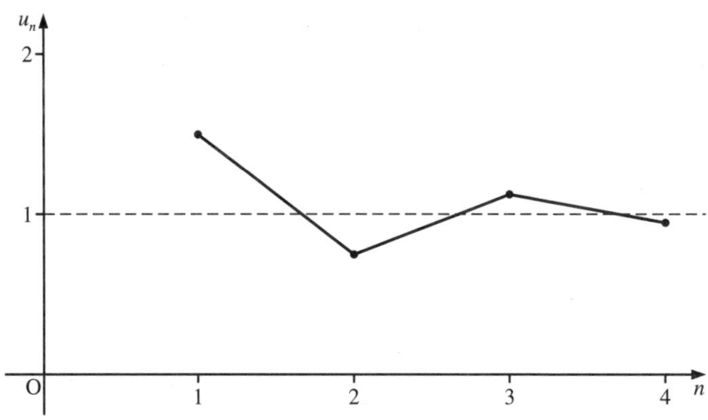

This is called an **oscillating sequence**. In this particular case, the sequence is converging to a limit of 1.

A sequence can oscillate but at the same time diverge. For example, the sequence

$$-2.5, 5, -10, 20, -40, \ldots, 5(-2)^{n-1}, \ldots$$

is clearly oscillating and is not converging to any particular value. We can see this on a graph of u_n against n.

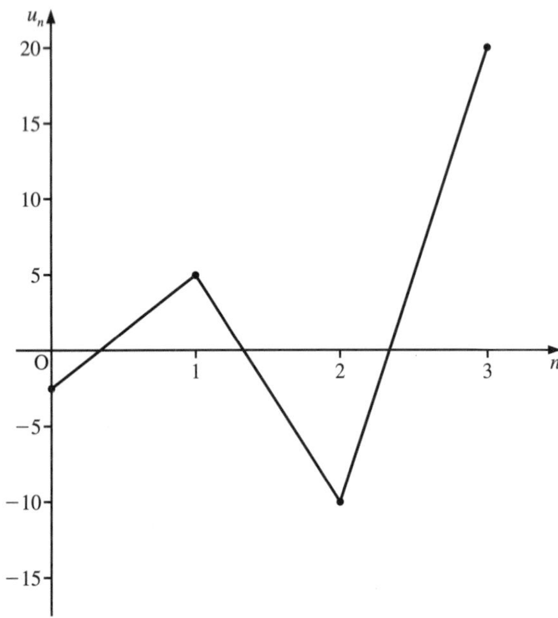

Periodic sequences

Consider the sequence

$$1, 3, 1, 3, \ldots, 2+(-1)^n, \ldots$$

As the number of terms increases, the sequence comprises only the terms 1 and 3. We say that the sequence is **periodic**. In this case, it is of period 2.

Example 5 A sequence is defined by

$$u_1 = 1, \quad u_n = \frac{1}{u_{n-1}} + 3 \quad \text{for } n > 1$$

Write down the first five terms of the sequence and determine whether the sequence is convergent or divergent.

SOLUTION

Let $n = 2$, then $\quad u_2 = \frac{1}{1} + 3 = 4$

Let $n = 3$, then $\quad u_3 = \frac{1}{4} + 3 = 3.25$

Let $n = 4$, then $\quad u_4 = \dfrac{1}{3.25} + 3 = 3.307\,69$

Let $n = 5$, then $\quad u_5 = \dfrac{1}{u_4} + 3 = 3.302\,33$

The sequence is clearly convergent and a graph of u_n against n shows this fact.

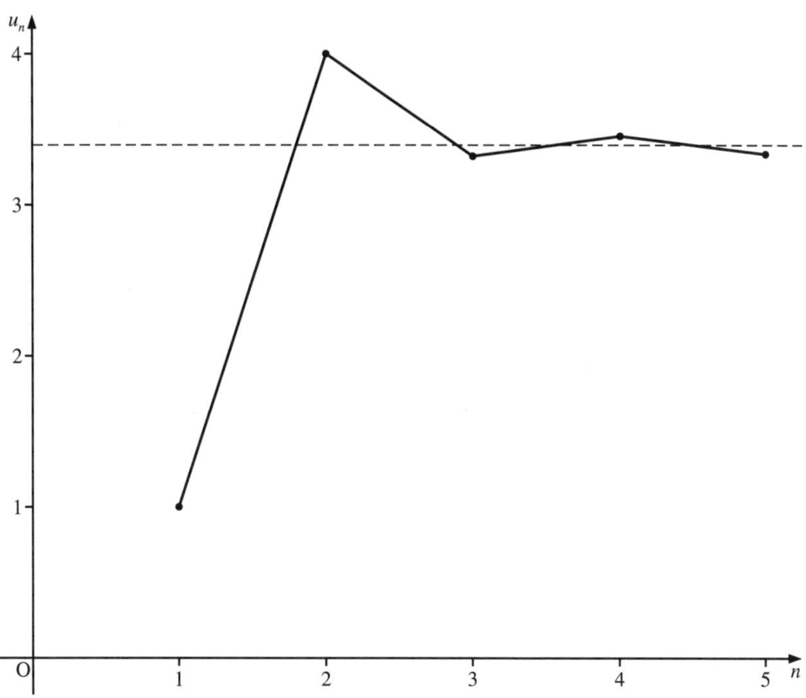

Example 6 A sequence is defined by the recurrence relation

$$u_n = 2u_{n-1} + u_{n-2} \quad \text{for } n > 1$$

Given that $u_5 = 99$, show that $u_7 - 2u_6 = 99$. Given further that $u_8 = 1393$, find the terms u_6 and u_7.

SOLUTION

Let $n = 7$, then $\quad u_7 = 2u_6 + u_5$

$$= 2u_6 + 99$$

$$\therefore \quad u_7 - 2u_6 = 99 \qquad [1]$$

Let $n = 8$, then $\quad u_8 = 2u_7 + u_6$

$$\therefore \quad 1393 = 2u_7 + u_6 \qquad [2]$$

Solving [1] and [2] simultaneously gives $u_6 = 239$ and $u_7 = 577$.

Exercise 9A

1 Write down the first six terms of the following sequences, and determine which of the sequences are convergent and which are divergent. For those which are convergent, determine the limiting value to which they are tending.

a) $u_n = 2n + 1$ **b)** $u_n = 3n - 2$ **c)** $u_n = 5 - 2n$

d) $u_n = n^2 + 3$ **e)** $u_n = \dfrac{1}{n}$ **f)** $u_n = \dfrac{n}{n+1}$

g) $u_n = \dfrac{1}{n^2+1}$ **h)** $u_n = 3 + \dfrac{1}{n(n+1)}$ **i)** $u_n = n(n+1)(n+2)$

j) $u_n = 2^n$ **k)** $u_n = (-1)^n n$ **l)** $u_n = \dfrac{(-1)^{n+1}}{n^2}$

2 Write down an expression, in terms of n, for the nth term of each of these sequences.

a) 4, 8, 12, 16, 20, ... **b)** 5, 7, 9, 11, 13, ... **c)** 4, 9, 14, 19, 24, ...

d) 8, 11, 14, 17, 20, ... **e)** $\frac{1}{2}, \frac{1}{3}, \frac{1}{4}, \frac{1}{5}, \frac{1}{6}, \ldots$ **f)** $\frac{1}{3}, \frac{1}{6}, \frac{1}{9}, \frac{1}{12}, \frac{1}{15}, \ldots$

g) $\frac{2}{5}, \frac{2}{8}, \frac{2}{11}, \frac{2}{14}, \frac{2}{17}, \ldots$ **h)** $\frac{1}{2}, \frac{2}{3}, \frac{3}{4}, \frac{4}{5}, \frac{5}{6}, \ldots$ **i)** $\frac{2}{1}, \frac{3}{4}, \frac{4}{7}, \frac{5}{10}, \frac{6}{13}, \ldots$

j) $\frac{3}{5}, \frac{5}{11}, \frac{7}{17}, \frac{9}{23}, \frac{11}{29}, \ldots$ **k)** $\frac{12}{7}, \frac{11}{12}, \frac{10}{17}, \frac{9}{22}, \frac{8}{27}, \ldots$ **l)** $\frac{4}{5}, \frac{1}{12}, -\frac{2}{19}, -\frac{5}{26}, -\frac{8}{33}, \ldots$

3 Write down an expression, in terms of n, for the nth term of each of these sequences.

a) 2, 4, 8, 16, 32, ... **b)** 10, 20, 40, 80, 160 ... **c)** 5, 10, 20, 40, 80, ...

d) 4, 12, 36, 108, 324, ... **e)** 2, −6, 18, −54, 162, ... **f)** $1, -\frac{1}{2}, \frac{1}{4}, -\frac{1}{8}, \frac{1}{16}, \ldots$

g) 1, 4, 9, 16, 25, ... **h)** $\frac{1}{4}, \frac{2}{9}, \frac{3}{16}, \frac{4}{25}, \frac{5}{36}, \ldots$ **i)** −2, 6, −12, 20, −30, ...

j) $\frac{2}{3}, \frac{3}{8}, \frac{4}{15}, \frac{5}{24}, \frac{6}{35}, \ldots$ **k)** $0, \frac{1}{4}, -\frac{2}{9}, \frac{3}{14}, -\frac{4}{19}, \ldots$ **l)** 1, 4, 27, 256, 3125, ...

4 Write down the first six terms of each of the following sequences, and determine which of the sequences are convergent, which are divergent, and which are periodic. For those which are convergent, determine the limiting value to which they are tending.

a) $u_{n+1} = 2 + u_n,\ u_1 = 5$ **b)** $u_{n+1} = 6 + u_n,\ u_1 = 3$ **c)** $u_{n+1} = 3 - u_n,\ u_1 = 2$

d) $u_{n+1} = 1 + 2u_n,\ u_1 = 3$ **e)** $u_{n+1} = 8 - 3u_n,\ u_1 = 3$ **f)** $u_{n+1} = 10 - u_n,\ u_1 = 5$

g) $u_{n+1} = \dfrac{1}{u_n},\ u_1 = 7$ **h)** $u_{n+1} = \dfrac{2}{u_n^2},\ u_1 = 1$ **i)** $u_{n+1} = u_n^2 - 3,\ u_1 = 2$

j) $u_{n+1} = u_n(u_n - 2),\ u_1 = 1$ **k)** $u_{n+1} = 1 - \dfrac{1}{u_n},\ u_1 = 2$ **l)** $u_{n+1} = u_n + \dfrac{1}{u_n},\ u_1 = 1$

5 Write down the first six terms of each of these sequences.

a) $u_{n+1} = 2u_n + u_{n-1},\ u_1 = 2,\ u_2 = 1$ **b)** $u_{n+1} = u_n + u_{n-1},\ u_1 = 3,\ u_2 = 4$

c) $u_{n+1} = 4u_n - 3u_{n-1},\ u_1 = 2,\ u_2 = 1$ **d)** $u_{n+1} = u_{n-1} - u_n,\ u_1 = 3,\ u_2 = 3$

e) $u_{n+1} = u_{n-1} - 4u_n,\ u_1 = -1,\ u_2 = 0$ **f)** $u_{n+1} = 3u_{n-1} - 4u_n,\ u_1 = 5,\ u_2 = 4$

g) $u_{n+1} = u_n \times u_{n-1},\ u_1 = 1,\ u_2 = 2$ **h)** $u_{n+1} = u_n(u_{n-1} - 2),\ u_1 = 3,\ u_2 = 1$

i) $u_{n+1} = u_n(2u_{n-1} - 3),\ u_1 = 2,\ u_2 = 3$ **j)** $u_{n+1} = u_n^2 - u_{n-1},\ u_1 = 8,\ u_2 = -3$

k) $u_{n+1} = \dfrac{u_n}{u_{n-1}},\ u_1 = 4,\ u_2 = 2$ **l)** $u_{n+1} = \dfrac{u_{n-1} - 1}{u_n},\ u_1 = 7,\ u_2 = 3$

6 A Fibonacci sequence is defined by the recurrence relation $u_{n+1} = u_n + u_{n-1}$, where $u_1 = 1$ and $u_2 = 2$.

a) Write down the first seven terms of the sequence.
b) Given that $u_{15} = 987$ and $u_{12} = 233$, deduce the equations

$$u_{14} + u_{13} = 987 \quad \text{and} \quad u_{14} - u_{13} = 233$$

c) Hence find the value of u_{14}.

7 The sequence $\{u_n\}$ is defined by the recurrence relation $u_{n+1} = u_n + 2u_{n-1}$.

a) Given that $u_{11} = 683$ and $u_8 = 85$, deduce the equations

$$u_{10} + 2u_9 = 683 \quad \text{and} \quad u_{10} - u_9 = 170$$

b) Hence find the value of u_9.

8 A sequence is defined by $v_{n+1} = 3v_n - v_{n-1}$.

a) Given that $v_{10} = 1597$ and $v_7 = 89$, deduce the equations

$$3v_9 - v_8 = 1597 \quad \text{and} \quad 3v_8 - v_9 = 89$$

b) Hence find the values of v_8 and v_9.

9 A sequence is given by $u_n = 2u_{n-1} + u_{n-2}$, where $u_6 = 41$ and $u_9 = 577$.

a) Show that $2u_8 + u_7 = 577$ and $u_8 - 2u_7 = 41$.
b) Hence find the values of u_7 and u_8.

10 For the sequence $u_1, u_2, u_3, \ldots$ the terms are related by $u_{n+1} = u_n - u_{n-1}$, where $u_1 = 1$ and $u_2 = 3$.

a) Show that the sequence is periodic.
b) Find the values of u_{13}, u_{63}, and u_{89}.

11 A sequence is defined by the recurrence relation $u_{n+1} = 3u_n - u_{n-1}$.

a) Rearrange this relation to give an expression for u_{n-1} in terms of u_{n+1} and u_n.
b) Given that $u_{11} = 589$ and $u_{10} = 225$, find the values of u_9, u_8 and u_1.

12 For the sequence $u_1, u_2, u_3, \ldots$ the terms are related by $u_{n+1} = 2u_n + u_{n-1}$, where $u_9 = 338$ and $u_{10} = 816$. Show that $u_1 = u_3 = u_5$.

13 A sequence is defined by $u_{n+1} = 2u_n - u_{n-1}$, where $u_1 = 1$ and $u_2 = 2$. Find an expression for u_n in terms of n.

14 The sequence $\{v_n\}$ is defined by $v_n = 3v_{n-1} - 2v_{n-2}$, where $v_1 = 1$ and $v_2 = 2$. Find an expression for v_n in terms of n.

***15** Show that $u_n = 3 + 2^n$ satisfies the recurrence relation $u_{n+1} = 3u_n - 2u_{n-1}$, where $u_1 = 5$ and $u_2 = 7$. Hence find the value of u_{16}.

***16 a)** Show that the sequence $u_{n+1} = \sqrt{2u_n + 5}$, where $u_1 = 3$, converges to a limit, u, and find the value of u correct to 2 decimal places.
b) Explain why u satisfies the quadratic equation $u^2 - 2u - 5 = 0$.
c) By choosing another value for u_1 and taking the negative square root in part **a**, find the other solution to the equation $u^2 - 2u - 5 = 0$.

Series and sigma notation

A **series** is the sum of the terms of a sequence. We write the sum of the first n terms of a sequence as S_n, where

$$S_n = u_1 + u_2 + u_3 + \ldots + u_n$$

This is an example of a finite series since there is a finite number of terms. It can be expressed more concisely using sigma ($\sum$) notation, as follows:

$$u_1 + u_2 + u_3 + \ldots + u_n = \sum_{r=1}^{n} u_r$$

For example, the finite series $7 + 11 + 15 + 19$ can be written as

$$\sum_{r=1}^{4} (4r + 3)$$

The infinite series $1 + 4 + 9 + 16 + \ldots$ can be written as

$$\sum_{r=1}^{\infty} r^2$$

Example 7 Find the sum of the first four terms of the sequence defined by

$$u_r = (-1)^r\, 3^{r+1} \quad \text{for } r \geqslant 1$$

SOLUTION

We want S_4, the sum of the first four terms, which is given by

$$S_4 = \sum_{r=1}^{4} (-1)^r\, 3^{r+1}$$

$$= (-1)\, 3^2 + (-1)^2\, 3^3 + (-1)^3\, 3^4 + (-1)^4\, 3^5$$

$$= -3^2 + 3^3 - 3^4 + 3^5$$

$$\therefore \quad S_4 = 180$$

The sum of the first four terms is 180.

Example 8 Write each of the following series in $\sum$ notation.

a) $-1 + 2 + 7 + 14 + 23$ **b)** $6 - 7 + 8 - 9 + \ldots$

SOLUTION

a) Consider the series whose terms are defined by $u_r = r^2$:

$$1 + 4 + 9 + 16 + 25$$

It is clear that we want the series whose rth term is given by $u_r = r^2 - 2$. Therefore, the given series can be written as

$$\sum_{r=1}^{5} (r^2 - 2)$$

b) Notice that this is an infinite series. Ignoring the alternating sign, we see that the terms increase by 1. The rth term of the series $6 + 7 + 8 + 9 + \ldots$ is given by $u_r = r + 5$.

Therefore, the given series has its rth term defined by

$$u_r = (-1)^{r+1}(r + 5)$$

and can be written as

$$\sum_{r=1}^{\infty} (-1)^{r+1}(r + 5)$$

The term $(-1)^{r+1}$ simply gives us the alternating sign. When $(r + 1)$ is even, we get a positive term. When $(r + 1)$ is odd, we get a negative term. Notice that this series could also be written as

$$\sum_{r=5}^{\infty} (-1)^{r+1}(r + 1)$$

Example 9 The sum of the first n terms of a series is given by

$$S_n = \frac{1}{2}(5n^2 + n) \quad \text{for } n \geqslant 1$$

Find u_r, an expression for the rth term of the series.

SOLUTION

The rth term can be found by finding the difference between the sum of the first r terms and the sum of the first $(r - 1)$ terms:

$$u_r = (u_1 + u_2 + \ldots u_{r-1} + u_r) - (u_1 + u_2 + \ldots + u_{r-1})$$

$$\therefore \quad u_r = S_r - S_{r-1}$$

Substituting for S_r and S_{r-1} gives

$$u_r = \frac{1}{2}(5r^2 + r) - \frac{1}{2}[5(r - 1)^2 + (r - 1)]$$

$$= \frac{1}{2}[5r^2 + r - 5(r - 1)^2 - (r - 1)]$$

$$= \frac{1}{2}[5r^2 + r - 5(r^2 - 2r + 1) - r + 1]$$

$$= \frac{1}{2}(10r - 4)$$

$$\therefore \quad u_r = 5r - 2$$

The rth term of the series is given by $u_r = 5r - 2$.

Convergent and divergent series

We have seen that a sequence can converge, diverge, oscillate and converge, or oscillate and diverge.

In exactly the same way, a series can also follow one of these patterns. For example, the series

$$1 + \frac{1}{2} + \frac{1}{4} + \frac{1}{8} + \dots$$

converges. This can be seen simply using a calculator to add on the next term of the sequence. This gives

n	1	2	3	4	5	6	7
S_n	1	1.5	1.75	1.875	1.9375	1.968 75	1.984 375

Drawing a graph of S_n against n gives

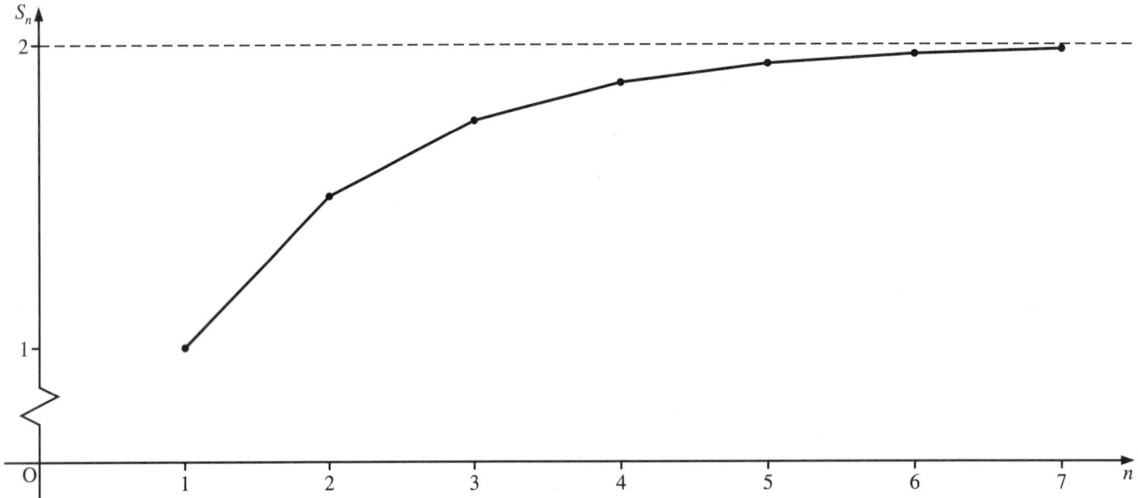

It is clear from the graph that the series is converging to the value 2. In other words, the limit of the series is 2.

However, the series

$$1^3 + 2^3 + 3^3 + \dots$$

does not converge to a particular value and is therefore a divergent series.

To see that a series can also oscillate, consider

$$-1 + \frac{1}{4} - \frac{1}{16} + \frac{1}{64} + \dots$$

Drawing the graph of S_n against n gives

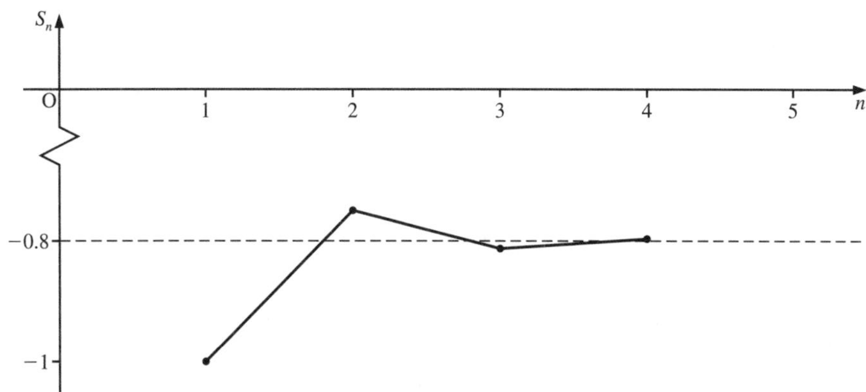

It is clear from the graph that the series oscillates and converges to −0.8.

Example 10 The terms of a series are defined by $u_r = 3^{(1-r)}$, for $r \geqslant 1$. Draw a graph of S_n against n for $n = 1, 2, 3, 4, 5, 6$. Determine whether the series is convergent or divergent.

SOLUTION

The series is

$$1 + \frac{1}{3} + \frac{1}{9} + \frac{1}{27} + \dots$$

Constructing a table of values for S_n (to three decimal places) against n gives

n	1	2	3	4	5	6
S_n	1	1.333	1.444	1.481	1.494	1.498

Drawing a graph of S_n against n gives

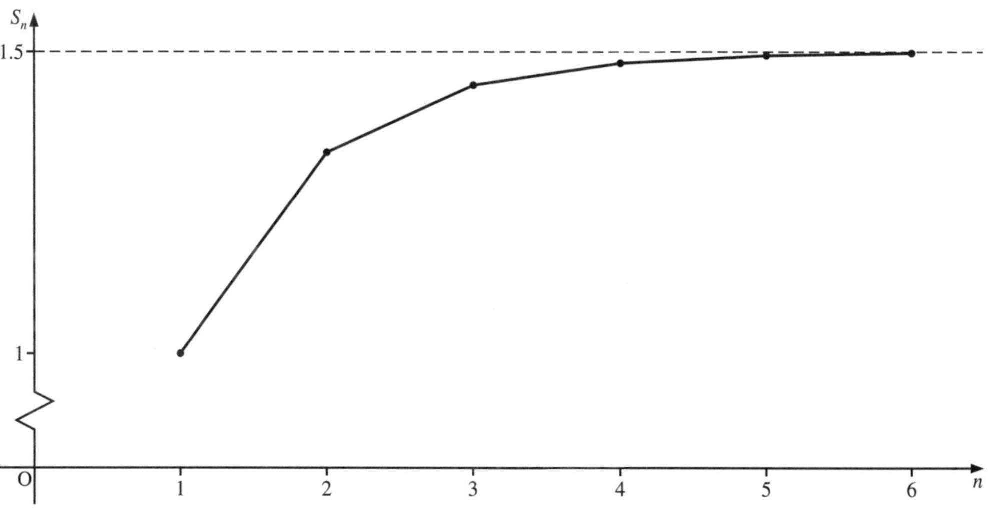

It is clear from both the table of values and the graph that the series is converging to 1.5.

Exercise 9B

1 Write down all the terms in each of these series.

a) $\sum_{r=1}^{5} r^2$ **b)** $\sum_{r=1}^{6} (3r-1)$ **c)** $\sum_{r=1}^{4} (2r^2+3)$ **d)** $\sum_{r=1}^{5} r(r+1)$

e) $\sum_{r=3}^{6} r^3$ **f)** $\sum_{r=5}^{10} r(r-3)$ **g)** $\sum_{r=0}^{5} (2r+1)^2$ **h)** $\sum_{r=1}^{n} \frac{1}{r}$

i) $\sum_{r=4}^{8} r^r$ **j)** $\sum_{r=1}^{7} \frac{(-1)^{r-1}}{r}$ **k)** $\sum_{r=1}^{10} [1-(-1)^r]r^2$ **l)** $\sum_{r=1}^{5} 3$

2 Write each of these series in $\sum$ notation.

a) $1+2+3+4+5$

b) $1^3+2^3+3^3+4^3+5^3+6^3+7^3$

c) $7+10+13+16+19+22+25$

d) $\frac{1}{3}+\frac{1}{4}+\frac{1}{5}+\ldots+\frac{1}{20}$

e) $5\times 6+6\times 7+7\times 8+\ldots+18\times 19$

f) $3^4+4^4+5^4+\ldots+n^4$

g) $1-2+3-4+5-6+7$

h) $4-8+16-32+64-128+256-512+1024$

i) $\frac{5}{5^2-1}+\frac{6}{6^2-1}+\frac{7}{7^2-1}+\ldots+\frac{n}{n^2-1}$

j) $\frac{1}{2\times 3}+\frac{2}{3\times 4}+\frac{3}{4\times 5}+\ldots+\frac{n}{(n+1)(n+2)}$

k) $1\times 4-3\times 7+5\times 10-\ldots+29\times 46$

l) $\frac{2}{3}+\frac{5}{9}+\frac{8}{27}+\frac{11}{81}+\frac{14}{243}+\frac{17}{729}$

***3** **a)** Sketch the graph of $y=\frac{1}{x^2}$, for $x>0$.

b) By comparing the area under your graph, for values of x between $x=r-1$ and $x=r$, with the area of an appropriate rectangle over the same interval, show that

$$\frac{1}{r^2}<\int_{r-1}^{r}\frac{1}{x^2}\,dx \quad \text{where } r\geqslant 2$$

c) Deduce that

$$\sum_{r=2}^{N}\frac{1}{r^2}<\int_{1}^{N}\frac{1}{x^2}\,dx$$

d) Hence show that

$$\sum_{r=1}^{N}\frac{1}{r^2}<2 \quad \text{for all } N$$

Arithmetic progressions

Consider the sequence of numbers 1, 3, 5, 7, ... Each term can be obtained from the previous term by adding 2. This sequence is an example of an arithmetic progression.

An **arithmetic progression** (AP) is a sequence of numbers in which any term can be obtained from the previous term by adding a certain number called the **common difference**.

- The first term of an AP is denoted by a.
- Its common difference is denoted by d.

In our example, $a = 1$ and $d = 2$.

Generally, the terms of an AP are given by

$a,$	$a + d,$	$a + 2d, \ldots,$	$a + (n - 1)d, \ldots$
1st	2nd	3rd	nth

where $a + (n - 1)d$ is the nth term.

Result I: Sum of the first *n* terms of an AP

The sum of the first n terms of an AP is given by

$$S_n = \frac{n}{2}[2a + (n-1)d]$$

$$\text{or} \quad S_n = \frac{n}{2}(a + l)$$

where l is the last term.

Proof

Consider the sum of the first n terms of an AP:

$$S_n = a + [a + d] + \ldots + [a + (n-1)d] \qquad [1]$$

Writing the terms on the right in reverse order gives

$$S_n = [a + (n-1)d] + [a + (n-2)d] + \ldots + a \qquad [2]$$

Adding [1] and [2] gives

$$\begin{aligned} S_n &= \{a + [a + (n-1)d]\} + \{(a + d) + [a + (n-2)d]\} + \ldots + \{[a + (n-1)d] + a\} \\ &= [2a + (n-1)d] + [2a + (n-1)d] + \ldots + [2a + (n-1)d] \\ &= n[2a + (n-1)d] \end{aligned}$$

$$\therefore \quad S_n = \frac{n}{2}[2a + (n-1)d]$$

as required.

Alternatively,

$$S_n = \frac{n}{2}[a + a + (n-1)d]$$

$$= \frac{n}{2}(\text{1st term} + n\text{th term})$$

$$\therefore \quad S_n = \frac{n}{2}(a + l)$$

where l is the nth term.

Example 11 The first four terms of an AP are 5, 11, 17 and 23. Find the 30th term and the sum of the first 30 terms.

SOLUTION

It is clear that for this AP we have $a = 5$ and $d = 6$.

The 30th term is given by

$$u_{30} = 5 + (30-1)6 = 5 + (29)6$$

$$\therefore \quad u_{30} = 179$$

The 30th term is 179.

The sum of the first 30 terms is given by

$$S_{30} = \frac{30}{2}[2(5) + (30-1)6] = 15[10 + (29)6]$$

$$\therefore \quad S_{30} = 2760$$

The sum of the first 30 terms is 2760.

Example 12 An AP has a first term of 2 and an nth term of 32. Given that the sum of the first n terms is 357, find n and the common difference of the AP.

SOLUTION

Since the nth term is 32, we have

$$a + (n-1)d = 32$$

We also know that $a = 2$. Therefore,

$$2 + (n-1)d = 32$$

$$\therefore \quad (n-1)d = 30 \qquad [1]$$

Since the sum of the first n terms is 357, we have

$$\frac{n}{2}[2a + (n-1)d] = 357$$

We also know that $a = 2$. Therefore,

$$\frac{n}{2}[2(2) + (n-1)d] = 357$$

$$\therefore \quad n[4 + (n-1)d] = 714 \qquad [2]$$

Substituting [1] into [2] gives

$$n(4+30) = 714$$

$$\therefore \quad 34n = 714$$

$$\therefore \quad n = 21$$

Substituting $n = 21$ into [1] gives

$$(21-1)d = 30$$

$$\therefore \quad d = \frac{3}{2}$$

The value of n is 21 and the common difference is $\frac{3}{2}$.

Example 13 The sum of the first five terms of an AP is $\frac{65}{2}$. Also, five times the 7th term is the same as six times the 2nd term. Find the first term and common difference of the AP.

SOLUTION

Since the sum of the first five terms is $\frac{65}{2}$, we have

$$\frac{5}{2}[2a+(5-1)d] = \frac{65}{2}$$

$$\therefore \quad 2a+4d = 13 \qquad [1]$$

We also know that five times the 7th term is the same as six times the 2nd term. That is,

$$5(a+6d) = 6(a+d)$$

$$\therefore \quad a-24d = 0 \qquad [2]$$

Solving [1] and [2] simultaneously gives $a = 6$ and $d = \frac{1}{4}$.

The first term of the AP is 6 and the common difference is $\frac{1}{4}$.

Example 14 The 3rd, 5th and 8th terms of an AP are $3x+8$, $x+24$ and x^3+15 respectively. Find the value of x and hence the common difference of the AP.

SOLUTION

Since the 3rd, 5th and 8th terms of the AP are $3x+8$, $x+24$ and x^3+15, we have

$$a+2d = 3x+8 \qquad [1]$$

$$a+4d = x+24 \qquad [2]$$

$$\text{and} \quad a+7d = x^3+15 \qquad [3]$$

Subtracting [1] from [2] gives

$$2d = -2x+16$$

$$\therefore \quad d = -x+8 \qquad [4]$$

Subtracting [1] from [3] gives

$$5d = x^3-3x+7 \qquad [5]$$

Substituting [4] into [5] gives

$$5(-x+8) = x^3 - 3x + 7$$
$$\therefore \quad -5x + 40 = x^3 - 3x + 7$$
$$\therefore \quad x^3 + 2x - 33 = 0$$
$$\therefore \quad (x-3)(x^2 + 3x + 11) = 0$$

Solving gives $x = 3$ or $x^2 + 3x + 11 = 0$. Since $x^2 + 3x + 11 = 0$ has no real solutions, we have $x = 3$. The common difference of the AP is given by $d = -x + 8$. Therefore, $d = -3 + 8 = 5$.

Exercise 9C

1 Decide which of the following series are APs. For those which are, write down the value of the common difference.

a) $8 + 11 + 14 + 17 + 20 + 23$
b) $83 + 72 + 61 + 50 + 39 + 28$
c) $1 + 2 + 4 + 8 + 16 + 32$
d) $1 + 1.1 + 1.11 + 1.111 + 1.1111 + 1.11111$
e) $1 + 1.1 + 1.2 + 1.3 + 1.4 + 1.5$
f) $\frac{1}{2} + \frac{4}{5} + \frac{11}{10} + \frac{7}{5} + \frac{17}{10} + 2$
g) $1 + \frac{1}{2} + \frac{1}{3} + \frac{1}{4} + \frac{1}{5}$
h) $1 - 2 + 3 - 4 + 5 - 6$
i) $-1 - 2 - 3 - 4 - 5 - 6$
j) $3\frac{1}{8} + 4\frac{1}{2} + 5\frac{7}{8} + 7\frac{1}{4} + 8\frac{5}{8} + 10$
k) $-9 - 7\frac{5}{6} - 6\frac{2}{3} - 5\frac{1}{2} - 4\frac{1}{3} - 3\frac{1}{6}$
l) $3a + 4a + 5a + 6a + 7a + 8a$

2 Write down the term indicated in square brackets in each of the following APs.

a) $1 + 5 + 9 + \dots$ [10th term]
b) $7 + 9 + 11 + \dots$ [30th term]
c) $20 + 17 + 14 + \dots$ [16th term]
d) $-6 - 11 - 16 - \dots$ [12th term]
e) $1.2 + 1.4 + 1.6 + \dots$ [14th term]
f) $81 + 77 + 73 + \dots$ [nth term]
g) $0.1 - 0.2 - 0.5 - \dots$ [25th term]
h) $1\frac{1}{2} + 2 + 2\frac{1}{2} + \dots$ [100th term]
i) $\frac{1}{6} + \frac{1}{3} + \frac{1}{2} + \dots$ [nth term]
j) $5\frac{1}{4} + 6\frac{1}{2} + 7\frac{3}{4} + \dots$ [16th term]
k) $a + 3a + 5a + \dots$ [nth term]
l) $\frac{5}{4}x + \frac{3}{2}x + \frac{7}{4}x + \dots$ [nth term]

3 Find the sum, as far as the term indicated in square brackets, of each of these APs.

a) $1 + 2 + 3 + \dots$ [10th term]
b) $5 + 7 + 9 + \dots$ [25th term]
c) $4 + 9 + 14 + \dots$ [18th term]
d) $60 + 55 + 50 + \dots$ [12th term]
e) $9 + 5 + 1 + \dots$ [20th term]
f) $7 + 10 + 13 + \dots$ [nth term]
g) $9 - 1 - 11 - \dots$ [25th term]
h) $-2 - \frac{1}{2} + 1 + \dots$ [30th term]
i) $4 + 3\frac{1}{3} + 2\frac{2}{3} + \dots$ [100th term]
j) $100 + 98 + 96 + \dots$ [101st term]
k) $b + 7b + 13b + \dots$ [nth term]
l) $14c + 4c - 6c - \dots$ [nth term]

4 Find the number of terms in each of these APs.

a) $5 + 6 + 7 + \dots + 15$
b) $10 + 20 + 30 + \dots + 210$
c) $5 + 8 + 11 + \dots + 302$
d) $-8 - 6 - 4 - \dots + 78$
e) $97 + 85 + 73 + \dots + 13$
f) $46 + 42 + 38 + \dots - 26$
g) $9 - 11 - 31 - \dots - 571$
h) $2.1 + 3.2 + 4.3 + \dots + 31.8$
i) $\frac{1}{2} + 1\frac{1}{6} + 1\frac{5}{6} + \dots + 19\frac{5}{6}$
j) $11\frac{1}{2} + 9 + 6\frac{1}{2} + \dots - 23\frac{1}{2}$
k) $7 + 9 + \dots + (2n + 1)$
l) $-9 - 4 + 1 + \dots + (5n - 4)$

5 Find the sum of each of these APs.

a) $1 + 2 + 3 + \ldots + 100$

b) $6 + 8 + 10 + \ldots + 30$

c) $9 + 13 + 17 + \ldots + 41$

d) $62 + 60 + 58 + \ldots + 38$

e) $8 + 3 - 2 - \ldots - 42$

f) $1.3 + 1.6 + 1.9 + \ldots + 4.6$

g) $3\frac{1}{3} + 4 + 4\frac{2}{3} + \ldots + 12\frac{2}{3}$

h) $9\frac{1}{5} + 8\frac{4}{5} + 8\frac{2}{5} + \ldots + 3\frac{3}{5}$

i) $7\frac{1}{2} + 6\frac{1}{4} + 5 + \ldots - 15$

j) $1 + 2 + 3 + \ldots + n$

k) $1 + 2 + 3 + \ldots + 2n$

l) $1 + 3 + 5 + \ldots + (2n - 1)$

6 **a)** Prove that $\sum_{r=1}^{n} r = \frac{n}{2}(n + 1)$.

b) Use your answer to part **a** to deduce the following.

i) $\sum_{r=1}^{n} (3r - 1) = \frac{n}{2}(3n + 1)$

ii) $\sum_{r=1}^{n} (5r - 3) = \frac{n}{2}(5n - 1)$

iii) $\sum_{r=0}^{n} (r + 5) = \frac{1}{2}(n + 1)(n + 10)$

iv) $\sum_{r=3}^{n} (2r + 3) = (n + 6)(n - 2)$

v) $\sum_{r=6}^{n} (4 + 3r) = \frac{1}{2}(n - 5)(3n + 26)$

vi) $\sum_{r=n+1}^{2n} r = \frac{n}{2}(3n + 1)$

7 In an AP, the 1st term is 13 and the 15th term is 111. Find the common difference and the sum of the first 20 terms.

8 In an arithmetic series, the 3rd term is 4 and the 8th term is 49. Find the 1st term, the common difference and the sum of the first ten terms.

9 The 2nd term of an AP is 7 and the 7th term is –8. Find the 1st term, the common difference and the sum of the first 14 terms.

10 The 5th term of an arithmetic series is 7 and the common difference is 4. Find the 1st term and the sum of the first ten positive terms.

11 The sum of the first ten terms of an AP is 95, and the sum of the first 20 terms of the same AP is 290. Calculate the 1st term and the common difference.

12 Given that both the sum of the first ten terms of an AP and the sum of the 11th and 12th terms of the same AP are equal to 60, find the 1st term and the common difference.

13 The 17th term of an AP is 22, and the sum of the first 17 terms is 102. Find the 1st term, the common difference and the sum of the first 30 terms.

14 The 8th term of an arithmetic series is 5 and the sum of the first 16 terms is 84. Calculate the sum of the first ten terms.

15 An AP has 1st term 2 and common difference 5. Given that the sum of the first n terms of the progression is 119, calculate the value of n.

16 The 1st term of an arithmetic series is 38 and the tenth term is 2. Given that the sum of the first n terms of the series is 72, calculate the possible values of n.

17 Find how many terms of the AP $3 + 8 + 13 + \ldots$ should be taken in order that the total should exceed 200.

18 An AP has 1st term 6 and common difference 12. Calculate how many terms should be taken in order that the total should exceed 500.

19 A child is collecting conkers. He collects six conkers on the first day of the month and stores them in a box. On the second day of the month he collects another ten conkers, and adds them to his box. He continues in this way, each day collecting four conkers more than he collected on the previous day. Find the day of the month on which the number of conkers in his box will first exceed 1000.

20 The sum of the first n terms of a series is $n(n+2)$. Find the first three terms of the series.

21 The sum of the first n terms of a series is $\frac{n}{2}(n+8)$. Find the 1st, 2nd and 10th terms.

22 An AP has a common difference of 3. Given that the nth term is 32, and the sum of the first n terms is 185, calculate the value of n.

23 The 1st, 2nd and 3rd terms of an AP are $8 - x$, $3x$ and $4x + 1$, respectively. Calculate the value of x, and find the sum of the first eight terms of the progression.

24 Given that the 2nd, 3rd and 4th terms of an AP are $16 - x$, $3x - 2$ and $2x$, respectively, calculate the value of x, and find the 1st term of the progression.

25 The 1st, 2nd and 4th terms of an arithmetic series are $11 - x$, $2x + 1$ and $3x + 1$, respectively. Calculate the value of x, and find the sum of the first 12 terms of the series.

26 **a)** Find the sum of the integers from 1 to 100.
b) Find the sum of the integers from 1 to 100 which are divisible by 3.
c) Hence find the sum of the integers from 1 to 100 which are not divisible by 3.

27 Find the sum of the integers from 1 to 200 which are not divisible by 5.

***28** In an AP the nth term is 11, the sum of the first n terms is 72, and the first term is $\frac{1}{n}$. Find the value of n.

***29** Given that a^2, b^2 and c^2 are in arithmetic progression show that $\frac{1}{b+c}$, $\frac{1}{c+a}$ and $\frac{1}{a+b}$ are also in arithmetic progression.

***30** Given that

$$\sum_{r=n+3}^{2n} r = 312$$

find the value of n.

Geometric progressions

Consider the sequence of numbers 2, 6, 18, 54, ... Each term of the sequence can be obtained from the previous term by multiplying by 3. This is an example of a geometric progression.

A **geometric progression** (GP) is a sequence of numbers in which any term can be obtained from the previous term by multiplying by a certain number called the **common ratio**.

- The first term of a GP is denoted by a.
- Its common ratio is denoted by r.

In our example, $a = 2$ and $r = 3$.

Generally, the terms of a GP are given by

$$a, ar, ar^2, ar^3, \ldots, ar^{n-1}, \ldots$$

where ar^{n-1} is the nth term.

Result II: Sum of the first *n* terms of a GP

The sum of the first n terms of a GP is given by

$$S_n = a\left(\frac{1-r^n}{1-r}\right) = a\left(\frac{r^n-1}{r-1}\right)$$

Proof

The sum of the first n terms is

$$S_n = a + ar + ar^2 + \ldots + ar^{n-1} \qquad [1]$$

Multiplying throughout by r gives

$$rS_n = \quad ar + ar^2 + ar^3 + \ldots + ar^n \qquad [2]$$

Subtracting [2] from [1] gives

$$S_n - rS_n = (a + ar + \ldots + ar^{n-1}) - (ar + ar^2 + \ldots + ar^n)$$

$$\therefore \quad S_n(1-r) = a - ar^n$$

$$\therefore \quad S_n = \frac{a(1-r^n)}{1-r} = a\left(\frac{1-r^n}{1-r}\right) \qquad [3]$$

as required.

Multiplying both the numerator and the denominator of [3] by -1 gives

$$S_n = a\left(\frac{r^n-1}{r-1}\right)$$

which is an alternative form.

Example 15 A GP has a 1st term of 1 and a common ratio of $\frac{1}{4}$. Find the sum of the first four terms and show that the nth term is given by $4^{(1-n)}$.

SOLUTION

The sum of the first four terms is given by

$$S_4 = 1\left[\frac{1-(\frac{1}{4})^4}{1-(\frac{1}{4})}\right] \quad \therefore \quad S_4 = \frac{85}{64}$$

The sum of the first four terms is $\frac{85}{64}$.

The nth term is given by

$$\begin{aligned} u_n &= ar^{n-1} \\ &= 1(\tfrac{1}{4})^{n-1} = (4^{-1})^{n-1} \\ \therefore \quad u_n &= 4^{(1-n)} \end{aligned}$$

Example 16 The sum of the 2nd and 3rd terms of a GP is 12. The sum of the 3rd and 4th terms is -36. Find the first term and the common ratio.

SOLUTION

Since the sum of the 2nd and 3rd terms is 12, we have

$$\begin{aligned} ar + ar^2 &= 12 \\ \therefore \quad ar(1+r) &= 12 \\ \therefore \quad 1 + r &= \frac{12}{ar} \qquad [1] \end{aligned}$$

Since the sum of the 3rd and 4th terms is -36, we have

$$\begin{aligned} ar^2 + ar^3 &= -36 \\ \therefore \quad ar^2(1+r) &= -36 \qquad [2] \end{aligned}$$

Substituting [1] into [2] gives

$$\begin{aligned} ar^2\left(\frac{12}{ar}\right) &= -36 \\ \therefore \quad 12r &= -36 \\ \therefore \quad r &= -3 \end{aligned}$$

From [1] we obtain

$$a = \frac{12}{r(1+r)}$$

Substituting $r = -3$ gives

$$a = \frac{12}{(-3)(1-3)} = 2$$

The 1st term of the GP is 2 and the common ratio is -3.

Example 17 Show that there are two possible GPs in each of which the 1st term is 8 and the sum of the first three terms is 14. For the GP with positive common ratio find, in term of n, an expression for the sum of the first n terms.

SOLUTION

Since the 1st term is 8 and the sum of the first three terms is 14, we have

$$8 + 8r + 8r^2 = 14$$

$$\therefore \quad 8r^2 + 8r - 6 = 0$$

$$\therefore \quad 2(2r - 1)(2r + 3) = 0$$

Solving gives $r = \frac{1}{2}$ or $r = -\frac{3}{2}$.

Hence, there are two GPs which have a first term of 8 and have the sum of their first three terms equal to 14, namely, one with a common ratio of $\frac{1}{2}$ and a second with a common ratio of $-\frac{3}{2}$.

To find the sum of the first n terms of the GP with positive common ratio, we use

$$S_n = a\left(\frac{1 - r^n}{1 - r}\right) \quad \text{with } a = 8 \text{ and } r = \tfrac{1}{2}$$

This gives

$$S_n = 8\left[\frac{1 - (\frac{1}{2})^n}{1 - (\frac{1}{2})}\right] = 16[1 - (2^{-1})^n]$$

$$\therefore \quad S_n = 16(1 - 2^{-n})$$

Exercise 9D

1 Decide which of the following series are GPs. For those which are, write down the value of the common ratio.

a) $2 + 6 + 18 + 54 + 162 + 486$
b) $3 - 6 + 12 - 24 + 48 - 96$
c) $3 + 9 + 15 + 21 + 27 + 33$
d) $1 + 1.1 + 1.11 + 1.111 + 1.1111 + 1.111\,11$
e) $1 + 1.2 + 1.44 + 1.728 + 2.0736 + 2.488\,32$
f) $1 + 2 - 4 - 8 + 16 + 32$
g) $1 + \frac{1}{3} + \frac{1}{6} + \frac{1}{9} + \frac{1}{12} + \frac{1}{15}$
h) $1 + \frac{1}{2} + \frac{1}{4} + \frac{1}{8} + \frac{1}{16} + \frac{1}{32}$
i) $-1\frac{9}{32} + 2\frac{9}{16} - 5\frac{1}{8} + 10\frac{1}{4} - 20\frac{1}{2} + 41$
j) $1 - 2 + 3 - 4 + 5 - 6$
k) $3 - 3 + 3 - 3 + 3 - 3$
l) $1 + a + a^2 + a^3 + a^4 + a^5$

2 Write down the term indicated in square brackets in each of the following GPs.

a) $2 + 4 + 8 + \ldots$ [10th term]
b) $1 - 3 + 9 + \ldots$ [7th term]
c) $5 + 10 + 20 + \ldots$ [8th term]
d) $2 + 3 + 4\frac{1}{2} + \ldots$ [9th term]
e) $81 - 54 + 36 - \ldots$ [8th term]
f) $2 + \frac{2}{5} + \frac{2}{25} + \ldots$ [5th term]
g) $1 + \frac{1}{2} + \frac{1}{4} + \ldots$ [12th term]
h) $1 - \frac{1}{3} + \frac{1}{9} - \ldots$ [6th term]
i) $36 + 24 + 16 + \ldots$ [7th term]
j) $7 - 7 + 7 - \ldots$ [100th term]
k) $x + x^2 + x^3 + \ldots$ [nth term]
l) $a - ar + ar^2 + \ldots$ [nth term]

3 Find the sum, as far as the term indicated in square brackets, of each of the following GPs.

a) $3 + 6 + 12 + \ldots$ [10th term]
b) $3 - 6 + 12 - \ldots$ [10th term]
c) $3 - 6 + 12 - \ldots$ [11th term]
d) $5 + 10 + 20 + \ldots$ [8th term]
e) $-2 + 8 - 32 + \ldots$ [6th term]
f) $1 + 10 + 100 + \ldots$ [7th term]
g) $1 + \frac{1}{3} + \frac{1}{9} + \ldots$ [7th term]
h) $\frac{8}{9} + \frac{4}{3} + 2 + \ldots$ [6th term]
i) $\frac{1}{2} - \frac{1}{4} + \frac{1}{8} + \ldots$ [12th term]
j) $1 + 1.1 + 1.21 + \ldots$ [6th term]
k) $\frac{1}{2} + \frac{1}{4} + \frac{1}{8} + \ldots$ [nth term]
l) $x + x^2 + x^3 + \ldots$ [nth term]

4 Find the number of terms in each of these geometric progressions.

a) $2 + 10 + 50 + \ldots + 1250$
b) $3 + 6 + 12 + \ldots + 768$
c) $2 + 6 + 18 + \ldots + 1458$
d) $1 - 2 + 4 - \ldots + 1024$
e) $4 - 12 + 36 - \ldots - 972$
f) $5 + 20 + 80 + \ldots + 5120$
g) $54 + 18 + 6 + \ldots + \frac{2}{27}$
h) $64 + 32 + 16 + \ldots + \frac{1}{8}$
i) $0.01 - 0.03 + 0.09 - \ldots + 65.61$
j) $0.03 + 0.12 + 0.48 + \ldots + 491.52$
k) $1\frac{11}{16} + 1\frac{1}{8} + \frac{3}{4} + \ldots + \frac{16}{243}$
l) $1\frac{1}{2} - 2\frac{1}{4} + 3\frac{3}{8} + \ldots - 25\frac{161}{256}$

5 Find the sum of each of these GPs.

a) $3 + 6 + 12 + \ldots + 384$
b) $2 + 6 + 18 + \ldots + 1458$
c) $4 - 12 + 36 - \ldots - 972$
d) $7 - 14 + 28 - \ldots + 448$
e) $36 + 12 + 4 + \ldots + \frac{4}{27}$
f) $20 + 10 + 5 + \ldots + \frac{5}{16}$
g) $\frac{1}{4} + \frac{1}{16} + \frac{1}{64} + \ldots + \frac{1}{4096}$
h) $\frac{1}{3} - \frac{1}{9} + \frac{1}{27} - \ldots - \frac{1}{729}$
i) $1 + 1.1 + 1.21 + \ldots + 1.771\,561$
j) $8 - 12 + 18 + \ldots - 136.6875$
k) $1 + \frac{1}{2} + \frac{1}{4} + \ldots + \frac{1}{2^n}$
l) $5 - 10 + 20 - \ldots + 5 \times (-2)^{n-1}$

6 A GP has 3rd term 75 and 4th term 375. Find the common ratio and the first term.

7 In a GP the 2nd term is -12 and the 5th term is 768. Find the common ratio and the first term.

8 The 4th term of a geometric series is 48, and the 6th term is 12. Find the possible values of the common ratio and the corresponding values of the 1st term.

9 A GP has 3rd term 7 and 5th term 847. Find the possible values of the common ratio, and the corresponding values of the 4th term.

10 Find the sum of the first ten terms of a GP which has 3rd term 20 and 8th term 640.

11 In a GP the 2nd term is 15 and the 5th term is -405. Find the sum of the first eight terms.

12 A GP has common ratio -3. Given that the sum of the first nine terms of the progression is 703, find the 1st term.

13 Find the 1st term of the geometric series in which the common ratio is 2 and the sum of the first ten terms is 93.

14 The common ratio of a GP is -5 and the sum of the first seven terms of the progression is 449. Find the first three terms.

15 A GP has 1st term $\frac{1}{11}$ and common ratio 2. Given that the sum of the first n terms is 93, calculate the value of n.

16 Find how many terms of the GP $5 - 10 + 20 - \ldots$ should be taken in order that the total should equal 215.

17 In a geometric series the 1st term is 8 and the sum of the first three terms is 104. Calculate the possible values of the common ratio, and, in each case, write down the corresponding first three terms of the series.

18 Given that the 1st term of a GP is 5 and the sum of the first three terms is 105, find the possible values of the common ratio, and, in each case, write down the corresponding values of the first three terms of the series.

19 In a GP the sum of the 2nd and 3rd terms is 12, and the sum of the 3rd and 4th terms is 60. Find the common ratio and the 1st term.

20 A GP is such that the sum of the 4th and 5th terms is -108, and the sum of the 5th and 6th terms is 324. Calculate the common ratio and the value of the 1st term.

21 Find the first five terms in the geometric series which is such that the sum of the 1st and 3rd terms is 50, and the sum of the 2nd and 4th terms is 150.

22 In the geometric series $u_1 + u_2 + u_3 + \ldots$, $u_1 + u_3 = 26$ and $u_3 + u_5 = 650$. Find the possible values of u_4.

23 The sum of the 1st and 4th terms of a GP is 430, and the sum of the 2nd and 5th terms of the same GP is -2580. Find the common ratio and the 1st term.

24 In a GP in which all the terms are positive and increasing, the difference between the 7th and 5th terms is 192, and the difference between the 4th and 2nd terms is 24. Find the common ratio and the 1st term.

25 In the geometrical series $\sum_{r=1}^{n} u_r$, $u_5 - u_2 = 156$, and $u_7 - u_4 = 1404$. Find the possible values of the common ratio and the corresponding values of u_1.

26 A child tries to negotiate a new deal for her pocket money for the 30 days of the month of June. She wants to be paid 1 p on the 1st of the month, 2 p on the 2nd of the month, and, in general, (2^{n-1}) p on the nth day of the month. Calculate how much she would get, in total, if this were accepted.

27 A man, who started work in 1990, planned an investment for his retirement in 2030 in the following way. On the first day of each year, from 1990 to 2029 inclusive, he is to place £100 in an investment account. The account pays 10% compound interest per annum, and interest is added on the 31 December of each year of the investment. Calculate the value of his investment on 1 January 2030.

28 A woman borrows £50 000 in order to buy a house. Compound interest at the rate of 12% per annum is charged on the loan. She agrees to pay back the loan in 25 equal instalments, at yearly intervals, the first repayment being made exactly one year after the loan is taken out. Calculate the value of each instalment.

29 The 3rd, 4th and 5th terms of a GP are $x-2$, $x+1$, and $x+7$. Calculate the value of x and write down the first three terms of the progression.

30 Given that the 4th, 5th and 6th terms of a GP are $x+1$, $x-1$, and $2x-5$, write down the possible values of the first three terms of the progression.

31 The first three terms of a GP are $x-1$, $x+2$ and $3x$. Calculate the possible values of x and write down the corresponding values of the first three terms.

***32** Given that $x-5$, $x-2$ and $3x$ are the 1st, 2nd and 4th terms of a GP, find the three possible values of x.

***33** Given that a, b, c and d are in geometric progression prove that

$$(b-c)^2+(c-a)^2+(d-b)^2=(a-d)^2$$

Infinite geometric progressions

On page 238, we looked at the series

$$1+\frac{1}{2}+\frac{1}{4}+\frac{1}{8}+\ldots+\left(\frac{1}{2}\right)^{n-1}+\ldots$$

which can be written as

$$\sum_{n=1}^{\infty}\left(\frac{1}{2}\right)^{n-1}$$

We are now in a position to identify this series as the sum of an infinite GP in which $a=1$ and $r=\frac{1}{2}$. Therefore,

$$S_n=\left[\frac{1-(\frac{1}{2})^n}{1-\frac{1}{2}}\right]=2\left[1-\left(\frac{1}{2}\right)^n\right]$$

If we look at S_n for $n=2$, 10, 20 and 30, we have

n	2	10	20	30
S_n	1.5	1.998	1.999 998 093	1.999 999 998

As $n\to\infty$, the term $(\frac{1}{2})^n\to 0$, therefore $S_n\to 2$. We called this the **limit** of the series. That is,

$$\lim_{n\to\infty} S_n=2$$

Since the sequence is a GP, we call this limit the **sum to infinity** of the GP.

Result III: Sum to infinity of a GP

The sum to infinity of a GP is given by

$$S_\infty = \sum_{n=1}^{\infty} ar^{n-1} = \frac{a}{1-r}$$

where $-1 < r < 1$.

Proof

The sum of the first n terms of a GP is given by

$$S_n = a\left(\frac{1-r^n}{1-r}\right)$$

If $-1 < r < 1$ then as $n \longrightarrow \infty$, $r^n \longrightarrow 0$. Therefore, as $n \longrightarrow \infty$ we have

$$S_n \longrightarrow a\left(\frac{1-0}{1-r}\right) = \frac{a}{1-r}$$

Notice that the proof of this result hangs on the fact that $-1 < r < 1$. If this is not the case, the sum to infinity does not exist.

The sum to infinity of a GP in which $-1 < r < 1$ is given by

$$S_\infty = \frac{a}{1-r}$$

as required.

Example 18 Calculate the sum to infinity of the series $2 + \frac{1}{2} + \frac{1}{8} + \frac{1}{32} + \ldots$.

SOLUTION

This is a GP with $a = 2$ and $r = \frac{1}{4}$. Therefore,

$$S_\infty = \frac{2}{1-\frac{1}{4}} = \frac{8}{3}$$

Example 19 Write the recurring decimal 0.3232... as the sum of a GP. Hence write this recurring decimal as a rational number.

SOLUTION

Now

$$0.323232\ldots = \frac{32}{100} + \frac{32}{10\,000} + \frac{32}{1\,000\,000} + \ldots$$

This is a GP with $a = \frac{32}{100}$ and $r = \frac{1}{100}$. Since $-1 < r < 1$ the sum to infinity exists and is given by

$$S_\infty = \frac{\left(\frac{32}{100}\right)}{\left(1-\frac{1}{100}\right)} = \frac{32}{99}$$

The recurring decimal $0.\dot{3}\dot{2}$ can be written as $\frac{32}{99}$.

Example 20 The sum to infinity of a GP is 7 and the sum of the first two terms is $\frac{48}{7}$. Show that the common ratio, r, satisfies the equation

$$1 - 49r^2 = 0$$

Hence find the first term of the GP with positive common ratio.

SOLUTION

Since the sum to infinity is 7, we have

$$\frac{a}{1-r} = 7$$

$$\therefore \quad a = 7(1-r) \qquad [1]$$

The sum of the first two terms is $\frac{48}{7}$, therefore,

$$a + ar = \frac{48}{7}$$

$$\therefore \quad a(1+r) = \frac{48}{7} \qquad [2]$$

Substituting [1] into [2] gives

$$7(1-r)(1+r) = \frac{48}{7}$$

$$\therefore \quad 49(1-r^2) = 48$$

$$\therefore \quad 1 - 49r^2 = 0$$

as required.

Solving gives $r = \frac{1}{7}$ or $r = -\frac{1}{7}$.

Since we require the GP with positive common ratio $r = \frac{1}{7}$, from [1] the first term is given by

$$a = 7(1-r)$$
$$= 7(1 - \tfrac{1}{7})$$
$$\therefore \quad a = 6$$

The first term of the GP with positive common ratio is 6.

Mixed example

Example 21 The 2nd, 3rd and 9th terms of an AP form a geometric progression. Find the common ratio of the GP.

SOLUTION

The 2nd, 3rd and 9th terms of an AP are given by $a+d$, $a+2d$ and $a+8d$ respectively. If these terms form a GP, the common ratio is given by

$$r=\frac{a+2d}{a+d} \qquad \text{or} \qquad r=\frac{a+8d}{a+2d}$$

Eliminating r gives

$$\frac{a+2d}{a+d}=\frac{a+8d}{a+2d}$$

$$\therefore \quad (a+2d)^2=(a+d)(a+8d)$$

$$\therefore \quad a^2+4ad+4d^2=a^2+9ad+8d^2$$

$$\therefore \quad 4d^2+5ad=0$$

$$\therefore \quad d(4d+5a)=0$$

Solving gives $d=0$ or $d=-\dfrac{5a}{4}$.

When $d=0$, all the terms in the AP are the same and

$$r=\frac{a+2(0)}{a+0}=1$$

In other words, all the terms in the GP are also all the same.

When $d=-\dfrac{5a}{4}$,

$$r=\frac{a+2\left(-\frac{5a}{4}\right)}{a+\left(-\frac{5a}{4}\right)}=\frac{a-\frac{5a}{2}}{a-\frac{5a}{4}}$$

$$\therefore \quad r=\frac{\left(-\frac{3}{2}\right)}{\left(-\frac{1}{4}\right)}=6$$

The common ratio of the GP is 6.

Exercise 9E

1 Work out each of the following.

a) $\sum_{r=0}^{\infty} \left(\frac{1}{2}\right)^r$ **b)** $\sum_{r=0}^{\infty} \left(\frac{1}{3}\right)^r$ **c)** $\sum_{r=1}^{\infty} \left(\frac{1}{5}\right)^r$

d) $\sum_{r=0}^{\infty} \left(-\frac{1}{4}\right)^r$ **e)** $\sum_{r=2}^{\infty} \left(-\frac{1}{8}\right)^r$ **f)** $\sum_{r=0}^{\infty} \left(\frac{1}{9}\right)^{r+1}$

g) $\sum_{r=1}^{\infty} (0.3)^{r+1}$ **h)** $\sum_{r=0}^{\infty} (-0.7)^{r+2}$ **i)** $\sum_{r=0}^{\infty} 4 \times \left(\frac{1}{3}\right)^r$

j) $\sum_{r=1}^{\infty} 3 \times \left(-\frac{1}{5}\right)^r$ **k)** $\sum_{r=0}^{\infty} a^r, \quad |a| < 1$ **l)** $\sum_{r=1}^{\infty} (3x)^{r+1}, \quad |x| < \frac{1}{3}$

2 Express each of the following recurring decimals as a fraction in simplest form.

a) $0.\dot{5}$ **b)** $0.\dot{8}$ **c)** $0.\dot{7}\dot{2}$
d) $0.\dot{1}0\dot{2}$ **e)** $2.\dot{4}$ **f)** $3.2\dot{8}1\dot{4}$

3 Find the 1st term of a GP that has a common ratio of $\frac{2}{5}$ and a sum to infinity of 20.

4 Find the 3rd term of a GP that has a common ratio of $-\frac{1}{3}$ and a sum to infinity of 18.

5 A GP has a 1st term of 6 and a sum to infinity of 60. Find the common ratio.

6 Find the common ratio of a GP that has a 1st term of 6 and a sum to infinity of 4.

7 Find the common ratio of a geometric series which has a 2nd term of 6 and a sum to infinity of 24.

8 A GP has a 2nd term of 6 and a sum to infinity of 27. Write down the possible values of the first three terms.

9 Given that $\sum_{r=0}^{\infty} 5 \times a^r = 15$, find the value of a.

10 In an AP the 1st, 2nd and 5th terms are in geometric progression. Find the common ratio of the GP.

11 The 1st, 2nd and 3rd terms of a GP are the 1st, 7th and 9th terms of an AP. Find the common ratio of the GP.

12 The 2nd, 4th and 5th terms of an AP are the first three terms of a GP. Find the ratio of the GP.

13 In an AP the 1st, 3rd and 7th terms are in geometric progression, and the sum of the 1st and the 3rd terms of the AP is 15. Find the first four terms of the AP.

14 Given that a, 10 and b are consecutive terms of an AP, and that 1, a and b are consecutive terms of a GP, find the possible values of a and b.

15 The numbers 2, p and q are consecutive terms of a GP, and the numbers p, 30 and q are consecutive terms of an AP. Find the possible values of p and q.

16 The first three terms of an AP are 31, x and y and the first three terms of a GP are y, 4 and x. Calculate the possible values of x and y.

17 Find the possible values of the constants a and c, given that a, 12, c are consecutive terms of a GP, and a, 20, c are consecutive terms of an AP.

***18** Given that the pth, qth and rth terms of an AP are in geometric progression, show that the common ratio of the GP is either $\dfrac{r-q}{q-p}$ or $\dfrac{p-q}{q-r}$.

***19** **a)** For $|x| < 1$ show that

$$\sum_{r=1}^{\infty} r \times x^r = \sum_{r=1}^{\infty}\left(x^r \times \sum_{s=0}^{\infty} x^s\right)$$

b) Deduce that

$$\sum_{r=1}^{\infty} r \times x^r = \frac{x}{(1-x)^2}$$

***20** **a)** Sketch the graph of $y = 2^x$ for $0 \leqslant x \leqslant 1$.

b) By comparing the area under your graph, for values of x between $\dfrac{r}{n}$ and $\dfrac{r+1}{n}$, with the areas of appropriate rectangles over the same interval, show that

$$\frac{2^{\frac{r}{n}}}{n} < \int_{\frac{r}{n}}^{\frac{r+1}{n}} 2^x \,\mathrm{d}x < \frac{2^{\frac{r+1}{n}}}{n}$$

c) Deduce that

$$\frac{1}{n}\sum_{r=0}^{n-1} 2^{\frac{r}{n}} < \int_0^1 2^x \,\mathrm{d}x < \frac{1}{n}\sum_{r=0}^{n-1} 2^{\frac{r+1}{n}}$$

d) By summing the appropriate GPs obtain the result

$$\frac{1}{n}\left(\frac{1}{2^{\frac{1}{n}}-1}\right) < \int_0^1 2^x \,\mathrm{d}x < \frac{1}{n}\left(\frac{2^{\frac{1}{n}}}{2^{\frac{1}{n}}-1}\right)$$

e) Hence, by substituting a suitably large value for n, deduce that, to three decimal places.

$$\int_0^1 2^x \,\mathrm{d}x = 1.443$$

Exercise 9F: Examination questions

1 The nth terms of two sequences are defined as follows:

a) $t_n = 1 - \dfrac{1}{n}$ **b)** $u_n = 1 - \dfrac{1}{u_{n-1}}$ where $u_1 = 2$

Decide in each case whether the sequence is convergent, divergent, oscillating or periodic, giving reasons for your answers. (UODLE)

2 The sequence $u_1, u_2, u_3, \ldots$, where u_1 is a given real number, is defined by $u_{n+1} = u_n^2 - 1$.

i) Describe the behaviour of the sequence for each of the cases $u_1 = 0$, $u_1 = 1$ and $u_1 = 2$.
ii) Given that $u_2 = u_1$, find exactly the two possible values of u_1.
iii) Given that $u_3 = u_1$, show that $u_1^4 - 2u_1^2 - u_1 = 0$. (UCLES)

3 The fourth term of an arithmetic progression is 9 and the sum of the first ten terms is 60. Find the first term and the common difference. (UODLE)

4 In an arithmetic progression, the 8th term is twice the 3rd term and the 20th term is 110.

a) Find the common difference.
b) Determine the sum of the first 100 terms. (AEB 92)

5 **i)** The tenth term of an arithmetic progression is 36, and the sum of the first ten terms is 180. Find the first term and the common difference.

ii) Evaluate $\sum_{r=1}^{1000} (3r - 1)$. (UCLES)

6 An arithmetic series has first term 5 and tenth term equal to 26.

a) Find the common difference.
b) Determine the least value of n for which the sum of the first n terms of the series exceeds 1000. (AEB 95)

7 Five numbers are in arithmetic progression and the sum of their squares is 147·5. The middle number is 5. Find the other four numbers. (WJEC)

8 John is given an interest-free loan to buy a second-hand car. He repays the loan in monthly instalments. He repays £20 the first month, £22 the second month and the repayments continue to rise by £2 per month until the loan is repaid. Given that the final monthly repayment is £114,

a) show that the number of months it will take John to repay the loan is 48,
b) find the amount, in pounds, of the loan. (EDEXCEL)

9 An employer offers the following schemes of salary payments over a five-year period:

Scheme X: 60 monthly payments, starting with £1000 and increasing by £6 each month [£1000, £1006, £1012, ...]
Scheme Y: 5 annual payments, starting with £12 000 and increasing by £d each year [£12 000, £$(12\,000 + d)$, ...].

a) Over the complete five-year period, find the total salary payable under Scheme X.
b) Find the value of d which gives the same total salary for both schemes over the complete five-year period. (EDEXCEL)

10 The training programme of a pilot requires him to fly 'circuits' of an airfield. Each day he flies three more circuits than the day before. On the fifth day he flew 14 circuits.
Calculate how many circuits he flew

i) on the first day
ii) in total by the end of the fifth day
iii) in total by the end of the nth day
iv) in total from the end of the nth day to the end of the $(2n)$th day. Simplify your answer. (MEI)

11 **a)** The first term of an arithmetic progression is 3; the sum of the second term and the fifth term is 26. Find the common difference and the sum of the first 10 terms of the arithmetic progression.

b) The third and fourth terms of a geometric progression are 32 and 64 respectively. Find the common ratio and the sum of the first 10 terms of the geometric progression. (WJEC)

12 The first term of a geometric series is 5 and the common ratio is 1.2. Find for this series

a) the 16th term, giving your answer to the nearest integer

b) the sum of the first 30 terms, giving your answer to the nearest integer.

c) Give a reason why this series has no sum to infinity. (EDEXCEL)

13 **a)** Find the sum of the whole numbers that are divisible by 3 and lie between 100 and 200.

b) The first and fourth terms of a geometric progression are 54 and 2 respectively. Find the common ratio and the sum to infinity of the geometric progression. (WJEC)

14 Use the formula for the sum of an infinite geometric series to determine the exact fractional value of the recurring decimal 0.733333... in its lowest terms. (NICCEA)

15 In a geometric progression, the sum of the first two terms is 9 and the third term is 12.

i) Find the two possible values of the common ratio r, and the corresponding values of the first term a.

ii) Find the sum to infinity of the series for which $|r| < 1$. (WJEC)

16 The first three terms of a geometric series are 1, p and q. Given also that 10, q and p are the first three terms of an arithmetic series, show that

$$2p^2 - p - 10 = 0$$

Hence find the possible values of p and q.

17 **a)** Find the third term of the geometric series whose first two terms are 3, 4.

b) Given that x, 4, $x + 6$ are consecutive terms of a geometric series, find

i) the possible values of x

ii) the corresponding values of the common ratio of the geometric series.

Given that x, 4, $x + 6$ are the sixth, seventh and eighth terms of a geometric series and that the sum to infinity of the series exists, find

iii) the first term

iv) the sum to infinity. (MEI)

18 If the sum of the infinite geometric series

$$x^2 + \frac{x^2}{1-x} + \frac{x^2}{(1-x)^2} + \frac{x^2}{(1-x)^3} + \dots$$

is 380, what are the two possible values of x? (NICCEA)

19 The first term of a geometric series is a, where $a \neq 0$, and the second term is $a^2 - 2a$.

a) Write down the common ratio of the series, in terms of a.

b) Find the set of values for a for which the series has a sum to infinity. (UODLE)

20 A pump is used to extract air from a bottle. The first operation of the pump extracts $56\,\text{cm}^3$ of air and subsequent extractions follow a geometric progression. The third operation of the pump extracts $31.5\,\text{cm}^3$ of air.
Determine the common ratio of the geometric progression and calculate the total amount of air that could be extracted from the bottle, if the pump were to extract air indefinitely. (AEB 95)

21 A company offers a ten-year contract to an employee. This gives a starting salary of £15 000 a year with an annual increase of 8% of the previous year's salary.

i) Show that the amounts of annual salary form a geometric sequence and write down its common ratio.
ii) How much does the employee expect to earn in the tenth year?
iii) Show that the total amount earned over the 10 years is nearly £217 500.

After considering the offer, the employee asks for a different scheme of payment. This has the same starting salary of £15 000 but with a fixed annual pay rise £d.

iv) Find d if the total amount paid out over 10 years is to be the same under the two schemes. (MEI)

22 A savings scheme pays 5% per annum compound interest. A deposit of £100 is invested in this scheme at the start of each year.

a) Show that at the start of the third year, after the annual deposit has been made, the amount in the scheme is £315.25.
b) Find the amount in the scheme at the start of the fortieth year, after the annual deposit has been made. (EDEXCEL)

23 When a child's ball is dropped from a height h metres on to a hard, flat floor, it rebounds to a height of $\frac{3}{5}h$ metres. The ball is dropped initially from a height of 1.2 m.

a) Find the maximum height to which the ball rises after two bounces.
b) Find the total distance that the ball has travelled when it hits the floor for the tenth time.
c) Assuming that the ball continues to bounce in the same way indefinitely, find the total distance that the ball travels. (UODLE)

10 Binomial expansions

Logic may explain mathematics but cannot prove it. The logical theory of mathematics is an exciting sophisticated speculation – like any scientific theory. The mathematician should not forget that his intuition is the final authority.
IMRE LAKATOS

Binomial theorem for a positive integral index

We know from usual algebraic multiplication that

$$
\begin{aligned}
(1+x) &= 1+x \\
(1+x)^2 = (1+x)(1+x) &= 1+2x+x^2 \\
(1+x)^3 = (1+x)^2(1+x) &= 1+3x+3x^2+x^3 \\
(1+x)^4 = (1+x)^3(1+x) &= 1+4x+6x^2+4x^3+x^4 \\
\ldots\ldots \qquad \ldots\ldots &\qquad \ldots\ldots
\end{aligned}
$$

The coefficients of the expansions form a triangle, namely Pascal's triangle.

$$
\begin{array}{ccccccccc}
 & & & 1 & & 1 & & & \\
 & & 1 & & 2 & & 1 & & \\
 & 1 & & 3 & & 3 & & 1 & \\
1 & & 4 & & 6 & & 4 & & 1 \\
\ldots & & \ldots & & \ldots & & \ldots & & \ldots
\end{array}
$$

The next row is obtained in the following way.

$$
\begin{array}{ccccccccccccc}
 & 1 & & 4 & & 6 & & 4 & & 1 & \\
1 & & (1+4) & & (4+6) & & (6+4) & & (4+1) & & 1 \\
1 & & 5 & & 10 & & 10 & & 5 & & 1
\end{array}
$$

The entry in the 5th row, 3rd position from the left is 10. Therefore, the coefficient of the x^2 term in the expansion of $(1+x)^5$ is 10.

The entry in the 3rd row, 2nd position from the left is 3. Therefore, the coefficient of the x term in the expansion of $(1+x)^3$ is 3.

Generally, we denote the entry in the nth row, $(r+1)$th position by

$$\binom{n}{r} = \frac{n!}{r!(n-r)!}$$

where

$$n! = n(n-1)(n-2)\ldots 3.2.1 \quad \text{(called } n \text{ factorial)}$$

and where, by definition, $0! = 1$.

Using this result, we can write down a general formula for the expansion of $(1+x)^n$:

$$(1+x)^n = 1 + \binom{n}{1}x + \binom{n}{2}x^2 + \binom{n}{3}x^3 + \ldots + x^n$$

This is known as the **binomial expansion** of $(1+x)^n$, for positive integer n. This can also be written using sigma notation:

$$(1+x)^n = \sum_{r=0}^{n} \binom{n}{r} x^r$$

Alternatively, the binomial expansion of $(1+x)^n$ can be written as

$$(1+x)^n = 1 + nx + \frac{n(n-1)}{2!}x^2 + \frac{n(n-1)(n-2)}{3!}x^3 + \ldots + x^n$$

Notice that this expansion terminates at the term x^n when n is a positive integer.

Example 1 Evaluate

a) $\binom{5}{1}$ **b)** $\binom{6}{4}$

SOLUTION

a)

$$\binom{5}{1} = \frac{5!}{1!(5-1)!} = \frac{5!}{1!\,4!} = \frac{5.4.3.2.1}{1 \times 4.3.2.1}$$

$$\therefore \quad \binom{5}{1} = 5$$

b)

$$\binom{6}{4} = \frac{6!}{4!(6-4)!} = \frac{6!}{4!\,2!} = \frac{6.5.4.3.2.1}{4.3.2.1 \times 2.1}$$

$$\therefore \quad \binom{6}{4} = 15$$

Example 2 Find the coefficients of the x^2 and x^3 terms in the expansion of $(1+x)^7$.

SOLUTION

The coefficient of the x^2 term is

$$\binom{7}{2} = \frac{7!}{2!\,(7-2)!} = 21$$

The coefficient of the x^2 term in the expansion of $(1+x)^7$ is 21.

The coefficient of the x^3 term is

$$\binom{7}{3} = \frac{7!}{3!\,(7-3)!} = 35$$

The coefficient of the x^3 term in the expansion of $(1+x)^7$ is 35.

Example 3 Expand $(3+y)^4$ in powers of y.

SOLUTION

In order to use the binomial expansion, we need to express $(3+y)^4$ in the form $(1+x)^4$. Now

$$(3+y)^4 = \left[3\left(1+\frac{y}{3}\right)\right]^4 = 3^4\left(1+\frac{y}{3}\right)^4$$

Hence, we can use the binomial expansion of $(1+x)^4$ with $x = \frac{y}{3}$, which gives

$$\begin{aligned}(3+y)^4 &= 3^4\left(1+\frac{y}{3}\right)^4 \\ &= 3^4\left[1+\binom{4}{1}\left(\frac{y}{3}\right)+\binom{4}{2}\left(\frac{y}{3}\right)^2+\binom{4}{3}\left(\frac{y}{3}\right)^3+\left(\frac{y}{3}\right)^4\right] \\ &= 81\left[1+4\left(\frac{y}{3}\right)+6\left(\frac{y}{3}\right)^2+4\left(\frac{y}{3}\right)^3+\left(\frac{y}{3}\right)^4\right] \\ &= 81\left(1+\frac{4y}{3}+\frac{6y^2}{9}+\frac{4y^3}{27}+\frac{y^4}{81}\right)\end{aligned}$$

$$\therefore\quad (3+y)^4 = 81+108y+54y^2+12y^2+y^4$$

Expansion of $(a+x)^n$

Since

$$(a+x)^n = \left[a\left(1+\frac{x}{a}\right)\right]^n = a^n\left(1+\frac{x}{a}\right)^n$$

we have

$$\begin{aligned}(a+x)^n &= a^n\left(1+\frac{x}{a}\right)^n \\ &= a^n\left[1+\binom{n}{1}\left(\frac{x}{a}\right)+\binom{n}{2}\left(\frac{x}{a}\right)^2+\binom{n}{3}\left(\frac{x}{a}\right)^3+\ldots+\left(\frac{x}{a}\right)^n\right] \\ &= a^n\left(1+\binom{n}{1}\frac{x}{a}+\binom{n}{2}\frac{x^2}{a^2}+\binom{n}{3}\frac{x^3}{a^3}+\ldots+\frac{x^n}{a^n}\right)\end{aligned}$$

This gives

$$(a+x)^n = a^n+\binom{n}{1}a^{n-1}x+\binom{n}{2}a^{n-2}x^2+\binom{n}{3}a^{n-3}x^3+\ldots+x^n$$

Example 4 Expand $(2-x)^6$ in powers of x.

SOLUTION

Using the general result which we have just derived together with the fact that

$$(2-x)^6 = [2+(-x)]^6$$

gives

$$(2-x)^6 = 2^6 + \binom{6}{1}2^5(-x) + \binom{6}{2}2^4(-x)^2 + \binom{6}{3}2^3(-x)^3 + \binom{6}{4}2^2(-x)^4 + \binom{6}{5}2(-x)^5 + x^6$$

$$\therefore \quad (2-x)^6 = 64 - 192x + 240x^2 - 160x^3 + 60x^4 - 12x^5 + x^6$$

Example 5 Given that $(1-2x)^5(2+x)^6 \equiv a + bx + cx^2 + dx^3 + \ldots$, find the values of the constants a, b, c and d.

SOLUTION

Expanding $(1-2x)^5$ using the binomial theorem gives

$$\begin{aligned}(1-2x)^5 &= 1 + \binom{5}{1}(-2x) + \binom{5}{2}(-2x)^2 + \binom{5}{3}(-2x)^3 + \ldots \\ &= 1 - 10x + 40x^2 - 80x^3 + \ldots\end{aligned}$$

Expanding $(2+x)^6$ using the binomial theorem gives

$$\begin{aligned}(2+x)^6 &= 2^6 + \binom{6}{1}2^5x + \binom{6}{2}2^4x^2 + \binom{6}{3}2^3x^3 + \ldots \\ &= 64 + 192x + 240x^2 + 160x^3 + \ldots\end{aligned}$$

Therefore, we have

$$\begin{aligned}(1-2x)^5(2+x)^6 &= (1 - 10x + 40x^2 - 80x^3 + \ldots)(64 + 192x + 240x^2 + 160x^3 + \ldots) \\ &= 64 - 448x + 880x^2 + 320x^3 + \ldots\end{aligned}$$

Therefore, the values of the constants a, b, c and d are

$$a = 64 \qquad b = -448 \qquad c = 880 \qquad d = 320$$

Example 6 Write down the expansion of $(1+y)^4$. Hence find the first four terms in the expansion of $(1+x+x^2)^4$.

SOLUTION

Using the binomial theorem gives

$$\begin{aligned}(1+y)^4 &= 1 + \binom{4}{1}y + \binom{4}{2}y^2 + \binom{4}{3}y^3 + y^4 \\ &= 1 + 4y + 6y^2 + 4y^3 + y^4 \qquad [1]\end{aligned}$$

Writing $(1+x+x^2)^4$ as $[1+(x+x^2)]^4$ and using [1] with $y = x + x^2$ give

$$\begin{aligned}[(1+(x+x^2)]^4 &= 1 + 4(x+x^2) + 6(x+x^2)^2 + 4(x+x^2)^3 + (x+x^2)^4 \\ &= 1 + 4(x+x^2) + 6x^2(1+x)^2 + 4x^3(1+x)^3 + x^4(1+x)^4\end{aligned}$$

Since we require only the first four terms of the expansion, we can ignore terms involving x^4 and higher powers of x. Therefore,

$$[1+(x+x^2)]^4 = 1+4(x+x^2)+6x^2(1+2x)+4x^3(1)+\dots$$

$$\therefore \quad (1+x+x^2)^4 = 1+4x+10x^2+16x^3+\dots$$

Example 7 Expand $(x+2y)^5$.

SOLUTION

Using the general result gives

$$(x+2y)^5 = x^5+\binom{5}{1}x^4(2y)+\binom{5}{2}x^3(2y)^2+\binom{5}{3}x^2(2y)^3+\binom{5}{4}x(2y)^4+(2y)^5$$

$$\therefore \quad (x+2y)^5 = x^5+10x^4y+40x^3y^2+80x^2y^3+80xy^4+32y^5$$

Example 8 Calculate the value of the constant a if the coefficient of the x^3 term in the expansion of $(a+2x)^4$ is 160.

SOLUTION

The x^3 term in the expansion of $(a+2x)^4$ is

$$\binom{4}{3}a(2x)^3 = 32ax^3$$

$$\therefore \quad 32a = 160$$

$$\therefore \quad a = 5$$

Approximations

Example 9 Expand $(1+4x)^{14}$ in ascending powers of x, up to and including the 4th term. Hence evaluate $(1.0004)^{14}$, correct to four decimal places.

SOLUTION

Using the binomial expansion gives

$$(1+4x)^{14} = 1+\binom{14}{1}(4x)+\binom{14}{2}(4x)^2+\binom{14}{3}(4x)^3+\dots$$

$$\therefore \quad (1+4x)^{14} = 1+56x+1456x^2+23\,296x^3+\dots$$

Since

$$(1.0004)^{14} = [1+4(0.0001)]^{14}$$

we can use the above expansion with $x = 0.0001$, which gives

$$\begin{aligned}(1.0004)^{14} &= 1+56(0.0001)+1456(0.0001)^2+23\,296(0.0001)^3\\ &= 1.005\,614\,583\\ &= 1.0056 \quad \text{to four decimal places}\end{aligned}$$

The next term in the expansion is

$$\binom{14}{4}(4x)^4 = 256\,256x^4$$

This term would contribute the value

$$256\,256(0.0001)^4 = 2.562\,56 \times 10^{-11}$$

to the approximation of $(1.0004)^{14}$. In other words, it would not effect the answer of 1.0056, which is given to four decimal places.

Exercise 10A

1 Simplify each of the following.

a) $\binom{5}{3}$ **b)** $\binom{6}{2}$ **c)** $\binom{9}{7}$ **d)** $\binom{6}{5}$
e) $\binom{5}{5}$ **f)** $\binom{12}{2}$ **g)** $\binom{7}{3}$ **h)** $\binom{100}{99}$

2 Expand

a) $(1+x)^4$ **b)** $(1+x)^5$ **c)** $(1+3x)^4$ **d)** $(1-x)^3$
e) $(1-2x)^4$ **f)** $(1-5x)^3$ **g)** $(1+\frac{1}{2}x)^4$ **h)** $(1-\frac{1}{5}x)^2$

3 Find the coefficient of the term indicated in square brackets in the expansion of each of the expressions below.

a) $(1+x)^7$ $[x^4]$ **b)** $(1+x)^9$ $[x^2]$ **c)** $(1+2x)^5$ $[x^3]$
d) $(1+5x)^8$ $[x^2]$ **e)** $(1-3x)^6$ $[x^3]$ **f)** $(1-6x)^7$ $[x]$
g) $(1-4x)^4$ $[x^2]$ **h)** $(1+2x)^5$ $[x^4]$ **i)** $(1-\frac{1}{2}x)^3$ $[x^2]$

4 Expand

a) $(2+x)^3$ **b)** $(3+x)^4$ **c)** $(6-5x)^3$ **d)** $(2+\frac{1}{2}x)^4$
e) $(3x+2y)^3$ **f)** $(2x-y)^5$ **g)** $(2x+5y)^3$ **h)** $(3x-4y)^4$

5 Find the coefficient of the term indicated in square brackets in the expansion of each of the expressions below.

a) $(2+3x)^5$ $[x^3]$ **b)** $(5+2x)^8$ $[x^6]$ **c)** $(3+2x)^7$ $[x^5]$ **d)** $(7-4x)^5$ $[x^4]$
e) $(2-7x)^4$ $[x]$ **f)** $(5+2x)^6$ $[x^3]$ **g)** $(\frac{1}{3}+\frac{3}{2}x)^6$ $[x^3]$ **h)** $(\frac{2}{3}-\frac{2}{5}x)^3$ $[x]$

6 Expand each of the following in ascending powers of x, up to and including the term in x^3.

a) $(1-3x)^5$ **b)** $(1+2x)^{10}$ **c)** $(1-5x)^7$ **d)** $(2-3x)^5$
e) $(4-x)^5$ **f)** $(2+3x)^6$ **g)** $(1+\frac{1}{3}x)^9$ **h)** $(4+\frac{1}{4}x)^6$

7 Expand each of the following in ascending powers of x, up to and including the term in x^2.

a) $(2+x)(1+x)^5$ **b)** $(5-x)(1+x)^6$
c) $(5+4x)(1-2x)^7$ **d)** $(6+5x)(3+4x)^4$
e) $(5+x^2)(1-3x)^6$ **f)** $(7-2x^2)(3-x)^4$
g) $(2+3x+7x^2)(1+x)^6$ **h)** $(1+x+x^2)(2+x)^5$

8 Find the coefficient of the term indicated in square brackets in the expansion of each of these expressions.

a) $(1+2x)(1+x)^5$ $[x^2]$
b) $(1-4x)(1+x)^6$ $[x^2]$
c) $(3-x)(1+x)^7$ $[x]$
d) $(3-5x)(1+3x)^4$ $[x^3]$
e) $(2+x^2)(1-x)^4$ $[x^4]$
f) $(4-3x^3)(2+3x)^4$ $[x^3]$
g) $(1+2x-x^2)(4+x)^6$ $[x^6]$
h) $(2-3x+x^2)(1+2x)^4$ $[x^2]$

9 Expand

a) $(1+x^3)^4$
b) $(1+3x^2)^3$
c) $(3-2x^3)^3$
d) $(1+x+x^2)^2$
e) $(1+2x-x^2)^3$
f) $(2+3x-x^2)^2$
g) $(2+x-4x^2)^2$
h) $(3+2x+x^2)^2$

10 Find the coefficient of the term indicated in square brackets in the expansion of each of the expressions below.

a) $\left(3x-\dfrac{2}{x}\right)^5$ $[x^3]$
b) $\left(2x+\dfrac{5}{x}\right)^6$ $[x^4]$
c) $\left(x^2-\dfrac{2}{x}\right)^6$ [constant]
d) $\left(x^3+\dfrac{3}{x}\right)^4$ $[x^4]$
e) $\left(2x^3-\dfrac{3}{x^2}\right)^4$ $[x^2]$
f) $\left(x^3+\dfrac{7}{x}\right)^5$ $[x^7]$
g) $\left(\dfrac{3}{x^2}-5x\right)^4$ $\left[\dfrac{1}{x^2}\right]$
h) $\left(\dfrac{4}{x}+x^4\right)^{10}$ [constant]

11 a) Expand each of the following in ascending powers of x up to and including the term in x^2.

i) $(1+x)^4$ **ii)** $(1-2x)^4$

b) By first factorising the quadratic $1-x-2x^2$, deduce the first three terms in the binomial expansion of $(1-x-2x^2)^4$.

12 a) Expand each of the following in ascending powers of x up to and including the term in x^2.

i) $(1+3x)^6$ **ii)** $(1-4x)^6$

b) By first factorising the quadratic $1-x-12x^2$, deduce the first three terms in the binomial expansion of $(1-x-12x^2)^6$.

13 a) Expand each of the following in ascending powers of x up to and including the term in x^2.

i) $(3+2x)^5$ **ii)** $(1-3x)^5$

b) By first factorising the quadratic $3-7x-6x^2$, deduce the first three terms in the binomial expansion of $(3-7x-6x^2)^4$.

14 a) Expand $(1-x)^{10}$ in ascending powers of x up to and including the term in x^4.
b) Hence evaluate $(0.99)^{10}$ correct to six decimal places.

15 a) Expand $(1+2x)^{14}$ in ascending powers of x up to and including the term in x^3.
b) Hence evaluate $(1.02)^{14}$ correct to three decimal places.

16 a) Write down the first three terms in the binomial expansion of $(1-3x)^8$.
b) Hence evaluate $(0.997)^8$ correct to five decimal places.

17 a) Write down the first three terms in the binomial expansion of $(2+5x)^9$.
b) Hence evaluate $(2.005)^9$ correct to two decimal places.

18 By first expanding $(3-2x)^{12}$ in ascending powers of x up to and including the term in x^2, work out the value of $(2.998)^{12}$ correct to the nearest whole number.

19 By first expanding $(5-4x)^5$ in ascending powers of x up to and including the term in x^3, work out the value of $(4.96)^5$ correct to the nearest whole number.

20 Use binomial expansions to evaluate each of the following to the stated degree of accuracy.

a) $(1.01)^7$ correct to three decimal places.
b) $(0.99)^{10}$ correct to two decimal places.
c) $(2.04)^8$ correct to four decimal places.
d) $(3.998)^{12}$ correct to the nearest whole number.
e) $(5.999)^4$ correct to two decimal places.
f) $(6.03)^5$ correct to three decimal places.

21 Given that $(1+ax)^n \equiv 1+30x+375x^2+\dots$, find the values of the constants a and n.

22 Given that $(1+bx)^n \equiv 1-15x+90x^2+\dots$, find the values of the constants b and n.

23 When $(1+cx)^n$ is expanded as a series in ascending powers of x, the first three terms are given by $1+20x+150x^2$. Calculate the values of the constants c and n.

24 When $(1+ax)^n$ is expanded as a series in ascending powers of x, the first three terms are given by $1-8x+30x^2$. Calculate the values of a and n.

25 **a)** Show that $(x+y)^6+(x-y)^6 \equiv 2x^6+30x^4y^2+30x^2y^4+2y^6$.
b) Hence deduce that $(\sqrt{3}+\sqrt{2})^6+(\sqrt{3}-\sqrt{2})^6 = 970$.

26 Without using a calculator simplify each of the following.

a) $(\sqrt{5}+\sqrt{2})^4+(\sqrt{5}-\sqrt{2})^4$
b) $(\sqrt{2}+1)^5-(\sqrt{2}-1)^5$
c) $(\sqrt{7}+\sqrt{3})^6+(\sqrt{7}-\sqrt{3})^6$
d) $(3+\sqrt{3})^3-(3-\sqrt{3})^3$

***27** Prove the following results.

a) $\binom{n+1}{r} = \binom{n}{r} + \binom{n}{r-1}$
b) $\binom{n+2}{3} - \binom{n}{3} = n^2$

***28** Given $(1+x)^{2n} \equiv c_0+c_1x+c_2x^2+c_3x^3+c_4x^4+\dots+c_{2n}x^{2n}$ for constants $c_0, c_1, c_2, \dots, c_{2n}$, show that

a) $c_0+c_1+c_2+c_3+\dots+c_{2n} = 2^{2n}$
b) $c_0-c_1+c_2-c_3+\dots+c_{2n} = 0$
c) $c_0+c_2+c_4+c_6+\dots+c_{2n} = 2^{2n-1}$
d) $c_1+c_3+c_5+c_7+\dots+c_{2n-1} = 2^{2n-1}$
e) $c_0^2+c_1^2+c_2^2+c_3^2+\dots+c_{2n}^2 = \binom{4n}{2n}$

Binomial theorem when *n* is not a positive integer

From the previous section we know that

$$(1+x)^n = 1+nx+\frac{n(n-1)x^2}{2!}+\frac{n(n-1)(n-2)x^3}{3!}+\dots$$

with the expansion terminating at the term x^n when n is a positive integer.

However, if n is not a positive integer then the expansion does not terminate and is only valid for $-1 < x < 1$, i.e. $|x| < 1$. In this case, the expansion gives an approximation to $(1+x)^n$.

Example 10 Obtain the first four terms in the expansion of $\dfrac{1}{1+x}$.

SOLUTION

Now

$$\frac{1}{1+x} = (1+x)^{-1}$$

Using the binomial expansion gives

$$\begin{aligned}(1+x)^{-1} &= 1 + (-1)x + \frac{(-1)(-1-1)x^2}{2!} + \frac{(-1)(-1-1)(-1-2)x^3}{3!} + \dots \\ &= 1 - x + x^2 - x^3 + \dots\end{aligned}$$

$$\therefore \quad (1+x)^{-1} \approx 1 - x + x^2 - x^3 \quad \text{valid for} \quad |x| < 1$$

Example 11 Obtain the expansion of $\dfrac{(1+x)^3}{(2-x)}$ up to and including the term in x^3. Hence evaluate $(1.2)^3$ correct to two decimal places.

SOLUTION

Now

$$\frac{(1+x)^3}{(2-x)} = (1+x)^3(2-x)^{-1}$$

Expanding $(1+x)^3$ using the binomial theorem gives

$$(1+x)^3 = 1 + \binom{3}{1}x + \binom{3}{2}x^2 + x^3$$

$$\therefore \quad (1+x)^3 = 1 + 3x + 3x^2 + x^3 \quad \text{valid for all } x$$

Expanding $(2-x)^{-1}$ using the binomial theorem gives

$$\begin{aligned}(2-x)^{-1} &= 2^{-1}\left[1 + \left(-\frac{x}{2}\right)\right]^{-1} \\ &= \frac{1}{2}\left[1 + (-1)\left(-\frac{x}{2}\right) + \frac{(-1)(-1-1)}{2!}\left(-\frac{x}{2}\right)^2 + \frac{(-1)(-1-1)(-1-2)}{3!}\left(-\frac{x}{2}\right)^3 + \dots\right] \\ &= \frac{1}{2}\left(1 + \frac{x}{2} + \frac{x^2}{4} + \frac{x^3}{8} + \dots\right)\end{aligned}$$

$$\therefore \quad (2-x)^{-1} = \frac{1}{2} + \frac{x}{4} + \frac{x^2}{8} + \frac{x^3}{16} + \dots \quad \text{valid for } \left|\frac{x}{2}\right| < 1, \text{ i.e. } |x| < 2$$

Therefore, we have

$$(1+x)^3(2-x)^{-1} = (1+3x+3x^2+x^3)\left(\frac{1}{2}+\frac{x}{4}+\frac{x^2}{8}+\frac{x^3}{16}+\dots\right)$$

$$= \frac{1}{2}+\frac{7x}{4}+\frac{19x^2}{8}+\frac{27x^3}{16}+\dots$$

$$\therefore \quad \frac{(1+x)^3}{(2-x)} \approx \frac{1}{2}+\frac{7x}{4}+\frac{19x^2}{8}+\frac{27x^3}{16} \quad \text{valid for } |x|<2$$

Let $x = 0.2$ (which lies in the valid range), then

$$\frac{(1+0.2)^3}{(2-0.2)} \approx \frac{1}{2}+\frac{7(0.2)}{4}+\frac{19(0.2)^2}{8}+\frac{27(0.2)^3}{16}$$

$$\therefore \quad \frac{(1.2)^3}{1.8} \approx 0.9585$$

$$\therefore \quad (1.2)^3 \approx 1.8 \times 0.9585 = 1.73 \quad \text{(2 decimal places)}$$

Example 12 Given that x is so small that x^3 and higher powers of x can be neglected, show that

$$\frac{1}{\sqrt{1+x}} = 1 - \frac{x}{2} + \frac{3x^2}{8}$$

By letting $x = \frac{1}{4}$, find a rational approximation of $\sqrt{5}$.

SOLUTION

Since

$$\frac{1}{\sqrt{1+x}} = (1+x)^{-\frac{1}{2}}$$

we can use the binomial expansion of $(1+x)^{-\frac{1}{2}}$:

$$(1+x)^{-\frac{1}{2}} = 1 + \left(-\frac{1}{2}\right)x + \frac{\left(-\frac{1}{2}\right)\left(-\frac{1}{2}-1\right)x^2}{2!} + \dots$$

$$\approx 1 - \frac{x}{2} + \frac{3x^2}{8}$$

as required. This expansion is valid for $|x| < 1$.

Let $x = \frac{1}{4}$, then

$$\frac{1}{\sqrt{1+\frac{1}{4}}} \approx 1 - \frac{1}{2}\left(\frac{1}{4}\right) + \frac{3}{8}\left(\frac{1}{4}\right)^2$$

$$\therefore \quad \sqrt{\frac{4}{5}} \approx \frac{115}{128}$$

$$\therefore \quad \frac{2}{\sqrt{5}} \approx \frac{115}{128}$$

$$\therefore \quad \sqrt{5} \approx \frac{2(128)}{115} = \frac{256}{115}$$

A rational approximation for $\sqrt{5}$ is $\frac{256}{115}$. This is 2.226 086 956 52 and since $\sqrt{5} \approx 2.236\,067\,977\,5$ on the calculator, we see that the approximation is only accurate to one decimal place.

Exercise 10B

1 Obtain the first four terms in the binomial expansion of each of the following, and state the range of values of x for which each is valid.

a) $(1+x)^{-2}$ **b)** $(1+x)^{\frac{1}{2}}$ **c)** $(1+2x)^{-3}$ **d)** $(1-3x)^{-2}$

e) $(1-3x)^{-\frac{1}{3}}$ **f)** $(1+x^2)^{-3}$ **g)** $\dfrac{1}{1-3x}$ **h)** $\sqrt{1-6x}$

2 Obtain the expansion of each of the following up to and including the term in x^3, giving the range of values of x for which each is valid.

a) $(2+x)^{-1}$ **b)** $(4+x)^{\frac{1}{2}}$ **c)** $(9-4x)^{-\frac{1}{2}}$ **d)** $(8+3x)^{\frac{2}{3}}$

e) $\dfrac{1}{(2-x)^3}$ **f)** $\sqrt{4-x}$ **g)** $\dfrac{1}{6+x}$ **h)** $\dfrac{1}{(3-2x)^2}$

3 Given x is so small that x^3 and higher powers of x may be neglected, write down a quadratic approximation to each of the following.

a) $\dfrac{1+x}{1-x}$ **b)** $\dfrac{1+3x}{2+x}$ **c)** $\dfrac{x}{4-x}$ **d)** $\dfrac{2+5x}{(1-3x)^2}$

e) $(4-x^2)\sqrt{4-x}$ **f)** $\dfrac{7-3x}{\sqrt{1-x}}$ **g)** $(4-3x)^2\sqrt{1-6x}$ **h)** $\dfrac{(2-x^2)^2}{(4+x)^3}$

4 a) Expand each of the following in ascending powers of x up to and including the term in x^2.

i) $(1+2x)^4$ **ii)** $\sqrt{1-x}$

b) Hence obtain a quadratic approximation to $(1+2x)^4\sqrt{1-x}$ which is valid for small values of x.

5 a) Given x is small write down a quadratic approximation to each of the following.

i) $\sqrt{1+2x}$ **ii)** $\dfrac{1}{(1-x)^4}$

b) Hence obtain the first three terms in the series expansion of

$$\frac{\sqrt{1+2x}}{(1-x)^4}$$

stating the range of values of x for which the series is valid.

6 a) Expand each of the following in ascending powers of x up to and including the term in x^3.

i) $\dfrac{1}{1-x}$ **ii)** $\dfrac{1}{1-2x}$

b) Hence obtain a cubic approximation to

$$\frac{1}{(1-x)(1-2x)}$$

which is valid for small values of x.

7 **a)** Given x is small, write down a cubic approximation to each of the following.

i) $\sqrt[3]{1-3x}$ **ii)** $\dfrac{1}{1-4x}$

b) Hence obtain the first four terms in the series expansion of

$$\frac{\sqrt[3]{1-3x}}{1-4x}$$

stating the range of values of x for which the series is valid.

8 Given x is so small that x^4 and higher powers of x may be neglected, show that

$$\frac{(1-3x)^4}{\sqrt{4+x^2}} = \tfrac{1}{2} - 6x + 26\tfrac{15}{16}x^2 - 53\tfrac{1}{4}x^3$$

9 Given $|x| < \frac{1}{2}$, write down the binomial expansion of $\sqrt{\dfrac{1+2x}{1-2x}}$ in ascending powers of x up to and including the term in x^3.

10 When $(1+cx)^n$ is expanded in ascending powers of x the first three terms of the expansion are $1-2x+7x^2$. Find the values of the constants c and n.

11 Given that $(1+ax)^n \equiv 1 + \dfrac{x}{9} - \dfrac{5}{162}x^2 + \ldots$, calculate the values of the constants a and n.

12 Expand $(1+x)^{\frac{1}{2}}$ in ascending powers of x up to and including the term in x^3. Hence evaluate $\sqrt{1.01}$ correct to eight decimal places.

13 Expand $(1-2x)^{\frac{1}{4}}$ in ascending powers of x up to and including the term in x^2. Hence evaluate $\sqrt[4]{0.998}$ correct to six decimal places.

14 Expand $(4-3x)^{-\frac{1}{2}}$ in ascending powers of x up to and including the term in x^2. Hence evaluate $\dfrac{1}{\sqrt{3.97}}$ correct to four decimal places.

15 **a)** Show that $\left(1+\dfrac{x}{25}\right)^{\frac{1}{2}} \equiv 1 + \dfrac{x}{50} - \dfrac{x^2}{5000} + \dfrac{x^3}{250\,000} - \ldots$

b) By substituting $x = 1$ into your answer to part **a**, deduce that $\sqrt{26} \approx 5.099\,02$.

16 **a)** Write down the first four terms in the binomial expansion of $\sqrt{1+\dfrac{x}{100}}$.

b) By substituting a suitable value for x into your answer to part **a**, deduce that $\sqrt{102} \approx 10.099\,505$.

17 **a)** Show that $\left(1+\dfrac{x}{125}\right)^{\frac{1}{3}} \equiv 1 + \dfrac{x}{375} - \dfrac{x^2}{140\,625} + \ldots$

b) By substituting a suitable value of x into your answer to part **a**, deduce that $\sqrt[3]{126} \approx 5.013$.

18 **a)** Write down the first three terms in the binomial expansion of $\sqrt[4]{1+\dfrac{x}{16}}$.

b) Deduce that $\sqrt[4]{15} \approx 1.968$.

***19 a)** Show that $\sqrt{\dfrac{1+2x}{1-x}} \equiv 1 + \frac{3}{2}x + \frac{3}{8}x^2 + \ldots$

b) By substituting $x = 0.02$, deduce that $\sqrt{13} \approx 3.606$.

***20 a)** Show that $\dfrac{1}{(a+bx)^2} \equiv \dfrac{1}{a^2} - \dfrac{2b}{a^3}x + \dfrac{3b^2}{a^4}x^2 + \ldots$

b) Given that the coefficient of the x term is equal to the coefficient of the x^2 term, show that $3b + 2a = 0$.

c) Given also that the sum of the constant term and the coefficient of the x^2 term is 84, find the possible values of the constants a and b.

Exercise 10C: Examination questions

1 Determine the coefficient of x^3 in the binomial expansion of $(1 - 2x)^7$. (AEB Spec)

2

$$f(x) \equiv (1+x)^4 - (1-x)^4$$

Expand f(x) as a series in ascending powers of x. (EDEXCEL)

3 i) Show that $(2+x)^4 = 16 + 32x + 24x^2 + 8x^3 + x^4$, for all x.

ii) Find the values of x for which $(2+x)^4 = 16 + 16x + x^4$. (MEI)

4 When $(1 - \frac{3}{2}x)^p$ is expanded in ascending powers of x, the coefficient of x is -24.

a) Find the value of p.

b) Find the coefficient of x^2 in the expansion.

c) Find the coefficient of x^3 in the expansion. (EDEXCEL)

5 Find the first three terms in the expansion of $(3 + 4x + x^2)(1 - x)^5$ in ascending powers of x. (WJEC)

6 Given that

$$(1 + kx)^8 = 1 + 12x + px^2 + qx^3 + \ldots \quad \text{for all } x \in \mathbb{R}$$

a) find the value of k, the value of p and the value of q.

b) Using your values of k, p and q, find the numerical coefficient of the x^3 term in the expansion of $(1-x)(1+kx)^8$. (EDEXCEL)

7 Expand $(1 + ax)^8$ in ascending powers of x up to and including the term in x^2. The coefficients of x and x^2 in the expansion of $(1 + bx)(1 + ax)^8$ are 0 and -36 respectively. Find the values of a and b, given that $a > 0$ and $b < 0$. (EDEXCEL)

8 Expand $\left(x - \dfrac{1}{x}\right)^5$, simplifying the coefficients. (EDEXCEL)

9 Write down and simplify the first three terms in the binomial expansion of

$$\left(x^2 + \frac{1}{3x}\right)^9$$

in descending powers of x, and find the term which is independent of x. (WJEC)

10 The binomial expansion of $(1 + 3x)^{-4}$ is $1 + ax + bx^2 + cx^3 + \ldots$, where a, b, c are constants.

a) Determine the values of a, b and c.
b) State the range of values of x for which the expansion is valid. (AEB 95)

11 Find, in their simplest form, the first three terms in the expansion of $(1 + 3t)^{\frac{2}{3}}$ in ascending powers of t, where $|t| < \frac{1}{3}$. (NEAB)

12 a) Obtain the first four non-zero terms of the binomial expansion in ascending powers of x of $(1 - x^2)^{-\frac{1}{2}}$, given that $|x| < 1$.
b) Show that, when $x = \frac{1}{3}$, $(1 - x^2)^{-\frac{1}{2}} = \frac{3}{4}\sqrt{2}$.
c) Substitute $x = \frac{1}{3}$ into your expansion and hence obtain an approximation to $\sqrt{2}$, giving your answer to five decimal places. (EDEXCEL)

13 i) Write down the expansion of $(2 - x)^4$.
ii) Find the first four terms in the expansion of $(1 + 2x)^{-3}$ in ascending powers of x. For what range of values of x is this expansion valid?
iii) When the expansion is valid,

$$\frac{(2 - x)^4}{(1 + 2x)^3} = 16 + ax + bx^2 + \ldots$$

Find the values of a and b. (MEI)

14 i) Show that $\dfrac{1}{\sqrt{4 - x}} = \dfrac{1}{2}\left(1 - \dfrac{x}{4}\right)^{-\frac{1}{2}}$.

ii) Write down the first three terms in the binomial expansion of $\left(1 - \dfrac{x}{4}\right)^{-\frac{1}{2}}$ in ascending powers of x, stating the range of values of x for which this expansion is valid.

iii) Find the first three terms in the expansion of $\dfrac{2(1 + x)}{\sqrt{4 - x}}$ in ascending powers of x, for small values of x. (MEI)

11 Algebraic fractions

Mathematics is a dangerous profession – an appreciable proportion of us go mad.
J. E. LITTLEWOOD

The process which is used for adding and subtracting numerical fractions is also used for adding and subtracting algebraic fractions. To find the sum or difference of two numerical fractions, each fraction must be expressed in terms of the same denominator, called the **lowest common denominator**. For example, in calculating

$$\frac{1}{3}+\frac{1}{6}$$

the lowest common denominator is 6, since 3 is a factor of 6. Therefore,

$$\frac{1}{3}+\frac{1}{6}=\frac{2}{6}+\frac{1}{6}=\frac{3}{6}=\frac{1}{2}$$

When the only common factor of the denominators is 1, we can find their product and use the product as the common denominator. For example, in calculating

$$\frac{1}{3}+\frac{1}{7}$$

the lowest common denominator is $3 \times 7 = 21$, since 3 and 7 have no common factor other than 1. Therefore,

$$\frac{1}{3}+\frac{1}{7}=\frac{7}{21}+\frac{3}{21}=\frac{10}{21}$$

A numerical fraction whose numerator is greater than or equal to the denominator is called an **improper fraction**. For example, $\frac{7}{5}$ is an improper fraction which can be written as $1\frac{2}{5}$.

An algebraic fraction is called improper if the degree of the numerator is greater than or equal to the degree of the denominator. For example,

$$\frac{x^2+2x-8}{3x+7}$$

is an improper fraction because the degree of the numerator is two and the degree of the denominator is one. We look at improper algebraic fractions and the techniques used in their simplification on pages 283–4.

Example 1 Express $\dfrac{4}{x+6} - \dfrac{2}{x+7}$ as a single fraction.

SOLUTION

The lowest common denominator is $(x+6)(x+7)$. Therefore,

$$\frac{4}{x+6} - \frac{2}{x+7} \equiv \frac{4(x+7) - 2(x+6)}{(x+6)(x+7)}$$

$$\equiv \frac{4x+28-2x-12}{(x+6)(x+7)}$$

$$\equiv \frac{2x+16}{(x+6)(x+7)}$$

$$\therefore \quad \frac{4}{x+6} - \frac{2}{x+7} \equiv \frac{2(x+8)}{(x+6)(x+7)}$$

Example 2 Express $\dfrac{2x}{x^2+3x+2} + \dfrac{3}{x+1}$ as a single fraction.

SOLUTION

In this case, $(x^2+3x+2)(x+1)$ is a common denominator but there is a much simpler common denominator, namely (x^2+3x+2), since

$$x^2+3x+2 \equiv (x+1)(x+2)$$

In other words, $(x+1)$ is a factor of x^2+3x+2. Therefore,

$$\frac{2x}{x^2+3x+2} + \frac{3}{x+1} \equiv \frac{2x}{(x+1)(x+2)} + \frac{3}{x+1}$$

$$\equiv \frac{2x+3(x+2)}{(x+1)(x+2)}$$

$$\therefore \quad \frac{2x}{x^2+3x+2} + \frac{3}{x+1} \equiv \frac{5x+6}{(x+1)(x+2)}$$

Example 3 Express $x+7+\dfrac{1}{x-4} - \dfrac{5}{x+1}$ as a single fraction.

SOLUTION

This can be written as

$$\frac{x+7}{1} + \frac{1}{x-4} - \frac{5}{x+1}$$

Therefore, the lowest common denominator is $(x-4)(x+1)$, and

$$x+7+\frac{1}{x-4} - \frac{5}{x+1} \equiv \frac{(x+7)(x-4)(x+1) + (x+1) - 5(x-4)}{(x-4)(x+1)}$$

$$\equiv \frac{x^3+4x^2-25x-28+x+1-5x+20}{(x-4)(x+1)}$$

$$\therefore \quad x+7+\frac{1}{x-4} - \frac{5}{x+1} \equiv \frac{x^3+4x^2-29x-7}{(x-4)(x+1)}$$

Example 4 Express $3\frac{3}{5} + \dfrac{4}{x+7}$ as a single fraction. Hence solve the equation

$$3\tfrac{3}{5} + \frac{4}{x+7} = \frac{8}{5-x}$$

SOLUTION

Writing $3\frac{3}{5}$ as an improper fraction gives $\frac{18}{5}$. Therefore,

$$\begin{aligned} 3\tfrac{3}{5} + \frac{4}{x+7} &\equiv \frac{18}{5} + \frac{4}{x+7} \\ &\equiv \frac{18(x+7) + 4(5)}{5(x+7)} \\ &\equiv \frac{18x+146}{5(x+7)} \end{aligned}$$

Using this result, the equation

$$3\tfrac{3}{5} + \frac{4}{x+7} = \frac{8}{5-x}$$

becomes

$$\frac{18x+146}{5(x+7)} = \frac{8}{5-x}$$

Cross-multiplying and simplifying give

$$(18x+146)(5-x) = 40(x+7)$$

$$\therefore \quad 18x^2 + 96x - 450 = 0$$

Dividing by 6 and factorising give

$$3x^2 + 16x - 75 = 0$$

$$\therefore \quad (3x+25)(x-3) = 0$$

Solving gives $x = -\frac{25}{3}$ or $x = 3$.

Exercise 11A

1 Express each of these as a single fraction.

a) $\dfrac{3}{x-1} + \dfrac{2}{x+3}$ **b)** $\dfrac{2}{x+4} - \dfrac{1}{x-3}$ **c)** $\dfrac{4}{x-5} - \dfrac{2}{x+4}$ **d)** $\dfrac{x}{x-3} - \dfrac{2}{x+2}$

e) $\dfrac{2x}{x+2} - \dfrac{1}{x-2}$ **f)** $\dfrac{5}{2x-3} - \dfrac{3}{2x+5}$ **g)** $\dfrac{2x+1}{x-2} + \dfrac{3x}{x+4}$ **h)** $\dfrac{5x}{x^2+3} + \dfrac{6}{2x-5}$

2 Express each of these as a single fraction.

a) $\dfrac{x}{x^2+3x-4} + \dfrac{2}{x-1}$ **b)** $\dfrac{5}{x^2+6x+8} - \dfrac{4}{x+2}$

c) $x+3+\dfrac{1}{x-2} - \dfrac{3}{x+4}$ **d)** $2x+5-\dfrac{1}{x-7} - \dfrac{2}{x-4}$

e) $1+\dfrac{3}{x^2-3x+2}+\dfrac{2}{x-1}$

f) $x-1+\dfrac{4x}{x^2-4x+4}+\dfrac{3}{x-2}$

g) $\dfrac{3}{x^2+5x+6}-\dfrac{2}{x+3}-\dfrac{1}{x+2}$

h) $\dfrac{5x}{2x^2+3x-5}-\dfrac{1}{x-1}-\dfrac{2}{2x+5}$

3 Show that

$$\frac{3}{x-2}+\frac{4}{x-1}\equiv\frac{7x-11}{(x-2)(x-1)}$$

Hence solve the equation

$$\frac{3}{x-2}+\frac{4}{x-1}=\frac{7}{x}$$

4 Express $\dfrac{2}{x-3}+\dfrac{1}{2x+1}$ as a single fraction. Hence solve the equation

$$\frac{2}{x-3}+\frac{1}{2x+1}=\frac{5}{2x-3}$$

5 Express $\dfrac{3}{x-2}+\dfrac{x}{x+1}$ as a single fraction. Hence solve the equation

$$\frac{3}{x-2}+\frac{x}{x+1}=2\tfrac{1}{7}$$

6 Show that

$$1+\frac{2}{x-1}-\frac{3}{x+4}\equiv\frac{x^2+2x+7}{(x-1)(x+4)}$$

Hence solve the equation

$$1+\frac{2}{x-1}-\frac{3}{x+4}=2\tfrac{1}{2}$$

7 a) Factorise the expression $2x^3-3x^2-11x+6$.

b) Express $\dfrac{1}{x+1}+\dfrac{12}{(x+1)^2}$ as a single fraction.

c) Hence solve the equation

$$7-2x=\frac{1}{x+1}+\frac{12}{(x+1)^2}$$

8 a) Show that $6x^3-29x^2+45x-22\equiv(6x-11)(x-1)(x-2)$.

b) Express $\dfrac{2}{x-3}+\dfrac{3}{x+4}$ as a single fraction.

c) Hence solve the equation

$$\frac{2}{x-3}+\frac{3}{x+4}=\frac{6x}{7x-22}$$

9 Given

$$\frac{a}{6x-1} - \frac{1}{3x+1} \equiv \frac{b}{(6x-1)(3x+1)}$$

where a and b are both constants, find the values of a and b.

10 Find the values of the constants A and B for which

$$\frac{A}{x-5} + \frac{B}{x+5} \equiv \frac{2x}{x^2-25}$$

11 Given

$$\frac{6}{x+3} + \frac{P}{x-1} \equiv \frac{Qx}{(x+3)(x-1)}$$

where P and Q are both constants, find the values of P and Q.

12 Find the values of the constants a and b for which

$$\frac{a}{x+b} + \frac{2}{x-4} \equiv \frac{3x}{(x+b)(x-4)}$$

Partial fractions

In Example 1 (page 276) the algebraic fraction $\frac{4}{x+6} - \frac{2}{x+7}$ is expressed as a single fraction, namely $\frac{2(x+8)}{(x+6)(x+7)}$. We now want to look at the techniques involved in expressing $\frac{2(x+8)}{(x+6)(x+7)}$ as $\frac{4}{x+6} - \frac{2}{x+7}$

This reverse process is called expressing $\frac{2(x+8)}{(x+6)(x+7)}$ in **partial fractions.**

There are basically four different types of algebraic fraction which can be expressed in partial fractions.

Type I: Denominator with linear factors

Each linear factor $(ax+b)$ in the denominator has a corresponding partial fraction of the form

$$\frac{A}{(ax+b)}$$

where a, b and A are constants.

Example 5 Express $\dfrac{7x+8}{(x+4)(x-6)}$ in partial fractions.

SOLUTION

We begin by assuming that

$$\frac{7x+8}{(x+4)(x-6)} \equiv \frac{A}{x+4} + \frac{B}{x-6}$$

Multiplying throughout by $(x+4)(x-6)$ gives

$$7x+8 \equiv A(x-6) + B(x+4) \qquad [1]$$

There are two techniques for finding the constants A and B. First, by letting $x = 6$, the constant A will be eliminated from the equation, thereby allowing the constant B to be determined. This gives

$$7(6)+8 = A(6-6) + B(6+4)$$

$$\therefore \quad 50 = 10B$$

$$\therefore \quad B = 5$$

Then, by letting $x = -4$, the constant A can be determined:

$$7(-4)+8 = A(-4-6) + B(-4+4)$$

$$\therefore \quad -20 = -10A$$

$$\therefore \quad A = 2$$

Alternatively, expanding the right-hand side of identity [1] gives

$$7x+8 \equiv Ax - 6A + Bx + 4B$$

The coefficients of the x terms and the constants can now be compared to find the constants A and B.

Comparing coefficients of the x terms gives

$$7 = A + B \qquad [2]$$

Comparing constant terms gives

$$8 = -6A + 4B$$

$$\therefore \quad 4 = -3A + 2B \qquad [3]$$

Solving [2]and [3] simultaneously gives $A = 2$ and $B = 5$, as before.

Therefore,

$$\frac{7x+8}{(x+4)(x-6)} \equiv \frac{2}{x+4} + \frac{5}{x-6}$$

We will see later in Examples 7 to 10 (pages 282–4) that it is easier to use a combination of the two techniques.

Example 6 Express $\dfrac{9x^2+34x+14}{(x+2)(x^2-x-12)}$ in partial fractions.

SOLUTION

Although the denominator appears to have only one linear factor, this is not the case since

$$x^2-x-12 \equiv (x+3)(x-4)$$

Therefore,

$$\frac{9x^2+34x+14}{(x+2)(x^2-x-12)} \equiv \frac{9x^2+34x+14}{(x+2)(x+3)(x-4)}$$

We assume that

$$\frac{9x^2+34x+14}{(x+2)(x+3)(x-4)} \equiv \frac{A}{x+2}+\frac{B}{x+3}+\frac{C}{x-4}$$

Multiplying throughout by $(x+2)(x+3)(x-4)$ gives

$$9x^2+34x+14 \equiv A(x+3)(x-4)+B(x+2)(x-4)+C(x+2)(x+3)$$

Let $x=-2$: $\quad 9(-2)^2+34(-2)+14 = A(-2+3)(-2-4)$

$$\therefore \quad -18 = -6A$$

$$\therefore \quad A = 3$$

Let $x=-3$: $\quad 9(-3)^2+34(-3)+14 = B(-3+2)(-3-4)$

$$\therefore \quad -7 = 7B$$

$$\therefore \quad B = -1$$

Let $x=4$: $\quad 9(4)^2+34(4)+14 = C(4+2)(4+3)$

$$\therefore \quad 294 = 42C$$

$$\therefore \quad C = 7$$

Therefore,

$$\frac{9x^2+34x+14}{(x+2)(x+3)(x-4)} \equiv \frac{3}{x+2}-\frac{1}{x+3}+\frac{7}{x-4}$$

Type II: Denominator with an irreducible quadratic factor

Each quadratic factor (ax^2+bx+c) in the denominator which is irreducible, i.e. will not factorise, has a corresponding partial fraction of the form

$$\frac{Ax+B}{(ax^2+bx+c)}$$

where a, b, c, A and B are constants.

Example 7 Express $\dfrac{7x^2+2x-28}{(x-6)(x^2+3x+5)}$ in partial fractions.

SOLUTION

Since x^2+3x+5 cannot be factorised, we assume that

$$\frac{7x^2+2x-28}{(x-6)(x^2+3x+5)} \equiv \frac{A}{x-6}+\frac{Bx+C}{x^2+3x+5}$$

Multiplying throughout by $(x-6)(x^2+3x+5)$ gives

$$7x^2+2x-28 \equiv A(x^2+3x+5)+(Bx+C)(x-6)$$

Let $x=6$: $7(6)^2+2(6)-28 = A\left[(6)^2+3(6)+5\right]$

$$\therefore \quad 236 = 59A$$

$$\therefore \quad A = 4$$

Comparing coefficients of the x^2 terms gives

$$7 = A + B \qquad [1]$$

Substituting $A=4$ into [1] gives $B=3$.

Comparing constant terms gives

$$-28 = 5A - 6C \qquad [2]$$

Substituting $A=4$ into [2] gives $C=8$. Therefore,

$$\frac{7x^2+2x-28}{(x-6)(x^2+3x+5)} \equiv \frac{4}{x-6}+\frac{3x+8}{x^2+3x+5}$$

Type III: Denominator with a repeated factor

Each repeated linear factor $(ax+b)^2$ in the denominator has corresponding partial fractions of the form

$$\frac{A}{ax+b}+\frac{B}{(ax+b)^2}$$

where a, b, A and B are constants.

Example 8 Express $\dfrac{2x^2+29x-11}{(2x+1)(x-2)^2}$ in partial fractions

SOLUTION

Assume that

$$\frac{2x^2+29x-11}{(2x+1)(x-2)^2} \equiv \frac{A}{2x+1}+\frac{B}{x-2}+\frac{C}{(x-2)^2}$$

Multiplying throughout by $(2x+1)(x-2)^2$ gives

$$2x^2+29x-11 \equiv A(x-2)^2+B(2x+1)(x-2)+C(2x+1)$$

Let $x = -\frac{1}{2}$: $\quad 2(-\frac{1}{2})^2 + 29(-\frac{1}{2}) - 11 = A(-\frac{1}{2} - 2)^2$

$$\therefore \quad -25 = \frac{25}{4}A$$

$$\therefore \quad A = -4$$

Let $x = 2$: $\quad 2(2)^2 + 29(2) - 11 = C(4 + 1)$

$$\therefore \quad 55 = 5C$$

$$\therefore \quad C = 11$$

Comparing the coefficients of the x^2 terms gives

$$2 = A + 2B \qquad [1]$$

Substituting $A = -4$ into [1] gives $B = 3$. Therefore,

$$\frac{2x^2 + 29x - 11}{(2x + 1)(x - 2)^2} \equiv -\frac{4}{2x + 1} + \frac{3}{x - 2} + \frac{11}{(x - 2)^2}$$

Type IV: Improper fractions

On page 275, we saw that an improper algebraic fraction is one in which the degree of the numerator is greater than or equal to the degree of the denominator. To simplify an improper algebraic fraction, we divide the numerator (a polynomial) by the denominator (a polynomial).

- When a polynomial of degree n is divided by a polynomial also of degree n, the quotient is a constant.
- When a polynomial of degree n is divided by a polynomial of degree m, where $m < n$, the quotient is a polynomial of degree $n - m$.

We will use these two facts when expressing improper algebraic fractions in partial fractions.

Example 9 Express $\dfrac{5x^2 - 71}{(x + 5)(x - 4)}$ in partial fractions.

SOLUTION

The degree of $5x^2 - 71$ is 2 and the degree of $(x + 5)(x - 4)$ is also 2, therefore the quotient is a constant (written as A below) and we assume that

$$\frac{5x^2 - 71}{(x + 5)(x - 4)} \equiv A + \frac{B}{x + 5} + \frac{C}{x - 4}$$

Multiplying throughout by $(x + 5)(x - 4)$ gives

$$5x^2 - 71 \equiv A(x + 5)(x - 4) + B(x - 4) + C(x + 5)$$

Comparing coefficients of the x^2 terms gives $A = 5$.

Let $x = -5$: $\quad 5(-5)^2 - 71 = B(-5 - 4)$

$$\therefore \quad 54 = -9B$$

$$\therefore \quad B = -6$$

Let $x = 4$: $\quad 5(4)^2 - 71 = C(4+5)$

$$\therefore \quad 9 = 9C$$

$$\therefore \quad C = 1$$

Therefore, we have

$$\frac{5x^2 - 71}{(x+5)(x-4)} \equiv 5 - \frac{6}{x+5} + \frac{1}{x-4}$$

Example 10 Express $\dfrac{3x^4 + 7x^3 + 8x^2 + 53x - 186}{(x+4)(x^2+9)}$ in partial fractions.

SOLUTION

The degree of $3x^4 + 7x^3 + 8x^2 + 53x - 186$ is 4 and the degree of $(x+4)(x^2+9)$ is 3, therefore the quotient is a polynomial of degree $(4-3) = 1$ and so we assume that

$$\frac{3x^4 + 7x^3 + 8x^2 + 53x - 186}{(x+4)(x^2+9)} \equiv Ax + B + \frac{C}{x+4} + \frac{Dx+E}{x^2+9}$$

Multiplying throughout by $(x+4)(x^2+9)$ gives

$$3x^4 + 7x^3 + 8x^2 + 53x - 186 \equiv (Ax+B)(x+4)(x^2+9) + C(x^2+9) + (Dx+E)(x+4)$$

Comparing coefficients of the x^4 terms gives $A = 3$.

Comparing coefficients of the x^3 terms gives

$$7 = 4A + B \qquad [1]$$

Substituting $A = 3$ into [1] gives $B = -5$.

Let $x = -4$: $\quad 3(-4)^4 + 7(-4)^3 + 8(-4)^2 + 53(-4) - 186 = C\left[(-4)^2 + 9\right]$

$$\therefore \quad 50 = 25C$$

$$\therefore \quad C = 2$$

Comparing coefficients of the x^2 terms gives

$$8 = 9A + 4B + C + D \qquad [2]$$

Substituting $A = 3$, $B = -5$ and $C = 2$ into [2] gives $D = -1$.

Comparing constant terms gives

$$-186 = 36B + 9C + 4E \qquad [3]$$

Substituting $B = -5$ and $C = 2$ into [3] gives $E = -6$. Therefore,

$$\frac{3x^4 + 7x^3 + 8x^2 + 53x - 186}{(x+4)(x^2+9)} \equiv 3x - 5 + \frac{2}{x+4} + \frac{-x-6}{x^2+9}$$

or

$$\frac{3x^4 + 7x^3 + 8x^2 + 53x - 186}{(x+4)(x^2+9)} \equiv 3x - 5 + \frac{2}{x+4} - \frac{x+6}{x^2+9}$$

Exercise 11B

In Questions **1** to **29**, write each of the expressions in partial fractions.

1 ***Denominator with two distinct linear factors***

a) $\dfrac{3x-1}{(x+3)(x-2)}$ **b)** $\dfrac{5x+6}{(x+4)(x-3)}$

c) $\dfrac{2x+1}{(x+2)(x+1)}$ **d)** $\dfrac{9-8x}{(2x-1)(3-x)}$

e) $\dfrac{7x+16}{x^2+2x-8}$ **f)** $\dfrac{5x-1}{2x^2+x-10}$

g) $\dfrac{4x}{x^2-9}$ **h)** $\dfrac{2x-7}{x^2-x-2}$

2 ***Denominator with three distinct linear factors***

a) $\dfrac{2}{(x+3)(x+2)(x+1)}$ **b)** $\dfrac{x^2-9x+2}{(x+1)(x-1)(x-2)}$

c) $\dfrac{x+1}{(x+3)(x+2)(x-1)}$ **d)** $\dfrac{2x^2-7x+1}{(2x+1)(2x-1)(x-2)}$

e) $\dfrac{7x^2+39x+56}{(x+4)(x+3)(2x+5)}$ **f)** $\dfrac{2x^2+11x+3}{x(3x+1)(x+3)}$

g) $\dfrac{2(x^2-2x-6)}{x^3+x^2-4x-4}$ **h)** $\dfrac{13x+19}{x^3+2x^2-5x-6}$

3 ***Denominator with a quadratic factor***

a) $\dfrac{5x^2-3x+1}{(x^2+1)(x-2)}$ **b)** $\dfrac{6x+7}{(x^2+2)(x+3)}$

c) $\dfrac{6x^2-7x-11}{(x^2+1)(x-5)}$ **d)** $\dfrac{9x+7}{(2x^2+3)(x+2)}$

e) $\dfrac{6x^2-3x+14}{(3x^2+1)(4-x)}$ **f)** $\dfrac{x-6}{(x^2+3)(2x+1)}$

g) $\dfrac{x^2+5x+4}{(x^2+3x+1)(x+3)}$ **h)** $\dfrac{3x^2+4x-1}{x^3-1}$

4 ***Denominator with a repeated factor***

a) $\dfrac{2x+3}{(x+2)^2}$ **b)** $\dfrac{4x-9}{(x-3)^2}$ **c)** $\dfrac{3x-14}{x^2-8x+16}$ **d)** $\dfrac{5x+7}{(x+1)^2(x+2)}$

e) $\dfrac{6x^2-11x+13}{(x-2)^2(x+3)}$ **f)** $\dfrac{7+5x-6x^2}{(2x+1)^2(x+2)}$ **g)** $\dfrac{x^2-3x-9}{x^3-6x^2+9x}$ **h)** $\dfrac{2x^2+9x+24}{x^3+4x^2-3x-18}$

5 *Improper fractions*

a) $\dfrac{x^2+7x-14}{(x+5)(x-3)}$ **b)** $\dfrac{2x^2+x-5}{(x+2)(x+1)}$

c) $\dfrac{x^3+4x^2-x-17}{(x+3)(x-2)}$ **d)** $\dfrac{3x^3-10x^2+2x-1}{(3x-1)(x-3)}$

e) $\dfrac{x^3+7x^2-18}{(x+4)(x-1)(x-2)}$ **f)** $\dfrac{2x^3+3x^2-x-4}{x^2(x+1)}$

g) $\dfrac{3x^3-5x^2+5x-11}{(x^2+1)(x-3)}$ **h)** $\dfrac{x^4+x^3-19x^2-44x-21}{(x+3)(x+2)(x+1)}$

6 $\dfrac{4x+7}{(x+1)(x+3)}$ **7** $\dfrac{x^2+29x+2}{(3x+5)(2x+1)(x-3)}$ **8** $\dfrac{1-2x}{(x+1)^2}$

9 $\dfrac{4x^2-7x-5}{(x-3)^2(x+2)}$ **10** $\dfrac{3x^2+7x+11}{(x^2+3x+3)(x-1)}$ **11** $\dfrac{x^2+3x-34}{x^2+2x-15}$

12 $\dfrac{x+10}{(2x-1)(x-4)}$ **13** $\dfrac{x^2+9x+17}{(x+4)^2(x+3)}$ **14** $\dfrac{x^3+6x^2+10x+1}{(x+5)(x-1)}$

15 $\dfrac{2x+7}{(x+5)^2}$ **16** $\dfrac{2(4x^3-2x^2-5x+1)}{(2x+1)^2}$ **17** $\dfrac{x-2}{(x+3)(x+2)}$

18 $\dfrac{x^2-5x+40}{x^3-3x^2-6x+8}$ **19** $\dfrac{x^2+13x+8}{x^3+5x^2+7x+35}$ **20** $\dfrac{17-11x-4x^2}{(x+5)^2(2x+3)}$

21 $\dfrac{2(5x-13)}{(x+3)(x-4)(x-5)}$ **22** $\dfrac{4x^2-8x+15}{x^3-x^2-8x+12}$ **23** $\dfrac{3x^2+5x+6}{(x^2+3)(x+5)}$

24 $\dfrac{5x^2-5x+6}{x^3-3x+2}$ **25** $\dfrac{x^4+2x^3-2x^2-9x-7}{(x-1)^2(x+4)}$ **26** $\dfrac{9x^2-22x+9}{x^3-4x^2+3x}$

27 $\dfrac{3x+1}{(x+3)(x-5)}$ **28** $\dfrac{3x-10}{(x^2+x+5)(2x+5)}$ **29** $\dfrac{3x^2-4x+13}{(x^2+3)(x-1)}$

30 a) Express $\dfrac{1}{(x-5)(x-4)}$ in partial fractions.

b) Hence prove that

$$\frac{1}{(x-5)^2(x-4)^2} \equiv \frac{1}{(x-5)^2} + \frac{1}{(x-4)^2} - \frac{2}{x-5} + \frac{2}{x-4}$$

31 a) Express $\dfrac{7}{(x+5)(x-2)}$ in partial fractions.

b) Hence prove that

$$\frac{1}{(x+5)^2(x-2)^2} \equiv \frac{1}{343}\left[\frac{7}{(x+5)^2} + \frac{7}{(x-2)^2} + \frac{2}{x+5} - \frac{2}{x-2}\right]$$

32 a) Express each of the following in partial fractions.

i) $\dfrac{x-8}{(x-2)(x-4)}$ **ii)** $\dfrac{12}{(x-2)(x-4)}$

b) Hence prove that

$$\frac{(x-8)^2}{(x-2)^2(x-4)^2} \equiv \frac{9}{(x-2)^2} + \frac{4}{(x-4)^2} + \frac{6}{x-2} - \frac{6}{x-4}$$

33 a) Express each of the following in partial fractions.

i) $\dfrac{3x+13}{(x+3)(x+1)}$ **ii)** $\dfrac{2}{(x+3)(x+1)}$

b) Hence prove that

$$\frac{(3x+13)^2}{(x+3)^2(x+1)^2} \equiv \frac{4}{(x+3)^2} + \frac{25}{(x+1)^2} + \frac{10}{x+3} - \frac{10}{x+1}$$

Application of partial fractions to series expansions

Example 11 Express

$$\frac{2x+5}{(x+1)(x+2)}$$

in partial fractions and hence find the first four terms in the series expansion of

$$\frac{2x+5}{(x+1)(x+2)}$$

State the values of x for which the expansion is valid.

SOLUTION

Assume that

$$\frac{2x+5}{(x+1)(x+2)} \equiv \frac{A}{x+1} + \frac{B}{x+2}$$

Multiplying throughout by $(x+1)(x+2)$ gives

$$2x+5 \equiv A(x+2) + B(x+1)$$

Let $x=-1$: $\quad 2(-1)+5 = A(-1+2)$

$$\therefore \quad A = 3$$

Let $x=-2$: $\quad 2(-2)+5 = B(-2+1)$

$$\therefore \quad B = -1$$

Therefore,

$$\frac{2x+5}{(x+1)(x+2)} \equiv \frac{3}{x+1} - \frac{1}{x+2}$$

To obtain the series expansion, we need the series expansion of both $\dfrac{3}{x+1}$ and $\dfrac{1}{x+2}$.

Now $\dfrac{3}{x+1} = 3(1+x)^{-1}$ and applying the binomial theorem gives

$$3(1+x)^{-1} = 3\left[1+(-1)x+\frac{(-1)(-1-1)}{2!}x^2+\frac{(-1)(-1-1)(-1-2)}{3!}x^3+\ldots\right]$$
$$= 3(1-x+x^2-x^3+\ldots)$$
$$= 3-3x+3x^2-3x^2+\ldots$$

This expansion is valid for $-1 < x < 1$. [1]

Also, $\dfrac{1}{x+2} = (2+x)^{-1}$ and applying the binomial theorem gives

$$(2+x)^{-1} = \left[2\left(1+\frac{x}{2}\right)\right]^{-1}$$
$$= 2^{-1}\left(1+\frac{x}{2}\right)^{-1}$$
$$= \frac{1}{2}\left[1+(-1)\left(\frac{x}{2}\right)+\frac{(-1)(-1-1)}{2!}\left(\frac{x}{2}\right)^2+\frac{(-1)(-1-1)(-1-2)}{3!}\left(\frac{x}{3}\right)^3+\ldots\right]$$
$$= \frac{1}{2}\left(1-\frac{x}{2}+\frac{x^2}{4}-\frac{x^2}{8}+\ldots\right)$$
$$= \frac{1}{2}-\frac{x}{4}+\frac{x^2}{8}-\frac{x^3}{16}+\ldots$$

This expansion is valid for $-1 < \dfrac{x}{2} < 1$, in other words for $-2 < x < 2$. [2]

Therefore,

$$\frac{3}{x+1}-\frac{1}{x+2} = (3-3x+3x^2-3x^3+\ldots)-\left(\frac{1}{2}-\frac{x}{4}+\frac{x^2}{8}-\frac{x^3}{16}+\ldots\right)$$
$$= \frac{5}{2}-\frac{11}{4}x+\frac{23}{8}x^2-\frac{47}{16}x^3\ldots$$

This expansion is valid for those values of x which satisfy [1] and [2], namely $-1 < x < 1$.

Exercise 11C

1 Express each of the following in partial fractions. Hence obtain the series expansion for each expression, giving all terms up to and including the term in x^3, and stating the values of x for which the expansion is valid.

a) $\dfrac{2-x}{(1+x)(1-2x)}$ **b)** $\dfrac{1+7x}{(1+3x)(1-x)}$ **c)** $\dfrac{1+x}{(1+3x)(1+5x)}$

d) $\dfrac{7+x}{(1+x)(2-x)}$ **e)** $\dfrac{1}{(2+x)(3+x)}$ **f)** $\dfrac{4-x+3x^2}{(1+x^2)(1-x)}$

2 Express each of the following in partial fractions, and hence obtain the series expansion for each expression, stating all terms up to and including the term in x^3.

a) $\dfrac{3+2x^2}{(1+x+x^2)(1-2x)}$ **b)** $\dfrac{1-x-x^2}{(1+3x+x^2)(1+x)}$ **c)** $\dfrac{5+4x+2x^2}{(1-x+x^2)(1+2x)}$

3 Find the coefficient of the term in x^4 when the fraction $\dfrac{x}{(1+2x)(1+3x)}$ is expanded as a series.

4 Calculate the coefficient of the term in x^5 in the series expansion of the fraction

$$\frac{1+x}{(1+x^2)(1-x)}$$

5 Calculate the coefficient of the term in x^4 when the fraction

$$\frac{3x^2+2}{(1+x+x^2)(1+3x)}$$

is expanded as a series.

***6** Find an expression for the coefficient of the term in x^n in the series expansions of the following:

a) $\dfrac{2-5x}{(1-2x)(1-3x)}$ **b)** $\dfrac{9}{(1-4x)(1+5x)}$ **c)** $\dfrac{3-2x}{(1-x)^2}$

***7** Given

$$\frac{(1+x)^n(2-x+x^2)}{(1+x^2)(1-x)} \equiv 2+ax+bx^2-29x^3+\dots$$

for $|x|<1$ and where a and b are constants, find **a)** the value of n, **b)** the values of a and b.

Exercise 11D: Examination questions

1 Express $\dfrac{x^2+1}{x^2(x-1)}$ in partial fractions. (WJEC)

2 Find the values of the constants A, B, C and D for which

$$\frac{x^3+3x^2-2x+1}{(x-1)(x+2)} \equiv A+Bx+\frac{C}{x-1}+\frac{D}{x+2}$$

3 Express in partial fractions the functions f and g given by

i) $f(x)=\dfrac{5x-11}{2x^2+x-6}$ **ii)** $g(x)=\dfrac{x^2}{(x-1)(x-2)}$ (OCSEB)

4 Express $\dfrac{2}{(x-1)(x-3)}$ in partial fractions. Hence show that

$$\frac{4}{(x-1)^2(x-3)^2} \equiv \frac{1}{(x-3)^2}+\frac{1}{(x-1)^2}-\frac{1}{(x-3)}+\frac{1}{(x-1)}$$ (UODLE)

5 a) Express $\dfrac{1-x-x^2}{(1-2x)(1-x)^2}$ as the sum of three partial fractions.

b) Hence, or otherwise, expand this expression in ascending powers of x up to and including the term in x^3.

c) State the range of values of x for which the full expansion is valid. (NEAB)

6 Express $E(x) = \dfrac{10+9x}{(1+3x)(2-x)}$ in the form $\dfrac{A}{1+3x} + \dfrac{B}{2-x}$, where A and B are constants, and hence find the first three terms in the expansion of $E(x)$ in ascending powers of x.

(OCSEB)

7 a) Given that

$$\frac{2-6x+10x^2}{(1-3x)(1+x^2)} \equiv \frac{A}{1-3x} + \frac{Bx+C}{1+x^2}$$

find the values of the constants A, B and C.

b) Write down the series expansions, up to and including the term in x^3, of

$$\frac{1}{1-3x} \quad \text{and} \quad \frac{1}{1+x^2}$$

c) Deduce that, if x is small enough for terms in x^4 and higher powers of x to be ignored, then

$$\frac{2-6x+10x^2}{(1-3x)(1+x^2)} \equiv 2+ax^2+bx^3$$

where a and b are constants to be determined.

8 a) Express $\dfrac{5}{(1+x^2)(2+x)}$ in partial fractions.

b) Given that $|x| < 1$, determine the expansion of

$$\frac{5}{(1+x^2)(2+x)}$$

in ascending powers of x up to and including the term in x^2. (UODLE)

12 Proof

No amount of experimentation can ever prove me right but a single experiment might prove me wrong.
ALBERT EINSTEIN

In mathematics we use lots of results (formulae). For example, to find the area of a right-angled triangle ABC we use the result

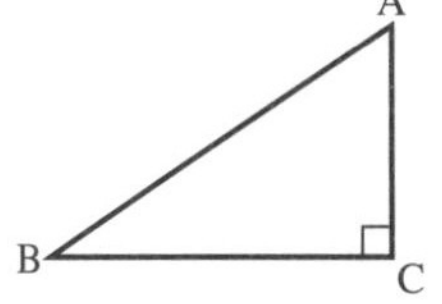

$$\text{Area} = \tfrac{1}{2}\text{BC.AC}$$

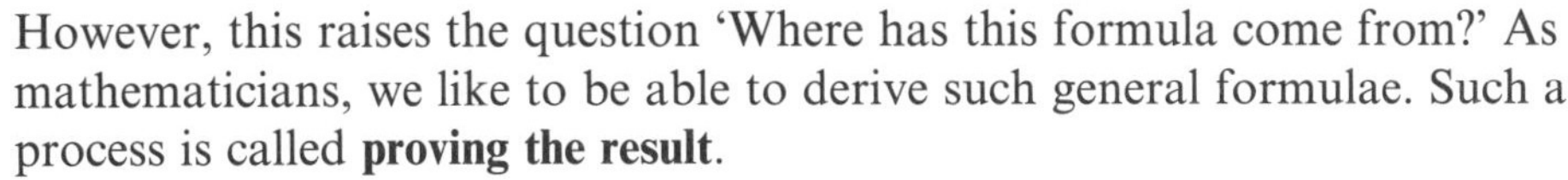

However, this raises the question 'Where has this formula come from?' As mathematicians, we like to be able to derive such general formulae. Such a process is called **proving the result**.

Throughout the book we see many examples of proof. For example, on pages 135–6 we prove that the distance between the two points $A(x_1, y_1)$ and $B(x_2, y_2)$ is given by

$$AB = \sqrt{(x_1 - x_2)^2 + (y_1 - y_2)^2}$$

Once the result is proved, we can use it over and over again.

The mechanics of a proof in mathematics consists of establishing statements by using certain facts which are already known to be true.

Consider the following result:

If $x - y = 1$, then $x^2 - y^2 = x + y$.

Now we could start by trying some numbers. For example, letting $x = 5$ and $y = 4$ gives $x - y = 1$ and if the result is true we should find that $x^2 - y^2 = x + y$. In fact, we have

$$\begin{aligned} x^2 - y^2 &= 5^2 - 4^2 \\ &= 25 - 16 \\ &= 9 \end{aligned}$$

and

$$\begin{aligned} x + y &= 5 + 4 \\ &= 9 \end{aligned}$$

which are the same.

We conclude that the result works when $x = 5$ and $y = 4$. This simply suggests that the result may be true for other values of x and y, and ultimately all values of x and y. To formally prove this result, we use algebraic manipulation. For example,

$$x^2 - y^2 = (x - y)(x + y)$$

Now using the fact that $x - y = 1$ gives

$$x^2 - y^2 = 1 \times (x + y)$$
$$= x + y$$

as required.

This is a mathematical proof using logical steps and algebraic manipulation. However, there exists a mathematical 'language' which we use when deducing one mathematical statement from another.

Implication between mathematical statements

If p and q are mathematical statements, then the compound statement

If p then q

is called the **implication of q by p** and is written $p \Rightarrow q$. In other words, the statement p is true implies that statement q is true.

For example, consider these two statements:

p: the triangle ABC is right-angled at B

q: the area of triangle ABC is given by $\frac{1}{2}$AB.BC

In this case, we have $p \Rightarrow q$. In other words, if the triangle is right-angled at B then the area of triangle ABC is given by $\frac{1}{2}$AB.BC.

Sometimes the statement p implies q and the statement q implies p. In such cases, we write $p \Leftrightarrow q$. We call $q \Rightarrow p$ the **converse statement** of $p \Rightarrow q$.

An example of such a case is p: $xy = 0$

q: $x = 0$ or $y = 0$

Here we have $xy = 0 \Rightarrow x = 0$ or $y = 0$

and $x = 0$ or $y = 0 \Rightarrow xy = 0$

This can be written $xy = 0 \Leftrightarrow x = 0$ or $y = 0$

Another way of writing $p \Leftrightarrow q$ is

p is true **if and only if** q is true

It is important to note that

- the part **if** corresponds to $q \Rightarrow p$
- the part **only if** corresponds to $p \Rightarrow q$.

Sometimes **if and only if** is written as **iff**.

Yet another way of writing $p \Leftrightarrow q$ is

A **necessary and sufficient** condition for p to be true is that q is true

In this case, it is important to note that

- the part **necessary** corresponds to $p \Rightarrow q$
- the part **sufficient** corresponds to $q \Rightarrow p$.

Consider the two statements $p\colon\ 2x + 1 = 7$

$q\colon\ x = 3$

Now it is obvious that $p \Leftrightarrow q$ and we can prove this in a number of ways.

First, consider the proof of $p \Rightarrow q$. This means that we start with statement p and, using logical arguments, we deduce statement q. So if

$$p\colon\ 2x + 1 = 7$$
$$\Rightarrow \quad 2x = 6$$
$$\Rightarrow \quad x = \frac{6}{2} = 3$$

That is, $x = 3$, which is statement q, as required.

Second, we must prove that $q \Rightarrow p$. This means that we start with statement q and deduce statement p. So if

$$q\colon\ x = 3$$
$$\Rightarrow \quad 2x = 6$$
$$\Rightarrow \quad 2x + 1 = 7$$

which is statement p, as required.

In fact, the converse of every step of the proof $p \Rightarrow q$ is also true and in this case the proof could have been written as

$$p\colon\ 2x + 1 = 7$$
$$\Leftrightarrow \quad 2x = 6$$
$$\Leftrightarrow \quad x = \frac{6}{2} = 3$$

That is, $x = 3$, which is statement q, as required.

However, it must be noted that when solving the equation $2x + 1 = 7$ we would not usually write a double implication sign, since we are starting with the equation $2x + 1 = 7$ and trying to deduce the value of x. Instead, we use a single implication sign.

Now consider the two statements $p\colon\ x > 3$

$q\colon\ x^2 > 9$

By inspection, we see that $p \Rightarrow q$, since the square of a number greater than 3 will always give a number greater than 9. To formally prove this result, we start with statement p, in other words

$$p\colon\ x > 3$$
$$\Rightarrow \quad x - 3 > 0$$

Multiplying both sides by $x+3$, which is allowed since $x>3$ and therefore $x+3$ will be positive,

$$\Rightarrow \quad (x-3)(x+3)>0$$
$$\Rightarrow \quad x^2-9>0$$
$$\Rightarrow \quad x^2>9$$

which is statement q, as required.

By inspection, we note that in this particular example the converse statement, namely $q \Rightarrow p$, is not true. We can see this by using the example $x=-4$, which satisfies $x^2>9$ but does not satisfy $x>3$. This example is being used to show that a result is not true and we call it a **counter-example**.

Knowing that $q \nRightarrow p$, we ask the question 'Where in our proof of $p \Rightarrow q$ does the double implication fail?' Starting with

$$q:\ x^2>9$$
$$\Rightarrow \quad x^2-9>0$$
$$\Rightarrow \quad (x-3)(x+3)>0$$

At this stage, we can only divide by $x+3$ if we know that $x+3>0$. That is, $x>-3$. Now in this case we cannot be guaranteed that $x>-3$, since the only statement we know to be true is $x^2>9$ and $x=-4$ would satisfy this but not $x>-3$.

In other words, we cannot make the deduction that

$$(x-3)(x+3)>0 \quad \Rightarrow \quad x-3>0$$

Example 1 Decide which of the signs $\Rightarrow$, $\Leftarrow$ or $\Leftrightarrow$ is appropriate in the compound statement

$$xy=y \ldots y=0$$

Justify your decision by proving your chosen result.

SOLUTION

By inspection, we see that if $y=0$ then $xy=y$. That is, $0=0$. However, if $x=1$ and $y=3$, we have

$$1\times 3=3$$
$$\text{that is} \quad xy=y$$

In other words, we can have $xy=y$ without $y=0$. This means that the compound statement should read

$$xy=y \quad \Leftarrow \quad y=0$$

It is worth noting that if $xy=y$ then we have the following sequence of deductions:

$$\Rightarrow \quad xy-y=0$$
$$\Rightarrow \quad y(x-1)=0$$
$$\Rightarrow \quad y=0 \quad \text{or} \quad x-1=0$$
$$\Rightarrow \quad y=0 \quad \text{or} \quad x=1$$

Example 2 Complete the following statement by inserting one of **if**, **only if** or **if and only if**, as appropriate.

The integer n is a multiple of 3 ...... it is a multiple of 6

SOLUTION

Now if n is a multiple of 3 it could be equal to 9. However, 9 is not a multiple of 6. We therefore have

n a multiple of 3 $\not\Rightarrow$ n is a multiple of 6

However, if n is a multiple of 6 then it must be a multiple of 3, since 3 divides 6. We therefore have

n is a multiple of 6 $\Rightarrow$ n is a multiple of 3

To conclude, we have

The integer n is a multiple of 3 **if** n is a multiple of 6

Example 3 Complete each of the following by inserting one of **necessary**, **sufficient** or **necessary and sufficient**, as appropriate.

a) A ...... condition that $xy = 0$ is that $x = 0$ and $y = 0$.

b) A ...... condition that all the sides of a triangle are equal is that all the angles are equal.

c) A ...... condition that $ax^2 + bx + c > 0$ is that $a > 0$.

SOLUTION

a) Now $xy = 0 \not\Rightarrow x = 0$ and $y = 0$. However, $x = 0$ and $y = 0 \Rightarrow xy = 0$. Therefore, we have

A **sufficient** condition that $xy = 0$ is that $x = 0$ and $y = 0$

b) Now

All the sides of a triangle are equal $\Rightarrow$ All the angles are equal

and

All the angles are equal $\Rightarrow$ All the sides of the triangle are equal

Therefore, we have

A **necessary and sufficient** condition that all the sides of a triangle are equal is that all the angles are equal

c) Now $ax^2 + bx + c > 0 \Rightarrow a > 0$, since the graph of $y = ax^2 + bx + c$ must be U-shaped if $y \geqslant 0$.

But $a > 0 \not\Rightarrow ax^2 + bx + c > 0$, since if $a = 4 > 0$ we have $y = 4x^2 + 8x + 3$, whose graph drops below the x-axis. In fact, the graph of

$$y = 4x^2 + 8x + 3 = (2x + 1)(2x + 3)$$

cuts the x-axis at $x = -\frac{1}{2}$ and $x = -\frac{3}{2}$ as sketched on the right. Therefore, we have

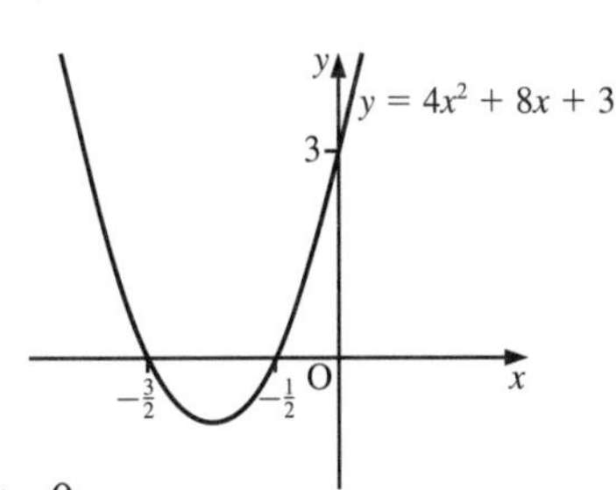

A **necessary** condition that $ax^2 + bx + c > 0$ is that $a > 0$.

Disproving statements

Consider the statement

If x is an odd number then x^2 is even.

If we let $x = 3$, an odd number, then $x^2 = 9$, which is **not** even. This example **disproves** the statement and is called a **counter-example**.

It is very important to notice that, although we cannot prove a result true using examples, we can disprove a result by using just one example – the counter-example.

Example 4 Decide which of the following statements are true and which are false. For those which are true, prove that they are true. For those which are false, find a counter-example to show that they are false.

a) $3^x - 3$ is always divisible by x, where x is a positive integer and $x > 1$.

b) The sum of two odd numbers is always even.

c) For any real numbers x and y, $\sqrt{x^2 + y^2} = x + y$.

d) For all positive integers n, $2^n > n^2$.

SOLUTION

a) Trying examples gives

$$x = 2 : 3^2 - 3 = 6, \text{ which is divisible by } 2$$

$$x = 3 : 3^3 - 3 = 24, \text{ which is divisible by } 3$$

$$x = 4 : 3^4 - 3 = 78, \text{ which is } \textbf{not} \text{ divisible by } 4$$

Although the first two examples suggest that the statement is true, the example $x = 4$ acts as a counter-example and disproves the statement.

In fact, $3^x - 3$ is always divisible by x when x is a prime. This is a special case of Fermat's theorem, which you do not need to know but may find interesting.

b) Trying examples gives

$$3 + 5 = 8, \text{ which is even}$$

$$7 + 13 = 20, \text{ which is even}$$

$$5 + 13 = 18, \text{ which is even}$$

These examples suggest that the sum of two odd numbers is always even, but we must prove it generally.

Proof Let n be a positive integer, then $2n$ will always be even and $2n + 1$ will always be odd.

Let $2n_1 + 1$ and $2n_2 + 1$ be two different odd numbers (hence the reason why n_1 and n_2 are different). Finding the sum of these two numbers gives

$$\begin{aligned}(2n_1 + 1) + (2n_2 + 1) &= 2n_1 + 2n_2 + 1 + 1 \\ &= 2(n_1 + n_2 + 1)\end{aligned}$$

Since $2(n_1 + n_2 + 1)$ is a multiple of 2, it is divisible by 2 and is even. Hence we have proved that the sum of any two odd numbers will always be even.

c) Trying the example $x = 2$, $y = 3$ gives

$$\sqrt{2^2 + 3^2} = \sqrt{13} \neq 2 + 3$$

This is a counter-example which disproves the statement.

d) Trying the example $n = 2$ we have

$$2^2 = 4 \not> (2)^2 = 4$$

This is a counter-example which disproves the statement.

Proof by contradiction

Suppose we wanted to prove the result

There does not exist a real value x such that $\dfrac{3}{1 - x^2} = 1$.

This can be proved using the method of proof by contradiction, which involves assuming the opposite of the given statement and arriving at a contradiction.

Therefore, if we assume that a real value of x does exist such that $\dfrac{3}{1 - x^2} = 1$ then

$$\frac{3}{1 - x^2} = 1$$
$$\Rightarrow \quad 3 = 1 - x^2$$
$$\Rightarrow \quad x^2 = -2$$
$$\Rightarrow \quad x = \pm\sqrt{-2}$$

Our original assumption is now contradicted since $\sqrt{-2}$ is not a real number. Therefore, the original statement must be true.

Example 5 Use the method of proof by contradiction to prove that, for all positive real values of x, $\dfrac{4x^2}{4 + x^2} \leqslant x$.

SOLUTION

We start by assuming the opposite of the given statement: namely, that there exists a positive real value of x such that $\dfrac{4x^2}{4 + x^2} > x$. We now try to find this value of x.

$$\frac{4x^2}{4 + x^2} > x$$
$$\Rightarrow \quad 4x^2 > x(4 + x^2) \quad (\text{since } 4 + x^2 > 0)$$
$$\Rightarrow \quad 4x^2 > 4x + x^3$$
$$\Rightarrow \quad x^3 - 4x^2 + 4x < 0$$
$$\Rightarrow \quad x(x^2 - 4x + 4) < 0$$
$$\Rightarrow \quad x(x - 2)^2 < 0 \qquad [1]$$

Now since $x > 0$ and $(x - 2)^2 \geqslant 0$, we have $x(x - 2)^2 \geqslant 0$, which means that [1] is a contradiction. Therefore, the original statement must be true.

Example 6 Use the method of proof by contradiction to prove the following result:

If x^2 is even then x is even, where x is an integer

SOLUTION

Here we have a compound statement made up of the two statements

p: x^2 is even

q: x is even

The compound statement we are given is $p \Rightarrow q$. To prove this result by contradiction, we suppose that statement p is true and that the opposite of statement q is true. We then proceed to show that this leads to a contradiction.

Proof Assume that x is odd (the opposite of statement q). In other words, $x = 2n + 1$, where n is an integer. We then have

$$\begin{aligned} x^2 &= (2n+1)^2 \\ &= 4n^2 + 4n + 1 \\ &= 2(2n^2 + 2n) + 1 \end{aligned}$$

Since x^2 can be written in the form $2k + 1$, where k is an integer, x^2 must be odd, which is a contradiction of statement p. We conclude that the result must be true.

Example 7 Use the method of proof by contradiction to prove that $\sqrt{2}$ is irrational.

SOLUTION

An irrational number is one which cannot be expressed as a fraction of integers. So we must prove that $\sqrt{2}$ cannot be expressed as a fraction. To prove this by contradiction, we start by supposing that $\sqrt{2}$ can be written as a fraction. That is,

$$\sqrt{2} = \frac{a}{b}$$

where a and b are integers with no common factor. Therefore,

$$\begin{aligned} & b\sqrt{2} = a \\ \Rightarrow \quad & 2b^2 = a^2 \end{aligned}$$

This tells us that a^2 must be even, since it is a multiple of 2. If a^2 is even then a must be even (by Example 6) and therefore $a = 2p$, where p is an integer. Therefore,

$$\begin{aligned} & a^2 = 4p^2 \\ \Rightarrow \quad & 2b^2 = 4p^2 \\ \Rightarrow \quad & b^2 = 2p^2 \quad \Rightarrow \quad b^2 \text{ is even} \end{aligned}$$

If b^2 is even then b must be even (by Example 6). However, we now have a even and b even, which means that a and b have a common factor of 2. This is a contradiction to the original supposition that $\sqrt{2}$ could be written as a fraction in its lowest terms.

Hence $\sqrt{2}$ is irrational.

Exercise 12A

1 In each of the following statements, insert one of the symbols $\Rightarrow$, $\Leftarrow$ or $\Leftrightarrow$ as appropriate.

a) $x - 4 = 0 \ldots\ldots x = 4$

b) $x^2 = 9 \ldots\ldots x = 3$

c) $x = 1 \ldots\ldots x^2 + x - 2 = 0$

d) $\dfrac{1}{x} = 5 \ldots\ldots x = \dfrac{1}{5}$

e) $4^x = 64 \ldots\ldots x = 3$

f) $x^2 - 6x + 8 = 0 \ldots\ldots x = 2$ or 4

g) $\text{p}(x)$ has a factor of $(x - 2) \ldots\ldots \text{p}(x) = x^3 - 3x^2 + 5x - 6$

h) n is an even integer …… n^2 is an even integer

i) $\sin x = \frac{1}{2} \ldots\ldots x = 30°$

j) $x > 12 \ldots\ldots x^2 > 144$

k) $y = 7 - 2x + x^2 \ldots\ldots y$ has a minimum at $(1, 6)$

l) $\text{F}(x) = \dfrac{1}{1 - x} + \dfrac{1}{1 + x} \ldots\ldots \text{F}(x) = \dfrac{2}{1 - x^2}$

2 Complete each of the following statements by inserting one of **if**, **only if** or **if and only if**, as appropriate.

a) $x - 2 \leqslant 0 \ldots\ldots x \leqslant 2$

b) $3x + 5 = 7 \ldots\ldots x = \frac{2}{3}$

c) $x - 5 = 0 \ldots\ldots x^2 - 25 = 0$

d) $x = \dfrac{4}{7} \ldots\ldots \dfrac{1}{x} = \dfrac{7}{4}$

e) $x^2 - 6x = 0 \ldots\ldots x = 6$

f) $x^2 - 7x + 10 = (x - 5)\text{q}(x) \ldots\ldots \text{q}(x) = (x - 2)$

g) $x^6 = 729 \ldots\ldots x = -3$

h) y has a point of inflexion at the origin …… $y = x^3$

i) $x = 45° \ldots\ldots \tan\ x = 1$

j) $\text{f}(x) = (x - 1)(x - 2)(x - 3) \ldots\ldots \text{f}(x) = x^3 - 6x^2 + 11x - 6$

k) $x > 3 \ldots\ldots x^2 > 9$

l) $N = \dfrac{1}{\sqrt{2} + 1} \ldots\ldots N = \sqrt{2} - 1$

3 Complete each of the following by inserting one of **necessary**, **sufficient** or **necessary and sufficient**, as appropriate.

a) A ...... condition that $5x < 20$ is that $x < 4$.

b) A ...... condition that $x^2 = 49$ is that $x = 7$.

c) A ...... condition that $x \geqslant 6$ is that $x^2 \geqslant 36$.

d) A ...... condition that $x^2 - 3x + 2 = 0$ is that $x = 2$.

e) A ...... condition that $\frac{1}{3}(2x + 3) = 4$ is that $2x = 9$.

f) A ...... condition that y has a maximum at the origin is that $\frac{\mathrm{d}y}{\mathrm{d}x} = 0$ at $x = 0$.

g) A ...... condition that $xy > 0$ is that either both $x > 0$ and $y > 0$ or both $x < 0$ and $y < 0$.

h) A ...... condition that a quadrilateral is a square is that all the sides of the quadrilateral are of equal length.

i) A ...... condition that a parallelogram is a rectangle is that at least two of the angles of the parallelogram are right-angles.

j) A ...... condition that $ab = ac$ is that $b = c$.

k) A ...... condition that $(x - k)$ is a factor of $ax^3 + bx^2 + cx + d$ is that $ak^3 + bk^2 + ck + d = 0$.

l) A ...... condition that $x^2 + bx + 1 = 0$ has two distinct real roots is that $b^2 > 4$.

4 Decide which of the following statements are true and which are false. For those which are true, prove that they are true. And for those which are false, find counter-examples to show that they are false.

a) If $a > b$ then $a^2 > b^2$.

b) If $p > q$ then $\frac{1}{p} < \frac{1}{q}$.

c) $n^2 + n$ is an even number for all positive integers n.

d) If n is an even number than $2n^2 + n - 1$ is odd.

e) If n is an odd number than $n^2 + 3n$ is odd.

f) If n is an even number than $n^2 + 2n$ is a multiple of four.

g) For positive real numbers a and b, $\sqrt{a+b} = \sqrt{a} + \sqrt{b}$.

h) $n(2n + 1)(7n + 1)$ is a multiple of 12 for all positive integers n.

i) If a and b are real numbers then $b^2 \geqslant 4a(b - a)$.

j) $n^3 - n$ is a multiple of 6 for all positive integral values of n.

k) $2^n - 1$ is prime for all positive integral values of n.

l) $n! + 1$ is prime for all positive integral values of n.

m) For non zero values of x and y, if $y = 1 + x$ then $\frac{1}{x} - \frac{1}{y} = \frac{1}{xy}$.

n) There exists an integer n such that $5n - n^2 > 7$.

o) $\sqrt{x^2 + y^2} \leqslant x + y$ for all real numbers x and y.

p) $\sqrt{x^2 + y^2} \leqslant x + y$ for all positive real numbers x and y.

q) The quadratic equation $x^2 + x + k = 0$ has two distinct real roots provided $k \leqslant \frac{1}{4}$.

r) If $\frac{\mathrm{d}^2y}{\mathrm{d}x^2} = 0$ when $x = 0$, then the curve $y = \mathrm{f}(x)$ has a point of inflexion at $x = 0$.

5 Prove the following results about odd and even numbers.

a) The sum of two odd numbers is always even.
b) The sum of two even numbers is always even.
c) The product of two odd numbers is always odd.
d) The product of two even numbers is always a multiple of four.
e) An odd number squared is always odd.
f) An odd number cubed is always odd.

6 For each of the following results about rational and irrational numbers decide whether they are true or false. For those which are true, prove that they are true. And for those which are false, find counter examples to show that they are false.

a) The sum of two rational numbers is always rational.
b) The sum of two irrational numbers is always irrational.
c) The product of two rational numbers is always rational.
d) The product of two irrational numbers is always irrational.

In questions **7** to **20**, use the method of proof by contradiction to prove each of the results.

7 If n is an integer and n^2 is odd, then n is odd.

8 if n is an integer and n^2 is even, then n is even.

9 The sum of a rational number and an irrational number is always irrational.

10 The product of a non-zero rational number and an irrational number is always irrational.

11 If x is a real number and $x > 0$, then $x + \dfrac{1}{x} \geqslant 2$.

12 For any two real numbers a and b it is always true that $a^2 + b^2 \geqslant 2ab$.

13 There does not exist a real value of x for which $\dfrac{1}{x^2+1} = 2$.

14 The equation $\dfrac{1}{1-x} = x$ has no real solution.

15 The equation $x = 1 + \dfrac{4x-5}{x+2}$ has no real solution.

16 For all positive real values of x, $\dfrac{2x^2}{1+x^2} \leqslant x$.

17 For all positive real values of x, $x \geqslant \dfrac{4x^2}{4x^2+1}$.

18 $\dfrac{x+y}{2} \geqslant \sqrt{xy}$ for all positive real values of x and y.

***19** $\sqrt{3}$ is irrational.

***20** There are infinitely many prime numbers.

Exercise 12B: Examination questions

1 a) Explain the difference between the term *necessary condition* and the term *sufficient condition.*

b) i) Prove that a necessary condition for n^2 to be a multiple of *eight* is that n should be even.

ii) Give an example to show that this is not a sufficient condition.

c) i) Prove that a sufficient condition for $x + y$ to be a multiple of *three* is that both x and y are multiples of *three*.

ii) Give an example to show that this is not a necessary condition.

2 One of the following statements is true and the other is false. For the one which is true, prove that it is true; and for the one which is false, give a counter-example to show that it is false.

I For $x \neq -1$, $\dfrac{x}{(x+1)^2} \leqslant \dfrac{1}{4}$

II For all real values of x and y, $x > y \quad \Rightarrow \quad (x-y)^2 < x^2 - y^2$

3 a) i) Show that $k^3 - 4k = k(k-2)(k+2)$.

ii) Given that k is even, explain why at least one of k, $k-2$ and $k+2$ is divisible by 4. Hence, prove that if k is even then 48 is always a factor of $k^3 - 4k$.

b) Decide whether the following statement is true or false.

If k is even then $n^2 + 2n + 4$ is divisible by 7.

If you decide it is true, prove the result. If you decide it is false, give a counter-example.

4 For each of the following statements determine whether it is true or false. For those which you decide to be true prove the result. For those which you decide to be false give a counter-example.

a) If $x > -1$ then $x^2 > 1$, for all real x.

b) If $x > 3$ then $\dfrac{1}{x-2} < 1$, for all real x.

c) $x^3 + y^3 \geqslant xy(x+y)$, for all real positive values of x and y.

d) There exist real numbers x and y such that

$$\cos(x+y) = \cos x + \cos y$$

e) All integers which are multiples of 4 are multiples of 12.

13 Differentiation II

How happy the lot of the mathematician. He is judged solely by his peers, and the standard is so high that no colleague or rival can ever win a reputation he does not deserve.
W. H. AUDEN

Function of a function

When we write, for example, $y = (x + 2)^4$, we say that y is a function of x. If we let $u = x + 2$ then we have

$$y = u^4 \quad \text{where } u = x + 2$$

In other words, y is now a function of u, and u is a function of x. The new variable, u, is the link between the two expressions.

In order to differentiate the expression $y = (x + 2)^4$ with respect to x, we would first need to expand the bracket. In this particular case, this would certainly be feasible. However, if we had an expression such as $y = (x + 2)^{12}$, this would involve more work. It is for this reason that we require a technique which will enable us to differentiate such expressions more easily.

Result I: The chain rule

If y is a function of u, and u is a function of x,

$$\frac{dy}{dx} = \frac{dy}{du}\frac{du}{dx}$$

Returning to our example of $y = (x + 2)^4$, we have

$$y = u^4 \quad \text{where } u = x + 2$$

Differentiating y with respect to u gives

$$\frac{dy}{du} = 4u^3$$

Differentiating u with respect to x gives

$$\frac{du}{dx} = 1$$

By the chain rule,

$$\frac{dy}{dx} = \frac{dy}{du}\frac{du}{dx} = (4u^3)(1)$$

$$\therefore \quad \frac{dy}{dx} = 4u^3$$

Substituting $u = x + 2$ gives

$$\frac{dy}{dx} = 4(x+2)^3$$

which is the required derivative.

Example 1 If $y = (3x-1)^7$ find $\frac{dy}{dx}$.

SOLUTION

By letting $u = 3x - 1$, we have $y = u^7$.

Differentiating each expression gives

$$\frac{dy}{du} = 7u^6 \quad \text{and} \quad \frac{du}{dx} = 3$$

By the chain rule,

$$\frac{dy}{dx} = \frac{dy}{du}\frac{du}{dx}$$

$$= (7u^6)(3) = 21u^6$$

$$\therefore \quad \frac{dy}{dx} = 21(3x-1)^6$$

Generally, using function notation, if $y = f(g(x))$, then

$$\frac{dy}{dx} = f'(g(x)).g'(x)$$

In Example 1, this can be interpreted as $y = (3x-1)^7$. Therefore,

$$\frac{dy}{dx} = 7(3x-1)^{7-1} \times \frac{d}{dx}(3x-1)$$

Simplifying gives

$$\frac{dy}{dx} = 7(3x-1)^6\,(3)$$

$$\therefore \quad \frac{dy}{dx} = 21(3x-1)^6$$

Notation We could have written $\frac{d}{dx}(3x-1)$ as $(3x-1)'$ and it is this form of notation that we will use throughout this chapter.

Example 2 Find $\frac{dy}{dx}$ for each of the following functions.

a) $y = 2(1-x)^5$ **b)** $y = (x^2+3)^4$

c) $y = \dfrac{1}{(3-7x)}$ **d)** $y = \sqrt{6x+1}$

SOLUTION

a) When $y = 2(1 - x)^5$,

$$\frac{dy}{dx} = 2 \times 5(1 - x)^4 \times (1 - x)'$$
$$= 10(1 - x)^4(-1)$$
$$\therefore \quad \frac{dy}{dx} = -10(1 - x)^4$$

b) When $y = (x^2 + 3)^4$,

$$\frac{dy}{dx} = 4(x^2 + 3)^3 \times (x^2 + 3)'$$
$$= 4(x^2 + 3)^3(2x)$$
$$\therefore \quad \frac{dy}{dx} = 8x(x^2 + 3)^3$$

c) When $y = \dfrac{1}{3 - 7x} = (3 - 7x)^{-1}$,

$$\frac{dy}{dx} = -(3 - 7x)^{-2} \times (3 - 7x)'$$
$$= -(3 - 7x)^{-2}(-7)$$
$$= 7(3 - 7x)^{-2}$$
$$\therefore \quad \frac{dy}{dx} = \frac{7}{(3 - 7x)^2}$$

d) When $y = \sqrt{6x + 1} = (6x + 1)^{\frac{1}{2}}$,

$$\frac{dy}{dx} = \tfrac{1}{2}(6x + 1)^{-\frac{1}{2}} \times (6x + 1)'$$
$$= \tfrac{1}{2}(6x + 1)^{-\frac{1}{2}}(6)$$
$$= 3(6x + 1)^{-\frac{1}{2}}$$
$$\therefore \quad \frac{dy}{dx} = \frac{3}{\sqrt{6x + 1}}$$

Inverse function of a function (Integration of $f'(x)[f(x)]^n$)

Consider the function $f(x) = (x^2 + 1)^4$.

$$f'(x) = 4(x^2 + 1)^3(x^2 + 1)'$$
$$\therefore \quad f'(x) = 8x(x^2 + 1)^3$$

Therefore, we can write

$$\int 8x(x^2 + 1)^3 \, dx = (x^2 + 1)^4 + c$$

However, we want to be able to recognise that $(x^2 + 1)^4$ is the integral of $8x(x^2 + 1)^3$. In order to do this, we must recognise that the integrand is of the form

$$f'(x)\,[f(x)]^n$$

In our example, $f(x) = x^2 + 1$, $f'(x) = 2x$ and $n = 3$.

Consider the integral $\int x(3x^2 - 2)^5 \, dx$. We first notice that the derivative of $(3x^2 - 2)$ is $6x$ and that we have an x term outside the main function, i.e. $(3x^2 - 2)^5$, of the integrand. This means that we consider $(3x^2 - 2)^6$, which when differentiated gives $36x(3x^2 - 2)^5$. Therefore,

$$\int x(3x^2 - 2)^5 \, dx = \frac{1}{36}(3x^2 - 2)^6 + c$$

Result II

$$\int f'(x)\,[f(x)]^n\,dx = \frac{[f(x)]^{n+1}}{n+1} + c$$

Example 3 Find each of the following integrals.

a) $\int (x-2)^2\,dx$ **b)** $\int x(3x^2+6)^4\,dx$ **c)** $\int 4x^2(x^3-3)^5\,dx$

d) $\int (x+2)(x^2+4x-1)^3\,dx$ **e)** $\int \frac{x}{\sqrt{x^2+3}}\,dx$

SOLUTION

a) To find $\int (x-2)^2\,dx$, consider $(x-2)^3$, which when differentiated gives $3(x-2)^2$. Therefore,

$$\int (x-2)^2\,dx = \frac{(x-2)^3}{3} + c$$

b) To find $\int x(3x^2+6)^4\,dx$, we notice that the derivative of $3x^2+6$ is $6x$ and we have an x term outside the main function in the integrand. This means that we consider $(3x^2+6)^5$, which when differentiated gives $30x(3x^2+6)^4$. Therefore,

$$\int x(3x^2+6)^4\,dx = \frac{(3x^2+6)^5}{30} + c$$

c) To find $\int 4x^2(x^3-3)^5\,dx$ we notice that the derivative of (x^3-3) is $3x^2$, and we have an x^2 term outside the main function in the integrand. This means that we consider $(x^3-3)^6$, which when differentiated gives $18x^2(x^3-3)^5$. Therefore,

$$\int 4x^2(x^3-3)^5\,dx = \frac{4}{18}(x^3-3)^6 + c$$

$$= \frac{2}{9}(x^3-3)^6 + c$$

d) To find $\int (x+2)(x^2+4x-1)^3\,dx$, we notice that the derivative of (x^2+4x-1) is $2x+4=2(x+2)$, and that we have the term $(x+2)$ outside the main function in the integrand. This means that we consider $(x^2+4x-1)^4$, which when differentiated gives

$$4(2x+4)(x^2+4x-1)^3 = 8(x+2)(x^2+4x-1)^3$$

Therefore,

$$\int (x+2)(x^2+4x-1)^3\,dx = \frac{1}{8}(x^2+4x-1)^4 + c$$

e) To find

$$\int \frac{x}{\sqrt{x^2+3}}\,dx = \int x(x^2+3)^{-\frac{1}{2}}\,dx$$

we notice that the derivative of (x^2+3) is $2x$ and we have an x outside the main function in the integrand. This means that we consider $(x^2+3)^{\frac{1}{2}}$, which when differentiated gives

$$2x \times \frac{1}{2}(x^2+3)^{-\frac{1}{2}} = x(x^2+3)^{-\frac{1}{2}}$$

as required. Therefore,

$$\int \frac{x}{\sqrt{x^2+3}}\,dx = (x^2+3)^{\frac{1}{2}} + c$$
$$= \sqrt{x^2+3} + c$$

Exercise 13A

1 Find $\frac{dy}{dx}$ for each of these.

a) $y=(2x-1)^3$ **b)** $y=(3x+4)^2$ **c)** $y=(5x-3)^4$ **d)** $y=(3-x)^5$
e) $y=(4-3x)^6$ **f)** $y=(x^2+1)^4$ **g)** $y=(x^3-6)^2$ **h)** $y=(1-2x^2)^3$
i) $y=(4-x^4)^2$ **j)** $y=(7-5x^3)^6$ **k)** $y=(6x^2-5)^4$ **l)** $y=(9-7x^2)^3$

2 Differentiate each of the following with respect to x.

a) $(2x-5)^{-3}$ **b)** $(3x+2)^{-1}$ **c)** $(x^2+3)^{-2}$ **d)** $(5-2x^3)^{-1}$
e) $\frac{1}{3+4x}$ **f)** $\frac{1}{4-x^2}$ **g)** $\frac{5}{3-2x}$ **h)** $\frac{3}{(x+1)^2}$
i) $\frac{7}{(2-x^2)^5}$ **j)** $-\frac{1}{(3x^2+8)}$ **k)** $(5x^3-4)^{-4}$ **l)** $\frac{1}{2(5-3x^4)^2}$

3 Find $f'(x)$ for each of these.

a) $f(x)=(2x-1)^{\frac{1}{2}}$ **b)** $f(x)=(6-x)^{\frac{1}{3}}$ **c)** $f(x)=(x^3-2)^{\frac{2}{3}}$ **d)** $f(x)=(4-x^5)^{-\frac{1}{5}}$
e) $f(x)=\sqrt{4x-5}$ **f)** $f(x)=\sqrt[3]{x^2+3}$ **g)** $f(x)=\frac{1}{\sqrt{5-2x}}$ **h)** $f(x)=\frac{6}{\sqrt[3]{x^2+5}}$
i) $f(x)=-\frac{3}{\sqrt[6]{4x-7}}$ **j)** $f(x)=(5-4\sqrt{x})^5$ **k)** $f(x)=\sqrt{3+\sqrt{x}}$ **l)** $f(x)=\frac{1}{4-\sqrt[3]{x}}$

4 Find $\frac{\mathrm{d}y}{\mathrm{d}x}$ for each of these

a) $y = (x^2 + x - 1)^4$ **b)** $y = \sqrt{x^3 - 6x}$ **c)** $y = \frac{1}{x^2 - 3x + 5}$ **d)** $y = \left(\frac{1}{\sqrt{x}} - 1\right)^4$

e) $y = (x^4 - 2x^2 + 3)^5$ **f)** $y = \frac{6}{\sqrt[3]{2 - x + x^3}}$ **g)** $y = \left(1 + \frac{3}{x}\right)^2$ **h)** $y = (2\sqrt{x} - x)^4$

i) $y = \sqrt[4]{6x - x^3}$ **j)** $y = (2 + 5x - 3x^4)^7$ **k)** $y = -\frac{3}{(1 + x - x^2)^4}$ **l)** $y = \sqrt{\frac{1}{x} + 3}$

5 Integrate each of the following with respect to x.

a) $(2x - 3)^4$ **b)** $(5x + 8)^2$ **c)** $(3x - 4)^5$ **d)** $3(x - 7)^2$

e) $(4 - x)^5$ **f)** $-(6 - 7x)^3$ **g)** $(3x - 4)^{-3}$ **h)** $6(5 - 9x)^{-2}$

i) $\frac{1}{(2x - 1)^7}$ **j)** $\frac{3}{(1 - x)^2}$ **k)** $\sqrt{2x - 3}$ **l)** $\frac{12}{\sqrt[3]{x - 4}}$

6 Find each of the following integrals.

a) $\int (2x - 7)^5\,\mathrm{d}x$ **b)** $\int \frac{1}{\sqrt{2x - 1}}\,\mathrm{d}x$ **c)** $\int x(x^2 + 2)^3\,\mathrm{d}x$ **d)** $\int 2x(4 - 3x^2)^5\,\mathrm{d}x$

e) $\int x^2(x^3 - 4)^2\,\mathrm{d}x$ **f)** $\int \frac{4x}{(3 - x^2)^2}\,\mathrm{d}x$ **g)** $\int x^3\sqrt{x^4 - 1}\,\mathrm{d}x$ **h)** $\int 4x\sqrt[3]{2 - 3x^2}\,\mathrm{d}x$

i) $\int x^{\frac{1}{3}}\left(x^{\frac{4}{3}} - 2\right)^2\mathrm{d}x$ **j)** $\int \frac{25x^4}{(3 - x^5)^2}\,\mathrm{d}x$ **k)** $\int x(x^2 + 1)^2\,\mathrm{d}x$ **l)** $\int (x^2 + 1)^2\,\mathrm{d}x$

***7** Given that $y = \sqrt{\sqrt{(1 + \sqrt{x})} - 1}$, show that

$$\frac{\mathrm{d}y}{\mathrm{d}x} = \frac{1}{8\sqrt{x(1 + \sqrt{x})(\sqrt{(1 + \sqrt{x})} - 1)}}$$

***8** Find an expression for $\frac{\mathrm{d}y}{\mathrm{d}x}$ given that

$$y = \frac{1}{\left(x + \sqrt{4 + (x^2 - 1)^3}\right)^4}$$

Product rule

Result III

If $y = uv$, where u and v are both functions of x, then

$$\frac{\mathrm{d}y}{\mathrm{d}x} = u\frac{\mathrm{d}v}{\mathrm{d}x} + v\frac{\mathrm{d}u}{\mathrm{d}x}$$

Proof

Let δx be a small increment in x and let δu, δv and δy be the resulting small increments in u, v and y, respectively. Then

$$y + \delta y = (u + \delta u)(v + \delta v)$$

$$\therefore \quad \delta y = uv + u\delta v + v\delta u + \delta u\delta v - y$$

But $y = uv$, therefore

$$\begin{aligned} \delta y &= uv + u\delta v + v\delta u + \delta u\delta v - uv \\ &= u\delta v + v\delta u + \delta u\delta v \end{aligned}$$

Dividing throughout by δx gives

$$\frac{\delta y}{\delta x} = u\frac{\delta v}{\delta x} + v\frac{\delta u}{\delta x} + \frac{\delta u\delta v}{\delta x}$$

As $\delta x \to 0$, we have $\delta u \to 0$, $\delta v \to 0$ and $\delta y \to 0$, with

$$\frac{\delta y}{\delta x} \to \frac{dy}{dx} \qquad \frac{\delta u}{\delta x} \to \frac{du}{dx} \qquad \frac{\delta v}{\delta x} \to \frac{dv}{dx} \qquad \frac{\delta u\delta v}{\delta x} \to 0$$

Therefore, we have

$$\frac{dy}{dx} = u\frac{dv}{dx} + v\frac{du}{dx}$$

as required.

This can also be expressed as

$$(uv)' = u'v + uv'$$

Example 4 If $y = x^2(x+2)^3$ find $\dfrac{dy}{dx}$.

SOLUTION

In this example, y is the product of two functions u and v, where

$$u = x^2 \quad \text{and} \quad v = (x+2)^3$$

Differentiating each with respect to x gives

$$\frac{du}{dx} = 2x \quad \text{and} \quad \begin{aligned} \frac{dv}{dx} &= 3(x+2)^2 \cdot (x+2)' \\ &= 3(x+2)^2 \end{aligned}$$

Using the product rule gives

$$\begin{aligned} \frac{dy}{dx} &= u\frac{dv}{dx} + v\frac{du}{dx} \\ &= x^2\left[3(x+2)^2\right] + (x+2)^3(2x) \end{aligned}$$

$$\therefore \quad \frac{dy}{dx} = 3x^2(x+2)^2 + 2x(x+2)^3$$

Factorising gives

$$\frac{dy}{dx} = x(x+2)^2\,[3x + 2(x+2)]$$

$$\therefore \quad \frac{dy}{dx} = x(x+2)^2(5x+4)$$

In all the examples which follow, we will fully factorise the derivative.

Example 5 Find $\frac{dy}{dx}$ for each of the following functions.

a) $y = (x+2)^3(1-x^2)^4$ **b)** $y = 7x^2\sqrt{x^2-1}$

SOLUTION

a) When $y = (x+2)^3(1-x^2)^4$, using the product rule gives

$$\frac{dy}{dx} = (x+2)^3\left[(1-x^2)^4\right]' + (1-x^2)^4\left[(x+2)^3\right]'$$

$$= (x+2)^3\left[4(1-x^2)^3(-2x)\right] + (1-x^2)^4\left[3(x+2)^2(1)\right]$$

$$\therefore \quad \frac{dy}{dx} = -8x(x+2)^3(1-x^2)^3 + 3(x+2)^2(1-x^2)^4$$

Factorising gives

$$\frac{dy}{dx} = (x+2)^2(1-x^2)^3\left[-8x(x+2) + 3(1-x^2)\right]$$

$$= (x+2)^2(1-x^2)^3(-8x^2 - 16x + 3 - 3x^2)$$

$$\therefore \quad \frac{dy}{dx} = (x+2)^2(1-x^2)^3(3 - 16x - 11x^2)$$

b) When $y = 7x^2\sqrt{x^2-1} = 7x^2(x^2-1)^{\frac{1}{2}}$, using the product rule gives

$$\frac{dy}{dx} = 7x^2\left[(x^2-1)^{\frac{1}{2}}\right]' + (x^2-1)^{\frac{1}{2}}(7x^2)'$$

$$= 7x^2\left[\tfrac{1}{2}(x^2-1)^{-\frac{1}{2}}(2x)\right] + (x^2-1)^{\frac{1}{2}}(14x)$$

$$\therefore \quad \frac{dy}{dx} = 7x^2\left[x(x^2-1)^{-\frac{1}{2}}\right] + 14x(x^2-1)^{\frac{1}{2}}$$

Factorising gives

$$\frac{dy}{dx} = 7x(x^2-1)^{-\frac{1}{2}}[x^2 + 2(x^2-1)]$$

$$= 7x(x^2-1)^{-\frac{1}{2}}(3x^2-2)$$

$$\therefore \quad \frac{dy}{dx} = \frac{7x(3x^2-2)}{\sqrt{x^2-1}}$$

Quotient rule

Result IV

If $y = \frac{u}{v}$, where u and v are functions of x, then

$$\frac{dy}{dx} = \frac{v\frac{du}{dx} - u\frac{dv}{dx}}{v^2}$$

Proof

Let δx be a small increment in x and let δu, δv and δy be the resulting small increments in u, v and y respectively. Then

$$y + \delta y = \frac{u + \delta u}{v + \delta v}$$

$$\therefore \quad \delta y = \frac{u + \delta u}{v + \delta v} - \frac{u}{v}$$

$$= \frac{v(u + \delta u) - u(v + \delta v)}{v(v + \delta v)}$$

$$= \frac{uv + v\delta u - uv - u\delta v}{v(v + \delta v)}$$

$$\therefore \quad \delta y = \frac{v\delta u - u\delta v}{v^2 + v\delta v}$$

Dividing throughout by δx gives

$$\frac{\delta y}{\delta x} = \frac{v\delta u - u\delta v}{v^2 + v\delta v} \times \frac{1}{\delta x}$$

$$\therefore \quad \frac{\delta y}{\delta x} = \frac{v\frac{\delta u}{\delta x} - u\frac{\delta v}{\delta x}}{v^2 + v\delta v}$$

As $\delta x \longrightarrow 0$, we have $\delta u \longrightarrow 0$, $\delta v \longrightarrow 0$ and $\delta y \longrightarrow 0$ with

$$\frac{\delta y}{\delta x} \longrightarrow \frac{dy}{dx} \qquad \frac{\delta u}{\delta x} \longrightarrow \frac{du}{dx} \qquad \frac{\delta v}{\delta x} \longrightarrow \frac{dv}{dx}$$

Therefore, we have

$$\frac{dy}{dx} = \frac{v\frac{du}{dx} - u\frac{dv}{dx}}{v^2}$$

as required.

This can also be expressed as

$$\left(\frac{u}{v}\right)' = \frac{u'v - uv'}{v^2}$$

Example 6 If $y = \dfrac{3x+2}{2x+1}$ find $\dfrac{dy}{dx}$.

SOLUTION

In this example, y is the quotient of two functions u and v, where

$$u = 3x + 2 \quad \text{and} \quad v = 2x + 1$$

Differentiating each with respect to x gives

$$u' = 3 \quad \text{and} \quad v' = 2$$

Using the quotient rule, we have

$$\frac{dy}{dx} = \frac{3 \times (2x+1) - (3x+2) \times 2}{(2x+1)^2}$$

$$\therefore \quad \frac{dy}{dx} = -\frac{1}{(2x+1)^2}$$

Example 7 Find $\dfrac{dy}{dx}$ for each of the following functions.

a) $y = \dfrac{x}{(x^2+4)^3}$ **b)** $y = \sqrt{\dfrac{x^3}{x^2-1}}$

SOLUTION

a) When $y = \dfrac{x}{(x^2+4)^3}$, using the quotient rule gives

$$\frac{dy}{dx} = \frac{(x^2+4)^3(x)' - x\left[(x^2+4)^3\right]'}{\left[(x^2+4)^3\right]^2}$$

$$= \frac{(x^2+4)^3(1) - x\left[3(x^2+4)^2(2x)\right]}{(x^2+4)^6}$$

$$\therefore \quad \frac{dy}{dx} = \frac{(x^2+4)^3 - 6x^2(x^2+4)^2}{(x^2+4)^6}$$

Factorising the numerator gives

$$\frac{dy}{dx} = \frac{(x^2+4)^2[(x^2+4) - 6x^2]}{(x^2+4)^6}$$

$$\therefore \quad \frac{dy}{dx} = \frac{(4-5x^2)}{(x^2+4)^4}$$

b) When $y = \sqrt{\dfrac{x^3}{x^2-1}} = \dfrac{x^{\frac{3}{2}}}{(x^2-1)^{\frac{1}{2}}}$, using the quotient rule gives

$$\frac{dy}{dx} = \frac{(x^2-1)^{\frac{1}{2}}\left(x^{\frac{3}{2}}\right)' - x^{\frac{3}{2}}\left[(x^2-1)^{\frac{1}{2}}\right]'}{\left[(x^2-1)^{\frac{1}{2}}\right]^2}$$

$$= \frac{(x^2-1)^{\frac{1}{2}}\left(\frac{3}{2}x^{\frac{1}{2}}\right) - x^{\frac{3}{2}}\left[\frac{1}{2}(x^2-1)^{-\frac{1}{2}}(2x)\right]}{(x^2-1)}$$

$$\therefore \quad \frac{dy}{dx} = \frac{\frac{3}{2}x^{\frac{1}{2}}(x^2-1)^{\frac{1}{2}} - x^{\frac{5}{2}}(x^2-1)^{-\frac{1}{2}}}{(x^2-1)}$$

Factorising the numerator gives

$$\frac{dy}{dx} = \frac{\frac{1}{2}x^{\frac{1}{2}}(x^2-1)^{-\frac{1}{2}}[3(x^2-1)-2x^2]}{(x^2-1)}$$

$$= \frac{\frac{1}{2}x^{\frac{1}{2}}(x^2-1)^{-\frac{1}{2}}(x^2-3)}{(x^2-1)}$$

$$\therefore \quad \frac{dy}{dx} = \frac{\sqrt{x}\,(x^2-3)}{2\sqrt{(x^2-1)^3}}$$

Exercise 13B

In each part of Questions **1** to **4**, use the product rule to differentiate the given function with respect to x.

1 **a)** $(x+3)(x-4)$ **b)** $(2x+5)(x-7)$ **c)** $(3x-4)(2x+5)$
d) $(6+x)(5-x)$ **e)** $(3-2x)(7+3x)$ **f)** $(x+4)(x^2-2)$
g) $(x^2-5)(4x-1)$ **h)** $(x+6)(x^2-3x+3)$ **i)** $(3x-5)(x^2-2x+7)$
j) $(x^2+1)(x^3-5)$ **k)** $(5x^3-3)(x^2+4x-1)$ **l)** $(x^6-2)(3x^3-x^2+4)$

2 **a)** $x^2(x+3)^4$ **b)** $x^3(2+x)^2$ **c)** $x^4(3x-1)^3$
d) $3x^2(2x+5)^2$ **e)** $x^3(4x^2-1)^3$ **f)** $5x^2(2-x^3)^2$
g) $3x^2(5x^2+1)^4$ **h)** $x^7(2-5x^3)^4$ **i)** $x^2(x^2+x-1)^3$
j) $x^3(4-x+2x^2)^4$ **k)** $2x^4(3x^2-6x+2)^3$ **l)** $5x^2(x^3-x+1)^3$

3 **a)** $(x+2)^2(x-5)^3$ **b)** $(2x-1)^3(x+4)^2$ **c)** $(5x+2)^4(4x-3)^3$
d) $(2-x)^6(5+2x)^4$ **e)** $(3+5x)^2(4-7x)^7$ **f)** $(x^2+1)^2(2x-3)^4$
g) $(5x+9)^3(x^2-2)^3$ **h)** $(2x^2-3)^5(4x-7)^6$ **i)** $(x^3-1)^3(4x^2+5)^2$
j) $(5-x^2)^4(6-5x^2)^6$ **k)** $(x^2-3x+1)^5(2x-3)^3$ **l)** $(5x^2-10x+12)^3(x^3-6)^5$

4 **a)** $x\sqrt{x+1}$ **b)** $2x\sqrt{3-x}$ **c)** $3x\sqrt{5+2x}$
d) $x^2\sqrt{x+3}$ **e)** $x^2\sqrt{3-4x}$ **f)** $(2x-1)\sqrt{x+3}$
g) $(1-3x)\sqrt{2x+5}$ **h)** $\sqrt{x}(5x-4)^3$ **i)** $(3x+5)^2\sqrt{x-2}$
j) $\sqrt{2x-3}\sqrt{4x+1}$ **k)** $\sqrt{6+x}\sqrt{3-2x}$ **l)** $\sqrt{x^2-2}\sqrt{6x-1}$

In each part of Questions **5** and **6**, use the quotient rule to differentiate the given function with respect to x.

5 **a)** $\dfrac{x}{x-2}$ **b)** $\dfrac{x+3}{x-1}$ **c)** $\dfrac{3-x}{4+x}$ **d)** $\dfrac{4x-3}{x+2}$

e) $\dfrac{2x-5}{x+4}$ **f)** $\dfrac{5x}{x+2}$ **g)** $\dfrac{1+3x}{2-5x}$ **h)** $\dfrac{4x+3}{2x-1}$

i) $\dfrac{x^2}{x+3}$ **j)** $\dfrac{x^2}{4-x}$ **k)** $\dfrac{x^3}{2x-3}$ **l)** $\dfrac{x^5}{3-x}$

6 **a)** $\dfrac{(3x-2)^2}{\sqrt{x}}$ **b)** $\dfrac{(5x+1)^3}{\sqrt{x}}$ **c)** $\dfrac{(x^2-4)^5}{\sqrt{x}}$ **d)** $\dfrac{\sqrt{x}}{2x-1}$

e) $\dfrac{3-\sqrt{x}}{(2+x)^2}$ **f)** $\dfrac{5+2\sqrt{x}}{(5-4x)^3}$ **g)** $\dfrac{(3x^2+2)^4}{\sqrt{2x-1}}$ **h)** $\dfrac{(2-3x)^2}{\sqrt{1-x^2}}$

i) $\sqrt{\dfrac{x-2}{x+1}}$ **j)** $\sqrt{\dfrac{x-3}{2x+5}}$ **k)** $\sqrt{\dfrac{3+x}{2-3x}}$ **l)** $\sqrt{\dfrac{x^2+1}{x^3-3}}$

7 Find $\dfrac{dy}{dx}$ for each of these.

a) $y = x^3(3-x)^2$ **b)** $y = \dfrac{x}{2x-1}$ **c)** $y = \sqrt{x}(5x-1)^2$

d) $y = \dfrac{2x}{\sqrt{x}+1}$ **e)** $y = (5x+3)^3(x-2)^2$ **f)** $y = \sqrt{\dfrac{3x-2}{x-3}}$

g) $y = x^3\sqrt{7-2x}$ **h)** $y = \dfrac{x^2}{2-x}$ **i)** $y = \dfrac{\sqrt{x}+1}{\sqrt{x}-1}$

j) $y = (3-x)^4(2+x)^5$ **k)** $y = \dfrac{x^2+1}{3x-1}$ **l)** $y = \sqrt{(x-1)(2x+1)}$

***8** Differentiate each of the following functions with respect to x.

a) $\dfrac{x(x-1)^3}{x-3}$ **b)** $\dfrac{x\sqrt{5-x^2}}{6-x}$ **c)** $\dfrac{x^3\sqrt{4-x^2}}{5-\sqrt{x}}$

***9** Given that $y = \left(\dfrac{x(x-3)^3}{(x+3)(x+5)^2}\right)^2$, show that

$$\frac{dy}{dx} = \frac{2x(x-3)^5(x^3+27x^2+69x-45)}{(x+3)^3(x+5)^5}$$

Applications

Example 8 Find the equation of the tangent to the curve $y = x^2(x+1)^4$ at the point P(1, 16).

SOLUTION

To find $\frac{dy}{dx}$ we use the product rule together with the chain rule, which gives

$$\begin{aligned}\frac{dy}{dx} &= x^2\left[(x+1)^4\right]' + (x+1)^4(x^2)' \\ &= x^2\left[4(x+1)^3\right] + (x+1)^4(2x) \\ &= 4x^2(x+1)^3 + 2x(x+1)^4 \\ &= 2x(x+1)^3[2x + (x+1)]\end{aligned}$$

$$\therefore \quad \frac{dy}{dx} = 2x(x+1)^3(3x+1)$$

At P(1, 16),

$$\left.\frac{dy}{dx}\right|_{x=1} = 2(8)(4) = 64$$

Therefore, the equation of the tangent line is of the form

$$y = 64x + c$$

Since the tangent line passes through P(1, 16), we have

$$16 = 64(1) + c$$

$$\therefore \quad c = -48$$

The equation of the tangent is $y = 64x - 48$.

Example 9 Find the coordinates of the stationary points on the curve $f(x) = \frac{x}{x^2+4}$. Show that

$$f''(x) = \frac{2x(x^2-12)}{(x^2+4)^3}$$

and hence determine the nature of the stationary points.

SOLUTION

At a stationary point, $f'(x) = 0$. Using the quotient rule gives

$$\begin{aligned}f'(x) &= \frac{(x^2+4)(1) - x(2x)}{(x^2+4)^2} \\ &= \frac{x^2+4-2x^2}{(x^2+4)^2}\end{aligned}$$

$$\therefore \quad f'(x) = \frac{4-x^2}{(x^2+4)^2}$$

When $f'(x) = 0$: $\quad \dfrac{4 - x^2}{(x^2+4)^2} = 0$

$$\therefore \quad 4 - x^2 = 0$$

$$\therefore \quad x = \pm 2$$

When $x = 2$: $\quad f(2) = \dfrac{2}{2^2 + 4} = \dfrac{1}{4}$

When $x = -2$: $\quad f(-2) = \dfrac{-2}{(-2)^2 + 4} = -\dfrac{1}{4}$

The coordinates of the stationary points on the curve are $\left(2, \frac{1}{4}\right)$ and $\left(-2, -\frac{1}{4}\right)$.

Since $f'(x) = \dfrac{4 - x^2}{(x^2+4)^2}$, use the quotient rule to find $f''(x)$, giving

$$\begin{aligned} f''(x) &= \frac{(x^2+4)^2(-2x) - (4 - x^2)[2(x^2+4)(2x)]}{(x^2+4)^4} \\ &= \frac{-2x(x^2+4)^2 - 4x(x^2+4)(4 - x^2)}{(x^2+4)^4} \\ &= \frac{2x(x^2+4)[-(x^2+4) - 2(4 - x^2)]}{(x^2+4)^4} \\ &= \frac{2x(x^2+4)(x^2 - 12)}{(x^2+4)^4} \end{aligned}$$

$$\therefore \quad f''(x) = \frac{2x(x^2 - 12)}{(x^2+4)^3}$$

as required.

When $x = 2$: $\quad f''(2) = \dfrac{-32}{512} = -\dfrac{1}{16} < 0$

Since $f''(2) < 0$, the stationary point $\left(2, \frac{1}{4}\right)$ is a maximum.

When $x = -2$: $\quad f''(-2) = \dfrac{32}{512} = \dfrac{1}{16} > 0$

Since $f''(-2) > 0$, the stationary point $\left(-2, -\frac{1}{4}\right)$ is a minimum.

Curve sketching

Very often we require a rough sketch of a curve without plotting a large number of points. Given below are five steps to follow when sketching a curve.

- **Zeros** Find the value of y when $x = 0$ and find (if possible) the value(s) of x when $y = 0$. This gives the points where the curve crosses the axes.
- **Infinities** Find the values of x for which y is not defined. In most cases these will be values of x which make the denominator of a rational function zero.
- **Sign** There are two places where a curve might change sign, either

  (i) at $y = 0$ or
  (ii) at $y = \infty$.

- **Turning points** Calculate $\dfrac{dy}{dx}$.
  - When $\dfrac{dy}{dx} > 0$, the curve slopes upwards from left to right.
  - When $\dfrac{dy}{dx} < 0$, the curve slopes downwards from left to right.
  - When $\dfrac{dy}{dx} = 0$, the curve has a turning point.
- **Asymptotes** Examine the behaviour of the function as $x \to +\infty$ and as $x \to -\infty$.

You can remember this checklist by remembering the word ZISTA – Zeros, Infinities, Sign, Turning points, Asymptotes.

Another useful point to remember is that if the equation of the curve involves only even powers of x, the curve will be symmetrical about the y-axis. For example, the curve $y = x^2$ is symmetrical about the y-axis.

Example 10 Sketch the curve $y = \dfrac{x+1}{2x-3}$.

SOLUTION

- Zeros: When $x = 0$, $y = -\frac{1}{3}$. When $y = 0$, $x = -1$. The curve cuts the x-axis at $(-1, 0)$ and the y-axis at $\left(0, -\frac{1}{3}\right)$.
- Infinities: y is not defined when $2x - 3 = 0$. That is, when $x = \frac{3}{2}$.
- Sign change: There are sign changes at $x = -1$ and $x = \frac{3}{2}$.
- Turning points: Using the quotient rule, we have

$$\frac{dy}{dx} = \frac{(2x-3) \times 1 - (x+1) \times 2}{(2x-3)^2}$$

$$\therefore \quad \frac{dy}{dx} = -\frac{5}{(2x-3)^2}$$

  We see that $\dfrac{dy}{dx}$ is always negative and never zero.

  Therefore, the curve slopes downwards from left to right and there are no turning points.

- Asymptotes: As $x \to +\infty$, we see that $y \to \frac{1}{2}$ from the positive direction.

  As $x \to -\infty$, we see that $y \to \frac{1}{2}$ from the negative direction.

  The curve is shown on the right.

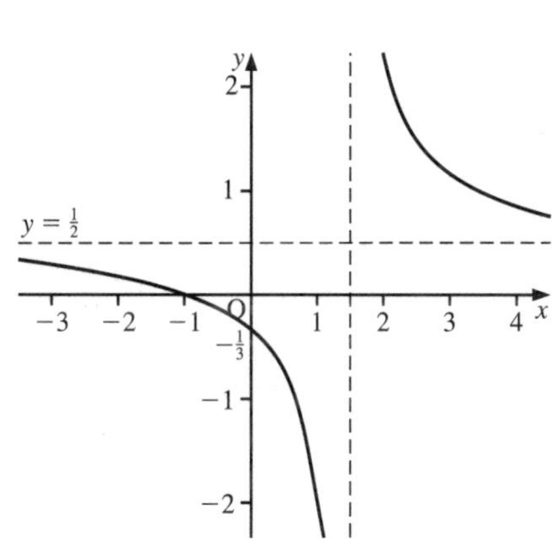

Example 11 Sketch the curve $y = \dfrac{2x}{x^2+1}$.

SOLUTION

- Zeros: When $x = 0$, $y = 0$. The curve passes through the origin.
- Infinities: y is defined for all x, since $x^2 + 1 > 0$.
- Sign: There is a change of sign at $x = 0$.
- Turning points: Using the quotient rule, we have

$$\frac{dy}{dx} = \frac{(x^2+1)\times 2 - 2x \times (2x)}{(x^2+1)^2}$$

$$= \frac{2-2x^2}{(x^2+1)^2}$$

$$\therefore \quad \frac{dy}{dx} = \frac{2(1-x^2)}{(x^2+1)^2}$$

We see that when $\dfrac{dy}{dx} = 0$,

$$2(1-x^2) = 0$$

$$\therefore \quad x^2 = 1 \quad \text{giving} \quad x = \pm 1$$

When $x = 1$, $y = 1$. When $x = -1$, $y = -1$. Therefore, there are two stationary points on the curve, namely $(1, 1)$ and $(-1, -1)$. Examining the gradient either side of each of the stationary points gives

x	0	1	2
$\dfrac{dy}{dx}$	2	0	$-\dfrac{6}{25}$
	/	—	\

x	-2	-1	0
$\dfrac{dy}{dx}$	$-\dfrac{6}{25}$	0	2
	\	—	/

Therefore, $(1, 1)$ is a maximum turning point and $(-1, -1)$ is a minimum turning point.

- Asymptotes: As $x \to +\infty$, $y \to 0$ from the positive direction.

As $x \to -\infty$, $y \to 0$ from the negative direction.

The curve is shown below.

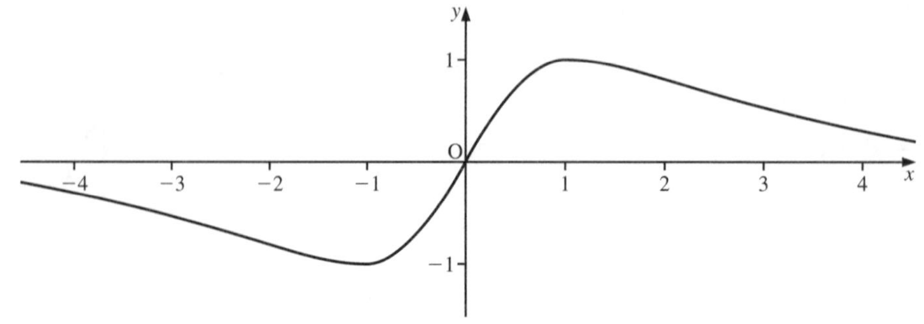

Exercise 13C

1 Find the equation of the tangent and the normal to the curve $y = x(4-x)^2$ at the point $(2, 8)$.

2 Find the equation of the tangent and the normal to the curve $y = \dfrac{2x}{x-1}$ at the point $(3, 3)$.

3 The tangent to the curve $y = x^3(x-2)^2$ at the point $(-1, -9)$, meets the normal to the same curve at the point $(1, 1)$, at the point P. Find the coordinates of P.

4 The tangent to the curve $y = 3x\sqrt{1+2x}$ at the point $(4, 36)$, meets the x-axis at P, and the y-axis at Q. Calculate the area of the triangle OPQ, where O is the origin.

5 Find the coordinates of the two points on the curve $y = \dfrac{x}{1+x}$ where the gradient is $\frac{1}{9}$.

6 Show there is just one point, P, on the curve $y = x(x-1)^3$, where the gradient is 7. Find the coordinates of P.

7 Given that $y = \dfrac{x^2}{2-x}$, show that

$$\frac{dy}{dx} = \frac{x(4-x)}{(2-x)^2}$$

Hence find the coordinates of the two points on the curve $y = \dfrac{x^2}{2-x}$ where the gradient of the curve is zero.

8 Given that $y = x\sqrt{3+2x}$, show that

$$\frac{dy}{dx} = \frac{3(1+x)}{\sqrt{3+2x}}$$

Hence find the point on the curve $y = x\sqrt{3+2x}$ where the gradient is zero.

9 Find the coordinates of the stationary points on the curve $y = (x+3)^2(2-x)$, and determine their nature.

10 Show that the curve $y = \dfrac{x+3}{(x+4)^2}$ has a single stationary point. Find the coordinates of that stationary point, and determine its nature.

11 Find the coordinates of the stationary point on the curve $y = \dfrac{x}{\sqrt{x-5}}$, and determine its nature.

12 Find and classify all the stationary values on the curve $y = (4x-1)(x^2-4)^2$.

13 Use ZISTA to sketch each of the following curves.

a) $y = \dfrac{1}{1+x}$ **b)** $y = \dfrac{2}{3+x}$ **c)** $y = \dfrac{4}{1-x}$ **d)** $y = \dfrac{x}{1+x}$

e) $y = \dfrac{x+4}{x-2}$ **f)** $y = \dfrac{x^2}{x+2}$ **g)** $y = \dfrac{x}{(x+1)^2}$ **h)** $y = \dfrac{2x-5}{x^2-4}$

i) $y = \dfrac{1}{1+x^2}$ **j)** $y = \dfrac{x}{1+x^2}$ **k)** $y = \dfrac{1}{1-x^2}$ **l)** $y = \dfrac{x}{1-x^2}$

14 The total profit, y thousand pounds, generated from the production and sale of x items of a particular product is given by the formula

$$y = \frac{300\sqrt{x}}{100 + x}$$

Calculate the value of x which gives a maximum profit, and determine that maximum profit.

15 A rectangle is drawn inside a semicircle of radius 5 cm, in such a way that one of its sides lies along the diameter of the semicircle, as in the diagram on the right.
Given that the width of the rectangle is x cm, show that the area of the rectangle is $\left[2x\sqrt{25 - x^2}\right]$ cm^2.
Calculate the maximum value of this area.

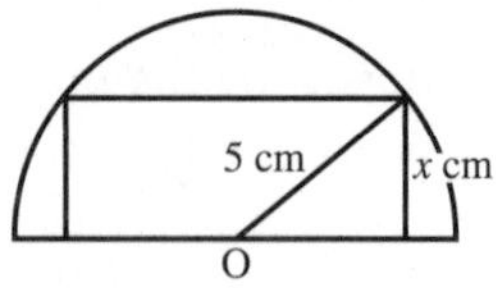

***16** **a)** Given that $y = \dfrac{x^2}{ax + b}$, $(a \neq 0, b \neq 0)$, show that $\dfrac{d^2y}{dx^2} = \dfrac{2b^2}{(ax + b)^3}$.

b) Given further that the curve $y = \dfrac{x^2}{ax + b}$ has a stationary point at (3, 3), find the values of the constants a and b, and show that the point (3, 3) is a minimum.

***17** The tangent to the curve $y = x^2 - 4$ at the point where $x = a$ meets the x- and y-axes at the points P and Q respectively.

a) Show that the area of the triangle OPQ, where O is the origin, is given by $\dfrac{(a^2 + 4)^2}{4a}$.

b) Hence find the minimum area of the triangle OPQ.

***18** A cylinder is inscribed inside a sphere of radius r. Given that the volume of the cylinder is a maximum, show that the ratio (volume of sphere) : (volume of cylinder) is $\sqrt{3} : 1$.

Exercise 13D: Examination questions

1 Express $\dfrac{13x + 16}{(x - 3)(3x + 2)}$ in partial fractions. Hence find the value of

$$\frac{d}{dx}\left[\frac{13x + 16}{(x - 3)(3x + 2)}\right]$$

when $x = 2$. (AEB 96)

2 Express $y = \dfrac{2x^2 - 6x - 1}{(x^2 + 1)(x - 2)}$ in partial fractions, and hence show that $\dfrac{dy}{dx} = \dfrac{13}{4}$ when $x = 0$.

3 Express

$$f(x) = \frac{3x}{(x - 1)(x - 4)}$$

in partial fractions. Use your result to write down an expression for $f'(x)$ and show that

$$f''(x) = \frac{-2}{(x - 1)^3} + \frac{8}{(x - 4)^3} \qquad \text{(WJEC)}$$

4 a) Differentiate $(1 + x^3)^{\frac{1}{2}}$ with respect to x.

b) Use the result from **a**, or an appropriate substitution, to find the value of

$$\int_0^2 \frac{x^2}{\sqrt{(1+x^3)}}\,dx \qquad \text{(AEB Spec)}$$

5 Prove that

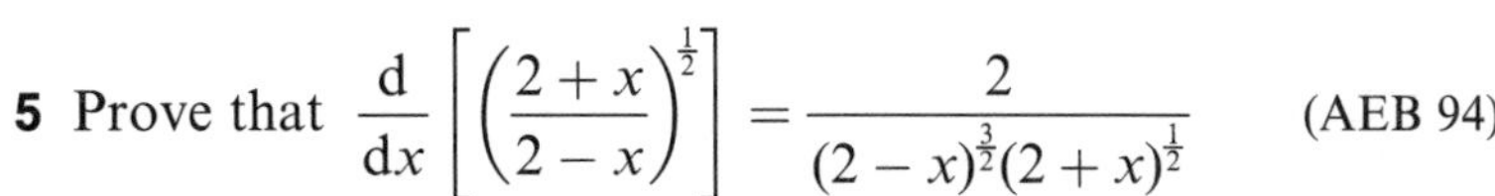

$$\frac{d}{dx}\left[\left(\frac{2+x}{2-x}\right)^{\frac{1}{2}}\right] = \frac{2}{(2-x)^{\frac{3}{2}}(2+x)^{\frac{1}{2}}} \qquad \text{(AEB 94)}$$

6 A sketch of the curve with equation $y = \dfrac{1+2x}{(1-x)^2}$ is shown on the right.

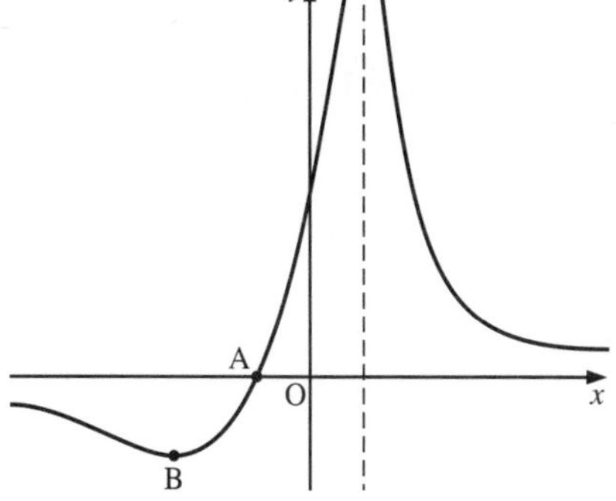

a) Write down the coordinates of A, and state the equation of the asymptote, shown as a broken line.

b) Find the equation of the tangent to the curve with equation

$$y = \frac{1+2x}{(1-x)^2}$$

at the point (0, 1). Hence determine the coordinates of the other point at which this tangent intersects the curve. (UODLE)

7 A curve has equation $f(x) = \dfrac{x+1}{\sqrt{x-1}}$. Find $f'(x)$ and hence find the coordinates of the turning point of the curve. Determine whether the turning point is a maximum or a minimum.

8 Given that $y = \dfrac{x^2+1}{x^2-2}$ find and simplify an expression for $\dfrac{dy}{dx}$. Hence determine the coordinates of the turning point on this curve, and determine its nature. (NICCEA)

9 Given that

$$y = \frac{4(x-2)}{4x^2+9}$$

find the maximum and minimum values of y and *sketch* the graph of y against x. (OCSEB)

14 Differentiation III

No mathematician can be a complete mathematician unless he is also something of a poet.
KARL WEIERSTRASS

Implicit functions

So far, we have looked at curves whose equations are in the form $y = \mathrm{f}(x)$. In other words, the variable y is given explicitly in terms of x. For example, $y = x^3 + 3x + 2$ is an **explicit function**.

Some curves are defined by **implicit functions**. That is, functions which are **not** expressed in the form $y = \mathrm{f}(x)$. For example, $x^2 + 3xy = 4$ is an implicit function. Notice that in this case, although we have an example of an implicit function, it can be rearranged to give an explicit function of x:

$$y = \frac{4 - x^2}{3x}$$

However, many implicit functions **cannot** be expressed in the form $y = \mathrm{f}(x)$. For example, the implicit function

$$x^2 + 3xy - 4y^3 = 7$$

cannot be expressed in the form $y = \mathrm{f}(x)$. It is for this reason that we must have a technique for differentiating implicit functions.

Implicit differentiation

Consider the implicit function

$$x^2 + y^2 = 2$$

Differentiating each term with respect to x gives

$$\frac{\mathrm{d}}{\mathrm{d}x}(x^2) + \frac{\mathrm{d}}{\mathrm{d}x}(y^2) = \frac{\mathrm{d}}{\mathrm{d}x}(2) \quad \text{or} \quad (x^2)' + (y^2)' = (2)' \qquad [1]$$

By the chain rule,

$$\frac{\mathrm{d}}{\mathrm{d}x}(y^2) = \frac{\mathrm{d}}{\mathrm{d}y}(y^2)\,\frac{\mathrm{d}y}{\mathrm{d}x} = 2y\frac{\mathrm{d}y}{\mathrm{d}x}$$

Therefore, [1] becomes

$$2x + 2y\,\frac{\mathrm{d}y}{\mathrm{d}x} = 0$$

Rearranging for $\dfrac{\mathrm{d}y}{\mathrm{d}x}$ gives

$$\frac{\mathrm{d}y}{\mathrm{d}x} = -\frac{2x}{2y} = -\frac{x}{y}$$

Example 1 Find $\dfrac{dy}{dx}$ for each of the following functions.

a) $x^2 - 6y^3 + y = 0$ **b)** $x^2y = 5x + 2$

c) $(x + y)^5 - 7x^2 = 0$ **d)** $\dfrac{x^3}{x + y} = 2$

SOLUTION

a) Differentiating each term of $x^2 - 6y^3 + y = 0$ with respect to x gives

$$2x - 18y^2\frac{dy}{dx} + \frac{dy}{dx} = 0$$

Rearranging for $\dfrac{dy}{dx}$ gives

$$\frac{dy}{dx}(1 - 18y^2) = -2x$$

$$\therefore \quad \frac{dy}{dx} = \frac{-2x}{1 - 18y^2} = \frac{2x}{18y^2 - 1}$$

b) Differentiating each term of $x^2y = 5x + 2$ with respect to x requires the use of the product rule. This gives

$$x^2\frac{dy}{dx} + y(2x) = 5$$

$$\therefore \quad x^2\frac{dy}{dx} = 5 - 2xy$$

$$\therefore \quad \frac{dy}{dx} = \frac{5 - 2xy}{x^2}$$

c) Differentiating each term of $(x + y)^5 - 7x^2 = 0$ with respect to x gives

$$5(x + y)^4\left(1 + \frac{dy}{dx}\right) - 14x = 0$$

$$\therefore \quad 1 + \frac{dy}{dx} = \frac{14x}{5(x + y)^4}$$

$$\therefore \quad \frac{dy}{dx} = \frac{14x}{5(x + y)^4} - 1$$

d) We have $\dfrac{x^3}{x + y} = 2$. Multiplying throughout by $(x + y)$ gives

$$x^3 = 2x + 2y$$

Differentiating each term with respect to x gives

$$3x^2 = 2 + 2\frac{dy}{dx}$$

$$\therefore \quad 2\frac{dy}{dx} = 3x^2 - 2$$

$$\therefore \quad \frac{dy}{dx} = \frac{3x^2 - 2}{2}$$

Example 2 Find the equations of the normals to the curve $x^2 + 3xy + 2y^2 = 10$ at the points where $x = -1$.

SOLUTION

To find the y coordinates of the points on the curve where $x = -1$, substitute $x = -1$ into $x^2 + 3xy + 2y^2 = 10$. That is,

$$(-1)^2 + 3(-1)y + 2y^2 = 10$$
$$\therefore \quad 1 - 3y + 2y^2 = 10$$
$$\therefore \quad 2y^2 - 3y - 9 = 0$$
$$\therefore \quad (2y + 3)(y - 3) = 0$$

Solving gives $y = -\frac{3}{2}$ and $y = 3$.

The points on the curve at which $x = -1$ are $\mathrm{P}\left(-1, -\frac{3}{2}\right)$ and $\mathrm{Q}(-1, 3)$. To find the equation of the normal at each point, we need the gradient of the curve at each point.

Since $x^2 + 3xy + 2y^2 = 10$, differentiating implicitly gives

$$2x + \left(3x\frac{dy}{dx} + 3y\right) + 4y\frac{dy}{dx} = 0$$
$$\therefore \quad \frac{dy}{dx}(3x + 4y) = -2x - 3y$$
$$\therefore \quad \frac{dy}{dx} = -\frac{2x + 3y}{3x + 4y}$$

At the point $\mathrm{P}\left(-1, -\frac{3}{2}\right)$,

$$\left.\frac{dy}{dx}\right|_{x=-1,\, y=-\frac{3}{2}} = -\frac{2(-1) + 3\left(-\frac{3}{2}\right)}{3(-1) + 4\left(-\frac{3}{2}\right)} = -\frac{13}{18}$$

At the point $\mathrm{Q}(-1, 3)$,

$$\left.\frac{dy}{dx}\right|_{x=-1,\, y=3} = -\frac{2(-1) + 3(3)}{3(-1) + 4(3)} = -\frac{7}{9}$$

The normal to the curve at point P has a gradient of

$$\frac{-1}{\left(-\dfrac{13}{18}\right)} = \frac{18}{13}$$

So, its equation has the form $y = \frac{18}{13}x + c_1$.

Since the normal passes through $\mathrm{P}\left(-1, -\frac{3}{2}\right)$, we have

$$-\frac{3}{2} = \frac{18}{13}(-1) + c_1$$
$$\therefore \quad c_1 = -\frac{3}{26}$$

The equation of the normal to the curve at P is

$$y = \frac{18}{13}x - \frac{3}{26} \quad \text{or} \quad 26y = 36x - 3$$

The normal to the curve Q has a gradient of

$$\frac{-1}{\left(-\frac{7}{9}\right)} = \frac{9}{7}$$

So, its equation has the form $y = \frac{9}{7}x + c_2$.

Since the normal passes through the point Q(−1, 3), we have

$$3 = \frac{9}{7}(-1) + c_2 \quad \therefore \quad c_2 = \frac{30}{7}$$

The equation of the normal to the curve at Q is

$$y = \frac{9}{7}x + \frac{30}{7} \quad \text{or} \quad 7y = 9x + 30$$

Second derivatives

Consider again the function $x^2 + y^2 = 2$ Differentiating it implicitly gives

$$2x + 2y\frac{dy}{dx} = 0 \quad \text{or} \quad 2x + 2yy' = 0$$

Differentiating again with respect to x gives

$$2 + 2y(y'') + 2y'(y') = 0 \qquad \text{(Using the product rule in } 2yy'\text{)}$$

$$\therefore \quad 2 + 2yy'' + 2(y')^2 = 0$$

where $y'' = \dfrac{d^2y}{dx^2}$.

Rearranging for y'' gives

$$y'' = \frac{-1 - (y')^2}{y} = -\frac{1 + (y')^2}{y}$$

We know that $\dfrac{dy}{dx} = -\dfrac{x}{y}$, Therefore,

$$y'' = -\frac{1 + \left(-\frac{x}{y}\right)^2}{y} = -\frac{(y^2 + x^2)}{y^3}$$

Since $x^2 + y^2 = 2$, we have

$$\frac{d^2y}{dx^2} = -\frac{2}{y^3}$$

Example 3 Find and classify the stationary points on the curve $x^2 + xy + y^2 = 27$.

SOLUTION

At a stationary point, $\frac{dy}{dx} = y' = 0$. Differentiating $x^2 + xy + y^2 = 27$ implicitly gives

$$2x + (xy' + y) + 2yy' = 0 \qquad [1]$$

Rearranging for y' gives

$$2yy' + xy' = -2x - y$$

$$\therefore \quad y'(2y + x) = -(2x + y)$$

$$\therefore \quad y' = -\frac{2x + y}{2y + x}$$

When $y' = 0$, we have

$$-\frac{2x + y}{2y + x} = 0$$

$$\therefore \quad 2x + y = 0$$

$$\therefore \quad y = -2x \qquad [2]$$

Substituting $y = -2x$ into $x^2 + xy + y^2 = 27$ gives

$$x^2 + x(-2x) + (-2x)^2 = 27$$

$$\therefore \quad 3x^2 = 27 \quad \text{giving} \quad x = \pm 3$$

Using [2], when $x = 3$, $y = -6$. When $x = -3$, $y = 6$. Therefore, the stationary points are $(3, -6)$ and $(-3, 6)$.

Differentiating [1] implicitly gives

$$2 + xy'' + y' + y' + 2yy'' + 2(y')^2 = 0$$

We know that $y' = 0$ at the stationary points. Therefore,

$$2 + xy'' + 2yy'' = 0$$

To determine the nature of the stationary points, we will examine the sign of y'' at each point.

At $(3, -6)$: $\quad 2 + 3y'' + 2(-6)y'' = 0$

$$\therefore \quad y'' = \frac{2}{9} > 0$$

Since $y'' > 0$, the stationary point $(3, -6)$ is a minimum.

At $(-3, 6)$: $\quad 2 + (-3)y'' + 2(6)y'' = 0$

$$\therefore \quad y'' = -\frac{2}{9} < 0$$

Since $y'' < 0$, the stationary point $(-3, 6)$ is a maximum.

Exercise 14A

1 For each of the following curves express $\dfrac{dy}{dx}$ in terms of x and y.

a) $x^2 - y^3 = 4$ **b)** $3xy - y^2 = 7$ **c)** $x^2y + xy^2 = 2$
d) $2x - y^3 = 3xy$ **e)** $x^4 - xy^2 = 6x$ **f)** $x^6 - 5xy^3 = 9xy$
g) $x^2(x - 3y) = 4$ **h)** $\dfrac{x^2}{x+y} = 2$ **i)** $\dfrac{y}{x^2 - 7y^3} = x^5$

2 In each part of this question find the gradient of the stated curve at the point specified.

a) $xy^2 - 6y = 8$ at $(2, -1)$ **b)** $x^4 - y^3 = 2$ at $(1, -1)$
c) $3y^4 - 7xy^2 - 12y = 5$ at $(-2, 1)$ **d)** $xy^3 - x^2y = 6$ at $(3, 2)$
e) $(x + y)^2 - 4x + y + 10 = 0$ at $(2, -3)$ **f)** $\dfrac{x^2}{x - y} = 8$ at $(4, 2)$
g) $\dfrac{2}{x} + \dfrac{5}{y} = 2xy$ at $\left(\frac{1}{2}, 5\right)$ **h)** $(x + 2y)^4 = 1$ at $(5, -2)$

3 Find the equation of the tangent to the curve $xy^2 + x^2y = 6$ at the point $(1, -3)$.

4 Find the equations of the tangent and the normal to the curve $xy^2 + 3x - 2y = 6$ at the point $(2, 1)$.

5 Find the equations of the tangent and the normal to the curve $xy = 2$ at the point $\left(6, \frac{1}{3}\right)$.

6 Find the equations of the tangents to the curve $x^2y - xy^2 = 12$ at the points where $y = 3$.

7 At what points are the tangents to the circle $x^2 + y^2 - 4x - 6y + 9 = 0$ parallel to the x-axis?

8 Find the equation of the tangents to the curve $x^2 + 3x - 2y^2 = 4$ at the points where the curve crosses the x-axis.

9 Find and classify the stationary values on each of the following curves.

a) $x^2 + y^2 - 4x + 6y + 12 = 0$ **b)** $3x^2 + y^2 - 6x + 4y + 6 = 0$
c) $3x^2 + x^2y + y = 2$ **d)** $xy + y - x^2 = 8$
e) $2xy + y^2 - x^2 = 2$ **f)** $4xy^3 = 1 + 4x^2y^2$

***10** Given that $x^n + y^n = 1$, show that $\dfrac{d^2y}{dx^2} = -\dfrac{(n-1)x^{n-2}}{y^{2n-1}}$.

***11** Given that $x^2 + xy + y^2 - 3x - y = 3$,

a) show that $\dfrac{dy}{dx} = \dfrac{3 - 2x - y}{x + 2y - 1}$

b) find, and classify, the maximum and minimum values of y

c) determine the coordinates of the points on the curve where the tangents to the curve are parallel to the y-axis.

Parametric equations

In some cases, y is defined as a function of x by expressing both y and x in terms of a third variable known as a parameter. Such equations are called **parametric equations**.

For example, the pair of equations

$$x = t + 1 \qquad [1] \qquad\qquad y = t^2 \qquad [2]$$

are parametric equations, with the parameter being t. In fact, these parametric equations define the parabola with equation

$$y = x^2 - 2x + 1$$

This can be seen by eliminating the parameter t between [1] and [2]. From [1] we have $t = x - 1$. Substituting into [2] gives

$$y = (x-1)^2$$

$$\therefore \quad y = x^2 - 2x + 1$$

Example 4 Find the cartesian equation for each of the following parametric forms.

a) $x = \sqrt{t},\ y = 2t^2 - 1$ **b)** $x = \dfrac{1}{t},\ y = 3t - 1$ **c)** $x = \dfrac{1}{2-t},\ y = \dfrac{3}{1+2t}$

SOLUTION

a) When $x = \sqrt{t}$, then $x^2 = t$ and $x^4 = t^2$. Substituting into $y = 2t^2 - 1$ gives

$$y = 2x^4 - 1$$

The cartesian equation is $y = 2x^4 - 1$.

b) When $x = \dfrac{1}{t}$, then $t = \dfrac{1}{x}$. Substituting into $y = 3t - 1$ gives

$$y = 3\left(\frac{1}{x}\right) - 1 = \frac{3}{x} - 1$$

The cartesian equation is $y = \dfrac{3}{x} - 1$ or, multiplying throughout by x, $xy + x = 3$.

c) When $x = \dfrac{1}{2-t}$, rearranging for t gives

$$x(2-t) = 1$$

$$\therefore \quad 2x - xt = 1$$

$$\therefore \quad xt = 2x - 1$$

$$\therefore \quad t = \frac{2x-1}{x}$$

Substituting into $y = \dfrac{3}{1+2t}$ gives

$$y = \frac{3}{1+2\left(\dfrac{2x-1}{x}\right)} = \frac{3x}{x+2(2x-1)}$$

$$\therefore \quad y = \frac{3x}{5x-2}$$

The cartesian equation is $y = \dfrac{3x}{5x-2}$.

However, it is not always possible to eliminate the parameter and it is for this reason that we need a technique for differentiating parametric equations.

Parametric differentiation

Consider the parametric equations

$$x = t + 1 \quad \text{and} \quad y = t^2$$

We can differentiate x with respect to t, and we can differentiate y with respect to t. This gives

$$\frac{dx}{dt} = 1 \quad \text{and} \quad \frac{dy}{dt} = 2t$$

By the chain rule,

$$\frac{dy}{dx} = \frac{dy}{dt}\,\frac{dt}{dx} = \frac{\left(\dfrac{dy}{dt}\right)}{\left(\dfrac{dx}{dt}\right)}$$

$$\therefore \quad \frac{dy}{dx} = 2t(1) = 2t$$

In this particular case, we know that $t = x - 1$. Therefore,

$$\frac{dy}{dx} = 2(x-1) = 2x - 2$$

Notice that we expect the derivative of $y = x^2 - 2x + 1$ to be $2x - 2$.

Example 5 Find $\dfrac{dy}{dx}$ in terms of the parameter for each of the following.

a) $y = 3t^2 + 2t,\ x = 1 - 2t$ **b)** $y = (1+2t)^3,\ x = t^3$

SOLUTION

a) When $y = 3t^2 + 2t$: $\dfrac{dy}{dt} = 6t + 2$

When $x = 1 - 2t$: $\dfrac{dx}{dt} = -2$

By the chain rule,

$$\frac{dy}{dx} = \frac{dy}{dt}\frac{dt}{dx}$$

$$= (6t + 2)\left(-\frac{1}{2}\right)$$

$$\therefore \quad \frac{dy}{dx} = -3t - 1$$

b) When $y = (1 + 2t)^3$: $\quad \frac{dy}{dt} = 3(1 + 2t)^2(2)$

$$\therefore \quad \frac{dy}{dt} = 6(1 + 2t)^2$$

When $x = t^3$: $\quad \frac{dx}{dt} = 3t^2$

By the chain rule,

$$\frac{dy}{dx} = \frac{dy}{dt}\frac{dt}{dx} = 6(1 + 2t)^2 \times \frac{1}{3t^2}$$

$$\therefore \quad \frac{dy}{dx} = \frac{2(1 + 2t)^2}{t^2}$$

Example 6 Find the equation of the tangent to the curve given parametrically by $x = \frac{2}{t}$ and $y = 3t^2 - 1$, at the point (2, 2).

SOLUTION

The value of the parameter t is found by substituting $x = 2$, $y = 2$ into the equations

$$x = \frac{2}{t} \quad \text{and} \quad y = 3t^2 - 1$$

This gives $t = 1$.

Differentiating parametrically gives

$$\frac{dy}{dt} = 6t \quad \text{and} \quad \frac{dx}{dt} = -\frac{2}{t^2}$$

By the chain rule

$$\frac{dy}{dx} = 6t\left(-\frac{t^2}{2}\right) \qquad \therefore \quad \frac{dy}{dx} = -3t^3$$

When $t = 1$,

$$\left.\frac{dy}{dx}\right|_{t=1} = -3$$

The gradient of the tangent is -3.

Hence, the equation of the tangent is of the form

$$y = -3x + c$$

Since the tangent line passes through the point (2, 2), we have

$$2 = -3(2) + c \quad \therefore \quad c = 8$$

The equation of the tangent is $y = -3x + 8$.

Second derivatives

Consider the parametric equations

$$x = t + 1 \quad \text{and} \quad y = t^3$$

Differentiating each with respect t gives

$$\frac{dx}{dt} = 1 \quad \text{and} \quad \frac{dy}{dt} = 3t^2.$$

Therefore,

$$\frac{dy}{dx} = \frac{dy}{dt}\,\frac{dt}{dx} = 3t^2 \qquad [1]$$

To find $\frac{d^2y}{dx^2}$, we differentiate every term of [1] with respect to x. That is,

$$\frac{d}{dx}\left(\frac{dy}{dx}\right) = \frac{d}{dx}(3t^2) \qquad [2]$$

Note that we cannot differentiate $3t^2$ with respect to x. Therefore, we use the chain rule, which gives

$$\begin{aligned}\frac{d}{dx}(3t^2) &= \frac{d}{dt}(3t^2)\,\frac{dt}{dx}\\ &= 6t\,\frac{dt}{dx}\end{aligned}$$

Therefore, we have from [2]

$$\frac{d^2y}{dx^2} = 6t\,\frac{dt}{dx}$$

We know that

$$\frac{dt}{dx} = \frac{1}{\left(\frac{dx}{dt}\right)} = 1$$

Therefore,

$$\frac{d^2y}{dx^2} = 6t(1) = 6t$$

Example 7 Find and classify the stationary point on the curve $x = 4 - t^3$ and $y = t^2 - 2t$.

SOLUTION

Differentiating parametrically gives

$$\frac{dx}{dt} = -3t^2 \quad \text{and} \quad \frac{dy}{dt} = 2t - 2$$

By the chain rule,

$$\frac{dy}{dx} = (2t-2)\left(-\frac{1}{3t^2}\right)$$

$$\therefore \quad \frac{dy}{dx} = \frac{2-2t}{3t^2} \qquad [1]$$

At a stationary point, $\frac{dy}{dx} = 0$. Therefore,

$$\frac{2-2t}{3t^2} = 0$$

$$\therefore \quad 2 - 2t = 0 \quad \text{giving} \quad t = 1$$

To find the coordinates of the point corresponding to the value $t = 1$, we substitute $t = 1$ into

$$x = 4 - t^3 \quad \text{and} \quad y = t^2 - 2t$$

This gives $x = 3$ and $y = -1$. The coordinates of the stationary point are $(3, -1)$.

To find $\frac{d^2y}{dx^2}$, we differentiate [1] with respect to x. That is,

$$\frac{d^2y}{dx^2} = \frac{d}{dx}\left(\frac{2-2t}{3t^2}\right) = \frac{d}{dt}\left(\frac{2-2t}{3t^2}\right)\frac{dt}{dx}$$

$$= \frac{(3t^2)(-2) - (2-2t)(6t)}{9t^4} \times \frac{dt}{dx}$$

We know that $\frac{dt}{dx} = -\frac{1}{3t^2}$. Therefore,

$$\frac{d^2y}{dx^2} = \frac{-6t^2 - 12t + 12t^2}{9t^4}\left(-\frac{1}{3t^2}\right)$$

$$= -\frac{6t(t + 2 - 2t)}{9t^4}\left(-\frac{1}{3t^2}\right)$$

$$\therefore \quad \frac{d^2y}{dx^2} = \frac{2(2-t)}{9t^5}$$

At the stationary point, $t = 1$. Therefore,

$$\left.\frac{d^2y}{dx^2}\right|_{t=1} = \frac{2(2-1)}{9(1)^5} = \frac{2}{9} > 0$$

Since $\frac{d^2y}{dx^2} > 0$, the stationary point $(3, -1)$ is a minimum.

Exercise 14B

1 Find the cartesian equation for each of the following parametric forms.

a) $x = t + 3,\ y = t^2$ **b)** $x = 2t - 1,\ y = 12t^2 - 5$ **c)** $x = 2\sqrt{t},\ y = 8t^2 + 5$

d) $x = t,\ y = \dfrac{4}{t}$ **e)** $x = \dfrac{2}{\sqrt{t}},\ y = \dfrac{3}{1+t}$ **f)** $x = 2 + 3t^2,\ y = 1 - t^2$

g) $x = \dfrac{1}{t+1},\ y = t(1+t)$ **h)** $x = \dfrac{1}{1+2t},\ y = \dfrac{t}{1+2t}$ **i)** $x = \dfrac{t}{1-3t},\ y = \dfrac{t}{1+2t}$

2 For the following curves, each of which is given in terms of a parameter t, find an expression for $\dfrac{dy}{dx}$ in terms of t.

a) $x = t^2,\ y = 4t - 1$ **b)** $x = 3t^4,\ y = 2t^2 - 3$ **c)** $x = 2\sqrt{t},\ y = 5t - 4$

d) $x = 6t - 5,\ y = (2t-1)^3$ **e)** $x = 4t(t-2),\ y = (t-1)^3$ **f)** $x = 4\sqrt{t} - t,\ y = t^2 - 2\sqrt{t}$

g) $x = \dfrac{1}{t},\ y = t^2 + 4t - 3$ **h)** $x = \dfrac{2}{3+\sqrt{t}},\ y = \sqrt{t}$ **i)** $x = \dfrac{2}{\sqrt[3]{3t-4}},\ y = \sqrt[3]{6t+1}$

3 In each part of this question find the gradient of the stated curve at the point defined by the given value of the parameter t.

a) $x = t + 5,\ y = t^2 - 2t$, where $t = 2$ **b)** $x = t^6,\ y = 6t^3 - 5$, where $t = -3$

c) $x = (t-2)^2,\ y = (3t+4)^3$, where $t = 0$ **d)** $x = (2t^3 - 5)^2,\ y = (1-3t)^3$, where $t = 1$

e) $x = \sqrt{t-1},\ y = \dfrac{1}{t}$, where $t = 10$ **f)** $x = \dfrac{1}{\sqrt{t-4}},\ y = \dfrac{3}{\sqrt{t^2-9}}$, where $t = 5$

g) $x = t(t^2 - 1),\ y = (4t+1)^3$, where $t = \frac{1}{2}$ **h)** $x = t^2(3t-1),\ y = \sqrt{3t+4}$, where $t = -1$

4 Find the equation of the tangent to the curve $x = 3t^2,\ y = 7 + 12t$, at the point where $t = 2$.

5 Find the equations of the tangents to the curve $x = t^2,\ y = 6t - 7$, at the points where $x = 1$.

6 Find the equation of the tangent and the normal to the curve $x = 6t^2,\ y = t^3 - 4t$, at the point where $t = -1$.

7 Find the equations of the tangents to the curve $x = \dfrac{4}{t},\ y = t^2 - 3t + 2$, at the points where the curve crosses the x-axis.

8 At which points are the tangents to the curve $x = 2t^2 - 3,\ y = t^3 - 6t^2 + 9t - 4$, parallel to the x-axis?

9 Find the equations of the normals to the curve $x = \dfrac{16}{t^2},\ y = 2t - 3$, at the points where the curve crosses the line $x = 1$.

10 A curve is given by $x = t^2,\ y = 4t$. The tangent at the point where $t = 2$, meets the tangent at the point where $t = -1$, at the point P. Find the coordinates of P.

11 a) Find the equation of the tangents to the curve $x = 8t + 1$, $y = 2t^2$, at the points $(9, 2)$ and $(-31, 32)$.

b) Show that these tangents are perpendicular, and find the coordinates of the point of intersection of the tangents.

12 a) Given the curve $x = \dfrac{t}{1+t}$, $y = \dfrac{t^2}{1+t}$, show that $\dfrac{dy}{dx} = t(t+2)$.

b) Hence find the points on the curve where the gradient is 15.

13 a) Given the curve $x = t^2(1 - 3t^2)$, $y = 5t^3(4 - t)$, show that

$$\frac{dy}{dx} = \frac{10t(t-3)}{6t^2 - 1}$$

b) Hence find the values of the parameter t at the points on the curve where the gradient is 8.

14 For the following curves, each of which is given in terms of a parameter t, find an expression for $\dfrac{dy}{dx}$ and for $\dfrac{d^2y}{dx^2}$ in terms of t.

a) $x = 3t + 1,\ y = t^3 + 5$ **b)** $x = 9t - 1,\ y = t^3 + 6t$ **c)** $x = t^2,\ y = 2t(1 - t)$

d) $x = 4t^3,\ y = 3t^4 - 6t^2$ **e)** $x = 2\sqrt{t},\ y = t^2 + 3t$ **f)** $x = \dfrac{1}{t},\ y = (t-2)^2$

g) $x = t^2 + 3t,\ y = t^2 - 4$ **h)** $x = t^2(2t + 3),\ y = 3t^4 + 4t^3 + 1$

i) $x = 4 - \dfrac{1}{t},\ y = (t+3)^2$ **j)** $x = \dfrac{1}{t+1},\ y = \dfrac{1}{t-1}$ **k)** $x = \dfrac{t+1}{t-1},\ y = \dfrac{1}{t-1}$

l) $x = \dfrac{1}{1+\sqrt{t}},\ y = \dfrac{1}{1-\sqrt{t}}$

15 a) Given the curve $x = 2t^2 - 5$, $y = t^2 - 8t + 3$, derive the results

i) $\dfrac{dy}{dx} = \dfrac{t-4}{2t}$ **ii)** $\dfrac{d^2y}{dx^2} = \dfrac{1}{2t^3}$

b) Hence find and classify any stationary values on the curve.

16 a) Given the curve $x = t^3 + 3t + 1$, $y = t^2 + 4t - 4$, derive the results

i) $\dfrac{dy}{dx} = \dfrac{2(t+2)}{3(t^2+1)}$ **ii)** $\dfrac{d^2y}{dx^2} = -\dfrac{2(t^2 + 4t - 1)}{9(t^2+1)^3}$

b) Hence find and classify any stationary values on the curve.

17 a) For the curve $x = t^2 + t$, $y = t^3 - 3t^2 - 9t + 20$, show that

i) $\dfrac{dy}{dx} = \dfrac{3(t+1)(t-3)}{2t+1}$ **ii)** $\dfrac{d^2y}{dx^2} = \dfrac{6(t^2 + t + 2)}{(2t+1)^3}$

b) Hence find and classify any stationary values on the curve.

18 a) For the curve $x = \frac{3t-1}{t}$, $y = \frac{t^2+4}{t}$, show that

i) $\frac{dy}{dx} = t^2 - 4$ **ii)** $\frac{d^2y}{dx^2} = 2t^3$

b) Hence find and classify any stationary values on the curve.

Rates of change

The derivative

$$\frac{dy}{dx} = \lim_{\delta x \to 0} \frac{\delta y}{\delta x}$$

is called the **rate of change of y with respect to x.**

It shows how changes in y are related to changes in x. For example, if $\frac{dy}{dx} = 3$, then y is increasing 3 times as fast as x.

In practical situations, letters other than x and y are used. For example, in mechanics, if s denotes the position of a particle at time t, then

- the velocity, v, of the particle is given by $\frac{ds}{dt}$, the rate of change of displacement with respect to time.
- the acceleration of the particle is given by $\frac{d^2s}{dt^2} = \frac{dv}{dt}$, the rate of change of velocity with respect to time.

Related rates of change

Many practical situations involve related rates of change. For example, suppose the side of a square piece of metal increases at a rate of 0.1 cm per second when it is heated. As a result, the area of the square surface of the metal also increases – but at what rate?

Let the square piece of metal have sides of length x cm. Then, its area is given by $A = x^2$.

We are given that $\frac{dx}{dt}$, the rate of increase of the length of a side with respect to time, is $0.1\,\text{cm}\,\text{s}^{-1}$. We want $\frac{dA}{dt}$, the rate of increase of the area with respect to time.

Now $A = x^2$, therefore $\frac{dA}{dx} = 2x$. We are given that $\frac{dx}{dt} = 0.1$, so

$$\frac{dA}{dt} = \frac{dA}{dx}\frac{dx}{dt} = 2x \times 0.1$$

$$\therefore \quad \frac{dA}{dt} = 0.2x$$

The rate of increase of the area is $0.2x\,\text{cm}^2\,\text{s}^{-1}$.

Example 8 A spherical balloon is inflated at a rate of $3\,\text{cm}^3\,\text{s}^{-1}$. Find the rate of increase of the radius when this radius is 2 cm.

SOLUTION

Let the balloon have radius r and volume V, then $V = \frac{4}{3}\pi r^3$. Therefore,

$$\frac{dV}{dr} = 4\pi r^2$$

We are given $\frac{dV}{dt} = 3$ and we know that

$$\frac{dV}{dt} = \frac{dV}{dr}\frac{dr}{dt}$$

$$\therefore \quad 3 = 4\pi r^2 \frac{dr}{dt}$$

$$\therefore \quad \frac{dr}{dt} = \frac{3}{4\pi r^2}$$

When $r = 2$,

$$\left.\frac{dr}{dt}\right|_{r=2} = \frac{3}{4\pi(2)^2} = \frac{3}{16\pi}$$

The rate of increase of the radius when the radius is 2 cm is $\frac{3}{16\pi}\,\text{cm}\,\text{s}^{-1}$.

Example 9 A container in the shape of a hollow cone of semi-vertical angle 30° is held with its vertex pointing downwards. Water is poured into the cone at the rate of $5\,\text{cm}^3\,\text{s}^{-1}$. Find the rate at which the depth of water in the cone is increasing when this depth is 10 cm.

SOLUTION

Let the depth of the water in the cone be x cm. The radius, r cm, of the cross-section of the water is given by

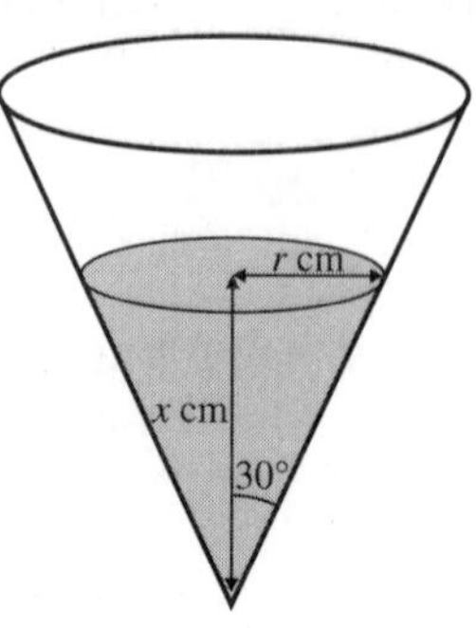

$$\tan 30° = \frac{r}{x}$$

$$\therefore \quad r = x\tan 30° = \frac{x}{\sqrt{3}}$$

The volume, $V\,\text{cm}^3$, of the water in the cone is given by

$$V = \frac{1}{3}\pi r^2 x = \frac{1}{3}\pi\left(\frac{x}{\sqrt{3}}\right)^2 x = \frac{1}{9}\pi x^3$$

Therefore,

$$\frac{dV}{dx} = \frac{1}{3}\pi x^2$$

We are given that $\frac{dV}{dt} = 5$ and we know that

$$\frac{dV}{dx} = \frac{dV}{dx}\frac{dx}{dt}$$

Therefore,

$$5 = \frac{1}{3}\pi x^2 \frac{dx}{dt}$$

$$\therefore \quad \frac{dx}{dt} = \frac{15}{\pi x^2}$$

When $x = 10$,

$$\frac{dx}{dt} = \frac{15}{\pi(10)^2} = \frac{3}{20\pi}$$

The rate of increase of the depth when the depth is 10 cm is, therefore, $\frac{3}{20\pi}$ cm s^{-1}

Exercise 14C

1 The side of a square is increasing at the rate of 3 cm s^{-1}. Find the rate of increase of the area when the length of a side is 4 cm.

2 The side of a square is increasing at the rate of 5 cm s^{-1}. Find the rate of increase of the area when the length of a side is 10 cm.

3 The radius of a circle is increasing at the rate of $\frac{1}{3}$ cm s^{-1}. Find the rate of increase of the area when this radius is 5 cm.

4 The area of a square is increasing at the rate of 7 cm^2 s^{-1}. Find the rate of increase of the length of a side when this area is 100 cm^2.

5 The area of a circle is increasing at the rate of (4π) cm^2 s^{-1}. Find the rate of increase of the radius when this radius is $\frac{1}{2}$ cm.

6 The volume of a cube is increasing at the rate of 18 cm^3 s^{-1}. Find the rate of increase of the length of a side when the volume is 125 cm^3.

7 The volume of a sphere is increasing at the rate of (12π) cm^3 s^{-1}. Find the rate of increase of the radius when this radius is 6 cm.

8 The surface area of a sphere is increasing at a rate of 2 cm^2 s^{-1}. Find the rate of increase of the radius when the surface area is (100π) cm^2.

9 The radius of a circle is increasing at a rate of $\frac{1}{5}$ cm s^{-1}. Find the rate of increase of the circumference.

10 The area of a square, of side x cm, is increasing at the rate of 6 cm^2 s^{-1}. Find an expression, in terms of x, for the rate of increase of the length of a side.

11 **a)** A circle has radius r cm, circumference C cm, and area A cm^2. Show that $\dfrac{dC}{dA} = \dfrac{1}{r}$.

b) The area of a circle is increasing at 2 cm^2 s^{-1}. Find the rate of increase of the circumference when the radius is 3 cm.

12 A boy is inflating a spherical balloon at the rate of 10 cm^3 s^{-1}. Find the rate of increase of surface area of the balloon when its radius is 5 cm.

13 A circular ink blot spreads at the rate of $\frac{1}{3}$ cm^2 s^{-1}. Find the rate of increase in the circumference of the ink blot when its radius is $\frac{1}{2}$ cm.

14 A container in the shape of a hollow cone of semi-vertical angle 45° is held with its vertex pointing downwards. Water drips into the container at a rate of 3 cm^3 per minute. Find the rate at which the depth of water in the cone is increasing when this depth is 2 cm.

15 The inside of a glass is in the shape of an inverted cone of depth 8 cm and radius 4 cm. Wine is poured into the glass at the rate of 4 cm^3 s^{-1}. Find the rate at which the depth of the wine in the glass is increasing when this depth is 6 cm.

16 A container in the shape of a hollow cone of depth 12 cm and radius 6 cm is held with its vertex pointing downwards. Water is poured into the container at a rate of 2 cm^3 s^{-1}. Find the rate at which the depth of water in the cone is increasing when this depth is $\frac{3}{4}$ cm.

17 A hollow cone of base radius 10 cm and height 10 cm is held with its vertex downwards. The cone is initially empty when water is poured into it at the rate of (4π) cm^3 s^{-1}. Find the rate of increase in the depth of the water in the cone 18 seconds after pouring has commenced.

18 A champagne glass is in the shape of an inverted cone of depth 9 cm and radius 3 cm. Champagne is poured into the glass at the rate of (2π) cm^3 s^{-1}. Find the rate at which the depth of the champagne in the glass is increasing 4 seconds after pouring has commenced.

***19** An empty drinking trough of length 5 m, has triangular cross-section in the shape of an equilateral triangle of side 80 cm. Water from a hose fills the trough at the rate of $100\sqrt{3}$ cm^3 s^{-1}. Find the rate of increase of the depth of the water in the trough when the hose has been running for 25 minutes.

***20** A hemispherical bowl of radius r cm is being filled with water at a constant rate.

a) Show that when the depth of water in the bowl is h cm, then the volume of water in the bowl is given by $\dfrac{\pi h^2(3r - h)}{3}$.

b) Show also that between the time when the water is half way to the top, and the time when the bowl is about to overflow, the rate at which the depth is rising has fallen by a quarter.

Exercise 14D: Examination questions

1 Find the gradient of the curve with equation

$$5x^2 + 5y^2 - 6xy = 13$$

at the point $(1, 2)$. (EDEXCEL)

2 The curve C has equation $x^2 + 4y^2 - 4x - 12y - 12 = 0$.

a) By differentiation find the gradient of C at the point $(6, 3)$.
b) Find an equation of the tangent to the curve C at the point $(6, 3)$. (EDEXCEL)

3 A curve is defined implicitly by the equation $x^2y + y^2 - 3x - 3 = 0$. The point A has coordinates $(1, 2)$ and the point B is where the curve crosses the x-axis.

a) Show that the point A lies on the curve.
b) Find the coordinates of the point B.
c) Calculate the gradient of the curve at the point A.
d) Find the equation of the normal to the curve at the point A. (AEB 95)

4 A curve has implicit equation $x^2 - 2xy + 4y^2 = 12$.

a) Find an expression for $\frac{dy}{dx}$ in terms of y and x.

Hence determine the coordinates of the points where the tangents to the curve are parallel to the x-axis.

b) Find the equation of the normal to the curve at the point $(2\sqrt{3}, \sqrt{3})$. (UODLE)

5 The curve C has parametric equations $x = t^2 + 1$, $y = 2 - t$.

a) Find the coordinates of the point A where C meets the x-axis.
b) Show that the normal to C at A has equation $4x - y - 20 = 0$.
c) Find the parameter of the point where this normal meets C again. (WJEC)

6 The diagram shows a sketch of the curve with parametric equations $x = 1 - 2t$, $y = t^2$. The tangent and normal at P are also shown.

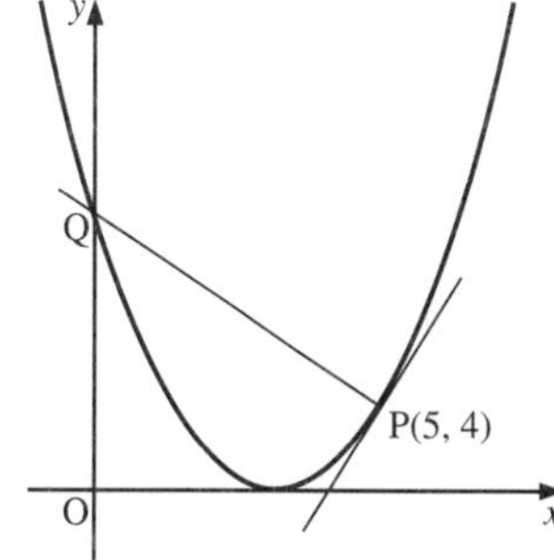

i) Show that the point P(5, 4) lies on the curve by stating the value of t corresponding to this point.

ii) Show that, at the point with parameter t, $\frac{dy}{dx} = -t$.

iii) Find the equation of the tangent at P.

iv) The normal at P cuts the curve again at Q. Find the coordinates of Q. (MEI)

7 A curve is defined parametrically by the equations

$$x = t^3 - \frac{3}{t} \qquad y = 2t^{\frac{3}{2}} \qquad (t > 0)$$

a) Show that $\frac{dy}{dx} = \frac{t^{\frac{5}{2}}}{t^4 + 1}$.

b) Find the equation of the normal to the curve at the point where $t = 1$. (AEB 95)

8 The curve C has parametric equations

$$x = t^2 \qquad y = 2 - \frac{1}{t} \qquad (t \neq 0)$$

a) Find the coordinates of the point A where C meets the x-axis.
b) Show that the tangent to C at A has equation $4x - y - 1 = 0$.
c) Find the coordinates of the point where this tangent meets C again. (WJEC)

9 The equation of a curve is given in terms of the parameter t by the equations $x = 2t$ and $y = \frac{2}{t}$, where t takes positive and negative values.

i) Sketch the curve.

At the point on the curve with parameter t,

ii) show that the gradient of the curve is $-t^{-2}$
iii) find and simplify the equation of the tangent.

P and Q are the points where a tangent to this curve crosses the x- and y-axes, and O is the origin.

iv) Show that the area of the triangle OPQ is independent of t. (MEI)

10 The curve C has parametric equations

$$x = at \qquad y = \frac{a}{t} \qquad t \in \mathbb{R},\ t \neq 0$$

where t is a parameter and a is a positive constant.

a) Sketch C.

b) Find $\frac{\mathrm{d}y}{\mathrm{d}x}$ in terms of t.

The point P on C has parameter $t = 2$.

c) Show that an equation of the normal to C at P is $2y = 8x - 15a$.

The normal meets C again at the point Q.

d) Find the value of t at Q. (EDEXCEL)

11 Air is pumped into a spherical balloon at the rate of $250\,\text{cm}^3$ per second. When the radius of the balloon is 15 cm, calculate

i) the rate at which its radius is increasing,
ii) the rate at which its surface area is increasing.

[The volume V and the surface area S of a sphere of radius r are given by $V = \frac{4}{3}\pi r^3$ and $S = 4\pi r^2$.] (WJEC)

15 Trigonometry I

Medicine makes people ill, mathematics makes them sad, and theology makes them sinful.
MARTIN LUTHER

Trigonometric functions

Graphs of $y = \sin\theta$, $y = \cos\theta$ and $y = \tan\theta$

Plotting values of y against θ, where θ is measured in degrees, gives the following graphs.

$y = \sin\theta$

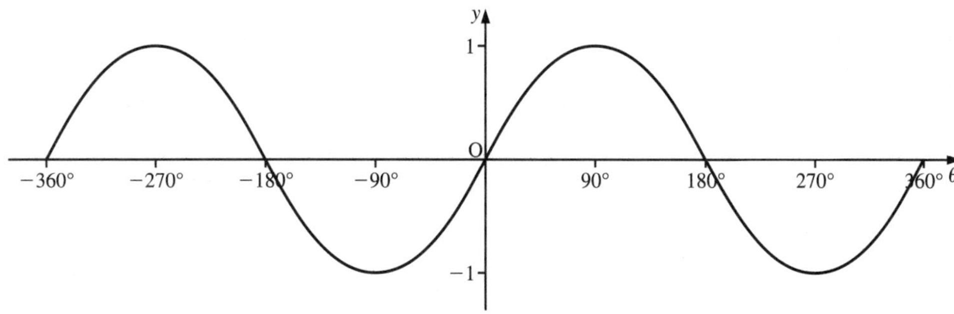

Properties of the sine function:

- The function $f(\theta) = \sin\theta$ is periodic, of period 360°. That is,

  $$\sin(\theta + 360°) = \sin\theta$$

- The function $f(\theta) = \sin\theta$ is an odd function since

  $$f(-\theta) = -\sin\theta = -f(\theta)$$

  In other words, the graph has rotational symmetry about the origin of order 2.
- The maximum value of $f(\theta)$ is 1 and its minimum value is -1. In other words, $-1 \leqslant f(\theta) \leqslant 1$.

$y = \cos\theta$

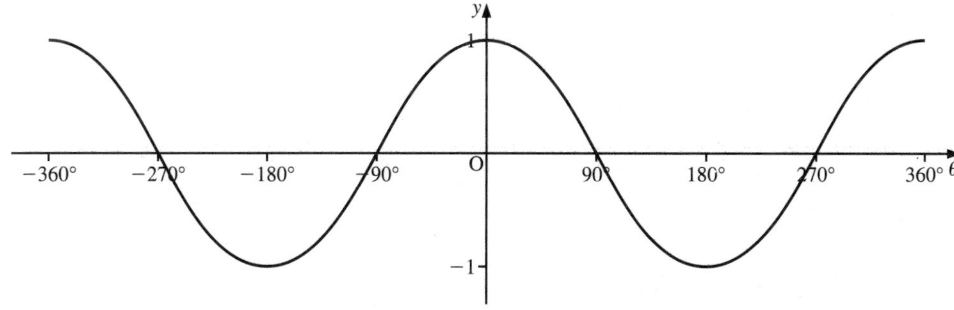

Properties of the cosine function:

- The function $f(\theta) = \cos\theta$ is periodic, of period 360°. That is,

$$\cos(\theta + 360°) = \cos\theta$$

- The function $f(\theta) = \cos\theta$ is an even function since

$$f(-\theta) = \cos\theta = f(\theta)$$

  In other words, the graph of $f(\theta)$ is symmetrical about the y-axis.
- The maximum value of $f(\theta)$ is 1 and its minimum value is -1. In other words, $-1 \leqslant f(\theta) \leqslant 1$.

$y = \tan\boldsymbol{\theta}$

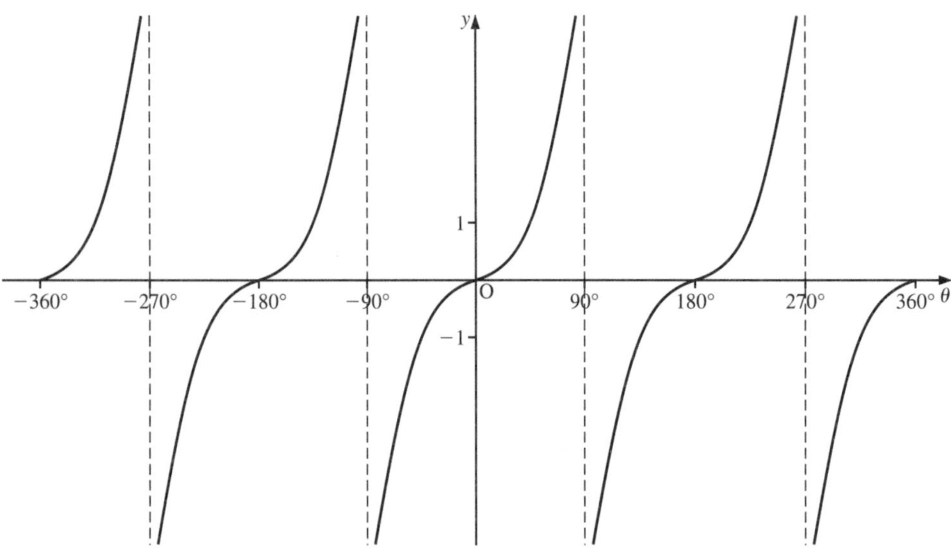

Properties of the tangent function:

- The function $f(\theta) = \tan\theta$ is periodic, of period 180°. That is,

$$\tan(\theta + 180°) = \tan\theta$$

- The function $f(\theta) = \tan\theta$ is an odd function since

$$f(-\theta) = -\tan\theta = -f(\theta)$$

  In other words, the graph of $f(\theta)$ has rotational symmetry about the origin of order 2.
- The function $f(\theta) = \tan\theta$ is not defined when $\theta = \pm 90°, \pm 270°, \ldots$

Graphs of $y = \operatorname{cosec}\theta$, $y = \sec\theta$ and $y = \cot\theta$

The trigonometric functions cosecant, secant and cotangent (abbreviated to cosec, sec and cot) are defined as:

$$\operatorname{cosec}\theta = \frac{1}{\sin\theta} \qquad \sec\theta = \frac{1}{\cos\theta} \qquad \cot\theta = \frac{1}{\tan\theta}$$

Plotting values of y against θ, where θ is measured in degrees, gives the following graphs.

$y = \operatorname{cosec} \theta$

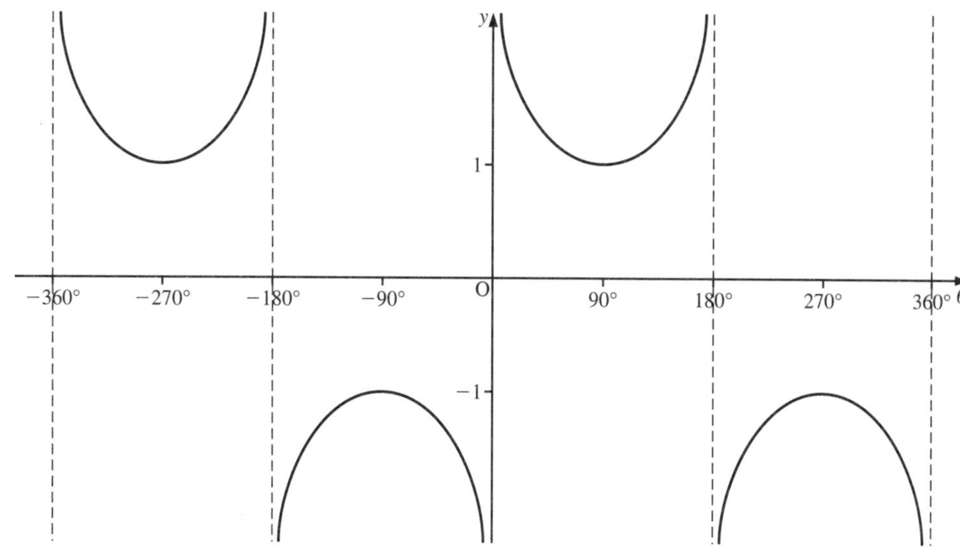

Properties of the cosec function:

- The function $f(\theta) = \operatorname{cosec} \theta$ is periodic, of period 360°. That is,

  $$\operatorname{cosec}(\theta + 360°) = \operatorname{cosec} \theta$$

- The function $f(\theta) = \operatorname{cosec} \theta$ is an odd function since

  $$f(-\theta) = -\operatorname{cosec} \theta = -f(\theta)$$

  In other words, the graph of $f(\theta)$ has rotational symmetry about the origin of order 2.
- $f(\theta) \geqslant 1$ or $f(\theta) \leqslant -1$.
- The function $f(\theta) = \operatorname{cosec} \theta$ is not defined when $\theta = 0°, \pm 180°, \pm 360°, \ldots$

$y = \sec \theta$

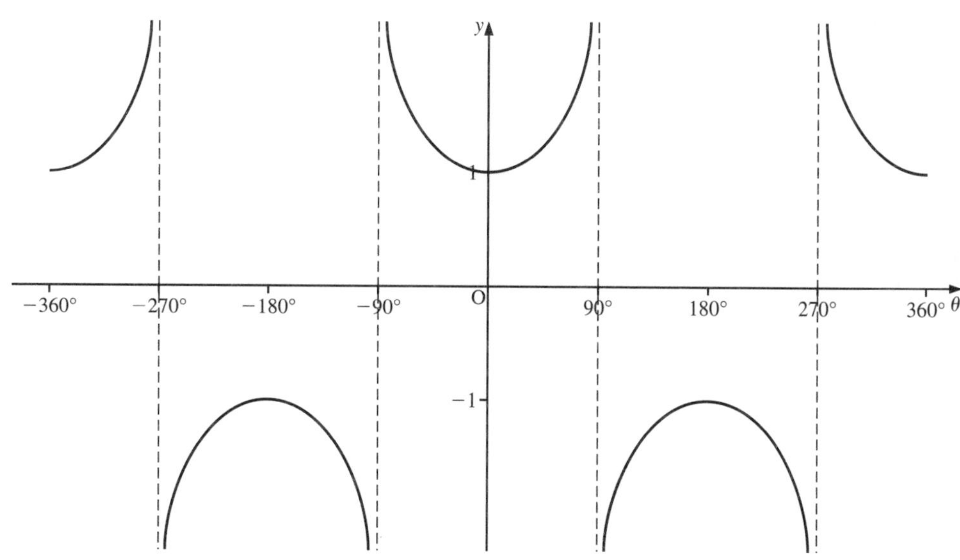

Properties of the secant function:

- The function $f(\theta) = \sec \theta$ is periodic, of period 360°. That is,

  $$\sec(\theta + 360°) = \sec \theta$$

- The function $f(\theta) = \sec\theta$ is an even function since

  $$f(-\theta) = \sec\theta = f(\theta)$$

  In other words, the graph of $f(\theta)$ is symmetrical about the y-axis.
- $f(\theta) \geqslant 1$ or $f(\theta) \leqslant -1$.
- The function $f(\theta) = \sec\theta$ is not defined when $\theta = \pm 90°, \pm 270°, \ldots$

$y = \cot\theta$

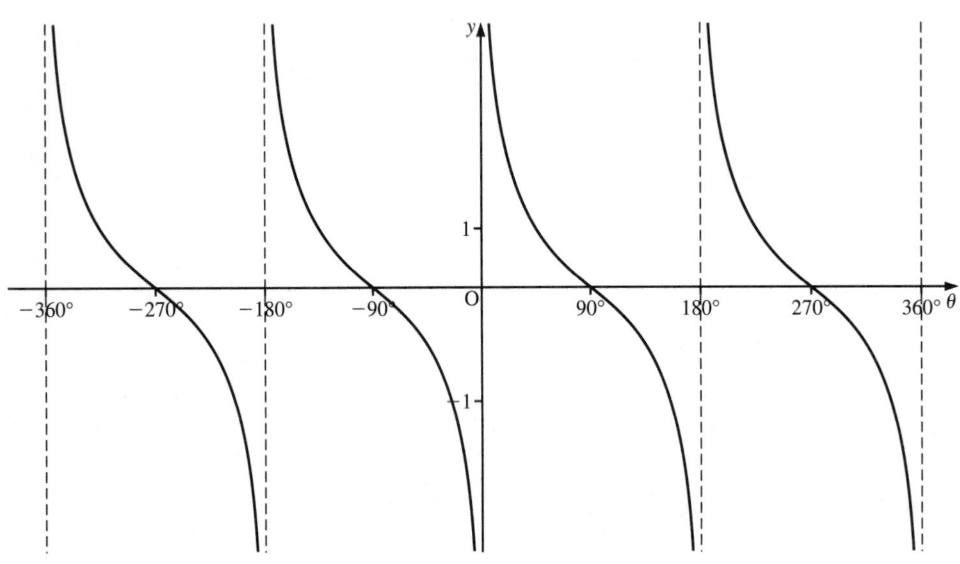

Properties of the cotangent function:

- The function $f(\theta) = \cot\theta$ is periodic, of period 180°. That is,

  $$\cot(\theta + 180°) = \cot\theta$$

- The function $f(\theta) = \cot\theta$ is an odd function since

  $$f(-\theta) = -\cot\theta = -f(\theta)$$

  In other words, the graph of $f(\theta)$ has rotational symmetry about the origin of order 2.
- The function $f(\theta) = \cot\theta$ is not defined when $\theta = 0°, \pm 180°, \pm 360°, \ldots$

Note As $\theta \to 90°$, $\tan\theta \to \infty$ and therefore $\cot\theta = \dfrac{1}{\tan\theta} \to 0$.

Curve sketching

In this section, we will be using the results on the transformation of graphs from pages 84–91.

Example 1 Sketch the graphs of each of the following functions for $-360° \leqslant \theta \leqslant 360°$.

a) $f(\theta) = 1 + \sin\theta$ **b)** $f(\theta) = \tan(\theta + 90°)$

c) $f(\theta) = 2\sin\theta$ **d)** $f(\theta) = \sec 2\theta$

For those functions which attain maximum and minimum values, write them down and state the values of θ in the range $0° \leqslant \theta \leqslant 360°$ for which they occur.

SOLUTION

a) To obtain the graph of $f(\theta) = 1 + \sin\theta$, we translate the graph of $\sin\theta$ by 1 unit parallel to the y-axis giving

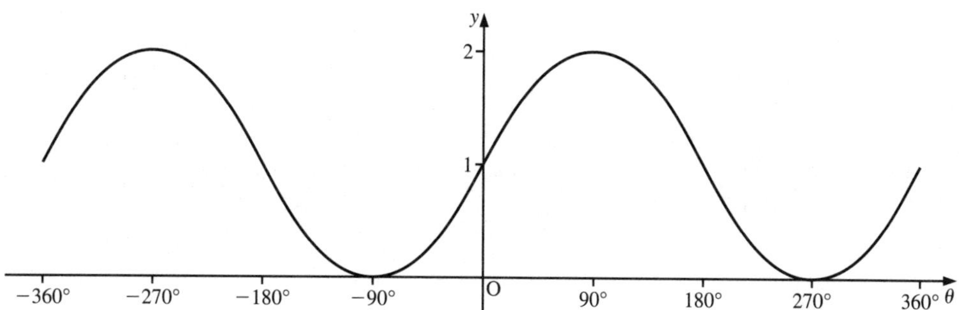

From the graph we see that the maximum value of $f(\theta)$ is 2 and this occurs when $\theta = 90°$. The minimum value of $f(\theta)$ is 0 and this occurs when $\theta = 270°$.

b) To obtain the graph of $f(\theta) = \tan(\theta + 90°)$, we translate the graph of $\tan\theta$ by $-90°$ parallel to the θ-axis giving

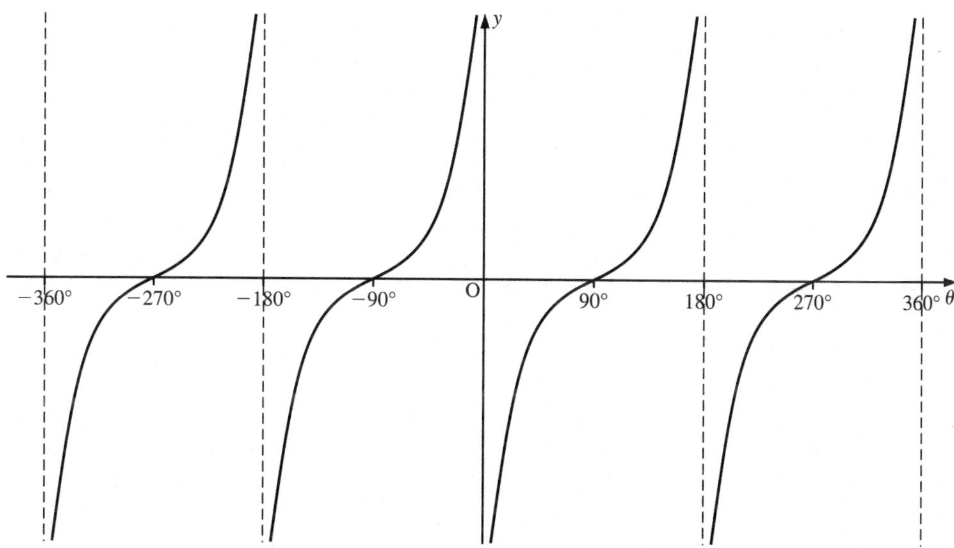

From the graph we see that $f(\theta)$ does not attain a maximum or minimum value.

c) To obtain the graph of $f(\theta) = 2\sin\theta$, we stretch the graph of $\sin\theta$ parallel to the y-axis by a scale factor of 2 giving

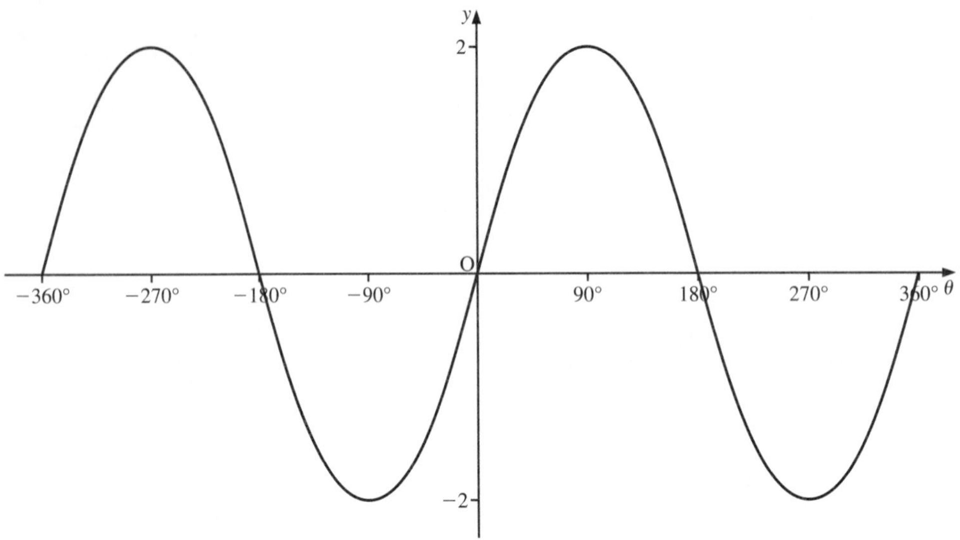

From the graph we see that the maximum value of $f(\theta)$ is 2 and this occurs when $\theta = 90°$. The minimum value of $f(\theta)$ is -2 and this occurs when $\theta = 270°$,

d) To obtain the graph of $f(\theta) = \sec 2\theta$, we stretch the graph of $\sec\theta$ parallel to the θ-axis by a scale factor of $\frac{1}{2}$ giving

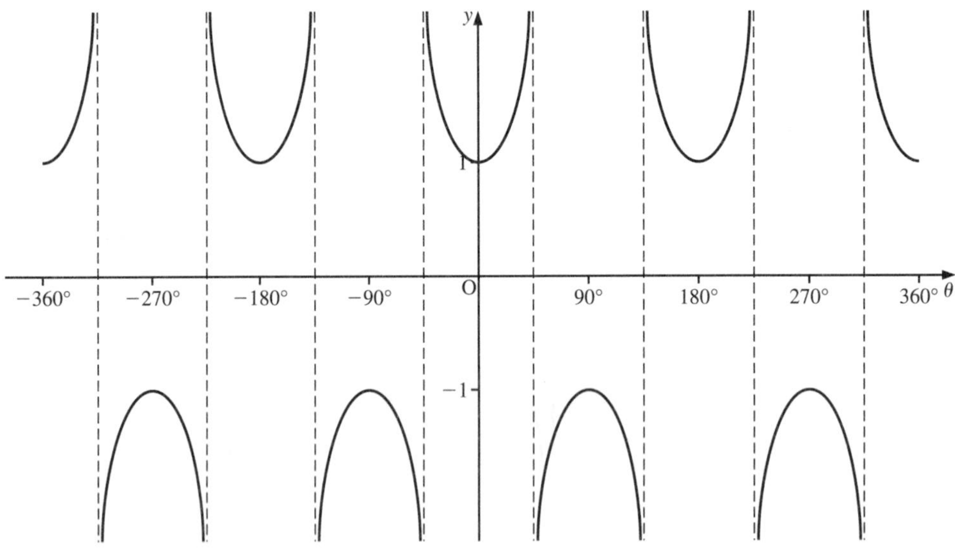

We see that the graph of $f(\theta)$ has turning points which are called **local maximum and minimum points**. We say that the function attains a local maximum value of -1 when $\theta = 90°$ and $270°$; and a local minimum value 1 when $\theta = 0°$, $180°$ and $360°$. We will call these the maximum and minimum values of the function since they correspond to turning points. Notice that when the entire domain of the function is considered the maximum and minimum values are ∞ and $-\infty$. We will call these the **greatest** and **least values** of the function.

Example 2 Express each of the following as the trigonometric ratio of an acute angle, using the same trigonometric function as given in the question.

a) $\sin 300^\circ$ **b)** $\cos(-350^\circ)$ **c)** $\tan 150^\circ$

SOLUTION

a) Sketching the sine graph gives

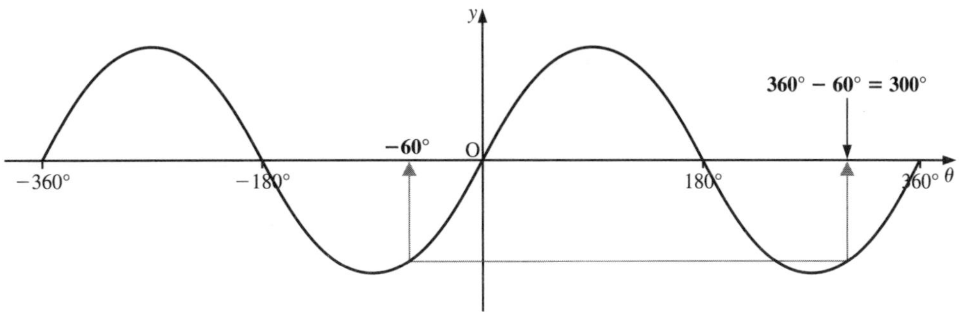

We see that

$$\sin 300^\circ = \sin(-60^\circ) = -\sin 60^\circ$$

since sine is an odd function. Therefore, $\sin 300^\circ = -\sin 60^\circ$.

b) Sketching the cosine graph gives

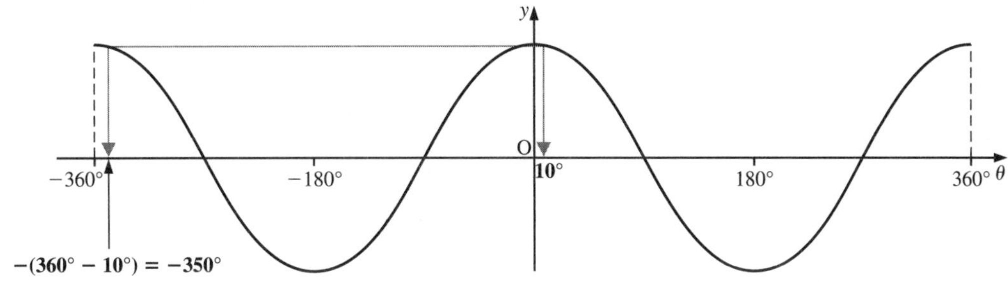

We see that $\cos(-350^\circ) = \cos 10^\circ$.

c) Sketching the tangent graph gives

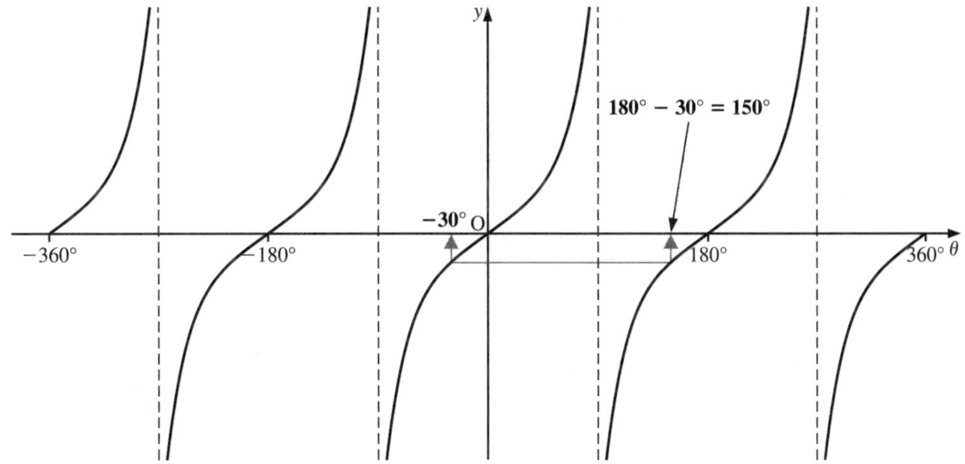

We see that $\tan 150^\circ = \tan(-30^\circ) = -\tan 30^\circ$.

Trigonometric equations

We have met trigonometric equations many times before. For example, $\cos\theta = \frac{1}{2}$ is a trigonometric equation in which the unknown is θ. We have met this type of equation when solving triangles. In this particular case, the acute angle θ which would satisfy this equation is 60°. However, from the periodic properties of the cosine graph, we know that there are other solutions to this equation.

Drawing the graph of $y = \cos\theta$ and $y = \frac{1}{2}$ on the same set of axes gives

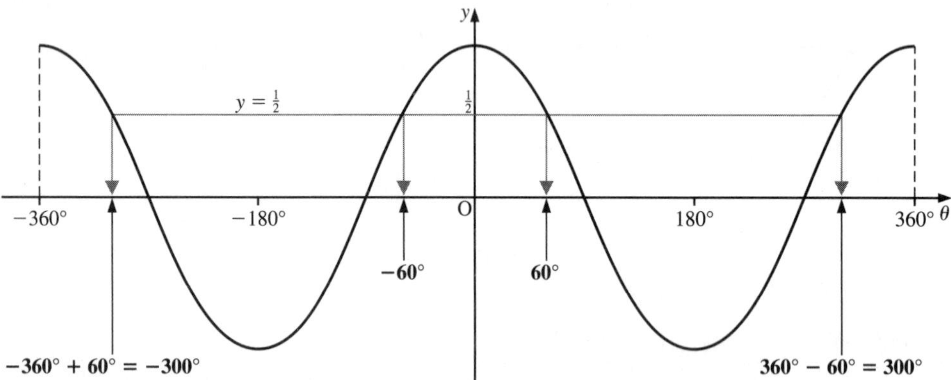

This shows that in the range $-360^\circ \leqslant \theta \leqslant 360^\circ$ there are actually **four** solutions to the equation $\cos\theta = \frac{1}{2}$. These are $\theta = \pm 60^\circ, \pm 300^\circ$.

It is clear that if no range for θ is stated, there is an infinite number of solutions to this equation. It is for this reason that trigonometric equations are always accompanied by a range for θ.

Example 3 Solve the following equations for θ, where $-360^\circ \leqslant \theta \leqslant 360^\circ$.

a) $\sin\theta = \dfrac{\sqrt{3}}{2}$ **b)** $\cos\theta = \dfrac{1}{3}$ **c)** $\tan\theta = -\dfrac{1}{4}$

SOLUTION

a) Drawing the graphs of $y = \sin\theta$ and $y = \dfrac{\sqrt{3}}{2}$ on the same set of axes for $-360^\circ \leqslant \theta \leqslant 360^\circ$ gives

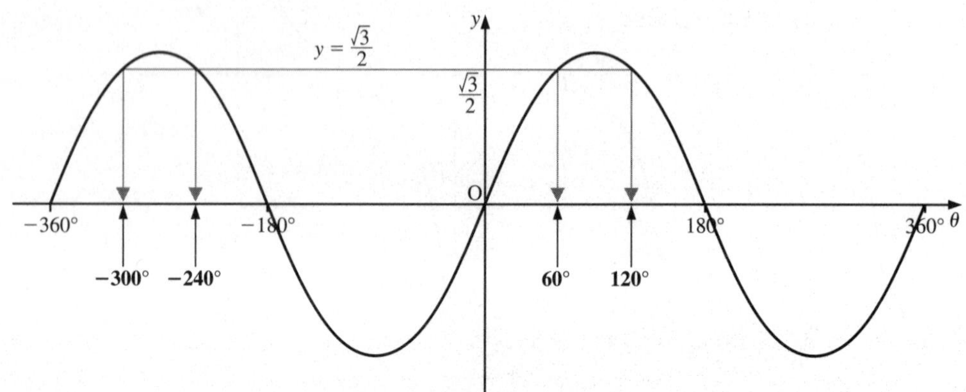

One solution is $\theta = 60°$. The other solutions in this range are

$$\theta = 180° - 60° = 120°$$

$$\theta = -180° - 60° = -240°$$

$$\theta = -360° + 60° = -300°$$

The solutions are $\theta = 60°$, $120°$, $-240°$ and $-300°$.

b) Drawing the graphs of $y = \cos\theta$ and $y = \frac{1}{3}$ on the same set of axes for $-360° \leqslant \theta \leqslant 360°$ gives

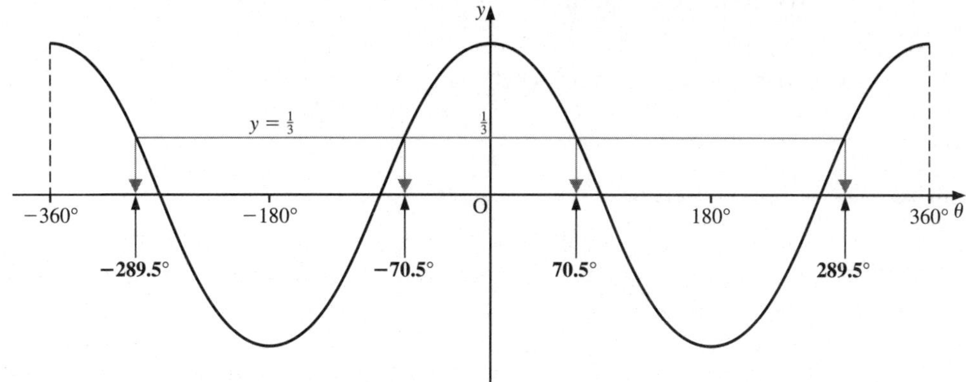

The calculator gives the value of θ as $\cos^{-1}\left(\frac{1}{3}\right) = 70.5°$, to one decimal place. The other positive solution in this range is $\theta = 360° - 70.5° = 289.5°$. Since the cosine graph is symmetrical about the y-axis the other solutions in this range are $\theta = -70.5°$ and $\theta = -289.5°$.

The solutions are $\theta = \pm 70.5°, \pm 289.5°$.

c) Drawing the graphs of $y = \tan\theta$ and $y = -\frac{1}{4}$ on the same set of axes for $-360° \leqslant \theta \leqslant 360°$ gives

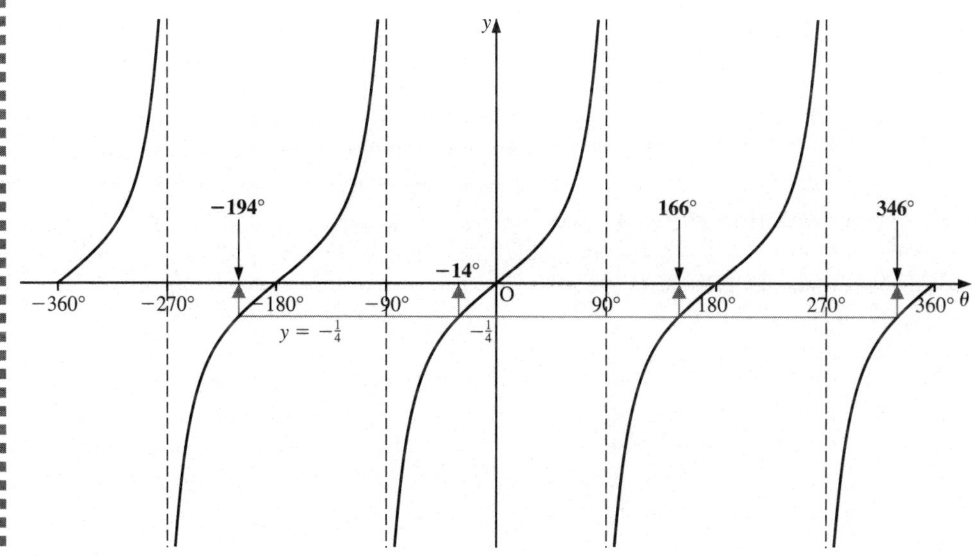

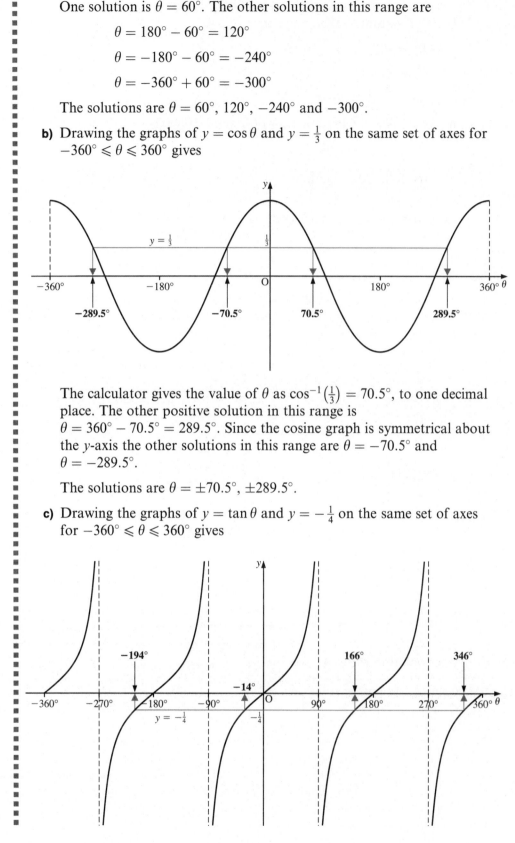

The calculator gives the value of θ as $\tan^{-1}\left(-\frac{1}{4}\right) = -14.0°$, to one decimal place. The other solutions in this range are

$$\theta = 180° - 14.0° = 166.0°$$

$$\theta = 360° - 14.0° = 346.0°$$

$$\theta = -180° - 14.0° = -194.0°$$

Therefore, the solutions are $\theta = 166.0°$, $346.0°$, $-14.0°$ and $-194.0°$.

Example 4 Solve each of the following equations for θ, where $0° \leqslant \theta \leqslant 360°$.

a) $\sqrt{3}\cot\theta = 1$ **b)** $\cos^2\theta = \frac{1}{2}$ **c)** $\operatorname{cosec}^2\theta - 4 = 0$

SOLUTION

a) By definition $\cot\theta = \dfrac{1}{\tan\theta}$. Therefore,

$\sqrt{3}\cot\theta = 1$ becomes

$$\frac{\sqrt{3}}{\tan\theta} = 1$$

$$\therefore \quad \tan\theta = \sqrt{3}$$

One solution is $\theta = 60°$. In the range $0 \leqslant \theta \leqslant 360°$, the other solution is $\theta = 180° + 60° = 240°$.

The solutions are $\theta = 60°$ and $240°$.

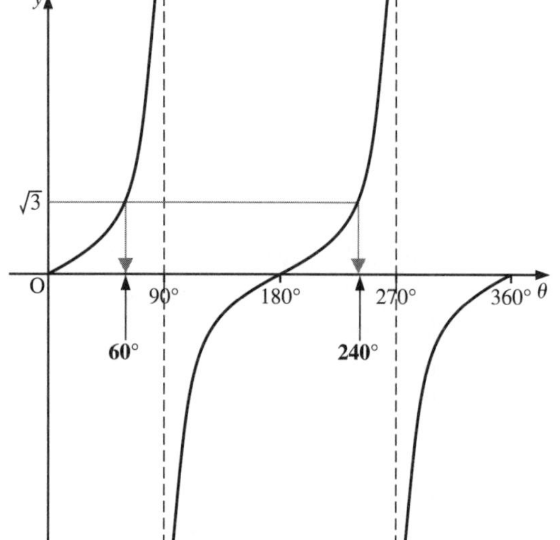

b) When $\cos^2\theta = \dfrac{1}{2}$: $\quad \cos\theta = \pm\sqrt{\dfrac{1}{2}}$

$$\therefore \quad \cos\theta = \pm\frac{1}{\sqrt{2}}$$

When $\cos\theta = \dfrac{1}{\sqrt{2}}$, one solution is $\theta = 45°$.

In the range $0° \leqslant \theta \leqslant 360°$, the other solution is $\theta = 360° - 45° = 315°$.

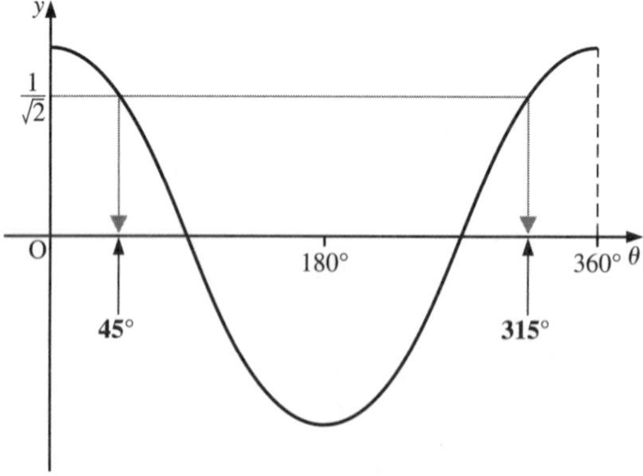

When $\cos\theta = -\dfrac{1}{\sqrt{2}}$, one solution is $\theta = 135°$.
In the range $0° \leqslant \theta \leqslant 360°$, the other solution is $\theta = 360° - 135° = 225°$.

The solutions are $\theta = 45°$, $135°$, $225°$ and $315°$.

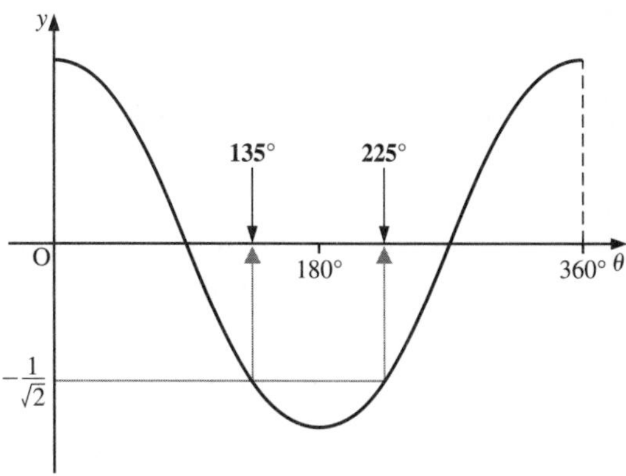

c) By definition $\operatorname{cosec}\theta = \dfrac{1}{\sin\theta}$. Therefore, $\operatorname{cosec}^2\theta - 4 = 0$ becomes

$$\frac{1}{\sin^2\theta} - 4 = 0$$

$$\therefore \quad \sin^2\theta = \frac{1}{4}$$

$$\therefore \quad \sin\theta = \pm\sqrt{\frac{1}{4}} = \pm\frac{1}{2}$$

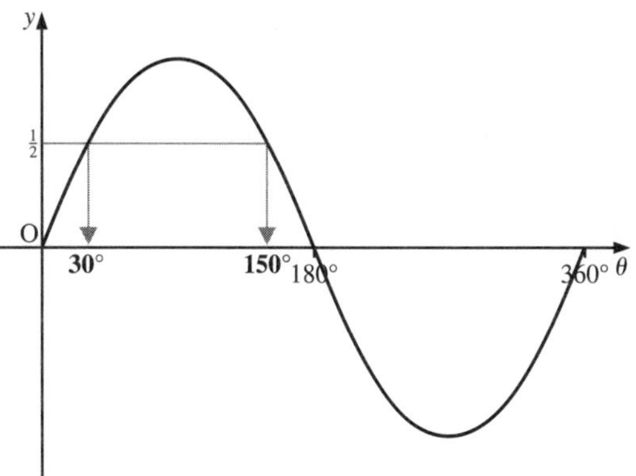

When $\sin\theta = \frac{1}{2}$, one solution is $\theta = 30°$. In the range $0° \leqslant \theta \leqslant 360°$, the other solution is $\theta = 180° - 30° = 150°$.

When $\sin\theta = -\frac{1}{2}$, one solution is $\theta = -30°$. However, this is not in the required range. Drawing the sine graph together with the line $y = -\frac{1}{2}$ gives the diagram on the right.

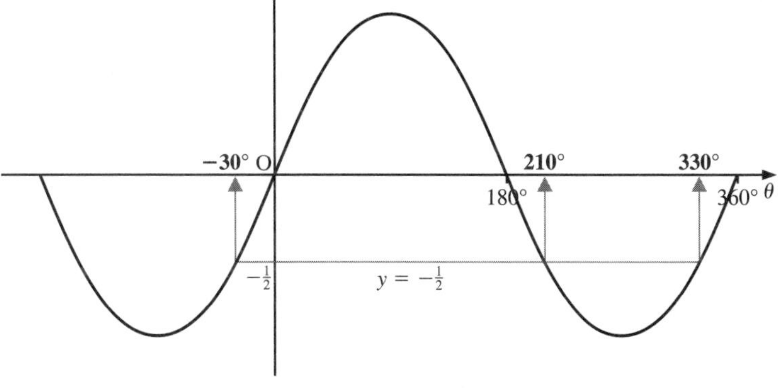

By the symmetry of the sine graph, we see that the required values of θ are $\theta = 180° + 30° = 210°$ and $\theta = 360° - 30° = 330°$.

The solutions are $\theta = 30°$, $150°$, $210°$ and $330°$.

Example 5 Solve each of the following equations for θ, where $-180° \leqslant \theta \leqslant 180°$.

a) $\tan^2\theta - \tan\theta = 0$ **b)** $2\cos^2\theta - \cos\theta - 1 = 0$ **c)** $\operatorname{cosec}\theta + \sin\theta + 2 = 0$

SOLUTION

a) To solve the equation $\tan^2\theta - \tan\theta = 0$, we notice that this is a quadratic equation in θ. Factorising gives

$$\tan\theta(\tan\theta - 1) = 0$$

Solving gives $\tan\theta = 0$ or $\tan\theta = 1$.

When $\tan\theta = 0$, $\theta = -180°$, $0°$ and $180°$ in the required range.

When $\tan\theta = 1$, one solution is $\theta = 45°$. The other solution is $\theta = -180° + 45° = -135°$.

The solutions are $\theta = -180°$, $-135°$, $0°$, $45°$ and $180°$.

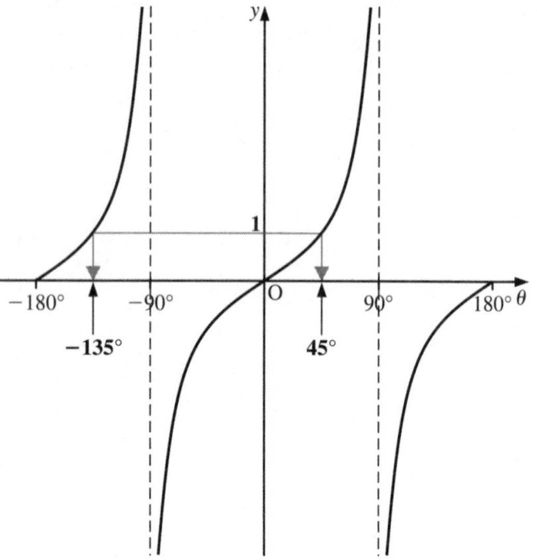

b) To solve $2\cos^2\theta - \cos\theta - 1 = 0$, we notice that this is a quadratic equation in $\cos\theta$. Factorising gives

$$(2\cos\theta + 1)(\cos\theta - 1) = 0$$

Solving gives $\cos\theta = -\frac{1}{2}$ or $\cos\theta = 1$.

When $\cos\theta = -\frac{1}{2}$, one solution is $\theta = -120°$. The other solution in the required range is $\theta = 120°$.

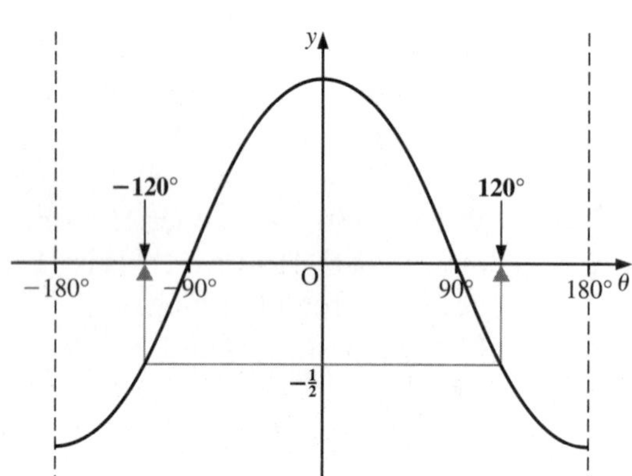

When $\cos\theta = 1$, the only solution in the required range is $0°$.

The solutions are $\theta = -120°$, $0°$ and $120°$.

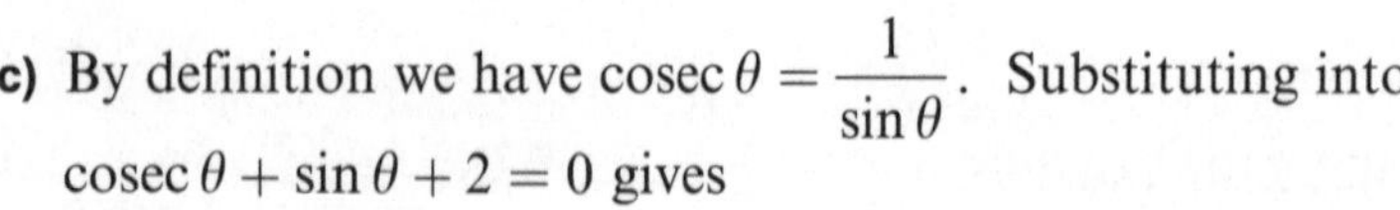

c) By definition we have $\operatorname{cosec}\theta = \dfrac{1}{\sin\theta}$. Substituting into $\operatorname{cosec}\theta + \sin\theta + 2 = 0$ gives

$$\frac{1}{\sin\theta} + \sin\theta + 2 = 0$$

Multiplying throughout by $\sin\theta$ and simplifying give

$$1 + \sin^2\theta + 2\sin\theta = 0$$

$$\therefore \quad \sin^2\theta + 2\sin\theta + 1 = 0$$

$$\therefore \quad (\sin\theta + 1)(\sin\theta + 1) = 0$$

Solving gives $\sin\theta = -1$. In the range $-180° \leqslant \theta \leqslant 180°$, $\theta = -90°$.

The solution is $\theta = -90°$.

Multiple angles

All the trigonometric equations we have looked at so far have involved solving $\sin\theta = k$, $\cos\theta = k$ or $\tan\theta = k$. However, we now look at equations which involve 2θ, 3θ, ..., etc.

Consider the equation $\sin 2\theta = \frac{1}{2}$, where $-180° \leqslant \theta \leqslant 180°$.

When we solve $\sin x = \frac{1}{2}$ in the range $-180° \leqslant x \leqslant 180°$, we obtain $x = 30°$ and $x = 150°$.

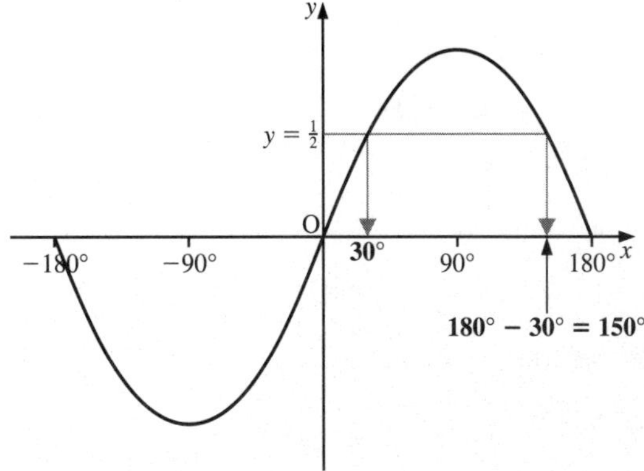

Since $x = 2\theta$, we have $\theta = 15°$ and $\theta = 75°$.

But we have, in fact, lost two other solutions to the equation $\sin 2\theta = \frac{1}{2}$, namely $\theta = -105°$ and $\theta = -165°$. These two other solutions have been missed because the range in which we have been working is for θ and not 2θ.

Therefore, to completely solve this equation, we must change the range to match the multiple angle. That is,

$$\sin 2\theta = \frac{1}{2} \qquad -360° \leqslant 2\theta \leqslant 360°$$

Now we have

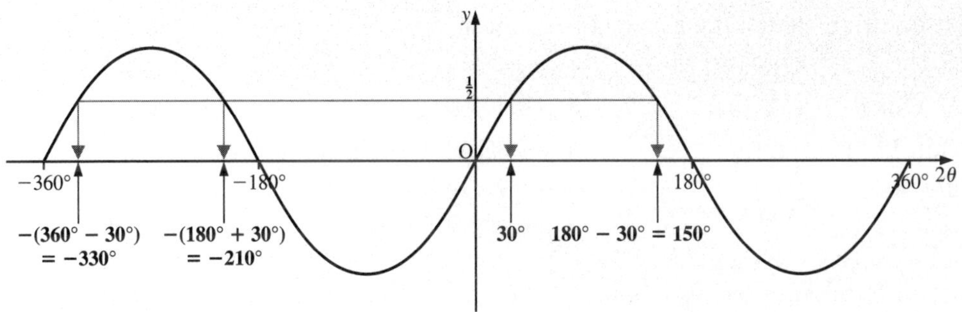

Therefore,

$$2\theta = -330°, -210°, 30° \text{ and } 150°$$

$$\therefore \quad \theta = -165°, -105°, 15° \text{ and } 75°$$

Example 6 Solve each of the following equations in the stated range.

a) $\cos^2 2\theta - 1 = 0, \quad -180° \leqslant \theta \leqslant 180°$

b) $\tan(2\theta + 45°) = \sqrt{3}, \quad -90° \leqslant \theta \leqslant 90°$

SOLUTION

a) Changing the range to 2θ gives $-360° \leqslant 2\theta \leqslant 360°$. We have

$$\cos^2 2\theta - 1 = 0$$
$$\therefore \quad \cos^2 2\theta = 1$$
$$\therefore \quad \cos 2\theta = \pm 1$$

When $\cos 2\theta = 1$: $\quad 2\theta = -360°,\ 0°$ and $360°$

$\therefore \quad \theta = -180°,\ 0°$ and $180°$

When $\cos 2\theta = -1$: $\quad 2\theta = -180°$ and $180°$

$\therefore \quad \theta = -90°$ and $90°$

The solutions are $\theta = -180°.\ -90°,\ 0°,\ 90°$ and $180°$.

b) Changing the range to $(2\theta + 45°)$ gives

$$-180° \leqslant 2\theta \leqslant 180°$$
$$\therefore \quad -180° + 45° \leqslant 2\theta + 45° \leqslant 180° + 45°$$
$$\therefore \quad -135° \leqslant 2\theta + 45° \leqslant 225°$$

We have $\tan(2\theta + 45°) = \sqrt{3}$. One solution is

$$2\theta + 45° = 60° \quad \therefore \quad 2\theta = 15° \quad \text{giving} \quad \theta = 7.5°$$

In the required range, the other solution is

$$2\theta + 45° = -120° \quad \therefore \quad 2\theta = -165° \quad \text{giving} \quad \theta = -82.5°$$

The solutions are $\theta = 7.5°$ and $-82.5°$.

Exercise 15A

1 In each part of this question sketch the graphs of the given functions on the same set of axes for $0° \leqslant \theta \leqslant 360°$.

a) $y = \sin\theta,\ y = 2 + \sin\theta,\ y = 3\sin\theta$
b) $y = \cos\theta,\ y = 1 - \cos\theta,\ y = 2\cos\theta - 1$
c) $y = \tan\theta,\ y = \tan 2\theta,\ y = \frac{1}{2}\tan\theta$
d) $y = \sin\theta,\ y = \sin 2\theta,\ y = \sin(2\theta - 90°)$
e) $y = \sec\theta,\ y = 1 + \sec\theta$
f) $y = \operatorname{cosec}\theta,\ y = 1 - \operatorname{cosec}\theta$
g) $y = \cot\theta,\ y = \cot(90° - \theta)$
h) $y = \operatorname{cosec} 2\theta,\ y = 1 + \operatorname{cosec}(2\theta - 180°)$

2 Write down the greatest and least values of each of the following expressions, and state the smallest non-negative value of θ for which they occur.

a) $\sin\theta$
b) $3 + \cos\theta$
c) $5 - 3\sin\theta$
d) $\sin(\theta + 20°)$
e) $3 - 4\cos(\theta - 40°)$
f) $3 + 7\sin(60° - \theta)$
g) $\dfrac{1}{3 + \sin\theta}$
h) $\dfrac{6}{2 - \cos 2\theta}$

3 Express each of the following as the trigonometric ratio of an **acute angle**, using the same trigonometric function as in the question.

a) $\sin 200°$ **b)** $\cos 240°$ **c)** $\tan 160°$ **d)** $\cos 310°$
e) $\tan 220°$ **f)** $\cos 490°$ **g)** $\sin(-20°)$ **h)** $\cos(-280°)$

4 Solve each of the following equations for $0° \leqslant \theta \leqslant 360°$, giving your answers correct to one decimal place.

a) $\sin\theta = 0.3$ **b)** $\cos\theta = 0.7$ **c)** $\tan\theta = 2$ **d)** $\cos\theta = -0.5$
e) $\sin\theta = -0.35$ **f)** $\tan\theta = -7$ **g)** $\sin\theta = 0.8$ **h)** $\sin\theta = -1$

5 Solve each of the following equations for $-180° \leqslant \theta \leqslant 180°$, giving your answers correct to one decimal place.

a) $\operatorname{cosec}\theta = 2$ **b)** $\sec\theta = 3$ **c)** $\cot\theta = 0.5$ **d)** $\cot\theta = -3$
e) $\sec\theta = 6$ **f)** $\operatorname{cosec}\theta = 5$ **g)** $\sec\theta = -1$ **h)** $\operatorname{cosec}\theta = -10$

6 Solve each of the following equations for $0° \leqslant \theta \leqslant 360°$, giving your answers correct to one decimal place.

a) $2\sin^2\theta - \sin\theta = 0$ **b)** $3\cos^2\theta = \cos\theta$ **c)** $5\sin\theta\cos\theta - \sin\theta = 0$
d) $\tan^2\theta + 4\tan\theta = 0$ **e)** $6\sin^2\theta - 5\sin\theta + 1 = 0$ **f)** $\cot^2\theta - 3\cot\theta + 2 = 0$
g) $\sec^2\theta + 4\sec\theta - 5 = 0$ **h)** $2\cot^2\theta - 7\cot\theta + 6 = 0$ **i)** $3\cos\theta + 4\sec\theta = 8$
j) $4\sin\theta + 1 = 3\operatorname{cosec}\theta$ **k)** $3\sec\theta + 11 = 4\cos\theta$ **l)** $(\tan\theta + 1)^2 = 9$

7 Solve each of the following equations for $0° \leqslant \theta \leqslant 180°$, giving your answers correct to one decimal place.

a) $\sin 2\theta = 0.3$ **b)** $\cot 3\theta = 4$ **c)** $4\sin 4\theta = 2$ **d)** $\cos 2\theta = -0.4$
e) $2 + 3\sin 2\theta = 0$ **f)** $5\cos 5\theta = 2$ **g)** $\sec 4\theta = -8$ **h)** $1 - \sin 5\theta = 0$

8 Solve each of the following equations for $0° \leqslant \theta \leqslant 360°$, giving your answers correct to one decimal place.

a) $\sin(\theta + 20°) = 0.4$ **b)** $\cos(\theta - 50°) = -0.3$
c) $1 + 5\sin(\theta - 100°) = 0$ **d)** $\tan(\theta - 162°) = 0.6$
e) $\sec(\theta - 62°) = 3$ **f)** $\cot(\theta + 17°) = -0.4$
g) $5\cot(\theta - 150°) = 3$ **h)** $\operatorname{cosec}(\theta + 210°) = 4$

9 Solve each of the following equations for $-90° \leqslant \theta \leqslant 90°$, giving your answers correct to one decimal place.

a) $\cos(2\theta - 25°) = 0.2$ **b)** $2\sin(3\theta + 48°) = 1$
c) $\tan(30° - \theta) = 2$ **d)** $\sec(4\theta + 67°) = -5$
e) $5\cot(2\theta + 40°) = 2$ **f)** $\operatorname{cosec}(60° - \theta) = 10$
g) $7 + 10\tan(5\theta - 200°) = 0$ **h)** $1 - 2\cos(47° - 2\theta) = 0$

Standard trigonometric identities

Result I

For any angle θ,

i) $\tan\theta \equiv \dfrac{\sin\theta}{\cos\theta}$

ii) $\sin^2\theta + \cos^2\theta \equiv 1$

To prove these two results for an acute angle θ, consider the right-angled triangle shown on the right.

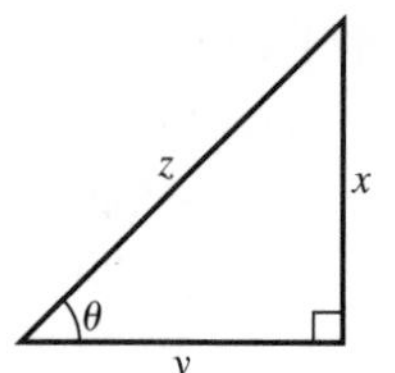

Proof of (i)

We know that $\sin\theta = \frac{x}{z}$ and $\cos\theta = \frac{y}{z}$. Therefore,

$$\frac{\sin\theta}{\cos\theta} = \frac{\left(\frac{x}{z}\right)}{\left(\frac{y}{z}\right)} = \frac{x}{y} = \tan\theta$$

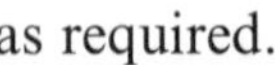
as required.

Proof of (ii)

Also we have

$$\sin^2\theta + \cos^2\theta = \left(\frac{x}{z}\right)^2 + \left(\frac{y}{z}\right)^2$$

$$= \frac{x^2 + y^2}{z^2}$$

By Pythagoras, we have $x^2 + y^2 = z^2$. Therefore,

$$\sin^2\theta + \cos^2\theta = \frac{z^2}{z^2} = 1 \text{ as required.}$$

as required.

Result II

For any angle θ,

i) $1 + \tan^2\theta \equiv \sec^2\theta$
ii) $1 + \cot^2\theta \equiv \operatorname{cosec}^2\theta$

Proof of (i)

Using the fact that $\tan\theta \equiv \frac{\sin\theta}{\cos\theta}$ for all θ, we have

$$\text{LHS} \equiv 1 + \tan^2\theta \equiv 1 + \left(\frac{\sin\theta}{\cos\theta}\right)^2$$

$$\equiv 1 + \frac{\sin^2\theta}{\cos^2\theta}$$

$$\equiv \frac{\cos^2\theta + \sin^2\theta}{\cos^2\theta}$$

$$\equiv \frac{1}{\cos^2\theta} \quad (\text{since } \sin^2\theta + \cos^2\theta \equiv 1)$$

$$\equiv \sec^2\theta \equiv \text{RHS}$$

as required.

Proof of (ii)

Using the definition $\cot\theta \equiv \dfrac{1}{\tan\theta} \equiv \dfrac{\cos\theta}{\sin\theta}$, we have

$$\begin{aligned}\text{LHS} &\equiv 1 + \cot^2\theta \equiv 1 + \frac{\cos^2\theta}{\sin^2\theta}\\ &\equiv \frac{\sin^2\theta + \cos^2\theta}{\sin^2\theta}\\ &\equiv \frac{1}{\sin^2\theta}\\ &\equiv \operatorname{cosec}^2\theta \equiv \text{RHS}\end{aligned}$$

as required.

These identities enable us to calculate all the trigonometric ratios if just one is known. Example 7 illustrates this fact.

Example 7 Given that θ is acute and that $\sin\theta = \dfrac{1}{\sqrt{3}}$, find the values of

a) $\cos\theta$ **b)** $\tan\theta$ **c)** $\operatorname{cosec}\theta$

SOLUTION

a) Consider the right-angled triangle shown on the right.

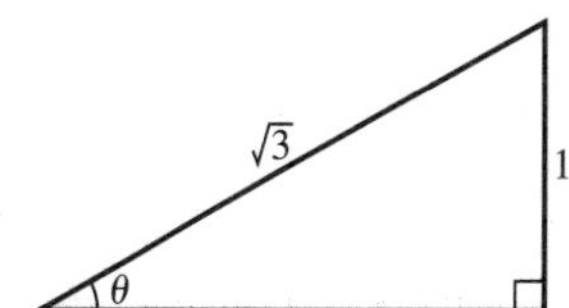

Using Pythagoras gives

$$x^2 = (\sqrt{3})^2 - 1^2 \quad \therefore \quad x = \sqrt{2}$$

From the right-angled triangle, we have

$$\cos\theta = \frac{\text{Adj}}{\text{Hyp}} = \frac{\sqrt{2}}{\sqrt{3}}$$

$$\therefore \quad \cos\theta = \sqrt{\frac{2}{3}}$$

b) Also from the right-angled triangle, we have

$$\tan\theta = \frac{\text{Opp}}{\text{Adj}} = \frac{1}{\sqrt{2}}$$

c)

$$\operatorname{cosec}\theta = \frac{\text{Hyp}}{\text{Opp}} = \frac{\sqrt{3}}{1}$$

$$\therefore \quad \operatorname{cosec}\theta = \sqrt{3}$$

Further trigonometric equations

In order to solve trigonometric equations such as

$$2\sin\theta - \cos\theta = 0 \quad \text{and} \quad 4\cos^2\theta + 3\sin\theta = 4$$

we must reduce them to one or more of the forms $\sin\theta = k$, $\cos\theta = k$ or $\tan\theta = k$ (where k is a constant), by using one or more of the identities from Results I and II. Examples 8 and 9 illustrate such techniques.

Example 8 Solve $2\sin\theta - \cos\theta = 0$, $0° \leqslant \theta \leqslant 180°$.

SOLUTION

$$2\sin\theta - \cos\theta = 0$$
$$\therefore \quad 2\sin\theta = \cos\theta$$

Since $\tan\theta = \dfrac{\sin\theta}{\cos\theta}$, divide by $\cos\theta$ to obtain

$$2\frac{\sin\theta}{\cos\theta} = \frac{\cos\theta}{\cos\theta}$$
$$\therefore \quad 2\tan\theta = 1$$
$$\therefore \quad \tan\theta = \tfrac{1}{2}$$

When $\tan\theta = \frac{1}{2}$, $\theta = 26.6°$, in the range $0° \leqslant \theta \leqslant 180°$.

Example 9 Solve each of the following equations in the stated range.

a) $4\cos^2\theta + 3\sin\theta = 4, \ 0° \leqslant \theta \leqslant 360°$
b) $3\sec^2\theta - 4\tan\theta - 2 = 0, \ -180° \leqslant \theta \leqslant 180°$
c) $5\cos^2 3\theta = 3(1 + \sin 3\theta), \ 0° \leqslant \theta \leqslant 180°$

SOLUTION

a) To solve the equation $4\cos^2\theta + 3\sin\theta = 4$, we notice that by replacing $\cos^2\theta$ with an expression in terms of $\sin^2\theta$ the original equation becomes a quadratic in $\sin\theta$.

We know that

$$\sin^2\theta + \cos^2\theta \equiv 1$$
$$\therefore \quad \cos^2\theta \equiv 1 - \sin^2\theta$$

Substituting this into the equation gives

$$4(1 - \sin^2\theta) + 3\sin\theta = 4$$
$$\therefore \quad 3\sin\theta - 4\sin^2\theta = 0$$
$$\therefore \quad \sin\theta(3 - 4\sin\theta) = 0$$

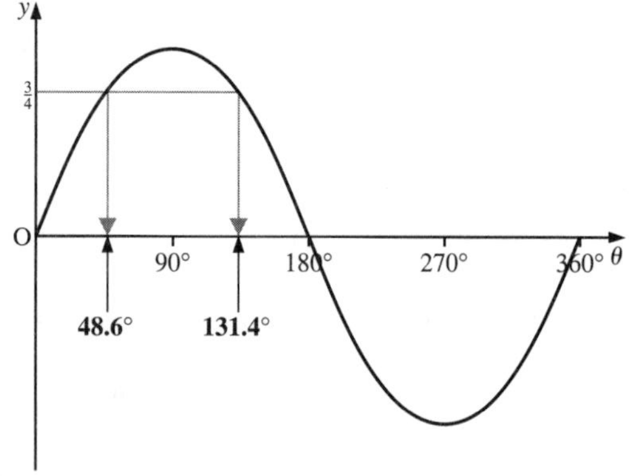

Solving gives $\sin\theta = 0$ or $\sin\theta = \frac{3}{4}$.

When $\sin\theta = 0$, $\theta = 0°$, $180°$ and $360°$ in the range $0° \leqslant \theta \leqslant 360°$.

When $\sin\theta = \frac{3}{4}$, one solution is $\theta = 48.6°$. In the required range, the other solution is $\theta = 180° - 48.6° = 131.4°$.

The solutions are $\theta = 0°$, $48.6°$, $131.4°$, $180°$ and $360°$.

b) To solve the equation $3\sec^2\theta - 4\tan\theta - 2 = 0$, we notice that $\sec^2\theta$ can be replaced by an expression involving $\tan^2\theta$ since

$$\sec^2\theta \equiv 1 + \tan^2\theta$$

Substituting this into the equation gives

$$3(1 + \tan^2\theta) - 4\tan\theta - 2 = 0$$

$$\therefore \quad 3\tan^2\theta - 4\tan\theta + 1 = 0$$

$$\therefore \quad (3\tan\theta - 1)(\tan\theta - 1) = 0$$

Solving gives $\tan\theta = \frac{1}{3}$ or $\tan\theta = 1$.

When $\tan\theta = \frac{1}{3}$, one solution is $\theta = 18.4°$. In the range $-180° \leqslant \theta \leqslant 180°$, the other solution is $\theta = -180° + 18.4° = -161.6°$.

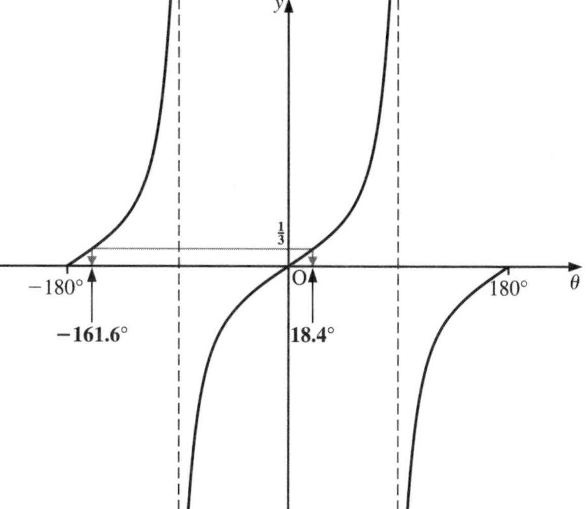

When $\tan\theta = 1$, one solution is $\theta = 45°$. In the range $-180° \leqslant \theta \leqslant 180°$, the other solution is $\theta = -180° + 45° = -135°$.

The solutions are $\theta = -135°, -161.6°, 18.4°$ and $45°$.

c) Replacing $\cos^2 3\theta$ by $1 - \sin^2 3\theta$ gives

$$5(1 - \sin^2 3\theta) = 3(1 + \sin 3\theta)$$

$$\therefore \quad 5 - 5\sin^2 3\theta = 3 + 3\sin 3\theta$$

$$\therefore \quad 5\sin^2 3\theta + 3\sin 3\theta - 2 = 0$$

$$\therefore \quad (5\sin 3\theta - 2)(\sin 3\theta + 1) = 0$$

Solving gives $\sin 3\theta = \frac{2}{5}$ or $\sin 3\theta = -1$.

Changing the range for θ gives $0° \leqslant 3\theta \leqslant 540°$. When $\sin 3\theta = \frac{2}{5}$, the other solutions in this range are

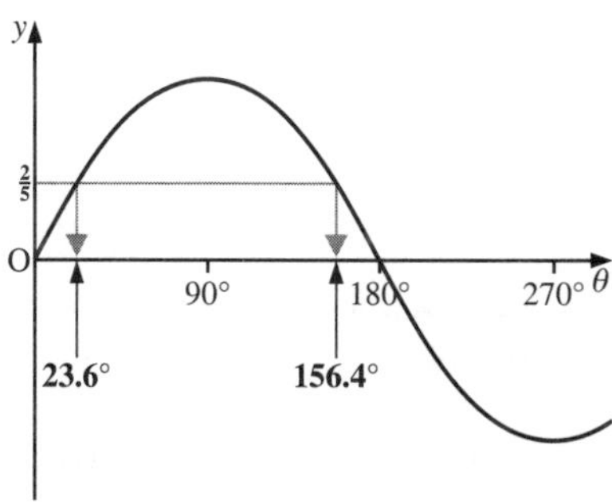

$$3\theta = 23.6°$$

$$3\theta = 180° - 23.6° = 156.4°$$

$$3\theta = 360° + 23.6° = 383.6°$$

$$3\theta = 540° - 23.6° = 516.4°$$

Therefore, $\theta = 7.9°, 52.1°, 127.9°$ and $172.1°$.

The solutions are $\theta = 7.9°, 52.1°, 90°, 127.9°$ and $172.1°$.

Exercise 15B

(For questions 1 and 2 you may want to refer to the table on page 50.)

1 Simplify each of the following, giving each of your answers as a fraction of integers or surds.

a) $\sin 135°$ **b)** $\cos 330°$ **c)** $\tan 210°$ **d)** $\sin 300°$
e) $\sec 150°$ **f)** $\cot 120°$ **g)** $\sin 270°$ **h)** $\tan 240°$
i) $\operatorname{cosec} 225°$ **j)** $\cos 390°$ **k)** $\sec 765°$ **l)** $\sin(-30°)$

2 Solve each of the following equations for $0° \leqslant \theta \leqslant 360°$.

a) $\sin\theta = \dfrac{\sqrt{3}}{2}$ **b)** $4\cos^2\theta = 1$ **c)** $\tan\theta = -1$

d) $\sin\theta = -\frac{1}{2}$ **e)** $\tan^2\theta = 3$ **f)** $\cos\theta = -\dfrac{\sqrt{3}}{2}$

g) $\cot 3\theta = 1$ **h)** $\tan 2\theta = \sqrt{3}$ **i)** $\cos 4\theta = -\frac{1}{2}$

j) $\tan(\theta + 48°) = \sqrt{3}$ **k)** $\sin(2\theta - 14°) = 1$ **l)** $\sec(3\theta + 12°) = -\dfrac{2}{\sqrt{3}}$

3 Given that θ is acute and that $\sin\theta = \frac{2}{3}$, express each of the following in surd form.

a) $\cos\theta$ **b)** $\tan\theta$ **c)** $\cot\theta$

4 Given that θ is acute and that $\cos\theta = \frac{1}{4}$, express each of the following in surd form.

a) $\sin\theta$ **b)** $\tan\theta$ **c)** $\operatorname{cosec}\theta$

5 Given that θ is acute and that $\tan\theta = 3$, express each of the following in surd form.

a) $\sin\theta$ **b)** $\sec\theta$ **c)** $\operatorname{cosec}\theta$

6 Given that θ is acute and that $\sec\theta = \frac{5}{3}$, express each of the following in surd form.

a) $\sin\theta$ **b)** $\tan\theta$ **c)** $\cot\theta$

7 Solve each of the following equations for $0° \leqslant \theta \leqslant 360°$, giving your answers correct to one decimal place.

a) $\sin\theta = 3\cos\theta$ **b)** $5\cos\theta = 3\sin\theta$ **c)** $\sin\theta + \cos\theta = 0$
d) $2\cos\theta - 3\sin\theta = 0$ **e)** $\sec\theta = 2\operatorname{cosec}\theta$ **f)** $5\sec\theta + 3\operatorname{cosec}\theta = 0$
g) $\sin^2\theta - 5\sin\theta\cos\theta = 0$ **h)** $3\cos^2\theta = 7\sin\theta\cos\theta$ **i)** $4\sin^2\theta = \cos^2\theta$
j) $\operatorname{cosec}^2\theta = 9\sec^2\theta$ **k)** $25\cos\theta = 16\sin\theta\tan\theta$ **l)** $\sec\theta\tan\theta = 8\operatorname{cosec}\theta\cot\theta$

8 Solve each of the following equations for $-180° \leqslant \theta \leqslant 180°$, giving your answers correct to one decimal place.

a) $2\cos^2\theta + 3\sin\theta - 3 = 0$ **b)** $3\sin^2\theta - 5\cos\theta - 1 = 0$
c) $8\sin^2\theta = 11 - 10\cos\theta$ **d)** $\sin^2\theta - 2 = 2\cos^2\theta - 4\sin\theta$
e) $4(2 + \cos^2\theta) = \sin\theta(11 + \sin\theta)$ **f)** $2\cos^3\theta - 5\cos^2\theta - 3\cos\theta = 0$
g) $2\cos^3\theta = 3\sin\theta\cos\theta$ **h)** $4\sin\theta\cos\theta(1 + \sin\theta) = 11\cos^3\theta - 7\cos\theta$

9 Solve each of the following equations for $0° \leqslant \theta \leqslant 360°$, giving your answers correct to one decimal place.

a) $\sec^2\theta = 7 + \tan\theta$ **b)** $2\operatorname{cosec}^2\theta + 7\cot\theta = 6$
c) $2\sec^2\theta + 11\tan\theta = 17$ **d)** $6\operatorname{cosec}^2\theta = 5(2 - \cot\theta)$
e) $5\operatorname{cosec}\theta(\operatorname{cosec}\theta - 1) = 3 - \cot^2\theta$ **f)** $4 + \tan^2\theta = \sec\theta(7 - \sec\theta)$
g) $3\tan^2\theta + \sec^2\theta = 5(1 - 3\tan\theta)$ **h)** $4\tan^3\theta - 4\tan^2\theta + \tan\theta = 0$
i) $\sec^3\theta + 2 = 3\sec\theta - 2\tan^2\theta$ **j)** $2\sin\theta\tan\theta = \sin\theta + \cos\theta$

10 Solve each of the following equations for $-180° \leqslant \theta \leqslant 180°$, giving your answers correct to one decimal place.

a) $6\sin^3\theta + 5\sin^2\theta - 2\sin\theta - 1 = 0$ **b)** $4\cos^3\theta - 4\cos^2\theta - \cos\theta + 1 = 0$
c) $\tan^3\theta - 4\tan^2\theta + \tan\theta + 6 = 0$ **d)** $\sec^3\theta - 2\sec^2\theta - 11\sec\theta + 12 = 0$
e) $3\sin^3\theta + 10\cos^2\theta + 9\sin\theta = 12$ **f)** $2\tan^3\theta = \sec^2\theta + 13\tan\theta + 5$

11 Solve each of the following equations for $0° \leqslant \theta \leqslant 180°$, giving your answers correct to one decimal place.

a) $\sin 5\theta = 3\cos 5\theta$ **b)** $12\sin^2 2\theta + \sin 2\theta = 1$
c) $3\tan 4\theta + 8 = 3\cot 4\theta$ **d)** $3\cot 3\theta = 4\cos 3\theta$
e) $11 + \tan^2 2\theta = 7\sec 2\theta$ **f)** $3\cos^2 4\theta = 4(1 - \sin 4\theta)$

***12** Given that

$$1 + 4\sin\theta(3\sin\theta - 4) = \operatorname{cosec}\theta(\operatorname{cosec}\theta - 4)$$

calculate the four possible values of $\sin\theta$.

***13** Given that

$$\cos\theta = \frac{20\sin^4\theta - 24\sin^2\theta + 6}{10\sin^3\theta - 7\sin\theta}$$

calculate the possible values of $\tan\theta$.

Proving trigonometric identities

Consider the following identity:

$$\tan\theta + \cot\theta \equiv \sec\theta\operatorname{cosec}\theta$$

Substituting different values of θ into the LHS and RHS will show this identity to be true for those particular values of θ. However, this does not **prove** the identity for all values of θ. Identities can be proved by using other simpler identities which we know to be true for all values of θ. For example, we know that the following identities are true for all values of θ.

- $\tan\theta \equiv \dfrac{\sin\theta}{\cos\theta}$
- $\sin^2\theta + \cos^2\theta \equiv 1$
- $1 + \tan^2\theta \equiv \sec^2\theta$
- $1 + \cot^2\theta \equiv \operatorname{cosec}^2\theta$

The general method for proving an identity is to choose either the LHS **or** the RHS (usually whichever is the more complicated) and show, by using known identities, that it can be manipulated into the form of the other. However, two alternative techniques are to show that

- $\text{LHS} - \text{RHS} \equiv 0$ or
- $\dfrac{\text{LHS}}{\text{RHS}} \equiv 1.$

Once the proof is completed, we write QED, which stands for the Latin ‘Quod Erat Demonstrandum’ – ‘which was to be proved’.

Example 10 Prove the identity $\tan\theta + \cot\theta \equiv \sec\theta \operatorname{cosec}\theta$.

SOLUTION

$$\begin{aligned}
\text{LHS} &\equiv \tan\theta + \cot\theta \\
&\equiv \frac{\sin\theta}{\cos\theta} + \frac{\cos\theta}{\sin\theta} \\
&\equiv \frac{\sin^2\theta + \cos^2\theta}{\sin\theta\cos\theta} \\
&\equiv \frac{1}{\sin\theta\cos\theta} \qquad (\text{since } \sin^2\theta + \cos^2\theta \equiv 1) \\
&\equiv \frac{1}{\sin\theta} \times \frac{1}{\cos\theta} \\
&\equiv \operatorname{cosec}\theta \sec\theta \equiv \text{RHS} \qquad \text{QED}
\end{aligned}$$

Example 11 Prove the identity

$$(1 - \sin\theta + \cos\theta)^2 \equiv 2(1 - \sin\theta)(1 + \cos\theta)$$

SOLUTION

$$\begin{aligned}
\text{LHS} &\equiv (1 - \sin\theta + \cos\theta)(1 - \sin\theta + \cos\theta) \\
&\equiv 1 - 2\sin\theta + 2\cos\theta + \sin^2\theta + \cos^2\theta - 2\sin\theta\cos\theta \\
&\equiv 2 - 2\sin\theta + 2\cos\theta - 2\sin\theta\cos\theta \quad (\text{since } \sin^2\theta + \cos^2\theta \equiv 1) \\
&\equiv 2(1 - \sin\theta + \cos\theta - \sin\theta\cos\theta) \\
&\equiv 2(1 - \sin\theta)(1 + \cos\theta) \equiv \text{RHS} \qquad \text{QED}
\end{aligned}$$

Example 12 Prove the identity

$$\frac{1 + \sin\theta}{1 - \sin\theta} \equiv (\tan\theta + \sec\theta)^2$$

SOLUTION

$$\begin{aligned}
\text{RHS} &\equiv \left(\frac{\sin\theta}{\cos\theta} + \frac{1}{\cos\theta}\right)^2 \\
&\equiv \left(\frac{\sin\theta + 1}{\cos\theta}\right)^2 \\
&\equiv \frac{(1 + \sin\theta)^2}{\cos^2\theta} \\
&\equiv \frac{(1 + \sin\theta)^2}{1 - \sin^2\theta} \\
&\equiv \frac{(1 + \sin\theta)^2}{(1 + \sin\theta)(1 - \sin\theta)} \\
&\equiv \frac{1 + \sin\theta}{1 - \sin\theta} \equiv \text{LHS} \qquad \text{QED}
\end{aligned}$$

Exercise 15C

Prove each of the following identities.

1 $\sin\theta\tan\theta + \cos\theta \equiv \sec\theta$

2 $\operatorname{cosec}\theta + \tan\theta\sec\theta \equiv \operatorname{cosec}\theta\sec^2\theta$

3 $\operatorname{cosec}\theta - \sin\theta \equiv \cot\theta\cos\theta$

4 $(\sin\theta + \cos\theta)^2 - 1 \equiv 2\sin\theta\cos\theta$

5 $(\sin\theta - \operatorname{cosec}\theta)^2 \equiv \sin^2\theta + \cot^2\theta - 1$

6 $(\sec\theta + \tan\theta)(\sec\theta - \tan\theta) \equiv 1$

7 $\tan^2\theta + \sin^2\theta \equiv (\sec\theta + \cos\theta)(\sec\theta - \cos\theta)$

8 $\sec^2\theta + \cot^2\theta \equiv \operatorname{cosec}^2\theta + \tan^2\theta$

9 $(\sin\theta + \cos\theta)(1 - \sin\theta\cos\theta) \equiv \sin^3\theta + \cos^3\theta$

10 $\tan^4\theta + \tan^2\theta \equiv \sec^4\theta - \sec^2\theta$

11 $\cos^4\theta - \sin^4\theta \equiv \cos^2\theta - \sin^2\theta$

12 $\sin\theta + \cos\theta \equiv \dfrac{1 - 2\cos^2\theta}{\sin\theta - \cos\theta}$

13 $\dfrac{\sin\theta}{1 + \cos\theta} + \dfrac{1 + \cos\theta}{\sin\theta} \equiv 2\operatorname{cosec}\theta$

14 $\dfrac{\operatorname{cosec}\theta}{\cot\theta + \tan\theta} \equiv \cos\theta$

15 $\dfrac{1}{1 + \tan^2\theta} + \dfrac{1}{1 + \cot^2\theta} \equiv 1$

16 $\dfrac{1 - \sin\theta}{\cos\theta} \equiv \dfrac{1}{\sec\theta + \tan\theta}$

17 $\dfrac{\tan\theta + \cot\theta}{\sec\theta + \operatorname{cosec}\theta} \equiv \dfrac{1}{\sin\theta + \cos\theta}$

18 $\sec^4\theta - \operatorname{cosec}^4\theta \equiv \dfrac{\sin^2\theta - \cos^2\theta}{\sin^4\theta\cos^4\theta}$

19 $\sqrt{(\sec^2\theta - 1)} + \sqrt{(\operatorname{cosec}^2\theta - 1)} \equiv \dfrac{1}{\sin\theta\cos\theta}$

20 $\dfrac{\sin\theta}{\sqrt{(1 + \cot^2\theta)}} + \dfrac{\cos\theta}{\sqrt{(1 + \tan^2\theta)}} \equiv 1$

21 $\sqrt{\dfrac{1 - \sin\theta}{1 + \sin\theta}} \equiv \sec\theta - \tan\theta$

22 $\dfrac{1 + \sin\theta + \cos\theta}{\cos\theta} \equiv \dfrac{1 - \sin\theta + \cos\theta}{1 - \sin\theta}$

***23** $\sqrt{\dfrac{\tan\theta + \sin\theta}{\cot\theta - \cos\theta}} \equiv \tan^2\theta\sqrt{\dfrac{1 + \sin\theta}{1 - \cos\theta}}$

***24** $\dfrac{\tan^3\theta}{1 + \tan^2\theta} + \dfrac{\cot^3\theta}{1 + \cot^2\theta} \equiv \dfrac{1 - 2\sin^2\theta\cos^2\theta}{\sin\theta\cos\theta}$

***25** $\dfrac{\sin^3\theta - \cos^3\theta}{\sin\theta - \cos\theta} \equiv 1 + \sin\theta\cos\theta$

***26** $\dfrac{\cot^2\theta(\sec\theta - 1)}{1 + \sin\theta} \equiv \dfrac{\sec^2\theta(1 - \sin\theta)}{1 + \sec\theta}$

Exercise 15D: Examination questions

1 The diagram shows part of the graph of $y = \sin x$, where x is measured in radians, and values α on the x-axis and k on the y-axis such that $\sin\alpha = k$. Write down, in terms of α,

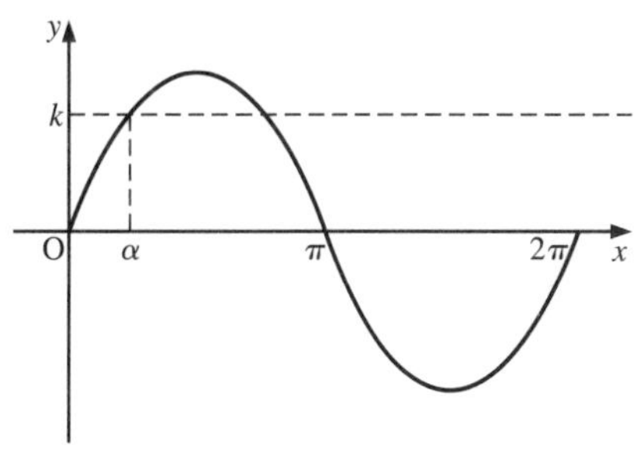

i) a value of x between $\frac{1}{2}\pi$ and π such that $\sin x = k$,

ii) two values of x between 3π and 4π such that $\sin x = -k$. (UCLES)

2 Solve the following equations for θ, giving your answers in degrees in the interval $0° \leqslant \theta \leqslant 360°$:

a) $\tan\theta = 0.4$ **b)** $\sin(2\theta) = 0.4$ (UODLE)

3 Given that $0 \leqslant x \leqslant \pi$, find the values of x for which

a) $\sin 3x = 0.5$ **b)** $\cot\left(x + \frac{\pi}{2}\right) = 1$ (EDEXCEL)

4 Given that $-90° < x < 90°$, find the values of x for which

a) $4\sin^2 x = 3$ **b)** $\sec(2x - 15°) = 2$ (EDEXCEL)

5 Find all solutions of the equation $2\cos^2 x + 3\sin x = 0$ in the interval $0° \leqslant x \leqslant 360°$. (AEB 95)

6 **a)** Find the values of $\cos x$ for which $6\sin^2 x = 5 + \cos x$.
b) Find all the values of x in the interval $180° < x < 540°$ for which

$$6\sin^2 x = 5 + \cos x$$ (EDEXCEL)

7 Determine, in radians, the solutions of the equation $3\cos^2 y + 8\sin y = 0$ for which $0 \leqslant y \leqslant 2\pi$, giving your answers to 2 decimal places. (EDEXCEL)

8 Find all the solutions in the interval $0° < x < 360°$ of the equation

$$3\tan x + 2\cos x = 0$$ (AEB 93)

9 Solve the equation $4\tan^2 x + 12\sec x + 1 = 0$, giving all solutions in degrees, to the nearest degree, in the interval $-180° < x < 180°$. (AEB Spec)

10 Show that $(x - 1)$ is a factor of $\text{f}(x) \equiv 2x^3 - 7x^2 + 2x + 3$ and find the three roots of the equation $\text{f}(x) = 0$.
Hence, or otherwise, solve the equation $2\sin^3\theta - 7\sin^2\theta + 2\sin\theta + 3 = 0$, giving all solutions in degrees in the interval $0° \leqslant \theta < 360°$. (UODLE)

11 **a)** Given that $\text{f}(x) = 2x^3 - 7x^2 + x + 1$, show that $(2x - 1)$ is a factor of $\text{f}(x)$ and find the remaining quadratic factor. Hence find the exact values of the three roots of the equation $\text{f}(x) = 0$.
b) Solve the equation $2\cos^3\theta - 7\cos^2\theta + \cos\theta + 1 = 0$ for $0° \leqslant \theta < 360°$, giving your answers to the nearest degree. (AEB 94)

12 Prove the identity $\operatorname{cosec} x - \sin x \equiv \cos x \cot x$. (UODLE)

13 Prove the identity $\dfrac{1}{1 + \cos\alpha} + \dfrac{1}{1 - \cos\alpha} \equiv 2\operatorname{cosec}^2\alpha$. (WJEC)

16 Trigonometry II

Mathematics, rightly viewed, possesses not only truth, but supreme beauty – a beauty cold and austere, like that of sculpture, without appeal to any part of our weaker nature, without the trappings of paintings or music, yet sublimely pure, and capable of stern perfection such as only the greatest art can show.
BERTRAND RUSSELL

Compound angles

Using substitution, it is clear to see that

$$\sin(A+B) \neq \sin A + \sin B$$
$$\cos(A+B) \neq \cos A + \cos B$$
$$\tan(A+B) \neq \tan A + \tan B$$

For example, let $A = B = 30^\circ$, then

$$\sin(30^\circ + 30^\circ) = \sin 60^\circ = \frac{\sqrt{3}}{2}$$

However,

$$\sin 30^\circ + \sin 30^\circ = \frac{1}{2} + \frac{1}{2} = 1 \neq \frac{\sqrt{3}}{2}$$

It is for this reason that it would be helpful if we had identities for each of $\sin(A+B)$, $\cos(A+B)$ and $\tan(A+B)$.

Result I

For any angles A and B,

$$\sin(A+B) \equiv \sin A \cos B + \cos A \sin B \quad \text{and} \quad \cos(A+B) \equiv \cos A \cos B - \sin A \sin B$$

Proof

We will prove the results for acute angles A and B. (To prove the results for any angles, we would need to use vectors and matrices.)
In the diagram, we see that

$$\begin{aligned}\sin(A+B) &= \frac{\text{PQ}}{\text{OP}}\\ &= \frac{\text{PT}+\text{TQ}}{\text{OP}}\\ &= \frac{\text{PT}+\text{RS}}{\text{OP}}\\ &= \frac{\text{PT}}{\text{OP}} + \frac{\text{RS}}{\text{OP}}\\ &= \frac{\text{PT}}{\text{PR}} \times \frac{\text{PR}}{\text{OP}} + \frac{\text{RS}}{\text{OR}} \times \frac{\text{OR}}{\text{OP}}\end{aligned}$$

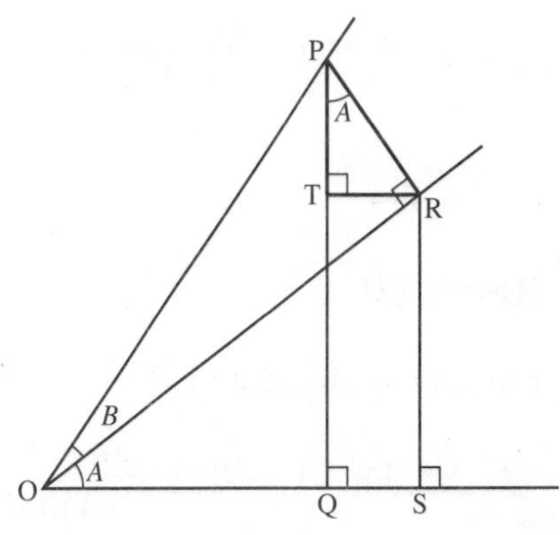

Therefore,

$$\sin(A+B) \equiv \cos A \sin B + \sin A \cos B$$

as required.

From the same diagram, we see that

$$\begin{aligned}\cos(A+B) &= \frac{\text{OQ}}{\text{OP}}\\ &= \frac{\text{OS}-\text{TR}}{\text{OP}}\\ &= \frac{\text{OS}}{\text{OP}}-\frac{\text{TR}}{\text{OP}}\\ &= \frac{\text{OS}}{\text{OR}}\times\frac{\text{OR}}{\text{OP}}-\frac{\text{TR}}{\text{PR}}\times\frac{\text{PR}}{\text{OP}}\end{aligned}$$

Therefore,

$$\cos(A+B) \equiv \cos A \cos B - \sin A \sin B$$

as required.

Result II

For any angles A and B,

$\sin(A-B) \equiv \sin A \cos B - \cos A \sin B$ and $\cos(A-B) \equiv \cos A \cos B + \sin A \sin B$

Proof

From Result I, we know that

$$\sin(A+B) \equiv \sin A \cos B + \cos A \sin B \quad \text{and} \quad \cos(A+B) \equiv \cos A \cos B - \sin A \sin B$$

Replacing B with $-B$ and noting that

- $\cos(-B) = \cos B$, since cosine is an even function
- $\sin(-B) = -\sin B$, since sine is an odd function

we have

$$\begin{aligned}\sin(A-B) \equiv \sin(A+(-B)) &\equiv \sin A \cos(-B) + \cos A \sin(-B)\\ &\equiv \sin A \cos B - \cos A \sin B\end{aligned}$$

as required.

We also have

$$\begin{aligned}\cos(A-B) \equiv \cos(A+(-B)) &\equiv \cos A \cos(-B) - \sin A \sin(-B)\\ &\equiv \cos A \cos B + \sin A \sin B\end{aligned}$$

as required.

Result III

For any angles A and B,

$$\tan(A \pm B) \equiv \frac{\tan A \pm \tan B}{1 \mp \tan A \tan B}$$

Proof

We know that $\tan\theta = \dfrac{\sin\theta}{\cos\theta}$, therefore,

$$\text{LHS} \equiv \tan(A \pm B) \equiv \frac{\sin(A \pm B)}{\cos(A \pm B)}$$
$$\equiv \frac{\sin A\cos B \pm \cos A\sin B}{\cos A\cos B \mp \sin A\sin B}$$

Because we want $1 \mp \tan A\tan B$ in the denominator, we divide the numerator and the denominator by $\cos A\cos B$, giving

$$\tan(A \pm B) \equiv \frac{\left(\dfrac{\sin A\cos B}{\cos A\cos B} \pm \dfrac{\cos A\sin B}{\cos A\cos B}\right)}{\left(\dfrac{\cos A\cos B}{\cos A\cos B} \mp \dfrac{\sin A\sin B}{\cos A\cos B}\right)}$$
$$\equiv \frac{\left(\dfrac{\sin A}{\cos A} \pm \dfrac{\sin B}{\cos B}\right)}{\left(1 \mp \dfrac{\sin A}{\cos A}\cdot\dfrac{\sin B}{\cos B}\right)}$$
$$\equiv \frac{\tan A \pm \tan B}{1 \mp \tan A\tan B}$$

as required.

Example 1 Given that A and B are acute angles and that $\sin A = \dfrac{1}{\sqrt{3}}$ and that $\cos B = \dfrac{3}{5}$, evaluate

a) $\sin(A+B)$ **b)** $\tan(A-B)$ **c)** $\operatorname{cosec}(A+B)$

SOLUTION

a) Now

$$\sin(A+B) \equiv \sin A\cos B + \cos A\sin B \qquad [1]$$

When $\sin A = \dfrac{1}{\sqrt{3}}$ (A acute), we have, from this right-angled triangle (shown right),

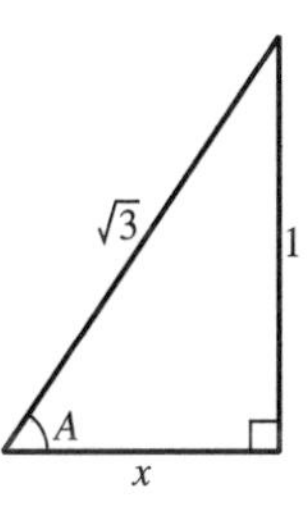

$$x^2 = (\sqrt{3})^2 - 1^2 \quad \therefore \quad x = \sqrt{2}$$

Therefore,

$$\cos A = \frac{\sqrt{2}}{\sqrt{3}} = \sqrt{\frac{2}{3}} \quad \text{and} \quad \tan A = \frac{1}{\sqrt{2}}$$

When $\cos B = \frac{3}{5}$ (B acute), we have, from this right-angled triangle (shown right), $y = 4$. Therefore,

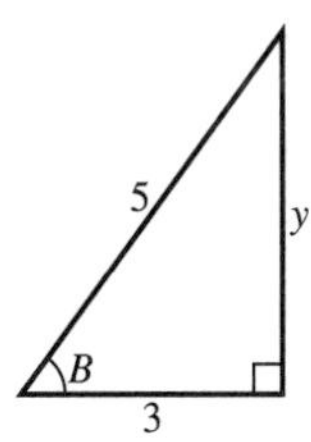

$$\sin B = \frac{4}{5} \quad \text{and} \quad \tan B = \frac{4}{3}$$

Substituting these values into identity [1] gives

$$\sin(A+B) = \frac{1}{\sqrt{3}} \times \frac{3}{5} + \frac{\sqrt{2}}{\sqrt{3}} \times \frac{4}{5} = \frac{3+4\sqrt{2}}{5\sqrt{3}}$$

b) Now

$$\tan(A-B) \equiv \frac{\tan A - \tan B}{1+\tan A \tan B} = \frac{\dfrac{1}{\sqrt{2}} - \dfrac{4}{3}}{1+\left(\dfrac{1}{\sqrt{2}}\right)\left(\dfrac{4}{3}\right)} = \frac{3-4\sqrt{2}}{3\sqrt{2}+4}$$

$$\therefore \quad \tan(A-B) = \frac{3-4\sqrt{2}}{4+3\sqrt{2}}$$

c) Now

$$\operatorname{cosec}(A+B) \equiv \frac{1}{\sin(A+B)} = \frac{1}{\left(\dfrac{3+4\sqrt{2}}{5\sqrt{3}}\right)}$$

$$\therefore \quad \operatorname{cosec}(A+B) = \frac{5\sqrt{3}}{3+4\sqrt{2}}$$

Example 2 Solve the equation $2\sin(30^\circ+\theta)+2\cos(60^\circ+\theta)=\sqrt{3}$, for $-180^\circ \leqslant \theta \leqslant 180^\circ$.

SOLUTION

Now, using the compound formula for $\sin(A+B)$, we have

$$\begin{aligned} 2\sin(30^\circ+\theta) &= 2(\sin 30^\circ \cos\theta + \cos 30^\circ \sin\theta) \\ &= 2\left(\frac{1}{2}\cos\theta + \frac{\sqrt{3}}{2}\sin\theta\right) \\ &= \cos\theta + \sqrt{3}\sin\theta \end{aligned}$$

And using the compound formula for $\cos(A+B)$, we also have

$$\begin{aligned} 2\cos(60^\circ+\theta) &= 2(\cos 60^\circ \cos\theta - \sin 60^\circ \sin\theta) \\ &= 2\left(\frac{1}{2}\cos\theta - \frac{\sqrt{3}}{2}\sin\theta\right) = \cos\theta - \sqrt{3}\sin\theta \end{aligned}$$

Therefore, the original equation becomes

$$\left(\cos\theta + \sqrt{3}\sin\theta\right) + \left(\cos\theta - \sqrt{3}\sin\theta\right) = \sqrt{3}$$

Simplifying and solving give

$$2\cos\theta = \sqrt{3} \quad \therefore \quad \cos\theta = \frac{\sqrt{3}}{2}$$

When $\cos\theta = \dfrac{\sqrt{3}}{2}$, $\theta = -30^\circ$ and 30°, in the required range.

The solutions are $\theta = \pm 30^\circ$.

Example 3 Prove the identity

$$\frac{\sin(A+B)}{\cos(A-B)} + 1 \equiv \frac{(1+\tan B)(1+\cot A)}{\cot A + \tan B}$$

SOLUTION

Expanding the LHS gives

$$\text{LHS} \equiv \frac{\sin A \cos B + \cos A \sin B}{\cos A \cos B + \sin A \sin B} + 1$$

Dividing both the numerator and the denominator of the fraction by $\sin A \cos B$ gives

$$\text{LHS} \equiv \frac{1 + \cot A \tan B}{\cot A + \tan B} + 1$$

$$\equiv \frac{1 + \cot A \tan B + \cot A + \tan B}{\cot A + \tan B}$$

$$\equiv \frac{(1+\tan B)(1+\cot A)}{\cot A + \tan B} \equiv \text{RHS} \qquad \text{QED}$$

Exercise 16A

Questions **1** to **9** should be answered without the use of a calculator.

1 Given that A and B are acute angles and that $\sin A = \frac{3}{5}$ and that $\cos B = \frac{5}{13}$, find the value of each of these.

a) $\sin(A+B)$ **b)** $\cos(A-B)$ **c)** $\tan(A-B)$

2 Given that C and D are acute angles and that $\cos C = \frac{12}{13}$ and that $\cos D = \frac{3}{5}$, find the value of each of these.

a) $\cos(C+D)$ **b)** $\cos(C-D)$ **c)** $\cot(C+D)$

3 Given that P and Q are acute angles and that $\tan P = \frac{7}{24}$ and that $\tan Q = 1$, find the value of each of these.

a) $\sin(P+Q)$ **b)** $\cos(P+Q)$ **c)** $\tan(P-Q)$

4 Given that $\tan(A-B) = \frac{1}{2}$ and that $\tan A = 3$, find the value of $\tan B$.

5 Given that $\tan(P+Q) = 5$ and that $\tan Q = 2$, find the value of $\tan P$.

6 Given that $\tan(\theta - 45°) = 4$, find the value of $\tan\theta$.

7 Given that $\tan(\theta + 60°) = 2$, find the value of $\cot\theta$.

8 Given that $\cot(30° - \theta) = 3$, find the value of $\cot\theta$.

9 In each part of the following question find the value of $\tan\theta$.

a) $\sin(\theta - 30^\circ) = \cos\theta$
b) $\sin(\theta + 45^\circ) = \cos\theta$
c) $\cos(\theta + 60^\circ) = \sin\theta$
d) $\sin(\theta + 60^\circ) = \cos(\theta - 60^\circ)$
e) $\cos(\theta + 60^\circ) = 2\cos(\theta + 30^\circ)$
f) $\sin(\theta + 60^\circ) = \cos(45^\circ - \theta)$

In questions **10** to **22**, prove each of the given identities.

10 $\sin(\theta + 90^\circ) \equiv \cos\theta$

11 $\cos(\theta + 90^\circ) \equiv -\sin\theta$

12 $\sin(180^\circ - \theta) \equiv \sin\theta$

13 $\cos(\theta - 180^\circ) \equiv -\cos\theta$

14 $\sin(A + B) + \sin(A - B) \equiv 2\sin A\cos B$

15 $\cos(A - B) - \cos(A + B) \equiv 2\sin A\sin B$

16 $\sin(A + B)\sin(A - B) \equiv \sin^2 A - \sin^2 B$

17 $\cos(A + B)\cos(A - B) \equiv -(\sin A + \cos B)(\sin A - \cos B)$

18 $\tan A - \tan B \equiv \dfrac{\sin(A - B)}{\cos A\cos B}$

19 $\cot A + \cot B \equiv \dfrac{\sin(A + B)}{\sin A\sin B}$

20 $\dfrac{\sin(A - B)}{\sin(A + B)} \equiv \dfrac{\tan A - \tan B}{\tan A + \tan B}$

21 $\dfrac{\cos(A - B)}{\cos(A + B)} \equiv \dfrac{\cot A\cot B + 1}{\cot A\cot B - 1}$

22 $\dfrac{\sin(A - B)}{\sin A\sin B} + \dfrac{\sin(B - C)}{\sin B\sin C} + \dfrac{\sin(C - A)}{\sin C\sin A} \equiv 0$

***23** Prove that if P, Q and R are the angles of a triangle, then $\dfrac{\tan P + \tan Q + \tan R}{\tan P\tan Q\tan R} \equiv 1$.

***24** Prove that $\tan 15^\circ = 2 - \sqrt{3}$

Double angles

Result IV

For any angle A,

$$\sin 2A \equiv 2\sin A\cos A$$

Proof

We know that

$$\sin(A + B) \equiv \sin A\cos B + \cos A\sin B$$

If we let $B = A$, then

$$\sin(A + A) \equiv \sin A\cos A + \cos A\sin A$$

$$\therefore \quad \sin 2A \equiv 2\sin A\cos A$$

as required.

Result V

For any angle A,

$$\begin{aligned}\cos 2A &\equiv \cos^2 A - \sin^2 A \\ &\equiv 2\cos^2 A - 1 \\ &\equiv 1 - 2\sin^2 A\end{aligned}$$

Proof

We know that

$$\cos(A+B) \equiv \cos A \cos B - \sin A \sin B$$

If we let $B = A$, then

$$\cos(A+A) \equiv \cos A \cos A - \sin A \sin A$$

$$\therefore \quad \cos 2A \equiv \cos^2 A - \sin^2 A \qquad [1]$$

as required.

We know that $\sin^2 A + \cos^2 A \equiv 1$. Rearranging this identity gives

$$\sin^2 A \equiv 1 - \cos^2 A \quad \text{and} \quad \cos^2 A \equiv 1 - \sin^2 A$$

Substituting $\sin^2 A \equiv 1 - \cos^2 A$ into [1] gives

$$\begin{aligned}\cos 2A &\equiv \cos^2 A - (1 - \cos^2 A) \\ &\equiv 2\cos^2 A - 1\end{aligned}$$

as required.

Substituting $\cos^2 A \equiv 1 - \sin^2 A$ into [1] gives

$$\begin{aligned}\cos 2A &\equiv (1 - \sin^2 A) - \sin^2 A \\ &\equiv 1 - 2\sin^2 A\end{aligned}$$

as required.

Result VI

For any angle A,

$$\tan 2A \equiv \frac{2\tan A}{1 - \tan^2 A}$$

Proof

We know that

$$\tan(A+B) \equiv \frac{\tan A + \tan B}{1 - \tan A \tan B}$$

If we let $B = A$, then

$$\begin{aligned}\tan(A+A) &\equiv \frac{\tan A + \tan A}{1 - \tan A \tan A} \\ &\equiv \frac{2\tan A}{1 - \tan^2 A}\end{aligned}$$

as required.

Example 4 Given that θ is acute and that $\tan\theta = \frac{1}{2}$, evaluate each of the following.

a) $\tan 2\theta$ **b)** $\sin 2\theta$ **c)** $\sec 2\theta$

SOLUTION

When $\tan\theta = \frac{1}{2}$, we have, from this right-angled triangle (shown right),

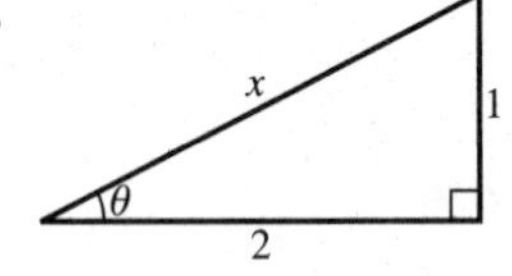

$$x^2 = 2^2 + 1^2 \quad \therefore \quad x = \sqrt{5}$$

Therefore, $\sin\theta = \dfrac{1}{\sqrt{5}}$ and $\cos\theta = \dfrac{2}{\sqrt{5}}$.

a) We know that

$$\tan 2\theta = \frac{2\tan\theta}{1-\tan^2\theta} = \frac{2\left(\frac{1}{2}\right)}{1-\left(\frac{1}{2}\right)^2}$$

$$\therefore \quad \tan 2\theta = \frac{4}{3}$$

b) We know that

$$\sin 2\theta = 2\sin\theta\cos\theta = 2\left(\frac{1}{\sqrt{5}}\right)\left(\frac{2}{\sqrt{5}}\right)$$

$$\therefore \quad \sin 2\theta = \frac{4}{5}$$

c) Now

$$\sec 2\theta = \frac{1}{\cos 2\theta} = \frac{1}{\cos^2\theta - \sin^2\theta} = \frac{1}{\left(\frac{2}{\sqrt{5}}\right)^2 - \left(\frac{1}{\sqrt{5}}\right)^2}$$

$$\therefore \quad \sec 2\theta = \frac{5}{3}$$

Example 5 Solve the equation $4\cos 2\theta - 2\cos\theta + 3 = 0$, for $0° \leqslant \theta \leqslant 360°$.

SOLUTION

From Result V, we know that $\cos 2\theta = 2\cos^2\theta - 1$. Substituting this into the equation will give a quadratic in $\cos\theta$, as follows:

$$4(2\cos^2\theta - 1) - 2\cos\theta + 3 = 0$$

$$\therefore \quad 8\cos^2\theta - 2\cos\theta - 1 = 0$$

$$\therefore \quad (4\cos\theta + 1)(2\cos\theta - 1) = 0$$

Solving gives $\cos\theta = -\frac{1}{4}$ or $\cos\theta = \frac{1}{2}$.

When $\cos\theta = -\frac{1}{4}$, one solution is $\theta = 104.5°$. In the range $0° \leqslant \theta \leqslant 360°$, the other solution is $\theta = 360° - 104.5° = 255.5°$.

When $\cos\theta = \frac{1}{2}$, one solution is $\theta = 60°$. In the range $0° \leqslant \theta \leqslant 360°$, the other solution is $\theta = 360° - 60° = 300°$.

The solutions are $\theta = 60°$, $104.5°$, $255.5°$ and $300°$.

Example 6 Prove that $\dfrac{\sin A + \sin 2A}{1 + \cos A + \cos 2A} \equiv \tan A$.

SOLUTION

Consider the left-hand side:

$$\text{LHS} \equiv \frac{\sin A + 2\sin A\cos A}{1 + \cos A + (2\cos^2 A - 1)} \equiv \frac{\sin A + 2\sin A\cos A}{\cos A + 2\cos^2 A}$$

$$\equiv \frac{\sin A(1 + 2\cos A)}{\cos A(1 + 2\cos A)}$$

$$\equiv \frac{\sin A}{\cos A}$$

$$\equiv \tan A \equiv \text{RHS} \qquad \text{QED}$$

Half-angle and other formulae

In the double-angle identities on pages 370–1, let $A = \dfrac{\theta}{2}$. We then get the following:

Result VII

For any angle θ,

i) $\sin\theta \equiv 2\sin\left(\dfrac{\theta}{2}\right)\cos\left(\dfrac{\theta}{2}\right)$

ii) $\cos\theta \equiv \cos^2\left(\dfrac{\theta}{2}\right) - \sin^2\left(\dfrac{\theta}{2}\right)$

$$\equiv 2\cos^2\left(\frac{\theta}{2}\right) - 1$$

$$\equiv 1 - 2\sin^2\left(\frac{\theta}{2}\right)$$

iii) $\tan\theta \equiv \dfrac{2\tan\left(\dfrac{\theta}{2}\right)}{1 - \tan^2\left(\dfrac{\theta}{2}\right)}$

Proof of (i)

We know that $\sin 2A = 2\sin A\cos A$. Now letting $A = \dfrac{\theta}{2}$ gives

$$\sin\left(2\left(\frac{\theta}{2}\right)\right) \equiv 2\sin\left(\frac{\theta}{2}\right)\cos\left(\frac{\theta}{2}\right)$$

$$\therefore \quad \sin\theta \equiv 2\sin\left(\frac{\theta}{2}\right)\cos\left(\frac{\theta}{2}\right)$$

as required.

Proof of (ii)

We know that

$$\cos 2A \equiv \cos^2 A - \sin^2 A$$
$$\equiv 2\cos^2 A - 1$$
$$\equiv 1 - 2\sin^2 A$$

Letting $A = \frac{\theta}{2}$ gives

$$\cos\theta \equiv \cos^2\left(\frac{\theta}{2}\right) - \sin^2\left(\frac{\theta}{2}\right)$$
$$\equiv 2\cos^2\left(\frac{\theta}{2}\right) - 1$$
$$\equiv 1 - 2\sin^2\left(\frac{\theta}{2}\right)$$

as required.

Proof of (iii)

We know that $\tan 2A \equiv \frac{2\tan A}{1-\tan^2 A}$.

Letting $A = \frac{\theta}{2}$ gives

$$\tan\theta \equiv \frac{2\tan\left(\frac{\theta}{2}\right)}{1-\tan^2\left(\frac{\theta}{2}\right)}$$

as required.

Example 7 Solve the equation $\sin\theta + \sin\frac{\theta}{2} = 0$, for $-360° \leqslant \theta \leqslant 360°$.

SOLUTION

From Result VII (i), we know that $\sin\theta \equiv 2\sin\frac{\theta}{2}\cos\frac{\theta}{2}$. Substituting this into the equation, we have

$$2\sin\frac{\theta}{2}\cos\frac{\theta}{2} + \sin\frac{\theta}{2} = 0$$
$$\therefore \quad \sin\frac{\theta}{2}(2\cos\frac{\theta}{2} + 1) = 0$$

Solving gives $\sin\frac{\theta}{2} = 0$ or $\cos\frac{\theta}{2} = -\frac{1}{2}$

When $\sin\frac{\theta}{2} = 0$, one solution is $\theta = 0°$. In the range $-360° \leqslant \theta \leqslant 360°$, the other solutions are $\frac{\theta}{2} = \pm 180°$. That is, $\theta = \pm 360°$.

When $\cos\frac{\theta}{2} = -\frac{1}{2}$, the solutions in the range $-360° \leqslant \theta \leqslant 360°$ are $\frac{\theta}{2} = \pm 120°$. That is, $\theta = \pm 240°$.

The solutions are $\theta = -360°, -240°, 0°, 240°$ and $360°$.

Example 8 Express $\sin 3\theta$ in terms of $\sin\theta$. Hence solve the equation

$$\sin^2\theta + \sin\theta = 1 - \sin 3\theta$$

for $0° \leqslant \theta \leqslant 360°$.

SOLUTION

We know that

$$\sin(A + B) \equiv \sin A \cos B + \cos A \sin B$$

If we let $A = \theta$ and $B = 2\theta$, then

$$\sin 3\theta \equiv \sin\theta \cos 2\theta + \cos\theta \sin 2\theta \qquad [1]$$

From Result I, we know that $\sin 2\theta \equiv 2\sin\theta\cos\theta$; and from Result II, we know that $\cos 2\theta \equiv 1 - 2\sin^2\theta$. Substituting each of these into [1] gives

$$\sin 3\theta \equiv \sin\theta(1 - 2\sin^2\theta) + \cos\theta(2\sin\theta\cos\theta)$$

$$\equiv \sin\theta - 2\sin^3\theta + 2\sin\theta\cos^2\theta$$

We know that $\cos^2\theta \equiv 1 - \sin^2\theta$. Therefore,

$$\sin 3\theta \equiv \sin\theta - 2\sin^3\theta + 2\sin\theta(1 - \sin^2\theta)$$

$$\equiv \sin\theta - 2\sin^3\theta + 2\sin\theta - 2\sin^3\theta$$

$$\therefore \quad \sin 3\theta \equiv 3\sin\theta - 4\sin^3\theta \qquad [2]$$

To solve the equation

$$\sin^2\theta + \sin\theta = 1 - \sin 3\theta$$

we substitute for $\sin 3\theta$ using [2]. This gives

$$\sin^2\theta + \sin\theta = 1 - (3\sin\theta - 4\sin^3\theta)$$

$$\sin^2\theta + \sin\theta = 1 - 3\sin\theta + 4\sin^3\theta$$

$$\therefore \quad 4\sin^3\theta - \sin^2\theta - 4\sin\theta + 1 = 0$$

This is a cubic equation in $\sin\theta$, which can be seen more clearly by letting $\sin\theta = x$. This gives the equation

$$4x^3 - x^2 - 4x + 1 = 0$$

Using the factor theorem, we see that $x = 1$ and $x = -1$ are two roots. That is, $(x - 1)$ and $(x + 1)$ are both factors of the cubic expression. So,

$$4x^3 - x^2 - 4x + 1 = (4x - 1)(x + 1)(x - 1)$$

which means that

$$4\sin^3\theta - \sin^2\theta - 4\sin\theta + 1 = 0$$

becomes

$$(4\sin\theta - 1)(\sin\theta + 1)(\sin\theta - 1) = 0$$

Solving gives $\sin\theta = \frac{1}{4}$, $\sin\theta = -1$ and $\sin\theta = 1$.

When $\sin\theta = \frac{1}{4}$, one solution is $\theta = 14.5°$. In the range $0° \leqslant \theta \leqslant 360°$, the other solution is $\theta = 180° - 14.5° = 165.5°$.

When $\sin\theta = -1$, $\theta = 270°$ in the required range. When $\sin\theta = 1$, $\theta = 90°$ in the required range.

The solutions are $\theta = 14.5°$, $90°$, $165.5°$ and $270°$.

Note The identity

$$\sin 3\theta \equiv 3\sin\theta - 4\sin^3\theta$$

is known as a **triple-angle formula**.

The other two triple-angle formulae, which are proved similarly, are

$$\cos 3\theta \equiv 4\cos^3\theta - 3\cos\theta$$

and $$\tan 3\theta \equiv \frac{3\tan\theta - \tan^3\theta}{1 - 3\tan^2\theta}$$

Exercise 16B

1 Given that θ is an acute angle and that $\sin\theta = \frac{4}{5}$, find the value of each of these.

a) $\sin 2\theta$ **b)** $\cos 2\theta$ **c)** $\tan 2\theta$

2 Given that θ is an acute angle and that $\cos\theta = \frac{5}{13}$, find the value of each of these.

a) $\cos 2\theta$ **b)** $\operatorname{cosec} 2\theta$ **c)** $\cot 2\theta$

3 Given that θ is an acute angle and that $\tan\theta = 2$, find the value of each of these.

a) $\tan 2\theta$ **b)** $\sin 2\theta$ **c)** $\sec 2\theta$

4 Solve each of the following equations for $0° \leqslant \theta \leqslant 360°$, giving your answers correct to one decimal place.

a) $3\sin 2\theta = \sin\theta$
b) $4\cos\theta = 3\sin 2\theta$
c) $\sin 2\theta + \cos\theta = 0$
d) $3\cos 2\theta - \cos\theta + 2 = 0$
e) $6\cos 2\theta - 7\sin\theta + 6 = 0$
f) $2\cos 2\theta = 1 - 3\sin\theta$
g) $\tan 2\theta + \tan\theta = 0$
h) $\sin 2\theta + \sin\theta - \tan\theta = 0$

5 Solve each of the following equations for $-360° \leqslant \theta \leqslant 360°$, giving your answers correct to one decimal place.

a) $\sin\theta = \sin\frac{\theta}{2}$
b) $3\cos\frac{\theta}{2} = 2\sin\theta$
c) $2\sin\theta = \tan\frac{\theta}{2}$
d) $2\cos\theta = 15\cos\frac{\theta}{2} + 2$
e) $3\tan\theta = 8\tan\frac{\theta}{2}$
f) $\sin\theta\operatorname{cosec}\frac{\theta}{2} - \cot\frac{\theta}{2} = 3\left(1 - 2\sin\frac{\theta}{2}\right)$

In Questions **6** to **21**, prove each of the given identities.

6 $2\cos^2\theta - \cos 2\theta \equiv 1$

7 $2\operatorname{cosec} 2\theta \equiv \operatorname{cosec}\theta\sec\theta$

8 $2\cos^3\theta + \sin 2\theta\sin\theta \equiv 2\cos\theta$

9 $\tan\theta + \cot\theta \equiv 2\operatorname{cosec} 2\theta$

10 $\cos^4\theta - \sin^4\theta \equiv \cos 2\theta$

11 $\dfrac{1-\cos 2\theta}{1+\cos 2\theta} \equiv \tan^2\theta$

12 $\cot\theta - \tan\theta \equiv 2\cot 2\theta$

13 $\cot 2\theta + \operatorname{cosec} 2\theta \equiv \cot\theta$

14 $\dfrac{\cos 2\theta}{\cos\theta + \sin\theta} \equiv \cos\theta - \sin\theta$

15 $\dfrac{\sin 2\theta}{1-\cos 2\theta} \equiv \cot\theta$

16 $\cos 2\theta \equiv \dfrac{1-\tan^2\theta}{1+\tan^2\theta}$

17 $\sin 2\theta \equiv \dfrac{2\tan\theta}{1+\tan^2\theta}$

18 $\dfrac{1}{\cos\theta + \sin\theta} + \dfrac{1}{\cos\theta - \sin\theta} \equiv \dfrac{2\cos\theta}{\cos 2\theta}$

19 $\dfrac{\sin\theta + \sin 2\theta}{1+\cos\theta + \cos 2\theta} \equiv \tan\theta$

20 $\dfrac{\sin A}{\sin B} - \dfrac{\cos A}{\cos B} \equiv \dfrac{2\sin(A-B)}{\sin 2B}$

21 $\dfrac{\cos A}{\sin B} - \dfrac{\sin A}{\cos B} \equiv \dfrac{2\cos(A+B)}{\sin 2B}$

22 **a)** Show that $\sin 3\theta \equiv 3\sin\theta - 4\sin^3\theta$
b) Hence solve the equation $1 - \sin 3\theta = 2\sin\theta(2\sin\theta - 1)$, for $0 \leqslant \theta \leqslant 360°$

23 **a)** Show that $\cos 3\theta \equiv 4\cos^3\theta - 3\cos\theta$
b) Hence solve the equation $1 + \cos 3\theta = \cos\theta(1+\cos\theta)$, for $0 \leqslant \theta \leqslant 360°$

***24** Prove that $\tan 3\theta \equiv \dfrac{3\tan\theta - \tan^3\theta}{1 - 3\tan^2\theta}$

***25** Prove that $\cos^5\theta \equiv \dfrac{\cos 5\theta + 5\cos 3\theta + 10\cos\theta}{16}$

***26** Find an expression for $\tan 22\frac{1}{2}°$ in the form $a + b\sqrt{2}$, where a and b are integers.

Factor formulae

Result VIII

For any angles P and Q,

i) $\sin P + \sin Q \equiv 2\sin\left(\dfrac{P+Q}{2}\right)\cos\left(\dfrac{P-Q}{2}\right)$

ii) $\sin P - \sin Q \equiv 2\cos\left(\dfrac{P+Q}{2}\right)\sin\left(\dfrac{P-Q}{2}\right)$

iii) $\cos P + \cos Q \equiv 2\cos\left(\dfrac{P+Q}{2}\right)\cos\left(\dfrac{P-Q}{2}\right)$

iv) $\cos P - \cos Q \equiv -2\sin\left(\dfrac{P+Q}{2}\right)\sin\left(\dfrac{P-Q}{2}\right)$

Proof of (i) and (ii)

We know that

$$\sin(A+B) \equiv \sin A \cos B + \cos A \sin B \qquad [1]$$

and

$$\sin(A-B) \equiv \sin A \cos B - \cos A \sin B \qquad [2]$$

Adding [1] and [2] gives

$$\sin(A+B) + \sin(A-B) \equiv 2\sin A \cos B \qquad [3]$$

Letting $A+B=P$ and $A-B=Q$, we have

$$A = \frac{P+Q}{2} \quad \text{and} \quad B = \frac{P-Q}{2}$$

Substituting into [3] gives

$$\sin P + \sin Q \equiv 2\sin\left(\frac{P+Q}{2}\right)\cos\left(\frac{P-Q}{2}\right)$$

as required for Result VIII (i).

In order to prove Result VIII (ii), we subtract [2] from [1], which gives

$$\sin(A+B) - \sin(A-B) \equiv 2\cos A \sin B$$

Substituting for A and B gives

$$\sin P - \sin Q \equiv 2\cos\left(\frac{P+Q}{2}\right)\sin\left(\frac{P-Q}{2}\right)$$

as required for Result VIII (ii).

Proof of (iii) and (iv)

Results VIII (iii) and (iv) are proved similarly, but starting with the identities for $\cos(A+B)$ and $\cos(A-B)$.

Example 9 Express each of the following as the product of two trigonometric functions.

a) $\sin 6\theta + \sin 4\theta$ **b)** $\cos 8\theta + \cos 4\theta$

SOLUTION

a) Using Result VIII (i), we have

$$\sin 6\theta + \sin 4\theta \equiv 2\sin\left(\frac{6\theta+4\theta}{2}\right)\cos\left(\frac{6\theta-4\theta}{2}\right)$$

$$\therefore \quad \sin 6\theta + \sin 4\theta \equiv 2\sin 5\theta \cos \theta$$

b) Using Result VIII (iii), we have

$$\cos 8\theta + \cos 4\theta \equiv 2\cos\left(\frac{8\theta+4\theta}{2}\right)\cos\left(\frac{8\theta-4\theta}{2}\right)$$

$$\therefore \quad \cos 8\theta + \cos 4\theta \equiv 2\cos 6\theta \cos 2\theta$$

Example 10 Find the exact value of

a) $\sin 105° - \sin 15°$ **b)** $\sin 105° \sin 15°$

SOLUTION

a) Using Result VIII (ii), we have

$$\sin 105° - \sin 15° = 2\cos\left(\frac{105° + 15°}{2}\right)\sin\left(\frac{105° - 15°}{2}\right)$$
$$= 2\cos 60° \sin 45°$$
$$= 2 \times \frac{1}{2} \times \frac{1}{\sqrt{2}}$$
$$\therefore \quad \sin 105° - \sin 15° = \frac{1}{\sqrt{2}}$$

b) Comparing $\sin 105° \sin 15°$ with

$$\cos A - \cos B \equiv -2\sin\left(\frac{A+B}{2}\right)\sin\left(\frac{A-B}{2}\right) \quad \text{(Result VIII (iv))}$$

we see that

$$\frac{A+B}{2} = 105° \quad \text{and} \quad \frac{A-B}{2} = 15°$$
$$\therefore \quad A + B = 210° \quad [1] \quad \text{and} \quad A - B = 30° \quad [2]$$

Solving [1] and [2] gives $A = 120°$ and $B = 90°$. Therefore,

$$-2\sin 105° \sin 15° = \cos 120° - \cos 90° = -\tfrac{1}{2} - 0$$
$$\therefore \quad \sin 105° \sin 15° = \tfrac{1}{4}$$

Example 11 Solve the equation $\cos 4\theta + \cos 2\theta = 0$, for $0° \leqslant \theta \leqslant 180°$.

SOLUTION

Using Result VIII (iii), we have

$$\cos 4\theta + \cos 2\theta \equiv 2\cos\left(\frac{4\theta + 2\theta}{2}\right)\cos\left(\frac{4\theta - 2\theta}{2}\right)$$
$$\equiv 2\cos 3\theta \cos\theta$$

The equation $\cos 4\theta + \cos 2\theta = 0$ then becomes

$$2\cos 3\theta \cos\theta = 0$$

Solving gives $\cos 3\theta = 0$ or $\cos\theta = 0$.

When $\cos\theta = 0$, $\theta = 90°$ in the required range.

Changing the range to 3θ gives $0° \leqslant 3\theta \leqslant 540°$. When $\cos 3\theta = 0$, $3\theta = 90°$, $270°$ and $450°$. That is, $\theta = 30°$, $90°$ and $150°$.

The solutions are $\theta = 30°$, $90°$ and $150°$.

Exercise 16C

1 Express each of the following as the product of two trigonometric functions.

a) $\sin 4\theta + \sin 2\theta$ **b)** $\cos 5\theta + \cos 3\theta$ **c)** $\cos 6\theta - \cos 2\theta$ **d)** $\sin 5\theta - \sin 3\theta$
e) $\cos 8\theta - \cos 4\theta$ **f)** $\cos 4\theta - \cos 3\theta$ **g)** $\cos 3\theta + \cos 7\theta$ **h)** $\sin 2\theta - \sin 8\theta$

2 Express each of the following as the sum or difference of two trigonometric functions.

a) $2\sin 4\theta \cos 2\theta$ **b)** $2\cos 3\theta \cos 2\theta$
c) $2\cos 4\theta \cos \theta$ **d)** $2\cos 6\theta \sin 3\theta$
e) $-2\sin 4\theta \sin \theta$ **f)** $2\cos 7\theta \cos 6\theta$
g) $2\cos\left(\frac{9\theta}{2}\right)\cos\left(\frac{7\theta}{2}\right)$ **h)** $-2\sin\left(\frac{5\theta}{2}\right)\sin\left(\frac{3\theta}{2}\right)$

3 Evaluate each of the following without using a calculator.

a) $\sin 75° - \sin 15°$ **b)** $\cos 105° - \cos 15°$ **c)** $\cos 105° - \cos 165°$ **d)** $\sin 255° - \sin 15°$
e) $\cos 75° \cos 15°$ **f)** $\sin 105° \sin 75°$ **g)** $\sin 82\frac{1}{2}° \cos 37\frac{1}{2}°$ **h)** $\cos 37\frac{1}{2}° \sin 7\frac{1}{2}°$

4 Solve each of the following equations for $0° \leqslant \theta \leqslant 180°$, giving your answers correct to one decimal place.

a) $\sin 7\theta + \sin 2\theta = 0$ **b)** $\cos 4\theta - \cos \theta = 0$ **c)** $\sin 6\theta + \sin 3\theta = 0$
d) $\cos 4\theta + \cos 2\theta = 0$ **e)** $\sin 7\theta = \sin 3\theta$ **f)** $\sin \theta + \sin 4\theta = 0$
g) $\cos(\theta + 10°) + \cos(\theta + 30°) = 0$

5 Solve each of the following equations for $0° \leqslant \theta \leqslant 180°$, giving your answers correct to one decimal place.

a) $\sin 4\theta + \sin 2\theta = \cos \theta$ **b)** $\cos 5\theta + \cos 3\theta = \cos 4\theta$
c) $\cos 5\theta - \cos \theta = \sin 2\theta$ **d)** $\sin \theta - \sin 4\theta + \sin 7\theta = 0$
e) $\sin 2\theta - \sin 5\theta + \sin 8\theta = 0$ **f)** $\sin 2\theta + \sin 10\theta + \cos 4\theta = 0$
g) $\sin 3\theta - \sin 2\theta = \sin 6\theta + \sin \theta$ **h)** $\cos 5\theta + \cos 4\theta = \sin 5\theta + \sin 4\theta$

In Questions **6** to **17**, prove each of the given identities.

6 $\sin 3\theta + \sin \theta \equiv 4\sin \theta \cos^2\theta$

7 $\cos 3\theta + \cos \theta \equiv 4\cos^3\theta - 2\cos \theta$

8 $\sin 7\theta - \sin 3\theta \equiv 4\sin \theta \cos \theta \cos 5\theta$

9 $\sin \theta - 2\sin 3\theta + \sin 5\theta \equiv 2\sin \theta(\cos 4\theta - \cos 2\theta)$

10 $\cos \theta + \cos 3\theta + \cos 5\theta + \cos 7\theta \equiv 4\cos \theta \cos 2\theta \cos 4\theta$

11 $\dfrac{\sin 5\theta + \sin \theta}{\sin 4\theta + \sin 2\theta} \equiv 2\cos \theta - \sec \theta$

12 $\dfrac{\sin \theta + \sin 2\theta + \sin 3\theta}{\cos \theta + \cos 2\theta + \cos 3\theta} \equiv \tan 2\theta$

13 $\sin(A+B) - \sin(A-B) \equiv 2\cos A \sin B$

14 $\sin A + \sin(A+B) + \sin(A+2B) \equiv \sin(A+B)[1+2\cos B]$

15 $\cos A - \cos(A+B) + \cos(A+3B) - \cos(A+4B) \equiv 2\sin B \sin(A+2B)[2\cos B - 1]$

16 $\dfrac{\sin A + \sin B}{\cos A + \cos B} \equiv \tan\left(\dfrac{A+B}{2}\right)$

17 $\dfrac{\sin A - \sin B}{\sin A + \sin B} \equiv \tan\left(\dfrac{A-B}{2}\right)\cot\left(\dfrac{A+B}{2}\right)$

***18** **a)** Show that $\cos 75^\circ = \dfrac{\sqrt{3}-1}{2\sqrt{2}}$.

b) Deduce that $\cos 37\tfrac{1}{2}^\circ = \sqrt{\dfrac{\sqrt{3}-1+2\sqrt{2}}{4\sqrt{2}}}$

Harmonic form

Expressions of the form $a\cos\theta + b\sin\theta$ arise in many practical situations. We will look at how to express the function

$$f(\theta) = a\cos\theta + b\sin\theta$$

in the form

$$R\cos(\theta \pm \alpha) \quad \text{or} \quad R\sin(\theta \pm \alpha)$$

where $R > 0$ is a constant and α is acute.

This alternative form will enable us to solve equations of the form

$$a\cos\theta + b\sin\theta = c$$

and to find the maximum and minimum values of such functions.

Example 12

a) Express $3\cos\theta - 4\sin\theta$ in the form $R\cos(\theta + \alpha)$.
b) Solve the equation $3\cos\theta - 4\sin\theta = 1$, for $0^\circ \leqslant \theta \leqslant 360^\circ$.

SOLUTION

a) Let

$$\begin{aligned} 3\cos\theta - 4\sin\theta &\equiv R\cos(\theta+\alpha) \\ &\equiv R(\cos\theta\cos\alpha - \sin\theta\sin\alpha) \\ &\equiv R\cos\theta\cos\alpha - R\sin\theta\sin\alpha \end{aligned}$$

and equate the corresponding coefficients of $\cos\theta$ and $\sin\theta$ to obtain

$$3 = R\cos\alpha \quad [1] \qquad \text{and} \qquad 4 = R\sin\alpha \quad [2]$$

Squaring each of [1] and [2] and adding give

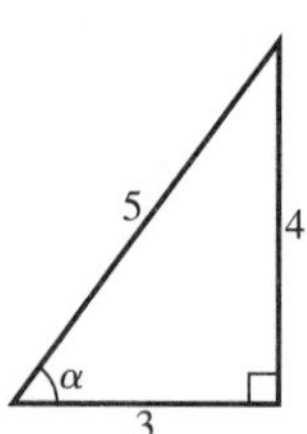

$$R^2\cos^2\alpha + R^2\sin^2\alpha = 3^2 + 4^2$$
$$R^2(\cos^2\alpha + \sin^2\alpha) = 25$$
$$\therefore \quad R^2 = 25 \quad (\text{since } \cos^2\alpha + \sin^2\alpha = 1)$$
$$\therefore \quad R = 5 \quad (\text{since } R > 0)$$

Dividing [2] by [1] gives

$$\frac{R\sin\alpha}{R\cos\alpha} = \frac{4}{3}$$
$$\therefore \quad \tan\alpha = \frac{4}{3}$$
$$\therefore \quad \alpha = 53.1°$$

Therefore, we have

$$3\cos\theta - 4\sin\theta = 5\cos(\theta + 53.1°)$$

b) The equation

$$3\cos\theta - 4\sin\theta = 1$$

becomes

$$5\cos(\theta + 53.1°) = 1$$
$$\therefore \quad \cos(\theta + 53.1°) = \frac{1}{5}$$

Changing the range for $\theta + 53.1°$ gives $53.1° \leqslant \theta + 53.1° \leqslant 413.1°$.

When $\cos(\theta + 53.1°) = \frac{1}{5}$, one solution in the required range is

$$\theta + 53.1° = 78.5° \quad \therefore \quad \theta = 25.4°$$

The other solution is

$$\theta + 53.1° = 360° - 78.5° \quad \therefore \quad \theta = 228.4°$$

The solutions are $\theta = 25.4°$ and $228.4°$.

Example 13 Find the maximum and minimum values of each of the following functions.

a) $f(\theta) = 5\cos\theta + 3\sin\theta$ **b)** $f(\theta) = \dfrac{1}{3 + \sin\theta - 2\cos\theta}$

SOLUTION

a) We start by expressing $f(\theta)$ in the form $R\cos(\theta - \alpha)$:

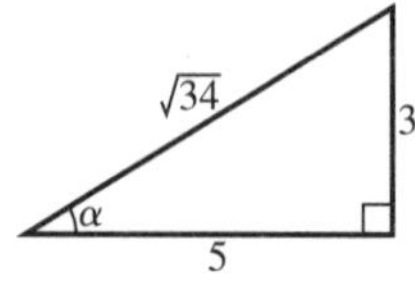

$$5\cos\theta + 3\sin\theta \equiv R\cos(\theta - \alpha)$$
$$\equiv R(\cos\theta\cos\alpha + \sin\theta\sin\alpha)$$
$$\equiv R\cos\theta\cos\alpha + R\sin\theta\sin\alpha$$

Equating the corresponding coefficients gives

$$R\cos\alpha = 5 \quad [1] \qquad \text{and} \qquad R\sin\alpha = 3 \quad [2]$$

Squaring each of [1] and [2] and adding give

$$R^2\cos^2\alpha + R^2\sin^2\alpha = 5^2 + 3^2$$

$$R^2(\cos^2\alpha + \sin^2\alpha) = 34$$

$$\therefore \quad R^2 = 34 \quad \text{giving} \quad R = \sqrt{34} \quad \text{(choosing the positive root)}$$

Dividing [2] by [1] gives

$$\frac{R\sin\alpha}{R\cos\alpha} = \frac{3}{5}$$

$$\therefore \quad \tan\alpha = \frac{3}{5} \quad \text{giving} \quad \alpha = 31.0^\circ$$

Therefore, we have

$$f(\theta) = \sqrt{34}\cos(\theta - 31.0^\circ)$$

The function f attains its maximum value when the cosine function attains its maximum value. Now $\cos(\theta - 31.0^\circ)$ has a maximum value of 1 when $\theta - 31^\circ = 0$. Therefore, the maximum value of f is $\sqrt{34}$ when $\theta = 31.0^\circ$. Similarly, the minimum value of f is $-\sqrt{34}$ when $\theta = 211.0^\circ$.

b) We start by expressing the denominator of $f(\theta)$ in the form $3 + R\sin(\theta - \alpha)$:

$$\sin\theta - 2\cos\theta \equiv R\sin(\theta - \alpha)$$

$$\equiv R(\sin\theta\cos\alpha - \cos\theta\sin\alpha)$$

$$\equiv R\sin\theta\cos\alpha - R\cos\theta\sin\alpha$$

Equating the corresponding coefficients gives

$$R\cos\alpha = 1 \quad [1] \qquad \text{and} \qquad R\sin\alpha = 2 \quad [2]$$

Squaring each of [1] and [2] and adding give

$$R^2\cos^2\alpha + R^2\sin^2\alpha = 1^2 + 2^2$$

$$\therefore \quad R^2(\cos^2\alpha + \sin^2\alpha) = 5$$

$$\therefore \quad R = \sqrt{5} \quad \text{(choosing the positive root)}$$

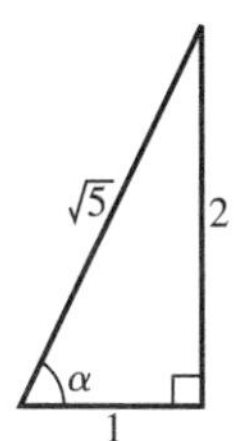

Dividing [2] by [1] gives

$$\frac{R\sin\alpha}{R\cos\alpha} = \frac{2}{1}$$

$$\therefore \quad \tan\alpha = 2 \quad \text{giving} \quad \alpha = 63.4^\circ$$

Therefore, we have

$$f(\theta) = \frac{1}{3 + \sqrt{5}\sin(\theta - 63.4^\circ)}$$

The graph of $f(\theta)$ attains a minimum when $\sin(\theta - 63.4^\circ)$ is a maximum. Now $\sin(\theta - 63.4^\circ)$ has a maximum value of 1 which occurs when

$$\theta - 63.4^\circ = 90^\circ \quad \therefore \quad \theta = 153.4^\circ$$

A minimum turning point on the graph of $f(\theta)$ is $\left(153.4^\circ, \dfrac{1}{3 + \sqrt{5}}\right)$.

Similarly, the graph of f attains a maximum value when $\sin(\theta - 63.4°)$ is a minimum. Now $\sin(\theta - 63.4°)$ has a minimum value of -1 which occurs when

$$\theta - 63.4° = 270° \quad \therefore \quad \theta = 333.4°$$

A maximum turning point on the graph of $f(\theta)$ is $\left(333.4°, \frac{1}{3-\sqrt{5}}\right)$.

Exercise 16D

1 Find the value of R and $\tan\alpha$ in each of the following identities.

a) $3\sin\theta + 4\cos\theta \equiv R\sin(\theta + \alpha)$
b) $5\sin\theta - 12\cos\theta \equiv R\sin(\theta - \alpha)$
c) $2\cos\theta + 5\sin\theta \equiv R\cos(\theta - \alpha)$
d) $2\cos\theta + 5\sin\theta \equiv R\sin(\theta + \alpha)$
e) $\cos\theta - \sin\theta \equiv R\cos(\theta + \alpha)$
f) $20\sin\theta - 15\cos\theta \equiv R\sin(\theta - \alpha)$
g) $\sqrt{3}\cos\theta + \sin\theta \equiv R\cos(\theta - \alpha)$
h) $2\cos\theta - 4\sin\theta \equiv R\cos(\theta + \alpha)$

2 Find the greatest and least values of each of the following expressions, and state, correct to one decimal place, the smallest non negative value of θ for which each occurs.

a) $12\sin\theta + 5\cos\theta$
b) $2\cos\theta + \sin\theta$
c) $7 + 3\sin\theta - 4\cos\theta$
d) $10 - 2\sin\theta + \cos\theta$
e) $\dfrac{1}{2 + \sin\theta + \cos\theta}$
f) $\dfrac{1}{7 - 2\cos\theta + \sqrt{5}\sin\theta}$
g) $\dfrac{3}{5\cos\theta - 12\sin\theta + 16}$
h) $\dfrac{2}{7 - 4\sqrt{3}\cos\theta + \sin\theta}$

3 Solve each of the following equations for $0° \leqslant \theta \leqslant 360°$, giving your answers correct to one decimal place.

a) $\sin\theta + \sqrt{3}\cos\theta = 1$
b) $4\sin\theta - 3\cos\theta = 2$
c) $\sin\theta + \cos\theta = \dfrac{1}{\sqrt{2}}$
d) $5\sin\theta + 12\cos\theta = 7$
e) $7\sin\theta - 4\cos\theta = 3$
f) $\cos\theta - 3\sin\theta = 2$
g) $5\cos\theta + 2\sin\theta = 4$
h) $9\cos 2\theta - 4\sin 2\theta = 6$
i) $\dfrac{\sqrt{5}}{2}\sec\theta - \tan\theta = 2$
j) $\cot\theta - \sqrt{13}\operatorname{cosec}\theta = 5$
k) $24 = 10\operatorname{cosec}\theta - 7\cot\theta$
l) $\sqrt{2}\tan 2\theta - \sqrt{3}\sec 2\theta = \sqrt{2}$

***4** Show that $1 - \sqrt{2} \leqslant 2\cos^2\theta + \sin 2\theta \leqslant 1 + \sqrt{2}$ for all values of θ.

***5** **a)** Given that $6\cos^2\theta - 8\sin\theta\cos\theta \equiv A + R\cos(2\theta + \alpha)$, find the values of the constants A, R and α.
b) Hence solve the equation $6\cos^2\theta - 8\sin\theta\cos\theta = 5$, for $0° \leqslant \theta \leqslant 360°$.
c) Deduce the solutions to the equation $6\sin^2\phi + 8\cos\phi\sin\phi = 5$, for $-90° \leqslant \phi \leqslant 270°$.

Radians

So far, we have given solutions to trigonometric equations in degrees. However, using the fact that

$$180° = \pi \text{ radians}$$

we can give the solutions of trigonometric equations in radians.

Example 14 Solve the equation $2\sin\theta - \cos\theta = 0$, for $-\pi \leqslant \theta \leqslant \pi$. Give your answer in radians.

SOLUTION

Rearranging gives

$$2\sin\theta = \cos\theta$$

$$\therefore \quad \frac{2\sin\theta}{\cos\theta} = 1$$

$$\therefore \quad \tan\theta = \tfrac{1}{2}$$

With the calculator in radian mode, this gives $\theta = 0.46$ rad. The other solution in the required range is $\theta = -\pi + 0.46 = -2.68$ rad.

The solutions are $\theta = 0.46$ rad and -2.68 rad.

Example 15 Solve the equation $2\cos\theta = \sec\theta$, for $-2\pi \leqslant \theta \leqslant 2\pi$. Giving your answer in radians

SOLUTION

Since $\sec\theta = \dfrac{1}{\cos\theta}$, we have

$$2\cos\theta = \frac{1}{\cos\theta}$$

$$\therefore \quad \cos^2\theta = \frac{1}{2}$$

$$\therefore \quad \cos\theta = \pm\frac{1}{\sqrt{2}}$$

In degrees, the solutions are $\theta = \pm 45°$. In radians, the solutions are $\pm\dfrac{\pi}{4}$. The other solutions in the required range are

$$\theta = \pm\left(2\pi - \frac{\pi}{4}\right) = \pm\frac{7\pi}{4}$$

The solutions are $\theta = \pm\dfrac{\pi}{4}$ and $\pm\dfrac{7\pi}{4}$.

Exercise 16E

1 Solve each of the following equations for $0 \leqslant \theta \leqslant 2\pi$ rad, giving your answers in radians correct to two decimal places.

a) $\cos\theta = 0.4$ **b)** $\sin\theta = 0.7$ **c)** $\tan\theta = 4$
d) $\cos\theta = -0.8$ **e)** $\sin\theta = -0.3$ **f)** $\tan\theta = 0.6$
g) $\operatorname{cosec}\theta = 7$ **h)** $\cot\theta = 6$ **i)** $\sec\theta = -2.5$
j) $\operatorname{cosec}\theta = -10$ **k)** $\sec\theta = -5$ **l)** $\cot\theta = 2.8$

2 Solve each of the following equations for $-\pi \leqslant \theta \leqslant \pi$, giving your answers in radians correct to two decimal places.

a) $\sin(\theta - 0.1) = 0.2$ **b)** $\cos(\theta + 0.2) = 0.6$ **c)** $\tan(\theta - 0.5) = 3$
d) $\sec(\theta - 0.3) = 5$ **e)** $\sin(\theta + 0.4) = -0.34$ **f)** $\operatorname{cosec}(\theta + 0.7) = 8$
g) $\cos 2\theta = 0.6$ **h)** $\sin 3\theta = -0.48$ **i)** $\sec 4\theta = 2.5$
j) $\cot 2\theta = -2.3$ **k)** $\sin(2\theta + 0.1) = 0.9$ **l)** $\cot(3\theta - 0.69) = 0.25$

3 Solve each of the following for $0 \leqslant \theta \leqslant 2\pi$, giving your answer in radians in terms of π.

a) $\operatorname{cosec} 2\theta = 2$ **b)** $\cos 3\theta + \cos\theta = 0$ **c)** $\cos 2\theta + \sin\theta = 0$
d) $\tan\left(2\theta - \dfrac{\pi}{6}\right) = 1$ **e)** $\cos^2\theta + \cos\theta = 0$ **f)** $\tan 2\theta + \tan\theta = 0$
g) $\operatorname{cosec}\left(3\theta - \dfrac{\pi}{3}\right) = -2$ **h)** $\sin 5\theta + \sin 3\theta = 0$ **i)** $\sin\theta - \cos\theta = \sqrt{\dfrac{3}{2}}$
j) $\sin^2\theta = 3\cos^2\theta$ **k)** $\sin 2\theta + \sin\theta = 0$ **l)** $\tan^2\theta = 1 + \sec\theta$

Exercise 16F: Examination questions

1 Given that $\tan A = \frac{2}{3}$ and $\cot B = \frac{7}{5}$, find, without using a calculator, the exact value of $\tan(A + B)$. (WJEC)

2 The acute angles A and B are such that $\sin A = \frac{4}{5}$ and $\cos B = \frac{15}{17}$. Without using a calculator find the exact value of $\sin(A + B)$. (WJEC)

3 The acute angle α is such that $\tan\left(\alpha + \dfrac{\pi}{4}\right) = 41$.

i) Find, without using a calculator, the exact value of $\tan\alpha$.
ii) Deduce that $\cos\alpha = \frac{21}{29}$.
iii) In triangle ABC, AB = 69 cm, AC = 29 cm and BAC = α. Find BC. (WJEC)

4 In the triangle ABC, $\cos B = \frac{8}{17}$, $\cos C = \frac{9}{41}$ and BC = 91 cm.

a) Find the exact value of $\sin(B + C)$.
b) Deduce, or obtain otherwise, the value of $\sin A$.
c) Find the length of AB. (WJEC)

5 Show that

$$\cos(A + B) + \sin(A - B) \equiv (\cos A + \sin A)(\cos B - \sin B)$$

Find all solutions to the equation

$$\cos 5x + \sin x = 0$$

for which $0° \leqslant x \leqslant 90°$.

6 Find, to the nearest degree, all the values of θ between 0° and 360° satisfying the equation

$$2\cos 2\theta + 5\cos\theta = 4 \qquad \text{(WJEC)}$$

7 Solve the equation $\sin\theta - 3\cos 2\theta = 2$, giving all the solutions between 0° and 360°. (WJEC)

8 Prove the identity $\operatorname{cosec} 2\theta - \cot 2\theta \equiv \tan \theta$. (WJEC)

9 Show that $\tan \theta + \cot \theta \equiv \dfrac{2}{\sin 2\theta}$.

Hence, or otherwise, solve the equation $\tan \theta + \cot \theta = 4$, giving all the values of θ between $0°$ and $360°$. (UCLES)

10 In the triangle ABC, AC = 3 cm, BC = 2 cm, $\angle$ BAC = θ and $\angle$ ABC = 2θ. Calculate the value of θ correct to the nearest tenth of a degree. Hence find the size of the angle ACB and, without further calculation, explain why the length of AB is greater than 2 cm. (NEAB)

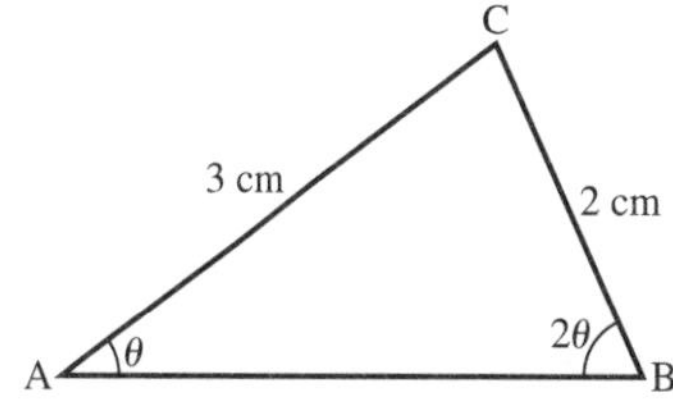

11 Simplify, as far as you can, each of the following expressions:

i) $\sin^2 x \operatorname{cosec} x$ **ii)** $\cos x \tan x$ **iii)** $\sin(\pi - x)$

iv) $\cos\left(\frac{3}{2}\pi + x\right)$ **v)** $\dfrac{\sin 2x}{2 \sin x}$ **vi)** $\dfrac{\sin x + \sin 5x}{\cos x + \cos 5x}$ (MEI)

12 Solve the equation $\sin 5\theta - \sin 3\theta = \sin \theta$, giving all the solutions in the range $0° \leqslant \theta \leqslant 180°$. (WJEC)

13 **i)** By writing $3\theta = \theta + 2\theta$ and using double angle formulae for the cosine and sine functions, establish the identity

$$\sin 3\theta \equiv \sin \theta(3 - 4 \sin^2 \theta)$$

ii) Hence, or otherwise, find the two values of θ in the range, $0° < \theta < 180°$, which satisfy the equation

$$\sin 3\theta + \sin \theta \cos 2\theta = 0 \qquad \text{(NICCEA)}$$

14 Express $4 \cos \theta - 3 \sin \theta$ in the form $r \cos(\theta + \alpha)$, where r is positive and α is an acute angle. Hence

i) state maximum and minimum values of the expression $4 \cos \theta - 3 \sin \theta$

ii) obtain all solutions of the equation $4 \cos \theta - 3 \sin \theta = 2$ in the interval $0° < \theta < 360°$. (MEI)

15 **i)** Express $35 \cos \theta + 12 \sin \theta$ in the form $R \cos(\theta - \alpha)$, where $R > 0$ and α is an acute angle.

ii) Find all the solutions of $35 \cos \theta + 12 \sin \theta = 20$ in the range $0° \leqslant \theta \leqslant 360°$. (WJEC)

16 Solve the equation $\sqrt{3} \tan \theta - \sec \theta = 1$, giving all solutions in the interval $0° < \theta < 360°$. (AEB 91)

17 The figure shows a rectangle PQSU containing a triangle PTR, right-angled at T.
Given that PT has length 6 cm, TR is of length 2 cm and angle UPT is denoted by θ, show that

$$\text{US} = (6\sin\theta + 2\cos\theta)\,\text{cm}$$

and express RQ in a similar form.

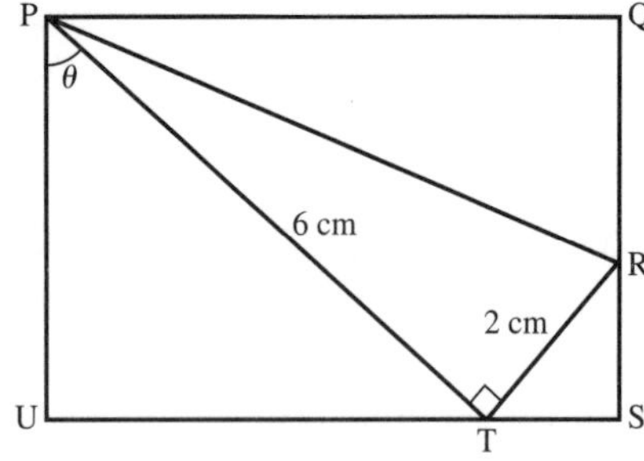

a) In the special case when the rectangle PQSU has perimeter 19 cm, show that

$$4\cos\theta + 3\sin\theta = 4.75$$

and hence find the two possible values of θ, to the nearest 0.1°.

b) Show that the triangle PQR has are $A\,\text{cm}^2$, where A can be written in the form

$$A = 8\sin 2\theta + 6\cos 2\theta$$

Hence determine the greatest value of A and the value of θ at which this occurs. (UODLE)

18 By expanding $\cos(\theta - 60°)$, show that $7\cos\theta + 8\cos(\theta - 60°)$ can be expressed in the form $13\sin(\theta + \alpha)$, where $0° < \alpha < 90°$, and state the value of α to the nearest 0.1°.
Hence find the solutions of the equation

$$7\cos\theta + 8\cos(\theta - 60°) = 6.5$$

in the interval $0° < \theta < 360°$, giving your answers to the nearest 0.1°. (AEB Spec)

19 The function f is defined for all real values of x by

$$\text{f}(x) = (\cos x - \sin x)(17\cos x - 7\sin x)$$

a) By first multiplying out the brackets, show that f(x) may be expressed in the form

$$5\cos 2x - 12\sin 2x + k$$

where k is a constant, and state the value of k.

b) Given that $5\cos 2x - 12\sin 2x \equiv R\cos(2x + \alpha)$, where $R > 0$ and $0 < \alpha < \frac{\pi}{2}$, state the value of R and find the value of α in radians to three decimal places.

c) Determine the greatest and least values of $\dfrac{39}{\text{f}(x) + 14}$ and state a value of x at which the greatest value occurs. (AEB 95)

20 Assuming the identities

$$\sin 3\theta \equiv 3\sin\theta - 4\sin^3\theta \quad \text{and} \quad \cos 3\theta \equiv 4\cos^3\theta - 3\cos\theta$$

prove that

$$\cos 5\theta \equiv 5\cos\theta - 20\cos^3\theta + 16\cos^5\theta$$

Find the set of all values of θ in the interval $0 < \theta < \pi$ for which $\cos 5\theta > 16\cos^5\theta$. (AEB 91)

17 Calculus with trigonometry

Most likely, logic is capable of justifying mathematics to no greater extent than biology is capable of justifying life.
Y. MANIN

Differentiation of $\sin x$ and $\cos x$

Result I

i) If $y = \sin x$, then $\dfrac{dy}{dx} = \cos x$

ii) If $y = \cos x$, then $\dfrac{dy}{dx} = -\sin x$

where x is measured in radians.

Proof of (i)

Let $P(x, y)$ and $Q(x + \delta x,\ y + \delta y)$ be two nearby points on the curve $y = \sin x$. Then

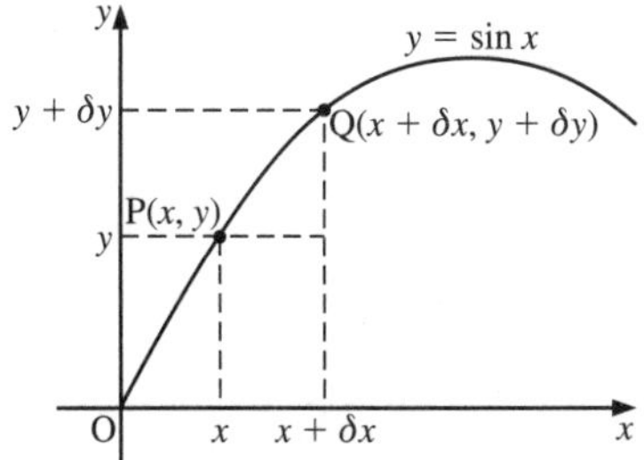

$$y + \delta y = \sin(x + \delta x)$$

$$\therefore \quad \delta y = \sin(x + \delta x) - y$$

$$\therefore \quad \delta y = \sin(x + \delta x) - \sin x$$

From Result VIII (ii) on page 377, we know that

$$\sin A - \sin B = 2\cos\left(\frac{A+B}{2}\right)\sin\left(\frac{A-B}{2}\right)$$

Therefore, we have

$$\delta y = 2\cos\left(\frac{(x+\delta x)+x}{2}\right)\sin\left(\frac{(x+\delta x)-x}{2}\right)$$

$$= 2\cos\left(\frac{2x+\delta x}{2}\right)\sin\left(\frac{\delta x}{2}\right)$$

$$= 2\cos\left(x+\frac{\delta x}{2}\right)\sin\left(\frac{\delta x}{2}\right)$$

$$\therefore \quad \frac{\delta y}{\delta x} = \frac{2\cos\left(x+\frac{\delta x}{2}\right)\sin\left(\frac{\delta x}{2}\right)}{\delta x}$$

$$= \frac{\cos\left(x+\frac{\delta x}{2}\right)\sin\left(\frac{\delta x}{2}\right)}{\left(\frac{\delta x}{2}\right)}$$

Now

$$\frac{dy}{dx} = \lim_{\delta x \to 0} \frac{\delta y}{\delta x}$$

Therefore, we have

$$\frac{dy}{dx} = \lim_{\delta x \to 0} \frac{\cos\left(x + \frac{\delta x}{2}\right) \sin\left(\frac{\delta x}{2}\right)}{\left(\frac{\delta x}{2}\right)}$$

$$= \lim_{\delta x \to 0} \cos\left(x + \frac{\delta x}{2}\right) \lim_{\delta x \to 0} \frac{\sin\left(\frac{\delta x}{2}\right)}{\left(\frac{\delta x}{2}\right)}$$

Now

$$\lim_{\delta x \to 0} \cos\left(x + \frac{\delta x}{2}\right) = \cos x$$

To find

$$\lim_{\delta x \to 0} \frac{\sin\left(\frac{\delta x}{2}\right)}{\left(\frac{\delta x}{2}\right)}$$

consider the arc AB of a circle centre O and radius r which subtends an angle $\alpha\left(< \frac{\pi}{2}\right)$ radians at O.

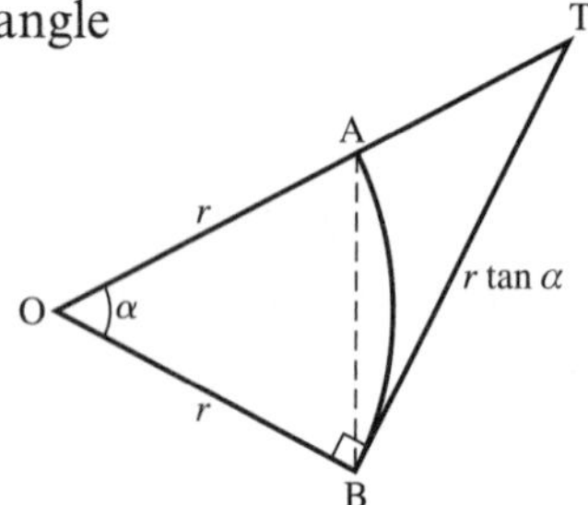

Area of triangle OAB $= \frac{1}{2} r^2 \sin\alpha$

Area of sector AOB $= \frac{1}{2} r^2 \alpha$

Area of triangle OBT $= \frac{1}{2}$OB $\times$ BT $= \frac{1}{2} r^2 \tan\alpha$

From the diagram we see that

Area of $\triangle$ OAB < Area of sector AOB < Area of $\triangle$ OBT

$\therefore \quad \frac{1}{2} r^2 \sin\alpha < \frac{1}{2} r^2 \alpha < \frac{1}{2} r^2 \tan\alpha$

$\therefore \quad \sin\alpha < \alpha < \tan\alpha$

Dividing throughout by $\sin\alpha$ (which is positive since α is acute) gives

$$1 < \frac{\alpha}{\sin\alpha} < \frac{1}{\cos\alpha}$$

or equivalently

$$1 > \frac{\sin\alpha}{\alpha} > \cos\alpha$$

Now as $\alpha \to 0$, $\cos\alpha \to 1$ and therefore the middle term of the inequality must also tend to 1. That is

$$\frac{\sin\alpha}{\alpha} \to 1$$

Therefore, we have

$$\lim_{\alpha\to 0} \frac{\sin\alpha}{\alpha} = 1$$

Using this result, we obtain

$$\lim_{\delta x\to 0} \frac{\sin\left(\frac{\delta x}{2}\right)}{\left(\frac{\delta x}{2}\right)} = 1$$

Therefore,

$$\frac{dy}{dx} = \cos x \times 1 = \cos x$$

as required.

Proof of (ii)

To differentiate $y = \cos x$, we note that

$$\cos x = \sin\left(x + \frac{\pi}{2}\right)$$

Differentiating with respect to x gives

$$\begin{aligned}\frac{d}{dx}(\cos x) &= \frac{d}{dx}\left[\sin\left(x + \frac{\pi}{2}\right)\right]\\ &= \cos\left(x + \frac{\pi}{2}\right)\\ &= -\sin x\end{aligned}$$

as required.

Differentiation of $\sin nx$ and $\cos nx$

To differentiate functions of the form $\sin nx$ and $\cos nx$, we use the chain rule. For example, if $y = \sin 4x$, then letting $u = 4x$ gives

$$y = \sin u \qquad \text{and} \qquad u = 4x$$

$$\therefore \quad \frac{dy}{du} = \cos u \qquad \text{and} \qquad \frac{du}{dx} = 4$$

By the chain rule,

$$\frac{dy}{dx} = \frac{dy}{du}\frac{du}{dx} = \cos u \times 4$$

$$\therefore \quad \frac{dy}{dx} = 4\cos 4x$$

In practice, we write

$$\frac{dy}{dx} = \cos 4x \times (4x)' = 4\cos 4x$$

Example 1 Find $\frac{dy}{dx}$ for each of the following functions.

a) $y = \cos 3x$ **b)** $y = \sin(x^2 + 2)$ **c)** $y = \cos\sqrt{x}$

SOLUTION

a) When $y = \cos 3x$: $\frac{dy}{dx} = -\sin 3x \times (3x)' = -3\sin 3x$

b) When $y = \sin(x^2 + 2)$: $\frac{dy}{dx} = \cos(x^2 + 2) \times (x^2 + 2)' = 2x\cos(x^2 + 2)$

c) When $y = \cos\sqrt{x}$: $\frac{dy}{dx} = -\sin\sqrt{x} \times \left(\sqrt{x}\right)' = -\frac{1}{2}x^{-\frac{1}{2}}\sin\sqrt{x}$

$$\therefore \quad \frac{dy}{dx} = -\frac{1}{2\sqrt{x}}\sin\sqrt{x}$$

Since we have

$$\frac{d(\sin x)}{dx} = \cos x \quad \text{and} \quad \frac{d(\cos x)}{dx} = -\sin x$$

we also have the following integrals:

- $\int \sin x \, dx = -\cos x + c$
- $\int \cos x \, dx = \sin x + c$

We can use these results together with inspection to integrate functions of the form

$$g'(x)f(g(x))$$

where f is a trigonometric function.

Example 2 Find each of the following integrals.

a) $\int \sin 5x \, dx$ **b)** $\int x^2 \cos(x^3 - 2) \, dx$ **c)** $\int \sin x \cos x \, dx$

SOLUTION

a) To find $\int \sin 5x \, dx$, we notice that the derivative of $5x$ is 5 and therefore

$$\int \sin 5x \, dx = -\tfrac{1}{5}\cos 5x + c$$

b) To find $\int x^2 \cos(x^3 - 2) \, dx$, we notice that the derivative of $x^3 - 2$ is $3x^2$ and that there is an x^2 term outside the main integrand. Therefore,

$$\int x^2 \cos(x^3 - 2) \, dx = \tfrac{1}{3}\sin(x^3 - 2) + c$$

c) To find $\int \sin x \cos x \, dx$, we can use the identity

$$\sin 2x \equiv 2 \sin x \cos x$$

$$\therefore \quad \sin x \cos x \equiv \tfrac{1}{2}\sin 2x$$

Therefore,

$$\int \sin x \cos x \, dx = \int \tfrac{1}{2}\sin 2x \, dx$$

$$= \tfrac{1}{2}\left(-\tfrac{1}{2}\cos 2x\right) + c = -\tfrac{1}{4}\cos 2x + c$$

Differentiation of $\sin^n x$ and $\cos^n x$

To differentiate functions of the form $\sin^n x$ and $\cos^n x$, we also use the chain rule. For example, if $y = \cos^2 x$, then this can be written as $y = (\cos x)^2$. Letting $u = \cos x$ gives

$$y = u^2 \qquad \text{and} \qquad u = \cos x$$

$$\therefore \quad \frac{dy}{du} = 2u \qquad \text{and} \qquad \frac{du}{dx} = -\sin x$$

By the chain rule,

$$\frac{dy}{dx} = \frac{dy}{du}\frac{du}{dx} = 2u(-\sin x) \quad \therefore \quad \frac{dy}{dx} = -2\cos x \sin x$$

In practice, we write

$$\frac{dy}{dx} = 2\cos x \times (\cos x)' = -2\cos x \sin x$$

Example 3 Find $\dfrac{dy}{dx}$ for each of the following functions.

a) $y = \sin^4 x$ **b)** $y = (\sin x + \cos x)^5$ **c)** $y = \cos^3 2x$

SOLUTION

a) When $y = \sin^4 x$:
$$\frac{dy}{dx} = 4\sin^3 x \times (\sin x)'$$
$$= 4\sin^3 x \cos x$$

b) When $y = (\sin x + \cos x)^5$:
$$\frac{dy}{dx} = 5(\sin x + \cos x)^4 \times (\sin x + \cos x)'$$
$$= 5(\sin x + \cos x)^4 (\cos x - \sin x)$$

c) When $y = \cos^3 2x = (\cos 2x)^3$:
$$\frac{dy}{dx} = 3(\cos 2x)^2 \times (\cos 2x)'$$
$$= 3(\cos 2x)^2 \times (-2\sin 2x)$$
$$\therefore \quad \frac{dy}{dx} = -6\cos^2 2x \sin 2x$$

Example 4 Find each of the following integrals.

a) $\displaystyle\int \cos x \sin^2 x \, dx$ **b)** $\displaystyle\int \frac{\cos x}{\sqrt{2 + \sin x}} \, dx$

SOLUTION

a) To find $\int \cos x \sin^2 x \, dx = \int \cos x (\sin x)^2 \, dx$, we notice that the derivative of $\sin x$ is $\cos x$, and that the function $\cos x$ is outside the main function of the integrand. Therefore,

$$\int \cos x \sin^2 x \; dx = \frac{(\sin x)^3}{3} + c$$
$$= \frac{\sin^3 x}{3} + c$$

b) To find

$$\int \frac{\cos x}{\sqrt{2 + \sin x}} \, dx = \int \cos x (2 + \sin x)^{-\frac{1}{2}} \, dx$$

we notice that the derivative of $2 + \sin x$ is $\cos x$ and that the function $\cos x$ is outside the main function of the integrand. Therefore,

$$\int \frac{\cos x}{\sqrt{2 + \sin x}} \, dx = \frac{(2 + \sin x)^{\frac{1}{2}}}{\left(\frac{1}{2}\right)} + c$$
$$= 2\sqrt{2 + \sin x} + c$$

Example 5 Find $\frac{dy}{dx}$ for each of these functions.

a) $y = x \sin x$ **b)** $y = \sin^2 x \cos 2x$ **c)** $y = \frac{\sin x}{\cos x}$

SOLUTION

a) $y = x \sin x$

Using the product rule, we have

$$\frac{dy}{dx} = x(\sin x)' + \sin x(x)'$$

$$= x \cos x + \sin x$$

b) $y = \sin^2 x \cos 2x$

Using the product rule, we have

$$\frac{dy}{dx} = \sin^2 x(\cos 2x)' + \cos 2x(\sin^2 x)'$$

$$= \sin^2 x(-2 \sin 2x) + \cos 2x(2 \sin x \cos x)$$

$$= -2 \sin^2 x \sin 2x + \cos 2x \sin 2x$$

$$\therefore \quad \frac{dy}{dx} = \sin 2x (\cos 2x - 2 \sin^2 x)$$

c) $y = \frac{\sin x}{\cos x}$

Using the quotient rule, we have

$$\frac{dy}{dx} = \frac{\cos x(\sin x)' - \sin x(\cos x)'}{(\cos x)^2}$$

$$= \frac{\cos^2 x + \sin^2 x}{\cos^2 x}$$

$$= \frac{1}{\cos^2 x}$$

$$\therefore \quad \frac{dy}{dx} = \sec^2 x$$

The solution to part **c** above is a standard result. Since $\tan x = \frac{\sin x}{\cos x}$, we have shown that

$$\frac{d}{dx}(\tan x) = \sec^2 x$$

We use this result on page 398.

Exercise 17A

1 Find $\frac{dy}{dx}$ for each of the following.

a) $y = \sin 3x$ **b)** $y = \cos 2x$ **c)** $y = \sin 5x$
d) $y = -\sin 6x$ **e)** $y = 2\cos 7x$ **f)** $y = -6\cos 5x$
g) $y = 8\sin \frac{1}{2}x$ **h)** $y = \cos(x+3)$ **i)** $y = \sin(x-4)$
j) $y = 3\sin\left(x + \frac{\pi}{4}\right)$ **k)** $y = -2\cos(4x-7)$ **l)** $y = 8\sin\left(\frac{3x-\pi}{2}\right)$

2 Differentiate each of the following with respect to x.

a) $\sin(x^2)$ **b)** $\cos(x^3)$ **c)** $2\cos(x^2-1)$
d) $3\sin(2x^3+3)$ **e)** $-4\sin(1-x^2)$ **f)** $6\cos(4-3x^4)$
g) $-\cos(x^2-2x)$ **h)** $\sin(x^3-3x^2)$ **i)** $\frac{1}{2}\sin(6x^2-4x+1)$
j) $-7\cos(2x-x^4)$ **k)** $6\sin\sqrt{x}$ **l)** $\cos\left(\frac{1}{x}\right)$

3 Find each of these integrals.

a) $\int 2\cos 2x\,dx$ **b)** $\int \sin 4x\,dx$ **c)** $\int 3\sin 5x\,dx$
d) $\int \cos(2x-1)\,dx$ **e)** $\int -6\sin(3x+2)\,dx$ **f)** $\int \sin\left(\frac{5x-\pi}{4}\right)dx$
g) $\int x\cos(x^2)\,dx$ **h)** $\int 8x^3\sin(x^4)\,dx$ **i)** $\int 3x\cos(x^2-7)\,dx$
j) $\int 2(x-2)\cos(x^2-4x)\,dx$ **k)** $\int (x^2-2x)\sin(3x^2-x^3)\,dx$ **l)** $\int \frac{\sin\sqrt{x}}{\sqrt{x}}\,dx$

4 Find $f'(x)$ for each of the following.

a) $f(x) = \sin^2 x$ **b)** $f(x) = \cos^3 x$ **c)** $f(x) = \sqrt{\cos x}$
d) $f(x) = \frac{1}{\cos^2 x}$ **e)** $f(x) = 2\sin^7 x$ **f)** $f(x) = -3\cos^6 x$
g) $f(x) = \sin^4 5x$ **h)** $f(x) = \cos^6 \frac{1}{2}x$ **i)** $f(x) = 2\sqrt{\cos 4x}$

5 Find $\frac{dy}{dx}$ for each of the following.

a) $y = (1+\sin x)^2$ **b)** $y = (3-\cos x)^4$ **c)** $y = (5+3\cos x)^6$
d) $y = (\sin x + \cos 2x)^3$ **e)** $y = \frac{1}{1+\cos x}$ **f)** $y = \sqrt{1-6\sin x}$
g) $y = -\frac{3}{1+\cos 3x}$ **h)** $y = \frac{4}{\sqrt{1-\sin 6x}}$ **i)** $y = (1+\sin^2 x)^3$

6 Find each of these integrals.

a) $\int 4\cos x \sin^3 x \, dx$ **b)** $\int \sin x \cos^2 x \, dx$ **c)** $\int \sin x (4 - \cos x)^5 \, dx$

d) $\int 2\cos x (3 + \sin x)^3 \, dx$ **e)** $\int \frac{\sin x}{(1 + \cos x)^2} \, dx$ **f)** $\int -\frac{\cos x}{\sqrt{4 - \sin x}} \, dx$

g) $\int 6\cos 3x \sin^5 3x \, dx$ **h)** $\int 8\sin 2x (5 - 2\cos 2x)^3 \, dx$ **i)** $\int 2\sin 4x \sqrt{6 + \cos 4x} \, dx$

j) $\int (1 - \cos x)(x - \sin x)^2 \, dx$ **k)** $\int \frac{x - \sin x}{\sqrt{x^2 + 2\cos x}} \, dx$ **l)** $\int \sin x \cos x \cos 2x \, dx$

7 Differentiate each of the following with respect to x.

a) $x \sin x$ **b)** $x^2 \cos x$ **c)** $x \cos 3x$ **d)** $x^3 \sin 6x$

e) $x \sin^5 x$ **f)** $3x^2 \cos^4 2x$ **g)** $\frac{x}{\sin x}$ **h)** $\frac{\cos 2x}{x + 1}$

i) $\frac{1}{1 + \sin x}$ **j)** $\frac{1 + \sin 2x}{\cos 2x}$ **k)** $\frac{x}{1 + \cos^2 x}$ **l)** $\frac{1 + \sin x}{1 + \cos x}$

8 Show that $\frac{d}{dx}\left(\frac{\cos x + \sin x}{\cos x - \sin x}\right) \equiv \frac{2}{1 - \sin 2x}$.

9 Given that $y = \sin x + 3\cos x$, show that $\cos x \frac{dy}{dx} + y \sin x \equiv 1$.

10 Given that $y = \sin^2 x - 2\cos x$, show that $\tan x \frac{dy}{dx} - y \equiv \sin^2 x + 2\sec x$.

11 Given that $y = \sin x (\sin x + 1)$, show that $\sin x \frac{dy}{dx} - y \cos x \equiv \sin^2 x \cos x$.

12 Given that $y = \cos 4x$, show that $\frac{d^2y}{dx^2} \equiv -16y$.

13 Given that $y = A \sin 3x + B \cos 3x$, where A and B are constants, show that $\frac{d^2y}{dx^2} + 9y \equiv 0$.

14 Given that $y = \sin x + 3\cos x$, show that $\frac{d^2y}{dx^2} - 3\frac{dy}{dx} + 2y \equiv 10 \sin x$.

15 Given that $y = x \sin 2x$, show that $x^2 \frac{d^2y}{dx^2} - 2x\frac{dy}{dx} + 2(2x^2 + 1)y \equiv 0$.

16 Given that $y = \frac{1}{1 + \sin x}$, show that $y\frac{d^2y}{dx^2} - 2\left(\frac{dy}{dx}\right)^2 \equiv y^3 \sin x$.

***17** **a)** Prove the following results

i) $\frac{d}{dx}(\sin^4 x - \cos^4 x) \equiv 2\sin 2x$

ii) $\frac{d}{dx}(\sin^4 x + \cos^4 x) \equiv -\sin 4x$

b) Deduce that $\frac{d}{dx}(\sin^8 x - \cos^8 x) \equiv 2\sin 2x - \sin^3 2x + \cos 2x \sin 4x$.

Differentiation of tan x, cosec x, sec x and cot x

Result II

i) If $y = \tan x$, then $\dfrac{dy}{dx} = \sec^2 x$

ii) If $y = \operatorname{cosec} x$, then $\dfrac{dy}{dx} = -\operatorname{cosec} x \cot x$

iii) If $y = \sec x$, then $\dfrac{dy}{dx} = \sec x \tan x$

iv) If $y = \cot x$, then $\dfrac{dy}{dx} = -\operatorname{cosec}^2 x$

The proofs of **(i)** to **(iv)** are left as exercises (see Exercise 17B, Questions 1 and 2).

Example 6 Find $\dfrac{dy}{dx}$ for each of these functions.

a) $y = \tan 3x$ **b)** $y = \sec (2x^2 - 1)$ **c)** $y = 4 \operatorname{cosec}^2 x$

SOLUTION

a) When $y = \tan 3x$: $\dfrac{dy}{dx} = \sec^2 3x \times (3x)' = 3 \sec^2 3x$

b) When $y = \sec (2x^2 - 1)$:
$$\frac{dy}{dx} = \sec (2x^2 - 1) \tan (2x^2 - 1) \times (2x^2 - 1)'$$
$$= 4x \sec (2x^2 - 1) \tan (2x^2 - 1)$$

c) When $y = 4 \operatorname{cosec}^2 x$:
$$\frac{dy}{dx} = 8 \operatorname{cosec} x \times (\operatorname{cosec} x)'$$
$$= 8 \operatorname{cosec} x (-\operatorname{cosec} x \cot x)$$
$$\therefore \quad \frac{dy}{dx} = -8 \operatorname{cosec}^2 x \cot x$$

These integrals follow from Result II:

- $\int \sec^2 x \, dx = \tan x + c$
- $\int \operatorname{cosec} x \cot x \, dx = -\operatorname{cosec} x + c$
- $\int \sec x \tan x \, dx = \sec x + c$
- $\int \operatorname{cosec}^2 x \, dx = -\cot x + c$

Example 7 Find each of these integrals.

a) $\int 2\sec 3x \tan 3x \,\mathrm{d}x$ **b)** $\int x\sec^2(1-x^2)\,\mathrm{d}x$ **c)** $\int \frac{\operatorname{cosec}^2\sqrt{x}}{\sqrt{x}}\,\mathrm{d}x$

SOLUTION

a) To find $\int 2\sec 3x \tan 3x \,\mathrm{d}x$, we notice that the derivative of $3x$ is 3. Therefore,

$$\int 2\sec 3x \tan 3x \,\mathrm{d}x = \tfrac{2}{3}\sec 3x + c$$

b) To find $\int x\sec^2(1-x^2)\,\mathrm{d}x$, we notice that the derivative of $1-x^2$ is $-2x$ and that there is an x term outside the main function of the integrand. Therefore,

$$\int x\sec^2(1-x^2)\,\mathrm{d}x = \frac{\tan(1-x^2)}{2} + c$$

c) To find $\int \frac{\operatorname{cosec}^2\sqrt{x}}{\sqrt{x}}\,\mathrm{d}x$, we notice that the derivative of $\sqrt{x}$ is $\frac{1}{2\sqrt{x}}$ and the function $\frac{1}{\sqrt{x}}$ is outside the main function of the integrand.

Therefore,

$$\int \frac{\operatorname{cosec}^2\sqrt{x}}{\sqrt{x}}\,\mathrm{d}x = -2\cot\sqrt{x} + c$$

Example 8 Find $\frac{\mathrm{d}y}{\mathrm{d}x}$ for each of the following functions.

a) $y = 3x\cot x$ **b)** $y = \frac{x}{\operatorname{cosec} 2x}$

SOLUTION

a) Using the product rule, we have

$$\frac{\mathrm{d}y}{\mathrm{d}x} = 3x(\cot x)' + \cot x(3x)'$$

$$= -3x\operatorname{cosec}^2 x + 3\cot x$$

$$\therefore \quad \frac{\mathrm{d}y}{\mathrm{d}x} = 3(\cot x - x\operatorname{cosec}^2 x)$$

b) Using the quotient rule, we have

$$\frac{\mathrm{d}y}{\mathrm{d}x} = \frac{\operatorname{cosec} 2x(x)' - x(\operatorname{cosec} 2x)'}{(\operatorname{cosec} 2x)^2}$$

$$= \frac{\operatorname{cosec} 2x - x(-2\operatorname{cosec} 2x\cot 2x)}{\operatorname{cosec}^2 2x}$$

$$\therefore \quad \frac{\mathrm{d}y}{\mathrm{d}x} = \frac{\operatorname{cosec} 2x(1+2x\cot 2x)}{\operatorname{cosec}^2 2x} = \frac{1+2x\cot 2x}{\operatorname{cosec} 2x}$$

Exercise 17B

1 Using the quotient rule together with the derivatives of $\sin x$ and $\cos x$, show that,

a) if $y = \tan x$, then $\dfrac{dy}{dx} = \sec^2 x$

b) if $y = \cot x$, then $\dfrac{dy}{dx} = -\operatorname{cosec}^2 x$

2 Using the chain rule together with the derivatives of $\sin x$ and $\cos x$, show that,

a) if $y = \operatorname{cosec} x$, then $\dfrac{dy}{dx} = -\operatorname{cosec} x \cot x$

b) if $y = \sec x$, then $\dfrac{dy}{dx} = \sec x \tan x$

3 Find $\dfrac{dy}{dx}$ for each of the following.

a) $y = \tan 2x$ **b)** $y = \sec 3x$ **c)** $y = \cot 6x$
d) $y = -\tan 4x$ **e)** $y = \operatorname{cosec} 5x$ **f)** $y = -2\cot 7x$
g) $y = 6\sec \frac{1}{3}x$ **h)** $y = \tan(x+2)$ **i)** $y = \operatorname{cosec}(x-1)$

4 Differentiate each of the following with respect to x.

a) $\sec(x^3)$ **b)** $\cot(x^2)$ **c)** $2\tan(x^4)$
d) $2\cot(3x^2)$ **e)** $-5\operatorname{cosec}(2x^4)$ **f)** $2\cot\sqrt{x}$
g) $\tan(x^2+3)$ **h)** $\operatorname{cosec}(1-x^3)$ **i)** $3\sec(6x^2+5)$
j) $-7\operatorname{cosec}(2-x^4)$ **k)** $2\tan(x^3-6x^2)$ **l)** $\frac{1}{2}\sec(4x-x^4)$

5 Find each of these integrals.

a) $\int 2\sec^2 2x\,dx$ **b)** $\int \operatorname{cosec} 3x \cot 3x\,dx$ **c)** $\int 4\operatorname{cosec}^2 8x\,dx$
d) $\int 3\sec 6x \tan 6x\,dx$ **e)** $\int -9\sec^2 3x\,dx$ **f)** $\int 12\operatorname{cosec} 4x \cot 4x\,dx$
g) $\int -2x\operatorname{cosec}^2(x^2)\,dx$ **h)** $\int 8x\sec x^2 \tan x^2\,dx$ **i)** $\int 3x^2\operatorname{cosec}^2 x^3\,dx$

6 Find $f'(x)$ for each of the following.

a) $f(x) = \tan^2 x$ **b)** $f(x) = \sec^3 x$ **c)** $f(x) = \cot^4 x$ **d)** $f(x) = -\operatorname{cosec}^3 x$
e) $f(x) = \sec^4 2x$ **f)** $f(x) = \cot^6 3x$ **g)** $f(x) = -\tan^2 5x$ **h)** $f(x) = \operatorname{cosec}^4 3x$

7 Find $\dfrac{dy}{dx}$ for each of the following.

a) $y = (1+\tan x)^2$ **b)** $y = (2-\operatorname{cosec} x)^4$ **c)** $y = (1+\sec x)^6$
d) $y = (3-\tan 2x)^3$ **e)** $y = -(1+\cot 3x)^4$ **f)** $y = (5-\operatorname{cosec} 2x)^3$
g) $y = \sqrt{1-\tan x}$ **h)** $y = \dfrac{1}{1+\operatorname{cosec} 3x}$ **i)** $y = \dfrac{3}{\sqrt{1-\cot 4x}}$
j) $y = \dfrac{2}{(1+\cot x)^3}$ **k)** $y = \sqrt[4]{1-\sec 8x}$ **l)** $y = -\dfrac{1}{(1+\sec^2 x)}$

8 Find each of the following integrals.

a) $\int 5\sec^2 x\tan^4 x\,dx$
b) $\int \operatorname{cosec}^2 x\cot^4 x\,dx$
c) $\int \sec^2 x(3+\tan x)^2\,dx$
d) $\int 2\operatorname{cosec}^2 x(1-\cot x)^4\,dx$
e) $\int \sec x\tan x(1+\sec x)^3\,dx$
f) $\int -\operatorname{cosec} x\cot x(3+\operatorname{cosec} x)^2\,dx$
g) $\int 4\operatorname{cosec}^2 2x\cot^2 2x\,dx$
h) $\int \frac{\operatorname{cosec} x\cot x}{(1+\operatorname{cosec} x)^2}\,dx$
i) $\int \operatorname{cosec}^2 5x\sqrt{2+\cot 5x}\,dx$
j) $\int \sec^5 x\tan x\,dx$

9 Differentiate each of the following with respect to x.

a) $x\tan x$
b) $x^3\operatorname{cosec} x$
c) $x^2\sec x$
d) $x\cot 6x$
e) $x^5\tan 3x$
f) $3x^2\sec^4 x$
g) $\sec 2x\tan 2x$
h) $\tan x(1+\sec x)$
i) $\frac{1}{1+\tan x}$
j) $\frac{x^2}{\sec 2x}$
k) $\frac{1+\operatorname{cosec} x}{x}$
l) $\frac{1}{\sec x+\tan x}$

10 Show that $\frac{d}{dx}\left(\frac{\tan x}{1+\sec x}\right)\equiv\frac{1}{1+\cos x}$.

11 Show that $\frac{d}{dx}\left(\frac{1+\cot x}{1-\cot x}\right)\equiv\frac{2}{\sin 2x-1}$.

12 Given that $y=\sec x+2\tan x$, show that $\cos x\frac{dy}{dx}+3\tan x\equiv 2y$.

13 Given that $y=\frac{x}{1+\tan x}$, show that $(1+\tan x)\frac{dy}{dx}+y\sec^2 x\equiv 1$.

14 Given that $y=\tan^2 x$, show that $\left(\frac{dy}{dx}\right)^2\equiv 4y(1+y)^2$.

15 Given that $y=\tan 3x$, show that $\frac{d^2y}{dx^2}\equiv 18y(1+y^2)$.

16 Given that $y=\cot 5x$, show that $\frac{d^2y}{dx^2}+10y\frac{dy}{dx}\equiv 0$.

17 Given that $y=x\tan x$, show that $x^2\frac{d^2y}{dx^2}\equiv 2(x^2+y^2)(1+y)$.

18 Given that $y=\sec 2x$, show that $\frac{d^2y}{dx^2}\equiv 4y(2y^2-1)$.

19 Given that $y=(\sec x+\tan x)^2$, show that $\cos x\frac{d^2y}{dx^2}-2\frac{dy}{dx}\equiv 2y\tan x$.

Applications

Example 9 Find the equation of the tangent to the curve $y = x + \tan x$ at the point where $x = \frac{\pi}{4}$.

SOLUTION

We need $\frac{dy}{dx}$ when $x = \frac{\pi}{4}$. Since $y = x + \tan x$, we have

$$\frac{dy}{dx} = 1 + \sec^2 x$$

When $x = \frac{\pi}{4}$: $\quad \left.\frac{dy}{dx}\right|_{x=\frac{\pi}{4}} = 1 + \sec^2\left(\frac{\pi}{4}\right)$

$$= 1 + \frac{1}{\cos^2\left(\frac{\pi}{4}\right)} = 1 + \frac{1}{\left(\frac{1}{2}\right)} = 3$$

The gradient of the tangent line is 3. Therefore, the equation of the tangent is of the form $y = 3x + c$.

When $x = \frac{\pi}{4}$, $y = \left(\frac{\pi}{4} + 1\right)$. Therefore, the tangent passes through the point $\left(\frac{\pi}{4}, \frac{\pi}{4} + 1\right)$. So,

$$\frac{\pi}{4} + 1 = 3\left(\frac{\pi}{4}\right) + c \quad \text{giving} \quad c = 1 - \frac{\pi}{2}$$

The equation of the tangent is

$$y = 3x + \left(1 - \frac{\pi}{2}\right) \quad \text{or} \quad 2y - 6x = 2 - \pi$$

Example 10 Find the area enclosed between the curve $y = \cos 2x$, the x-axis and the y-axis.

SOLUTION

The graph on the right shows the required area.

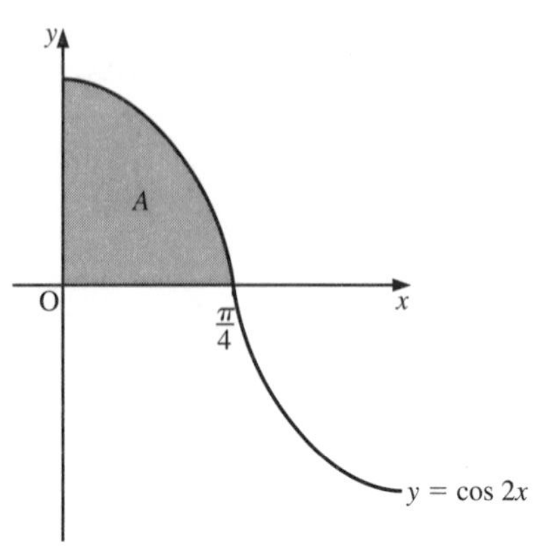

The required area A is given by

$$A = \int_0^{\frac{\pi}{4}} \cos 2x \, dx$$

$$= \left[\frac{\sin 2x}{2}\right]_0^{\frac{\pi}{4}}$$

$$= \frac{\sin 2\left(\frac{\pi}{4}\right)}{2} - \frac{\sin 2(0)}{2} = \frac{1}{2}$$

The required area is $\frac{1}{2}$.

Example 11 A curve is given by the equations $y = 2\sin^3 t$ and $x = 2\cos^3 t$.

Find the equation of the normal to the curve at the point where $t = \dfrac{\pi}{6}$.

SOLUTION

We first need to find $\dfrac{dy}{dx}$ by differentiating parametrically.

When $y = 2\sin^3 t$: $\quad \dfrac{dy}{dt} = 6\sin^2 t\cos t$

When $x = 2\cos^3 t$: $\quad \dfrac{dx}{dt} = 6\cos^2 t(-\sin t) = -6\cos^2 t\sin t$

By the chain rule,

$$\frac{dy}{dx} = \frac{dy}{dt} \times \frac{dt}{dx}$$

$$= -\frac{6\sin^2 t\cos t}{6\cos^2 t\sin t} = -\tan t$$

When $t = \dfrac{\pi}{6}$: $\quad \left.\dfrac{dy}{dx}\right|_{t=\frac{\pi}{6}} = -\tan\left(\dfrac{\pi}{6}\right) = -\dfrac{1}{\sqrt{3}}$

Therefore, the gradient of the normal is $\sqrt{3}$. The normal has equation of the form

$$y = \sqrt{3}x + c$$

When $t = \dfrac{\pi}{6}$: $\quad x = 2\cos^3\left(\dfrac{\pi}{6}\right) = \dfrac{3\sqrt{3}}{4} \quad$ and $\quad y = 2\sin^3\left(\dfrac{\pi}{6}\right) = \dfrac{1}{4}$

The normal passes through the point $\left(\dfrac{3\sqrt{3}}{4}, \dfrac{1}{4}\right)$. Therefore,

$$\frac{1}{4} = \sqrt{3}\left(\frac{3\sqrt{3}}{4}\right) + c \quad \text{giving} \quad \frac{1}{4} = \frac{9}{4} + c \quad \text{that is} \quad c = -2$$

The equation of the normal is $y = x\sqrt{3} - 2$.

Exercise 17C

This exercise revises the calculus techniques which were developed in earlier chapters, and applies these techniques to trigonometric functions.

Tangents and normals

1 Find the equation of the tangent to the curve $y = x + \sin x$ at the point where $x = \dfrac{\pi}{3}$.

2 Find the equation of the tangent and normal to the curve $y = x\cos x$ at the point where $x = \pi$.

3 The normals to the curve $y = \cos 2x$ at the points $A\left(\frac{\pi}{4}, 0\right)$ and $B\left(\frac{3\pi}{4}, 0\right)$ meet at the point C.

Find the coordinates of the point C, and the area of the triangle ABC.

4 Find the equation of the tangent and the normal to the curve $y = \dfrac{1}{1 + 2\sin x}$ at the point where $x = \dfrac{\pi}{6}$.

5 Find the coordinates of the two points on the curve $y = \sin x(2\cos x + 1)$, in the range $-\dfrac{\pi}{2} \leqslant x \leqslant \dfrac{\pi}{2}$, where the gradient is $-\frac{1}{2}$.

6 Show that there are two points on the curve $y = \dfrac{\sin x}{1 + \cos x}$, in the range $0 \leqslant x \leqslant 2\pi$, where the gradient is $\frac{2}{3}$. Find the coordinates of these points.

7 Find the coordinates of the points on the curve $y = 2\operatorname{cosec} x - \cot x$, in the range $0 \leqslant x \leqslant 2\pi$, where the gradient is $\frac{8}{3}$.

Stationary points

8 Given that $y = \sin x(1 - \cos x)$, show that

$$\frac{dy}{dx} = (1 + 2\cos x)(1 - \cos x)$$

Hence find the coordinates of the points on the curve $y = \sin x(1 - \cos x)$, in the range $0 \leqslant x \leqslant \pi$, where the gradient is zero.

9 Given that $y = \sec x + \operatorname{cosec} x$, show that

$$\frac{dy}{dx} = \frac{\sin^3 x - \cos^3 x}{\sin^2 x \cos^2 x}$$

Hence find the coordinates of the point on the curve $y = \sec x + \operatorname{cosec} x$, in the range $0 \leqslant x \leqslant \pi$, where the gradient is zero.

10 Given that $y = \dfrac{x - \sin x}{1 + \cos x}$, show that

$$\frac{dy}{dx} = \frac{x \sin x}{(1 + \cos x)^2}$$

Hence find the coordinates of the two points on the curve

$$y = \frac{x - \sin x}{1 + \cos x}$$

in the range $0 \leqslant x \leqslant 2\pi$, where the gradient is zero.

11 Find and classify the stationary values on each of the following curves in the range $0 \leqslant x \leqslant 2\pi$.

a) $y = x + 2\cos x$ **b)** $y = 2\cos x - \cos 2x$ **c)** $y = \dfrac{\sin x}{2 - \sin x}$ **d)** $y = \sin x \cos^3 x$

Areas and volumes of revolution

12 Evaluate the following definite integrals.

a) $\int_0^{\frac{\pi}{2}} (1 - \cos x)\,dx$ **b)** $\int_0^{\frac{\pi}{6}} \sin 3x\,dx$ **c)** $\int_{\frac{\pi}{6}}^{\frac{\pi}{4}} \cos x \sin^3 x\,dx$ **d)** $\int_{-\frac{\pi}{6}}^{\frac{\pi}{3}} \sec 2x \tan 2x\,dx$

13 Find the area between the curve $y = \sin x$ and the x-axis from $x = 0$ to $x = \pi$.

14 Find the area between the curve $y = 3\cos x + 2\sin x$ and the x-axis from $x = \frac{\pi}{6}$ to $x = \frac{\pi}{3}$.

15 In the interval $0 \leqslant x \leqslant \pi$, the line $y = \frac{1}{2}$ meets the curve $y = \sin x$ at the points A and B.

a) Find the coordinates of A and B.
b) Calculate the area enclosed between the curve and the line between A and B.

16 In the interval $0 \leqslant x \leqslant \pi$, the curve $y = \sin x$ meets the curve $y = \sin 2x$ at the origin and at the point P.

a) Find the coordinates of P.
b) Calculate the area enclosed by the two curves between the origin and P.

17 Find the volume of the solid of revolution formed by rotating about the x-axis the area between the curve $y = \sec 2x$ and the x-axis, from $x = 0$ to $x = \frac{\pi}{12}$.

Implicit differentiation

18 Find the equation of the tangent and the normal to the curve $\cos 2x + \cos y - 1 = 0$, at the point $\left(\frac{\pi}{6}, \frac{\pi}{3}\right)$.

19 Find the equation of the tangent and the normal to the curve $x \tan y = 6 - x^2$, at the point $\left(2, \frac{\pi}{4}\right)$.

20 Find the equation of the tangent to the curve $\frac{y}{x + \sin y} = 3$ at the point where $y = \pi$.

21 a) Given that $2\cos y - x^2 = 1$, deduce the following results.

i) $\frac{dy}{dx} = -\frac{x}{\sin y}$ **ii)** $\frac{d^2y}{dx^2} = -\left(\frac{x^2 \cos y + \sin^2 y}{\sin^3 y}\right)$

b) Hence find and classify any stationary values on the curve $2\cos y - x^2 = 1$, in the range $0 \leqslant y \leqslant \pi$.

22 A curve is given by the equation $\tan y = 2x + x^2$.

a) Deduce the following results.

i) $\frac{dy}{dx} = 2(x + 1)\cos^2 y$ **ii)** $\frac{d^2y}{dx^2} = 2\cos^2 y\left[1 - 2(x + 1)^2 \sin 2y\right]$

b) Hence find and classify any stationary values on the curve $\tan y = 2x + x^2$, in the range $0 \leqslant y \leqslant \pi$.

Parametric differentiation

23 Find the equation of the tangent to the curve $x = \sin^2 t$, $y = \cos t$, at the point where $t = \frac{\pi}{3}$.

24 Find the equation of the tangent and the normal to the curve $x = \cos 2t$, $y = \sin t$, at the point where $t = \frac{\pi}{6}$.

25 The tangent to the curve $x = 2\sin t$, $y = 2 + \cos 2t$, at the point (2, 1) meets the x-axis at P and the y-axis at Q. Find the distance PQ.

26 The curve C is given by the parametric equations $x = 3t + \sin t$ and $y = t + 2\cos t$.

a) Deduce the following result

i) $\dfrac{dy}{dx} = \dfrac{1 - 2\sin t}{3 + \cos t}$ **ii)** $\dfrac{d^2y}{dx^2} = \dfrac{\sin t - 6\cos t - 2}{(3 + \cos t)^3}$

b) Hence find and classify all stationary points on C in the range $0 \leqslant t \leqslant 2\pi$.

27 A curve is given by the parametric equations $x = 1 + 2\sin t$ and $y = \sin t + \cos t$.

a) Deduce the following results

i) $\dfrac{dy}{dx} = \dfrac{1 - \tan t}{2}$ **ii)** $\dfrac{d^2y}{dx^2} = -\dfrac{1}{4}\sec^3 t$

b) Hence find and classify all stationary points on the curve in the range $0 \leqslant t \leqslant 2\pi$.

Maxima and minima

28 A farmer has three pieces of fencing, each of length 5 metres, and wishes to enclose a pen as in the diagram. A long wall, which is already in place, will comprise the fourth side of the pen.

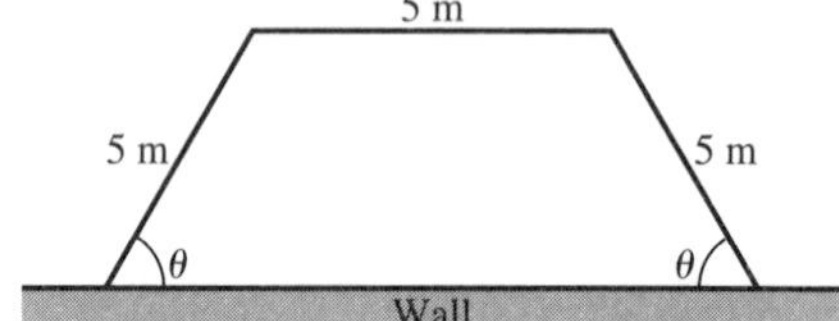

a) Show that the area of the pen is $[25\sin\theta(1 + \cos\theta)]\ \text{m}^2$.

b) Calculate the maximum value of this area.

29 A rectangle is drawn inside a semi-circle of base radius 10 cm, as in the diagram opposite.

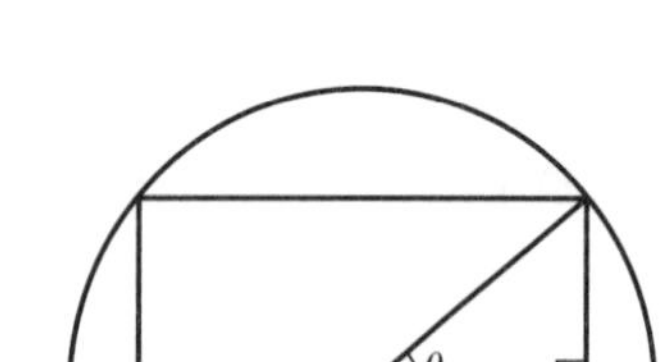

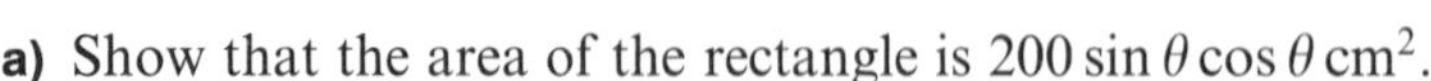

a) Show that the area of the rectangle is $200\sin\theta\cos\theta\ \text{cm}^2$.

b) Calculate the maximum value of this area.

30 A solid cylinder just fits under a hemispherical shell of radius a cm, as in the diagram.

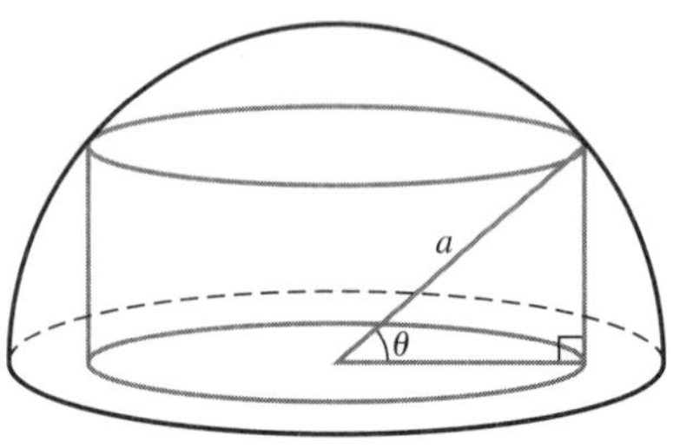

a) Show that the volume of the cylinder is $\pi a^3 \sin\theta\cos^2\theta$.

b) Hence find an expression, in terms of a, for the maximum value of this volume.

***31** A right-angled triangle just fits inside a semicircle of base radius r, as shown in the diagram.

a) Show that the area of the triangle is $2r^2 \sin\theta \cos^3\theta$.
b) Calculate the maximum value of this area.

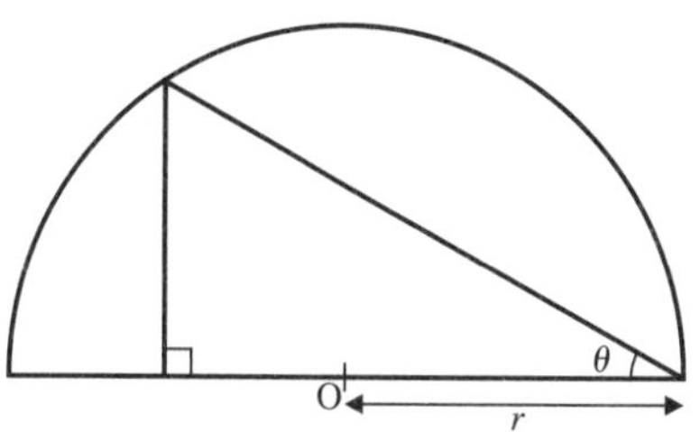

***32** A right circular cone of semi-vertical angle $\theta°$ just fits inside a sphere of radius 1 cm as shown opposite.

a) Show that the volume of the cone is given by the expression

$$\frac{\pi}{3}\sin^2 2\theta\,(1+\cos 2\theta)\,\text{cm}^3$$

b) Calculate the maximum value of this volume.

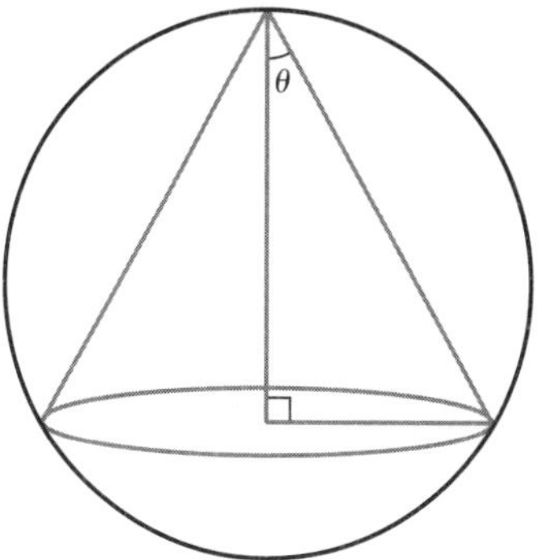

Exercise 17D: Examination questions

1 Given that $y = \cos 2x + \sin x$, $0 < x < 2\pi$, and x is in radians,
a) find, in terms of π, the values of x for which $y = 0$.

b) Find, to two decimal places, the values of x for which $\dfrac{dy}{dx} = 0$. (EDEXCEL)

2 a) i) Show that $(\cos x + \sin x)^2 = 1 + \sin 2x$, for all x.
ii) Hence, or otherwise, find the derivative of $(\cos x + \sin x)^2$.
b) i) By expanding $(\cos^2 x + \sin^2 x)^2$, find and simplify an expression for $\cos^4 x + \sin^4 x$ involving $\sin 2x$.
ii) Hence, or otherwise, show that the derivative of $\cos^4 x + \sin^4 x$ is $-\sin 4x$. (MEI)

3 a) Differentiate $\frac{1}{2}\tan 2x$ with respect to x.

b) Use the result found in part **a** to show that $\displaystyle\int_0^{\frac{\pi}{8}} \sec^2 2x\,dx = 0.5$, and hence evaluate

$\displaystyle\int_0^{\frac{\pi}{8}} \tan^2 2x\,dx.$ (WJEC)

4 a) Explain why $(1 - \cos x)$ is never negative.
b) Sketch the curve $y = 1 - \cos x$ in the interval $0 \leqslant x \leqslant \pi$.
c) Find the area bounded by the curve, the x-axis and the ordinate $x = \theta$, where $\theta > 0$.
d) Deduce that, for positive values of θ, $\theta > \sin\theta$. (NEAB)

5 The function f is defined by

$$f(x) = \frac{\sin^2 x}{1+\cos^2 x} \quad (x \in \mathbb{R})$$

a) Show that

$$f'(x) = \frac{2\sin 2x}{(1+\cos^2 x)^2}$$

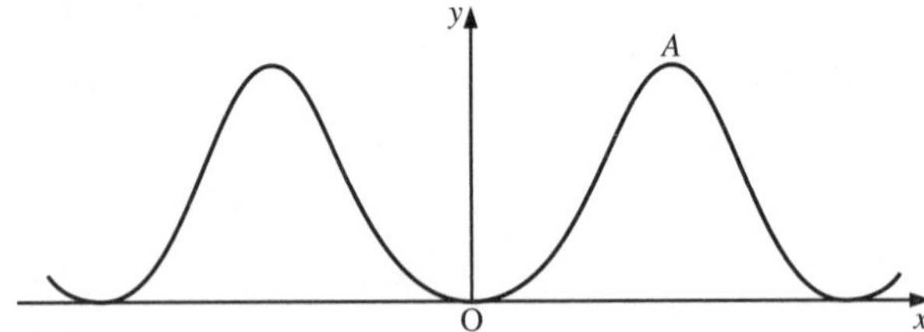

b) The figure shows a sketch of the graph of $y = f(x)$. Find the coordinates of the stationary point A and hence state the range of f.

c) Sketch the graph of

$$y = \frac{\cos^2 x}{1+\sin^2 x} \qquad \text{(AEB 92)}$$

6 The function f has domain $\left\{x : \frac{\pi}{8} \leqslant x \leqslant \frac{\pi}{6}\right\}$ and is defined by $f : x \mapsto \tan 2x$.

Sketch the graph of f and state the range of f.

The region bounded by the curve with equation $y = f(x)$, the x-axis and the lines $x = \frac{\pi}{8}$ and $x = \frac{\pi}{6}$ is rotated through 2π radians about the x-axis. Calculate the volume of the solid formed. (AEB 94)

7 The figure shows two circles of radii 2 cm and 4 cm with common centre O. The points P and Q lie on the circumference of the larger circle and angle POQ is θ radians $\left(0 < \theta < \frac{\pi}{2}\right)$.

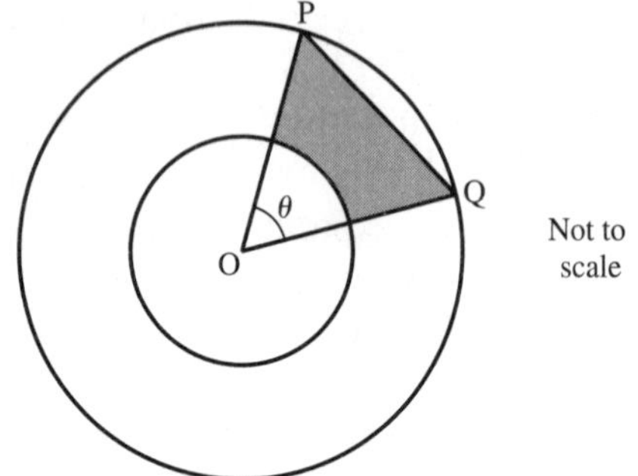

The shaded region bounded by OP, PQ, OQ and an arc of the smaller circle has area A cm^2. Show that

$$A = 8\sin\theta - 2\theta$$

Find, to two decimal places, the value of θ for which A is stationary, and show that this value gives the maximum area. (AEB 95)

8 The figure shows a conical container with slant height 3 cm and semi-vertical angle θ. It encloses a volume V cm^3, where

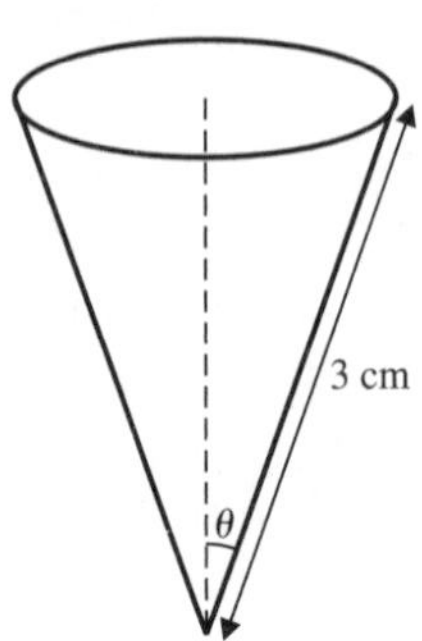

$$V = 9\pi \sin^2\theta \cos\theta$$

and θ varies so that $0 \leqslant \theta \leqslant \frac{\pi}{2}$.

a) Show that $\frac{dV}{d\theta} = 9\pi(2\sin\theta - 3\sin^3\theta)$.

b) Find $\frac{d^2V}{d\theta^2}$.

c) Determine the values of $\sin\theta$ and $\cos\theta$ for which V has a stationary value, leaving your answers in surd form.

d) Find the value of $\dfrac{d^2V}{d\theta^2}$ at the stationary point and hence show that this gives a maximum value of V. State the maximum value of V in terms of π.

e) Sketch the graph of V as θ varies from 0 to $\dfrac{\pi}{2}$. (UODLE)

9 A curve C is given by the equations

$$x = 2\cos t + \sin 2t \qquad y = \cos t - 2\sin 2t \qquad 0 \leqslant t < \pi$$

where t is a parameter.

a) Find $\dfrac{dx}{dt}$ and $\dfrac{dy}{dt}$ in terms of t.

b) Find the value of $\dfrac{dy}{dx}$ at the point P on C where $t = \dfrac{\pi}{4}$.

c) Find an equation of the normal to the curve at P. (EDEXCEL)

10 A function y of x is defined parametrically by $x = t - \sin t$, $y = 1 - \cos t$.

i) Find $\dfrac{dy}{dx}$.

ii) Show that $y^2\dfrac{d^2y}{dx^2} + 1 = 0$. (NICCEA)

18 Indices and logarithms

A theory has the alternative of being right or wrong. A model has a third possibility; it may be right but irrelevant.
MANFRED EIGEN

Indices

We know that for positive integers m and n:

- $x^m \times x^n = x^{m+n}$ First law of indices
- $x^m \div x^n = x^{m-n}$ Second law of indices
- $(x^m)^n = x^{mn}$ Third law of indices

We now assume that these results are true for all values of m and n.

Negative and fractional indices

x^0

We know that $3 \div 3 = 1$. Using the second law of indices, we have

$$3^1 \div 3^1 = 3^0$$

Therefore, $3^0 = 1$. Generally, we have

$$x^0 = 1 \qquad (x \neq 0)$$

x^{-m}

We also know that $3^0 \div 3^1 = 1 \div 3 = \frac{1}{3}$. Using the second law of indices, we have

$$3^0 \div 3^1 = 3^{-1}$$

Therefore, $3^{-1} = \frac{1}{3}$. Generally, we have

$$x^{-1} = \frac{1}{x} \qquad (x \neq 0)$$

We also know that

$$\left(x^{-1}\right)^m = \left(\frac{1}{x}\right)^m = \frac{1}{x^m}$$

and by the third law of indices we have

$$\left(x^{-1}\right)^m = x^{-m}$$

Therefore,

$$x^{-m} = \frac{1}{x^m}$$

$x^{\frac{1}{n}}$

By the first law of indices we have

$$5^{\frac{1}{2}} \times 5^{\frac{1}{2}} = 5^1$$

Therefore, $5^{\frac{1}{2}} = \sqrt{5}$.

Similarly,

$$5^{\frac{1}{3}} \times 5^{\frac{1}{3}} \times 5^{\frac{1}{3}} = 5^1$$

Therefore, $5^{\frac{1}{3}} = \sqrt[3]{5}$. Generally, we have

$$x^{\frac{1}{n}} = \sqrt[n]{x}$$

$x^{\frac{m}{n}}$

To interpret $x^{\frac{m}{n}}$ consider

$$x^{\frac{m}{n}} = \left(x^{\frac{1}{n}}\right)^m \quad \text{or} \quad x^{\frac{m}{n}} = (x^m)^{\frac{1}{n}}$$

$$= \left(\sqrt[n]{x}\right)^m \qquad\qquad = \sqrt[n]{x^m}$$

Usually we write

$$x^{\frac{m}{n}} = \sqrt[n]{x^m}$$

Example 1 Simplify each of these.

a) $4^{\frac{3}{2}}$ **b)** $\left(\frac{2}{3}\right)^{-3}$ **c)** $\left(\frac{1}{8}\right)^{-\frac{4}{3}}$ **d)** $\dfrac{4^{-1} \times 9^{\frac{1}{2}}}{8^{-2}}$

SOLUTION

a) Rewriting $4^{\frac{3}{2}}$ gives

$$4^{\frac{3}{2}} = \left(4^{\frac{1}{2}}\right)^3 = 2^3 = 8$$

b) Again we rewrite in a form which has no negative indices,

$$\left(\frac{2}{3}\right)^{-3} = \frac{1}{\left(\frac{2}{3}\right)^3} = \left(\frac{3}{2}\right)^3 = \frac{27}{8}$$

c) Rewriting we have

$$\left(\frac{1}{8}\right)^{-\frac{4}{3}} = \frac{1}{\left(\frac{1}{8}\right)^{\frac{4}{3}}} = 8^{\frac{4}{3}} = (\sqrt[3]{8})^4 = 2^4 = 16$$

d)

$$\frac{4^{-1} \times 9^{\frac{1}{2}}}{8^{-2}} = \frac{\frac{1}{4} \times \sqrt{9}}{\left(\frac{1}{8^2}\right)} = \frac{3}{4} \times 64 = 48$$

Example 2 Solve each of these.

a) $x^{\frac{1}{5}} = 3$ **b)** $x^{\frac{4}{3}} = 81$ **c)** $2x^{\frac{3}{4}} = x^{\frac{1}{2}}$ **d)** $x^{\frac{1}{3}} - 3 = 28x^{-\frac{1}{3}}$

SOLUTION

a) When $x^{\frac{1}{5}} = 3$: $\left(x^{\frac{1}{5}}\right)^5 = 3^5$

$\therefore \quad x = 243$

b) When $x^{\frac{4}{3}} = 81$: $\left(x^{\frac{4}{3}}\right)^{\frac{1}{4}} = 81^{\frac{1}{4}}$

$\therefore \quad x^{\frac{1}{3}} = 3$

$\therefore \quad \left(x^{\frac{1}{3}}\right)^3 = 3^3$

$\therefore \quad x = 27$

c) When $2x^{\frac{3}{4}} = x^{\frac{1}{2}}$: $\left(2x^{\frac{3}{4}}\right)^4 = \left(x^{\frac{1}{2}}\right)^4$

$\therefore \quad 2^4\left(x^{\frac{3}{4}}\right)^4 = x^2$

$\therefore \quad 16x^3 = x^2$

$\therefore \quad 16x^3 - x^2 = 0$

$\therefore \quad x^2(16x - 1) = 0$

Solving gives $x = 0$ or $16x - 1 = 0$. That is, $x = \frac{1}{16}$.

d) For $x^{\frac{1}{3}} - 3 = 28x^{-\frac{1}{3}}$ multiply throughout by $x^{\frac{1}{3}}$ to make all the indices positive. That is,

$$x^{\frac{1}{3}}\left(x^{\frac{1}{3}}\right) - 3x^{\frac{1}{3}} = x^{\frac{1}{3}}\left(28x^{-\frac{1}{3}}\right)$$

$$\therefore \quad x^{\frac{2}{3}} - 3x^{\frac{1}{3}} - 28 = 0$$

Letting $y = x^{\frac{1}{3}}$ gives

$$y^2 - 3y - 28 = 0$$

$$\therefore \quad (y + 4)(y - 7) = 0$$

Solving gives $y = -4$ and $y = 7$.

When $y = -4$: $x^{\frac{1}{3}} = -4$

$\therefore \quad \left(x^{\frac{1}{3}}\right)^3 = (-4)^3$ giving $x = -64$

When $y = 7$: $x^{\frac{1}{3}} = 7$

$\therefore \quad \left(x^{\frac{1}{3}}\right)^3 = 7^3$ giving $x = 343$

The solutions are $x = -64$ and $x = 343$.

Exercise 18A

1 Simplify each of these.

a) $x^5 \times x^4$ **b)** $p^3 \times p^{-1}$ **c)** $(3k^3)^2$ **d)** $y^{\frac{1}{2}} \times y^{\frac{1}{3}}$

e) $c^7 \div c^3$ **f)** $9h^2 \div 6h^{-4}$ **g)** $(4d^2)^2 \div (2d)^3$ **h)** $(6p^{-3})^4 \div (9p^{-4})^2$

2 Simplify each of these.

a) $4^{\frac{1}{2}}$ **b)** $27^{\frac{1}{3}}$ **c)** $9^{\frac{3}{2}}$ **d)** $8^{\frac{5}{3}}$

e) $125^{\frac{2}{3}}$ **f)** $49^{\frac{3}{2}}$ **g)** $\left(\frac{1}{25}\right)^{\frac{1}{2}}$ **h)** $\left(\frac{8}{27}\right)^{\frac{2}{3}}$

3 Simplify each of these.

a) 7^{-1} **b)** 3^{-2} **c)** $4^{-\frac{1}{2}}$ **d)** $25^{-\frac{3}{2}}$

e) $\left(\frac{2}{3}\right)^{-1}$ **f)** $\left(\frac{3}{11}\right)^{-2}$ **g)** $\left(\frac{343}{512}\right)^{-\frac{2}{3}}$ **h)** $\left(64^{\frac{5}{12}}\right)^{2}$

4 Solve each of the following equations for x.

a) $3x^3 = 375$ **b)** $98x^2 = 2$ **c)** $x^3 + 343 = 0$ **d)** $9x^{-1} = 5$

e) $x^{-3} = 8$ **f)** $\frac{1}{32}x^3 = 8x^{-1}$ **g)** $\frac{9}{25}x = \frac{5}{3}x^{-2}$ **h)** $\frac{2}{49}x^{-2} + 14x = 0$

5 Solve each of the following equations for x.

a) $x^{\frac{1}{2}} = 3$ **b)** $x^{\frac{1}{5}} = 2$ **c)** $7x^{\frac{1}{2}} + 2 = 0$ **d)** $x^{-\frac{1}{4}} = 4$

e) $4x^{\frac{1}{2}} = x^{-\frac{3}{2}}$ **f)** $5x^{\frac{2}{3}} = x^{-\frac{1}{3}}$ **g)** $4x^{-\frac{1}{3}} = 5x^{\frac{1}{6}}$ **h)** $6x^{\frac{3}{2}} - \frac{2}{3}x^{-\frac{1}{2}} = 0$

6 Solve each of the following equations for x.

a) $x^{\frac{2}{3}} = 9$ **b)** $x^{\frac{3}{2}} = 64$ **c)** $5x^{\frac{3}{4}} + 40 = 0$ **d)** $x^{-\frac{2}{3}} = 81$

e) $x^{-\frac{1}{4}} = 125x^{\frac{1}{2}}$ **f)** $8x^{-2} = 343x^{-\frac{1}{2}}$ **g)** $2x^{\frac{1}{3}} = \frac{81}{8}x^{-1}$ **h)** $49x^{-\frac{5}{8}} - \frac{8}{7}x^{\frac{7}{8}} = 0$

7 Solve each of the following equations for x.

a) $x^{\frac{2}{3}} - x^{\frac{1}{3}} - 2 = 0$ **b)** $x^{\frac{1}{2}} - 5x^{\frac{1}{4}} + 6 = 0$ **c)** $x - 10x^{\frac{1}{2}} + 24 = 0$ **d)** $2x^{\frac{2}{3}} - 5x^{\frac{1}{3}} = 12$

e) $3x^{\frac{2}{5}} = 2 - x^{\frac{1}{5}}$ **f)** $2x^{\frac{1}{4}} = 9 - 4x^{-\frac{1}{4}}$ **g)** $6x^{\frac{1}{3}} + 5 + x^{-\frac{1}{3}} = 0$ **h)** $x^3 + 8 = 9x^{\frac{3}{2}}$

***8** Solve each of the following equations for x.

a) $27^x = 9$ **b)** $8^x = 4^{1-x}$ **c)** $5^x = 25^{2x-5}$ **d)** $8 \times 2^x = \left(\frac{1}{4}\right)^{3x+5}$

Surds

We know that $\sqrt{16} = 4$ and that $\sqrt{\frac{1}{4}} = \frac{1}{2}$. These are examples of **rational** numbers. However, $\sqrt{2}$ cannot be expressed as a fraction of two integers. It is for this reason that $\sqrt{2}$ is an example of an **irrational number**.

Roots such as $\sqrt{2}$, $\sqrt{3}$, $\sqrt{5}$, ... are called **surds**. In this section, we will look at the simplification of expressions involving surds. In order to do this, we will use the following properties of surds.

- $\sqrt{a} \times \sqrt{b} = \sqrt{ab}$
- $\dfrac{\sqrt{a}}{\sqrt{b}} = \sqrt{\dfrac{a}{b}}$
- $a\sqrt{c} \pm b\sqrt{c} = (a \pm b)\sqrt{c}$

Example 3 Simplify each of the following.

a) $\sqrt{48}$ **b)** $3\sqrt{50} + 2\sqrt{18} - \sqrt{32}$

SOLUTION

a) In order to simplify $\sqrt{48}$, we notice that

$$48 = \underbrace{16}_{\text{Largest square factor of 48}} \times 3 = 4^2 \times 3$$

Therefore,

$$\sqrt{48} = \sqrt{4^2 \times 3} = 4\sqrt{3}$$

b) We have

$$\begin{aligned} 3\sqrt{50} + 2\sqrt{18} - \sqrt{32} &= 3\sqrt{25 \times 2} + 2\sqrt{9 \times 2} - \sqrt{16 \times 2} \\ &= 15\sqrt{2} + 6\sqrt{2} - 4\sqrt{2} \\ &= 17\sqrt{2} \end{aligned}$$

When surds appear in the denominator of a fraction, it is usual to eliminate them. This is called **rationalising the denominator**. For example, to rationalise the fraction $\dfrac{1}{\sqrt{3}}$, we multiply its numerator and its denominator by $\sqrt{3}$ (which is the same as multiplying by 1), giving

$$\frac{1}{\sqrt{3}} \times \frac{\sqrt{3}}{\sqrt{3}} = \frac{\sqrt{3}}{3}$$

To rationalise the fraction $\dfrac{1}{1+\sqrt{3}}$, we multiply its numerator and its denominator by $1 - \sqrt{3}$, giving

$$\begin{aligned} \frac{1}{(1+\sqrt{3})} \times \frac{1-\sqrt{3}}{(1-\sqrt{3})} &= \frac{1-\sqrt{3}}{1-\sqrt{3}+\sqrt{3}-3} \\ &= \frac{1-\sqrt{3}}{-2} \\ &= -\frac{1}{2} + \frac{1}{2}\sqrt{3} \end{aligned}$$

In general, to rationalise the fraction $\dfrac{1}{a \pm \sqrt{b}}$, we multiply its numerator and its denominator by $a \mp \sqrt{b}$.

Example 4 Express each of the following in the form $a+b\sqrt{c}$.

a) $\dfrac{3}{\sqrt{5}}$ **b)** $\dfrac{2+\sqrt{3}}{1-\sqrt{3}}$

SOLUTION

a) Multiplying numerator and denominator by $\sqrt{5}$ gives

$$\frac{3}{\sqrt{5}}\times\frac{\sqrt{5}}{\sqrt{5}}=\frac{3\sqrt{5}}{5}$$

(This is in the form $a+b\sqrt{c}$, where $a=0$, $b=\dfrac{3}{5}$ and $c=5$.)

b) Multiplying numerator and denominator by $1+\sqrt{3}$ gives

$$\frac{2+\sqrt{3}}{1-\sqrt{3}}\times\frac{1+\sqrt{3}}{1+\sqrt{3}}=\frac{2+2\sqrt{3}+\sqrt{3}+3}{1-3}$$

$$=\frac{5+3\sqrt{3}}{-2}=-\frac{5}{2}-\frac{3}{2}\sqrt{3}$$

(This is in the form $a+b\sqrt{c}$, where $a=-\frac{5}{2}$, $b=-\frac{3}{2}$ and $c=3$.)

Exercise 18B

1 Simplify each of these.

a) $\sqrt{12}$ **b)** $\sqrt{50}$

c) $\sqrt{112}$ **d)** $\sqrt{75}+2\sqrt{27}$

e) $5\sqrt{20}+2\sqrt{45}$ **f)** $2\sqrt{8}+\sqrt{200}-4\sqrt{18}$

g) $\sqrt{32}+\sqrt{128}-\sqrt{200}$ **h)** $7\sqrt{5}+3\sqrt{20}-\sqrt{80}$

2 Express each of the following in the form $\dfrac{a\sqrt{c}}{b}$, where a, b and c are integers.

a) $\dfrac{3}{\sqrt{2}}$ **b)** $\dfrac{5}{\sqrt{3}}$ **c)** $\dfrac{2}{\sqrt{6}}$ **d)** $\dfrac{\sqrt{7}}{\sqrt{2}}$

e) $\dfrac{10\sqrt{7}}{\sqrt{5}}$ **f)** $\dfrac{3\sqrt{5}}{2\sqrt{6}}$ **g)** $\dfrac{3\sqrt{50}}{5\sqrt{27}}$ **h)** $\dfrac{4\sqrt{45}}{5\sqrt{8}}$

3 Express each of the following in the form $\dfrac{a+b\sqrt{c}}{d}$, where a, b, c and d are integers.

a) $\dfrac{1}{2-\sqrt{3}}$ **b)** $\dfrac{1}{3+\sqrt{5}}$ **c)** $\dfrac{2}{5-\sqrt{7}}$ **d)** $\dfrac{3}{6+\sqrt{3}}$

e) $\dfrac{2+\sqrt{2}}{2-\sqrt{2}}$ **f)** $\dfrac{3+\sqrt{2}}{5+\sqrt{2}}$ **g)** $\dfrac{6+\sqrt{5}}{2-\sqrt{5}}$ **h)** $\dfrac{3+\sqrt{24}}{2+\sqrt{6}}$

***4** Simplify $\dfrac{(2+\sqrt{2})(3+\sqrt{5})(\sqrt{5}-2)}{(\sqrt{5}-1)(1+\sqrt{2})}$

Logarithms

A logarithm ('log' for short) is an index. To see this, consider the result

$$10^2 = 100$$

This can be written using logarithm notation as

$$\log_{10} 100 = 2$$

The number 10 is called the **base** of the logarithm.

Similarly, the result $2^3 = 8$ can be written as

$$\log_2 8 = 3$$

In this case, the base is 2.

Generally,

$$a^b = c \quad \text{is written as} \quad \log_a c = b$$

Most calculators only have the function keys $\log_{10}$ and $\log_e$, where $e = 2.718$ (to three decimal places). We usually write

$$\log_e x = \ln x$$

There will be more about the constant e on pages 429–31.

Result

i) $\log ab = \log a + \log b$

ii) $\log\left(\dfrac{a}{b}\right) = \log a - \log b$

iii) $\log a^n = n \log a$

These three results are true for any base.

Proof of (i)

Let $x = \log_c a$ and $y = \log_c b$. Then

$$c^x = a \quad \text{and} \quad c^y = b \qquad [1]$$

Therefore, from [1] we have $ab = c^x c^y = c^{x+y}$ and, by the definition of log,

$$\log_c(ab) = x + y$$
$$= \log_c a + \log_c b$$

as required.

Proof of (ii)

Using [1] again, we have

$$\frac{a}{b} = \frac{c^x}{c^y} = c^{x-y}$$

Therefore, by the definition of log,

$$\log_c\left(\frac{a}{b}\right) = x - y$$
$$= \log_c a - \log_c b$$

as required.

Proof of (iii)

Using [1] again, we have

$$a^n = (c^x)^n = c^{xn}$$

Therefore, by definition of log,

$$\log_c a^n = xn$$
$$= n \log_c a$$

as required.

Example 5 Express each of the following as a single logarithm.

a) $\log 3 + \log 5$ **b)** $\log 27 - \log 9$

c) $3 \log 2 + \log 4 - \log 8$ **d)** $2 \log x - 3 \log y + 2 \log xy$

SOLUTION

a)
$$\log 3 + \log 5 = \log (3 \times 5)$$
$$= \log 15$$

b)
$$\log 27 - \log 9 = \log\left(\frac{27}{9}\right)$$
$$= \log 3$$

c) Notice that $3 \log 2 = \log 2^3 = \log 8$. Therefore,

$$3 \log 2 + \log 4 - \log 8 = \log 8 + \log 4 - \log 8$$
$$= \log 4$$

d) Now

$$2 \log x = \log x^2 \qquad 3 \log y = \log y^3 \quad \text{and} \quad 2 \log xy = \log (xy)^2$$

Therefore,

$$2 \log x - 3 \log y + 2 \log xy = \log x^2 - \log y^3 + \log (xy)^2$$
$$= \log\left(\frac{x^2}{y^3}\right) + \log (xy)^2$$
$$= \log\left(\frac{x^2}{y^3} \times x^2 y^2\right)$$
$$= \log\left(\frac{x^4}{y}\right)$$

Example 6 Express each of the following in terms of $\log a$, $\log b$, $\log c$.

a) $\log\left(\dfrac{1}{a^2}\right)$ **b)** $\log\left(\dfrac{ab}{c}\right)$ **c)** $\log\sqrt{\dfrac{a}{bc^2}}$

SOLUTION

a)

$$\log\left(\frac{1}{a^2}\right) = \log a^{-2}$$

$$= -2\log a$$

b)

$$\log\left(\frac{ab}{c}\right) = \log(ab) - \log c$$

$$= \log a + \log b - \log c$$

c)

$$\log\sqrt{\frac{a}{bc^2}} = \log\left(\frac{a}{bc^2}\right)^{\frac{1}{2}}$$

$$= \frac{1}{2}\log\left(\frac{a}{bc^2}\right)$$

$$= \frac{1}{2}(\log a - \log bc^2)$$

$$= \frac{1}{2}(\log a - (\log b + 2\log c))$$

$$\therefore \quad \log\sqrt{\frac{a}{bc^2}} = \frac{1}{2}(\log a - \log b - 2\log c)$$

Example 7 Solve the following equations for x.

a) $3^x = 10$ **b)** $5^{4x-1} = 7^{x+2}$ **c)** $2^{2x} - 2^x = 6$

SOLUTION

a)

$$3^x = 10$$

Taking logarithms of both sides gives

$$\log 3^x = \log 10$$

$$\therefore \quad x\log 3 = \log 10$$

$$\therefore \quad x = \frac{\log 10}{\log 3} = 2.10$$

b)

$$5^{4x-1} = 7^{x+2}$$

Taking logarithms of both sides gives

$$\log 5^{4x-1} = \log 7^{x+2}$$

$$\therefore \quad (4x-1)\log 5 = (x+2)\log 7$$

$$\therefore \quad 4x\log 5 - x\log 7 = \log 5 + 2\log 7$$

$$\therefore \quad x(4\log 5 - \log 7) = \log 5 + 2\log 7$$

$$\therefore \quad x = \frac{\log 5 + 2\log 7}{4\log 5 - \log 7} = 1.22$$

c) Rearranging gives

$$2^{2x} - 2^x - 6 = 0$$

We notice that this is a quadratic equation in 2^x. That is,

$$(2^x)^2 - 2^x - 6 = 0$$

If we let $y = 2^x$, then we have

$$y^2 - y - 6 = 0$$

$$\therefore \quad (y+2)(y-3) = 0$$

Solving gives $y = -2$ or $y = 3$. Therefore,

$$2^x = -2 \quad \text{or} \quad 2^x = 3$$

Now when $2^x = -2$, taking logarithms gives

$$x \log 2 = \log(-2)$$

But $\log(-2)$ does not exist, since the log function is not defined for negative values. Therefore, this gives no solutions.

When $2^x = 3$, taking logarithms gives

$$x \log 2 = \log 3$$

$$\therefore \quad x = \frac{\log 3}{\log 2} = 1.58$$

Example 8 Solve the pair of simultaneous equations

$$\log(y - x) = 0 \quad \text{and} \quad 2\log y = \log(21 + x)$$

SOLUTION

If $\log(y - x) = 0$, then $y - x = 1$. Using the properties of logs, $2\log y = \log(21 + x)$ becomes

$$\log y^2 = \log(21 + x)$$

$$\therefore \quad y^2 = 21 + x$$

$$\therefore \quad y^2 - x = 21$$

The problem has been reduced to solving

$$y - x = 1 \qquad [1] \qquad \text{and} \qquad y^2 - x = 21 \qquad [2]$$

Subtracting [2] from [1] gives

$$y^2 - y = 20$$

$$\therefore \quad y^2 - y - 20 = 0$$

$$\therefore \quad (y-5)(y+4) = 0$$

Solving gives $y = 5$ and $y = -4$.

When $y = 5$, we have $5 - x = 1$. That is, $x = 4$. When $y = -4$, we have $-4 - x = 1$. That is, $x = -5$.

The solutions are $x = 4$, $y = 5$ or $x = -5$, $y = -4$.

Exercise 18C

1 Express each of the following in terms of $\log a$, $\log b$.

a) $\log(ab)$ **b)** $\log\left(\frac{a}{b}\right)$ **c)** $\log(a^2b)$ **d)** $\log(\sqrt{a})$

e) $\log\left(\frac{1}{a^2}\right)$ **f)** $\log(a\sqrt{b})$ **g)** $\log\left(\frac{a^3}{b}\right)$ **h)** $\log\left(\frac{a^2}{b^3}\right)$

i) $\log\left(\sqrt{\frac{a}{b}}\right)$ **j)** $\log\left(\frac{1}{ab^4}\right)$ **k)** $\log\left(\frac{1}{\sqrt{ab}}\right)$ **l)** $\log\left(\sqrt[6]{a^2b}\right)$

2 Express each of the following as a single logarithm.

a) $\log 3 + \log 4$ **b)** $\log 2 + \log 7$ **c)** $\log 15 - \log 3$
d) $\log 24 - \log 4$ **e)** $\log 2 + \log 3 + \log 5$ **f)** $\log 6 + \log 3 - \log 9$
g) $2\log 3 + \log 4 - \log 12$ **h)** $3\log 2 + 2\log 5 - \log 20$ **i)** $\frac{1}{2}\log 80 - \frac{1}{2}\log 5$
j) $\log 15 - \frac{1}{2}\log 9$ **k)** $2\log a - \log b - \log c$ **l)** $\log a + \frac{1}{2}\log b - 3\log c$

3 Solve each of the following equations for x, giving your answers correct to two decimal places.

a) $2^x = 5$ **b)** $3^x = 7$ **c)** $9^x = 28$ **d)** $12^x = 5$
e) $2^{x-1} = 3^{x+1}$ **f)** $5^{x-2} = 2^{x+3}$ **g)** $3^{2x-1} = 5^x$ **h)** $6^{1-x} = 2^{3x+1}$

4 Solve each of the following equations for x, giving your answers correct to two decimal places where necessary.

a) $2^{2x} - 5 \times 2^x + 4 = 0$ **b)** $3^{2x} - 30 \times 3^x + 81 = 0$ **c)** $4^{2x} - 5 \times 4^x + 6 = 0$
d) $2 \times 2^{2x} - 11 \times 2^x + 5 = 0$ **e)** $3 \times 5^{2x} - 14 \times 5^x + 8 = 0$ **f)** $2 \times 3^{2x} - 5 \times 3^x = 4$

5 **a)** Factorise $y^3 - 13y^2 + 39y - 27$.
b) Hence solve the equation $3^{3x} - 13 \times 3^{2x} + 39 \times 3^x - 27 = 0$.

6 **a)** Factorise $4a^3 - 29a^2 + 47a - 10$.
b) Hence solve the equation

$$4 \times 4^{3x} - 29 \times 4^{2x} + 47 \times 4^x - 10 = 0$$

giving your answers correct to two decimal places.

7 Solve the equation $5 \times 5^{3x} - 31 \times 5^{2x} + 31 \times 5^x - 5 = 0$.

8 Solve the equation

$$3 \times 2^{3x} - 22 \times 2^{2x} + 37 \times 2^x = 10$$

giving your answers correct to two decimal places.

9 Given x and y are both positive, solve the simultaneous equations

$$\log(xy) = 7 \qquad \log\left(\frac{x}{y}\right) = 1$$

10 Given that $\log(p - q + 1) = 0$ and $\log(pq) + 1 = 0$, show that $p = q = \frac{1}{\sqrt{10}}$.

***11** Solve the following equations for x

a) $2^{2x+1} - 9 \times 2^x + 4 = 0$
b) $5^{2x} - 6 \times 5^{x+1} + 125 = 0$
c) $4^x - 5 \times 2^{x+1} + 16 = 0$
d) $3^{3x} - 13 \times 3^{2x} + 13 \times 3^{x+1} - 27 = 0$

Modelling curves

When an experiment is performed and data obtained, we hope to establish a mathematical relationship between the variables in question. If the data falls on a straight line, then the relationship will be of the form $y = mx + c$, where m is the gradient and c is the y intercept.

Example 9 The table gives the mass (M) of a chemical that dissolved in water at each of six selected temperature levels (θ °C).

θ	5	10	15	20	25	30
M	26	27	29.5	30	31	33

It is suspected that the relationship between θ and M is of the form $M = a\theta + b$, where a and b are constants. Find estimates for the values of a and b.

SOLUTION

Plotting the data gives the graph on the right.

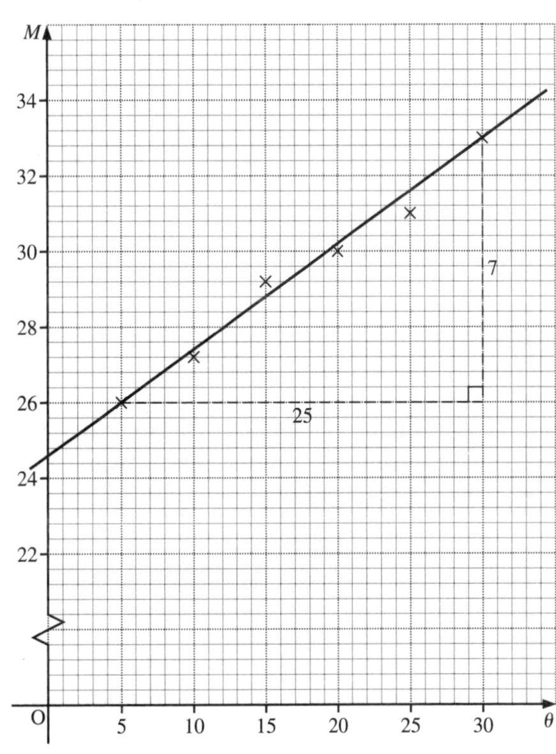

It can be seen that the data is linear and the gradient of the line is given by

$$\text{Gradient} = \frac{33 - 26}{30 - 5} = 0.3$$

The y intercept is 24.6. Therefore, the equation of the line is

$$M = 0.3\theta + 24.6$$

In many cases, the results of an experiment do not lie on a straight line but on a curve. If the relationship has one of the forms $y = kx^n$ or $y = ka^x$, then logarithms can be used to convert the curved graph into a straight line.

Example 10 The results of an experiment involving the variables a and b are shown in the table.

a	0.1	0.2	0.3	0.4	0.5
b	0.012	0.048	0.108	0.192	0.3

It is suspected that the relationship between a and b is of the form $b = ka^n$, where k and n are constants. Estimate the values of n and k.

SOLUTION

We assume that the relationship is of the form $b = ka^n$ and take logarithms of both sides, giving

$$\log b = \log ka^n$$

$$\therefore \quad \log b = \log k + \log (a^n)$$

$$\therefore \quad \log b = n \log a + \log k$$

This equation is of the form $y = mx + c$, with

$y = \log b \qquad m = n$ (the gradient) $\qquad x = \log a \qquad c = \log k$

Therefore, if we plot $\log b$ against $\log a$, the graph should be a straight line.

$\log a$	-1	-0.7	-0.5	-0.4	-0.3
$\log b$	-1.9	-1.3	-1.0	-0.7	-0.5

Plotting these values gives the graph on the right.

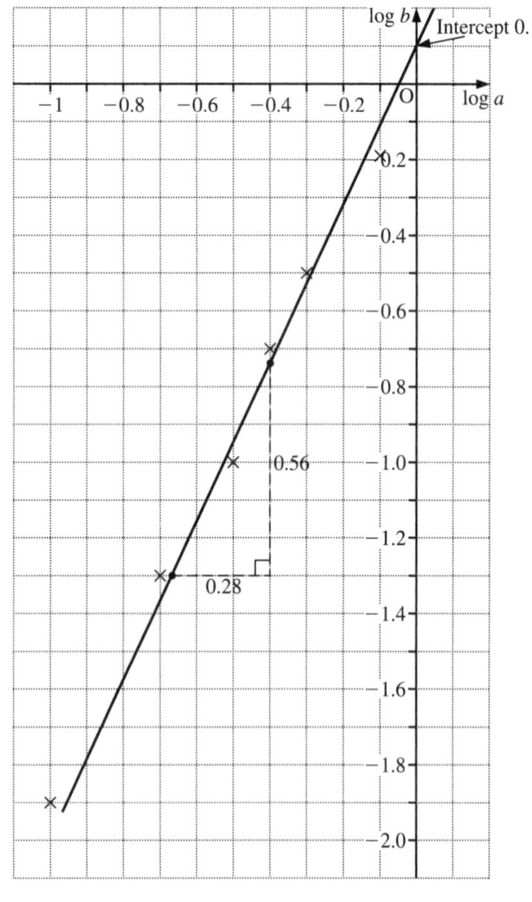

The graph is a straight line with gradient given by

$$\text{Gradient} = \frac{-1.3 + 0.74}{-0.68 + 0.4}$$

$$= \frac{-0.56}{-0.28} = 2$$

and intercept 0.1. This tells us that $b = ka^n$ is a good model for the relationship and that $n = 2$. Therefore,

$$\log k = 0.1$$

$$\therefore \quad k = 10^{0.1}$$

$$\therefore \quad k = 1.3$$

A model for the relationship is $b = 1.3a^2$.

Example 11 The temperature (y °C) of a body t minutes after being placed in a certain liquid is given in the table.

t	1	2	3	4	5	6
y	30	23.2	17	15.6	12.1	9.5

It is suspected that a model for the relationship between y and t is $y = ka^t$, where k and a are constants. Estimate the values of k and a, and comment on the suitability of the model.

SOLUTION

We assume the relationship is of the form $y = ka^t$ and take logarithms of both sides, giving

$$\log y = \log(ka^t)$$

$$\therefore \quad \log y = \log k + t \log a$$

$$\therefore \quad \log y = t \log a + \log k$$

This equation is of the form $Y = mX + c$ with

$$Y = \log y \qquad m = \log a \text{ (the gradient)} \qquad X = t \qquad c = \log k$$

Therefore, we use the values below to plot $\log y$ against t.

t	1	2	3	4	5	6
$\log y$	1.5	1.4	1.2	1.2	1.1	1.0

Plotting gives the graph on the right.

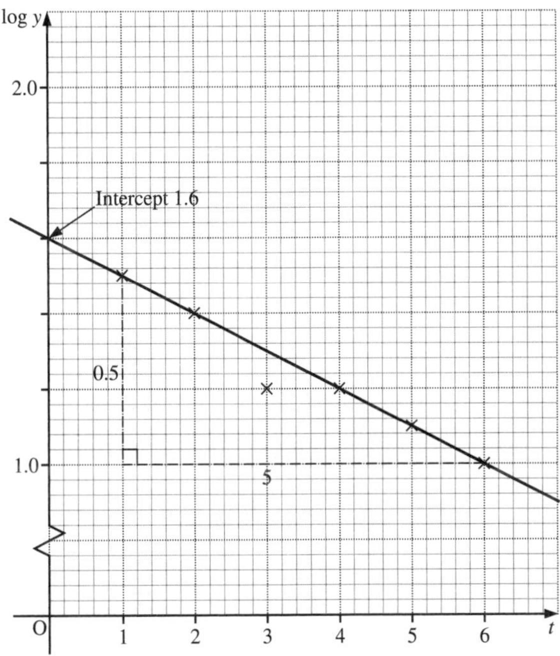

A straight line fits the data closely, which tells us that the relationship $y = ka^t$ is a good model. The gradient of the line is given by

$$\frac{1.0 - 1.5}{6 - 1} = -0.1$$

and the intercept is 1.6. Therefore, we have

$$\log a = -0.1 \qquad \text{and} \qquad \log k = 1.6$$
$$\therefore \quad a = 10^{-0.1} \qquad \therefore \quad k = 10^{1.6}$$
$$\therefore \quad a = 0.8 \qquad \therefore \quad k = 39.8$$

A model for the relationship is $y = 39.8 \times (0.8)^t$.

Exercise 18D

1 The following table shows the estimated total cost, £y, of producing x copies of a magazine.

Number of copies, x	100	500	1000	2000	3000	5000
Cost (£y)	450	650	900	1400	1900	2900

a) Represent these data on a graph.
b) Hence find an equation connecting x and y.

2 A cookery book gives the following table of times for roasting a turkey.

Mass of bird, x (kg)	2	4	6	9	12
Number of hours, y	1.5	2.5	3.5	5	6.5

a) Represent these data on a graph, plotting the mass on the x-axis and the time on the y-axis.
b) Hence find an equation connecting x and y.

3 Various weights, w newtons, are hung on an elastic string, and the corresponding length, l cm, of the stretched string is recorded.

Weight, w (N)	2	4	10	15	20	25
Length, l (cm)	26	31	52	63	81	120

a) Represent these data on a graph.
b) Show that, ignoring one pair of data, the rest are approximately linearly related, and find an equation connecting w and l in the form $l = \alpha w + \beta$.
c) Estimate the unstretched length of the string.

4 In a chemical experiment, an investigation is under way to determine a relationship between the amount, x grams, of a catalyst which is used, and the time, t seconds, until the reaction commences. The results of seven such experiments are recorded below.

Amount, x (grams)	5	10	15	20	30	50	100
Time, t (seconds)	300	292	278	266	245	202	106

a) Represent these data on a graph.
b) Find a linear relationship which approximately models the connection between x and t.
c) Criticise the model.

5 The following data are expected to follow a formula of the type $y = Ax^2 + B$, where A and B are constants to be determined.

x	2	4	6	8	10
y	22	26	34	46	60

a) Plot a graph of y against x^2.
b) Estimate the values of A and B.
c) Assuming that the model holds for larger values of x, estimate the value of y when $x = 20$.

6 Two variables h and t, are thought to be related by a formula of the type $h = A + B\sqrt{t}$, where A and B are constants to be determined.

t	10	20	30	40	50	60
h	9.3	11.9	14.0	15.6	17.1	18.5

a) By plotting a graph of h against $\sqrt{t}$, verify the claim, and find the values of A and B.
b) Hence estimate the value of t when $h = 25$.

7 A body is heated and then allowed to cool. The following table shows the temperature, T °C, at a time t minutes after cooling has commenced.

t (min)	10	16	20	40	50	100	200
T (°C)	120	83	70	45	40	30	25

a) By plotting a graph of T against $\frac{1}{t}$ find an equation connecting T and t.

b) Estimate the value which the temperature approaches after a long time.

8 The flow of water through pipes of various diameters is recorded below.

Diameter, d (cm)	1	2	3	4	5	6
Flow, f (litres min^{-1})	20	60	105	160	225	290

a) Find a rule connecting f and d of the form $f = A \times d^n$, where A and n are constants to be determined.
b) Hence estimate the size of pipe which should be used if a flow of 400 litres min^{-1} is required.

9 The table below shows the maximum load capacity for wires of varying radii:

Radius, x (mm)	3	6	9	12	15
Load, T (tonnes)	0.7	3.7	9.8	19.5	33.2

a) Show that x and T follow a rule of the form $T = A \times x^n$, where A and n are constants to be determined.
b) What size of wire should be used to lift a load of 25 tonnes?

10 The following data are expected to follow a formula of the type $y = A \times x^n$, where A and B are constants to be determined.

x	2	5	10	20	50	100
y	70	45	32	22	14	10

a) By plotting a graph of $\log y$ against $\log x$, verify the claim, and hence find the values of A and n.
b) Estimate the value of y when $x = 200$.

11 Show that the following data follow an approximate rule of the form $y = A \times x^n$, where A and n are constants to be determined.

x	10	20	30	40	50
y	600	75	20	10	5

12 **a)** For the following data plot a graph of $\log y$ against x.

x	5	10	15	20	25	30
y	10	25	60	150	380	950

b) Deduce that x and y are related by a formula $y = A \times B^x$, and write down the values of the constants A and B.

13 A researcher wishes to find a formula to fit the figures below.

n	10	20	30	35	40	45	50
P	3	165	9590	72 800	302 320	4 198 330	31 881 080

a) By plotting a suitable graph show that, ignoring one pair, these data are approximately related by a formula of the form $P = A \times B^n$.
b) Assuming that the values of n are correct, suggest a value of P for the incorrect reading.

14 The number of infected cells in a body is expected to follow a formula $N = A \times B^t$, where N is the number of cells at a time t hours from when the infection was introduced. The following table shows the result of an experiment in which attempts were made to estimate N for various values of t.

t	2	4	6	8	10	12
N	20	160	1460	13 120	118 100	1 062 900

Show that the readings do fit the proposed model, and find the values of the constants A and B.

Exercise 18E: Examination questions

1 Given that $p = t^{\frac{1}{2}} + t^{-\frac{1}{2}}$ and $q = t^{\frac{1}{2}} - t^{-\frac{1}{2}}$, find

$$p^2q^2 + 2$$

in terms of t, giving your answer in its simplest form. (EDEXCEL)

2 Given that $27^x = 9^{x-1}$, find the value of x. (EDEXCEL)

3 Given that $t^{\frac{1}{3}} = y$, $y \neq 0$,

a) express $6t^{-\frac{1}{3}}$ in terms of y.

b) Hence, or otherwise, find the values of t for which $6t^{-\frac{1}{3}} - t^{\frac{1}{3}} = 5$. (EDEXCEL)

4 a) Solve the equation $2x^{\frac{1}{3}} = x^{-\frac{2}{3}}$.

b) Use your calculator to find the value of y, correct to three significant figures, where $3^y = 6$. (MEI)

5 a) Write down the exact value of x given that $4^x = 8$.

b) Use logarithms to find y, correct to 3 decimal places, when $5^y = 10$. (UODLE)

6 An athlete plans a training schedule which involves running 20 km in the first week of training; in each subsequent week the distance is to be increased by 10% over the previous week. Write down an expression for the distance to be covered in the nth week according to this schedule, and find in which week the athlete would first cover more than 100 km. (UCLES)

7 The variable x satisfies the equation $3^x.4^{2x+1} = 6^{x+2}$.

By taking logarithms of both sides, show that $x = \dfrac{\log 9}{\log 8}$. (WJEC)

8 Express as a single logarithm in its simplest form

$$\log 2 + 2\log 18 - \tfrac{3}{2}\log 36$$ (UODLE)

9 Solve the pair of simultaneous equations

$$\log(x + y) = 0$$

$$2\log x = \log(y - 1)$$ (NICCEA)

10 Show that $x + 1$ is a factor of $x^3 - 3x^2 + 4$. Hence solve the equation

$$x^3 - 3x^2 + 4 = 0$$

Given that $x = 3^y$, show that

$$x^3 - 3x^2 + 4 = 3^{3y} - 3^{2y+1} + 4$$

Hence, solve the equation

$$3^{3y} - 3^{2y+1} + 4 = 0$$

giving your answer to three significant figures. (AEB 95)

11 The variables x and y are believed to satisfy an equation of the form

$$\frac{1}{y} = \frac{1}{x} + \frac{1}{a}$$

where a is a constant. For four chosen values of x, the corresponding approximate values of y, rounded to three significant figures, are obtained experimentally. The results are given in the following table.

x	2.5	5	7.5	10
y	4.35	28.5	−33.0	−14.9

By drawing a suitable linear graph, obtain an estimate for the value of a. (NEAB)

12 At time t seconds the displacement, s metres, of a particle, from a fixed origin, is given by the formula $s = ut + \frac{1}{2}at^2$, where u and a are constants. The values of s are measured after each second and results shown in the table below.

t	1	2	3	4	5	6
s	60	105	130	135	125	95

a) By drawing a graph of $\left(\frac{s}{t}\right)$ against t show that the formula above is approximately valid.

b) Use your graph to estimate values for u and a.

13 A zoo keeps an official record of the mass, x kg, and the average daily food intake, y kg, of each adult animal. Four selected pairs of values of x and y are given in the table below.

Animal	Cheetah	Deer	Rhinoceros	Hippopotamus
x	40	170	1500	3000
y	1.8	5.0	33.2	50.0

Show, by drawing a suitable linear graph, that these values are approximately consistent with a relationship between x and y of the form

$$y = ax^m$$

where a and m are constants.

Use your linear graph to find an estimate of the value of m.

The zoo has a bear with mass 500 kg. Assuming that this animal's food intake and mass conform to the relationship mentioned above, indicate a point on your linear graph corresponding to the bear and estimate the bear's average daily food intake. (NEAB)

19 Calculus with exponentials and logarithms

A first-rate theory predicts; a second-rate theory forbids; and a third-rate theory explains after the event.
A.I. KITAIGORODSKII

Exponential functions

On pages 416–21, we met equations which contained expressions such as 2^x, 3^x, that is, expressions of the form a^x. These are examples of **exponential functions**.

The graph of $y = 2^x$ is given on the right.

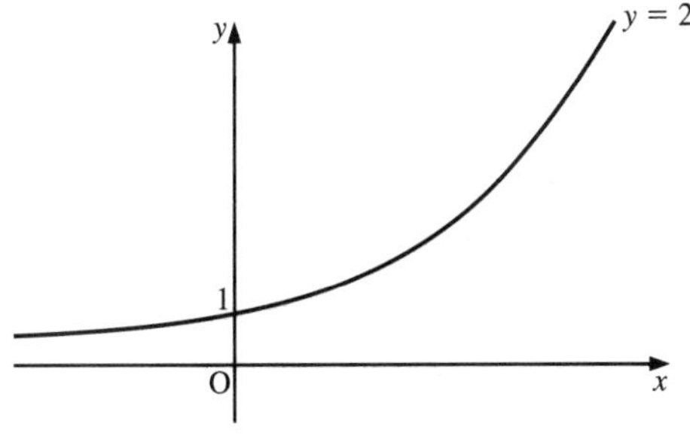

Let P(x, y) and Q($x + \delta x$, $y + \delta y$) be two nearby points on the curve $y = 2^x$. Then

$$y + \delta y = 2^{x+\delta x}$$

$$\therefore \quad \delta y = 2^{x+\delta x} - y$$

$$= 2^{x+\delta x} - 2^x$$

$$\therefore \quad \delta y = 2^x(2^{\delta x} - 1)$$

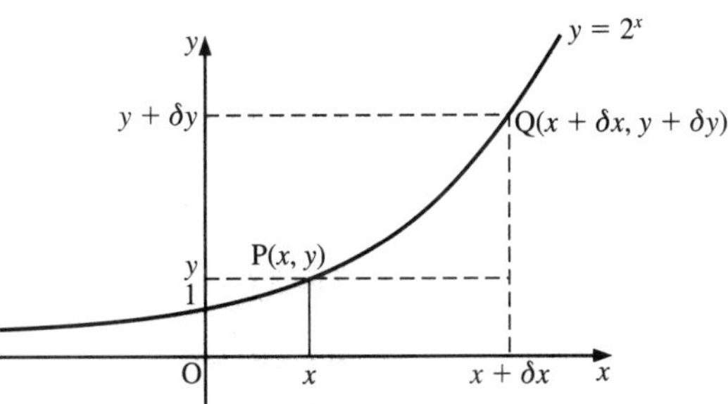

This gives

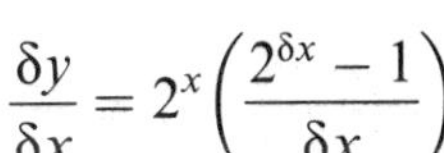

$$\frac{\delta y}{\delta x} = 2^x\left(\frac{2^{\delta x} - 1}{\delta x}\right)$$

Now

$$\frac{dy}{dx} = \lim_{\delta x \to 0}\left[\frac{dy}{dx}\right]$$

Therefore, we have

$$\frac{dy}{dx} = \lim_{\delta x \to 0}\left[2^x\left(\frac{2^{\delta x} - 1}{\delta x}\right)\right]$$

$$= 2^x \times \lim_{\delta x \to 0}\left(\frac{2^{\delta x} - 1}{\delta x}\right)$$

Taking small values for δx and using a calculator, it appears that

$$\lim_{\delta x \to 0}\left(\frac{2^{\delta x} - 1}{\delta x}\right) \approx 0.693$$

$$\therefore \quad \frac{d}{dx}(2^x) \approx 0.693 \times 2^x$$

A similar analysis of $y = 3^x$ gives

$$\frac{d}{dx}(3^x) = 3^x \times \lim_{\delta x \to 0}\left(\frac{3^{\delta x} - 1}{\delta x}\right)$$

Taking small values for δx and again using a calculator, it appears that

$$\lim_{\delta x \to 0}\left(\frac{3^{\delta x}-1}{\delta x}\right) \approx 1.099$$

$$\therefore \quad \frac{\mathrm{d}}{\mathrm{d}x}(3^x) \approx 1.099 \times 3^x$$

From the above, we notice that in the case of $y = 2^x$ the gradient of the curve is less than $y = 2^x$, whereas in the case of $y = 3^x$ the gradient of the curve is greater than $y = 3^x$. In fact, there exists an exponential function $y = a^x$ such that the gradient function equals $y = a^x$.

From our investigation above, we see that $2 < a < 3$. In fact, $a = 2.718\,28$ to five decimal places. This value of a is denoted by the symbol e and is irrational.

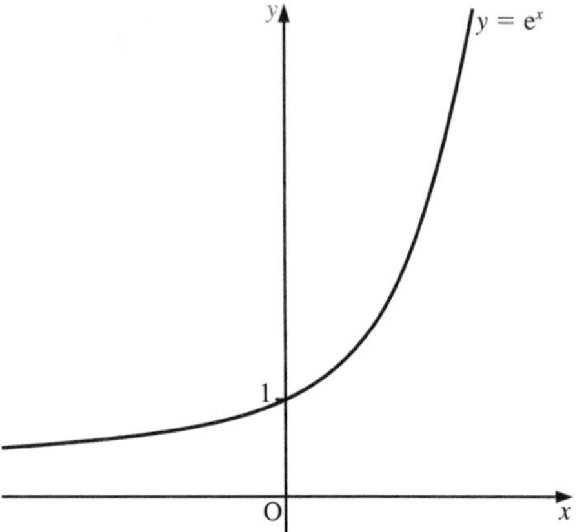

The function $y = \mathrm{e}^x$ is called the **exponential function**, and

$$\frac{\mathrm{d}y}{\mathrm{d}x} = \mathrm{e}^x$$

The graph of $y = \mathrm{e}^x$ is shown on the right.

In general

$$\frac{\mathrm{d}}{\mathrm{d}x}(\mathrm{e}^{ax}) = a\mathrm{e}^{ax}$$

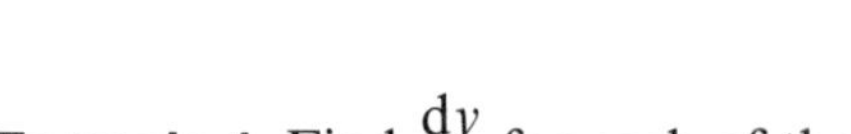

Example 1 Find $\frac{\mathrm{d}y}{\mathrm{d}x}$ for each of the following.

a) $y = \mathrm{e}^{2x}$ **b)** $y = 5\mathrm{e}^{\frac{1}{x}}$

SOLUTION

a) Using the chain rule gives

$$\frac{\mathrm{d}y}{\mathrm{d}x} = \mathrm{e}^{2x}(2x)' = 2\mathrm{e}^{2x}$$

b) Using the chain rule gives

$$\frac{\mathrm{d}y}{\mathrm{d}x} = 5\mathrm{e}^{\frac{1}{x}}\left(\frac{1}{x}\right)' = 5\mathrm{e}^{\frac{1}{x}}\left(-\frac{1}{x^2}\right)$$

$$\therefore \quad \frac{\mathrm{d}y}{\mathrm{d}x} = -\frac{5}{x^2}\,\mathrm{e}^{\frac{1}{x}}$$

Example 2 Find $\frac{dy}{dx}$ for each of the following.

a) $y = (1 - e^x)^4$ **b)** $y = \frac{2}{3 + e^{3x}}$

SOLUTION

a) Using the chain rule gives

$$\frac{dy}{dx} = 4(1 - e^x)^3(1 - e^x)' = -4e^x(1 - e^x)^3$$

b) We can write $y = 2(3 + e^{3x})^{-1}$ and then use the chain rule to obtain

$$\frac{dy}{dx} = -2(3 + e^{3x})^{-2}(3 + e^{3x})'$$

$$= -2(3 + e^{3x})^{-2}(3e^{3x})$$

$$\therefore \quad \frac{dy}{dx} = \frac{-6e^{3x}}{(3 + e^{3x})^2}$$

Using these techniques, we can now integrate functions of the form $g'(x)f(g(x))$.

Example 3 Find each of the following integrals.

a) $\int 2e^{-x}\,dx$ **b)** $\int (1 - e^{-3x})^2\,dx$ **c)** $\int 5xe^{x^2}\,dx$

SOLUTION

a) To find $\int 2e^{-x}\,dx$, we notice that the derivative of e^{-x} is $-e^{-x}$. Therefore,

$$\int 2e^{-x}\,dx = -2e^{-x} + c$$

b) First expand the bracket, obtaining

$$\int (1 - e^{-3x})^2\,dx = \int (1 - 2e^{-3x} + e^{-6x})\,dx$$

Therefore,

$$\int (1 - e^{-3x})^2\,dx = x + \frac{2}{3}e^{-3x} - \frac{e^{-6x}}{6} + c$$

c) To find $\int 5xe^{x^2}\,dx$, we notice that the derivative of e^{x^2} is $2xe^{x^2}$ and that we have e^{x^2} multiplied by an x term. Therefore,

$$\int 5xe^{x^2}\,dx = \frac{5}{2}e^{x^2} + c$$

Example 4 Find $\dfrac{dy}{dx}$ for each of the following.

a) $y = x^2e^x$ **b)** $y = \dfrac{e^{3x}}{x}$

SOLUTION

a) Using the product rule, we have

$$\frac{dy}{dx} = x^2(e^x)' + e^x(x^2)' = x^2e^x + 2xe^x$$

$$\therefore \quad \frac{dy}{dx} = xe^x(x+2)$$

b) Using the quotient rule, we have

$$\frac{dy}{dx} = \frac{x(e^{3x})' - e^{3x}(x)'}{x^2} = \frac{3xe^{3x} - e^{3x}}{x^2}$$

$$\therefore \quad \frac{dy}{dx} = \frac{e^{3x}(3x-1)}{x^2}$$

Exercise 19A

1 Find $\dfrac{dy}{dx}$ for each of the following.

a) $y = e^{3x}$ **b)** $y = 2e^{5x}$ **c)** $y = e^{x^2}$ **d)** $y = 4e^{\sqrt{x}}$

e) $y = e^{\frac{5}{x}}$ **f)** $y = e^{2x-3}$ **g)** $y = e^{x^2+1}$ **h)** $y = \dfrac{3}{e^{x^4}}$

2 Integrate each of the following with respect to x.

a) e^{4x} **b)** e^{6x} **c)** e^{-2x} **d)** $6e^{3x}$ **e)** $10e^{-2x}$

f) $2xe^{x^2}$ **g)** $x^2e^{x^3}$ **h)** $2x^3e^{x^4}$ **i)** $\dfrac{e^{\sqrt{x}}}{\sqrt{x}}$ **j)** $\dfrac{x^2}{e^{x^3}}$

3 Find $f'(x)$ for each of the following.

a) $f(x) = (1 + e^x)^2$ **b)** $f(x) = (1 - e^{-3x})^4$ **c)** $f(x) = \dfrac{1}{e^x + 1}$ **d)** $f(x) = \sqrt{1 - 2e^{4x}}$

e) $f(x) = (2 + e^{x^3})^4$ **f)** $f(x) = \dfrac{1}{3 - e^{4x^2}}$ **g)** $f(x) = (x + e^{2x})^3$ **h)** $f(x) = (3e^x + e^{3x})^4$

4 Find each of these integrals.

a) $\displaystyle\int e^x(3 + e^x)^2\,dx$ **b)** $\displaystyle\int 2e^x(e^x - 4)^3\,dx$ **c)** $\displaystyle\int \frac{4e^{-2x}}{(1 + e^{-2x})^2}\,dx$

d) $\displaystyle\int \frac{(e^{-x} + 7)^2}{e^x}\,dx$ **e)** $\displaystyle\int e^x\sqrt{4 + e^x}\,dx$ **f)** $\displaystyle\int e^{5x}\sqrt{e^{5x} + 2}\,dx$

g) $\displaystyle\int \frac{e^{-x}}{(1 - e^{-x})^2}\,dx$ **h)** $\displaystyle\int \frac{e^{3x}}{\sqrt{e^{3x} - 1}}\,dx$ **i)** $\displaystyle\int \frac{1}{2e^x\sqrt{1 - e^{-x}}}\,dx$

5 Differentiate each of the following with respect to x.

a) xe^{2x} **b)** x^2e^{4x} **c)** $2x^3e^{-3x}$

d) $e^{2x}(1+e^x)^2$ **e)** $e^{3x}(1-2e^{-x})^3$ **f)** $e^{-x}(1+3e^x)^4$

g) $\dfrac{e^{2x}}{x}$ **h)** $\dfrac{e^{3x}}{x^2}$ **i)** $\dfrac{e^{2x}}{1+e^{2x}}$

j) $\dfrac{2e^{4x}}{1-e^x}$ **k)** $\dfrac{e^{3x}}{1+e^{2x}}$ **l)** $\dfrac{1+e^x}{1-e^x}$

6 Given that $y = xe^{2x}$, show that $x\dfrac{dy}{dx} \equiv (2x+1)y$.

7 Given that $y = \dfrac{e^x}{e^x+1}$, show that $(1+e^x)\dfrac{dy}{dx} \equiv y$.

8 Given that $y = \dfrac{e^{x^2}}{x}$, show that $x\dfrac{dy}{dx} + y \equiv 2xe^{x^2}$.

9 Given that $y = e^x - e^{-x}$, show that $\left(\dfrac{dy}{dx}\right)^2 \equiv y^2 + 4$.

10 Given that $y = Ae^{4x} + Be^{-4x}$, where A and B are constants, show that $\dfrac{d^2y}{dx^2} - 16y \equiv 0$.

11 Given that $y = 2e^{3x} + e^{-2x}$, show that $\dfrac{d^2y}{dx^2} - 2\dfrac{dy}{dx} - 3y \equiv 5e^{-2x}$.

12 Given that $y = (1-3x)e^{2x}$, show that $\dfrac{d^2y}{dx^2} - 4\dfrac{dy}{dx} + 4y \equiv 0$.

***13** Given that $y = \dfrac{e^x - e^{-x}}{e^x + e^{-x}}$, show that $\dfrac{d^2y}{dx^2} + 2y\dfrac{dy}{dx} \equiv 0$.

***14** Given that $y = \dfrac{2}{e^x + e^{-x}}$, show that $\dfrac{d^2y}{dx^2} \equiv y - 2y^3$.

***15** Find $\displaystyle\int e^{(x+e^x)}\,dx$.

Natural logarithms

Logarithms to the base e are called **natural logarithms**. The notation $\ln x$ is used as the standard abbreviation for $\log_e x$. We will use $\ln x$ from this point onwards.

The function $\ln x$ is the inverse function of e^x. Notice that

$$\ln e^x = x \ln e = x \log_e e = x(1)$$

Therefore, we have

$$\ln(e^x) = x \quad \text{and} \quad e^{\ln x} = x$$

The graph of $y = \ln x$ is shown on the right.

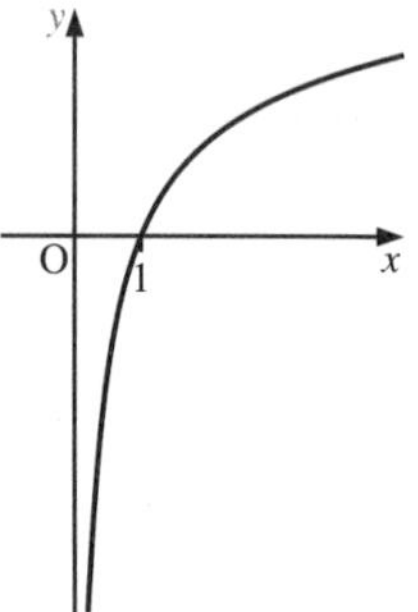

Notice further that since $\ln x$ is the inverse function of e^x, the graph of $y = \ln x$ is a reflection, in the line $y = x$, of the graph of e^x.

Result I

If $y = \ln x$, then

$$\frac{dy}{dx} = \frac{1}{x}$$

Proof

If $y = \ln x$, then by definition we have $e^y = x$.

Differentiating e^y with respect to y gives

$$e^y = \frac{dx}{dy}$$

$$\therefore \quad \frac{dy}{dx} = \frac{1}{e^y}$$

$$\therefore \quad \frac{dy}{dx} = \frac{1}{x}$$

as required.

Example 5 Find $\frac{dy}{dx}$ for each of the following.

a) $y = \ln 3x$ **b)** $y = \ln(x^2 - 1)$

SOLUTION

a) Using the chain rule, we have

$$\frac{dy}{dx} = \frac{1}{3x} \times (3x)'$$

$$\therefore \quad \frac{dy}{dx} = \frac{1}{x}$$

b) Using the chain rule, we have

$$\frac{dy}{dx} = \frac{1}{x^2 - 1} \times (x^2 - 1)'$$

$$\therefore \quad \frac{dy}{dx} = \frac{2x}{x^2 - 1}$$

Result II

$$\int \frac{f'(x)}{f(x)}\,dx = \ln(f(x)) + c$$

Proof

If $y = \ln(f(x))$, then

$$\frac{dy}{dx} = \frac{1}{f(x)} \times f'(x) = \frac{f'(x)}{f(x)}$$

as required.

Example 6 Find each of these integrals.

a) $\int \frac{1}{4x+1}\,dx$ **b)** $\int \frac{x^2+1}{x^3+3x}$ **c)** $\int \frac{1}{x \ln x}$

SOLUTION

a) We rewrite the integral with the derivative of the denominator as the numerator and compensate by introducing a constant outside the integral. We thus obtain

$$\int \frac{1}{4x+1}\,dx = \frac{1}{4}\int \frac{4}{4x+1}\,dx = \frac{1}{4}\ln(4x+1) + c$$

b) Here we note that the derivative of $x^3 + 3x$ is $3x^2 + 3 = 3(x^2 + 1)$. Therefore,

$$\int \frac{x^2+1}{x^3+3x}\,dx = \frac{1}{3}\int \frac{3x^2+3}{x^3+3x}\,dx = \frac{1}{3}\ln(x^3+3x) + c$$

c) Rewriting the integral gives

$$\int \frac{1}{x\ln x}\,dx = \int \frac{\left(\frac{1}{x}\right)}{\ln x}\,dx = \ln(\ln x) + c$$

Definite integrals involving logarithms

We know that

$$\int \frac{1}{x}\,\mathrm{d}x = \ln x + c \qquad [1]$$

However, we also know that the function $\ln x$ is only valid provided $x > 0$. Therefore, the integral [1] is only valid when $x > 0$. We therefore have a problem, since $\frac{1}{x}$ exists for negative values of x.

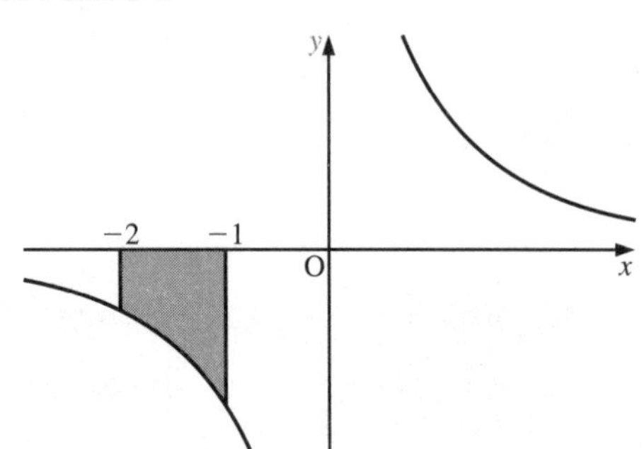

Consider the area between the curve $y = \frac{1}{x}$, the x-axis and the ordinates $x = -1$ and $x = -2$.

Clearly, this area exists and is identical to the area between the curve, the x-axis and the ordinates $x = 1$ and $x = 2$. Therefore, for both positive and negative values of x, we have

$$\int \frac{1}{x}\,\mathrm{d}x = \ln|x| + c \quad \text{and} \quad \int \frac{\mathrm{f}'(x)}{\mathrm{f}(x)}\,\mathrm{d}x = \ln|\mathrm{f}(x)| + c$$

It is usual practice to write only the modulus sign in the case of definite integrals.

Since the function $y = \frac{1}{x}$ is not defined when $x = 0$, the definite integral

$$\int_a^b \frac{1}{x}\,\mathrm{d}x$$

is invalid if the interval $[a, b]$ includes $x = 0$. In other words, a and b must both have the same sign for the integral to be valid. Note that it is **not** possible to integrate across a discontinuity.

Example 7 Evaluate $\displaystyle\int_1^2 \frac{2}{1-3x}\,\mathrm{d}x$

SOLUTION

We have

$$\int_1^2 \frac{2}{1-3x}\,\mathrm{d}x = -\frac{2}{3}\int_1^2 \frac{-3}{1-3x}\,\mathrm{d}x$$

$$= -\frac{2}{3}\Big[\ln|1-3x|\Big]_1^2 = -\frac{2}{3}\left(\ln|-5| - \ln|-2|\right)$$

$$\therefore \quad \int_1^2 \frac{2}{1-3x}\,\mathrm{d}x = -\frac{2}{3}\ln\left(\frac{5}{2}\right)$$

Example 8 Find $\dfrac{\mathrm{d}y}{\mathrm{d}x}$ for each of the following

a) $y = (2 - 3\ln x)^3$ **b)** $y = \dfrac{1}{\sqrt{1+\ln x}}$ **c)** $y = x\ln x$ **d)** $y = \dfrac{\ln 2x}{x^3}$

SOLUTION

a) Using the chain rule gives

$$\frac{\mathrm{d}y}{\mathrm{d}x} = 3(2-3\ln x)^2 \times (2-3\ln x)'$$

$$= 3(2-3\ln x)^2 \times -\frac{3}{x}$$

$$\therefore \quad \frac{\mathrm{d}y}{\mathrm{d}x} = -\frac{9(2-3\ln x)^2}{x}$$

b) We can write $y = \dfrac{1}{\sqrt{1+\ln x}}$ in the form $y = (1+\ln x)^{-\frac{1}{2}}$ and then use the chain rule. This gives

$$\frac{\mathrm{d}y}{\mathrm{d}x} = -\frac{1}{2}(1+\ln x)^{-\frac{3}{2}} \times (1+\ln x)'$$

$$= -\frac{1}{2}(1+\ln x)^{-\frac{3}{2}} \times \frac{1}{x}$$

$$\therefore \quad \frac{\mathrm{d}y}{\mathrm{d}x} = -\frac{1}{2x(1+\ln x)^{\frac{3}{2}}}$$

c) Using the product rule, we have

$$\frac{\mathrm{d}y}{\mathrm{d}x} = x(\ln x)' + \ln x\,(x)' = x\left(\frac{1}{x}\right) + \ln x$$

$$\therefore \quad \frac{\mathrm{d}y}{\mathrm{d}x} = 1 + \ln x$$

d) Using the quotient rule, we have

$$\frac{\mathrm{d}y}{\mathrm{d}x} = \frac{(\ln 2x)'x^3 - (\ln 2x)(x^3)'}{x^6}$$

$$= \frac{\left(\dfrac{2}{2x}\right)x^3 - (\ln 2x)3x^2}{x^6}$$

$$= \frac{x^2 - 3x^2\ln 2x}{x^6}$$

$$\therefore \quad \frac{\mathrm{d}y}{\mathrm{d}x} = \frac{1-3\ln 2x}{x^4}$$

Exercise 19B

1 Find $\frac{dy}{dx}$ for each of the following.

a) $y = \ln(1 + 2x)$ **b)** $y = \ln(1 - 4x)$ **c)** $y = \ln(1 + x^2)$ **d)** $y = \ln(x^3 - 2)$

e) $y = \ln(x^3 - 3x)$ **f)** $y = \ln(e^x + 4)$ **g)** $y = \ln(1 + e^{6x})$ **h)** $y = \ln(\sqrt{x})$

2 Integrate each of the following with respect to x.

a) $\frac{1}{1+x}$ **b)** $\frac{3}{2+3x}$ **c)** $\frac{2}{x}$ **d)** $\frac{2}{5+x}$

e) $\frac{4}{2x-1}$ **f)** $\frac{1}{4-x}$ **g)** $\frac{3}{5+6x}$ **h)** $\frac{5}{2-3x}$

3 Integrate each of the following with respect to x.

a) $\frac{2x}{x^2+1}$ **b)** $\frac{3x^2}{x^3+4}$ **c)** $\frac{4x}{2-x^2}$ **d)** $\frac{x^2}{3+x^3}$

e) $\frac{2x-1}{x^2-x}$ **f)** $\frac{x-3}{x^2-6x+1}$ **g)** $\frac{e^x}{1+e^x}$ **h)** $\frac{e^{5x}}{e^{5x}+1}$

4 Differentiate each of the following with respect to x.

a) $(1 + \ln x)^2$ **b)** $(3 - 2\ln x)^3$ **c)** $\frac{1}{1+\ln x}$ **d)** $x^2 \ln x$

e) $x^3 \ln x$ **f)** $x \ln(1 + x)$ **g)** $x \ln(2 - x)$ **h)** $x^2 \ln(3 + 2x)$

i) $x \ln(1 + x^2)$ **j)** $\frac{x}{\ln x}$ **k)** $\frac{2-\ln x}{x}$ **l)** $\frac{1+\ln x}{x^2}$

5 Given that $y = \frac{\ln(1+x)}{x^2}$, show that $x^2\frac{dy}{dx} + 2xy \equiv \frac{1}{1+x}$.

6 Given that $y = \ln\left(\frac{1+x}{1-x}\right)$, show that $(1 - x^2)\frac{dy}{dx} \equiv 2$.

7 Given that $y = \ln(\ln x)$, show that $(\ln x)\frac{d^2y}{dx^2} + \frac{1}{x}\frac{dy}{dx} + \frac{1}{x^2} \equiv 0$.

8 Given that $y = \ln(1 + e^x)$, show that $\frac{d^2y}{dx^2} \equiv e^x\left(1 - \frac{dy}{dx}\right)^2$.

***9** Show that $\frac{d}{dx}\left[\ln(x + \sqrt{x^2+1})\right] = \frac{1}{\sqrt{x^2+1}}$.

***10 a)** Given $y = \dfrac{(x+4)^5(x-3)^3}{(x-2)^8}$, show that

$$\ln y = A\ln(x+4) + B\ln(x-3) + C\ln(x-2)$$

where A, B and C are constants to be determined.

b) Deduce that $\dfrac{d}{dx}\left[\dfrac{(x+4)^5(x-3)^3}{(x-2)^8}\right] = \dfrac{3(34-9x)(x+4)^4(x-3)^2}{(x-2)^9}$.

***11** Show that for any positive constant a,

a) $a^x = e^{x\ln a}$ **b)** $\dfrac{d}{dx}[a^x] = a^x \ln a$

***12** Find $\displaystyle\int \frac{2e^x}{e^x + e^{-x}}\,dx$.

Applications

Example 9 Find the equation of the tangent and the normal to the curve $y = \ln\left(\dfrac{x-1}{x+1}\right)$ at the point P where $x = 3$.

SOLUTION

Differentiating $y = \ln\left(\dfrac{x-1}{x+1}\right)$ gives

$$\frac{dy}{dx} = \frac{1}{\left(\dfrac{x-1}{x+1}\right)}\left[\frac{(x+1)-(x-1)}{(x+1)^2}\right]$$

$$= \left(\frac{x+1}{x-1}\right)\left[\frac{2}{(x+1)^2}\right]$$

$$\therefore \quad \frac{dy}{dx} = \frac{2}{x^2-1}$$

When $x = 3$: $\quad y = \ln\left(\dfrac{3-1}{3+1}\right) = \ln\left(\dfrac{1}{2}\right) = -\ln 2$

At P$(3, -\ln 2)$: $\quad \dfrac{dy}{dx} = \dfrac{2}{3^2-1} = \dfrac{1}{4}$

The tangent is of the form $y = \frac{1}{4}x + c$. Using P$(3, -\ln 2)$, we have

$$-\ln 2 = \frac{3}{4} + c \quad \therefore \quad c = -\ln 2 - \frac{3}{4}$$

The equation of the tangent line is

$$y = \frac{1}{4}x - \ln 2 - \frac{3}{4}$$

or alternatively,

$$x - 4y = 4\ln 2 + 3$$

When the gradient of the tangent at P is $\frac{1}{4}$, the gradient of the normal at P is -4. Therefore, the normal is of the form $y = -4x + c$. Using P$(3, -\ln 2)$ gives

$$-\ln 2 = -4(3) + c \quad \therefore \quad c = 12 - \ln 2$$

The equation of the normal is $y = -4x + 12 - \ln 2$.

Example 10 Find and classify the stationary points on the curve $y = x^2e^x$. Hence sketch the curve.

SOLUTION

At stationary points, $\frac{dy}{dx} = 0$. Using the product rule, we have

$$\frac{dy}{dx} = x^2e^x + 2xe^x$$

When $\frac{dy}{dx} = 0$,

$$x^2e^x + 2xe^x = 0$$

$$\therefore \quad xe^x(x + 2) = 0$$

Solving gives $x = 0$ and $x = -2$.

When $x = 0$, $y = 0$. When $x = -2$, $y = 4e^{-2}$.

The points $(0, 0)$ and $(-2, 4e^{-2})$ are stationary points. To determine their nature, we consider $\frac{d^2y}{dx^2}$.

Now

$$\frac{d^2y}{dx^2} = x^2e^x + 4xe^x + 2e^x$$

$$= e^x(x^2 + 4x + 2)$$

When $x = 0$: $\left.\frac{d^2y}{dx^2}\right|_{x=0} = 2 > 0$

Therefore, $(0, 0)$ is a minimum.

When $x = -2$: $\left.\frac{d^2y}{dx^2}\right|_{x=-2} = -2e^{-2} < 0$

Therefore, $(-2, 4e^{-2})$ is a maximum.

The sketch of $y = x^2e^x$ is shown on the right.

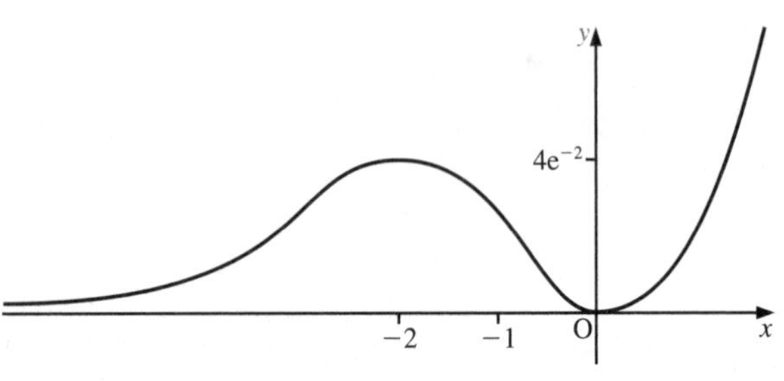

Exercise 19C

This exercise revises the calculus techniques which were developed in earlier chapters, and applies the techniques to the exponential and logarithmic functions.

Tangents and normals

1 Find the equation of the tangent to the curve $y = x + e^{2x}$ at the point where $x = 0$.

2 Find the equation of the tangent and the normal to the curve $y = \ln(1 + x)$ at the point where $x = 2$.

3 Find the equation of the tangent and the normal to the curve $y = xe^x$ at the point where $x = 1$.

4 The tangent to the curve $y = x \ln x$ at the point (e, e) meets the x-axis at A and the y-axis at B. Find the distance AB.

5 Find the equation of the tangent and the normal to the curve $y = e^x \ln x$ at the point where $x = 1$.

6 Find the coordinates of the points on the curve $y = x^2 + \ln x$ where the gradient is 3.

7 Show that there are two points on the curve $y = \ln(1 + x^2)$ where the gradient is $\frac{5}{13}$. Find the coordinates of these points.

8 Find the coordinates of the point on the curve $y = \ln(e^x + e^{-x})$ where the gradient is $\frac{3}{5}$.

Stationary points

9 Given that $y = x^2e^{-x}$, show that $\frac{dy}{dx} = x(2 - x)e^{-x}$. Hence find the coordinates of the two points on the curve $y = x^2e^{-x}$ where the gradient is zero.

10 Given that $y = \frac{\ln x}{x}$ for $x > 0$, show that $\frac{dy}{dx} = \frac{1 - \ln x}{x^2}$. Hence find the coordinates of the point on the curve $y = \frac{\ln x}{x}$ where the gradient is zero.

11 Given that $y = \frac{e^x}{x^2 - 3}$, show that $\frac{dy}{dx} = \frac{e^x(x + 1)(x - 3)}{(x^2 - 3)^2}$. Hence find the coordinates of the two points on the curve $y = \frac{e^x}{x^2 - 3}$ where the gradient is zero.

12 Find and classify the stationary values on each of the following curves.

a) $y = 2\ln(1 + x) - \ln x \quad (x > 0)$

b) $y = \frac{e^x}{x^3}$

c) $y = x(3 - \ln x) \quad (x > 0)$

d) $y = e^x(x - 1)^2$

Areas and volumes of revolution

13 Find the area between the curve $y = e^{2x}$ and the x-axis from $x = 0$ to $x = 3$.

14 Find the area between the curve $y = \dfrac{2}{x+3}$ and the x-axis from $x = 2$ to $x = 7$.

15 Find the area between the curve $y = 4e^{2x} - 3e^{x}$ and the x-axis from $x = 1$ to $x = 2$.

16 The line $y = \frac{1}{3}$ meets the curve $y = \dfrac{1}{x+1}$ at the point P.

a) Find the coordinates of P.
b) Calculate the area bounded by the line, the curve and the y-axis.

17 The line $y = x + 1$ meets the curve $y = \dfrac{8}{5-x}$ at the points P and Q

a) Find the coordinates of P and Q.
b) Show that the area enclosed between the curve and the line between P and Q is $6 - 8\ln 2$.

18 The region bounded by the curve $y = e^{x} + 1$, the x-axis, the line $x = 0$ and the line $x = 2$ is rotated through 360° about the x-axis. Calculate the volume of the solid generated.

19 The region R is bounded by the curve $y = 3 + \dfrac{2}{x+1}$, the x-axis, the y-axis and the line $x = 4$.

a) Show that the area of R is $12 + 2\ln 5$.

R is rotated through 360° about the x-axis.

b) Show that the volume of the solid generated is $\dfrac{4\pi}{5}[49 + 15\ln 5]$.

Implicit differentiation

20 Find the equations of the tangent and the normal to the curve $e^{x+y} = 1 + x^2 - y^2$, at the point $(3, -3)$.

21 Find the equations of the tangent and the normal to the curve $\ln(x^2 - y + 1) = 8x - y^2$, at the point $(2, 4)$.

22 Show that the tangent to the curve $e^{y} + x^2 = 2e^2$, at the point $(e, 2)$, passes through the point $(0, 4)$.

23 Given $x\ln y + 2y = 3$, show that $\dfrac{dy}{dx} \equiv \dfrac{y(2y-3)}{x(2y+x)}$.

24 A curve is given by the equation $e^{xy} + x = 4$. Deduce these results.

a) $e^{xy}\left(y + x\dfrac{dy}{dx}\right) + 1 = 0$ **b)** $x\dfrac{d^2y}{dx^2} + x^2\left(\dfrac{dy}{dx}\right)^2 + 2(1 + xy)\dfrac{dy}{dx} + y^2 \equiv 0$

Parametric differentiation

25 Find the equations of the tangent and the normal to the curve $x = e^{t} + t$, $y = e^{3t} - 2t$, at the point where $t = 0$.

26 Find the equations of the tangent and the normal to the curve $x = 2 + \ln t$, $y = t^3$, at the point $(2, 1)$.

27 Find the equations of the tangents to the curve $x = t\ln t$, $y = 3t - t^2$, at the points where $y = 2$.

28 A curve is given by the parametric equations $x = \ln(1-t)$, $y = \ln(1-t^2)$, where $-1 \leqslant t \leqslant 1$.

a) Deduce the following results.

i) $\dfrac{dy}{dx} = \dfrac{2t}{1+t}$ **ii)** $\dfrac{d^2y}{dx^2} = \dfrac{2(t-1)}{(1+t)^2}$

b) Hence find and classify any stationary points on the curve.

29 A curve is given by the parametric equations $x = te^{-2t}$, $y = t^2e^{-2t}$.

a) Deduce the following results.

i) $\dfrac{dy}{dx} = \dfrac{2t(1-t)}{1-2t}$ **ii)** $\dfrac{d^2y}{dx^2} = \dfrac{2(1-2t+2t^2)e^{2t}}{(1-2t)^3}$

b) Hence find and classify all the stationary values on the curve.

Exponentials, logarithms and trigonometric functions combined

30 Show that $\dfrac{d}{dx}[\ln(\sec x + \tan x)] = \sec x$.

31 Work out the following integrals.

a) $\displaystyle\int \frac{\cos x}{1+\sin x}$ **b)** $\displaystyle\int \frac{\sin x - \cos x}{\sin x + \cos x}\,dx$ **c)** $\displaystyle\int \tan x\,dx$

d) $\displaystyle\int \frac{\sec^2 x}{1+\tan x}\,dx$ **e)** $\displaystyle\int \frac{\sec x \tan x}{1+\sec x}\,dx$

32 Find and classify all stationary values on the curve $y = e^x \cos x$, in the range $0 \leqslant x \leqslant 2\pi$.

33 Given that $y = e^{3x}\sin 2x$, show that $\dfrac{d^2y}{dx^2} - 6\dfrac{dy}{dx} + 13y \equiv 0$.

***34** Show that $\dfrac{d}{dx}[\ln(\tan \frac{1}{2}x)] \equiv \operatorname{cosec} x$.

***35** Find $\displaystyle\int \frac{2\cos x}{\sin x + \cos x}\,dx$.

Exercise 19D: Examination questions

1 A curve has the equation $y = (2x+1)e^{-2x}$.

a) Find $\dfrac{dy}{dx}$.

b) Calculate the coordinates of the turning point of the curve and determine its nature.

(UODLE)

2 a) Show that, if $y = xe^{-x}$, then $\frac{dy}{dx} = (1 + Ax)e^{-x}$, where A is a constant to be determined.

b) Find the coordinates of the turning point on the graph of $y = xe^{-x}$. (You do not need to determine its nature.)

c) Find the values of $\frac{dy}{dx}$ and $\frac{d^2y}{dx^2}$ when $x = 0$.

d) Explain, with the aid of a sketch, the information given about the shape of the graph of $y = xe^{-x}$ near $x = 0$ by the values of $\frac{dy}{dx}$ and $\frac{d^2y}{dx^2}$ when $x = 0$. (UODLE)

3 The curve with equation $y = e^{3x} + 1$ meets the line $y = 8$ at the point $(h, 8)$.

a) Find h, giving your answer in terms of natural logarithms.

b) Show that the area of the finite region enclosed by the curve with equation $y = e^{3x} + 1$, the x-axis, the y-axis and the line $x = h$ is $2 + \frac{1}{3}\ln 7$. (EDEXCEL)

4

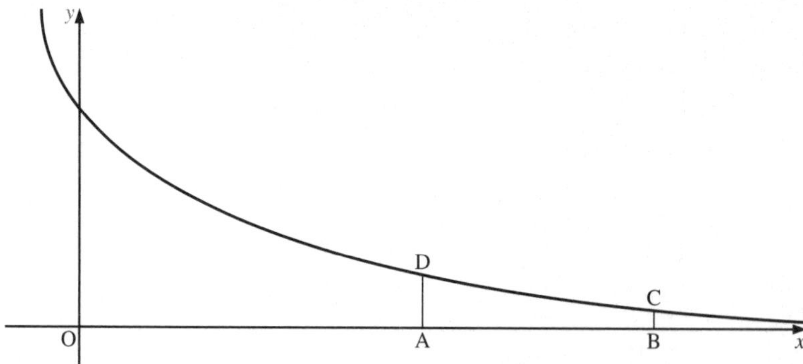

The diagram shows a sketch of the graph of $y = e^{-x}$. The points A and B have coordinates $(n, 0)$ and $(n + 1, 0)$ respectively, and the points C and D on the curve are such that AD and BC are parallel to the y-axis.

i) Show that B lies on the tangent to the curve at D.

ii) Find the area of the region ABCD under the curve, and show that the line BD divides this region into two parts whose areas are in the ratio e : e – 2. (UCLES)

5 Without using a calculator, show that $\int_{\frac{2}{3}}^{3} \frac{1}{3x - 1}\, dx = \ln 2$. (WJEC)

6 Show that $\int_{2}^{14} \frac{1}{2x - 1}\, dx = \ln 3$. (WJEC)

7 Given $y = \ln(x^2 - 4x + 5)$, find an expression for $\frac{dy}{dx}$.

Hence find $\int_{3}^{4} \frac{x - 2}{x^2 - 4x + 5}\, dx$.

8 The region R in the first quadrant is bounded by the curve $y = e^{-x}$, the x-axis, the y-axis, and the line $x = 2$. Show that the volume of the solid formed when R is completely rotated about the x-axis is

$$\pi \int_0^2 e^{-2x}\, dx$$

and evaluate this volume, giving your answer in terms of e and π. (UCLES)

9 The diagram shows part of the curve $y = \dfrac{1}{x-1}$ together with the line $y = \frac{1}{2}x$.

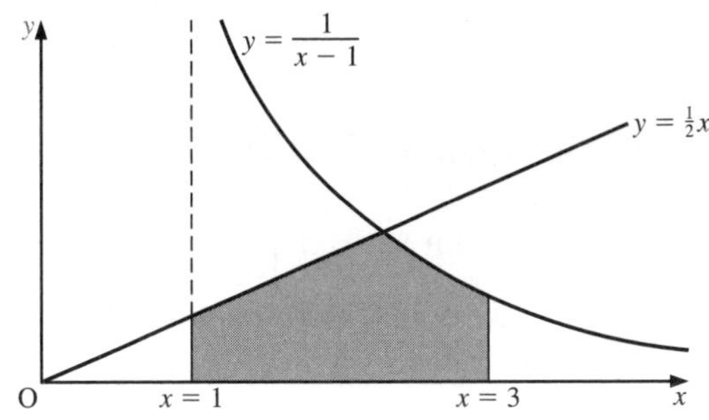

Show that the area, A, of the shaded region is given by

$$A = \frac{3}{4} + \int_2^3 \frac{1}{x-1}\, dx$$

Hence find A.

10 The figure shows a sketch of the curve with equation

$$y = \frac{3\cos x}{2 - \sin x} \text{ for } 0 \leqslant x \leqslant \pi.$$

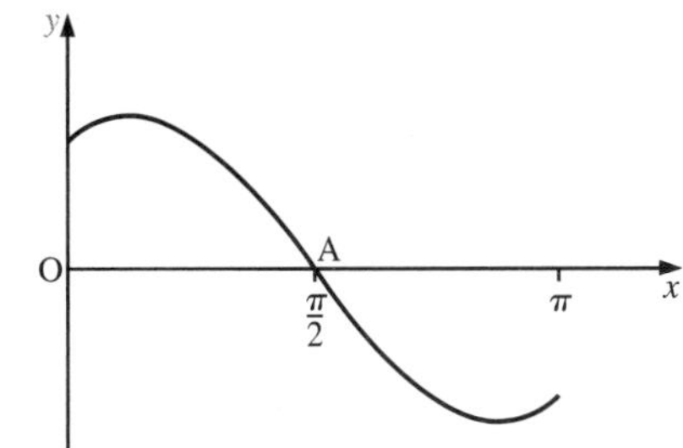

a) Find the values of x, in the interval $0 \leqslant x \leqslant \pi$, for which $\dfrac{dy}{dx} = 0$, giving your answers in radians.

b) Determine the range of values taken by y.

c) Determine the equation of the normal to the curve at the point $A\left(\dfrac{\pi}{2}, 0\right)$.

d) Calculate the area of the finite region bounded by the curve, the y-axis and the normal at A. (AEB 92)

20 Integration II

I have only to touch mathematics, and I forget everything else on earth.
SONYA KOVALEVSKY

Change of variable

Suppose we want to find $\int x(2x+1)^3\,\mathrm{d}x$. This can be done by either expanding the brackets and writing $x(2x+1)^3$ as a polynomial, or using a change of variable.

We can change the variable to u by letting $u = 2x + 1$. Then the integral becomes

$$\int xu^3\,\mathrm{d}x = \int xu^3\frac{\mathrm{d}x}{\mathrm{d}u}\,\mathrm{d}u$$

However, we must change to the variable u entirely. Therefore, the x and $\frac{\mathrm{d}x}{\mathrm{d}u}$ must be replaced.

Since $u = 2x + 1$, we have $\frac{\mathrm{d}u}{\mathrm{d}x} = 2$, i.e. $\frac{\mathrm{d}x}{\mathrm{d}u} = \frac{1}{2}$, and $x = \frac{u-1}{2}$. Therefore, the integral becomes

$$\int\frac{(u-1)}{2}u^3\,\frac{1}{2}\,\mathrm{d}u$$

Expanding and integrating give

$$\begin{aligned}\int x(2x+1)^3\,\mathrm{d}x &= \int\frac{u^4-u^3}{4}\,\mathrm{d}u\\ &= \frac{u^5}{20}-\frac{u^4}{16}+c\\ &= \frac{u^4}{80}(4u-5)+c\end{aligned}$$

Since $u = 2x + 1$, we have

$$\begin{aligned}\int x(2x+1)^3\,\mathrm{d}x &= \frac{(2x+1)^4}{80}[4(2x+1)-5]+c\\ &= \frac{(2x+1)^4}{80}(8x-1)+c\end{aligned}$$

Example 1 Find each of the following integrals by using the substitution suggested.

a) $\int (x+1)(3x-2)^5\,\mathrm{d}x, \quad u = 3x-2$

b) $\int x\sqrt{x-1}\,\mathrm{d}x, \quad u = x-1$

c) $\int \frac{3x-4}{\sqrt{2x+1}}\,\mathrm{d}x, \quad u = \sqrt{2x+1}$

SOLUTION

a) Given $u = 3x-2$, then $\frac{\mathrm{d}u}{\mathrm{d}x} = 3$ and $x = \frac{u+2}{3}$. Substituting gives

$$I = \int (x+1)(3x-2)^5\,\mathrm{d}x = \int (x+1)(3x-2)^5 \frac{\mathrm{d}x}{\mathrm{d}u}\,\mathrm{d}u$$

$$= \int \left(\frac{u+2}{3}+1\right)u^5\,\frac{1}{3}\,\mathrm{d}u$$

$$= \int \left(\frac{u^6}{9}+\frac{5u^5}{9}\right)\mathrm{d}u$$

$$\therefore \quad I = \frac{u^7}{63}+\frac{5u^6}{54}+c = \frac{u^6}{378}(6u+35)+c$$

Since $u = 3x-2$, we have

$$I = \frac{(3x-2)^6}{378}[6(3x-2)+35]+c$$

$$\therefore \quad I = \frac{(3x-2)^6}{378}(18x+23)+c$$

b) Given $u = x-1$, then $\frac{\mathrm{d}u}{\mathrm{d}x} = 1$ and $x = u+1$. Substituting gives

$$I = \int x\sqrt{x-1}\,\mathrm{d}x = \int x\sqrt{x-1}\,\frac{\mathrm{d}x}{\mathrm{d}u}\,\mathrm{d}u$$

$$= \int (u+1)u^{\frac{1}{2}}\,\mathrm{d}u$$

$$= \int u^{\frac{3}{2}}+u^{\frac{1}{2}}\,\mathrm{d}u$$

$$\therefore \quad I = \frac{2u^{\frac{5}{2}}}{5}+\frac{2u^{\frac{3}{2}}}{3}+c = \frac{2u^{\frac{3}{2}}}{15}(3u+5)+c$$

Since $u = x-1$, we have

$$I = \frac{2(x-1)^{\frac{3}{2}}}{15}[3(x-1)+5]+c$$

$$\therefore \quad I = \frac{2(x-1)^{\frac{3}{2}}}{15}(3x+2)+c$$

c) Given $u = \sqrt{2x+1}$, then

$$\frac{du}{dx} = \frac{1}{2}(2x+1)^{-\frac{1}{2}} \times 2 = \frac{1}{(2x+1)^{\frac{1}{2}}} = \frac{1}{u}$$

Also we have

$$u^2 = 2x + 1 \quad \therefore \quad x = \frac{u^2 - 1}{2}$$

Therefore,

$$\begin{aligned} I &= \int \frac{3x-4}{\sqrt{2x+1}} \frac{dx}{du}\, du \\ &= \int \frac{3x-4}{u} u\, du \\ &= \int \left[3\left(\frac{u^2-1}{2}\right) - 4\right] du \\ &= \int \left(\frac{3u^2}{2} - \frac{11}{2}\right) du \end{aligned}$$

$$\therefore \quad I = \frac{u^3}{2} - \frac{11u}{2} + c = \frac{u}{2}(u^2 - 11) + c$$

Since $u = (2x+1)^{\frac{1}{2}}$, we have

$$I = \frac{(2x+1)^{\frac{1}{2}}}{2}[(2x+1) - 11] + c = \frac{(2x+1)^{\frac{1}{2}}}{2}(2x - 10) + c$$

$$\therefore \quad I = (x-5)\sqrt{2x+1} + c$$

Example 2 Evaluate $\displaystyle\int_3^4 \frac{3x}{\sqrt{x-2}}\, dx$ using the substitution $u = \sqrt{x-2}$.

SOLUTION

Given $u = (x-2)^{\frac{1}{2}}$, then

$$\frac{du}{dx} = \frac{1}{2}(x-2)^{-\frac{1}{2}} = \frac{1}{2(x-2)^{\frac{1}{2}}} = \frac{1}{2u}$$

The limits must be changed from x limits to u limits. This is done by calculating the value of u when $x = 3$ and when $x = 4$.

When $x = 3$, $u = \sqrt{3-2} = 1$. When $x = 4$, $u = \sqrt{4-2} = \sqrt{2}$.

Also we have

$$u^2 = x - 2 \quad \therefore \quad x = u^2 + 2$$

Therefore,

$$\int_3^4 \frac{3x}{\sqrt{x-2}}\, dx = \int_1^{\sqrt{2}} \frac{3(u^2+2)}{u} 2u\, du = \int_1^{\sqrt{2}} (6u^2 + 12)\, du$$

$$= \Big[2u^3 + 12u\Big]_1^{\sqrt{2}} = 16\sqrt{2} - 14$$

Standard forms

Result I

i) $\displaystyle\int \frac{\mathrm{d}x}{a^2+x^2} = \frac{1}{a}\tan^{-1}\left(\frac{x}{a}\right) + c$

ii) $\displaystyle\int \frac{\mathrm{d}x}{\sqrt{a^2-x^2}} = \sin^{-1}\left(\frac{x}{a}\right) + c$

Proof of (i)

Let $x = a\tan\theta$, then $\dfrac{\mathrm{d}x}{\mathrm{d}\theta} = a\sec^2\theta$. Therefore,

$$\int \frac{\mathrm{d}x}{a^2+x^2} = \int \frac{1}{a^2+(a\tan\theta)^2}\frac{\mathrm{d}x}{\mathrm{d}\theta}\,\mathrm{d}\theta = \int \frac{1}{a^2(1+\tan^2\theta)}\,a\sec^2\theta\,\mathrm{d}\theta$$

$$\therefore \quad I = \int \frac{\mathrm{d}\theta}{a} \qquad (\text{since } 1+\tan^2\theta \equiv \sec^2\theta)$$

$$= \frac{\theta}{a} + c$$

Since $x = a\tan\theta$, we have $\tan\theta = \dfrac{x}{a}$, which gives $\theta = \tan^{-1}\left(\dfrac{x}{a}\right)$. Therefore,

$$I = \frac{1}{a}\tan^{-1}\left(\frac{x}{a}\right) + c$$

as required.

Proof of (ii)

Let $x = a\sin\theta$, then $\dfrac{\mathrm{d}x}{\mathrm{d}\theta} = a\cos\theta$. Therefore,

$$\int \frac{\mathrm{d}x}{\sqrt{a^2-x^2}} = \int \frac{1}{\sqrt{a^2-a^2\sin^2\theta}}\frac{\mathrm{d}x}{\mathrm{d}\theta}\,\mathrm{d}\theta = \int \frac{1}{\sqrt{a^2(1-\sin^2\theta)}}\,a\cos\theta\,\mathrm{d}\theta$$

$$\therefore \quad I = \int \frac{\cos\theta}{\sqrt{\cos^2\theta}}\,\mathrm{d}\theta \qquad (\text{since } 1-\sin^2\theta \equiv \cos^2\theta)$$

$$= \theta + c$$

Since $x = a\sin\theta$, we have $\sin\theta = \dfrac{x}{a}$, which gives $\theta = \sin^{-1}\left(\dfrac{x}{a}\right)$. Therefore,

$$I = \sin^{-1}\left(\frac{x}{a}\right) + c$$

as required.

Example 3 Find each of the following integrals.

a) $\displaystyle\int \frac{1}{\sqrt{16-x^2}}\,\mathrm{d}x$ **b)** $\displaystyle\int \frac{1}{\sqrt{36-4x^2}}\,\mathrm{d}x$ **c)** $\displaystyle\int \frac{1}{1+25x^2}\,\mathrm{d}x$

SOLUTION

a)
$$\int \frac{1}{\sqrt{16-x^2}}\,\mathrm{d}x = \int \frac{1}{\sqrt{4^2-x^2}}\,\mathrm{d}x$$
$$= \sin^{-1}\left(\frac{x}{4}\right) + c$$

b)
$$\int \frac{1}{\sqrt{36-4x^2}}\,\mathrm{d}x = \int \frac{1}{\sqrt{4(9-x^2)}}\,\mathrm{d}x = \frac{1}{2}\int \frac{1}{\sqrt{3^2-x^2}}\,\mathrm{d}x$$
$$= \frac{1}{2}\sin^{-1}\left(\frac{x}{3}\right) + c$$

c)
$$\int \frac{1}{1+25x^2}\,\mathrm{d}x = \int \frac{1}{25\left(\frac{1}{25}+x^2\right)}\,\mathrm{d}x = \frac{1}{25}\int \frac{1}{\left(\frac{1}{5}\right)^2+x^2}\,\mathrm{d}x$$
$$= \frac{1}{25} \times \frac{1}{\left(\frac{1}{5}\right)}\tan^{-1}\left[\frac{x}{\left(\frac{1}{5}\right)}\right] + c$$
$$= \frac{1}{5}\tan^{-1} 5x + c$$

Example 4 Evaluate each of these.

a) $\displaystyle\int_1^2 \frac{x-4}{x^2+9}\,\mathrm{d}x$ **b)** $\displaystyle\int_0^{\sqrt{2}} \frac{x}{\sqrt{16-x^4}}\,\mathrm{d}x$

SOLUTION

a)
$$\int_1^2 \frac{x-4}{x^2+9}\,\mathrm{d}x = \int_1^2 \frac{x}{x^2+9}\,\mathrm{d}x - \int_1^2 \frac{4}{x^2+9}\,\mathrm{d}x$$
$$= \frac{1}{2}\int_1^2 \frac{2x}{x^2+9}\,\mathrm{d}x - \int_1^2 \frac{4}{x^2+9}\,\mathrm{d}x$$
$$= \frac{1}{2}\Big[\ln(x^2+9)\Big]_1^2 - 4\left[\frac{1}{3}\tan^{-1}\left(\frac{x}{3}\right)\right]_1^2$$
$$= \frac{1}{2}(\ln 13 - \ln 10) - 4\left[\frac{1}{3}\tan^{-1}\left(\frac{2}{3}\right) - \frac{1}{3}\tan^{-1}\left(\frac{1}{3}\right)\right]$$
$$= \frac{1}{2}\ln\left(\frac{13}{10}\right) - \frac{4}{3}\left[\tan^{-1}\left(\frac{2}{3}\right) - \tan^{-1}\left(\frac{1}{3}\right)\right]$$

b) Let $u = x^2$, then $\dfrac{du}{dx} = 2x$. The new limits are found by substituting $x = 0$ and $x = \sqrt{2}$ into $u = x^2$.

When $x = 0$, $u = 0$. When $x = \sqrt{2}$, $u = 2$. Therefore, we have

$$\int_0^{\sqrt{2}} \frac{x}{\sqrt{16 - x^4}}\,dx = \int_0^2 \frac{x}{\sqrt{16 - (x^2)^2}} \frac{dx}{du}\,du$$

$$= \int_0^2 \frac{x}{\sqrt{16 - u^2}} \frac{1}{2x}\,du$$

$$= \frac{1}{2}\int_0^2 \frac{1}{\sqrt{16 - u^2}}\,du$$

$$= \frac{1}{2}\left[\sin^{-1}\left(\frac{u}{4}\right)\right]_0^2$$

$$= \frac{1}{2}\left[\sin^{-1}\left(\frac{1}{2}\right) - \sin^{-1}(0)\right] = \frac{\pi}{12}$$

Example 5 Use the substitution $t = e^x$ to find

$$\int \frac{e^x}{e^x + e^{-x}}\,dx$$

SOLUTION

Let $t = e^x$, then $\dfrac{dt}{dx} = e^x = t$. Therefore,

$$\int \frac{e^x}{e^x + e^{-x}}\,dx = \int \frac{t}{t + \frac{1}{t}} \frac{dx}{dt}\,dt$$

$$= \int \frac{t}{t + \frac{1}{t}} \frac{1}{t}\,dt$$

$$= \int \frac{t}{t^2 + 1}\,dt$$

$$= \tfrac{1}{2}\ln(1 + t^2) + c = \tfrac{1}{2}\ln(1 + e^{2x}) + c$$

Exercise 20A

1 Find the following integrals, in each case using the suggested substitution.

a) $\int x(x - 3)^2\,dx, \quad u = x - 3$

b) $\int x(x + 4)^3\,dx, \quad u = x + 4$

c) $\int (x - 4)(x - 1)^3\,dx, \quad u = x - 1$

d) $\int x(2x - 3)^2\,dx, \quad u = 2x - 3$

e) $\int (3x+1)(2x-5)^2\,\mathrm{d}x, \quad u = 2x-5$

f) $\int \frac{x}{x+3}\,\mathrm{d}x, \quad u = x+3$

g) $\int \frac{x}{(x+1)^2}\,\mathrm{d}x, \quad u = x+1$

h) $\int \frac{x+2}{(2x-3)^3}\,\mathrm{d}x, \quad u = 2x-3$

2 Find the following integrals, in each case using the suggested substitution.

a) $\int x\sqrt{x+1}\,\mathrm{d}x, \quad u = x+1$

b) $\int x\sqrt{x-1}\,\mathrm{d}x, \quad u = \sqrt{x-1}$

c) $\int (x-4)\sqrt{x+5}\,\mathrm{d}x, \quad u = x+5$

d) $\int (3x-2)\sqrt{1-2x}\,\mathrm{d}x, \quad u = \sqrt{1-2x}$

e) $\int \frac{x}{\sqrt{x+1}}\,\mathrm{d}x, \quad u = x+1$

f) $\int \frac{x}{\sqrt{x-3}}\,\mathrm{d}x, \quad u = \sqrt{x-3}$

g) $\int \frac{x-2}{\sqrt{x-4}}\,\mathrm{d}x, \quad u = x-4$

h) $\int \frac{x+3}{\sqrt{5-x}}\,\mathrm{d}x, \quad u = \sqrt{5-x}$

3 Integrate each of the following with respect to x.

a) $x(x+3)^3$

b) $\frac{x+2}{x-1}$

c) $x\sqrt{5-x}$

d) $\frac{x-3}{(x+2)^2}$

e) $\frac{x}{\sqrt{2x+1}}$

f) $(x-3)(5-2x)^4$

g) $\frac{2x-1}{x+7}$

h) $\frac{x}{\sqrt{(x+1)^3}}$

i) $\frac{x+2}{\sqrt{4-x}}$

j) $\frac{x+3}{(3-x)^2}$

k) $x^2(x-1)^4$

l) $x\sqrt{(1-x)^3}$

4 Evaluate each of these.

a) $\int_3^4 \frac{x}{x-2}\,\mathrm{d}x$

b) $\int_3^5 x(x-3)^2\,\mathrm{d}x$

c) $\int_1^6 x\sqrt{x+3}\,\mathrm{d}x$

d) $\int_1^3 \frac{x^2}{2x-1}\,\mathrm{d}x$

e) $\int_4^7 \frac{5-x}{\sqrt{x-3}}\,\mathrm{d}x$

f) $\int_1^3 (3x+1)(2-x)^4\,\mathrm{d}x$

5 Find each of the following integrals.

a) $\int \frac{1}{\sqrt{4-x^2}}\,\mathrm{d}x$

b) $\int \frac{1}{\sqrt{25-x^2}}\,\mathrm{d}x$

c) $\int \frac{1}{\sqrt{1-9x^2}}\,\mathrm{d}x$

d) $\int \frac{6}{\sqrt{1-36x^2}}\,\mathrm{d}x$

e) $\int \frac{1}{\sqrt{4-9x^2}}\,\mathrm{d}x$

f) $\int \frac{1}{\sqrt{25-16x^2}}\,\mathrm{d}x$

g) $\int \frac{4}{4+x^2}\,\mathrm{d}x$

h) $\int \frac{1}{100+x^2}\,\mathrm{d}x$

i) $\int \frac{1}{1+9x^2}\,\mathrm{d}x$

j) $\int \frac{1}{1+25x^2}\,\mathrm{d}x$

k) $\int \frac{1}{9+4x^2}\,\mathrm{d}x$

l) $\int \frac{14}{49+16x^2}\,\mathrm{d}x$

6 Evaluate each of these.

a) $\displaystyle\int_0^{\frac{1}{2}} \frac{1}{\sqrt{1-x^2}}\,dx$ **b)** $\displaystyle\int_0^1 \frac{1}{1+x^2}\,dx$ **c)** $\displaystyle\int_{\sqrt{3}}^3 \frac{1}{9+x^2}\,dx$

d) $\displaystyle\int_2^4 \frac{3}{\sqrt{16-x^2}}\,dx$ **e)** $\displaystyle\int_{-\sqrt{3}}^{\sqrt{3}} \frac{1}{\sqrt{4-x^2}}\,dx$ **f)** $\displaystyle\int_{-6}^6 \frac{24}{36+x^2}\,dx$

7 a) By using the substitution $t = e^x$, show that $\displaystyle\int \frac{e^{2x}}{1+e^x}\,dx = \int \frac{t}{1+t}\,dt.$

b) Deduce that $\displaystyle\int \frac{e^{2x}}{1+e^x}\,dx = e^x - \ln(1+e^x) + c.$

8 a) By using the substitution $u = x^3$, show that

$$\int \frac{3x^5}{x^3-1}\,dx = \int \frac{u}{u-1}\,du$$

b) Deduce that $\displaystyle\int \frac{3x^5}{x^3-1}\,dx = x^3 + \ln(x^3-1) + c.$

9 a) By using the substitution $t = e^x$, show that

$$\int \frac{e^x}{1+e^{2x}}\,dx = \int \frac{1}{1+t^2}\,dt$$

b) Deduce that $\displaystyle\int \frac{e^x}{1+e^{2x}}\,dx = \tan^{-1}(e^x) + c.$

10 By using the substitution $u = \sqrt{1+x^2}$, show that

$$\int_0^{\sqrt{3}} x^3\sqrt{1+x^2}\,dx = 3\tfrac{13}{15}$$

11 By using the substitution $x = 2\cos\theta$, show that

$$\int_1^2 \frac{1}{x^2\sqrt{4-x^2}}\,dx = \frac{\sqrt{3}}{4}$$

12 The diagram shows part of the curve with equation $y = \dfrac{x}{4-x}$, together with the line $y = x$.

The curve and the line intersect at O and P.

a) Find the coordinates of the point P.

b) Calculate the area of the region bounded by the curve and the line between O and P.

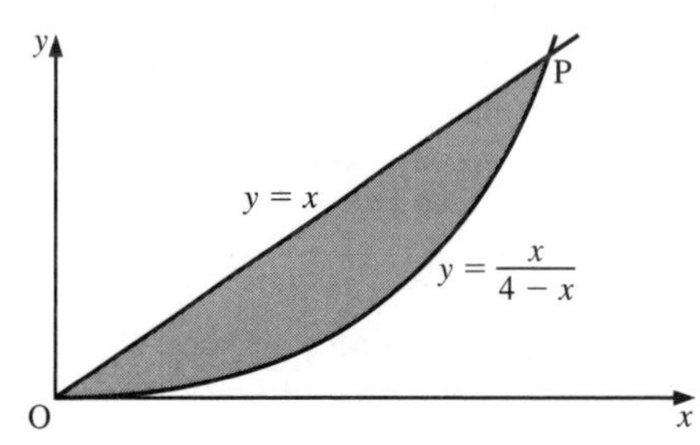

13 The diagram shows part of the curve with equation $y = x\sqrt{3-x}$, together with a line segment, OA. The curve has a maximum at A, and crosses the x-axis at B.

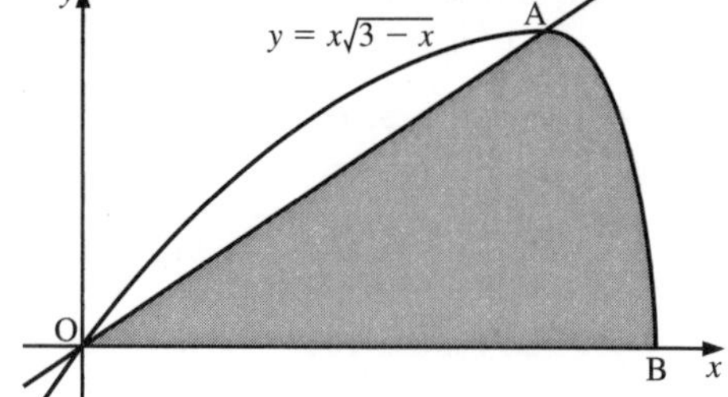

a) Find the coordinates of the points A and B.
b) Find the area of the shaded region bounded by the line segment OA, the arc of the curve AB, and the x-axis.

14 The diagram shows part of the curve with equation $y = \dfrac{x^2}{4-x}$, together with the line $y = 3x$. The curve and the line meet at the origin and at the point A.

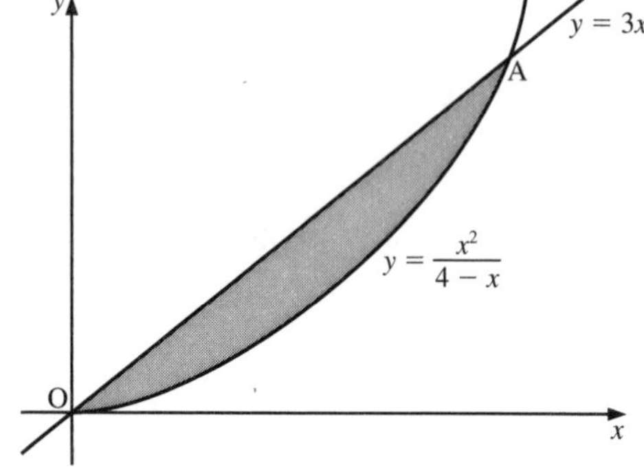

a) Find the coordinates of the point A.
b) Calculate the area of the shaded region.

15 a) Sketch the curve with equation $y = (x-1)^4(2-x)$.
b) Calculate the areas of each of the two finite regions bounded by the curve and the x-axis.

16 The curve $y = \dfrac{x}{x+1}$ meets the curve $y = \dfrac{2}{5-x}$ at the points A and B.

a) Find the coordinates of the points A and B.
b) Sketch the two curves on the same set of axes.
c) Calculate the area of the finite region bounded by the two curves.

***17 a)** Given that $x^2 + 2x + 5 \equiv (x+a)^2 + b^2$, find the values of the constants a and b.
b) By substituting $\tan u = x + 1$, deduce that $\displaystyle\int \frac{1}{x^2+2x+5}\,\mathrm{d}x = \frac{1}{2}\tan^{-1}\left(\frac{x+1}{2}\right) + c$.

***18 a)** By using the substitution $x = \sin\theta$, show that

$$\int \sqrt{1-x^2}\,\mathrm{d}x = \int \frac{1+\cos 2\theta}{2}\,\mathrm{d}\theta$$

b) Deduce that $\displaystyle\int \sqrt{1-x^2}\,\mathrm{d}x = \tfrac{1}{2}(x\sqrt{1-x^2} + \sin^{-1}x) + c$.

***19 a)** Show that $\displaystyle\int_0^a \mathrm{f}(x)\,\mathrm{d}x = \int_0^a \mathrm{f}(a-x)\,\mathrm{d}x$.
b) Deduce that $\displaystyle\int_0^\pi x\sin x\,\mathrm{d}x = \frac{\pi}{2}\int_0^\pi \sin x\,\mathrm{d}x$.
c) Hence evaluate $\displaystyle\int_0^\pi x\sin x\,\mathrm{d}x$.

Using partial fractions

Example 6 Find $\int \frac{2x^2 + x - 5}{x - 4}\,dx$.

SOLUTION

Using techniques developed on pages 279–90, we know that

$$\frac{2x^2 + x - 5}{x - 4} \equiv Ax + B + \frac{C}{x - 4}$$

Multiplying throughout by $x - 4$ gives

$$2x^2 + x - 5 = (Ax + B)(x - 4) + C$$

Comparing coefficients of x^2 gives $A = 2$.

Letting $x = 4$ gives $2(4)^2 + 4 - 5 = C$.

$$\therefore \quad C = 31.$$

Comparing constant terms gives $-5 = -4B + C$.

Since $C = 31$, we have $B = 9$. Therefore,

$$\int \frac{2x^2 + x - 5}{x - 4}\,dx = \int \left(2x + 9 + \frac{31}{x - 4}\right) dx$$
$$= x^2 + 9x + 31\ln(x - 4) + c$$

Example 7 Find $\int \frac{2x - 5}{(4x - 1)(x + 2)}\,dx$.

SOLUTION

To resolve into partial fractions, we let

$$\frac{2x - 5}{(4x - 1)(x + 2)} \equiv \frac{A}{4x - 1} + \frac{B}{x + 2}$$

Multiplying throughout by $(4x - 1)(x + 2)$ gives

$$2x - 5 = A(x + 2) + B(4x - 1)$$

Let $x = -2$: $\quad -9 = -9B \quad \therefore \quad B = 1$

Let $x = \frac{1}{4}$: $\quad -\frac{9}{2} = \frac{9}{4}A \quad \therefore \quad A = -2$

We have

$$\int \frac{2x - 5}{(4x - 1)(x + 2)}\,dx = \int \frac{-2}{4x - 1}\,dx + \int \frac{1}{x + 2}\,dx$$
$$= \frac{1}{2}\int \frac{-4}{4x - 1}\,dx + \int \frac{1}{x + 2}\,dx$$
$$= -\frac{1}{2}\ln(4x - 1) + \ln(x + 2) + c$$

Exercise 20B

Revision in the methods of partial fractions may be useful before tackling this exercise. Examples are to be found in Exercise 11B, pages 285–7.

1 Find each of the following integrals.

a) $\displaystyle\int \frac{x^2-7}{x-2}\,\mathrm{d}x$ **b)** $\displaystyle\int \frac{x^2+x-11}{x+4}\,\mathrm{d}x$

c) $\displaystyle\int \frac{2x^2-x-9}{x-3}\,\mathrm{d}x$ **d)** $\displaystyle\int \frac{4x^2+3}{2x-1}\,\mathrm{d}x$

e) $\displaystyle\int \frac{x^3-2x^2-x}{x-2}\,\mathrm{d}x$ **f)** $\displaystyle\int \frac{3x^3+x^2-2x+1}{x+1}\,\mathrm{d}x$

g) $\displaystyle\int \frac{3x^3-5x^2-18x+19}{x-3}\,\mathrm{d}x$ **h)** $\displaystyle\int \frac{x^4}{x-1}\,\mathrm{d}x$

2 Integrate each of the following with respect to x.

a) $\dfrac{x-3}{(x-2)(x-1)}$ **b)** $\dfrac{3x+2}{(x+3)(2x-1)}$

c) $\dfrac{1}{x^2-4}$ **d)** $\dfrac{x-6}{(x^2+3)(2x+1)}$

e) $\dfrac{5x-4}{(x-2)^2(x+1)}$ **f)** $\dfrac{x+2}{(x+1)^2(2x+1)}$

g) $\dfrac{2x-13}{(5-x)^2(2-x)}$ **h)** $\dfrac{2x^2-5x+4}{(x-1)(2x-3)}$

3 Evaluate each of these.

a) $\displaystyle\int_5^7 \frac{1}{(x-3)(x-4)}\,\mathrm{d}x$ **b)** $\displaystyle\int_{-1}^2 \frac{2x+1}{(2x+3)^2}\,\mathrm{d}x$

c) $\displaystyle\int_4^6 \frac{x^3-4x+2}{x(x-2)}\,\mathrm{d}x$ **d)** $\displaystyle\int_4^5 \frac{4}{(x-1)^2(x-3)}\,\mathrm{d}x$

e) $\displaystyle\int_0^2 \frac{1}{(x+4)(x+2)}\,\mathrm{d}x$ **f)** $\displaystyle\int_{-3}^3 \frac{x^2+11x+29}{(x+6)(x+5)}\,\mathrm{d}x$

g) $\displaystyle\int_0^3 \frac{x+2}{(x+3)^2(x+1)}\,\mathrm{d}x$ **h)** $\displaystyle\int_5^9 \frac{x^3-5x^2+6x-5}{(x-1)(x-4)}\,\mathrm{d}x$

4 The following integrals may look similar, but they involve a variety of methods. In each case, select an appropriate method and find the integral.

a) $\displaystyle\int \frac{1}{1+x}\,\mathrm{d}x$ **b)** $\displaystyle\int \frac{x}{1-x}\,\mathrm{d}x$ **c)** $\displaystyle\int \frac{x}{1+x^2}\,\mathrm{d}x$ **d)** $\displaystyle\int \frac{1}{\sqrt{1-x^2}}\,\mathrm{d}x$

e) $\displaystyle\int \frac{x}{(1-x)^2}\,\mathrm{d}x$ **f)** $\displaystyle\int \frac{1}{\sqrt{1+x}}\,\mathrm{d}x$ **g)** $\displaystyle\int \frac{1}{1+x^2}\,\mathrm{d}x$ **h)** $\displaystyle\int \frac{1}{(1-x)^2}\,\mathrm{d}x$

i) $\displaystyle\int \frac{x}{1+x}\,\mathrm{d}x$ **j)** $\displaystyle\int \frac{1}{1-x^2}\,\mathrm{d}x$ **k)** $\displaystyle\int \frac{x}{\sqrt{1-x^2}}\,\mathrm{d}x$ **l)** $\displaystyle\int \frac{x}{\sqrt{1+x}}\,\mathrm{d}x$

5 a) By using the substitution $t = e^x$, show that

$$\int \frac{1}{e^x - e^{-x}}\,dx = \int \frac{1}{t^2 - 1}\,dt$$

b) Deduce that $\int \frac{1}{e^x - e^{-x}}\,dx = \frac{1}{2}\ln\left(\frac{e^x - 1}{e^x + 1}\right) + c$.

6 The diagram shows part of the curve with equation $y = \frac{1}{9 - x^2}$, together with the line $y = \frac{1}{5}$. The curve and the line meet at the points A and B.

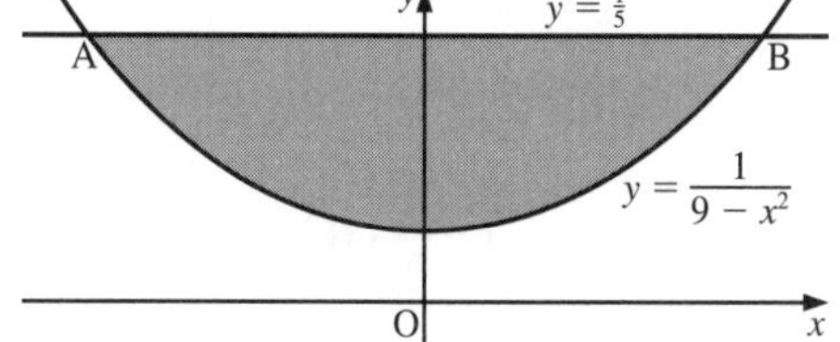

a) Find the coordinates of A and B.

b) Calculate the area of the shaded region.

7 The diagram shows part of the curve with equation $y = \frac{x}{(4 - x)^2}$, together with the line $y = x$.
The curve and the line meet at the origin and at the point P.

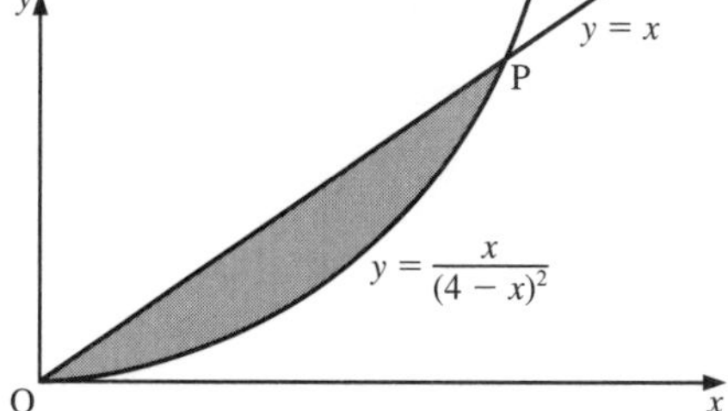

a) Find the coordinates of the point P.

b) Calculate the area of the shaded region.

8 a) Sketch the curve with equation $y = \frac{4}{(5 - x)(x - 1)}$.

b) Calculate the area of the finite region bounded by the curve, the x-axis, the line $x = 2$ and the line $x = 4$.

9 The line $y = 4x$ meets the curve $y = \frac{9x^2}{25 - x^2}$ at the origin and at the points P and Q.

a) Find P and Q.

b) Sketch the line and the curve on the same set of axes.

c) Show that the area of the region, in the positive quadrant, bounded by the line and the curve is given by $68 - 45\ln 3$.

***10 a)** By using the substitution $t = \tan x$, show that $\int \frac{1}{1 - \tan x}\,dx = \int \frac{1}{(1 + t^2)(1 - t)}\,dt$.

b) Deduce that $\int \frac{2}{1 - \tan x}\,dx = x - \ln(\cos x - \sin x) + c$.

***11** Find $\int \frac{1}{1 - x^3}\,dx$.

Integration by parts

This method is used to integrate the product of two functions. We already know that

$$\frac{\mathrm{d}(uv)}{\mathrm{d}x} = u\frac{\mathrm{d}v}{\mathrm{d}x} + v\frac{\mathrm{d}u}{\mathrm{d}x}$$

Integrating both sides with respect to x gives

$$uv = \int u\frac{\mathrm{d}v}{\mathrm{d}x}\,\mathrm{d}x + v\frac{\mathrm{d}u}{\mathrm{d}x}\,\mathrm{d}x$$

$$\therefore \quad \int u\frac{\mathrm{d}v}{\mathrm{d}x}\,\mathrm{d}x = uv - \int v\frac{\mathrm{d}u}{\mathrm{d}x}\,\mathrm{d}x$$

or

$$\int uv' = uv - \int vu'$$

The product to be integrated comprises two parts:

- the function u, which is differentiated
- the function $v' = \frac{\mathrm{d}v}{\mathrm{d}x}$, which is integrated.

Example 8 Find $\int x(x+3)^3\,\mathrm{d}x$.

SOLUTION

We write the integral as

$$\int x(x+3)^3\,\mathrm{d}x$$

Then the function x becomes simpler when differentiated. Therefore, we let $u = x$ and $v' = (x+3)^3$.

When $u = x$, then $u' = 1$.

When $v' = (x+3)^3$, then $v = \frac{1}{4}(x+3)^4$.

This is illustrated in the table on the right.

$u = x:$	$v = \frac{1}{4}(x+3)^4$
$u' = 1:$	$v' = (x+3)^3$

Therefore,

$$\begin{aligned}\int x(x+3)^3\,\mathrm{d}x &= \frac{x(x+3)^4}{4} - \int \frac{(x+3)^4}{4} \times 1\,\mathrm{d}x \\ &= \frac{x(x+3)^4}{4} - \frac{(x+3)^5}{20} + c \\ &= \frac{(x+3)^4}{20}[5x - (x+3)] + c \\ &= \frac{1}{20}(x+3)^4(4x-3) + c\end{aligned}$$

Example 9 Find $\int xe^x \, dx$.

SOLUTION

In this case, both x and e^x can be easily integrated but x becomes simpler when differentiated. Therefore, we let $u = x$ and $v' = e^x$, which gives the table on the right.

$u = x:$	$v = e^x$
$u' = 1:$	$v' = e^x$

Therefore,

$$\int xe^x \, dx = xe^x - \int e^x \times 1 \, dx$$
$$= xe^x - e^x + c$$

Example 10 Find $\int x^3 \ln x \, dx$.

SOLUTION

In this case, we do not know how to integrate $\ln x$. Therefore, we let $u = \ln x$ and $v' = x^3$, which gives the table on the right.

$u = \ln x:$	$v = \frac{1}{4}x^4$
$u' = \dfrac{1}{x}:$	$v' = x^3$

Therefore

$$\int x^3 \ln x \, dx = \frac{x^4}{4} \ln x - \int \frac{x^4}{4} \left(\frac{1}{x}\right) dx$$
$$= \frac{x^4}{4} \ln x - \int \frac{x^3}{4} \, dx$$
$$= \frac{x^4}{4} \ln x - \frac{x^4}{16} + c$$
$$= \frac{x^4}{16} (4 \ln x - 1) + c$$

In some cases, the process of integration by parts needs to be performed more than once. The following example illustrates such a case.

Example 11 Find $\int x^2 \cos x \, dx$.

SOLUTION

In this case, we let $u = x^2$ and $v' = \cos x$. This is because the term x^2 becomes a constant, namely 2, when differentiated twice.

$u = x^2:$	$v = \sin x$
$u' = 2x:$	$v' = \cos x$

Therefore,

$$I = \int x^2 \cos x \, dx = x^2 \sin x - \int 2x \sin x \, dx$$

At this point, we see that the process of integrating by parts needs to be applied to $\int 2x \sin x \, dx$.

We let $u = 2x$ and $v' = \sin x$, since $2x$ becomes simpler when differentiated.

Therefore,

$u = 2x:$	$v = -\cos x$
$u' = 2:$	$v' = \sin x$

$$\int 2x \sin x \, dx = 2x(-\cos x) - \int 2 \times (-\cos x) \, dx$$
$$= -2x \cos x + \int 2 \cos x \, dx$$
$$= -2\cos x + 2 \sin x + c$$

Therefore,

$$I = x^2 \sin x - (-2x \cos x + 2 \sin x) + c$$
$$= x^2 \sin x + 2x \cos x - 2 \sin x + c$$
$$= (x^2 - 2) \sin x + 2x \cos x + c$$

So far, we have not found a way of integrating $\ln x$. However, we can find $\int \ln x \, dx$ using integration by parts.

Example 12 Find $\int \ln x \, dx$.

SOLUTION

We write $\int \ln x \, dx$ as $\int 1 \times \ln x \, dx$ and then let $u = \ln x$ and $v' = 1$.

Therefore,

$u = \ln x:$	$v = x$
$u' = \dfrac{1}{x}:$	$v' = 1$

$$\int \ln x \, dx = x \ln x - \int x\left(\frac{1}{x}\right) dx$$
$$= x \ln x - \int 1 \, dx$$
$$= x \ln x - x + c$$

Exercise 20C

All questions in this exercise should be tackled by integration by parts.

1 Find each of the following integrals.

a) $x(x-1)^2$ **b)** $x(x+1)^3$ **c)** $x(4-x)^3$ **d)** $x(2x+3)^5$

e) $(x-1)(x+2)^2$ **f)** $(x+3)(x-4)^5$ **g)** $(3x-1)(2x+3)^2$ **h)** $(2-5x)(4-x)^4$

2 Find each of these integrals.

a) $\displaystyle\int \frac{x}{(x-1)^2}\,\mathrm{d}x$ **b)** $\displaystyle\int \frac{x}{(x+1)^2}\,\mathrm{d}x$ **c)** $\displaystyle\int \frac{x-2}{(2x-3)^2}\,\mathrm{d}x$

d) $\displaystyle\int \frac{3x-4}{(x+2)^4}\,\mathrm{d}x$ **e)** $\displaystyle\int \frac{x}{\sqrt{2x-3}}\,\mathrm{d}x$ **f)** $\displaystyle\int \frac{x+4}{\sqrt{3x-2}}\,\mathrm{d}x$

g) $\displaystyle\int \frac{3x+1}{\sqrt{1-2x}}\,\mathrm{d}x$ **h)** $\displaystyle\int x\sqrt{4-x}\,\mathrm{d}x$ **i)** $\displaystyle\int (2-5x)\sqrt{3-2x}\,\mathrm{d}x.$

3 Integrate each of the following with respect to x.

a) $x\cos x$ **b)** $x\sin 2x$ **c)** xe^{3x} **d)** xe^{-x}

e) $(6x-1)\cos 3x$ **f)** $x\ln x$ **g)** $x^2\ln x$ **h)** $\sqrt{x}\ln x$

4 Evaluate each of the following.

a) $\displaystyle\int_0^2 x(x-3)^2\,\mathrm{d}x$ **b)** $\displaystyle\int_0^{\frac{\pi}{2}} x\sin 2x\,\mathrm{d}x$ **c)** $\displaystyle\int_{-1}^{1} (x+1)e^x\,\mathrm{d}x$

d) $\displaystyle\int_3^6 \frac{x}{\sqrt{x-2}}\,\mathrm{d}x$ **e)** $\displaystyle\int_2^4 x^3\ln x\,\mathrm{d}x$ **f)** $\displaystyle\int_{\frac{\pi}{6}}^{\frac{\pi}{3}} x\cos 3x\,\mathrm{d}x$

5 Integrate each of the following with respect to x.

a) $x^2\sin x$ **b)** $x^2(x+3)^3$ **c)** x^2e^x

d) $x^2\cos 2x$ **e)** x^2e^{-2x} **f)** $(x+1)^2\sin x$

6 Evaluate each of the following.

a) $\displaystyle\int_0^1 x^2e^{2x}\,\mathrm{d}x$ **b)** $\displaystyle\int_0^{\pi} x^2\sin x\,\mathrm{d}x$ **c)** $\displaystyle\int_{-1}^{1} x^2(x+3)^3\,\mathrm{d}x$

d) $\displaystyle\int_0^{\frac{\pi}{4}} x^2\cos 2x\,\mathrm{d}x$ **e)** $\displaystyle\int_3^6 \frac{x^2}{\sqrt{x-2}}\,\mathrm{d}x$ **f)** $\displaystyle\int_0^{\pi} (x-\pi)^2\sin x\,\mathrm{d}x$

7 The diagram shows part of the curve with equation $y = xe^{-x}$, together with a line segment OA, where A is a local maximum of the curve.

a) Find the coordinates of the point A.
b) Calculate the area of the shaded region.

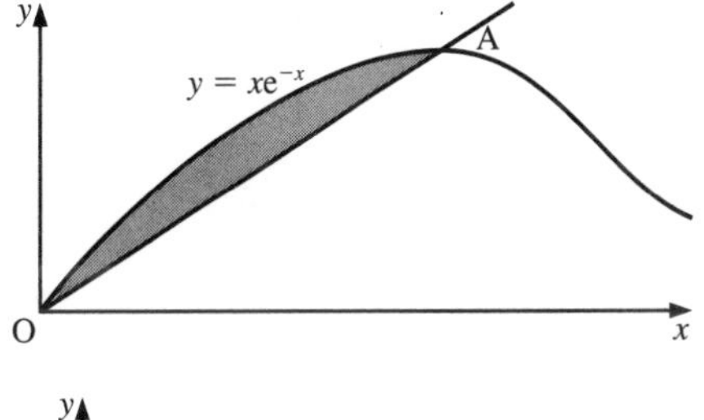

8 The diagram shows the curve with equation $y = x\sin x$ for values of x between 0 and 2π. The curve cuts the x-axis at O, P and Q.

a) Find the coordinates of P and Q.
b) Calculate the area of each of the shaded regions, marked A_1 and A_2.

9 Find the area of the finite region bounded by the curve $y = x\ln x$, the lines $x = 1$ and $x = e$.

10 a) Sketch the curve with equation $y = (1 - x)e^x$.

b) Calculate the area of the region in the positive quadrant bounded by the curve and the x- and y-axes.

***11** Find each of the following integrals.

a) $\int (\ln x)^2\,dx$ **b)** $\int x^3 e^{x^2}\,dx$ **c)** $\int e^x \sin x\,dx$

Integrals of tan x, cosec x, sec x and cot x

Result II

i) $\int \tan x\,dx = \ln(\sec x) + c$

ii) $\int \operatorname{cosec} x\,dx = -\ln(\operatorname{cosec} x + \cot x) + c$

iii) $\int \sec x\,dx = \ln(\sec x + \tan x) + c$

iv) $\int \cot x\,dx = \ln(\sin x) + c$

Proofs

These are most easily proved by differentiation. For example, take case (i):

$$\frac{d}{dx}\left[\ln(\sec x)\right] = \frac{1}{\sec x} \times \sec x\,\tan x$$

$$= \tan x$$

as required.

The remaining results are proved similarly.

Example 13 Find $\int x\cot(x^2)\,dx$.

SOLUTION

We notice that the derivative of x^2 is $2x$ and that there is an x term outside the main integrand. Therefore, we have

$$\int x\cot(x^2)\,dx = \frac{1}{2}\ln[\sin(x^2)] + c$$

Example 14 Find each of these integrals.

a) $\displaystyle\int \frac{1+\sin x}{\cos x}\,dx$ **b)** $\displaystyle\int \frac{1}{\sin x \cos x}\,dx$

SOLUTION

a) We have

$$\frac{1+\sin x}{\cos x} = \frac{1}{\cos x} + \frac{\sin x}{\cos x}$$
$$= \sec x + \tan x$$

Therefore,

$$\int \frac{1+\sin x}{\cos x}\,dx = \int \sec x\,dx + \int \tan x\,dx$$
$$= \ln(\sec x + \tan x) + \ln(\sec x) + c$$
$$= \ln[\sec x(\sec x + \tan x)] + c$$

b) We know that $\sin 2x = 2\sin x \cos x$. Therefore, $\sin x \cos x = \frac{1}{2}\sin 2x$.
We have

$$\int \frac{1}{\sin x \cos x}\,dx = \int \frac{1}{\frac{1}{2}\sin 2x}\,dx$$
$$= \int 2\operatorname{cosec} 2x\,dx$$
$$= -\frac{2\ln(\operatorname{cosec} 2x + \cot 2x)}{2} + c$$
$$= -\ln(\operatorname{cosec} 2x + \cot 2x) + c$$

Further results

We already know that

$$\int \sin x\,dx = -\cos x + c$$
$$\int \cos x\,dx = \sin x + c$$

We also know that

$$\int \sec^2 x\,dx = \tan x + c \qquad \int \sec x \tan x\,dx = \sec x + c$$
$$\int \operatorname{cosec} x \cot x\,dx = -\operatorname{cosec} x + c \qquad \int \operatorname{cosec}^2 x\,dx = -\cot x + c$$

Odd powers of sin x and cos x

Suppose we want to find $I = \int \sin^3 x \, dx$.

We write

$$\int \sin^3 x \, dx = \int \sin x \sin^2 x \, dx$$

and use the identity $\sin^2 x \equiv 1 - \cos^2 x$ to obtain

$$\begin{aligned} I &= \int \sin x(1 - \cos^2 x) \, dx \\ &= \int \sin x \, dx - \int \sin x \cos^2 x \, dx \\ &= -\cos x - \left(-\frac{\cos^3 x}{3}\right) + c \\ &= -\cos x + \frac{\cos^3 x}{3} + c \end{aligned}$$

Odd powers of $\cos x$ are treated similarly, taking out a single factor of $\cos x$ and using the identity $\cos^2 x \equiv 1 - \sin^2 x$ to express what remains in terms of $\sin x$.

Even powers of sin x and cos x

To integrate $\sin^2 x$ and $\cos^2 x$, we must express each in a different form using trigonometric identities.

To find $\int \sin^2 x \, dx$, we use the identity $\cos 2x \equiv 1 - 2\sin^2 x$. That is,

$$\sin^2 x \equiv \frac{1}{2} - \frac{\cos 2x}{2} \qquad [1]$$

Therefore,

$$\begin{aligned} \int \sin^2 x \, dx &= \int \left(\frac{1}{2} - \frac{\cos 2x}{2}\right) dx \\ &= \frac{x}{2} - \frac{\sin 2x}{4} + c \end{aligned}$$

In order to find $\int \cos^2 x \, dx$, we use the identity $\cos 2x \equiv 2\cos^2 x - 1$. That is,

$$\cos^2 x \equiv \frac{1}{2} + \frac{\cos 2x}{2} \qquad [2]$$

Therefore,

$$\begin{aligned} \int \cos^2 x \, dx &= \int \left(\frac{1}{2} + \frac{\cos 2x}{2}\right) dx \\ &= \frac{x}{2} + \frac{\sin 2x}{4} + c \end{aligned}$$

To find the integrals $\int \sin^n x \, dx$ and $\int \cos^n x \, dx$, where n is an even positive integer and relatively small, we can still use the above identities. For example, to find $\int \sin^4 x \, dx$, we can use the identity

$$\sin^4 x \equiv \tfrac{1}{4}(1 - \cos 2x)^2$$

Therefore,

$$I = \int \sin^4 x \, dx = \int \frac{1}{4}(1 - \cos 2x)^2 \, dx$$

$$= \frac{1}{4}\int 1 - 2\cos 2x + \cos^2 2x \, dx = \frac{1}{4}\int 1 - 2\cos 2x + \frac{1 + \cos 4x}{2} \, dx$$

$$= \frac{1}{4}\left(x - \frac{2\sin 2x}{2} + \frac{x}{2} + \frac{\sin 4x}{8}\right) + c$$

$$= \frac{3x}{8} - \frac{\sin 2x}{4} + \frac{\sin 4x}{32} + c$$

However, it is clear that an alternative method is required when n is a much larger integer. It is for this reason that we derive a reduction formula for such integrals, but this is beyond the scope of the present book.

Powers of tan x and cot x

To find $\int \tan^2 x \, dx$, we use the identity $\sec^2 x \equiv 1 + \tan^2 x$. That is,

$$\tan^2 x \equiv \sec^2 x - 1$$

Therefore,

$$\int \tan^2 x \, dx = \int (\sec^2 x - 1) \, dx$$

$$= \tan x - x + c$$

To find the integral $\int \tan^n x \, dx$, where n is a positive integer, we write it in the form

$$\int \tan^2 x \tan^{n-2} x \, dx$$

and then use the identity $\tan^2 x \equiv \sec^2 x - 1$.

For example, to find $\int \tan^4 x \, dx$, we write

$$\int \tan^2 x \tan^2 x \, dx = \int (\sec^2 x - 1) \tan^2 x \, dx$$

$$= \int \sec^2 x \tan^2 x \, dx - \int \tan^2 x \, dx$$

Now by inspection we see that

$$\int \sec^2 x \tan^2 x \, dx = \frac{\tan^3 x}{3} + c$$

Therefore, we have

$$\int \tan^4 x \, dx = \frac{\tan^3 x}{3} - \tan x + x + c$$

Likewise, to find $\int \cot^n x \, dx$, where n is a positive integer, we write it in the form

$$\int \cot^2 x \cot^{n-2} x \, dx$$

ad use the identity $\cot^2 x \equiv \operatorname{cosec}^2 x - 1$.

For example, to find $\int \cot^3 x \, dx$ we write

$$\int \cot^3 x \, dx = \int \cot^2 x \cot x \, dx$$

$$= \int (\operatorname{cosec}^2 x - 1) \cot x \, dx$$

$$= \int \operatorname{cosec}^2 x \cot x \, dx - \int \cot x \, dx$$

Now by inspection we see that

$$\int \operatorname{cosec}^2 x \cot x \, dx = -\frac{\cot^2 x}{2} + c$$

Therefore, we have

$$\int \cot^3 x \, dx = -\frac{\cot^2 x}{2} - \ln(\sin x) + c$$

Even powers of sec x and cosec x

To find $I = \int \sec^4 x \, dx$, we write

$$\int \sec^4 x \, dx = \int \sec^2 x \sec^2 x \, dx$$

and use the identity $\sec^2 x \equiv 1 + \tan^2 x$ to obtain

$$I = \int \sec^2 x (1 + \tan^2 x) \, dx$$

$$= \int \sec^2 x \, dx + \int \sec^2 x \tan^2 x \, dx$$

$$= \tan x + \left(\frac{\tan^3 x}{3}\right) + c = \tan x + \frac{\tan^3 x}{3} + c$$

Even powers of $\operatorname{cosec} x$ are treated similarly, taking out a factor of $\operatorname{cosec}^2 x$, and using the identity $\operatorname{cosec}^2 x \equiv 1 + \cot^2 x$ to express what remains in terms of $\cot x$.

Odd powers of sec x and cosec x

To find $I = \int \sec^3 x\,dx$, we write

$$\int \sec^3 x\,dx = \int \sec^2 x \sec x\,dx$$

and use integration by parts with $u = \sec x$ and $v' = \sec^2 x$. The table is shown on the the right.

$u = \sec x$:	$v = \tan x$
$u' = \sec x \tan x$:	$v' = \sec^2 x$

This gives

$$I = \sec x \tan x - \int (\sec x \tan x) \tan x\,dx$$

$$= \sec x \tan x - \int \sec x \tan^2 x\,dx$$

Using the identity $\tan^2 x \equiv \sec^2 x - 1$, we get

$$I = \sec x \tan x - \int \sec x\,(\sec^2 x - 1)\,dx$$

$$= \sec x \tan x - \int \sec^3 x\,dx + \int \sec x\,dx$$

$$= \sec x \tan x + I + \ln(\sec x + \tan x) + c$$

$$\therefore \quad 2I = \sec x \tan x + \ln(\sec x + \tan x)$$

$$\therefore \quad I = \tfrac{1}{2}[(\sec x \tan x + \ln(\sec x + \tan x)] + c$$

Odd powers of cosec x are treated similarly, taking out a factor of $\operatorname{cosec}^2 x$ and again using integration by parts.

Exercise 20D

1 Find each of these integrals.

a) $\int \sec 2x\,dx$ **b)** $\int \tan 3x\,dx$ **c)** $\int 2\cot 5x\,dx$

d) $\int \operatorname{cosec} 2x\,dx$ **e)** $\int \cot 3x\,dx$ **f)** $\int 4\operatorname{cosec}(\tfrac{1}{2}x)\,dx$

g) $\int 2x\tan(x^2)\,dx$ **h)** $\int x\operatorname{cosec}(x^2)\,dx$ **i)** $\int x^2\cot(x^3)\,dx$

j) $\int 3x^4\sec(x^5)\,dx$ **k)** $\int (x^2-1)\tan(x^3-3x)\,dx$ **l)** $\int \frac{\operatorname{cosec}(\sqrt{x})}{\sqrt{x}}\,dx$

In questions **2** to **6**, integrate each of the given expressions with respect to x.

2 ***Odd powers of sin x and cos x***

a) $\sin x \cos^5 x$ **b)** $\cos^3 x$ **c)** $\sin^3 2x$ **d)** $\cos^3 4x$

e) $\sin^3 x \cos^2 x$ **f)** $\cos^3 x \sin^4 x$ **g)** $\cos^3 2x \sin^2 2x$ **h)** $\sin 2x \sin^2 x$

3 ***Even powers of sin x and cos x***

a) $\sin^2 x$ **b)** $\cos^2 2x$ **c)** $\cos^2 3x$ **d)** $\sin^2 4x$

e) $\cos^2 6x$ **f)** $\sin^2(\frac{1}{2}x)$ **g)** $\cos^4 x$ **h)** $\sin^4 2x$

4 ***Powers of tan x and cot x***

a) $\cot^2 x$ **b)** $\tan^2 2x$ **c)** $\sec^2 x \tan^3 x$ **d)** $\operatorname{cosec}^2 2x \cot^4 2x$

e) $\tan^3 x$ **f)** $\cot^4 3x$ **g)** $\tan^4 5x$ **h)** $\tan^5 2x$

5 ***Powers of sec x and cosec x***

a) $\operatorname{cosec}^2 x$ **b)** $\sec^2 3x$ **c)** $\operatorname{cosec}^2(\frac{1}{3}x)$ **d)** $\sec^4 x$

e) $\operatorname{cosec}^4 5x$ **f)** $\sec^4 3x$ **g)** $\sec^6 x$ ***h)** $\operatorname{cosec}^5 x$

6 ***Mixed***

a) $\sin^3 4x$ **b)** $\cos^2 5x$ **c)** $\tan^4 3x$ **d)** $\cot 2x$

e) $\sec 4x$ **f)** $\operatorname{cosec}^3 5x \cot 5x$ **g)** $\cot^2 3x$ **h)** $\sin^5 6x$

i) $\tan(\frac{1}{2}x)$ **j)** $\tan^2 5x$ **k)** $\operatorname{cosec} 4x$ **l)** $\sin^3 3x \cos^4 3x$

7 Evaluate each of these.

a) $\displaystyle\int_{-\frac{\pi}{3}}^{\frac{\pi}{3}} \sec x \, dx$ **b)** $\displaystyle\int_{0}^{\frac{\pi}{6}} \sin^3 2x \, dx$ **c)** $\displaystyle\int_{\frac{\pi}{12}}^{\frac{\pi}{4}} \cos^2 x \, dx$

d) $\displaystyle\int_{-\frac{\pi}{6}}^{\frac{\pi}{3}} \sec^3 x \tan x \, dx$ **e)** $\displaystyle\int_{0}^{\frac{\pi}{12}} \sec^4 3x \, dx$ **f)** $\displaystyle\int_{\frac{\pi}{6}}^{\frac{\pi}{3}} \tan^3 x \, dx$

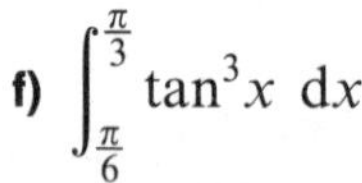

8 The diagram shows the two curves with equations $y = \sin x$ and $y = \sin 2x$ for values of x between 0 and π. The curves meet at the origin, and at the points P and Q.

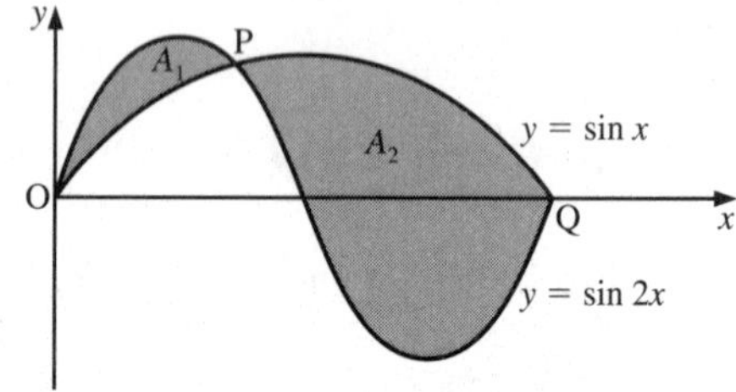

a) Find P and Q.
b) Find the areas of the shaded regions A_1 and A_2.

9 Show that the area of the finite region in the positive quadrant bounded by the curve $y = \cos^3 x$, the curve $y = \sin^3 x$, and the y-axis, is given by $\dfrac{5\sqrt{2} - 4}{6}$.

10 a) Sketch the curves with equations $y = \cos x$ and $y = \cos 2x$, on the same set of axes, for $0 \leqslant x \leqslant \pi$, labelling the two points of intersection as A and B.
b) Find A and B.
c) Calculate the area of the finite region bounded by the two curves between A and B.

***11 a)** By using the substitution $x = \tan\theta$, show that $\displaystyle\int \frac{1}{\sqrt{1+x^2}} \, dx = \int \sec\theta \, d\theta$.

b) Deduce that

$$\int \frac{1}{\sqrt{1+x^2}} \, dx = \ln(x + \sqrt{1+x^2}) + c$$

***12** Find $\int \sqrt{1-x^2}\,dx$.

***13** **a)** Show that, for $n \geqslant 2$,

$$\int_0^{\frac{\pi}{4}} \tan^n x\,dx = \frac{1}{n-1} - \int_0^{\frac{\pi}{4}} \tan^{n-2} x\,dx$$

b) Hence show that $\int_0^{\frac{\pi}{4}} \tan^{10} x\,dx = \frac{263}{315} - \frac{\pi}{4}$.

Exercise 20E

The following questions require a variety of methods. Part of the satisfaction in solving each question involves the selection of an appropriate method.

1 Integrate each of the following with respect to x.

a) $x(x+3)^2$ **b)** $\frac{x+2}{x(x+1)}$ **c)** $\tan 5x$

d) $x \sin x$ **e)** $\frac{4}{(2-x)^2}$ **f)** $\frac{1+x}{\sqrt{1-x^2}}$

g) $\frac{x}{x^2-4}$ **h)** $\sin x\, e^{\cos x}$ **i)** $\sec 4x$

j) $\frac{x^2+2}{x^3}$ **k)** $\frac{x-5}{x^2-1}$ **l)** $\ln x$

m) $\frac{x}{3-x}$ **n)** $x\sqrt{x^2+1}$ **o)** xe^{2x}

p) $(x-5)^3$ **q)** $\sin x \cos^3 x$ **r)** $\frac{x^2+1}{(x+3)(x-2)}$

s) $\frac{e^x}{e^x+1}$ **t)** $\sin^3 3x$ **u)** $x\sqrt{2-x}$

v) $\frac{x^2}{\sqrt{x^3+2}}$ **w)** $\sin^2 2x$ **x)** $\tan^3 6x$

2 Integrate each of the following with respect to x.

a) $\cos^2 6x$ **b)** $\frac{x-1}{x+2}$ **c)** $x^2 \operatorname{cosec}(x^3)$

d) $\frac{1}{(x+3)(x+4)}$ **e)** $x^3(x^4-5)^4$ **f)** $\frac{2}{5-x}$

g) xe^{-2x} **h)** $(x+2)(x-3)$ **i)** $\frac{x^2-9x+5}{(x-1)^2(x+2)}$

j) $\frac{x}{x+4}$ **k)** $\sec^2 3x \tan^2 3x$ **l)** $\frac{x+3}{\sqrt{1-x^2}}$

m) e^{2x} **n)** $x \cos 2x$ **o)** $\sqrt{5x-1}$

p) $\dfrac{x^2-x+1}{x(x^2+1)}$ **q)** $\dfrac{3x^2+5}{x(x^2+5)}$ **r)** $\operatorname{cosec} x \cot x \, e^{\operatorname{cosec} x}$

s) $\dfrac{\sin x}{1-\cos x}$ **t)** $x\sqrt{x+2}$ **u)** $\dfrac{x+1}{(x^2+2x-5)^3}$

v) $\cos^3 2x$ **w)** $x^5 \ln x$ **x)** $\cot^3 3x$

3 Integrate each of the following with respect to x.

a) $\sin^5 2x$ **b)** $\dfrac{x}{x-5}$ **c)** $\dfrac{\sec^2 x}{1+\tan x}$

d) $(x^2+1)(x^3+3x-2)^2$ **e)** $\dfrac{1}{x^2-9}$ **f)** $\sin^3 x \cos^2 x$

g) xe^{5x} **h)** $\dfrac{x}{(x^2+1)^2}$ **i)** $\dfrac{4x^2-x-2}{x^2(x+1)}$

j) $\sec 5x$ **k)** xe^{x^2} **l)** $x(2x+3)^3$

m) $\dfrac{x-1}{\sqrt{x}}$ **n)** $x \sin 3x$ **o)** $\dfrac{x+1}{x^2+1}$

p) $x \sin(x^2)$ **q)** $\dfrac{1}{2x+1}$ **r)** $\dfrac{4x}{(x+1)(x-3)}$

s) $\dfrac{3}{\sqrt{2x+1}}$ **t)** $\sec^3 x \tan x$ **u)** $\dfrac{x}{(2-x)^2}$

v) xe^{-2x^2} **w)** $\ln(\sqrt{x})$ **x)** $\cos^2 5x$

***4** Find each of these.

a) $\displaystyle\int \frac{1}{1+\sqrt{x}}\,dx$ **b)** $\displaystyle\int \frac{1}{x^{\frac{1}{2}}+x^{\frac{1}{3}}}\,dx$ **c)** $\displaystyle\int \cos(\ln x)\,dx$

d) $\displaystyle\int 2^x\,dx$ **e)** $\displaystyle\int \frac{1}{1+\sin 2x}\,dx$

Differential equations

An equation which involves only a first-order derivative, such as $\dfrac{dy}{dx}$, is called a **first-order differential equation**. One which involves a second-order derivative, such as $\dfrac{d^2y}{dx^2}$, is called a **second-order differential equation**.

Suppose $y = x^2 + 1$, then $\dfrac{dy}{dx} = 2x$. The equation

$$\frac{dy}{dx} = 2x$$

is a first-order differential equation and can be solved by integrating. That is,

$$y = \int 2x\,\mathrm{d}x \quad \therefore \quad y = x^2 + c$$

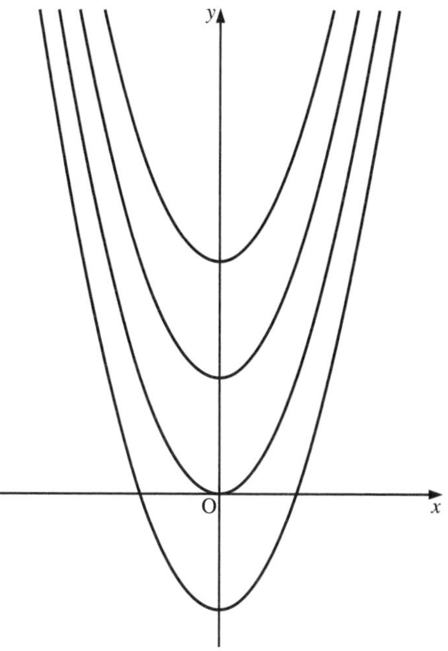

We see that solving the differential equation does not give a unique solution. In fact, we get a whole family of solutions for different values of the constant c, as shown on the right. The solution $y = x^2 + c$ is called the **general solution** of the differential equation.

If we also know that $y = 2$ when $x = 1$, then we have

$$2 = 1^2 + c \quad \therefore \quad c = 1$$

Therefore, we have a **particular solution** of the differential equation, namely $y = x^2 + 1$.

In the above example, the differential equation is of the form

$$\frac{\mathrm{d}y}{\mathrm{d}x} = \mathrm{f}(x)$$

In other words, the right-hand side is a function of x only. However, we now look at differential equations of the form

$$\frac{\mathrm{d}y}{\mathrm{d}x} = \mathrm{f}(x, y) = \mathrm{h}(x)\mathrm{g}(y)$$

in which the variables are separable.

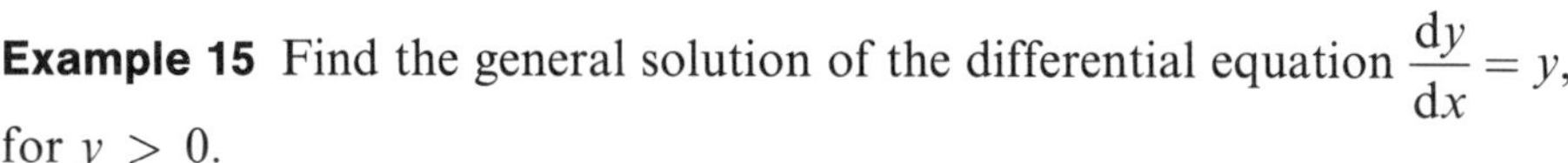

Example 15 Find the general solution of the differential equation $\dfrac{\mathrm{d}y}{\mathrm{d}x} = y$, for $y > 0$.

SOLUTION

The equation may be written as

$$\frac{1}{y}\frac{\mathrm{d}y}{\mathrm{d}x} = 1$$

Integrating both sides with respect to x gives

$$\int \frac{1}{y}\,\mathrm{d}y = \int 1\,\mathrm{d}x$$

$$\therefore \quad \ln y = x + c$$

We now need y in terms of x:

$$\mathrm{e}^{\ln y} = \mathrm{e}^{x+c}$$

$$\therefore \quad y = \mathrm{e}^x \mathrm{e}^c$$

$$\therefore \quad y = A\mathrm{e}^x$$

where $A = \mathrm{e}^c$ is a constant.

Example 16 Find the particular solution of each of the following differential equations.

a) $\dfrac{dy}{dx}(x-1) = y$, such that $y = 5$ when $x = 2$.

b) $\dfrac{dy}{dx} = e^{2x-y}$, such that $y = \ln\left(\dfrac{3}{2}\right)$ when $x = 0$.

c) $\dfrac{dy}{dx} = 3x\sin^2 y$, such that $y = -\dfrac{\pi}{4}$ when $x = \dfrac{1}{\sqrt{3}}$.

SOLUTION

a) After rearranging, we have

$$\frac{1}{y}\frac{dy}{dx} = \frac{1}{x-1}$$

Integrating both sides with respect to x gives

$$\int \frac{1}{y}\,dy = \int \frac{1}{x-1}\,dx$$

$$\therefore \quad \ln y = \ln(x-1) + c$$

$$\therefore \quad y = e^{\ln(x-1)+c}$$

$$\therefore \quad y = e^{\ln(x-1)}e^{c}$$

Hence

$$y = A(x-1)$$

where $A = e^c$.

This is the general solution of the differential equation. To find the particular solution, we know that $y = 5$ when $x = 2$. Substituting these values gives

$$5 = A(2-1) \quad \therefore \quad A = 5$$

The particular solution is $y = 5(x-1)$.

b) After rearranging, we have

$$\frac{dy}{dx} = e^{2x}e^{-y}$$

$$\therefore \quad e^{y}\frac{dy}{dx} = e^{2x}$$

Integrating both sides with respect to x gives

$$\int e^{y}\,dy = \int e^{2x}\,dx$$

$$\therefore \quad e^{y} = \tfrac{1}{2}e^{2x} + c$$

$$\therefore \quad y = \ln(\tfrac{1}{2}e^{2x} + c)$$

This is the general solution of the differential equation. To find the particular solution, we know that $y = \ln(\frac{3}{2})$ when $x = 0$. Substituting these values gives

$$\ln(\tfrac{3}{2}) = \ln(\tfrac{1}{2}e^0 + c) = \ln(\tfrac{1}{2} + c)$$

$$\therefore \quad c = 1$$

The particular solution is $y = \ln(\frac{1}{2}e^{2x} + 1)$.

c) After rearranging, we have

$$\frac{1}{\sin^2 y}\frac{dy}{dx} = 3x$$

$$\therefore \quad \operatorname{cosec}^2 y\frac{dy}{dx} = 3x$$

Integrating both sides with respect to x gives

$$\int \operatorname{cosec}^2 y\, dy = \int 3x\, dx$$

$$\therefore \quad -\cot y = \frac{3x^2}{2} + c$$

To find the particular solution, we know that $y = -\dfrac{\pi}{4}$ when $x = \dfrac{1}{\sqrt{3}}$.

Substituting gives

$$-\cot\left(-\frac{\pi}{4}\right) = \frac{3}{2}\left(\frac{1}{\sqrt{3}}\right)^2 + c$$

$$1 = \frac{1}{2} + c$$

$$\therefore \quad c = \frac{1}{2}$$

The particular solution is given by

$$-\cot y = \frac{3x^2}{2} + \frac{1}{2}$$

$$\therefore \quad \cot y = -\frac{3x^2 + 1}{2}$$

$$\therefore \quad y = \cot^{-1}\left(-\frac{3x^2 + 1}{2}\right)$$

Example 17 Find the equation of a curve given that it passes through the point (1,0) and that its gradient at any point (x, y) is equal to $x(y-1)^2$.

SOLUTION

If the gradient equals $x(y-1)^2$ at any point (x, y) then

$$\frac{dy}{dx} = x(y-1)^2$$

$$\therefore \quad \frac{1}{(y-1)^2}\frac{dy}{dx} = x$$

Integrating both sides with respect to x gives

$$\int \frac{1}{(y-1)^2}\,dy = \int x\,dx$$

$$\therefore \quad -\frac{1}{y-1} = \frac{x^2}{2} + c$$

We know that the curve passes through (1, 0). Therefore,

$$1 = \frac{1}{2} + c \quad \therefore \quad c = \frac{1}{2}$$

The equation of the curve is

$$-\frac{1}{y-1} = \frac{x^2}{2} + \frac{1}{2} = \frac{x^2+1}{2}$$

$$\therefore \quad -\frac{2}{x^2+1} = y - 1$$

Hence

$$y = 1 - \frac{2}{x^2+1} \quad \text{giving} \quad y = \frac{x^2-1}{x^2+1}$$

Exercise 20F

1 Find the general solution to each of the following differential equations.

a) $\dfrac{dy}{dx} = \dfrac{x+1}{y}$ **b)** $\dfrac{dy}{dx} = \dfrac{3x^2-2}{2y}$ **c)** $y^2\dfrac{dy}{dx} = 2x+3$

d) $\dfrac{dy}{dx} = 3x^2\sqrt{y}$ **e)** $(x-1)^2\dfrac{dy}{dx} + 2\sqrt{y} = 0$ **f)** $\dfrac{dy}{dx} = 2x\sqrt{2y-1}$

g) $\dfrac{dy}{dx} = 3x^2y$ **h)** $e^y\dfrac{dy}{dx} = x$ **i)** $\dfrac{dy}{dx} = \cos x \tan y$

j) $\dfrac{dy}{dx} = \sin x\, e^y$ **k)** $\dfrac{dy}{dx} = 4x^3\cos^2 y$ **l)** $\cos y\dfrac{dy}{dx} + \sin x = 1$

2 Find the particular solution to each of the following differential equations.

a) $\dfrac{dy}{dx} = 4 - 3x^2, \quad y = 5 \text{ at } x = 1$

b) $\dfrac{dy}{dx} = \dfrac{x+1}{y}, \quad y = 3 \text{ at } x = -2$

c) $(y-3)\dfrac{dy}{dx} = x + 3, \quad y = 4 \text{ at } x = 0$

d) $3y^2\dfrac{dy}{dx} + 2x = 1, \quad y = 2 \text{ at } x = 4$

e) $x^3\dfrac{dy}{dx} = 2y^2, \quad y = -\frac{4}{3} \text{ at } x = 2$

f) $\dfrac{dy}{dx} = \sqrt{\dfrac{y}{x+1}}, \quad y = 9 \text{ at } x = 3$

g) $\dfrac{dy}{dx} = 2xy, \quad y = 7 \text{ at } x = 0$

h) $\cos y\dfrac{dy}{dx} = \sin^3 x, \quad y = \dfrac{\pi}{2} \text{ at } x = 0$

i) $x^2\dfrac{dy}{dx} = \operatorname{cosec} y, \quad y = \dfrac{\pi}{3} \text{ at } x = 4$

j) $x(x-1)\dfrac{dy}{dx} = y, \quad y = 1 \text{ at } x = 2$

k) $e^y\dfrac{dy}{dx} + \sin x = 0, \quad y = 0 \text{ at } x = \dfrac{\pi}{3}$

l) $(x^3 - 7)\dfrac{dy}{dx} - 3x^2\cos^2 y = 0, \quad y = \dfrac{\pi}{4} \text{ at } x = 2$

3 The gradient of a curve at the point (x, y) is given by the expression $2x(y+1)$. Given that the curve passes through the point (3, 0), find an expression for y in terms of x.

4 The gradient of a curve at the point (x, y) is given by the expression $2\cos x\sqrt{y+3}$.

Given that the curve passes through the point $\left(\dfrac{\pi}{2}, 1\right)$, find an expression for y in terms of x.

***5** Find the general solution to the equation

$$\left(\frac{dy}{dx}\right)^2 + \sin x\cos^2 x\frac{dy}{dx} - \sin^4 x = 0$$

***6 a)** By substituting $y = ux$, where u is a function of x, show that the equation

$$x(1+x)\frac{dy}{dx} = y(1+x+y)$$

may be written as $(1+x)\dfrac{du}{dx} = u^2$.

b) Hence solve the equation

$$x(1+x)\frac{dy}{dx} = y(1+x+y)$$

given that $y = \frac{1}{2}$ when $x = -\frac{1}{2}$.

***7 a)** By substituting $p = \dfrac{dy}{dx}$ show that the equation $\dfrac{d^2y}{dx^2} - \dfrac{dy}{dx} = 1$ may be written as $\dfrac{dp}{dx} = p + 1$.

b) Hence solve the equation $\dfrac{d^2y}{dx^2} - \dfrac{dy}{dx} = 1$, given that $y = 4$ and $\dfrac{dy}{dx} = 2$ when $x = 0$.

Applications to exponential laws of growth and decay

Laws of growth and decay can be expressed in the form of differential equations. For example,

- if the rate of growth of x is proportional to x, then

  $$\frac{\mathrm{d}x}{\mathrm{d}t} = kx$$

  where k is a positive constant;

- if the rate of decay of x is proportional to x, then

  $$\frac{\mathrm{d}x}{\mathrm{d}t} = -kx$$

  where k is a positive constant.

One particular example of 'decay' is Newton's law of cooling, which states that the rate at which the temperature of a body falls is proportional to the amount by which its temperature exceeds that of its surroundings. This can be expressed in the form

$$\frac{\mathrm{d}\theta}{\mathrm{d}t} = -k\theta$$

where θ is the amount by which the temperature of the body exceeds the temperature of its surroundings.

The following example illustrates the solution of a differential equation which has been derived from a 'decay' situation.

Example 18 At time t minutes, the rate of change of temperature of a cooling body is proportional to the temperature T °C of that body at that time. Initially, $T = 72$ °C. Show that

$$T = 72\,\mathrm{e}^{-kt}$$

where k is a constant.

Given also that $T = 32$ °C when $t = 10$, find how much longer it will take the body to cool to 27 °C under these conditions.

SOLUTION

We can form the differential equation

$$\frac{\mathrm{d}T}{\mathrm{d}t} = -kT$$

Separating the variables and solving give

$$\int_{72}^{T} \frac{\mathrm{d}T}{T} = -\int_{0}^{t} k\,\mathrm{d}t$$

$$\therefore \quad \Big[\ln T\Big]_{72}^{T} = \Big[-kt\Big]_{0}^{t}$$

$$\therefore \quad \ln T - \ln 72 = -kt$$

$$\therefore \quad \ln\left(\frac{T}{72}\right) = -kt$$

$$\therefore \quad \frac{T}{72} = \mathrm{e}^{-kt}$$

$$\therefore \quad T = 72\mathrm{e}^{-kt} \qquad [1]$$

as required.

When $t = 10$, $T = 32\,^{\circ}\mathrm{C}$. Substituting into [1] gives the value of the constant k.

$$32 = 72\mathrm{e}^{-10k}$$

$$\therefore \quad \mathrm{e}^{-10k} = \frac{32}{72} = \frac{4}{9}$$

$$\therefore \quad -10k = \ln\left(\frac{4}{9}\right)$$

$$\therefore \quad k = -\frac{1}{10}\ln\left(\frac{4}{9}\right)$$

Now if t is the time it takes the body to cool to $27\,^{\circ}\mathrm{C}$, we have

$$27 = 72\mathrm{e}^{-kt} \qquad [2]$$

$$\therefore \quad \mathrm{e}^{-kt} = \frac{27}{72} = \frac{3}{8}$$

Hence

$$-kt = \ln\left(\frac{3}{8}\right) \qquad [3]$$

From [2] and [3] we have

$$\frac{1}{10}\ln\left(\frac{4}{9}\right)t = \ln\left(\frac{3}{8}\right)$$

$$\therefore \quad t = 10\frac{\ln\left(\frac{3}{8}\right)}{\ln\left(\frac{4}{9}\right)} = 12.1$$

So it will take the body $12.1 - 10 = 2.1$ minutes longer to cool to $27\,^{\circ}\mathrm{C}$.

Exercise 20G

1 The rate, in $cm^3 s^{-1}$, at which air is escaping from a balloon at time t seconds is proportional to the volume of air, $V cm^3$, in the balloon at that instant. Initially $V = 1000$.

a) Show that $V = 1000e^{-kt}$, where k is a positive constant.

Given also that $V = 500$ when $t = 6$,

b) show that $k = \frac{1}{6}\ln 2$

c) calculate the value of V when $t = 12$.

2 At time t minutes the rate of change of temperature of a cooling liquid is proportional to the temperature, T°C, of that liquid at that time. Initially $T = 80$.

a) Show that $T = 80e^{-kt}$, where k is a positive constant.

Given also that $T = 20$ when $t = 6$,

b) show that $k = \frac{1}{3}\ln 2$

c) calculate the time at which the temperature will reach 10 °C.

3 The value of a car depreciates in such a way that when it is t years old the rate of decrease in its value is proportional to the value, £V, of the car at that time. The car cost £12 000 when new.

a) Show that $V = 12000e^{-kt}$.

When the car is three years old its value has dropped to £4000.

b) Show that $k = \frac{1}{3}\ln 3$.

The owner decides to sell the car when its value reaches £2000.

c) Calculate, to the nearest month, the age of the car at that time.

4 A lump of a radioactive substance is decaying. At time t hours the rate of decay of the mass of the substance is proportional to the mass M grams of the substance at that time. At $t = 0$, $M = 72$; and at $t = 2$, $M = 50$.

a) Show that $M = 72e^{-t\ln(6/5)}$.

b) Sketch a graph of M against t.

5 By treatment with certain chemicals a scientist is able to control a killer virus. The rate of decrease in the number of viruses is found to be proportional to n, the number of viruses present. Using these chemicals it is found that the virus population is halved in six days. Show that the virus population is reduced to 1% of its original value after approximately 40 days.

6 A population is growing in such a way that, at time t years, the rate at which the population is increasing is proportional to the size, x, of that population at that time. Initially the size of the population is 2.

a) Show that $x = 2e^{kt}$, where k is a positive constant.

After 6 years the population size is 100.

b) Show that $k = \frac{1}{6}\ln 50$.

c) Calculate an estimate, to the nearest 1000, of the population size after 20 years.

7 During the initial stages of the spread of a disease in a body, the rate of increase of the number of infected cells in the body is proportional to the number, n, of infected cells present in the body at that time. Initially, n_0 infected cells are introduced to the body, and one day later the number of infected cells has risen to $2n_0$.

a) Show that $n = n_0 e^{t \ln 2}$.

b) Show that five days after the infection was introduced the number of infected cells has risen to $32n_0$.

***8** A patient is required to take a course of a certain drug. The rate of decrease in the concentration of the drug in the bloodstream at time t hours is proportional to the amount, x mg, of the drug in the bloodstream at that time.

a) Show that $x = x_0 e^{-kt}$, where k is a positive constant and x_0 mg is the size of the dose.

The patient repeats the dose of x_0 mg at regular intervals of T hours.

b) Show that the amount of the drug in the bloodstream will never exceed $\left(\dfrac{x_0}{1 - e^{-kT}}\right)$ mg.

Exercise 20H: Examination questions

1 Using the substitution $y = 2x - 1$, evaluate the integral $\displaystyle\int_1^2 \frac{x\,dx}{(2x-1)^3}$. (WJEC)

2 By using the substitution $x = 1 + u$, or otherwise, show that

$$\int_2^5 \frac{4x^2}{(x-1)^2}\,dx = 15 + 16\ln 2 \qquad \text{(NICCEA)}$$

3 **a)** By substituting $u = 1 + x$ or otherwise, find

i) $\displaystyle\int (1+x)^3 dx$ **ii)** $\displaystyle\int_{-1}^{+1} x(1+x)^3 dx$

b) By substituting $t = 1 + x^2$ or otherwise, evaluate the integral $\displaystyle\int_0^1 x\sqrt{1+x^2}\,dx$. (MEI)

4 By use of the substitution $2x = \sin\theta$, or otherwise, find the exact value of

$$\int_0^{\frac{1}{4}} \frac{1}{\sqrt{1-4x^2}}\,dx \qquad \text{(NEAB)}$$

5 Use the substitution $x = 2\cos\theta$, or otherwise, to evaluate $\displaystyle\int_1^{\sqrt{2}} \frac{1}{x^2\sqrt{(4-x^2)}}\,dx$,

giving the answer in surd form. (UODLE)

6 Given that $f(x) \equiv \dfrac{1}{3\cos^2 x + \sin^2 x}$,

a) find $\displaystyle\int f(x)\,dx$ **b)** evaluate $\displaystyle\int_0^{\frac{\pi}{4}} f(x)\,dx$ (EDEXCEL)

7 By use of the substitution $u^2 = 1 + x$, or otherwise, find the exact value of

$$\int_3^8 \frac{dx}{x\sqrt{(1+x)}} \qquad \text{(EDEXCEL)}$$

8 a) Integrate the following expressions with respect to x:

i) $\dfrac{1}{\sqrt{(3-x^2)}}$ **ii)** $\dfrac{x}{1+x^2}$

b) i) One method of finding

$$I = \int \frac{x}{(1+x^2)\sqrt{(3-x^2)}}\, dx$$

involves substituting $3 - x^2 = u^2$ in the integral. Use this substitution to obtain

$$I = \int \frac{1}{u^2 - 4}\, du$$

ii) Write $\dfrac{1}{u^2-4}$ in partial fractions.

iii) Complete the integration to find I in terms of x. (MEI)

9 Express $\dfrac{x}{(x-2)^2}$ in the form $\dfrac{A}{(x-2)} + \dfrac{B}{(x-2)^2}$, where A and B are constants.

Hence, or otherwise, evaluate $\displaystyle\int_3^4 \frac{x}{(x-2)^2}\, dx$. (AEB 94)

10 Express $\dfrac{7}{(2-x)(1+3x)}$ in partial fractions.

Hence find $\displaystyle\int_0^1 \frac{7}{(2-x)(1+3x)}\, dx$, giving your answer as a multiple of $\ln 2$. (AEB 95)

11 Express as the sum of partial fractions

$$\frac{2}{x(x+1)(x+2)}$$

Hence show that

$$\int_2^4 \frac{2}{x(x+1)(x+2)}\, dx = 3\ln 3 - 2\ln 5 \qquad \text{(EDEXCEL)}$$

12 Express $f(x) \equiv \dfrac{3x^2 - 7x + 6}{(x-3)^2(x+1)}$ in the form

$$\frac{A}{(x+1)} + \frac{B}{(x-3)} + \frac{C}{(x-3)^2}$$

where A, B and C are constants to be determined.

Hence evaluate $\displaystyle\int_1^2 f(x)\,dx$ giving your answer in the form $a + \ln b$, where a and b are rational numbers.

13 $f(x) \equiv \dfrac{x^2 + 6x + 7}{(x+2)(x+3)}, x \in \mathbb{R}$

Given that $f(x) \equiv A + \dfrac{B}{x+2} + \dfrac{C}{x+3}$,

a) find the values of the constants A, B and C

b) show that $\displaystyle\int_0^2 f(x)\,dx = 2 + \ln\left(\frac{25}{18}\right)$. (EDEXCEL)

14 Given that

$$\frac{7x - x^2}{(2-x)(x^2+1)} \equiv \frac{A}{(2-x)} + \frac{Bx + C}{(x^2+1)}$$

determine the values of A, B and C.

A curve has equation $y = \dfrac{7x - x^2}{(2-x)(x^2+1)}$.

Determine the equation of the normal to the curve at the point (1, 3).

Prove that the area of the region bounded by the curve, the x-axis and the line $x = 1$ is $\dfrac{7}{2}\ln 2 - \dfrac{\pi}{4}$. (AEB 87)

15 Use integration by parts to show that

$$\int_2^4 x \ln x\,dx = 7\ln 4 - 3 \qquad \text{(EDEXCEL)}$$

16 The diagram shows the graph of $y = xe^{-x}$.

i) Differentiate xe^{-x}.

ii) Find the coordinates of the point A, the maximum point on the curve.

iii) Find the area between the curve and the x-axis for $0 \leqslant x \leqslant 2$. (MEI)

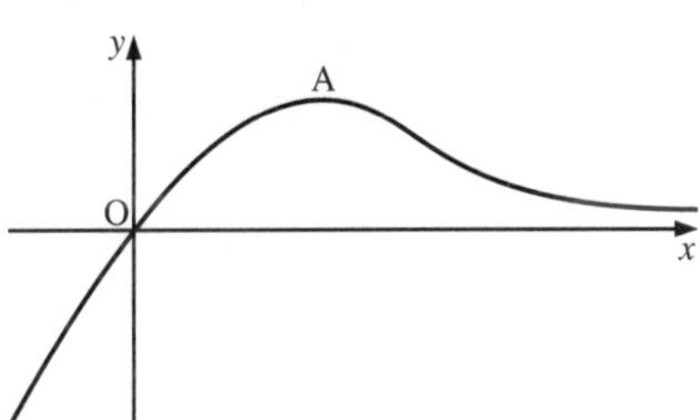

17 The figure shows the curve with equation $y = (4 - 2x)e^{-x}$ which cuts the x-axis at the point A and the y-axis at the point B. The gradient at the point C on the curve is zero.

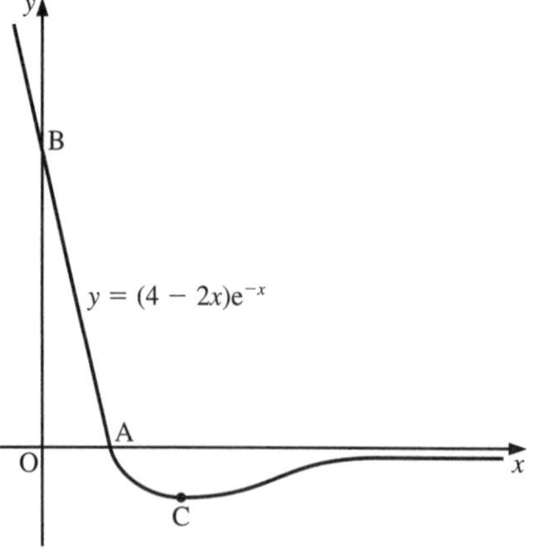

a) Write down the coordinates of the points A and B.

b) Find $\frac{dy}{dx}$ and hence obtain the coordinates of the point C.

c) Show that the area of the finite region bounded by the lines OA and OB and the curve is $2e^{-2} + 2$. (EDEXCEL)

18 The diagram shows the curve with equation $y = \frac{\ln x}{x^2}$ for $x > 0$.

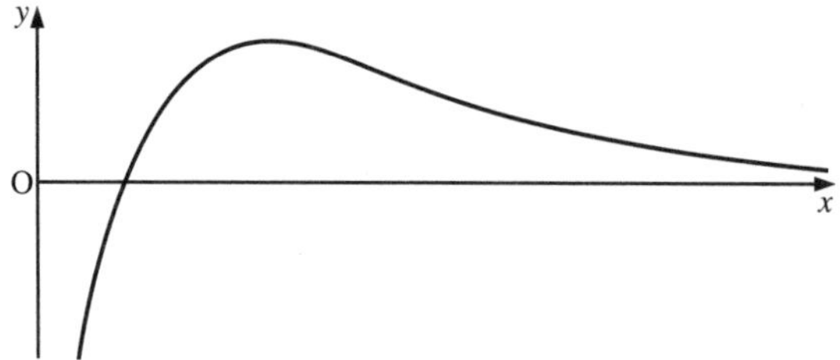

a) State the x coordinate of the point where the curve crosses the x-axis.

b) Show that

$$\frac{dy}{dx} = \frac{1 - 2\ln x}{x^3}$$

c) Find the coordinates of the maximum point of the curve and calculate the value of $\frac{d^2y}{dx^2}$ there.

d) The finite region bounded by the curve, the x-axis and the line $x = 2$ has area equal to

$$\int_1^2 \frac{\ln x}{x^2}\, dx$$

Evaluate this integral, leaving your answer in terms of natural logarithms. (AEB 94)

19 **i)** Show that

$$\frac{d}{dx}(2x - \sin 2x) = 4\sin^2 x$$

ii) Use integration by parts to show that

$$\int_0^{\frac{\pi}{2}} x\sin^2 x\, dx = \frac{1}{16}(\pi^2 + 4) \qquad \text{(NICCEA)}$$

20 **a)** Differentiate $(1 + x^2)^{\frac{3}{2}}$ with respect to x.

b) Solve the differential equation

$$\frac{dy}{dx} = x(1 + x^2)^{\frac{1}{2}} y^2$$

given that $y = 1$ when $x = 0$. (AEB 94)

21 **a)** Find $\int \frac{x+3}{x}\,\mathrm{d}x$.

b) Solve the differential equation

$$\frac{\mathrm{d}y}{\mathrm{d}x} = \frac{(x+3)\sqrt{y}}{x}$$

given that $y = 1$ when $x = 1$. (AEB 95)

22 Solve the first-order differential equation $\frac{\mathrm{d}y}{\mathrm{d}x} = y\mathrm{e}^{-x}$, given that $y = 1$ when $x = 0$. Express your final answer in the form $y = \mathrm{f}(x)$. (UODLE)

23 If $x < 1$, solve the differential equation

$$(1-x)\frac{\mathrm{d}y}{\mathrm{d}x} = (1+x)y$$

given that $y = 2$ when $x = 0$. (NICCEA)

24 Use integration by parts to find $\int x\sec^2 x\,\mathrm{d}x$.

By separating the variables, solve the differential equation

$$\cos^2 x\frac{\mathrm{d}y}{\mathrm{d}x} = xy^2 \qquad \left(0 \leqslant x < \frac{\pi}{2}\right)$$

given that $y = 1$ when $x = 0$. (AEB 95)

25 A bottle is shaped so that when the depth of water is x cm, the volume of water in the bottle is $(x^2 + 4x)\,\mathrm{cm}^3$, $x \geqslant 0$. Water is poured into the bottle so that at time t s after pouring commences, the depth of water is x cm and the rate of increase of the volume of the water is $(x^2 + 25)\,\mathrm{cm}^3\,\mathrm{s}^{-1}$.

a) Show that $\frac{\mathrm{d}x}{\mathrm{d}t} = \frac{x^2+25}{2x+4}$.

Given that the bottle was empty at $t = 0$,

b) solve this differential equation to obtain t in terms of x. (ULEAC)

26 The rate, in $\mathrm{cm}^3\,\mathrm{s}^{-1}$, at which oil is leaking from an engine sump at any time t seconds is proportional to the volume of oil, $V\,\mathrm{cm}^3$, in the sump at that instant. At time $t = 0$, $V = A$.

a) By forming and integrating a differential equation, show that

$$V = A\mathrm{e}^{-kt}$$

where k is a positive constant.

b) Sketch a graph to show the relation between V and t.

Given further that $V = \frac{1}{2}A$ at $t = T$,

c) show that $kT = \ln 2$. (EDEXCEL)

27 At time t hours the rate of decay of the mass of a radioactive substance is proportional to the mass x kg of the substance at that time. At time $t = 0$ the mass of the substance is A kg.

a) By forming and integrating a differential equation, show that $x = Ae^{-kt}$, where k is a constant.

It is observed that $x = \frac{1}{3}A$ at time $t = 10$.

b) Find the value of t when $x = \frac{1}{2}A$, giving your answer to 2 decimal places. (EDEXCEL)

21 Numerical methods

A modern computer, which according to the popular notion is so clever that it can do almost anything, is in fact a sub-moron.

C.S. OGILVY

Error

Many mathematical calculations involve rounding, which causes an error in the answer. Suppose a calculation involved the rational number 1.435, then using 1.44 (to two decimal places) causes a round-off error given by

$$1.435 - 1.44 = -0.005$$

However, rounding 1.435 to two significant figures gives 1.4. In this case, the round-off error is given by

$$1.435 - 1.4 = 0.035$$

Notice that the two round-off errors are different.

If x represents an approximation to the value X and e is the error in this approximation, then the numerical value of e, denoted $|e|$ (modulus of e) is called the **absolute error** and $\left|\frac{e}{X}\right|$ is called the **relative error**.

In the first of the two examples above, we have $e = -0.005$. Therefore, the absolute error is given by

$$|-0.005| = 0.005$$

The relative error is given by

$$\frac{0.005}{1.435} = \frac{1}{287}$$

The relative error gives an indication of whether the error is relatively large or small. For example, if in measuring a distance of 100 metres an error of 1 cm is introduced, the relative error is

$$\frac{1}{10\,000} = 0.0001$$

However, if in measuring a length of 10 cm an error of 1 cm is introduced, the relative error is

$$\frac{1}{10} = 0.1$$

In other words, it is far more important.

Example 1 Round off each of the following rational numbers in the stated way and find **i)** the round-off error, **ii)** the absolute round-off error, **iii)** the relative round-off error.

a) 32.678 94, to three decimal places
b) 1.702 165 1, to four significant figures

SOLUTION

a) $32.678\,94 = 32.679$ (3 dp). Therefore,

i) $$\text{round-off error} = 32.678\,94 - 32.679 = -0.000\,06$$

ii) $\text{absolute error} = 0.000\,06$

iii) $$\text{relative error} = \frac{0.000\,06}{32.678\,94} = 0.000\,001\,8 \quad (7 \text{ dp})$$

b) $1.702\,165\,1 = 1.702$ (4 sf). Therefore,

i) $$\text{round-off error} = 1.702\,165\,1 - 1.702 = -0.000\,165\,1$$

ii) $\text{absolute error} = 0.000\,165\,1$

iii) $$\text{relative error} = \frac{0.000\,165\,1}{1.702\,165\,1} = 0.000\,10 \quad (5 \text{ dp})$$

The percentage error is given by

$$\text{Percentage error} = \text{Relative error} \times 100\%$$

Example 2 Using your calculator state, as accurately as possible, the value of $\sqrt{5}$. Find the percentage error involved in approximating this value to one decimal place.

SOLUTION

Now $\sqrt{5} = 2.236\,067\,977$ on the calculator, which is 2.2 to one decimal place. Therefore,

$$\text{round-off error} = e = 2.236\,067\,977 - 2.2 = 0.036\,067\,977$$

Therefore, the percentage error is

$$\frac{0.036\,067\,977}{2.236\,067\,977} \times 100\% = 1.61\%$$

Suppose the exact answer to a calculation is $A = 2.346$. Rounding this to one decimal place gives 2.3. Now suppose we are given the information the opposite way round. That is, we are told that the answer to a calculation is 2.3, to one decimal place, and asked to give the bounds of A.

We know that if $A = 2.25$, this would be 2.3, to one decimal place. We also know that if A has any value less than 2.35, this would be 2.3, to one decimal place. So, we have the bounds of A as 2.35 and 2.25.

We write $A_u = 2.35$, for the upper bound of A; and $A_l = 2.25$, for the lower bound of A.

Example 3 In each of the following cases, X is given to a stated level of accuracy. State the upper and lower bounds of X.

a) 4.7, to one decimal place
b) 0.1842, to four significant figures

SOLUTION

a) We are given $X = 4.7$ (1 dp). Therefore, $X_u = 4.75$ and $X_l = 4.65$.

b) We are given $X = 0.1842$ (4 sf). Therefore, $X_u = 0.18425$ and $X_l = 0.18415$.

Example 4 The sides of a rectangle are measured as 4.24 m and 8.38 m, to the nearest centimetre.

a) Calculate the upper bound of the perimeter.
b) Calculate the lower bound of the area.

SOLUTION

a) The upper bound of the perimeter, P_u, is obtained by summing the upper bounds of the sides. We have

$$\begin{aligned} P_u &= 4.245 + 8.385 + 4.245 + 8.385 \\ &= 25.26\,\text{m} \end{aligned}$$

b) The lower bound of the area, A_l, is obtained by multiplying the lower bounds of the sides. We have

$$\begin{aligned} A_l &= 4.235 \times 8.375 \\ &= 35.468\,125\,\text{m}^2 \end{aligned}$$

Example 5 In mechanics, the formula $\omega = \sqrt{\dfrac{9.8}{l}}$ is used for calculating the angular velocity of a simple pendulum of string length l. In an experiment, the measured value of l is 5.62, correct to two decimal places.

a) Write down the upper and lower bounds of l.

b) Hence, find the upper and lower bounds of ω, correct to four decimal places.

SOLUTION

a) We are given $l = 5.62$ (1 dp). Therefore, $l_u = 5.625$ and $l_l = 5.615$.

b) Since we are dividing by l in the formula $\omega = \sqrt{\frac{9.8}{l}}$, the upper bound of ω can be found by using the lower bound for l. Therefore,

$$\omega_u = \sqrt{\frac{9.8}{5.615}}$$

$$= 1.321\,107\,498$$

Therefore, $\omega_u = 1.3211$, correct to four decimal places.

Now the lower bound of ω can be found by using the upper bound of l. Therefore,

$$\omega_l = \sqrt{\frac{9.8}{5.625}}$$

$$= 1.319\,932\,658$$

Therefore, $\omega_l = 1.3199$, correct to four decimal places.

Exercise 21A

1 Round off each of the following rational numbers in the stated way and find, correct to three decimal places, the percentage round-off error.

a) 3.896 234 to three decimal places

b) 10.0956 to two significant figures

c) $\frac{3}{7}$ to four decimal places

2 A circle has a radius of 3 cm.

a) Find the area of the circle using:

i) the calculator value of π

ii) π as 3.142

b) Find, correct to three decimal places, the percentage error in the area calculation using π as 3.142.

3 In each of the following cases, Y is given to a stated level of accuracy. State the upper and lower bounds of Y.

a) 8.34, to two decimal places

b) 61.234, to three decimal places

c) 0.005, to three decimal places

d) 3.467, to four significant figures

e) 6, to one significant figure

f) 0.462, to three significant figures

4 The sides of a rectangle are measured as 1.3 cm and 4.7 cm to the nearest mm.

a) Calculate the upper and lower bounds of the perimeter.

b) Calculate the upper and lower bounds of its area.

5 The side of square is measured as 4.00 m to the nearest cm. Calculate the lower bound of its area.

6 The radius of a circle is measured as 2.34 m to the nearest cm. Calculate the upper bound of its area, correct to four decimal places.

7 It is known that two variables p and q are related by the formula $p = \dfrac{1}{q+1}$. In an experiment, the value of q was found to be 0.207 to three significant figures. Calculate the upper and lower bounds of p, correct to three significant figures.

8 The variables x, y and z are related by the formula

$$z = \frac{x^2}{3y} + 2$$

In an experiment, the values of x and y were found to be 3.1 and 10.8 respectively, each correct to one decimal place. Calculate the upper and lower bounds of z, correct to four decimal places.

***9** The side of a square was measured as 1.8 cm. The maximum percentage error in measuring the side was 3%. Calculate, to four decimal places, the upper and lower bounds of the area of the square.

***10** In quantum theory, the formula

$$E = \frac{120}{r} - 2r^2$$

is used for calculating the energy E of an electron moving in an orbit of radius r. In an experiment, the value of r was found to be 2.13, correct to two decimal places. Calculate the upper and lower bounds of E, correct to three decimal places.

Numerical solution of equations

In previous chapters, we have needed to solve equations, but in most cases this was possible using algebraic methods. We now look at some numerical ways of solving equations.

Suppose we want to solve the equation $x^4 - 2x^3 - 2 = 0$.
Constructing a table of values for $y = x^4 - 2x^3 - 2$, $-2 \leqslant x \leqslant 3$ gives

x	−2	−1	0	1	2	3
y	30	1	−2	−3	−2	25

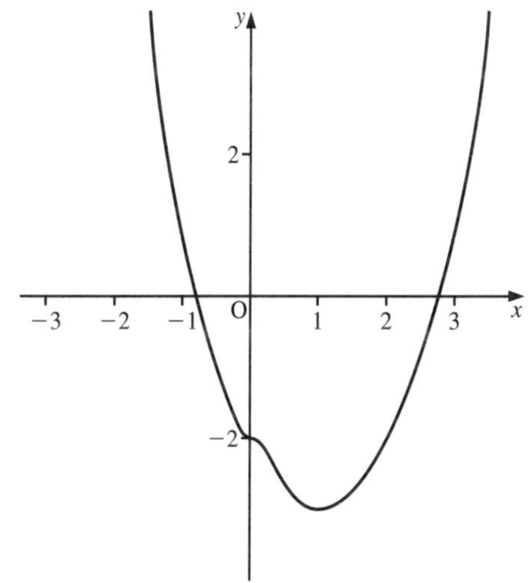

We see that the curve must cross the x-axis between −1 and 0, and between 2 and 3 since

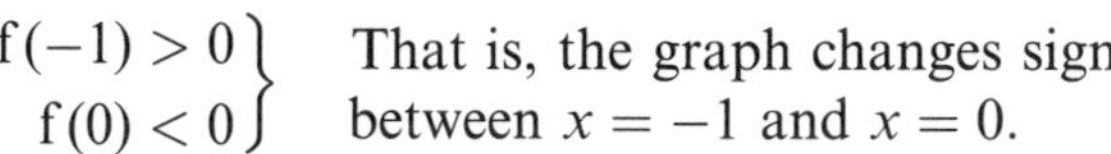

$\left.\begin{array}{r} f(-1) > 0 \\ f(0) < 0 \end{array}\right\}$ That is, the graph changes sign between $x = -1$ and $x = 0$.

and

$\left.\begin{array}{r} f(2) < 0 \\ f(3) > 0 \end{array}\right\}$ That is, the graph changes sign between $x = 2$ and $x = 3$.

and we also know that the graph of $y = x^4 - 2x^3 - 2$ is continuous.

Consider the negative root first. The method we will use is the **bisection method**, whereby we continually bisect the interval in which we know the root occurs, until we obtain the root to the required number of decimal places. In this case, we bisect the interval between -1 and 0. This process gives the following results.

$f(-0.5) = -1.6875 < 0$ so the root lies between -1 and -0.5
$f(-0.75) = -0.8398 < 0$ so the root lies between -1 and -0.75
$f(-0.82) = -0.4451 < 0$ so the root lies between -1 and -0.82
$f(-0.91) = 0.1929 > 0$ so the root lies between -0.91 and -0.82
$f(-0.87) = -0.1101 < 0$ so the root lies between -0.91 and -0.87

So the negative root is -0.9, to one decimal place.

We also know that there is also a root between 2 and 3. Applying the same process, we have

$f(2.5) = 5.8125 > 0$ so the root lies between 2 and 2.5
$f(2.25) = 0.8477 > 0$ so the root lies between 2 and 2.25
$f(2.13) = -0.7437 < 0$ so the root lies between 2.13 and 2.25
$f(2.19) = -0.0043 < 0$ so the root lies between 2.19 and 2.25

Therefore, the positive root is 2.2, to one decimal place.

This method is called the **interval bisection** method. In the example above, we mentioned the fact that the graph of $y = f(x)$ is continuous. This bisection method will only work when the graph of $y = f(x)$ is continuous over the interval in question. For example, consider $y = \dfrac{x-1}{x+1}$, whose graph is shown on the right.

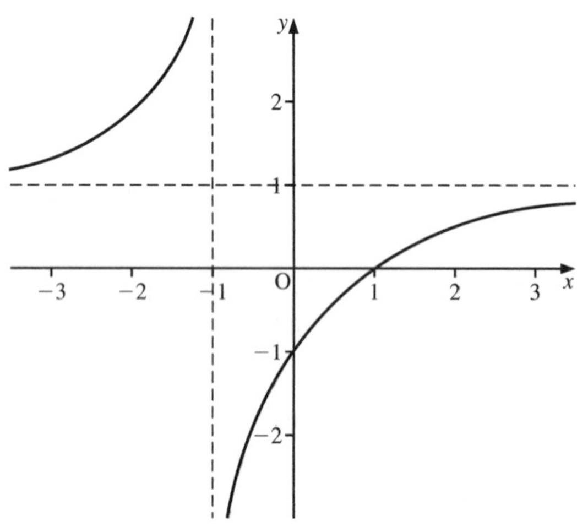

We see that $f(0) < 0$ and $f(-2) > 0$, but there is no root between 0 and -2. The method fails because the graph is not continuous at $x = -1$.

Graphical methods

Example 6 Draw the graphs of $y = x^2 + 1$ and $y = \frac{1}{x}$ on the same set of axes for $-3 \leqslant x \leqslant 4$.

Hence solve the equation $x^3 + x - 1 = 0$, giving your answer to one decimal place.

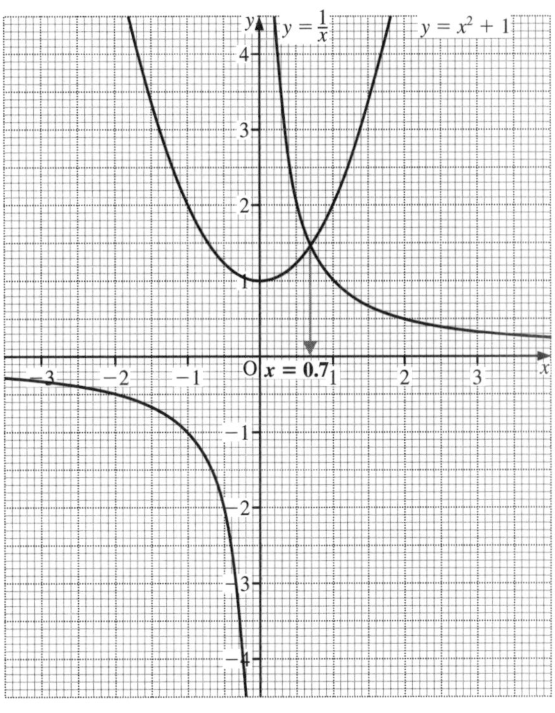

SOLUTION

Drawing the graphs gives the diagram shown on the right.

At the point of intersection, we have

$$x^2 + 1 = \frac{1}{x}$$

$$\therefore \quad x^3 + x = 1$$

$$\therefore \quad x^3 + x - 1 = 0$$

The x coordinate of the point of intersection is a solution of the equation $x^3 + x - 1 = 0$. In fact, it is the only real solution, since the two curves only intersect once. So, the solution is $x = 0.7$.

We could plot the graph of $y = x^3 + x - 1$, as shown on the right.

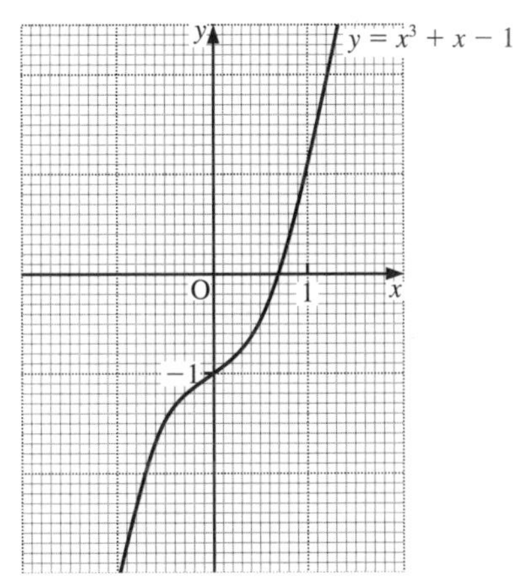

This also demonstrates that there is only one solution to the equation $x^3 + x - 1 = 0$, since the curve cuts the x-axis just once.

Iterative methods

If we want to solve the equation $f(x) = 0$ by an iterative method, we need a relationship

$$x_{r+1} = F(x_r)$$

where x_{r+1} is a better approximation to the solution of $f(x) = 0$ than is x_r. To find such a relationship, we need to rearrange $f(x) = 0$ into the form

$$x = F(x)$$

Suppose that the graphs of $y = x$ and $y = \mathrm{F}(x)$ are as shown below.

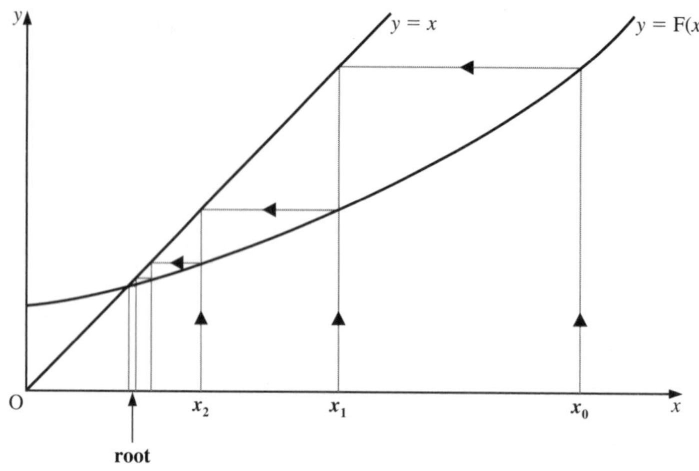

If our initial guess for the root of the equation is x_0, then $x_1 = \mathrm{F}(x_0)$ gives a better approximation. And $x_2 = \mathrm{F}(x_1)$ gives a better approximation than x_1, and so on.

By repeating the process, we can obtain an approximation to the root of the equation.

Suppose we want to solve the equation $x^2 + 4x - 1 = 0$ using iterative methods. Two of the forms into which the equation can be rearranged are

$$\text{I:} \quad x(x+4) - 1 = 0 \quad \text{which gives} \quad x = \frac{1}{x+4}$$

$$\text{II:} \quad x^2 = 1 - 4x \quad \text{which gives} \quad x = \frac{1}{x} - 4$$

giving the iterative formulae

$$\text{I:} \quad x_{r+1} = \frac{1}{x_r + 4} \quad \text{and} \quad \text{II:} \quad x_{r+1} = \frac{1}{x_r} - 4$$

Using $x_{r+1} = \dfrac{1}{x_r + 4}$ together with an initial guess of $x_0 = 1$ gives

$$x_1 = \frac{1}{1+4} = 0.2$$

$$x_2 = \frac{1}{0.2+4} = 0.238\,095\,2381$$

$$x_3 = 0.235\,955\,0562$$

$$x_4 = 0.236\,074\,2706$$

This gives one of the solutions as 0.2361 to four decimal places. The graphs of $y = \dfrac{1}{x+4}$ and $y = x$, shown opposite, illustrate how x_r is converging to the required root.

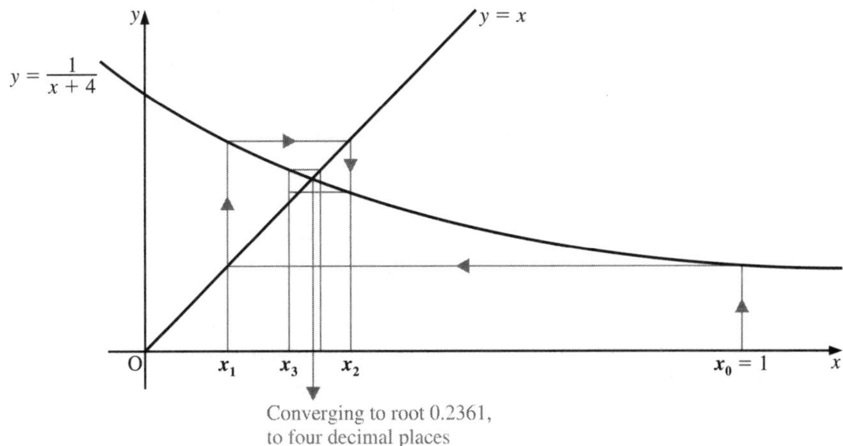

Using $x_{r+1} = \dfrac{1}{x_r} - 4$ together with an initial guess of $x_0 = 1$ gives

$$x_1 = \frac{1}{1} - 4 = -3$$

$$x_2 = \frac{1}{-3} = \frac{1}{-3} - 4 = -4.\dot{3}$$

$$x_3 = -4.230\,769\,231$$

$$x_4 = -4.236\,363\,636$$

$$x_5 = -4.236\,051\,502$$

$$x_6 = -4.236\,068\,896$$

This gives the other solution as −4.2361, to four decimal places. Again, the graphs show how x_r is converging to the required root.

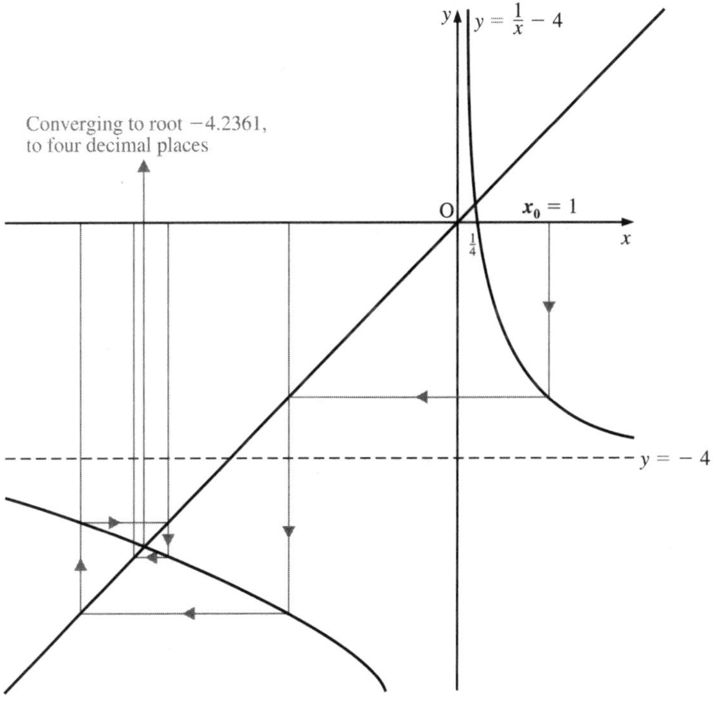

Graphics calculator It should be noted that with a graphics calculator iteration is made very simple. In the above example, we would first key

1 EXE

(the x_1 value). Then we would enter the iterative formula in the form

1 ÷ ANS – 4 EXE

This gives the value of x_2 as -3. Repeatedly pressing EXE gives the values $x_3, x_4, \ldots$.

This example shows two possible arrangements of the original equation, both of which gave iterative formulae which converged to a root of the original equation. However, this is not always the case. For example, the equation $x^2 + 4x + 2 = 0$ has roots between 0 and -1 and between -3 and -4.

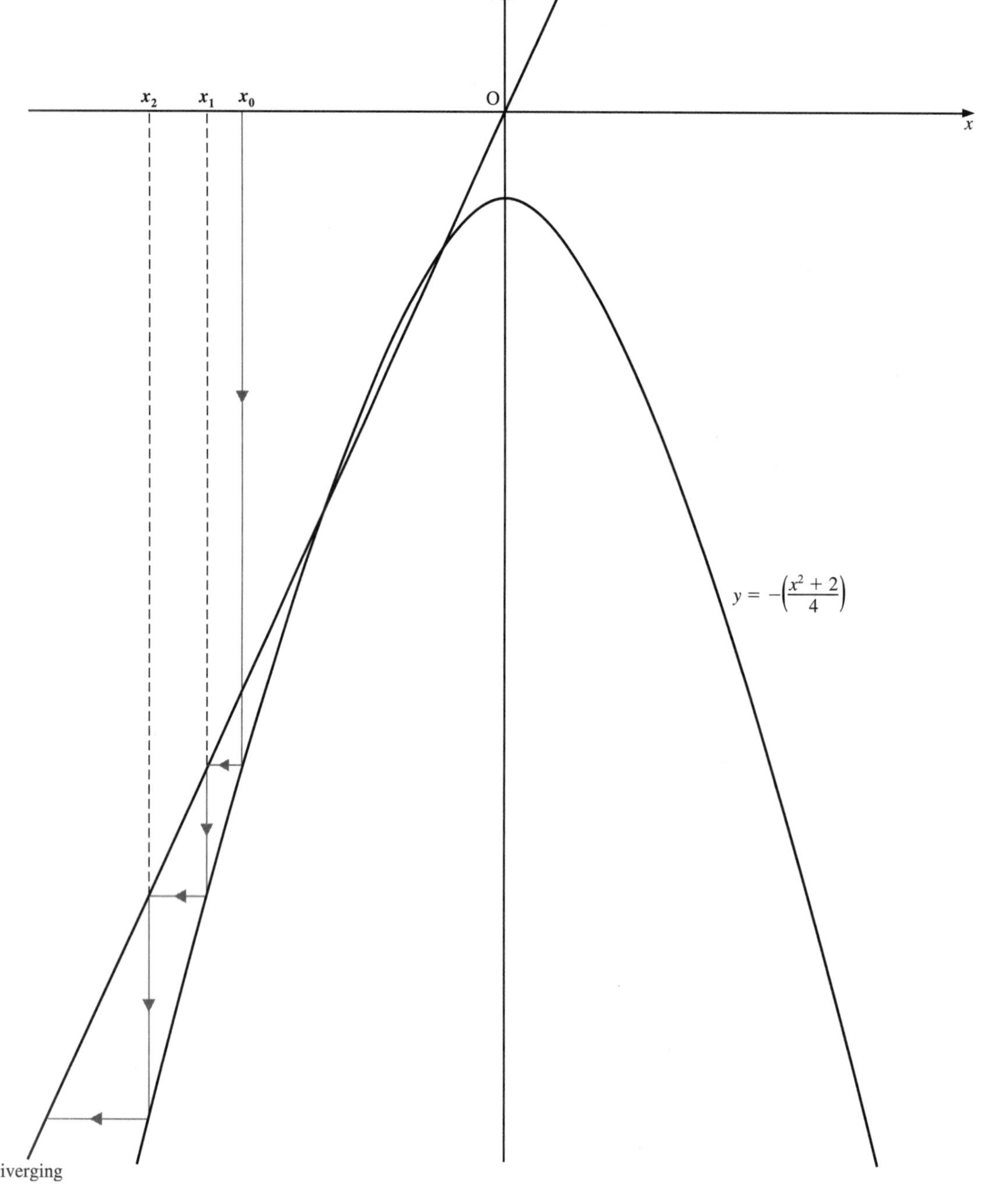

Diverging

One rearrangement of this equation is $x = -\left(\frac{x^2+2}{4}\right)$, which gives the iterative formula

$$x_{r+1} = -\left(\frac{x_r^{\,2}+2}{4}\right)$$

Letting $x_0 = -4$ gives

$$x_1 = -4.5$$
$$x_2 = -5.5625$$
$$x_3 = -8.235\,351\,563$$
$$x_4 = -17.455\,253\,84$$

It is clear that these values are getting further away from the required root. We say that the sequence $x_0, x_1, x_2, \ldots$ is diverging. The diagram illustrates geometrically what is happening in this case.

Notice that if we had chosen x_0 to be -2, then it would have produced a converging sequence, namely $-1.5, -1.0625, -0.782\,226\,5625, \ldots, -0.585\,786\,4376$.

Example 7

a) Show that the equation $x^3 + 7x - 2 = 0$ has a root between 0 and 1.
b) Show that the equation $x^3 + 7x - 2 = 0$ can be rearranged in the form

$$x = \frac{2}{x^2+7}$$

c) Use an iteration based on this rearrangement with an initial value $x_0 = 1$ to find this root correct to three decimal places.

SOLUTION

a) Letting $f(x) = x^3 + 7x - 2$, then $f(0) = -2$ and $f(1) = 6$. Since $f(0) < 0$ and $f(1) > 0$ and $f(x)$ is continuous, we know that the graph of $y = f(x)$ intersects the x-axis between $x = 0$ and $x = 1$.

b) Given $x^3 + 7x - 2 = 0$, rearranging leads to

$$x(x^2+7) - 2 = 0$$
$$\therefore \quad x = \frac{2}{x^2+7}$$

c) This gives the iterative formula

$$x_{r+1} = \frac{2}{x_r^{\,2}+7}$$

Letting $x_0 = 1$ gives

$$x_1 = 0.25$$
$$x_2 = 0.283\,185\,8407$$
$$x_3 = 0.282\,478\,1267$$
$$x_4 = 0.282\,494\,0995$$

The root is 0.282, to three decimal places.

Exercise 21B

1 For each of the following equations show that there is a root between the values stated, and using the bisection method find the root correct to one decimal place.

a) $x^3 + 2x - 7 = 0, \quad x = 1, x = 2$
b) $2x^3 + x^2 - 7x + 1 = 0, \quad x = -2, x = -3$
c) $x^4 + 2x^3 - x - 1 = 0, \quad x = 0, x = 1$
d) $4x^4 + x^2 - 4 = 0, \quad x = 0, x = -1$
e) $x^5 + x^2 = 1, \quad x = 0, x = 1$
f) $2x^5 = 7 - x, \quad x = 1, x = 2$

2 In each of the following cases:
i) Show how the iterative formula is derived from the given equation.
ii) Determine whether the iterative formula converges to a root of the equation. If the formula converges, state the root of the equation correct to four decimal places.

a) $x^2 - 6x + 1 = 0; \quad x_{r+1} = \dfrac{1}{6 - x_r}, \; x_0 = 5$
b) $x^2 + 10x - 3 = 0; \quad x_{r+1} = \dfrac{3 - x_r{}^2}{10}, \; x_0 = 2$
c) $3x^2 - 6x + 1 = 0; \quad x_{r+1} = \dfrac{1}{6 - 3x_r}, \; x_0 = 1$
d) $2x^3 - x^2 + 1 = 0; \quad x_{r+1} = \dfrac{2x_r{}^3 + 1}{x_r}, \; x_0 = 0.5$
e) $x^4 + x - 3 = 0; \quad x_{r+1} = \dfrac{3}{x_r{}^3 + 1}, \; x_0 = 1$

3 a) Starting with $x_0 = 2.5$ use the iterative formula below to find $x_1, x_2, \ldots, x_5$

$$x_{r+1} = \frac{5}{x_r} + 1$$

b) Find the equation which is solved by this iterative formula.

4 a) Starting with $x_0 = 1$, use the iterative formula

$$x_{r+1} = \frac{1}{x_r + 4}$$

to find x_1, x_2, x_3 and x_4.
b) Find the equation which is solved by this iterative formula.

5 a) Starting with $x_0 = 2$, use the iterative formula

$$x_{r+1} = \sqrt{\left(\frac{x_r + 7}{3}\right)}$$

to find $x_1, x_2, \ldots, x_5$.
b) Find the equation which is solved by this iterative formula.

6 a) Find the equation which is solved by the iterative formula

$$x_{r+1} = 2 + \frac{1}{x_r{}^2}$$

b) Starting with $x_0 = 2.5$, find a solution of this equation to four decimal places.

7 An iterative formula for finding the square root of any positive number N is

$$x_{r+1} = \frac{1}{2}\left(x_r + \frac{N}{x_r}\right)$$

a) Explain why the formula works.

b) Use this iterative formula to find the square root of each of the following, correct to three decimal places.

i) 7 **ii)** 15 **iii)** 38

8 Show that the equation $x^3 + 7x^2 - 2x - 2 = 0$ has a root between $x = 0$ and $x = 1$. By using a suitable iterative formula, find this root correct to three decimal places.

9 Show that the equation $2x^4 - x - 5 = 0$ has a root between $x = 1$ and $x = 2$. By using a suitable iterative formula, find this root correct to three decimal places.

10 Show that the equation $xe^x - 1 = 0$ has a root between $x = 0$ and $x = 1$. By using a suitable iterative formula, find this root correct to three decimal places.

11 The graph of $y = x$ intersects the graph of $y = e^x - 2$ at the point P in the positive quadrant. By deriving a suitable iterative formula, determine, correct to three decimal places, the coordinates of P.

***12** Shown on the right is a cylindrical container of radius r cm with a hemispherical base also of radius r cm. Given that the volume of the container is $50\,\text{cm}^3$ and that the total height of the container is 10 cm show that

$$30\pi r^2 - \pi r^3 - 150 = 0$$

Using a suitable iterative formula, find, correct to three decimal places, the value of r.

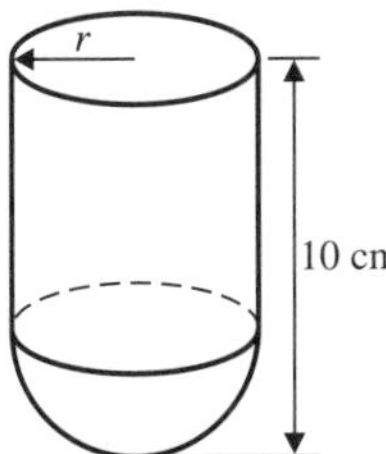

Trapezium rule

Suppose we want to find the area under the curve $y = x^3 + 1$ between $x = 0$ and $x = 2$.

The obvious way would be to find $\int_0^2 (x^3 + 1)\,dx$.

However, if the function were more complicated, then integration might not be such an easy option. It is for this reason that we use numerical techniques to find areas under curves.

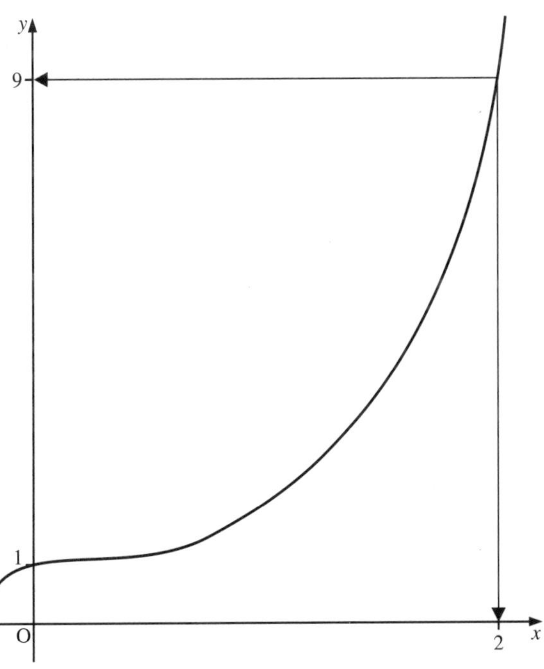

To find an approximation of the area indicated above, we fit trapeziums under the curve, as shown in the diagram below left.

The lines $x = 1$ and $x = 2$ are called **ordinates** and we let y_i be the value of y at the ith ordinate.

An estimate for the required area is given by finding the areas of both trapeziums and adding them together. We have

$$\left.\begin{aligned} \text{Area of OABC} &= \frac{h}{2}(y_0 + y_1) = \frac{1}{2}(1+2) = \frac{3}{2} \\ \text{Area of BCDE} &= \frac{h}{2}(y_1 + y_2) = \frac{1}{2}(2+9) = \frac{11}{2} \end{aligned}\right\} \text{giving a total area of 7.}$$

We can get a better approximation by dividing the required area into smaller trapeziums, as shown in the diagram below right.

An estimate of the area is given by

$$\begin{aligned} A &\approx \frac{0.5}{2}(y_0 + y_1) + \frac{0.5}{2}(y_1 + y_2) + \frac{0.5}{2}(y_2 + y_3) + \frac{0.5}{2}(y_3 + y_4) \\ &= 0.25\,[y_0 + 2(y_1 + y_2 + y_3) + y_4] \\ &= 0.25\left[1 + 2\left(\tfrac{9}{8} + 2 + \tfrac{35}{8}\right) + 9\right] \\ &= 6.25 \end{aligned}$$

An even better approximation can be obtained by dividing the area into trapeziums of width 0.25.

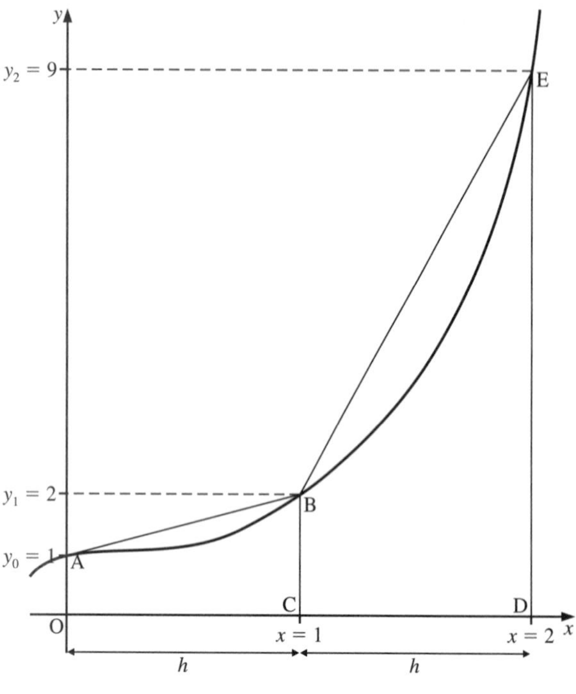

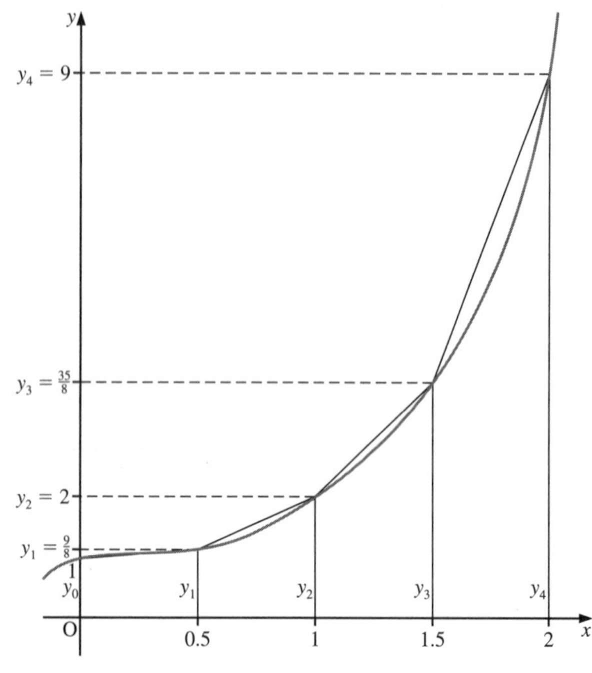

Generally, the area A between the curve $y = \mathrm{f}(x)$, the ordinates a and b and the x-axis is given by

$$A \approx \frac{h}{2}\,[y_0 + 2(y_1 + y_2 + \ldots + y_{n-2}) + y_{n-1}]$$

where $h = \dfrac{b-a}{n-1}$ and $y_i = y(a + ih)$.

Notice that n is the number of ordinates, therefore $(n - 1)$ is the number of trapeziums or strips.

This is known as the **trapezium rule**.

Example 8 Using the trapezium rule with five ordinates, find an estimate for the area under the curve $y = 2x^2 + 3$ between $x = 1$ and $x = 3$.

SOLUTION

Now $h = \dfrac{3-1}{5-1} = \dfrac{1}{2}$.

Therefore,

$$A \approx \frac{0.5}{2}\,[y_0 + 2(y_1 + y_2 + y_3) + y_4]$$

$$\therefore \qquad A \approx 0.25\,[5 + 2(7.5 + 11 + 15.5) + 21] = 23.5$$

An estimate of the area is 23.5.

Example 9 Using the trapezium rule with three ordinates, find an estimate for the area under the curve $y = \sin x$ between $x = 0$ and $x = \dfrac{\pi}{2}$.

SOLUTION

Now $h = \dfrac{\frac{\pi}{2} - 0}{2} = \dfrac{\pi}{4}$.

Therefore,

$$A \approx \frac{\left(\frac{\pi}{4}\right)}{2}\,(y_0 + 2y_1 + y_2)$$

$$\therefore \qquad A \approx \frac{\pi}{8}\,(0 + 1.414 + 1) \approx 0.95$$

An estimate of the area is 0.95.

Simpson's rule

Consider the area between the curve $y = f(x)$, the x-axis and the ordinates a and b, as shown on the right.

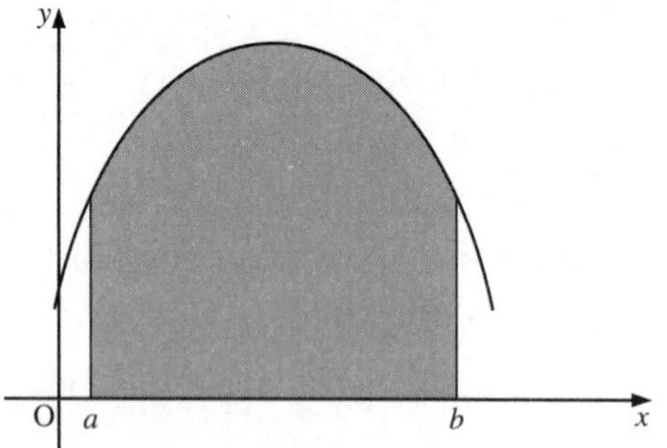

Using the trapezium rule with a small number of strips to find this area will not be very accurate.

For example, we see in the second diagram on the right that the total area covered by the trapeziums is smaller than the actual area under the curve.

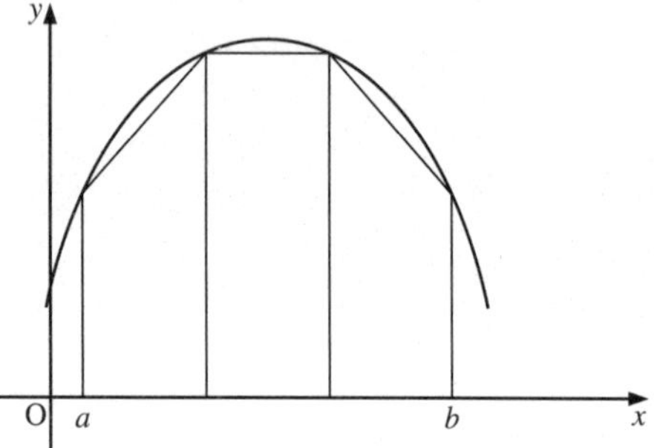

Such a curve can be approximated by a parabola. This approximation leads to Simpson's rule, which states that the area A between the curve $y = f(x)$, the ordinates a and b and the x-axis is given by

$$A \approx \tfrac{1}{3}h(y_0 + 4y_1 + 2y_2 + 4y_3 + 2y_4 + \ldots + 4y_{n-2} + y_{n-1})$$

where n is odd, $h = \dfrac{b-a}{n-1}$ and $y_i = y(a + ih)$.

The proof of Simpson's rule is left as question 16 of Exercise 21C (page 502).

Example 10 Using Simpson's rule with two strips, find an approximation for the area under the curve $y = \dfrac{1}{x}$ between $x = 1$ and $x = 2$.

SOLUTION

Two strips include three ordinates. Therefore,

$$h = \frac{2-1}{3-1} = \frac{1}{2}$$

Now

x_i	1	1.5	2
y_i	1	$0.\dot{6}$	0.5

By Simpson's rule we have

$$A \approx \frac{1}{3}\left(\frac{1}{2}\right)[1 + 4(0.\dot{6}) + 0.5]$$

$$A \approx 0.694 \quad \text{(3 decimal places)}$$

An estimate of the area is 0.694.

Example 11 Using Simpson's rule with 5 ordinates, find an approximation for the area under the curve $f(x) = e^{-2x}$ between $x = 1$ and $x = 3$.

SOLUTION

Now $h = \frac{3-1}{5-1} = \frac{2}{4} = \frac{1}{2}$ and we have

x_i	1	1.5	2	2.5	3
y_i	e^{-2}	e^{-3}	e^{-4}	e^{-5}	e^{-6}

By Simpson's rule we have

$$A \approx \frac{1}{3}\left(\frac{1}{2}\right)(e^{-2} + 4e^{-3} + 2e^{-4} + 4e^{-5} + e^{-6})$$

$$A \approx 0.067 \quad \text{(3 decimal places)}$$

An estimate of the area is 0.067.

Exercise 21C

Unless stated otherwise, the answers in this exercise should be given correct to three significant figures.

1 Given that $y = f(x)$ and that for the given values of x the corresponding values of y are shown in the table below, use the trapezium rule with 6 strips to find an approximate value for $\int_1^{19} f(x)\,dx$.

x	1	4	7	10	13	16	19
$f(x)$	2	7	15	21	26	31	36

2 a) Copy and complete the following table giving values correct to three decimal places:

x	1	2	3	4	5
$\sqrt{1+x^2}$					

b) Plot the graph of $y = \sqrt{1+x^2}$, for $0 \leqslant x \leqslant 5$.

c) Use the trapezium rule with 4 strips to find an approximate value for $\int_1^5 \sqrt{1+x^2}\,dx$.

3 Use the trapezium rule with 6 strips to find an approximate value for $\int_4^{16} \frac{1}{1+\sqrt{x}}\,dx$.

4 Find an approximate value for $\int_0^1 \sqrt{1-x^3}\,dx$ using the trapezium rule with 6 ordinates.

5 Use the trapezium rule with 4 strips to find an approximate value for $\int_1^9 \sqrt{\ln x}\,dx$.

6 Find an approximate value for $\int_0^{\frac{\pi}{3}} \frac{1}{1+\tan x}\,dx$ using the trapezium rule with 5 ordinates.

7 Given that $y = \mathrm{f}(x)$ and that for the given values of x, the corresponding values of y are shown in the table below, use Simpson's rule with 4 strips to find an approximate value for $\int_0^8 \mathrm{f}(x)\,\mathrm{d}x$.

x	0	2	4	6	8
$\mathbf{f}(x)$	1	3	9	15	30

8 Use Simpson's rule with 4 strips to find an approximate value for $\int_0^2 \frac{1}{\sqrt{1+x^3}}\,\mathrm{d}x$.

9 Use Simpson's rule with 6 strips to find an approximate value for $\int_0^6 \sqrt{36-x^2}\,\mathrm{d}x$.

10 Find an approximate value for $\int_1^2 \mathrm{e}^{x^2}\,\mathrm{d}x$ using Simpson's rule with 5 ordinates.

11 Find an approximate value for $\int_{\frac{\pi}{6}}^{\frac{\pi}{2}} \sqrt{\cos x}\,\mathrm{d}x$ using Simpson's rule with 3 ordinates.

12 Use Simpson's rule with 4 strips to find an approximate value for $\int_1^{21} \frac{1}{1+\ln x}\,\mathrm{d}x$.

13 **a)** Sketch the graph of $y = \frac{1}{x}$ for $x > 0$.

b) Use Simpson's rule with 7 ordinates to estimate the value of $\int_1^7 \frac{1}{x}\,\mathrm{d}x$.

c) Calculate, to three decimal places, the percentage error involved when taking the answer to part **b** as an approximation to $\ln 7$.

14 **a)** Given that $I = \int_0^{\pi} \sin x\,\mathrm{d}x$,

i) estimate, to three decimal places, the value of I using the trapezium rule with 4 strips,
ii) estimate, to three decimal places, the value of I using Simpson's rule with 2 strips.

b) Calculate the exact value of I.

c) Calculate, to one decimal place, the percentage error involved with each of the estimates in part **a**.

***15** By evaluating $\int_0^1 \frac{1}{1+x^2}\,\mathrm{d}x$, and using Simpson's rule with 11 ordinates, show that $\pi \approx 3.141\,593$.

***16** **a)** Given that the parabola $y = ax^2 + bx + c$ passes through the points $(-h, y_0)$, $(0, y_1)$ and (h, y_2), show that

i) $c = y_2$

ii) $a = \frac{y_0 - 2y_1 + y_2}{2h^2}$

b) Deduce that $\int_{-h}^{h} (ax^2 + bx + c)\,\mathrm{d}x = \frac{h}{3}(y_0 + 4y_1 + y_2)$.

Exercise 21D: Examination questions

1 The figure shows an arc AB of a circle, radius r cm, which subtends an angle θ radians at the centre O of the circle. Given that $r = 9$, measured to the nearest integer, and $\theta = 1.2$, measured to 1 decimal place, calculate

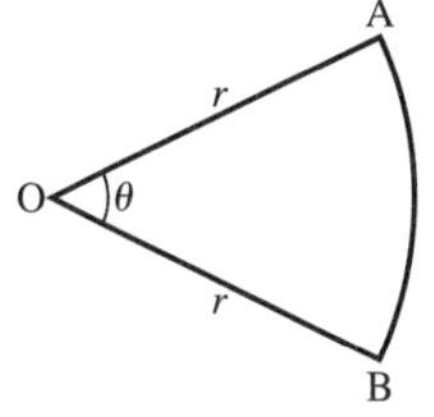

a) the least possible length of the arc AB of the circle.

The area of the sector of the circle bounded by the arc AB and the radii OA and OB is calculated using $r = 9$ and $\theta = 1.2$.

b) Calculate the greatest absolute error possible in this evaluation of the area. (EDEXCEL)

2 **a)** On the same diagram sketch the graphs of $y = \ln x$ and $y = 3 - x$ for x in the interval $1 \leqslant x \leqslant 3$.

b) Show that the equation $\ln x = 3 - x$ has a root between $x = 2$ and $x = 3$.

c) Showing the values of your intermediate approximations, use an iterative method to find this root correct to 3 decimal places.

d) Demonstrate that your answer has the required degree of accuracy. (UODLE)

3 The equation $x^3 - 2x - 5 = 0$ has one real root. Show that this root lies between 2 and 3. This root is to be found using the iterative formula $x_{r+1} = f(x)$. Show that the equation can be written as $x = \mathrm{f}(x)$ in the three forms:

1) $x = \frac{1}{2}(x^3 - 5)$ **2)** $x = 5(x^2 - 2)^{-1}$ **3)** $x = (2x + 5)^{1/3}$

Choose the form for which the iteration converges and hence find the root correct to five decimal places. (OCSEB)

4 **a)** Show that the equation $x^3 - 5x + 1 = 0$ has a root α which lies between 0 and 1.

b) Rearrangements of the above equation give the following iterative formulae

$$x_{n+1} = \frac{1}{5}\left(x_n^3 + 1\right) \qquad x_{n+1} = \sqrt[3]{5x_n - 1}$$

Using $x_0 = 0.5$, show that only one of these formulae will enable you to find α and determine the value of α correct to four decimal places. Record the values of x_1, x_2, $x_3, \ldots$ as accurately as your calculator will allow. (WJEC)

5 **a)** Show the equation $x^3 - 3x + 1 = 0$ has a root α lying between 1.5 and 1.6.
Given that x_0 is an approximate solution to this equation, a better approximation x_1 is sought using the iterative formula

$$x_1 = \sqrt{\left(\frac{3x_0 - 1}{x_0}\right)}$$

b) Take 1.5 as a first approximation to α, and apply this iterative formula twice to obtain two further approximations to α. Hence state the value of α as accurately as your working justifies. (EDEXCEL)

6 Show that the equation $x^5 - 5x - 6 = 0$ has a root in the interval (1, 2).
Stating the values of the constants p, q and r, use an iteration of the form

$$x_{n+1} = (px_n + q)^{\frac{1}{r}}$$

the appropriate number of times to calculate this root of the equation $x^5 - 5x - 6 = 0$ correct to 3 decimal places. Show sufficient working to justify your final answer. (EDEXCEL)

7 A golden rectangle has one side of length 1 unit and a shorter side of length ψ units, where ψ is called the golden section.

ψ can be found using the iterative formula

$$x_{n+1} = \sqrt[3]{x_n(1 - x_n)}$$

Choosing a suitable value for x_1 and showing intermediate values, use this iterative formula to obtain the value of ψ to 2 decimal places. (NEAB)

8 The sequence given by the iteration formula

$$x_{n+1} = 2(1 + e^{-x_n})$$

with $x_1 = 0$, converges to α. Find α correct to 3 decimal places, and state an equation of which α is a root. (UCLES)

9 The shaded region R, as shown in the figure, is bounded by the lines $x = 2$, $x = 4$, $y = 0$ and the curve with equation $y = \ln x$. Use the trapezium rule, with three equally spaced ordinates, to find an approximation for the area of R. Give your answer to 3 significant figures. (EDEXCEL)

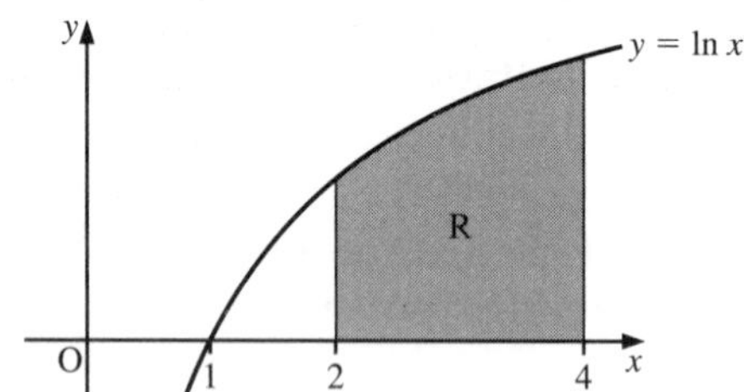

10 Use the trapezium rule with 5 ordinates and interval width 0.25 to evaluate approximately the integral

$$\int_1^2 \ln(1 + x^2)\,dx$$

Show your working and give your answer correct to 2 decimal places. (WJEC)

11 The diagram shows part of the graph of $y = \sqrt{x^2 - 1}$.

a) Use the trapezium rule with 6 ordinates and interval 0.2 to find an approximate value for the area of the shaded region R.

b) Find, in terms of π, the volume generated when R is rotated through 360° about the x-axis. (WJEC)

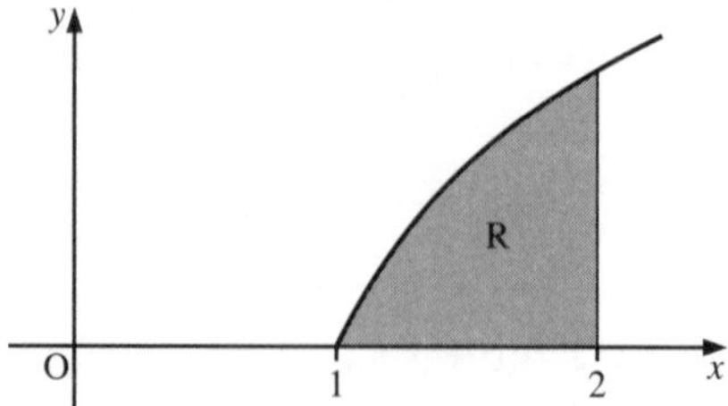

12 The table shows three values of x with the corresponding values of $f(x)$.

x	-1	2	5
$f(x)$	8	26	206

Use Simpson's rule with 3 ordinates to find an approximate value for

$$\int_{-1}^{5} f(x)\,dx$$

(EDEXCEL)

13 Showing all relevant working, use Simpson's rule with 5 equally spaced ordinates to find an estimate for

$$\int_{1}^{9} \ln(1 + x^3)\,dx$$

giving your final answer to 2 decimal places. (EDEXCEL)

14 The finite region R in the figure is bounded by the curve with equation $y = \dfrac{1}{1 + \sqrt{x}}$, the x-axis, the y-axis and the line with equation $x = 1$. Use Simpson's rule with 5 equally spaced ordinates to find an approximate value for the area of R, giving your answer to 2 decimal places. (EDEXCEL)

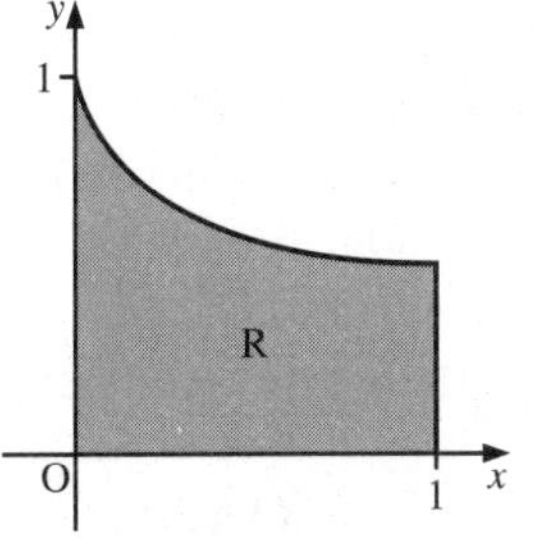

15 Use Simpson's rule with 5 equally spaced ordinates to find an approximate value for

$$\int_{0}^{2} \sin(1 + \sqrt{x})\,dx$$

giving your answer to three decimal places. (EDEXCEL)

22 Vectors

Physics is far too hard for physicists.
DAVID HILBERT

A quantity which is specified by a magnitude and a direction is called a **vector**. For example, displacement and velocity are both specified by a magnitude and a direction and are therefore examples of vector quantities.

A quantity which is specified by just its magnitude is called a **scalar**. For example, distance and speed are both fully specified by a magnitude and are therefore examples of scalar quantities.

In two dimensions, a vector is represented by a straight line with an arrowhead. In the diagram on the right, the line OA represents a vector $\overrightarrow{OA}$.

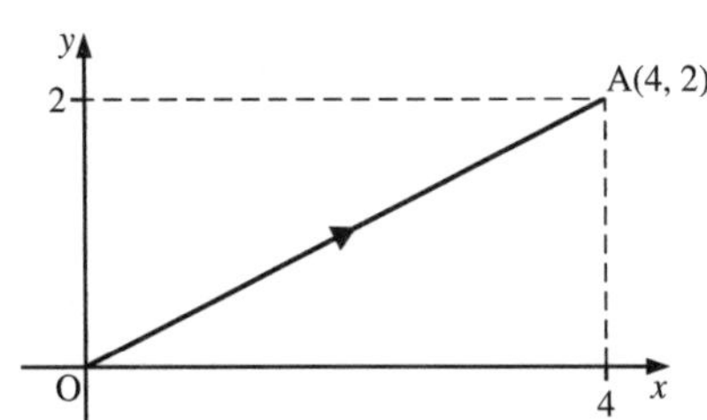

One way of writing this vector is

$$\overrightarrow{OA} = \begin{pmatrix} 4 \\ 2 \end{pmatrix}$$

which means that to go from O to A we move 4 units in the positive x direction and 2 units in the positive y direction. This is called a **column vector**.

The magnitude or modulus of the vector $\overrightarrow{OA}$ is represented by the length OA and is denoted by $|\overrightarrow{OA}|$. In this example, we have

$$|\overrightarrow{OA}|^2 = 4^2 + 2^2 = 20$$

$$\therefore \quad |\overrightarrow{OA}| = \sqrt{20} = 2\sqrt{5}$$

A **unit vector** is a vector of length **1**. The standard unit vectors in two dimensions are

$$\mathbf{i} = \begin{pmatrix} 1 \\ 0 \end{pmatrix} \quad \text{and} \quad \mathbf{j} = \begin{pmatrix} 0 \\ 1 \end{pmatrix}$$

They can be represented diagramatically as shown on the right

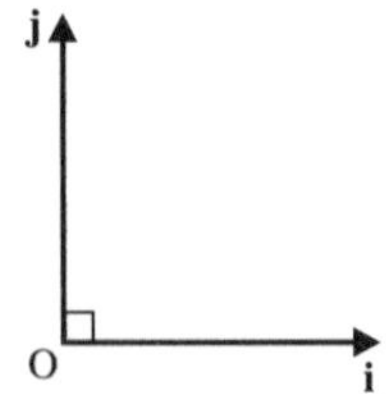

Notice that **i** and **j** are in bold face. This is to indicate that they are vectors. You will indicate a vector by placing a line beneath the letter representing it: for example, $\underline{i}$ and $\underline{j}$.

The vector $\overrightarrow{OA}$ can be written as

$$\overrightarrow{OA} = 4\mathbf{i} + 2\mathbf{j}$$

In three dimensions, the standard unit vectors are

$$\mathbf{i} = \begin{pmatrix} 1 \\ 0 \\ 0 \end{pmatrix} \quad \mathbf{j} = \begin{pmatrix} 0 \\ 1 \\ 0 \end{pmatrix} \quad \mathbf{k} = \begin{pmatrix} 0 \\ 0 \\ 1 \end{pmatrix}$$

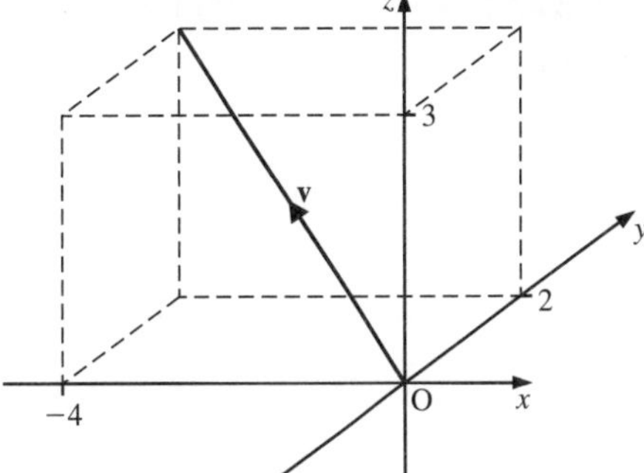

For example, the column vector $\mathbf{v} = \begin{pmatrix} -4 \\ 2 \\ 3 \end{pmatrix}$ can be written as $\mathbf{v} = -4\mathbf{i} + 2\mathbf{j} + 3\mathbf{k}$. The magnitude of $\mathbf{v}$ is given by

$$|\mathbf{v}| = \sqrt{(-4)^2 + 2^2 + 3^2} = \sqrt{29}$$

Example 1 Find the unit vector in the direction of $\mathbf{v} = 5\mathbf{i} - 2\mathbf{j} + 4\mathbf{k}$.

SOLUTION

The magnitude of $\mathbf{v}$ is given by

$$|\mathbf{v}| = \sqrt{5^2 + (-2)^2 + 4^2} = \sqrt{45} = 3\sqrt{5}$$

Therefore, using $\hat{\mathbf{v}}$ to denote a unit vector in the direction of $\mathbf{v}$, we have

$$\hat{\mathbf{v}} = \frac{5}{3\sqrt{5}}\mathbf{i} - \frac{2}{3\sqrt{5}}\mathbf{j} + \frac{4}{3\sqrt{5}}\mathbf{k}$$

The vector has length 1 and is in the direction of $\mathbf{v}$.

Note that $\hat{\mathbf{v}}$ could be written as $\frac{1}{3\sqrt{5}}\begin{pmatrix} 5 \\ -2 \\ 4 \end{pmatrix}$.

Addition and subtraction of vectors

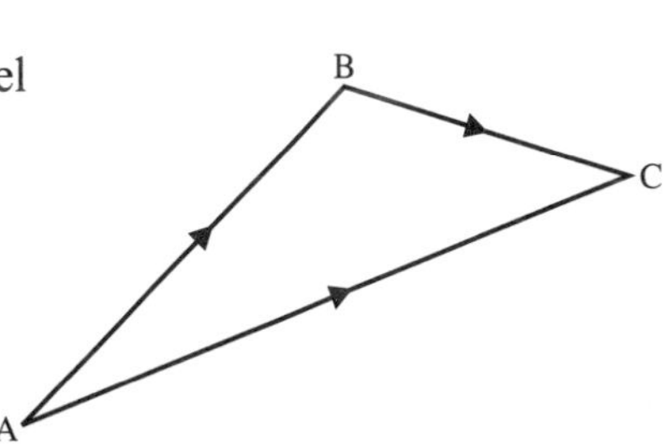

The diagram shows two possible paths that could be taken to travel from A to C.

One route is A to B then B to C. The other route is to go directly from A to C. This can be written as a vector equation:

$$\overrightarrow{AC} = \overrightarrow{AB} + \overrightarrow{BC}$$

The vector $\overrightarrow{AC}$ is called the **resultant** of vectors $\overrightarrow{AB}$ and $\overrightarrow{BC}$.

Since vectors may also be written as single letters in bold type, if we let $\mathbf{u} = \overrightarrow{AB}$, $\mathbf{v} = \overrightarrow{BC}$ and $\mathbf{w} = \overrightarrow{AC}$, then we have

$$\mathbf{w} = \mathbf{u} + \mathbf{v}$$

Since $\overrightarrow{AB} = \mathbf{u}$, we have $\overrightarrow{BA} = -\mathbf{u}$. In other words, the vector $\overrightarrow{AB}$ has the same magnitude as $\overrightarrow{BA}$ but is in the opposite direction.

Example 2 Given that $\overrightarrow{AB} = 3\mathbf{i} + 5\mathbf{j} - 4\mathbf{k}$ and $\overrightarrow{BC} = -\mathbf{i} + 4\mathbf{j} - \mathbf{k}$, find $\overrightarrow{AC}$.

SOLUTION

We have

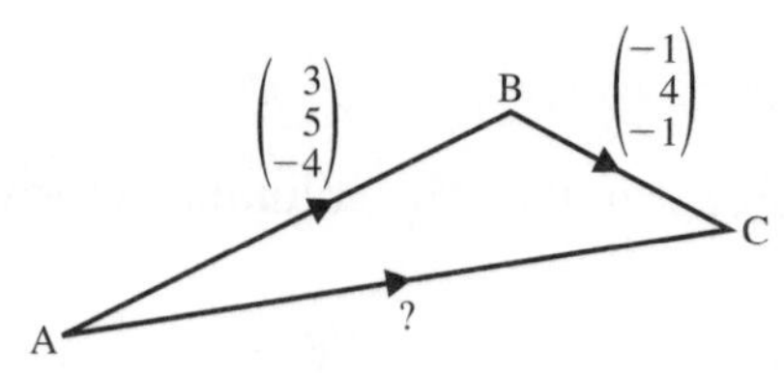

$$\overrightarrow{AC} = \overrightarrow{AB} + \overrightarrow{BC}$$

$$= (3\mathbf{i} + 5\mathbf{j} - 4\mathbf{k}) + (-\mathbf{i} + 4\mathbf{j} - \mathbf{k})$$

$$\therefore \quad \overrightarrow{AC} = 2\mathbf{i} + 9\mathbf{j} - 5\mathbf{k}$$

Example 3 Given that $\overrightarrow{BC} = 7\mathbf{i} - 2\mathbf{j} + \mathbf{k}$ and $\overrightarrow{AC} = \mathbf{i} - 6\mathbf{k}$, find $\overrightarrow{BA}$.

SOLUTION

We have

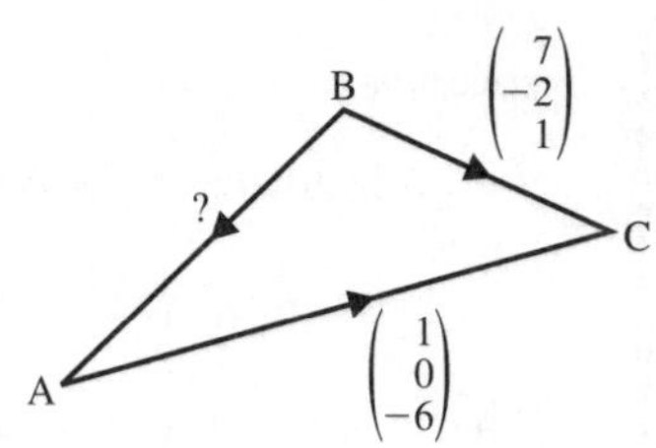

$$\overrightarrow{BA} = \overrightarrow{BC} + \overrightarrow{CA}$$

$$= \overrightarrow{BC} - \overrightarrow{AC}$$

$$= (7\mathbf{i} - 2\mathbf{j} + \mathbf{k}) - (\mathbf{i} - 6\mathbf{k})$$

$$\therefore \quad \overrightarrow{BA} = 6\mathbf{i} - 2\mathbf{j} + 7\mathbf{k}$$

Example 4 Two vectors are given by $\mathbf{a} = 2\mathbf{i} - \mathbf{j} - \mathbf{k}$ and $\mathbf{b} = -\mathbf{i} + 3\mathbf{j} + 4\mathbf{k}$.

a) Find $\mathbf{a} + \mathbf{b}$ and $\mathbf{a} - \mathbf{b}$.

b) Draw a diagram showing $\mathbf{a} + \mathbf{b}$ and another showing $\mathbf{a} - \mathbf{b}$.

SOLUTION

a) Adding the two vectors gives

$$\mathbf{a} + \mathbf{b} = (2\mathbf{i} - \mathbf{j} - \mathbf{k}) + (-\mathbf{i} + 3\mathbf{j} + 4\mathbf{k})$$

$$= \mathbf{i} + 2\mathbf{j} + 3\mathbf{k}$$

Subtracting the two vectors gives

$$\mathbf{a} - \mathbf{b} = (2\mathbf{i} - \mathbf{j} - \mathbf{k}) - (-\mathbf{i} + 3\mathbf{j} + 4\mathbf{k})$$

$$= 3\mathbf{i} - 4\mathbf{j} - 5\mathbf{k}$$

b) Adding **a** to **b** gives

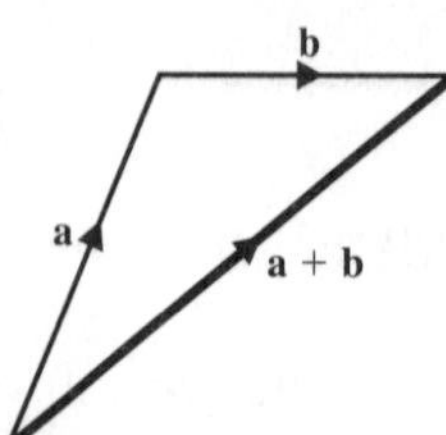

To illustrate $\mathbf{a} - \mathbf{b}$, we add $-\mathbf{b}$ to $\mathbf{a}$ giving

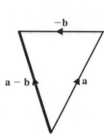

Example 5 The position vectors of the points, A, B and C are $2\mathbf{i} - \mathbf{j} + \mathbf{k}$, $3\mathbf{i} + 2\mathbf{j} - \mathbf{k}$ and $6\mathbf{i} + 11\mathbf{j} - 7\mathbf{k}$, respectively. Show that A, B and C are collinear.

SOLUTION

Let $\mathbf{a} = 2\mathbf{i} - \mathbf{j} + \mathbf{k}$, $\mathbf{b} = 3\mathbf{i} + 2\mathbf{j} - \mathbf{k}$ and $\mathbf{c} = 6\mathbf{i} + 11\mathbf{j} - 7\mathbf{k}$.

Therefore

$$\overrightarrow{AB} = \mathbf{b} - \mathbf{a}$$

$$= \begin{pmatrix} 3 \\ 2 \\ -1 \end{pmatrix} - \begin{pmatrix} 2 \\ -1 \\ 1 \end{pmatrix}$$

$$\therefore \quad \overrightarrow{AB} = \begin{pmatrix} 1 \\ 3 \\ -2 \end{pmatrix}$$

and

$$\overrightarrow{BC} = \mathbf{c} - \mathbf{b}$$

$$= \begin{pmatrix} 6 \\ 11 \\ -7 \end{pmatrix} - \begin{pmatrix} 3 \\ 2 \\ -1 \end{pmatrix} = \begin{pmatrix} 3 \\ 9 \\ -6 \end{pmatrix} = 3\begin{pmatrix} 1 \\ 3 \\ -2 \end{pmatrix}$$

$$\therefore \quad \overrightarrow{BC} = 3\overrightarrow{AB}$$

It is clear that since $\overrightarrow{BC}$ is a multiple of $\overrightarrow{AB}$, $\overrightarrow{BC}$ and $\overrightarrow{AB}$ are parallel. But $\overrightarrow{AB}$ and $\overrightarrow{BC}$ have a common point, namely B. Therefore, A, B and C are collinear.

Exercise 22A

1 Find the magnitude of each of these vectors.

a) $4\mathbf{i} + 3\mathbf{j}$ **b)** $5\mathbf{i} - 7\mathbf{j}$ **c)** $2\mathbf{i} - 2\mathbf{j} + \mathbf{k}$ **d)** $6\mathbf{i} - 3\mathbf{j} + 4\mathbf{k}$

e) $\begin{pmatrix} 12 \\ 5 \end{pmatrix}$ **f)** $\begin{pmatrix} 2 \\ -4 \end{pmatrix}$ **g)** $\begin{pmatrix} -9 \\ 7 \end{pmatrix}$ **h)** $\begin{pmatrix} 5 \\ -7 \\ 3 \end{pmatrix}$

2 Given $\mathbf{v} = \alpha\mathbf{i} + 5\mathbf{j} - \sqrt{7}\mathbf{k}$ and $|\mathbf{v}| = 9$, find the possible values of the constant α.

3 Given that $|2\mathbf{i} + \beta\mathbf{j} - 4\mathbf{k}| = 6$, find the possible values of the constant β.

4 Find the possible values of the constant δ such that $|\delta\mathbf{i} + 4\delta\mathbf{j} + 4\mathbf{k}| = 13$.

5 Find a unit vector in the direction of the vector $8\mathbf{i} - 6\mathbf{j}$.

6 Find a unit vector in the direction of $\mathbf{v} = 5\mathbf{i} - 8\mathbf{j}$.

7 Find a unit vector in the direction of the vector $\begin{pmatrix} -7 \\ 9 \end{pmatrix}$.

8 Find a unit vector in the direction of $\mathbf{v} = 3\mathbf{i} - 2\mathbf{j} + 5\mathbf{k}$.

9 Find a unit vector in the direction of the vector $\mathbf{i} - 3\mathbf{j} + 2\mathbf{k}$.

10 Find a unit vector in the direction of the vector $\begin{pmatrix} -3 \\ 12 \\ -4 \end{pmatrix}$.

11 Find a vector of magnitude 14 in the direction of the vector $6\mathbf{i} - 3\mathbf{j} + 2\mathbf{k}$.

12 Find a vector of magnitude $\sqrt{5}$ in the direction of $\mathbf{v} = 4\mathbf{i} - 8\mathbf{k}$.

13 Find a vector of magnitude $\sqrt{7}$ in the direction of the vector $\begin{pmatrix} 5 \\ -3 \\ 1 \end{pmatrix}$.

14 Given that $\overrightarrow{AB} = 2\mathbf{i} - 4\mathbf{j} + 5\mathbf{k}$ and $\overrightarrow{BC} = 3\mathbf{i} + 6\mathbf{j} - 2\mathbf{k}$ find $\overrightarrow{AC}$.

15 Given that $\overrightarrow{PQ} = -2\mathbf{i} + 3\mathbf{j} - 6\mathbf{k}$ and $\overrightarrow{QR} = 5\mathbf{i} - 7\mathbf{k}$ find $\overrightarrow{PR}$.

16 Given that $\overrightarrow{RS} = \begin{pmatrix} 3 \\ 6 \\ -8 \end{pmatrix}$ and $\overrightarrow{ST} = \begin{pmatrix} 5 \\ -5 \\ 0 \end{pmatrix}$ find $\overrightarrow{RT}$.

17 Given that $\overrightarrow{AB} = 5\mathbf{i} - 7\mathbf{j} - 2\mathbf{k}$ and $\overrightarrow{AC} = 2\mathbf{i} + 3\mathbf{j} - 2\mathbf{k}$ find $\overrightarrow{BC}$.

18 Given that $\overrightarrow{ML} = 10\mathbf{i} - 4\mathbf{j} - 6\mathbf{k}$ and $\overrightarrow{MN} = 7\mathbf{j} - 5\mathbf{k}$ find $\overrightarrow{NL}$.

19 Given that $\overrightarrow{PQ} = \begin{pmatrix} 5 \\ 2 \\ -8 \end{pmatrix}$ and $\overrightarrow{PR} = \begin{pmatrix} -2 \\ 5 \\ -6 \end{pmatrix}$ find $\overrightarrow{QR}$.

20 Given $\overrightarrow{AB} = \alpha\mathbf{i} + 6\mathbf{j} + 4\mathbf{k}$, $\overrightarrow{BC} = 4\mathbf{i} + \beta\mathbf{j} - 3\mathbf{k}$ and $\overrightarrow{AC} = -3\mathbf{i} + \gamma\mathbf{k}$, find the values of the constants α, β and γ.

21 Points P, Q and R have position vectors $\begin{pmatrix} 5 \\ 4 \\ 1 \end{pmatrix}$, $\begin{pmatrix} 7 \\ 5 \\ 4 \end{pmatrix}$ and $\begin{pmatrix} 11 \\ 7 \\ 10 \end{pmatrix}$, respectively.

a) Find $\overrightarrow{PQ}$ and $\overrightarrow{QR}$.
b) Deduce that P, Q and R are collinear and find the ratio PQ : QR.

22 The points A, B and C have coordinates (1, −5, 6), (3, −2, 10) and (7, 4, 18) respectively. Show that A, B and C are collinear.

23 Show that the points P(5, 4, −3), Q(3, 8, −1) and R(0, 14, 2) are collinear.

24 Given that A(2, 13, −5), B(3, β, −3) and C(6, −7, γ) are collinear, find the values of the constants β and γ.

Position vectors

The position vector of a point P with respect to a fixed origin O is the vector $\overrightarrow{OP}$. This is not a free vector, since O is a fixed point. We usually write

$$\overrightarrow{OP} = \mathbf{p}$$

Suppose the position vectors of two points P and Q are **p** and **q** respectively with respect to the origin O. The diagram illustrates the two vectors.

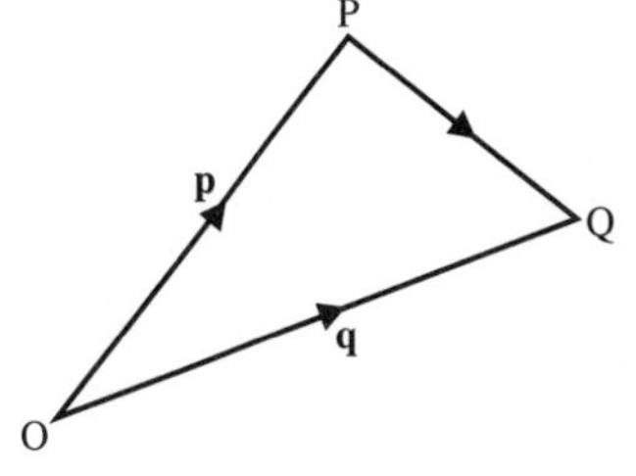

Then

$$\overrightarrow{PQ} = \overrightarrow{PO} + \overrightarrow{OQ}$$

$$= -\mathbf{p} + \mathbf{q}$$

$$\therefore \quad \overrightarrow{PQ} = \mathbf{q} - \mathbf{p}$$

We also have $\overrightarrow{QP} = \mathbf{p} - \mathbf{q}$.

Example 6 The points A, B and C have position vectors **a**, **b** and **c**. Point P is the mid-point of AB and point Q is the mid-point of BC. Find

a) the position vectors of P and Q **b)** $\overrightarrow{PQ}$

SOLUTION

a) The diagram illustrates the situation.

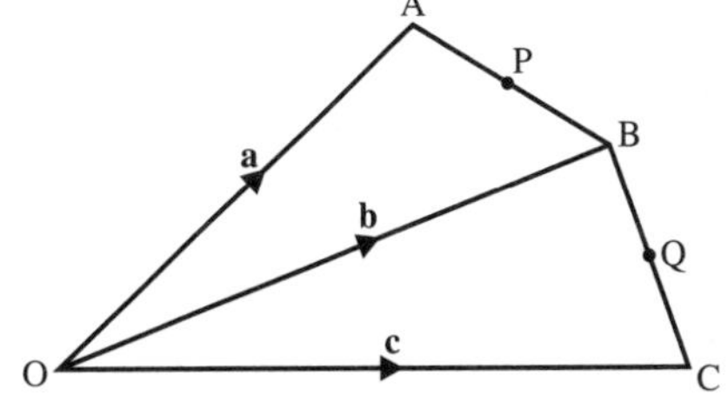

The position vector of P is given by

$$\overrightarrow{OP} = \overrightarrow{OA} + \overrightarrow{AP}$$

$$= \overrightarrow{OA} + \tfrac{1}{2}\overrightarrow{AB}$$

$$= \mathbf{a} + \tfrac{1}{2}(\mathbf{b} - \mathbf{a})$$

$$= \tfrac{1}{2}\mathbf{a} + \tfrac{1}{2}\mathbf{b}$$

$$\therefore \quad \overrightarrow{OP} = \tfrac{1}{2}(\mathbf{a} + \mathbf{b})$$

The position vector of Q is given by

$$\overrightarrow{OQ} = \overrightarrow{OB} + \overrightarrow{BQ}$$

$$= \overrightarrow{OB} + \tfrac{1}{2}\overrightarrow{BC}$$

$$= \mathbf{b} + \tfrac{1}{2}(\mathbf{c} - \mathbf{b})$$

$$= \tfrac{1}{2}\mathbf{b} + \tfrac{1}{2}\mathbf{c}$$

$$\therefore \quad \overrightarrow{OQ} = \tfrac{1}{2}(\mathbf{b} + \mathbf{c})$$

b) The vector $\overrightarrow{PQ}$ is given by

$$\overrightarrow{PQ} = \overrightarrow{OQ} - \overrightarrow{OP}$$

$$= \tfrac{1}{2}(\mathbf{b} + \mathbf{c}) - \tfrac{1}{2}(\mathbf{a} + \mathbf{b})$$

$$= \tfrac{1}{2}\mathbf{c} - \tfrac{1}{2}\mathbf{a}$$

$$\therefore \quad \overrightarrow{PQ} = \tfrac{1}{2}(\mathbf{c} - \mathbf{a})$$

Example 7 The point O is the centre of the regular hexagon ABCDEF. Given that $\overrightarrow{OA} = \mathbf{a}$ and $\overrightarrow{AB} = \mathbf{b}$, find

a) $\overrightarrow{OB}$ **b)** $\overrightarrow{BD}$ **c)** $\overrightarrow{CF}$

SOLUTION

a) The diagram illustrates the hexagon ABCDEF.

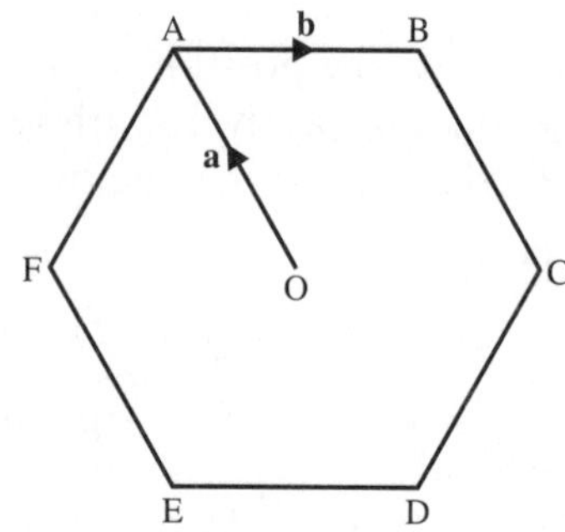

$$\overrightarrow{OB} = \overrightarrow{OA} + \overrightarrow{AB}$$
$$= \mathbf{a} + \mathbf{b}$$

b) We see that $\overrightarrow{BD} = \overrightarrow{BC} + \overrightarrow{CD}$. We also have $\overrightarrow{BC} = \overrightarrow{AO}$, since AO and BC are parallel and the same length. Also $\overrightarrow{CD} = \overrightarrow{BO}$, since CD and BO are parallel and the same length. Therefore, we have

$$\overrightarrow{BD} = -\mathbf{a} - (\mathbf{a} + \mathbf{b})$$
$$= -2\mathbf{a} - \mathbf{b}$$

c) We see that $\overrightarrow{CF} = \overrightarrow{CD} + \overrightarrow{DE} + \overrightarrow{EF}$. We also know that

$$\overrightarrow{CD} = \overrightarrow{BO} = -(\mathbf{a} + \mathbf{b})$$

and

$$\overrightarrow{DE} = \overrightarrow{BA} = -\mathbf{b}$$

and

$$\overrightarrow{EF} = \overrightarrow{OA} = \mathbf{a}$$

Therefore,

$$\overrightarrow{CF} = -(\mathbf{a} + \mathbf{b}) - \mathbf{b} + \mathbf{a}$$
$$= -2\mathbf{b}$$

This tells us that FC is parallel to AB and twice the length of AB.

Example 8 In the rectangle OABC, $\overrightarrow{OA} = \mathbf{a}$ and $\overrightarrow{OC} = \mathbf{c}$. M is the mid-point of AB and N is the mid-point of BC. ON meets MC at the point P.

a) Find an expression for OP in terms of **a** and **c**.

b) Show that OP : PN = 4 : 1.

SOLUTION

a) Now $\overrightarrow{OP} = \overrightarrow{OA} + \overrightarrow{AM} + \overrightarrow{MP}$, but this requires us to know the ratio MP : PC, which we do not have. We can find $\overrightarrow{OP}$ by finding $\overrightarrow{ON}$ and $\overrightarrow{MC}$ first. We have

$$\overrightarrow{ON} = \overrightarrow{OC} + \overrightarrow{CN}$$
$$= \mathbf{c} + \tfrac{1}{2}\mathbf{a}$$
$$= \tfrac{1}{2}\mathbf{a} + \mathbf{c}$$

We also have

$$\overrightarrow{MC} = \overrightarrow{MB} + \overrightarrow{BC}$$
$$= \tfrac{1}{2}\mathbf{c} + (-\mathbf{a})$$
$$\therefore \quad \overrightarrow{MC} = \tfrac{1}{2}\mathbf{c} - \mathbf{a}$$

Now

$$\overrightarrow{OP} = \lambda\overrightarrow{ON}$$
$$= \lambda(\tfrac{1}{2}\mathbf{a} + \mathbf{c}) \qquad [1]$$

and

$$\overrightarrow{OP} = \overrightarrow{OA} + \overrightarrow{AM} + \overrightarrow{MP}$$
$$= \mathbf{a} + \tfrac{1}{2}\mathbf{c} + \mu(\tfrac{1}{2}\mathbf{c} - \mathbf{a})$$
$$\therefore \quad \overrightarrow{OP} = \mathbf{a}(1-\mu) + \mathbf{c}(\tfrac{1}{2} + \tfrac{1}{2}\mu) \qquad [2]$$

At the point P we have [1] equal to [2] giving

$$\mathbf{a}\frac{\lambda}{2} + \mathbf{c}\lambda = \mathbf{a}(1-\mu) + \mathbf{c}\left(\frac{1}{2} + \frac{1}{2}\mu\right)$$

Therefore,

$$\frac{\lambda}{2} = 1 - \mu \quad \text{and} \quad \lambda = \frac{1}{2}(1+\mu)$$

Solving simultaneously gives $\lambda = \frac{4}{5}$ and $\mu = \frac{3}{5}$. Therefore,

$$\overrightarrow{OP} = \frac{4}{5}\left(\frac{1}{2}\mathbf{a} + \mathbf{c}\right)$$
$$= \frac{2}{5}\mathbf{a} + \frac{4}{5}\mathbf{c}$$

b) We have $\overrightarrow{OP} = \frac{4}{5}\overrightarrow{ON}$. Therefore, the ratio OP : PN = 4 : 1.

Exercise 22B

1 OAB is a triangle with $\overrightarrow{OA} = \mathbf{a}$ and $\overrightarrow{OB} = \mathbf{b}$. P and Q are the mid-points of OA and AB, respectively.

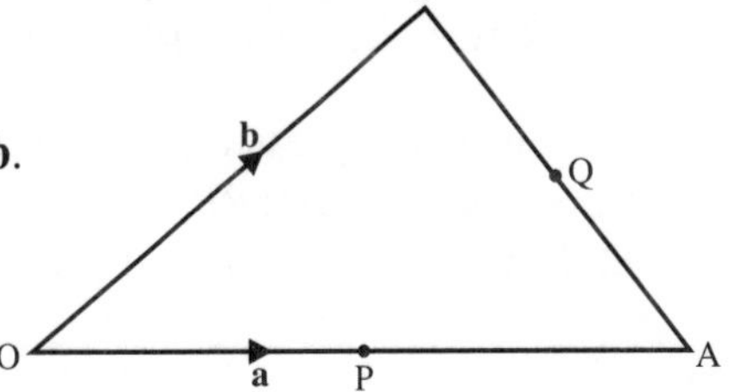

a) Express each of the following in terms of **a** or **b** or **a** and **b**.

i) $\overrightarrow{PA}$ **ii)** $\overrightarrow{AB}$ **iii)** $\overrightarrow{AQ}$ **iv)** $\overrightarrow{PQ}$

b) State two geometrical relationships connecting $\overrightarrow{OB}$ and $\overrightarrow{PQ}$.

2 OABC is a rectangle with $\overrightarrow{OA} = \mathbf{a}$ and $\overrightarrow{OB} = \mathbf{b}$. M is the mid-point of OC, and N is the point on CB such that CN : NB = 2 : 1. Express each of the following in terms of **a** or **b** or **a** and **b**.

a) $\overrightarrow{OC}$ **b)** $\overrightarrow{ON}$ **c)** $\overrightarrow{MO}$ **d)** $\overrightarrow{MN}$

3 OABC is a parallelogram with $\overrightarrow{OA} = \mathbf{a}$ and $\overrightarrow{OC} = \mathbf{c}$. S is the point on AB such that AS : SB = 3 : 1, and T is the point on BC such that BT : TC = 1 : 3.

a) Express each of the following in terms of **a** or **c** or **a** and **c**.

i) $\overrightarrow{AC}$ **ii)** $\overrightarrow{SB}$ **iii)** $\overrightarrow{BT}$ **iv)** $\overrightarrow{ST}$

b) Explain why $\overrightarrow{ST}$ and $\overrightarrow{AC}$ are parallel, and state the value of the ratio ST : AC.

4 OABC is a trapezium with $\overrightarrow{OA} = \mathbf{a}$ and $\overrightarrow{OB} = \mathbf{b}$. OA is parallel to, and twice as long as CB. Express each of the following in terms of **a** or **b** or **a** and **b**.

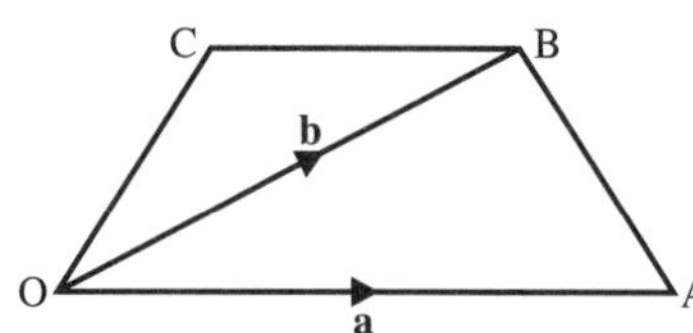

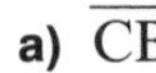

a) $\overrightarrow{CB}$ **b)** $\overrightarrow{BA}$ **c)** $\overrightarrow{CA}$ **d)** $\overrightarrow{CO}$

5

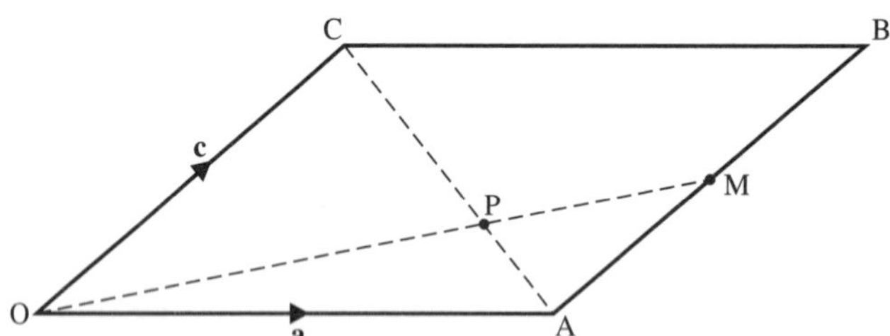

In the parallelogram OABC, $\overrightarrow{OA} = \mathbf{a}$, $\overrightarrow{OC} = \mathbf{c}$ and M is the mid-point of AB. OM meets AC at the point P.

a) Find an expression for $\overrightarrow{OP}$ in terms of **a** and **c**.
b) Find the value of the ratio OP : PM.
c) Show also that P is a point of trisection of AC, (i.e. that AP : PC = 1 : 2).

6 In the rectangle OABC, $\overrightarrow{OA} = \mathbf{a}$ and $\overrightarrow{OC} = \mathbf{c}$. R is the point on AB such that AR : RB = 1 : 2 and S is the point on BC such that BS : SC = 3 : 1. AS meets OR at P.

a) Find an expression for $\overrightarrow{OP}$ in terms of **a** and **c**.
b) Show that OP : PR = 4 : 1.
c) Find also the value of the ratio AP : PS.

7 In the triangle OAB, $\overrightarrow{OA} = \mathbf{a}$ and $\overrightarrow{OB} = \mathbf{b}$. M is the mid-point of AB and N is the point on OB such that ON : NB = 1 : 4. OM meets AN at P.

a) Find an expression for $\overrightarrow{OP}$ in terms of **a** and **b**.
b) Deduce that AP : PN = 5 : 1.

8 In the trapezium OABC, $\overrightarrow{OA} = \mathbf{a}$, $\overrightarrow{OC} = \mathbf{c}$ and $\overrightarrow{CB} = 3\mathbf{a}$. T is the point on BC such that BT : TC = 1 : 2. OT meets AC at P.

a) Find an expression for $\overrightarrow{OP}$ in terms of **a** and **c**.
b) Deduce that P is a point of trisection of both AC and OT.

9 In the rectangle OABC, M is the midpoint of OA and N is the mid-point of AB. OB meets MC at P and NC at Q. Show that OP = PQ = QB.

10 In the parallelogram OABC, P is the point on OA such that OP : PA = 1 : 2 and Q is the point on AB such that AQ : QB = 1 : 3. OB meets PC at L and QC at M. Show that OL : LM : MB = 7 : 9 : 12.

***11** ABCD is any quadrilateral. P, Q, R and S are respectively the mid-points of AB, BC, CD and DA. Prove that PQRS is a parallelogram.

***12** In the triangle OAB, M is the mid-point of AB and N is the point on OB such that ON : NB = $\lambda : \mu$. OM meets AN at P. Deduce the following results.

a) AP : PN = $(\lambda + \mu) : \mu$ **b)** OP : PM = $2\lambda : \mu$

Scalar product

The scalar product **a . b** of two vectors **a** and **b** is defined by

$$\mathbf{a} \cdot \mathbf{b} = |\mathbf{a}|\,|\mathbf{b}|\cos\theta$$

where θ is the angle between the vectors.

- When the two vectors **a** and **b** are perpendicular, $\theta = 90°$ and $\cos 90° = 0$. Therefore, $\mathbf{a} \cdot \mathbf{b} = 0$.
- When the angle between the vectors **a** and **b** is acute, $\cos\theta > 0$ and therefore $\mathbf{a} \cdot \mathbf{b} > 0$.

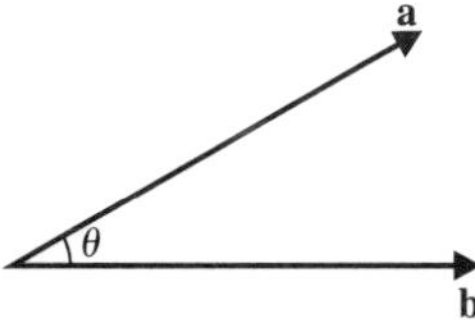

- When the angle between the vectors **a** and **b** is between 90° and 180°, $\cos\theta < 0$ and therefore $\mathbf{a} \cdot \mathbf{b} < 0$.

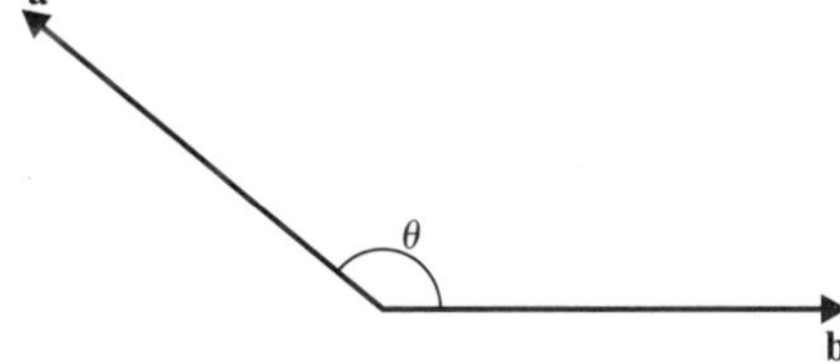

To see how we calculate the scalar product **a . b**, let $\mathbf{a} = a_1\mathbf{i} + a_2\mathbf{j}$ and $\mathbf{b} = b_1\mathbf{i} + b_2\mathbf{j}$. Then

$$\begin{aligned}\mathbf{a} \cdot \mathbf{b} &= (a_1\mathbf{i} + a_2\mathbf{j}) \cdot (b_1\mathbf{i} + b_2\mathbf{j}) \\ &= a_1b_1\mathbf{i} \cdot \mathbf{i} + a_1b_2\mathbf{i} \cdot \mathbf{j} + a_2b_1\mathbf{i} \cdot \mathbf{j} + a_2b_2\mathbf{j} \cdot \mathbf{j}\end{aligned}$$

Now $\mathbf{i}\,.\,\mathbf{i} = \mathbf{j}\,.\,\mathbf{j} = 1$, and since **i** and **j** are perpendicular to each other $\mathbf{i}\,.\,\mathbf{j} = \mathbf{j}\,.\,\mathbf{i} = 0$. Therefore,

$$\mathbf{a}\,.\,\mathbf{b} = a_1b_1 + a_2b_2$$

In three dimensions, the product is given by

$$\mathbf{a}\,.\,\mathbf{b} = a_1b_1 + a_2b_2 + a_3b_3$$

Example 9 Find the scalar product of each of the following pairs of vectors

a) $2\mathbf{i} + 3\mathbf{j}$ and $\mathbf{i} - 6\mathbf{j}$ **b)** $4\mathbf{i} - 2\mathbf{j} + \mathbf{k}$ and $2\mathbf{i} + \mathbf{j} - 3\mathbf{k}$

c) $\begin{pmatrix} 2 \\ -1 \\ 4 \end{pmatrix}$ and $\begin{pmatrix} -3 \\ 1 \\ 5 \end{pmatrix}$

SOLUTION

a) $(2\mathbf{i} + 3\mathbf{j})\,.\,(\mathbf{i} - 6\mathbf{j}) = (2 \times 1) + (3 \times -6) = -16$

b) $(4\mathbf{i} - 2\mathbf{j} + \mathbf{k})\,.\,(2\mathbf{i} + \mathbf{j} - 3\mathbf{k}) = (4 \times 2) + (-2 \times 1) + (1 \times -3) = 3$

c) $\begin{pmatrix} 2 \\ -1 \\ 4 \end{pmatrix} . \begin{pmatrix} -3 \\ 1 \\ 5 \end{pmatrix} = (2 \times -3) + (-1 \times 1) + (4 \times 5) = 13$

Example 10 Find the angle between the vectors $\mathbf{a} = 2\mathbf{i} + \mathbf{j} + \mathbf{k}$ and $\mathbf{b} = \mathbf{i} - \mathbf{j} + 3\mathbf{k}$.

SOLUTION

We know that

$$\mathbf{a}\,.\,\mathbf{b} = |\mathbf{a}|\,|\mathbf{b}|\cos\theta \qquad [1]$$

Now

$$\mathbf{a}\,.\,\mathbf{b} = (2 \times 1) + (1 \times -1) + (1 \times 3) = 4$$

We also have

$$|\mathbf{a}| = \sqrt{2^2 + 1^2 + 1^2} = \sqrt{6}$$

and

$$|\mathbf{b}| = \sqrt{1^2 + (-1)^2 + 3^2} = \sqrt{11}$$

Substituting into [1] gives

$$4 = \sqrt{6}\,\sqrt{11}\cos\theta$$

$$\therefore \quad \cos\theta = \frac{4}{\sqrt{66}} \quad \text{giving} \quad \theta = 60.5^\circ$$

Example 11 Given that the two vectors $\mathbf{a} = (3t+1)\mathbf{i} + \mathbf{j} - \mathbf{k}$ and $\mathbf{b} = (t+3)\mathbf{i} + 3\mathbf{j} - 2\mathbf{k}$ are perpendicular, find the possible values of the constant t.

SOLUTION

a and **b** are perpendicular, so $\mathbf{a}\,.\,\mathbf{b} = 0$. That is,

$$(3t+1)(t+3) + (1 \times 3) + (-1 \times -2) = 0$$
$$3t^2 + 10t + 8 = 0$$
$$(3t+4)(t+2) = 0$$

Solving gives $t = -\frac{4}{3}$ and $t = -2$, so the possible values of t are $-\frac{4}{3}$ and -2.

Exercise 22C

1 Given $\mathbf{a} = 3\mathbf{i} + 4\mathbf{j}$, $\mathbf{b} = \mathbf{i} - 3\mathbf{j}$ and $\mathbf{c} = 2\mathbf{i} + 5\mathbf{j}$, evaluate each of the following.

a) $\mathbf{a}\,.\,\mathbf{b}$ **b)** $\mathbf{b}\,.\,\mathbf{a}$ **c)** $\mathbf{a}\,.\,\mathbf{c}$ **d)** $\mathbf{c}\,.\,\mathbf{b}$ **e)** $\mathbf{a}\,.\,\mathbf{a}$ **f)** $\mathbf{c}\,.\,(\mathbf{a}+\mathbf{b})$

2 Given $\mathbf{x} = 2\mathbf{i} - 3\mathbf{j} + \mathbf{k}$, $\mathbf{y} = 5\mathbf{i} + 2\mathbf{j} - 7\mathbf{k}$ and $\mathbf{z} = \mathbf{i} - 4\mathbf{j} - 2\mathbf{k}$, evaluate each of the following.

a) $\mathbf{x}\,.\,\mathbf{y}$ **b)** $\mathbf{y}\,.\,\mathbf{x}$ **c)** $\mathbf{x}\,.\,\mathbf{z}$ **d)** $\mathbf{z}\,.\,\mathbf{z}$ **e)** $\mathbf{x}\,.\,(\mathbf{y}+\mathbf{z})$ **f)** $\mathbf{y}\,.\,(\mathbf{z}-\mathbf{x})$

3 Given

$$\mathbf{p} = \begin{pmatrix} -2 \\ 3 \end{pmatrix} \quad \mathbf{q} = \begin{pmatrix} 1 \\ 1 \end{pmatrix} \quad \text{and} \quad \mathbf{r} = \begin{pmatrix} 5 \\ -2 \end{pmatrix}$$

evaluate each of the following.

a) $\mathbf{p}\,.\,\mathbf{q}$ **b)** $\mathbf{q}\,.\,\mathbf{r}$ **c)** $\mathbf{r}\,.\,\mathbf{q}$ **d)** $\mathbf{q}\,.\,\mathbf{q}$ **e)** $\mathbf{r}\,.\,(\mathbf{q}+\mathbf{p})$ **f)** $\mathbf{p}\,.\,(\mathbf{q}-\mathbf{r})$

4 Given

$$\mathbf{c} = \begin{pmatrix} 3 \\ 1 \\ -4 \end{pmatrix} \quad \mathbf{d} = \begin{pmatrix} -5 \\ -2 \\ 7 \end{pmatrix} \quad \text{and} \quad \mathbf{e} = \begin{pmatrix} 0 \\ 4 \\ -5 \end{pmatrix}$$

evaluate each of the following

a) $\mathbf{c}\,.\,\mathbf{d}$ **b)** $\mathbf{d}\,.\,\mathbf{e}$ **c)** $\mathbf{c}\,.\,\mathbf{e}$ **d)** $\mathbf{d}\,.\,(\mathbf{e}-\mathbf{c})$ **e)** $\mathbf{c}\,.\,(\mathbf{c}+\mathbf{d})$ **f)** $\mathbf{e}\,.\,(2\mathbf{c}-\mathbf{d})$

5 Decide which of the following pairs of vectors are perpendicular, which are parallel, and which are neither perpendicular nor parallel.

a) $2\mathbf{i} + 8\mathbf{j}$ and $4\mathbf{i} - \mathbf{j}$
b) $3\mathbf{i} + 5\mathbf{j}$ and $6\mathbf{i} + 10\mathbf{j}$
c) $6\mathbf{i} - 8\mathbf{j} + 2\mathbf{k}$ and $9\mathbf{i} - 12\mathbf{j} + 3\mathbf{k}$
d) $5\mathbf{i} - 6\mathbf{j} + 2\mathbf{k}$ and $3\mathbf{i} + 2\mathbf{j} + \mathbf{k}$
e) $\begin{pmatrix} -3 \\ 1 \end{pmatrix}$ and $\begin{pmatrix} 6 \\ -2 \end{pmatrix}$
f) $\begin{pmatrix} 12 \\ 6 \end{pmatrix}$ and $\begin{pmatrix} 1 \\ -2 \end{pmatrix}$
g) $\begin{pmatrix} 3 \\ -1 \\ 4 \end{pmatrix}$ and $\begin{pmatrix} 9 \\ -3 \\ 12 \end{pmatrix}$
h) $\begin{pmatrix} 1 \\ 2 \\ 3 \end{pmatrix}$ and $\begin{pmatrix} 3 \\ 2 \\ 1 \end{pmatrix}$

6 Find the angle between each of the following pairs of vectors, giving your answers correct to one decimal place.

a) $3\mathbf{i} - 4\mathbf{j}$ and $12\mathbf{i} + 5\mathbf{j}$
b) $\mathbf{i}$ and $\mathbf{i} + \mathbf{j}$
c) $2\mathbf{i} + \mathbf{j} - 2\mathbf{k}$ and $4\mathbf{i} - 3\mathbf{j} + 12\mathbf{k}$
d) $3\mathbf{i} - 5\mathbf{j} - 2\mathbf{k}$ and $\mathbf{i} - 6\mathbf{k}$
e) $\begin{pmatrix} 2 \\ -1 \end{pmatrix}$ and $\begin{pmatrix} 6 \\ 3 \end{pmatrix}$
f) $\begin{pmatrix} 3 \\ 5 \end{pmatrix}$ and $\begin{pmatrix} 2 \\ -3 \end{pmatrix}$
g) $\begin{pmatrix} -2 \\ 1 \\ 3 \end{pmatrix}$ and $\begin{pmatrix} 4 \\ -3 \\ 3 \end{pmatrix}$
h) $\begin{pmatrix} 3 \\ 0 \\ -1 \end{pmatrix}$ and $\begin{pmatrix} 2 \\ 5 \\ 0 \end{pmatrix}$

7 $\mathbf{a} = 4\mathbf{i} + 5\mathbf{j}$, $\mathbf{b} = \lambda\mathbf{i} - 8\mathbf{j}$ and $\mathbf{c} = \mathbf{i} + \mu\mathbf{j}$.

a) Find the value of the constant λ given that **a** and **b** are perpendicular.
b) Find the value of the constant μ given that **a** and **c** are parallel.

8 $\mathbf{p} = 6\mathbf{i} - \mathbf{j}$, $\mathbf{q} = \lambda\mathbf{i} + 2\mathbf{j}$ and $\mathbf{r} = 2\mathbf{i} + \mu\mathbf{j}$.

a) Find the value of the constant λ given that **p** and **q** are parallel.
b) Find the value of the constant μ given that **p** and **r** are perpendicular.

9 Given that the vectors $2\mathbf{i} + t\mathbf{j} - 4\mathbf{k}$ and $\mathbf{i} - 3\mathbf{j} + (t - 4)\mathbf{k}$ are perpendicular, find the value of the constant t.

10 Given that $\begin{pmatrix} \lambda \\ 2 + \lambda \\ 3 \end{pmatrix}$ and $\begin{pmatrix} -1 \\ 3 \\ 4 - \lambda \end{pmatrix}$ are perpendicular vectors, find the value of the constant λ.

11 Find the possible values of the constant α, given that the vectors

$$\alpha\mathbf{i} + 8\mathbf{j} + (3\alpha + 1)\mathbf{k} \quad \text{and} \quad (\alpha + 1)\mathbf{i} + (\alpha - 1)\mathbf{j} - 2\mathbf{k}$$

are perpendicular.

12 Given that the vectors $\begin{pmatrix} t \\ 4 \\ 2t + 1 \end{pmatrix}$ and $\begin{pmatrix} t + 2 \\ 1 - t \\ -1 \end{pmatrix}$ are perpendicular, find the possible values of the constant t.

***13** Find a unit vector which is perpendicular to both of the vectors $4\mathbf{i} + 2\mathbf{j} - 3\mathbf{k}$ and $2\mathbf{i} - 3\mathbf{j} + \mathbf{k}$.

***14** In the triangle OAB, $\overrightarrow{OA} = \mathbf{a}$ and $\overrightarrow{OB} = \mathbf{b}$.

a) Show that $(\mathbf{a} - \mathbf{b}) \cdot (\mathbf{a} - \mathbf{b}) = \mathbf{a} \cdot \mathbf{a} + \mathbf{b} \cdot \mathbf{b} - 2\mathbf{a} \cdot \mathbf{b}$
b) Hence prove the cosine rule.

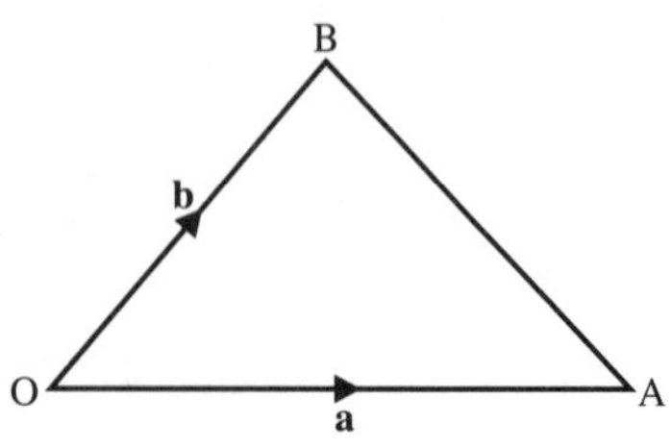

Vector equation of a line

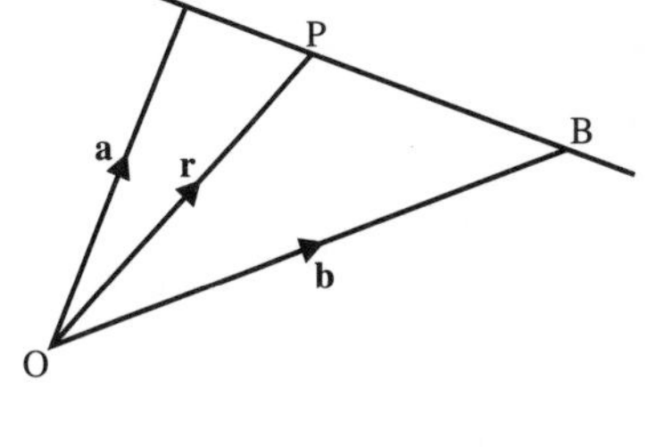

Let **a** and **b** be the position vectors of two points A and B with respect to an origin O. Let **r** be the position vector of a point P on the line AB.

We have

$$\overrightarrow{OP} = \overrightarrow{OA} + \overrightarrow{AP}$$
$$= \overrightarrow{OA} + t\,\overrightarrow{AB}$$

where t is a scalar. Therefore,

$$\mathbf{r} = \mathbf{a} + t(\mathbf{b} - \mathbf{a}) \quad \text{or} \quad \mathbf{r} = (1 - t)\mathbf{a} + t\mathbf{b}$$

This is the vector equation of the line AB. The vector $(\mathbf{b} - \mathbf{a})$ is the direction vector of the line.

Each value of the parameter t corresponds to a point on the line AB.

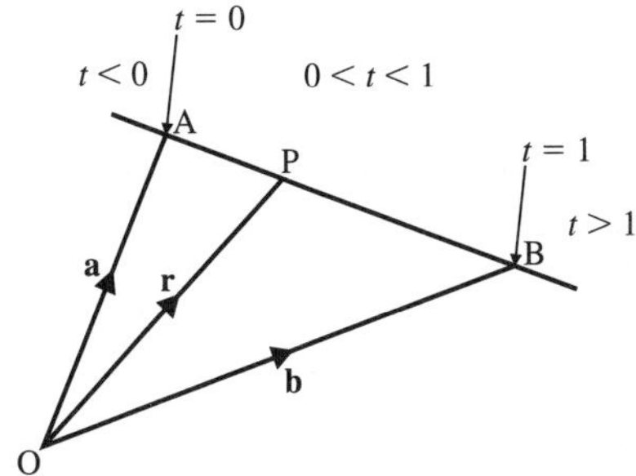

- When $t < 0$, point P is on the line BA produced.
- When $t = 0$, $\mathbf{r} = \mathbf{a}$, i.e. P = A
- When $0 < t < 1$, point P is between A and B.
- When $t = 1$, $\mathbf{r} = \mathbf{b}$, i.e. P = B
- When $t > 1$, point P is on the line AB produced.

Example 12 Find the vector equation of the line passing through A(1, 3, 2) and B(0, −1, 4). Does the point P(−2, 9, 1) line on the line AB?

SOLUTION

The position vectors of A and B are given by

$$\mathbf{a} = \begin{pmatrix} 1 \\ 3 \\ 2 \end{pmatrix} \quad \text{and} \quad \mathbf{b} = \begin{pmatrix} 0 \\ -1 \\ 4 \end{pmatrix}$$

The vector equation of the line is given by

$$\mathbf{r} = \mathbf{a} + t(\mathbf{b} - \mathbf{a})$$
$$= \begin{pmatrix} 1 \\ 3 \\ 2 \end{pmatrix} + t\left(\begin{pmatrix} 0 \\ -1 \\ 4 \end{pmatrix} - \begin{pmatrix} 1 \\ 3 \\ 2 \end{pmatrix}\right)$$

$$\therefore \quad \mathbf{r} = \begin{pmatrix} 1 \\ 3 \\ 2 \end{pmatrix} + t\begin{pmatrix} -1 \\ -4 \\ 2 \end{pmatrix}$$

This could be written in the form

$$\mathbf{r} = (1 - t)\mathbf{i} + (3 - 4t)\mathbf{j} + (2 + 2t)\mathbf{k}$$

If the point P(−2, 9, 1) lies on the line AB, there will exist a unique value of t for which

$$\begin{pmatrix}-2\\9\\1\end{pmatrix}=\begin{pmatrix}1-t\\3-4t\\2+2t\end{pmatrix}$$

If $-2=1-t$, then $t=3$. However, we see that $t=3$ does not satisfy $9=3-4t$. Therefore, the point P(−2, 9, 1) does not lie on the line AB.

Example 13 Two lines l and m have vector equations

$$\mathbf{r}_l=(2-3\lambda)\mathbf{i}+(1+\lambda)\mathbf{j}+4\lambda\mathbf{j} \quad \text{and}$$

$$\mathbf{r}_m=(-1+3\mu)\mathbf{i}+3\mathbf{j}+(7-\mu)\mathbf{k}$$

respectively. Find

a) the position vector of their common point

b) the angle between the lines.

SOLUTION

a) Rewriting the vector equations of the two lines in column form gives

$$\mathbf{r}_l=\begin{pmatrix}2-3\lambda\\1+\lambda\\4\lambda\end{pmatrix} \quad \text{and} \quad \mathbf{r}_m=\begin{pmatrix}-1+3\mu\\3\\7-\mu\end{pmatrix}$$

At the point common to l and m, we have

$$\mathbf{r}_l=\mathbf{r}_m$$

$$\therefore \quad \begin{pmatrix}2-3\lambda\\1+\lambda\\4\lambda\end{pmatrix}=\begin{pmatrix}-1+3\mu\\3\\7-\mu\end{pmatrix}$$

Equating **i**, **j** and **k** to coefficients, we get

$$\mathbf{i}: \quad 2-3\lambda=-1+3\mu \qquad [1]$$

$$\mathbf{j}: \quad 1+\lambda=3 \qquad [2]$$

$$\mathbf{k}: \quad 4\lambda=7-\mu \qquad [3]$$

From [2], $\lambda=2$. Substituting into [1] gives

$$2-3(2)=-1+3\mu$$

$$\therefore \quad \mu=-1$$

We notice that $\lambda=2$, $\mu=-1$ also satisfies [3]. So, at the common point, we have

$$\mathbf{r}_l=\begin{pmatrix}2-3(2)\\1+2\\4(2)\end{pmatrix}=\begin{pmatrix}-4\\3\\8\end{pmatrix}$$

The position vector of the common point is $-4\mathbf{i}+3\mathbf{i}+8\mathbf{k}$.

b) We know that

$$\mathbf{r}_l = \begin{pmatrix} 2-3\lambda \\ 1+\lambda \\ 4\lambda \end{pmatrix} = \begin{pmatrix} 2 \\ 1 \\ 0 \end{pmatrix} + \lambda \begin{pmatrix} -3 \\ 1 \\ 4 \end{pmatrix}$$

and

$$\mathbf{r}_m = \begin{pmatrix} -1+3\mu \\ 3 \\ 7-\mu \end{pmatrix} = \begin{pmatrix} -1 \\ 3 \\ 7 \end{pmatrix} + \mu \begin{pmatrix} 3 \\ 0 \\ -1 \end{pmatrix}$$

Therefore, the direction vector of line l is $-3\mathbf{i}+\mathbf{j}+4\mathbf{k}$ and the direction vector of line m is $3\mathbf{i}-\mathbf{k}$. The angle between the lines l and m is the angle between these two direction vectors.

Let $\mathbf{a} = -3\mathbf{i}+\mathbf{j}+4\mathbf{k}$ and $\mathbf{b} = 3\mathbf{i}-\mathbf{k}$. Then we have

$$|\mathbf{a}| = \sqrt{(-3)^2+1^2+4^2} = \sqrt{26}$$

and

$$|\mathbf{b}| = \sqrt{3^2+(-1)^2} = \sqrt{10}$$

Using the scalar product, we have

$$\mathbf{a}\,.\,\mathbf{b} = |\mathbf{a}|\,|\mathbf{b}|\cos\theta$$

$$\therefore \quad \begin{pmatrix} -3 \\ 1 \\ 4 \end{pmatrix} . \begin{pmatrix} 3 \\ 0 \\ -1 \end{pmatrix} = \sqrt{26}\sqrt{10}\cos\theta$$

$$-9-4 = \sqrt{260}\cos\theta$$

$$\therefore \quad \cos\theta = -\frac{13}{\sqrt{260}}$$

$$\therefore \quad \theta = 143.7^\circ$$

The angle between the lines l and m is 143.7°.

Exercise 22D

1 Find a vector equation for the line passing through the point (4, 3) and parallel to the vector $\mathbf{i}-2\mathbf{j}$.

2 Find a vector equation for the line passing through the point (5, −2, 3) and parallel to the vector $4\mathbf{i}-3\mathbf{j}+\mathbf{k}$.

3 Find a vector equation for the line passing through the point (5, −1) and perpendicular to the vector $\mathbf{i}+\mathbf{j}$.

4 Find a vector equation for the line joining the points (2, 6) and (5, −2).

5 Find a vector equation for the line joining the points (−1, 2, −3) and (6, 3, 0).

6 Points A and B have coordinates (4, 1) and (2, −5), respectively. Find a vector equation for the line which passes through the point A, and which is perpendicular to the line AB.

7 Points P and Q have coordinates (3, 5) and (–3, –7), respectively. Find a vector equation for the line which passes through the point P, and which is perpendicular to the line PQ.

8 Find a vector equation for the perpendicular bisector of the points (6, 3) and (2, –5).

9 Find a vector equation for the perpendicular bisector of the points (7, –1) and (3, –3).

10 Points A, B and C have position vectors $2\mathbf{i} + 3\mathbf{j}$, $4\mathbf{i} - \mathbf{j}$ and $6\mathbf{i} + 3\mathbf{j}$, respectively.

a) Find a vector equation for the line, l_1, which is the perpendicular bisector to the points A and B.
b) Find a vector equation for the line, l_2, which is the perpendicular bisector to the points B and C.
c) Hence find the position vector of the point where l_1 and l_2 meet.

11 Points P, Q and R have position vectors $4\mathbf{i} - 4\mathbf{j}$, $2\mathbf{i} + 2\mathbf{j}$ and $8\mathbf{i} + 6\mathbf{j}$, respectively.

a) Find a vector equation for the line, l_1, which is the perpendicular bisector to the points P and Q.
b) Find a vector equation for the line, l_2, which is the perpendicular bisector to the points Q and R.
c) Hence find the position vector of the point where l_1 and l_2 meet.

12 Three lines, l_1, l_2 and l_3, have equations

$$l_1: \quad \begin{pmatrix} x \\ y \end{pmatrix} = \begin{pmatrix} 1 \\ 1 \end{pmatrix} + \lambda \begin{pmatrix} 3 \\ 1 \end{pmatrix}$$

$$l_2: \quad \begin{pmatrix} x \\ y \end{pmatrix} = \begin{pmatrix} 6 \\ -4 \end{pmatrix} + \mu \begin{pmatrix} 1 \\ 2 \end{pmatrix}$$

and

$$l_3: \quad \begin{pmatrix} x \\ y \end{pmatrix} = \begin{pmatrix} 12 \\ -8 \end{pmatrix} + \nu \begin{pmatrix} -1 \\ 6 \end{pmatrix}$$

Show that l_1, l_2 and l_3 are concurrent, and find the position vector of their point of intersection.

13 Show that the lines $\mathbf{r}_1 = (6 - 2\lambda)\mathbf{i} + (\lambda - 5)\mathbf{j}$, $\mathbf{r}_2 = \mu\mathbf{i} + 3(1 - \mu)\mathbf{j}$ and $\mathbf{r}_3 = (5 - \nu)\mathbf{i} + (2\nu - 9)\mathbf{j}$ are concurrent, and find the position vector of their point of intersection.

14 Two lines, l_1 and l_2, have equations

$$l_1: \quad \begin{pmatrix} x \\ y \\ z \end{pmatrix} = \begin{pmatrix} 0 \\ -1 \\ -3 \end{pmatrix} + \lambda \begin{pmatrix} 1 \\ 3 \\ 6 \end{pmatrix}$$

and

$$l_2: \quad \begin{pmatrix} x \\ y \\ z \end{pmatrix} = \begin{pmatrix} -2 \\ 1 \\ 1 \end{pmatrix} + \mu \begin{pmatrix} 1 \\ 1 \\ 2 \end{pmatrix}$$

a) Show that l_1 and l_2 are concurrent, and find the position vector of their point of intersection.
b) Find also the angle between l_1 and l_2.

15 Two lines, l and m, have vector equations

$$\mathbf{r}_l = (3 - \lambda)\mathbf{i} + (4\lambda - 5)\mathbf{j} + (3\lambda - 1)\mathbf{k}$$

and

$$\mathbf{r}_m = (2\mu - 5)\mathbf{i} + \mu\mathbf{j} + (2\mu - 1)\mathbf{k}$$

a) Show that l and m are concurrent, and find the position vector of their point of intersection.
b) Find also the angle between l and m.

16 Points A, B and C have coordinates (0, 5), (9, 8) and (4, 3), respectively.

a) Find a vector equation for the line, l, joining A and B.
b) Find a vector equation for the line, p, which passes through the point C and which is perpendicular to the line AB.
c) Find the coordinates of the point of intersection of the lines l and p.
d) Deduce that the perpendicular distance from the point C to the line AB is $\sqrt{10}$.

17 Points A, B and C have coordinates (−1, −2), (5, 10) and (0, 5) respectively.

a) Find a vector equation for the line, l, joining A and B.
b) Find a vector equation for the line, p, which passes through the point C and which is perpendicular to the line AB.
c) Find the coordinates of the point of intersection of the lines l and p.
d) Deduce that the perpendicular distance from the point C to the line AB is $\sqrt{5}$.

18 Points P, Q and R have coordinates (−1, 1), (4, 6) and (7, 3) respectively.

a) Show that the perpendicular distance from the point R to the line PQ is $3\sqrt{2}$.
b) Deduce that the area of the triangle PQR is 15 units^2.

19 The points A, B and C have position vectors $4\mathbf{i} + \mathbf{j} - 4\mathbf{k}$, $3\mathbf{i} + 2\mathbf{j} - 3\mathbf{k}$ and $2\mathbf{i} + 3\mathbf{j} - 5\mathbf{k}$, respectively.

a) Given that the angle between $\overrightarrow{AB}$ and $\overrightarrow{AC}$ is θ,
  i) find the value of $\cos\theta$
  ii) deduce that $\sin\theta = \dfrac{\sqrt{6}}{3}$
b) Hence show that the perpendicular distance from the point C to the line AB is $\sqrt{6}$.

20 The points P, Q and R have position vectors $2\mathbf{i} + 5\mathbf{j} - 3\mathbf{k}$, $\mathbf{i} + 4\mathbf{j} - 2\mathbf{k}$ and $3\mathbf{i} + 3\mathbf{j} - 2\mathbf{k}$, respectively.

a) Given that the angle between $\overrightarrow{PQ}$ and $\overrightarrow{PR}$ is θ,
  i) find the value of $\cos\theta$
  ii) deduce that $\sin\theta = \dfrac{\sqrt{7}}{3}$
b) Hence show that the perpendicular distance from the point R to the line PQ is $\dfrac{\sqrt{42}}{3}$.

21 Points A, B and C have position vectors $-\mathbf{i} + 3\mathbf{j} + 5\mathbf{k}$, $5\mathbf{i} + 6\mathbf{j} - 4\mathbf{k}$ and $4\mathbf{i} + 7\mathbf{j} + 5\mathbf{k}$ respectively. P is the point on AB such that $\overrightarrow{AP} = \lambda\overrightarrow{AB}$.

a) Find $\overrightarrow{AB}$. **b)** Find $\overrightarrow{CP}$.
c) By considering the scalar product $\overrightarrow{AB} \cdot \overrightarrow{CP}$ find the position vector of the point on the line AB which is closest to C.
d) Deduce that the perpendicular distance from the point C to the line AB is $3\sqrt{3}$.

22 Points A, B and C have position vectors $3\mathbf{i} - 2\mathbf{j} + 5\mathbf{k}$, $7\mathbf{i} + 6\mathbf{j} + \mathbf{k}$ and $8\mathbf{i} + 6\mathbf{j} + 8\mathbf{k}$ respectively. P is the point on AB such that $\overrightarrow{AP} = \lambda\overrightarrow{AB}$.

a) Find $\overrightarrow{AB}$.

b) Find $\overrightarrow{CP}$.

c) By considering the scalar product $\overrightarrow{AB} \cdot \overrightarrow{CP}$ find the position vector of the point on the line AB which is closest to C.

d) Deduce that the perpendicular distance from the point C to the line AB is $2\sqrt{11}$.

***23** Show that the perpendicular distance from the point (h, k) to the line $ax + by + c = 0$ is

$$\left| \frac{ah + bk + c}{\sqrt{a^2 + b^2}} \right|$$

Exercise 22E Examination questions

1 The vertices A, B of the triangle OAB have position vectors **a**, **b** relative to O. C and D are the mid-points of OA and AB respectively.

a) Show that the position vector of M, the mid-point of CD, is $\frac{1}{2}\mathbf{a} + \frac{1}{4}\mathbf{b}$.

b) Write down, in terms of **a**, **b** and λ, the position vector of the point which divides BM in the ratio $\lambda : 1 - \lambda$. Hence find the position vector of the point of intersection of BM and OA.

(WJEC)

2 In the diagram OABCDEFG is a cube in which the length of each edge is 2 units. Unit vectors **i**, **j**, **k** are parallel to $\overrightarrow{OA}$, $\overrightarrow{OC}$, $\overrightarrow{OD}$ respectively. The mid-points of AB and FG are M and N, respectively.

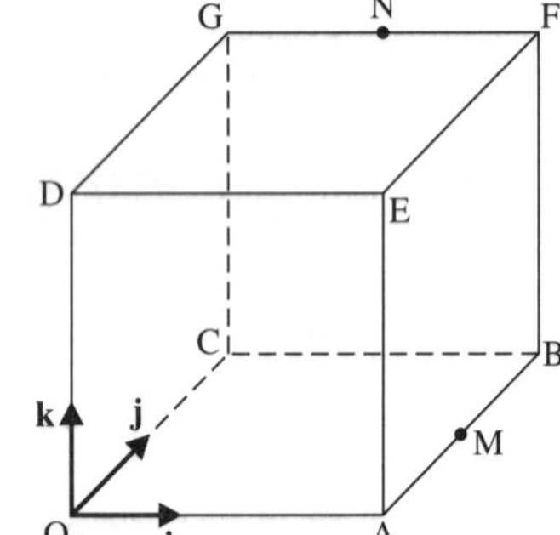

i) Express each of the vectors $\overrightarrow{ON}$ and $\overrightarrow{MG}$ in terms of **i**, **j** and **k**.

ii) Show that the acute angle between the directions of $\overrightarrow{ON}$ and $\overrightarrow{MG}$ is 63.6°, correct to the nearest 0.1°.

(UCLES)

3 Three points P, Q and R have position vectors, **p**, **q** and **r** respectively, where

$$\mathbf{p} = 7\mathbf{i} + 10\mathbf{j} \qquad \mathbf{q} = 3\mathbf{i} + 12\mathbf{j} \qquad \mathbf{r} = -\mathbf{i} + 4\mathbf{j}$$

i) Write down the vectors $\overrightarrow{PQ}$ and $\overrightarrow{RQ}$, and show that they are perpendicular.

ii) Using a scalar product, or otherwise, find the angle PRQ.

iii) Find the position vector of S, the mid-point of PR.

iv) Show that $|\overrightarrow{QS}| = |\overrightarrow{RS}|$. Using your previous results, or otherwise, find the angle PSQ.

(MEI)

4 The points A, B, C have position vectors

$$\mathbf{a} = 2\mathbf{i} + \mathbf{j} - \mathbf{k} \qquad \mathbf{b} = 3\mathbf{i} + 4\mathbf{j} - 2\mathbf{k} \qquad \mathbf{c} = 5\mathbf{i} - \mathbf{j} + 2\mathbf{k}$$

respectively, relative to a fixed origin O.

a) Evaluate the scalar product $(\mathbf{a} - \mathbf{b}) \cdot (\mathbf{c} - \mathbf{b})$.
Hence calculate the size of angle ABC, giving your answer to the nearest 0.1°.

b) Given that ABCD is a parallelogram:

i) determine the position vector of D

ii) calculate the area of ABCD.

c) The point E lies on BA produced so that $\overrightarrow{BE} = 3\overrightarrow{BA}$. Write down the position vector of E. The line CE cuts the line AD at X. Find the position vector of X. (UODLE)

5 The points A and B have position vectors $\mathbf{i} + 2\mathbf{j} + 2\mathbf{k}$ and $4\mathbf{i} + 3\mathbf{j}$ respectively, relative to an origin O.

a) Find the lengths of OA and OB.

b) Find the scalar product of **OA** and **OB** and hence find angle AOB.

c) Find the area of the triangle AOB, giving your answer correct to 2 decimal places.

d) The point C divides AB in the ratio $\lambda : 1 - \lambda$.

i) Find an expression for **OC**.

ii) Show that $OC^2 = 14\lambda^2 + 2\lambda + 9$.

iii) Find the position vectors of the two points on AB whose distance from O is $\sqrt{21}$.

iv) Show that the perpendicular distance of O from AB is approximately 2.99. (WJEC)

6 With respect to an origin O, the position vectors of the points L and M are $2\mathbf{i} - 3\mathbf{j} + 3\mathbf{k}$ and $5\mathbf{i} + \mathbf{j} + c\mathbf{k}$ respectively, where c is a constant. The point N is such that OLMN is a rectangle.

a) Find the value of c.

b) Write down the position vector of N.

c) Find, in the form $\mathbf{r} = \mathbf{p} + t\mathbf{q}$, an equation of the line MN. (EDEXCEL)

7 The position vectors of A and B are defined by

$$\overrightarrow{OA} = \begin{pmatrix} 4 \\ -3 \\ 2 \end{pmatrix} \quad \text{and} \quad \overrightarrow{OB} = \begin{pmatrix} 2 \\ -2 \\ -7 \end{pmatrix}$$

i) Show that $\overrightarrow{OA}$ and $\overrightarrow{OB}$ are perpendicular.

ii) Find a vector equation for the line AB. Show that this line intersects the line

$$\mathbf{r} = \begin{pmatrix} 3 \\ -1 \\ 2 \end{pmatrix} + s\begin{pmatrix} -1 \\ 1 \\ -3 \end{pmatrix}$$

and find the coordinates of the point of intersection. (NEAB)

8 With respect to a fixed origin O, the points A, B, and C have position vectors $\mathbf{i} + \mathbf{j} + 8\mathbf{k}$, $\mathbf{i} + 2\mathbf{j} + 6\mathbf{k}$ and $3\mathbf{i} + 12\mathbf{j} + 6\mathbf{k}$, respectively.

i) Show that $\overrightarrow{OC}$ and $\overrightarrow{AB}$ are perpendicular.

ii) Show also that the line through O and C intersects the line through A and B, and find the position vector of the point where they intersect. (NICCEA)

9 A line l_1 passes through the point A, with position vector $5\mathbf{i} + 3\mathbf{j}$, and the point B, with position vector $-2\mathbf{i} - 4\mathbf{j} + 7\mathbf{k}$.

a) Write down an equation of the line l_1.

A second line l_2 has equation

$$\mathbf{r} = \mathbf{i} - 3\mathbf{j} - 4\mathbf{k} + \mu(\mathbf{i} + 2\mathbf{j} + 3\mathbf{k})$$

where μ is a parameter.

b) Show that l_1 and l_2 are perpendicular to each other.
c) Show that the two lines meet, and find the position vector of the point of intersection.

The point C has position vector $2\mathbf{i} - \mathbf{j} - \mathbf{k}$.

d) Show that C lies on l_2.

The point D is the image of C after reflection in the line l_1.

e) Find the position vector of D. (EDEXCEL)

10 The lines l_1, l_2 have vector equations

$$l_1: \quad \mathbf{r} = 2\mathbf{i} + 3\mathbf{j} + 5\mathbf{k} + \mu(\mathbf{i} + \mathbf{j} + 2\mathbf{k})$$

$$l_2: \quad \mathbf{r} = 4\mathbf{j} + 6\mathbf{k} + \mu(-\mathbf{i} + 2\mathbf{j} + 3\mathbf{k})$$

a) Show that l_1 and l_2 intersect and find the position vector of the point of intersection.
b) Find the acute angle between l_1 and l_2, giving your answer correct to the nearest degree.
(AEB 95)

11 The point A has coordinates (7, −1, 3) and the point B has coordinates (10, −2, 2). The line l has vector equation

$$\mathbf{r} = \mathbf{i} + \mathbf{j} + \mathbf{k} + \lambda(3\mathbf{i} - \mathbf{j} + \mathbf{k})$$

where λ is a real parameter.

a) Show that the point A lies on the line l.
b) Find the length of AB.
c) Find the size of the acute angle between the line l and the line segment AB, giving your answer to the nearest degree.
d) Hence, or otherwise, calculate the perpendicular distance from B to the line l, giving your answer to 2 significant figures. (EDEXCEL)

12 a) Write down, in vector form, an equation of the line l which passes through L(−3, 1, −7) and M(5, 3, 5).
b) Find the position vector of the point P on the line for which OP is perpendicular to l, where O is the origin.
c) Hence find the shortest distance from O to the line l. (EDEXCEL)

13 The points P and Q have position vectors $\mathbf{p} = 3\mathbf{i} - \mathbf{j} + 2\mathbf{k}$, $\mathbf{q} = 4\mathbf{i} - 2\mathbf{j} - \mathbf{k}$ respectively, relative to a fixed origin O.

a) Determine a vector equation of the line l_1, passing through P and Q in the form $\mathbf{r} = \mathbf{a} + s\mathbf{b}$, where s is a scalar parameter.
b) The line l_2 has a vector equation $\mathbf{r} = 2\mathbf{i} - 2\mathbf{j} - 3\mathbf{k} + t(2\mathbf{i} - \mathbf{j} - 2\mathbf{k})$. Show that l_1 and l_2 intersect and find the position vector of the point of intersection V.
c) Show that PV has length $3\sqrt{11}$.
d) The acute angle between l_1 and l_2 is θ. Show that $\cos\theta = \dfrac{3}{\sqrt{11}}$.
e) Calculate the perpendicular distance from P to l_2. (AEB 92)

14 The points A(24, 6, 0), B(30, 12, 12) and C(18, 6, 36) are referred to cartesian axes, origin O.

a) Find a vector equation for the line passing through the points A and B.
The point P lies on the the line passing through A and B.

b) Show that $\overrightarrow{CP}$ can be expressed as

$$(6 + t)\mathbf{i} + t\mathbf{j} + (2t - 36)\mathbf{k}, \text{ where } t \text{ is a parameter.}$$

c) Given that $\overrightarrow{CP}$ is perpendicular to $\overrightarrow{AB}$, find the coordinates of P.
d) Hence, or otherwise, find the area of the triangle ABC, giving your answer to three significant figures. (EDEXCEL)

15 A pyramid has a rectangular base OABC and vertex D. The position vectors of A, B, C, and D with reference to the fixed origin O are $\mathbf{a} = 8\mathbf{i}$, $\mathbf{b} = 8\mathbf{i} + 4\mathbf{j}$, $\mathbf{c} = 4\mathbf{j}$, $\mathbf{d} = 4\mathbf{i} + 2\mathbf{j} + 6\mathbf{k}$ respectively.

a) Express the vector $\overrightarrow{AD}$ in terms of **i**, **j** and **k**.
b) Find the cosine of the angle ODA and hence show that the triangular face ODA has area $8\sqrt{10}$.
c) By using the result from (**b**), or otherwise, find the perpendicular distance from O to the line AD.
d) Determine a vector equation of the straight line through the points B and D.
e) By considering an appropriate scalar product, or otherwise, find the position vector of the point on the line BD which is closest to O. (AEB 94)

23 Probability

If a man who cannot count finds a four-leaf clover, is he entitled to happiness?
LEE

Suppose two fair dice are rolled. Then there are 36 possible outcomes, namely,

[1, 1] [1, 2] [1, 3] [1, 4] [1, 5] [1, 6]
[2, 1] [2, 2] [2, 3] [2, 4] [2, 5] [2, 6]
[3, 1] [3, 2] [3, 3] [3, 4] [3, 5] [3, 6]
[4, 1] [4, 2] [4, 3] [4, 4] [4, 5] [4, 6]
[5, 1] [5, 2] [5, 3] [5, 4] [5, 5] [5, 6]
[6, 1] [6, 2] [6, 3] [6, 4] [6, 5] [6, 6]

where, for example, [1, 2] denotes the outcome of getting a *1* on the first die and a *2* on the second die.

Suppose that we want to know the likelihood of getting a *total score of five* when the two dice are rolled. There are four ways in which such a total can be achieved, namely,

[1, 4] [2, 3] [3, 2] [4, 1]

Therefore, the probability of obtaining a *total score of five* is $\frac{4}{36} = \frac{1}{9}$.

Generally, the list of all possible outcomes is called the **sample space**, and those outcomes which meet the particular requirement are called the **event**.

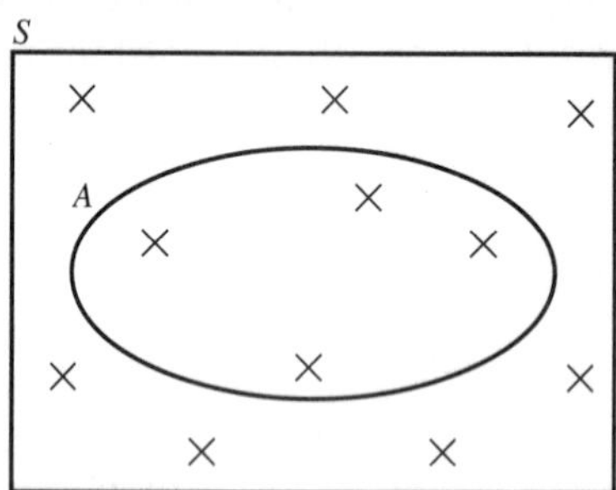

Let the sample space be donated by S and the event by A. The probability of A is defined by

$$P(A) = \frac{n(A)}{n(S)}$$

where $n(A)$ denotes the number of elements in the set A, and $n(S)$ denotes the number of elements in the sample space, S. That is,

$$\text{The probability of } A = \frac{\text{Number of ways in which } A \text{ can happen}}{\text{Number of possible outcomes}}$$

Returning to the initial example, A is the event that the *total score is five*, so $n(A) = 4$ and $n(S) = 36$. Therefore,

$$\mathrm{P}(A) = \frac{n(A)}{n(S)} = \frac{4}{36} = \frac{1}{9}$$

as shown earlier.

Combined events

Now consider two events, A and B. Two possible outcomes are

$A \cap B$, which means that A and B **both** occur

and $A \cup B$, which means that A occurs **or** B occurs.

It should be noted that $A \cup B$ includes the case when A and B both occur.

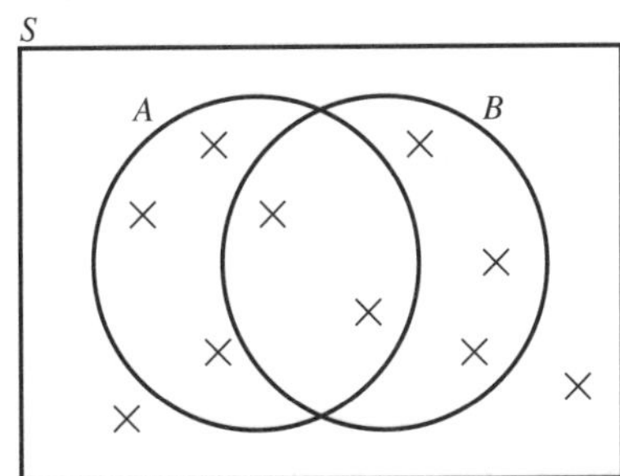

Using elementary set theory, we know that

$$n(A \cup B) = n(A) + n(B) - n(A \cap B)$$

Therefore, we have

$$\mathrm{P}(A \cup B) = \mathrm{P}(A) + \mathrm{P}(B) - \mathrm{P}(A \cap B)$$

Example 1 A card is selected at random from an ordinary pack of 52 cards. Find the probability that the card is

a) a king **b)** a heart **c)** the king of hearts **d)** either a king or a heart

Let K denote the event that the card is a king, and let H denote the event that the card is a heart.

a) $\mathrm{P}(K) = \frac{4}{52} = \frac{1}{13}$

b) $\mathrm{P}(H) = \frac{13}{52} = \frac{1}{4}$

c) Noting that the *king of hearts* is denoted by the event $K \cap H$, we can write

$$\mathrm{P}(K \cap H) = \frac{1}{52}$$

d) Noting that the *king* or a *heart* is denoted by the event $K \cup H$, and using

$$P(K \cup H) = P(K) + P(H) - P(K \cap H)$$

we get

$$P(K \cup H) = \frac{4}{52} + \frac{13}{52} - \frac{1}{52}$$

$$= \frac{16}{52} = \frac{4}{13}$$

Mutually exclusive events

Two events, A and B, are said to be **mutually exclusive** when they have **no outcome in common**. In other words, when $n(A \cap B) = 0$. The Venn diagram below illustrates such a case.

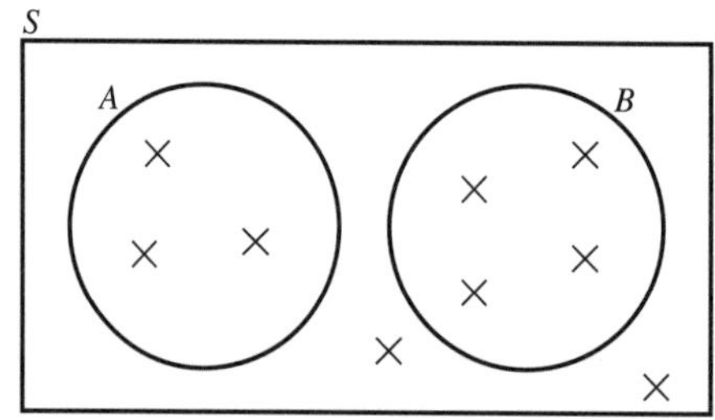

When A and B are mutually exclusive, we have

$$P(A \cup B) = P(A) + P(B)$$

Example 2 Given that the events A and B are mutually exclusive with $P(A) = \frac{3}{10}$ and $P(B) = \frac{2}{5}$, find the value of $P(A \cup B)$.

Using

$$P(A \cup B) = P(A) + P(B) - P(A \cap B)$$

with $P(A \cap B) = 0$, we get

$$P(A \cup B) = \frac{3}{10} + \frac{2}{5} = \frac{7}{10}$$

Exhaustive events

Two events, A and B, are said to be **exhaustive** if together they include **all possible outcomes** in the sample space. In other words, when $A \cup B = S$. The Venn diagram below illustrates such a case.

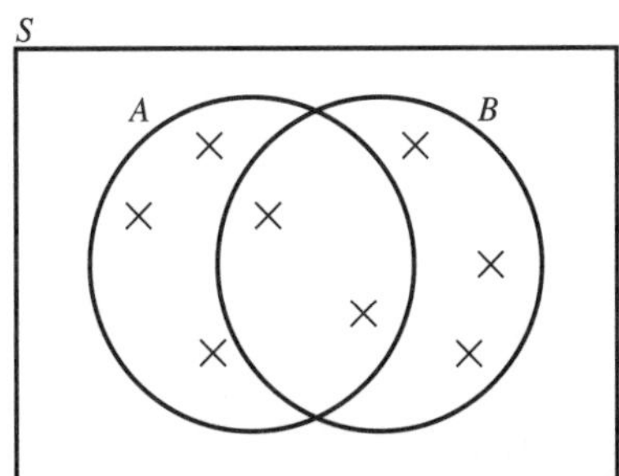

When A and B are exhaustive, we have

$$P(A \cup B) = 1$$

Example 3 Given $P(X) = \frac{4}{5}$, $P(Y) = \frac{1}{2}$ and $P(X \cap Y) = \frac{3}{10}$, show that the events X and Y are exhaustive.

We know that

$$P(X \cup Y) = P(X) + P(Y) - P(X \cap Y)$$

Therefore,

$$P(X \cup Y) = \frac{4}{5} + \frac{1}{2} - \frac{3}{10} = 1$$

Hence X and Y are exhaustive.

Complementary events

Given an event A, the complement of A (written A') consists of all outcomes in the sample space which are not contained in A. Noting that A and A' are both mutually exclusive and exhaustive, we have

$$P(A' \cup A) = P(A') + P(A)$$

by mutual exclusivity, and

$$P(A' \cup A) = 1$$

by exhaustivity. Therefore,

$$P(A') + P(A) = 1$$

$$\therefore \quad P(A') = 1 - P(A)$$

Example 4 Given $P(A) = 0.55$, $P(A \cup B) = 0.7$ and $P(A \cap B) = 0.2$, find $P(B')$.

Using

$$P(A \cup B) = P(A) + P(B) - P(A \cap B)$$

we have

$$P(B) = 0.7 + 0.2 - 0.55 = 0.35.$$

We also know that $P(B') = 1 - P(B)$. Therefore,

$$P(B') = 1 - 0.35 = 0.65.$$

Example 5 Given $P(G') = 5x$, $P(H) = \frac{3}{5}$, $P(G \cup H) = 8x$ and $P(G \cap H) = 3x$, find the value of x.

Since $P(G) = 1 - P(G')$, we have $P(G) = 1 - 5x$.

Using

$$P(G \cup H) = P(G) + P(H) - P(G \cap H)$$

we have

$$8x = (1 - 5x) + \frac{3}{5} - 3x$$

$$\therefore \quad 16x = \frac{8}{5}$$

$$\therefore \quad x = \frac{1}{10}$$

Exercise 23A

1 A card is selected at random from a pack of 52 cards. Find the probability that the card is

a) black

b) an honour [Aces, kings, queens and jacks are honours.]

c) a black honour

d) either black or an honour.

2 In a bag are 100 discs numbered 1 to 100. A disc is selected at random from the bag. Find the probability that the number on the selected disc is

a) even

b) a multiple of five

c) a multiple of ten

d) either even or a multiple of five.

3 Two fair dice are thrown. Find the probability that

a) at least one of the dice shows a *four*

b) the sum of the scores on the two dice is *nine*

c) one of the dice shows a *four* and the other shows a *five*

d) either at least one of the dice shows a *four* or the total of the scores on the two dice is *nine*.

4 In a class half the pupils study Mathematics, a third study English and a quarter study both Mathematics and English. Find the probability that a pupil selected at random from the class studies either Mathematics or English.

5 Given $\mathrm{P}(A) = \frac{3}{5}$, $\mathrm{P}(B) = \frac{2}{3}$ and $\mathrm{P}(A \cap B) = \frac{1}{2}$, find the value of $\mathrm{P}(A \cup B)$.

6 Given $\mathrm{P}(X) = 0.37$, $\mathrm{P}(Y) = 0.48$ and $\mathrm{P}(X \cup Y) = 0.69$, find the value of $\mathrm{P}(X \cap Y)$.

7 Given $\mathrm{P}(A) = \frac{7}{10}$, $\mathrm{P}(A \cup B) = \frac{9}{10}$ and $\mathrm{P}(A \cap B) = \frac{3}{20}$, find the value of $\mathrm{P}(B)$.

8 Given $\mathrm{P}(F) = 4x$, $\mathrm{P}(G) = \frac{1}{3}$, $\mathrm{P}(F \cap G) = x$ and $\mathrm{P}(F \cup G) = 8x$, find the value of x.

9 Given that the events X and Y are mutually exclusive with $\mathrm{P}(X) = \frac{4}{7}$ and $\mathrm{P}(Y) = \frac{1}{3}$, find the value of $\mathrm{P}(X \cup Y)$.

10 Given $\mathrm{P}(S) = 0.34$, $\mathrm{P}(T) = 0.49$ and $\mathrm{P}(S \cup T) = 0.83$, show that the events S and T are mutually exclusive.

11 Given that the events M and N are mutually exclusive with $\mathrm{P}(M) = 3x$, $\mathrm{P}(N) = 4x$ and $\mathrm{P}(M \cup N) = 1 - x$, find the value of x.

12 Given that the events A and B are exhaustive with $\mathrm{P}(A) = \frac{2}{3}$ and $\mathrm{P}(B) = \frac{3}{4}$, find the value of $\mathrm{P}(A \cap B)$.

13 Given $\mathrm{P}(X) = \frac{5}{8}$, $\mathrm{P}(Y) = \frac{11}{12}$ and $\mathrm{P}(X \cap Y) = \frac{13}{24}$, show that the events X and Y are exhaustive.

14 Given that the events S and T are exhaustive with $\mathrm{P}(S) = x$, $\mathrm{P}(T) = 3x$ and $\mathrm{P}(S \cap T) = 1 - 5x$, find the value of x.

15 When a roulette wheel is spun, the score will be a number from 0 to 36 inclusive. Each score is equally likely. Find the probability that the score is

a) an even number

b) a multiple of 3

c) a multiple of 6

d) an odd number which is not a multiple of 3.

16 As a result of a survey of the households in a town, it is found that 80% have a video recorder and 24% have satellite television. Given that 15% have both a video recorder and satellite television, find the proportion of households with neither a video recorder nor satellite television.

17 The children at a party were asked about their pets. Two-thirds had a dog and three-quarters had a cat. Given that half the children had both a cat and a dog, calculate the probability that a child selected at random is found to have neither a cat nor a dog.

18 Given $P(C) = 0.44$, $P(C \cap D) = 0.21$ and $P(C \cup D) = 0.83$ find $P(D')$.

19 Given $P(G') = 3x$, $P(H) = 4x$, $P(G \cap H) = \frac{1}{4}$ and $P(G \cup H) = 9x$, find the value of x.

***20** Given $P(A) = 0.6$, $P(A \cap B') = 0.4$ and $P(A \cup B) = 0.85$ find the value of $P(B)$.

***21** Show that

$$P(A \cup B \cup C) = P(A) + P(B) + P(C) - P(A \cap B) - P(B \cap C) - P(C \cap A) + P(A \cap B \cap C)$$

Conditional probability

Suppose we are considering two related events, A and B, and we are told that B has occurred. This information may influence the likelihood of A occuring. For example, if we select a card at random from a pack of 52, then the probability that it will be a heart is $\frac{13}{52} = \frac{1}{4}$.

However, if we are given the additional information that the card selected is red, then this probability is increased to $\frac{13}{26} = \frac{1}{2}$.

We formalise this by stating that the probability that event A will occur given that event B has already occurred is given by

$$\frac{n(A \cap B)}{n(B)}$$

This probability, which is denoted by $P(A|B)$, is illustrated in the Venn diagram below.

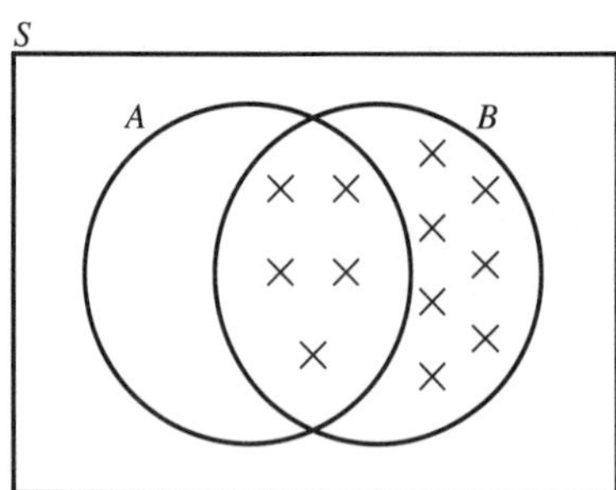

We therefore have

$$P(A|B) = \frac{P(A \cap B)}{P(B)}$$

In the above example, A is the event *obtain a heart* and B is the event *obtain a red card*, so $A \cap B$ is the event *obtain a heart* (all hearts being red).

Now $P(A \cap B) = \frac{13}{52}$ and $P(B) = \frac{26}{52}$. Therefore,

$$P(A|B) = \frac{\frac{13}{52}}{\frac{26}{52}} = \frac{1}{2}$$

Example 6 Two fair dice are thrown. Find the probability that one of the dice shows a *four* given that the total on the two dice is *ten*.

Let F denote the event that one of the dice shows a four and T denote the event that the total on the two dice is ten.

[1, 1]	[1, 2]	[1, 3]	[1, 4]	[1, 5]	[1, 6]
[2, 1]	[2, 2]	[2, 3]	[2, 4]	[2, 5]	[2, 6]
[3, 1]	[3, 2]	[3, 3]	[3, 4]	[3, 5]	[3, 6]
[4, 1]	[4, 2]	[4, 3]	[4, 4]	[4, 5]	[4, 6]
[5, 1]	[5, 2]	[5, 3]	[5, 4]	[5, 5]	[5, 6]
[6, 1]	[6, 2]	[6, 3]	[6, 4]	[6, 5]	[6, 6]

Then $F \cap T$ denotes [4, 6] or [6, 4], and T denotes [4, 6] or [5, 5] or [6, 4].

Therefore,

$$P(F \cap T) = \frac{2}{36} \quad \text{and} \quad P(T) = \frac{3}{36}$$

Using $P(F|T) = \dfrac{P(F \cap T)}{P(T)}$, we have

$$P(F|T) = \frac{\frac{2}{36}}{\frac{3}{36}} = \frac{2}{3}$$

Example 7 Given $P(A) = \frac{1}{2}$, $P(A|B) = \frac{1}{4}$ and $P(A \cup B) = \frac{2}{3}$, find $P(B)$.

Using $P(A|B) = \dfrac{P(A \cap B)}{P(B)}$, we have

$$\frac{1}{4} = \frac{P(A \cap B)}{P(B)}$$

$$\therefore \quad P(A \cap B) = \frac{1}{4}P(B) \qquad [1]$$

Using $P(A \cup B) = P(A) + P(B) - P(A \cap B)$, we have

$$\frac{2}{3} = \frac{1}{2} + P(B) - P(A \cap B)$$

$$\therefore \quad \frac{1}{6} = P(B) - P(A \cap B) \qquad [2]$$

Eliminating $P(A \cap B)$ between [1] and [2] gives

$$\frac{1}{6} = P(B) - \frac{1}{4}P(B)$$

$$\therefore \quad \frac{3}{4}P(B) = \frac{1}{6}$$

$$\therefore \quad P(B) = \frac{2}{9}$$

Independence

Two events A and B are said to be **independent** if

$$P(A|B) = P(A)$$

In this case we have

$$\frac{P(A \cap B)}{P(B)} = P(A)$$

$$\therefore \quad P(A \cap B) = P(A) \times P(B)$$

Example 8 A card is picked at random from a pack of 52 and a fair die is thrown. Find the probability that the card is the ace of spades and the die shows a three.

We could choose to list all of the 52×6 possible outcomes, but it is easier to use independence and write

$$\text{P(Ace of spades and a two)} = \text{P(Ace of spades)} \times \text{P(Two)}$$

$$= \frac{1}{52} \times \frac{1}{6}$$

$$= \frac{1}{312}$$

Tree diagrams

Sometimes, rather than list all the possible outcomes, it is easier to view probabilities on a **tree diagram**. The following example illustrates how this is done.

Example 9 A bag contains four red discs and five blue discs. Three discs are selected at random. Find the probability that two are red and the other is blue.

First, without using a tree diagram, and just using independence to multiply probabilities, and mutual exclusivity to add them, we have

$$\begin{aligned}\text{P(Two red and one blue)} &= \text{P(RRB)} + \text{P(RBR)} + \text{P(BRR)} \\ &= \frac{4}{9}\times\frac{3}{8}\times\frac{5}{7}+\frac{4}{9}\times\frac{5}{8}\times\frac{3}{7}+\frac{5}{9}\times\frac{4}{8}\times\frac{3}{7} \\ &= \frac{60}{504}+\frac{60}{504}+\frac{60}{504} \\ &= \frac{5}{14}\end{aligned}$$

Alternatively, using a tree diagram gives all the possible outcomes and probabilities as shown below.

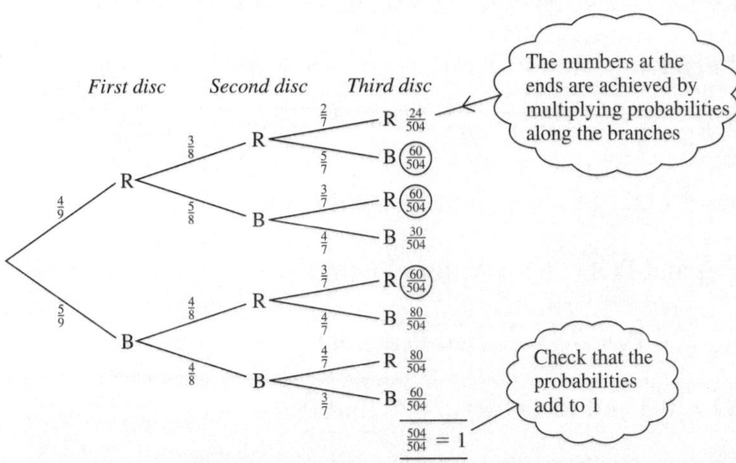

We want all those combinations of branches which include two red discs and one blue disc: for example, RRB.

To calculate P(RRB), we multiply the probabilities along the branches. That is,

$$\text{P(RRB)} = \frac{4}{9}\times\frac{3}{8}\times\frac{5}{7} = \frac{60}{504}$$

It can now easily be seen that the probability we want is the sum of those circled on the diagram: RRB, RBR and BRR. That is,

$$\frac{60}{504}+\frac{60}{504}+\frac{60}{504}=\frac{5}{14}$$

Exercise 23B

1 Two fair dice are rolled. Find the probability that one of the dice shows a *two* given that the total on the two dice is *six*.

2 Two fair dice are rolled. Find the probability that the total on the two dice is *eight* given that neither die shows a *five*.

3 Three fair coins are tossed. Find the probability that all three show *heads* given that there is an odd number of *heads* showing.

4 A bag contains three red discs and four green discs. Three discs are selected at random. Find the probability that they are green given that they are all of the same colour.

5 A box contains eleven balls numbered 1 to 11. Two balls are selected at random from the box. Given the sum of the numbers of the two selected balls is even, find the probability that the numbers on each of the selected balls are both odd.

6 Each of the pupils in two classes sits the same examination. In class A, 15 out of 20 pupils pass; in class B, 8 out of 12 pupils pass. A pupil is selected at random from the 32 pupils who have taken the examination. Given that she passed, find the probability that she was in class A.

7 In a given week, 200 people take a driving test in a given centre. Of the 120 who passed, 50 were taking the test for the first time; and of those who failed, 60 were taking the test for the first time. Estimate the probability of passing the driving test at a first attempt at this centre.

8 Given $P(A \cap B) = \frac{1}{3}$ and $P(B) = \frac{3}{5}$, find $P(A|B)$.

9 Given $P(A|B) = \frac{1}{5}$ and $P(B) = \frac{1}{2}$, find $P(A \cap B)$.

10 Given $P(A|B) = \frac{5}{6}$, $P(A) = \frac{3}{4}$ and $P(B) = \frac{2}{3}$, find $P(A \cup B)$.

11 Given $P(A) = \frac{2}{5}$, $P(B) = \frac{7}{10}$ and $P(A \cup B) = \frac{4}{5}$, find $P(A|B)$.

12 Given $P(A) = \frac{2}{3}$, $P(B) = \frac{4}{9}$ and $P(A|B) = \frac{3}{4}$, find $P(A \cup B)$.

13 Given $P(A) = 0.2$, $P(A|B) = 0.3$ and $P(A \cup B) = 0.4$, find $P(B)$.

14 Given $P(A) = \frac{1}{3}$, $P(B) = \frac{3}{8}$ and $P(A \cup B) = \frac{7}{12}$, show that A and B are independent.

15 Given $P(A) = x$, $P(B) = 2x$, $P(A \cup B) = 1 - x$ and $P(A|B) = \frac{1}{5}$, find the value of x.

16 A bag contains six white cards and four black cards. Three cards are selected at random. Draw a tree diagram and calculate the probability that two of the selected cards are white and the other is black.

17 Two sacks, A and B, each contain a mixture of plastic and leather rugby balls. Sack A contains four plastic balls and two leather balls, and sack B contains three plastic balls and five leather balls. A sack is selected at random and a ball is taken from it. Draw a tree diagram and calculate the probability that **a)** the ball is leather, **b)** the ball came from sack A, given it is leather.

18 A girl cannot decide what to wear for a party. She will wear either trousers or a skirt. She has two wardrobes. In wardrobe A are two pairs of trousers and three skirts, and in wardrobe B are five pairs of trousers and one skirt. She decides to select a wardrobe at random and then randomly select one item from that wardrobe. Draw a tree diagram and find the probability that

a) she wears trousers

b) she selected wardrobe A, given that she wears trousers.

19 An eccentric mathematics teacher decides to award a prize to a pupil selected from one of his three A-level classes. Class 1 has five boys and seven girls; class 2 has eight boys and two girls; and class 3 has three boys and three girls. He decides to award the prize by selecting a class at random and then randomly selecting a pupil from that class. Use a tree diagram to find the probability that

a) the selected pupil is a boy

b) the selected pupil is from class 3, given the selected pupil is a boy.

***20** My neighbour's cat has two kittens, and she tells me that at least one of them is female. What is the probability that both are female?

***21** In a game of bridge, each player is randomly dealt a hand of 13 cards from a pack of 52 cards. Find the probability that a randomly selected hand consists of three cards of each of three suits and four cards of the remaining suit.

Exercise 23C Examination questions

1 Two unbiased cubical dice are thrown simultaneously. Calculate the probability that

i) the scores on both dice are at least 3

ii) the scores on the two dice differ by 2. (WJEC)

2 A bag contains 4 red balls, 4 blue balls and 2 green balls. Four balls are drawn at random without replacement from the bag. Find the probability that the balls drawn are all of the same colour. (NEAB)

3 A committee consists of 6 men and 4 women. A sub-committee of 4 members is to be formed and it is decided to select these 4 members at random. Find the probability that the sub-committee contains at least 1 member of each sex. (WJEC)

4 Two events A and B are such that $P(A) = 0.6$, $P(B) = 0.3$ and $P(A \cup B) = 0.8$.

a) Determine whether or not A and B are mutually exclusive.
b) Calculate $P(A \cap B)$.
c) Find $P(A \cap B')$. (WJEC)

5 Two events A and B are such that $P(A) = 0.4$, $P(B) = 0.7$, $P(A \text{ or } B) = 0.8$. Calculate

a) $P(A \text{ and } B)$
b) the conditional probability $P(A \mid B)$. (AEB Spec)

6 The events A and B are such that $P(A) = 0.45$, $P(B) = 0.35$ and $P(A \cup B) = 0.7$.

a) Find the value of $P(A \cap B)$.
b) Explain why the events A and B are not independent.
c) Find the value of $P(A \mid B)$. (ULEAC)

7 Three events, A, B and C, are such that $P(A) = 0.5$, $P(B) = 0.4$, $P(C) = 0.3$ and $P(A \cup B) = 0.62$. The events B and C are independent. The events A and C are mutually exclusive.

i) Find $P(B \cap C)$.
ii) Find $P(A \cap B)$.
iii) Find $P(A \mid B)$.
iv) Show that B cannot occur without A or C also occurring. (NEAB)

8 An unbiased cubical die has one face lettered A, two faces lettered B and three faces lettered C. Three boxes are labelled A, B, C respectively. Box A contains 4 red and 2 white balls; box B contains 3 red and 3 white balls; box C contains 2 red and 4 white balls. The die is rolled and 3 balls are drawn at random without replacement from the box labelled with the letter uppermost on the die.

a) Show that the probability that 1 red ball and 2 white balls will be drawn from box A is $\frac{1}{30}$.
b) Show that the probability that 1 red ball and 2 white balls will be drawn is $\frac{29}{60}$.
c) Given that 1 red ball and 2 white balls were drawn, find the conditional probability that they were drawn from box A. (WJEC)

9 It is known that 1% of the population suffer from a certain disease. A diagnostic test for the disease gives a positive response with probability 0.98 if the disease is present. If the disease is not present, the probability of a positive response is 0.005.

The test is applied to a randomly selected member of the population.

i) Show that the probability of obtaining a positive response is 0.01475.
ii) Given that a positive response is obtained, calculate the probability that the person has the disease. (WJEC)

Answers

Exercise 1A

1 a) 6 **b)** 7 **c)** -2 **d)** 5 **e)** 2 **f)** $\frac{11}{2}$ **g)** 3 **h)** -2 **i)** $\frac{13}{4}$ **j)** $-\frac{1}{10}$ **k)** -3 **l)** $-\frac{1}{3}$ **2 a)** 2 **b)** $-\frac{4}{15}$ **c)** 2 **d)** $\frac{29}{11}$ **e)** $-\frac{3}{4}$
2 f) $-\frac{27}{2}$ **g)** -1 **h)** $\frac{14}{9}$ **i)** $\frac{5}{2}$ **j)** 11 **k)** $\frac{17}{2}$ **l)** 0 **3 a)** 7 **b)** 3 **c)** $\frac{5}{13}$ **d)** $-\frac{11}{9}$ **e)** -4 **f)** $\frac{31}{11}$ **g)** $\frac{16}{13}$ **h)** $-\frac{17}{4}$ **i)** $-\frac{1}{2}$
3 j) $\frac{17}{2}$ **k)** $\frac{21}{2}$ **l)** $\frac{49}{6}$ **4 a)** 2 **b)** 3 **c)** -1 **d)** $\frac{1}{2}$ **e)** $-\frac{1}{3}$ **f)** $\frac{17}{7}$ **g)** $-\frac{24}{23}$ **h)** -1 **i)** $\frac{43}{3}$ **j)** $-\frac{17}{30}$ **k)** $\frac{7}{10}$ **l)** $\frac{35}{26}$ **5 a)** 4
5 b) $-\frac{7}{2}$ **c)** $\frac{16}{7}$ **d)** -2 **e)** 27 **f)** $-\frac{15}{13}$ **g)** $\frac{7}{2}$ **h)** $\frac{52}{41}$ **i)** $\frac{13}{4}$ **j)** $\frac{31}{6}$ **k)** -2 **l)** 1 **6 a)** $-\frac{1}{2}$ **b)** -2 **c)** $-\frac{15}{8}$ **d)** $-\frac{11}{18}$ **e)** $\frac{13}{9}$
6 f) $\frac{1}{2}$ **g)** $-\frac{1}{4}$ **h)** $\frac{7}{9}$ **i)** $\frac{2}{5}$ **j)** $\frac{5}{4}$ **k)** $-\frac{7}{2}$ **l)** $\frac{14}{13}$ **7** $\frac{1}{3}$ **8** $-\frac{16}{29}$ **9** $-\frac{5}{8}$ **10** $\frac{5}{3}$ **11** -2 **12** $\frac{18}{7}$ **13** $\frac{1}{2}$ **14** $\frac{28}{11}$ **15** -5
16 $\frac{7}{2}$ **17** $-\frac{4}{5}$ **18** -8 **19** $\frac{22}{21}$ **20** $\frac{18}{7}$ **21** 10 **22** $\frac{13}{7}$ **23** 0 **24** $\frac{5}{2}$ **25** 10 **26** $\frac{2}{3}$ **27** $\frac{1}{3}$ **28** 12 **29** $-\frac{3}{4}$ **30** $-\frac{22}{37}$

Exercise 1B

1 a) $x > 4$ **b)** $x \geqslant -4$ **c)** $x \geqslant \frac{11}{2}$ **d)** $x > \frac{15}{4}$ **e)** $x < 5$ **f)** $x \geqslant \frac{2}{5}$ **g)** $x > 2$ **h)** $x \leqslant -\frac{6}{7}$ **i)** $x \geqslant \frac{5}{3}$ **j)** $x < \frac{5}{8}$ **k)** $x > -2$
1 l) $x \leqslant \frac{7}{2}$ **2 a)** $x < 4$ **b)** $x < 2$ **c)** $x > 9$ **d)** $x \leqslant 0$ **e)** $x > \frac{19}{9}$ **f)** $x \geqslant -1$ **g)** $x < -\frac{9}{13}$ **h)** $x > -\frac{2}{9}$ **i)** $x > -\frac{4}{3}$ **j)** $x \geqslant -\frac{28}{5}$
2 k) $x < 17$ **l)** $x \geqslant -18$ **3 a)** $x < 10$ **b)** $x \leqslant 7$ **c)** $x \leqslant -19$ **d)** $x > -\frac{19}{9}$ **e)** $x > \frac{6}{5}$ **f)** $x \leqslant 8$ **g)** $x \leqslant -6$ **h)** $x > 2$
3 i) $x > \frac{19}{11}$ **j)** $x < -\frac{4}{7}$ **k)** $x \leqslant \frac{42}{29}$ **l)** $x \leqslant \frac{5}{7}$ **4 a)** 1, 2, 3 **b)** 1 **c)** 2, 3, 4, 5 **d)** −3, −2, −1 **e)** 1, 2 **f)** 2, 3, 4, 5, 6 **7** 3

Exercise 1C

1 a) $(2,-1)$ **b)** $(1,2)$ **c)** $(4,1)$ **d)** $(-1,-2)$ **e)** $(3,-4)$ **f)** $(3,-1)$ **g)** $(2,-1)$ **h)** $(3,2)$ **i)** $(-3,2)$ **j)** $(1,-1)$ **k)** $(7,6)$
1 l) $\left(2,-\frac{3}{2}\right)$ **2 a)** $(4,2)$ **b)** $(3,4)$ **c)** $(2,3)$ **d)** $(-2,-1)$ **e)** $\left(\frac{1}{2},-\frac{3}{2}\right)$ **f)** $(3,2)$ **g)** $\left(\frac{1}{3},-\frac{2}{3}\right)$ **h)** $\left(\frac{1}{2},\frac{1}{4}\right)$ **i)** $\left(\frac{3}{4},-\frac{5}{4}\right)$
2 j) $(2,0)$ **k)** $\left(\frac{10}{3},1\right)$ **l)** $\left(\frac{3}{2},-\frac{1}{3}\right)$ **3 a)** $(2,5)$ **b)** $(6,2)$ **c)** $(-1,7)$ **d)** $(-2,6)$ **e)** $(-1,-4)$ **f)** $(6,2)$ **g)** $(-3,-2)$ **h)** $(0,2)$
3 i) $(-2,4)$ **j)** $(-3,3)$ **k)** $(-3,-4)$ **l)** $(1,4)$

Exercise 1D

1 a) $2,3$ **b)** $-1,4$ **c)** $2,5$ **d)** $-3,-2$ **e)** $2,4$ **f)** $-1,6$ **g)** ± 3 **h)** $-4,2$ **i)** $-3,4$ **j)** $4,5$ **k)** $0,4$ **l)** $-1,8$ **2 a)** $-2,-\frac{1}{2}$
2 b) $\frac{1}{3},2$ **c)** $-1,\frac{5}{2}$ **d)** $-3,\frac{1}{5}$ **e)** $-1,-\frac{1}{4}$ **f)** $\frac{1}{3},\frac{1}{2}$ **g)** $-\frac{2}{3},4$ **h)** $-3,\frac{5}{2}$ **i)** $\pm\frac{3}{4}$ **j)** $-\frac{5}{3},2$ **k)** $-3,\frac{2}{5}$ **l)** $\frac{1}{4},\frac{3}{2}$ **3 a)** $-5,1$
3 b) $-4,2$ **c)** $-\frac{1}{2},5$ **d)** $-12,2$ **e)** $-5,-1$ **f)** $-6,\frac{1}{3}$ **g)** $\frac{1}{7},3$ **h)** ± 1 **i)** 2 **j)** $-1,-\frac{1}{3}$ **k)** $0,-1$ **l)** $-3,0$ **4 a)** $-6,1$
4 b) $-4,2$ **c)** $-2,\frac{3}{5}$ **d)** $-7,3$ **e)** $-1,2$ **f)** $-\frac{1}{3},\frac{1}{2}$ **g)** ± 2 **h)** $\frac{11}{2},3$ **i)** $-3,-1$ **j)** $-\frac{1}{4},3$ **k)** $\frac{2}{5},2$ **l)** $-\frac{5}{2},2$
5 5 cm × 12 cm **6** $\frac{2}{3}$ **7** $\frac{3}{4}$ **8 b)** $x^2 - 6x + 9 = 0, 3$ **9** 120 **10** 10

Exercise 1E

1 a) $2 \pm \sqrt{5}$ **b)** $-3 \pm \sqrt{7}$ **c)** $1 \pm \sqrt{2}$ **d)** $4 \pm \sqrt{19}$ **e)** $-\frac{1}{2} \pm \frac{\sqrt{5}}{2}$ **f)** $-\frac{3}{2} \pm \frac{\sqrt{5}}{2}$ **g)** $\frac{5}{2} \pm \frac{\sqrt{33}}{2}$ **h)** $\frac{1}{2} \pm \frac{\sqrt{13}}{2}$ **i)** $-\frac{5}{2} \pm \frac{\sqrt{21}}{2}$

1 j) $-6 \pm \sqrt{31}$ **k)** $\frac{9}{2} \pm \frac{\sqrt{41}}{2}$ **l)** $\frac{1}{4} \pm \frac{\sqrt{5}}{4}$ **2 a)** $\frac{3}{4} \pm \frac{\sqrt{33}}{4}$ **b)** $1 \pm \sqrt{\frac{2}{3}}$ **c)** $-\frac{1}{2} \pm \sqrt{\frac{3}{2}}$ **d)** $-\frac{5}{6} \pm \frac{\sqrt{37}}{6}$ **e)** $-\frac{1}{10} \pm \frac{\sqrt{61}}{10}$

2 f) $\frac{3}{4} \pm \frac{\sqrt{17}}{4}$ **g)** $\frac{1}{4} \pm \frac{\sqrt{17}}{4}$ **h)** $-\frac{3}{8} \pm \frac{\sqrt{41}}{8}$ **i)** $1 \pm \frac{\sqrt{14}}{7}$ **j)** $-\frac{1}{3} \pm \sqrt{\frac{11}{18}}$ **k)** $2 \pm \sqrt{\frac{3}{5}}$ **l)** $-\frac{9}{2} \pm \frac{\sqrt{39}}{2}$ **3** $(x+2)^2 + 3$

4 $5(x-3)^2 + 2$

Exercise 1F

1 a) $-2.41, 0.41$ **b)** $-3.41, -0.59$ **c)** $-1.19, 4.19$ **d)** $0.63, 6.37$ **e)** $-4.19, 1.19$ **f)** $-9.10, 1.10$ **g)** $-1.62, 0.62$ **h)** $-1.19, 4.19$
1 i) $-4.30, -0.70$ **j)** $1.27, 4.73$ **k)** $1.84, 8.16$ **l)** $-13.48, 1.48$ **2 a)** $-2.35, 0.85$ **b)** $-1.18, 0.85$ **c)** $-2.09, 0.84$
2 d) $-2.78, -0.72$ **e)** $-0.69, 0.29$ **f)** $-1.24, 0.40$ **g)** $-2.27, 1.77$ **h)** $-0.73, 0.23$ **i)** $-1.77, -0.57$ **j)** $-1.39, 2.89$

2 k) $-1.23, -0.27$ **l)** $-1.27, 0.47$ **3 a)** $-1.65, 3.65$ **b)** $-9.22, 0.22$ **c)** $-2.59, 1.84$ **d)** $-1.13, 4.63$ **e)** $-13.29, -2.71$

3 f) $0.22, 2.28$ **g)** $-4.55, 0.88$ **h)** $-3.32, -0.48$ **i)** $-3.36, 1.69$ **j)** $-0.59, 5.09$ **k)** $-12.92, -0.08$ **l)** $1.53, 3.27$

4 -71 **5** -16 **10** 4 or 16 **11** $-\frac{2}{9}$ or 2 **12** $-\frac{5}{13}$ or 1 **13** $-\frac{50}{31}$ or 2 **15** $a = -2$, roots $0, 2$; $a = 6$, roots $-4, -2$

Exercise 1G

1 a) $\pm 2, \pm 3$ **b)** $\pm\sqrt{3}$ **c)** $1, 3$ **d)** $-2, \sqrt[3]{3}$ **e)** $4, 9$ **f)** $1, 25$ **g)** $\pm\sqrt{3}$ **h)** $1, 9$ **i)** $\pm 1, \pm 2$ **j)** $-\sqrt[3]{2}, \sqrt[3]{4}$ **k)** $9, 25$ **l)** $\pm\frac{1}{\sqrt{2}}, \pm 2$

2 a) $\pm\sqrt{2}$ **b)** $1, -2$ **c)** $25, 49$ **d)** $\pm\sqrt{2}, \pm 2$ **e)** $4, 25$ **f)** $\pm\sqrt{3}$ **g)** $\pm\sqrt{2}, \pm 2$ **h)** $9, 25$ **i)** $\pm 1, \pm\sqrt{2}$ **j)** $\pm\sqrt{\frac{3}{2}}, \pm 2$

2 k) $4, \frac{25}{4}$ **l)** $-\sqrt[3]{2}, 2$ **3** $1, -2$ **4** $-2, \frac{2}{3}$ **5 a)** $2, 5$ **b)** $\pm 1, \pm 2$ **6 a)** $-2, 7$ **b)** $-1, 2$ **7** $-4, -2, 1, 3$ **8** $-2, -\frac{1}{2}, 1$

Exercise 1H

1 a) $y = 2$ at $x = -2$ **b)** $y = 4$ at $x = 3$ **c)** $y = 15$ at $x = 5$ **d)** $y = -6$ at $x = -1$ **e)** $y = 5\frac{3}{4}$ at $x = -1\frac{1}{2}$

1 f) $y = 2\frac{3}{4}$ at $x = 3\frac{1}{2}$ **g)** $y = -17\frac{1}{2}$ at $x = -2\frac{1}{2}$ **h)** $y = 11$ at $x = -1$ **i)** $y = -7\frac{1}{16}$ at $x = -\frac{1}{8}$ **j)** $y = 8\frac{2}{3}$ at $x = \frac{1}{3}$

1 k) $y = 4\frac{23}{24}$ at $x = -\frac{1}{12}$ **l)** $y = -7\frac{4}{5}$ at $x = \frac{1}{5}$ **2 a)** $y = 4$ at $x = -1$ **b)** $y = 9$ at $x = 2$ **c)** $y = 9$ at $x = 1$ **d)** $y = 29$ at $x = 3$

2 e) $y = 6\frac{1}{4}$ at $x = -1\frac{1}{2}$ **f)** $y = 2$ at $x = 2$ **g)** $y = 6\frac{1}{4}$ at $x = -\frac{1}{2}$ **h)** $y = 3\frac{1}{2}$ at $x = \frac{1}{2}$ **i)** $y = 7\frac{9}{16}$ at $x = -\frac{3}{8}$

2 j) $y = 14\frac{1}{12}$ at $x = \frac{1}{6}$ **k)** $y = 9\frac{1}{8}$ at $x = -\frac{5}{4}$ **l)** $y = 12\frac{7}{8}$ at $x = \frac{1}{4}$ **4 b)** $100\,\text{m}^2$ **5** $x(40 - 2x)\,\text{m}^2$ **6** $133\frac{1}{3}\,\text{m}^2$ **7** 45 m, 3 s

10 b) $2x(1 - x)\,\text{m}^2$

Exercise 1I

1 a) $x < 3$ or $x > 5$ **b)** $-4 < x < 3$ **c)** $-3 \leqslant x \leqslant -1$ **d)** $-\frac{3}{2} < x < 1$ **e)** $x \leqslant -\frac{2}{3}$ or $x \geqslant \frac{1}{2}$ **f)** $x < -3$ or $x > 1$

1 g) $-2 < x < 3$ **h)** $x \leqslant -\frac{4}{3}$ or $x \geqslant \frac{5}{2}$ **i)** $x < -3$ or $x > 6$ **j)** $x \leqslant -5$ or $x \geqslant -\frac{4}{5}$ **k)** $-3 \leqslant x \leqslant 3$ **l)** $0 < x < \frac{5}{6}$

2 a) $x < -2$ or $x > 4$ **b)** $-3 \leqslant x \leqslant 5$ **c)** $-4 < x < -3$ **d)** $x \leqslant -\frac{5}{2}$ or $x \geqslant 1$ **e)** $-\frac{6}{5} \leqslant x \leqslant -1$ **f)** $x = 4$ **g)** $-3 < x < 2$

2 h) $x < -\frac{2}{3}$ or $x > 1$ **i)** $-\frac{2}{3} \leqslant x \leqslant 5$ **j)** $x < -4$ or $x > 4$ **k)** $-\frac{1}{5} \leqslant x \leqslant \frac{1}{2}$ **l)** $x \in \mathbb{R}, x \neq \frac{3}{2}$ **3 a)** $x \leqslant -5$ or $x \geqslant 2$

3 b) $3 < x < 7$ **c)** $x < 1$ or $x > 4$ **d)** $3 \leqslant x \leqslant 8$ **e)** $x < -3$ or $x > \frac{1}{2}$ **f)** $\frac{1}{2} < x < \frac{4}{5}$ **g)** $x \leqslant -\frac{3}{2}$ or $x \geqslant 4$ **h)** $-5 < x < 5$

3 i) $-\frac{1}{3} < x < \frac{1}{2}$ **j)** $-8 \leqslant x \leqslant \frac{1}{2}$ **k)** No solution **l)** $x \leqslant 0$ or $x \geqslant 5$ **4 a)** $x < 3$ or $x \geqslant 5$ **b)** $-1 < x < \frac{3}{2}$ **c)** $x < \frac{5}{2}$ or $x \geqslant \frac{8}{3}$

4 c) $x < \frac{5}{2}$ or $x \geqslant \frac{8}{3}$ **d)** $x < -\frac{1}{3}$ or $x > -\frac{1}{5}$ **e)** $2 < x \leqslant \frac{7}{3}$ **f)** $x < \frac{1}{6}$ or $x > \frac{1}{2}$ **g)** $-\frac{1}{2} \leqslant x < 2$ **h)** $-\frac{14}{3} < x < -3$

4 i) $1 < x < 2$ **j)** $x \leqslant \frac{4}{3}$ or $x > \frac{3}{2}$ **k)** $x < -1$ or $x > -\frac{3}{5}$ **l)** $x < -2$ or $x > 0$ **5 a)** $\frac{1}{2} \leqslant x \leqslant 2$ or $x > 4$

5 b) $2 < x < \frac{5}{2}$ or $x > 3$

Exercise 1J

1 a) $(2, 2), (5, 11)$ **b)** $(\frac{1}{2}, 3), (3, 13)$ **c)** $(-\frac{1}{3}, 3), (4, 16)$ **d)** $(-3, 37), (-\frac{5}{2}, \frac{63}{2})$ **e)** $(\frac{1}{2}, -\frac{5}{2}), (2, -1)$ **f)** $(-4, 11), (\frac{1}{2}, 2)$ **g)** $(2, 3), (3, 2)$

1 h) $(-6, -\frac{1}{3}), (\frac{1}{2}, 4)$ **i)** $(\pm 2, \pm 1), (\pm 1, \pm 2)$ **j)** $(\pm 4, \pm 2), (\pm 2, \pm 4)$ **k)** $(-1, 2), (3, -2)$ **l)** $(-1, 0), (3, 4)$ **2** $(6, 2), (-2, 10); (6, 2)$

3 12 cm × 4 cm **4** $(10, 26), (48, 7), (10, 26)$ **5** $(2, 4), (4, 2)$ **6 a)** $4, (-1, 0), (0, 3), (3, 0)$ **b)** $2, (0, 3)$ **c)** $(0, 3), (2, 3)$

7 a) $9, (-1, 0), (0, 5), (5, 0)$ **b)** $4, (0, 5)$ **c)** $(0, 5), (3, 8)$

Exercise 1K

1 20 **2** $3\frac{1}{2}$ **3** 15 N **4** £1.80 **5** $A = 7B$, $\frac{20}{7}$ **6** $F = \frac{3}{4}x$, $\frac{2}{3}$ cm **7** 45 **8** $\sqrt{10}$ **9** 112.5 **10** $A = 2\sqrt{B}$, 8

11 a) $t = \frac{1}{2}\sqrt{h}$ **b)** 81 m **12** 3.6 **13** 5 **14** 750 **15** 1.875 **16** $P = \frac{200}{\sqrt{Q}}$, 100

Exercise 1L

1 i) $(x + 3)^2 + 7$ **ii)** 7 at $x = -3$ **iii)** $\frac{1}{7}$ **2** $(x - 9)^2 + 13$ **3 b)** $a = 100, b = 10\,000$

4 a) $p = 3, q = 2, r = -7$ **b)** -7 **c)** $-3.5, -0.5$ **5** $-4 < x < 4$ **6** $x < -3$ or $x > 4$ **7** $x < -6$ or $x > -1$

8 $-8 < x < -4$ **9** $x < -1$ or $x > 1$ **10** $x(6 - x) \geqslant 5$, $1 \leqslant x \leqslant 5$ **11** $8.4 < x < 9$ **12** $-4 < k < 1$ **13** $k < 2$ or $k > 10$

14 -2 or 4 **15** $(1, -1)$ or $(4, 2)$ **16** $(3, 1)$ or $(-\frac{11}{5}, -\frac{8}{5})$ **17** $(3, -1)$ or $(-\frac{1}{3}, \frac{7}{3})$

Exercise 2A

1 $(12+4\sqrt{2}+4\sqrt{3})$ cm **2** $(12+6\sqrt{2}+6\sqrt{6})$ cm **3 a)** $4\sqrt{2}$ cm **b)** $4\sqrt{3}$ cm **c)** 35.3° **d)** 60° **4 a)** 13 cm **b)** $\sqrt{185}$ cm

4 c) 72.9° **d)** 28.1° **e)** 61.9° **5 a)** $6\sqrt{2}$ cm **b)** 9 cm **c)** 19.5° **d)** $3\sqrt{5}$ cm **e)** 78.5° **6 a)** $8\sqrt{2}$ cm **b)** 9 cm **c)** 51.1°

6 d) 63.6° **e)** 67.9° **7 a)** $6\sqrt{2}$ cm **b)** $3\sqrt{7}$ cm **c)** 61.9° **d)** 41.4° **e)** 97.2° **8 a)** 10 cm **b)** 13 cm **c)** 67.4° **d)** 72.1°

8 e) 36.9° **9 a)** 114 m **b)** 288 m **c)** 264 m **10 a)** 10 m **b)** 4.5 m **c)** 42° **d)** 97.3° **11 b)** $400\sqrt{3}$ m **c)** 150°

12 $\frac{20}{\sqrt{3}-1}$ m **13** 91.6 m

Exercise 2B

1 a) $x=8.96, y=8.65$ **b)** $x=6.60, \theta=39.9$ **c)** $x=10.6, y=9.13$ **d)** $x=8.01, \theta=27.3$

1 e) $x=7.67, \theta=60.1$ or $x=1.69, \theta=119.9$ **f)** $x=7.62, y=7.25$ **g)** $x=7.18, \theta=38.7$

1 h) $x=3.80, \theta=69.1$ or $x=1.23, \theta=110.9$ **i)** $x=3.84, y=10.1$ **j)** $x=13.8, y=11.7$

1 k) $x=5.50, \theta=102.8$ or $x=8.38, \theta=77.2$ **l)** $x=5.84, \theta=20.7$ **2 a)** 109.5° **b)** 13.8 **c)** 3.91 **d)** 52.2° **e)** 2.33 **f)** 8.94

2 g) 2.18 **h)** 7.43 **i)** 5.68 **j)** 53.6° **k)** 163.2° **l)** 4.18 **3 a)** 6.19 cm **b)** 82.0° **4 a)** 74.8° **b)** 49.3° **5 a)** 146.8°

5 b) 3.33 cm **6** 3.93 cm or 9.93 cm **7 b)** 48 km **c)** 320.1° **8** 1139 m at 345.8° **9 b)** 94.3 m at 328° **c)** 154.0 m at 144.9°

10 a) 4.14 km **b)** 2.93 km **11 a)** $\sqrt{89}$ cm **b)** $\sqrt{61}$ cm **c)** 10 cm **d)** 62.6° **12 a)** 67.1° **b)** 73.9° **c)** $\sqrt{74}$ cm **d)** 97.0°

13 a) $2\sqrt{41}$ cm **b)** $9\sqrt{2}$ cm **c)** 71.7° **d)** 66.9° **e)** 98.9° **14 a)** 72.5° **b)** 63.3° **c)** $3\sqrt{13}$ cm **d)** 102.3° **15** 34.8°

Exercise 2C

1 $2(9\pi-4\sqrt{3})\,\text{cm}^2$ **3 b)** 12.4 cm^2 **c)** 61.9 cm^2 **4** 166 cm^2 **5** 391 cm^2 **6 a)** 6.10 cm^2 **b)** 3.21 cm **7** 16.8 cm^2, 5.26 cm

8 33.2 cm^2, 9.21 cm **9 a)** 12.6 m^2 **b)** 12.4 m^2 **c)** £132.50 **10 a)** 64.6 mm^2 **b)** 93.1 mm^2 **c)** £2.52 **11 a)** 90 m^2 **b)** 164 m^2

11 c) 5000

Exercise 2D

1 a) $\frac{\pi}{6}$ rad **b)** $\frac{\pi}{2}$ rad **c)** $\frac{2\pi}{3}$ rad **d)** $\frac{\pi}{18}$ rad **e)** $\frac{4\pi}{9}$ rad **f)** $\frac{5\pi}{3}$ rad **g)** $\frac{\pi}{5}$ rad **h)** $\frac{4\pi}{3}$ rad **i)** $\frac{2\pi}{5}$ rad **j)** 2π rad **k)** $\frac{19\pi}{10}$ rad

1 l) $\frac{\pi}{180}$ rad **2 a)** 180° **b)** 45° **c)** 540° **d)** 30° **e)** 144° **f)** 15° **g)** 300° **h)** 180° **i)** 75° **j)** 2° **k)** 270° **l)** 210°

3 a) 229.2° **b)** 11.5° **c)** 246.4° **d)** 28.6° **e)** 40.1° **f)** 171.9° **g)** 297.9° **h)** 120.3° **i)** 286.5° **j)** 2.3° **k)** 916.7° **l)** 57.3°

4 a) $\frac{3\pi}{2}$ cm **b)** $\frac{15\pi}{4}\,\text{cm}^2$ **5 a)** 6π cm **b)** $27\pi\,\text{cm}^2$ **6 a)** $\left(12+\frac{5\pi}{2}\right)$ cm **b)** $\frac{15\pi}{2}\,\text{cm}^2$ **7 a)** $\left(14+\frac{5\pi}{2}\right)$ cm **b)** $\frac{35\pi}{4}\,\text{cm}^2$

8 a) $\frac{4\pi}{3}$ cm **b)** $\frac{16\pi}{3}\,\text{cm}^2$ **9 a)** $(24+3\pi)$ cm **b)** $18\pi\,\text{cm}^2$ **10** $\frac{25\pi}{4}\,\text{cm}^2$ **11** $\frac{50\pi}{3}\,\text{cm}^2$ **12 a)** $36(\pi-2)\,\text{cm}^2$ **b)** $6(\pi+2\sqrt{2})$ cm

12 b) $6(\pi+2\sqrt{2})$ cm **13 a)** 1.30 cm^2 **b)** 9.97 cm **14 a)** 17.0 cm^2 **b)** 24.5 cm **15 a)** $16\sqrt{3}\,\text{cm}^2$ **b)** $\frac{32\pi}{3}\,\text{cm}^2$

15 c) $(32\pi-48\sqrt{3})\,\text{cm}^2$ **16 a)** $\frac{9\pi}{4}\,\text{m}^2$ **b)** 9 m^2 **c)** $\frac{9}{2}(\pi-2)\,\text{m}^2$ **d)** $\frac{3\sqrt{2}\pi}{4}$ m **e)** $6(\sqrt{2}-1)$ m **f)** $\frac{3}{2}\left[\sqrt{2}(\pi+4)-4\right]$ m

Exercise 2E

1 $159\pi\,\text{cm}^3$ **2** 500 m^3 **3 a)** $\frac{1280\pi}{3}\,\text{cm}^3$ **b)** $\frac{20\pi}{3}\,\text{cm}^3$ **c)** $420\pi\,\text{cm}^3$ **4 a)** $6\sqrt{3}$ m **b)** $\frac{\pi}{3}$ rad **c)** $3(2\pi+9\sqrt{3})\,\text{m}^3$

4 d) $360(2\pi+9\sqrt{3})\,\text{m}^3$ **5** 2 cm **6** $\frac{9}{4}$ cm

Exercise 2F

1 a) 4.11 cm **b)** 46.9° **2 i)** 1.6 m **ii)** 93° **3 ii)** 217 cm **iii)** 30° **4 a)** 208.3 m **b)** 188.8 m **5 a)** 60° **b)** 31.2°

6 a) 53° **b)** 60° **c)** 44° **7 a)** 40.3° **b)** 79.6° **c)** 9.15 cm **8 a)** 43.6° **b)** 11.5 cm **c)** 70.8°, 7.56 cm **9** 24.6 m

10 a) 1.25 rad **b)** 2.41 cm^2 **11 a)** 8.4 cm **b)** 21.2 cm^2 **12 a)** $\frac{2\pi}{3}$ **b)** 170 000 m^3 **13 a)** 9.8 m **b)** 4.32 m^2 **14 a)** 22.2°

14 b) 1.63 cm **c)** 3.38 cm^2 **16 c)** $\frac{(3\sqrt{3}-\pi)r^2}{6}\,\text{cm}^2$ **17 a) i)** $\left[5-2r+\frac{\pi}{6}(5+2r)\right]$ cm **ii)** $\frac{\pi}{24}(25-4r^2)\,\text{cm}^2$ **b)** 0.94 cm

Exercise 3A

14 Stretch by a factor of 2 parallel to the y-axis, followed by translation $\begin{pmatrix}0\\5\end{pmatrix}$

Or Stretch by a factor of $\frac{1}{2}$ parallel to the x-axis, followed by translation $\begin{pmatrix}0\\3\end{pmatrix}$

15 a) $g(x) = (x+1)^2 + 7$ **b)** Reflection in the x-axis followed by translation $\begin{pmatrix}-1\\7\end{pmatrix}$ **16 b)** $g(x) = -x^2$ $h(x) = 2 - x^2$

17 b) $g(x) = \dfrac{3}{x+4}$ $h(x) = -\dfrac{3}{x+4}$ **18** $f(x-p)+q$

Exercise 3B

1 a) one-to-one **b)** two-to-one **c)** one-to-one **d)** two-to-one **e)** two-to-one **f)** one-to-one **g)** one-to-one

1 h) one-to-one **i)** two-to-one **j)** one-to-one **k)** two-to-one **l)** two-to-one

2 a) $f(x) \in \mathbb{R}, 4 < f(x) < 9$ **b)** $f(x) \in \mathbb{R}, f(x) \geqslant 7$ **c)** $f(x) \in \mathbb{R}, 1 < f(x) \leqslant 9$ **d)** $f(x) \in \mathbb{R}, \frac{1}{18} \leqslant f(x) \leqslant \frac{1}{3}$ **e)** $f(x) \in \mathbb{R}, f(x) \geqslant 9$

2 f) $f(x) \in \mathbb{R}, 4 < f(x) < 134$ **g)** $f(x) \in \mathbb{R}, -9 \leqslant f(x) \leqslant 0$ **h)** $f(x) \in \mathbb{R}, \frac{1}{10} < f(x) \leqslant \frac{1}{2}$ **i)** $f(x) \in \mathbb{R}, -4 < f(x) < \infty$

2 j) $f(x) \in \mathbb{R}, 2 \leqslant f(x) \leqslant 5$ **k)** $f(x) \in \mathbb{R}, 0 < f(x) \leqslant 20$ **l)** $f(x) \in \mathbb{R}, 0 < f(x) \leqslant \frac{1}{3}$ **3 a)** $4 \leqslant f(x) \leqslant 36$ **b)** $0 \leqslant g(x) \leqslant 9$

3 c) $0 \leqslant h(x) \leqslant 7$ **d)** $0 \leqslant f(x) \leqslant 4$ **e)** $1 \leqslant g(x) \leqslant 4$ **f)** $0 \leqslant h(x) \leqslant 16$ **4 a)** $f(x) \in \mathbb{R}, f(x) \geqslant 7$ **b)** $f(x) \in \mathbb{R}, f(x) \geqslant 4$

c) $f(x) \in \mathbb{R}, 0 < f(x) \leqslant \frac{1}{3}$

Exercise 3C

1 a) $-2, 6$ **b)** $-9, 1$ **c)** $-3, 9$ **d)** $2, 6$ **e)** $-\frac{5}{3}, 1$ **f)** $-\frac{4}{5}, 2$ **g)** 1 **h)** 5 **i)** $\frac{1}{3}, 1$ **j)** $-\frac{5}{4}, \frac{3}{2}$ **k)** $-3, \frac{7}{5}$ **l)** $-2, -\frac{1}{3}$

2 $-1, \frac{7}{3}$ **3** $-1, 2$ **4** $2, 10$ **5** 4 **6** $-6 < x < 1$ **7** $x < -\frac{3}{4}$ or $x > \frac{3}{2}$ **8** $x \leqslant -24$ or $x \geqslant 0$ **9** $-\frac{8}{3} \leqslant x \leqslant 8$

11 a) $x > 1$ **b)** $x \leqslant -\frac{1}{2}$ **c)** $-\frac{1}{3} \leqslant x \leqslant 7$ **d)** $x \leqslant -\frac{7}{2}$ or $x \geqslant \frac{3}{4}$ **e)** $-6 < x < -\frac{4}{3}$ **f)** $x < \frac{7}{2}$ **g)** $-1 \leqslant x \leqslant \frac{2}{3}$ **h)** $x < \frac{7}{3}$ or $x > 7$

11 i) $x < 0$ or $x > 2$ **j)** $-9 \leqslant x \leqslant -\frac{1}{3}$ **k)** $x \leqslant 1$ or $x \geqslant 5$ **l)** $-\frac{1}{3} < x < 4$ **13 a)** $-\frac{2}{3} \leqslant x \leqslant \frac{4}{3}$ **b)** $x < -1$ or $x > \frac{1}{3}$

Exercise 3D

1 a) Even **b)** Odd **c)** Neither **d)** Even **e)** Neither **f)** Even **g)** Neither **h)** Neither **i)** Even **j)** Neither **k)** Odd **l)** Odd

2 a) Odd **b)** Even **c)** Even **d)** Neither **e)** Even **f)** Odd **g)** Even **h)** Odd **i)** Even **j)** Even **k)** Odd **l)** Odd

Exercise 3E

1 a) 7 **b)** 4 **c)** $\frac{1}{4}$ **d)** 19 **e)** 9 **f)** $\frac{1}{4}$ **g)** $\frac{1}{9}$ **h)** 23 **i)** 81 **j)** 12 **k)** $\frac{3}{2}$ **l)** $\frac{1}{33}$ **2 a)** $3x^2 - 1$ **b)** $(3x-1)^2$ **c)** $\dfrac{6}{x} - 1$ **d)** $\dfrac{2}{x^2}$

2 e) x^4 **f)** $9x - 4$ **3 a) i)** $x^2 + 10x + 28$ **ii)** $x^2 + 8$ **b)** -2 **4 a)** $\dfrac{3}{x+5}, -2$ **b)** $\dfrac{3}{x} + 5, 3$ **5 a)** $-4, -1$ **b)** $\frac{3}{2}$

6 $2x^2 + 5, 1 \leqslant x \leqslant 5, 7 \leqslant gf(x) \leqslant 55$ **7** $9x^4 + 6x^2 - 1, -1 \leqslant qp(x) \leqslant 167$ **8** $\dfrac{1}{x^2+1}, 0 < gf(x) \leqslant 1$

9 a) $x^2 - 2, hg(x) \in \mathbb{R}, hg(x) \geqslant -2$ **b)** $x^4 + 6x^2 + 12, gg(x) \in \mathbb{R}, gg(x) \geqslant 12$ **10 a)** $\sqrt{x^2+1}, fg(x) \in \mathbb{R}, fg(x) > 1$

10 b) $x + 1, gf(x) \in \mathbb{R}, gf(x) > 1$ **11 a)** $9x + 20, x$ **b)** $-\frac{5}{2}$ **12** $-2, 1$ **13** -2 **14** 2 **15** $\pm(3x+1)$ **16 a)** $a = 4, b = 1$

16 b) $4^n x + \dfrac{4^n - 1}{3}$

Exercise 3F

1 a) $\dfrac{x-2}{3}$ **b)** $\dfrac{1+x}{5}$ **c)** $\dfrac{4-x}{3}$ **d)** $\dfrac{2}{x}, x \neq 0$ **e)** $\dfrac{3+x}{x}, x \neq 0$ **f)** $\dfrac{2x-5}{3x}, x \neq 0$ **g)** $\dfrac{2x}{1-x}, x \neq 1$ **h)** $\dfrac{5x}{2+x}, x \neq -2$

1 i) $\dfrac{x}{3-2x}, x \neq \dfrac{3}{2}$ **j)** $\dfrac{1}{x-1}, x \neq 1$ **k)** $\dfrac{3-x}{x-4}, x \neq 4$ **l)** $\dfrac{4x-5}{2-x}, x \neq 2$ **2 a)** $\sqrt{x}, x \in \mathbb{R}, x > 4$ **b)** $\dfrac{1-2x}{x}, x \in \mathbb{R}, 0 < x < \frac{1}{2}$

2 c) $2 + x^2, x \in \mathbb{R}, x > 1$ **d)** $\sqrt{\dfrac{x+1}{3}}, x \in \mathbb{R}, 2 < x < 47$ **e)** $\dfrac{x^2-3}{2}, x \in \mathbb{R}, x \geqslant 5$ **f)** $\dfrac{1}{x+3}, x \in \mathbb{R}, -\dfrac{14}{5} < x < -\dfrac{5}{2}$

2 g) $-2 + \sqrt{x-3}, x \in \mathbb{R}, x \geqslant 3$ **h)** $\sqrt[3]{x-1}, x \in \mathbb{R}$ **i)** $\dfrac{2x+1}{x}, x \in \mathbb{R}, 0 < x < 1$ **j)** $3 - x^2, x \in \mathbb{R}, x \geqslant 1$

2 k) $3 + \sqrt{x-5}, x \in \mathbb{R}, 6 \leqslant x \leqslant 14$ **l)** $(x-5)^2 - 3, x \in \mathbb{R}, x \leqslant 5$ **3 a)** $\dfrac{x+4}{3}, x \in \mathbb{R}$ **c)** 2 **4 b)** $\dfrac{10-x}{2}, x \in \mathbb{R}, x \leqslant 10$ **c)** $\dfrac{10}{3}$

5 a) $\sqrt{x+6}, x \in \mathbb{R}, x > -6$ **c)** 3 **6 b)** $2+\sqrt{x}, x \in \mathbb{R}, x > 0$ **c)** 4 **7 a)** 2 **b)** $\dfrac{x+5}{2}, x \in \mathbb{R}$; $\dfrac{7-x}{4}, x \in \mathbb{R}$

7 c) -1, from **a)** $f(2) = g(2) = -1$ **8 a)** $0 < h(x) \leqslant \frac{4}{3}$ **b)** $h^{-1}(x) = \dfrac{4}{x} - 3, x \in \mathbb{R}, x \neq 0$ **c)** 1 **9 a)** $3x - 5$

9 b) i) $\dfrac{x-1}{3}, x \in \mathbb{R}$ **ii)** $x+2, x \in \mathbb{R}$ **iii)** $\dfrac{x+5}{3}, x \in \mathbb{R}$ **10 a)** $g(x) \geqslant -3$ **b)** g^{-1} exists because g is one-to-one **c)** $x \geqslant -\frac{1}{9}$

11 a) $\dfrac{x-3}{2}, x \in \mathbb{R}$ **b)** $\dfrac{1}{2(x+1)}, x \in \mathbb{R}, x \neq -1$ **c)** $\pm\sqrt{2}$ **12 a) i)** $1 - \dfrac{1}{x}, x \in \mathbb{R}, x \neq 0$ **ii)** $x, x \in \mathbb{R}$ **iii)** $\dfrac{1}{1-x}, x \in \mathbb{R}, x \neq 1$

12 b) i) $1 - \dfrac{1}{x}, x \in \mathbb{R}, x \neq 0$ **ii)** $\dfrac{1}{1-x}, x \in \mathbb{R}, x \neq 1$ **iii)** $x, x \in \mathbb{R}$ **13 a)** $3 + \dfrac{5}{x}, x \in \mathbb{R}, x \neq 0$ **b)** $\dfrac{5}{x^2+1}, x \in \mathbb{R}, x > 0$ **c)** $\frac{1}{3}$, 3

14 b) $\dfrac{10x-4}{3x-1}, x \in \mathbb{R}, x \neq \dfrac{1}{3}$

Exercise 3G

6 $x < \frac{1}{2}$ **7** $2 < x < 6$ **8 a)** $(\frac{1}{3}, \frac{1}{3}), (1, 1)$ **b)** $x < \frac{1}{3}$ or $x > 1$ **9 a)** ± 2 or ± 4 **b)** 3, 4 **10 i)** 3 **iii)** $f(x) \geqslant 3$

10 iv) f not one-to-one; $x \geqslant -2$ **11 a)** $g(x) \geqslant 0$ **b)** 0, 8 **c)** 2, 6 **12 a)** $fg : x \to 4x^2 + 4x + 4, x \in \mathbb{R}$ **b)** $fg(x) \geqslant 3$ **c)** 3

13 a) $f(x) \in \mathbb{R}\ f(x) \geqslant -16$ **b)** -2 **c)** $f^{-1} : x \to 4 - \sqrt{x+16}, x \in \mathbb{R}, x \geqslant -16$ **14 a)** $g(x) \in \mathbb{R}\ g(x) \geqslant 1$ **b)** 0, 1 **c)** $-\frac{7}{3}$, 3

14 d) $-\frac{3}{2}$ **15 a)** $f(x) \in \mathbb{R}, 0 < f(x) \leqslant 4$ **b)** $f^{-1}(x) = \dfrac{8-2x}{x}, x \in \mathbb{R}, 0 < x \leqslant 4$ **c)** 2 **16 a)** $-1, \frac{3}{2}$ **b)** $g(x) \in \mathbb{R}, g(x) \geqslant 1$

16 c) $\dfrac{3}{2x^2+1}$; $fg(x) \in \mathbb{R}, 0 < fg(x) \leqslant 3$ **17 i)** x **ii)** $\dfrac{x}{x-1}$ **iii)** $f(x) \in \mathbb{R}, f(x) \neq 1$ **18 b)** $x < a$ or $x > a + \frac{1}{2}$

Exercise 4A

1 a) 3 **b)** 2 **c)** 1 **d)** 12 **e)** 2 **f)** 4 **2 a)** 14 **b)** 25 **c)** -4 **d)** 70 **e)** -1 **f)** 47 **3 a)** $3x^3 + x^2 + 2x - 2$

3 b) $4x^3 - 2x^2 - x - 5$ **c)** $x^3 + x^2 + x - 1$ **d)** $4x^3 + 7x^2 + 17x - 15$ **e)** $9x^4 - 2x^3 - x^2 + 13x - 20$ **f)** $9x^5 + 2x^4 + 23x^2 - 20$

4 a) $x^5 - 3x^4 + 4x^3 - 7x^2 + 7x - 2$ **b)** $x^5 + 8x^4 + 10x^3 - 27x^2 - 6x + 10$ **c)** $x^6 + 5x^5 + 15x^3 - 11x^2 - 10x + 6$

4 d) $6x^5 + 21x^4 - 5x^3 - 11x^2 - 13x + 2$ **e)** $2x^7 + 9x^5 - x^4 + 7x^3 + 4x^2 - 3x$ **f)** $6x^6 + 11x^5 + 19x^4 + 6x^3 + 47x^2 + 13x - 6$

5 a) $4x^3 - 2x^2 + 23x - 12$ **b)** $3x^3 - 5x^2 - 3x$ **c)** $x^5 - 4x^4 - 3x^3 + 6x^2 - 12x + 8$ **d)** $3x^3 - x^2 - 11x + 7$

5 e) $3x^4 - 10x^3 + 12x^2 + 2x + 9$ **f)** $2x^3 + 7x^2 + 8x + 3$ **6 a)** 21 **b)** -5 **c)** -13 **d)** -14 **e)** 26 **f)** -13

Exercise 4B

1 a) $x + 4, -3$ **b)** $x - 5, 8$ **c)** $2x + 1, -6$ **d)** $2x - 5, 8$ **e)** $3x - 2, 4$ **f)** $3x + 1, 8$ **2 a)** $x^2 + 5x + 8, 17$ **b)** $x^2 + 2x - 12, 39$

2 c) $2x^2 + 5x + 16, 65$ **d)** $2x^2 + 5x + 7, 0$ **e)** $x^2 + x + 3, -1$ **f)** $2x^3 - 3x^2 + 3x - 4, 14$ **3 a)** 10 **b)** 24 **c)** 1 **d)** 5

3 e) -4 **f)** 6 **4** 4 **5** 8 **6** -2 **7** 1 **8** -20 **9** $a = 5$ rem 13 **10** $b = -52$ rem -56 **11** $-13, 5$ **12** $1, -7$ **13** 1, 6

14 $3, -5$ **15** $-8, 12, (x-2)^2(x+3)$ **16** $3, -4; (x-2)(x+2)(x+3)$ **17** $3, -1$

Exercise 4C

1 a) $(x-3)(x-1)(x+3)$ **b)** $(x+1)(x+2)(x+3)$ **c)** $(x-2)(x^2+2)$ **d)** $(x-4)(x-1)(x+1)$ **e)** $(x-3)(x-1)(x+2)$

1 f) $x(x-7)(x+3)$ **g)** $(x-2)^2(x-1)$ **h)** $(x-3)^3$ **i)** $(x-2)(x+3)(x+5)$ **j)** $(x^2+4)(x+1)$ **k)** $(x+1)^2(x-3)$

1 l) $(x-1)(x+1)(x+7)$ **2 a)** $(x-1)(2x-1)(x+1)$ **b)** $(x-2)(2x+1)(x+1)$ **c)** $(x-1)(3x-1)(x+2)$ **d)** $(2x-1)(x+1)(x+3)$

2 e) $(x^2+1)(3x-1)$ **f)** $(5x-1)(x+1)(x+2)$ **g)** $(x-1)^2(x+1)(x+2)$ **h)** $(x-2)(x-1)(x+2)(x+3)$ **i)** $(x+1)^2(x-1)(x+2)$

2 j) $(x-1)^3(x-4)$ **k)** $(4x-1)(x-1)(x+1)(x+3)$ **l)** $(x^2+4)(x-2)(x+2)$ **3 a)** $-4, 1, 2$ **b)** 2 **3 c)** $-2, -1, 3$ **d)** $-1, 2$

3 e) $-5, -3, 1$ **f)** $-\frac{3}{2}, 1, 2$ **g)** $-\frac{1}{3}, 3, 4$ **h)** $-\frac{5}{2}, -1, 3$ **i)** $-\frac{1}{5}, 1, 6$ **j)** $-2, -6.16, 0.162$ **k)** 3, 1.70, 5.30 **l)** 1, 1.19, -4.19

4 $(x-3)(x-1)(x+4), -4, 1, 3$ **5** $(x+1)^2(x-5), -1, 5$ **6** 5 **7** $a = 5, b = 6, c = -1, -5, -3, 2$ **8** $c = -2, d = 5, e = 10$ (or $c = 5, d = -2, e = 10$), $-5, 1, 2$ **9** $a = -1, b = 14, c = -8$; 1, 2, 4 **10** $a = 5, b = -4, c = 5$; 4 **11** $b = 5, c = -3$, 3

12 a) $-3 \leqslant x \leqslant -1$ or $x \geqslant 2$ **b)** $-1 < x < 2$ or $x > 4$ **c)** $1 < x < 2$ or $x > 3$ **d)** $x \leqslant -1$ or $2 \leqslant x \leqslant 5$ **e)** $x < -5$

12 f) $-3 < x < \frac{1}{2}$ or $x > 2$ **g)** $\frac{1}{3} \leqslant x \leqslant 2$ or $x \geqslant 3$ **h)** $x < -\frac{3}{2}$ or $1 < x < 4$ **i)** $x > 4$ **j)** $x \leqslant -6$ **k)** $x < 2$

12 l) $x < -\frac{2}{3}$ or $\frac{1}{2} < x < 3$ **13** $(2x+1)(x-2)(x-5), -\frac{1}{2} \leqslant x \leqslant 2$ or $x \geqslant 5$ **14** $(x+2)^2(x-3), x = -2$ or $x \geqslant 3$

15 $a = 3, b = -2, c = 60, x < -5$ or $3 < x < 4$ **16** $c = 3, d = 1, e = -27, -3 < x < -\frac{3}{2}$ or $x > 4$

17 $a = 4, b = 2, c = -3, x \leqslant -\frac{1}{3}$ or $\frac{1}{4} \leqslant x \leqslant \frac{1}{2}$

Exercise 4D

1 $(2x+1)(x-2)(x-3)$ **2 a)** 2 **b)** $x+6,\ x-2,\ x-2$ **3 ii)** $2x+1,\ 3x-2,\ x-2$ **4** $a=-5,\ b=19$
5 b) $(2x+1)(x+2)(x-2)$ **c)** $-2,\ -\frac{1}{2},\ 2$ **6 b)** $-4,\ -1,\ 3$ **7** $A=-13;\ -3,\ -1,\ 4$ **8 i)** $0,\ -21;\ (x-1)$
8 ii) $p=-1,\ q=-2,\ r=-1$ **iii)** $1,\ 1\pm\sqrt{2}$ **9 i)** $30,\ 0;\ x-3$ **ii)** $p=2,\ q=-15$ **iii)** $-5,\ 2,\ 3$
10 b) $(2x-1)(x+2)(x-3)$ **c)** $x\leqslant-2$ or $\frac{1}{2}\leqslant x\leqslant 3$ **11** $-1<x<2$ or $x>5$ **12** $x<2$ or $3<x<4$

Exercise 5A

1 a) 5 **b)** 13 **c)** 10 **d)** 17 **e)** 20 **f)** $4\sqrt{2}$ **g)** $\sqrt{29}$ **h)** 10 **i)** $5\sqrt{5}$ **j)** $3\sqrt{5}$ **k)** $2\sqrt{10}$ **l)** $3\sqrt{2}$ **3** 4 **9** 15 **11** 15
12 -3 or 7 **13** -5 or 9 **14** 3 **15** 3 or 7 **16** 1 or -8

Exercise 5B

1 a) $(5,3)$ **b)** $(5,6)$ **c)** $(1,2)$ **d)** $(4,2)$ **e)** $\left(2\frac{1}{2},-4\right)$ **f)** $(-5,-5)$ **g)** $(5,0)$ **h)** $\left(5\frac{1}{2},\frac{1}{2}\right)$ **i)** $(0,0)$ **j)** $(4,-5)$ **k)** $\left(3\frac{1}{2},2\right)$
1 l) $\left(3\frac{1}{2},\frac{1}{2}\right)$ **2** $(10,7)$ **3** $(9,-1)$ **4** $(-3,-2)$ **5** $\left(-\frac{1}{2},-2\frac{1}{2}\right)$ **8 a)** P(4, 3), Q(8, 4), R(9, 3), S(5, 2) **9 a)** $\left(\frac{1}{2}(a+c),b\right)$
9 b) $\left(\frac{1}{2}(a+c),\ b+2\right)$ **c)** $A=c-a$

Exercise 5C

1 a) 2 **b)** 3 **c)** $\frac{1}{2}$ **d)** $\frac{3}{4}$ **e)** $-\frac{1}{4}$ **f)** $\frac{4}{7}$ **g)** $\frac{1}{5}$ **h)** 0 **i)** 3 **j)** $-\frac{8}{7}$ **k)** $-\frac{11}{4}$ **l)** Infinity **2 a)** 76.0° **b)** 55.0° **c)** 63.4°
2 d) (−)45° **e)** 36.9° **f)** 0° **g)** (−)26.6° **h)** (−)35.0° **i)** 63.4° **j)** 54.5° **k)** (−)24.0° **l)** 90° **5** $12,\ -\frac{5}{2},\ \frac{2}{5}$ **6** 1 **7** 3, 11
8 -2, 5 **9** 20 **13 a)** Square **14 b)** 5, 10, 11, 2

Exercise 5D

1 a) 5 **b)** 3 **c)** -4 **d)** 7 **e)** 5 **f)** $\frac{1}{4}$ **g)** $\frac{5}{2}$ **h)** $-\frac{2}{7}$ **i)** $-\frac{5}{2}$ **j)** $-\frac{5}{4}$ **k)** $-\frac{3}{2}$ **l)** $-\frac{3}{5}$ **2 a)** $y=2x-7$ **b)** $y=5x-6$
2 c) $y=-2x+9$ **d)** $y=\frac{1}{2}x-5$ **e)** $4y-x=18$ **f)** $3y+2x+2=0$ **3 a)** $y=3x-9$ **b)** $y=-2x+7$ **c)** $y=\frac{1}{3}x-4$
3 d) $y=\frac{4}{3}x+11$ **e)** $2y+5x=5$ **f)** $3x-10y=3$ **4 a)** $y=2x-1$ **b)** $y=\frac{1}{2}x-1$ **c)** $y=-9x+15$ **d)** $y=4x+4$
4 e) $y=-12x+43$ **f)** $y=9$ **5 a)** $y=-2x+12$ **b)** $y=-x+6$ **c)** $2y+x+3=0$ **d)** $4y+2x=11$ **e)** $2y-8x+1=0$
5 f) $y=-7x-2$ **6** $y=x+3$ **7** $y=-x+2$ **8** $y=-2x+9,\ 20\frac{1}{4}$ **9** $16y-6x+13=0,\ \ 3y+8x-60=0$
11 a) $y=-3x+25,\ \ y=-3x+5$ **b)** $104\frac{1}{6}$ **c)** $4\frac{1}{6}$ **12 a)** $y=-2x+5$ **b)** 1.25 **c)** 5

Exercise 5E

1 a) $(1,5)$ **b)** $(2,8)$ **c)** $(-1,2)$ **d)** $(1,4)$ **e)** $(-1,5)$ **f)** $\left(\frac{1}{2},\frac{1}{2}\right)$ **g)** $(-1,2)$ **h)** $(3,2)$ **i)** $(2,-1)$ **j)** $\left(2,\frac{1}{2}\right)$ **k)** $\left(3\frac{1}{2},-\frac{1}{2}\right)$
1 l) $\left(4\frac{6}{7},-3\frac{1}{7}\right)$ **2 a)** 26.6° **b)** 15.3° **c)** 18.4° **d)** 54.5° **e)** 53.1° **f)** 8.1° **g)** 27.9° **h)** 90° **i)** 2.7° **j)** 63.4° **k)** 63.0°
2 l) 33.7° **3 a)** $(3,1)$ **5** $y=3x+7$ **6 a)** $y=5x-6$ **b)** $(3,9)$ **7 a)** $(2,4)$ **b)** $(14,0)$ **c)** 28 **8 a)** (9, 0) **b)** (0, 12)
8 c) (6, 4), (3, 8) **9** $\left(\frac{6}{5},\frac{9}{5}\right),\ \left(-\frac{1}{2},-\frac{3}{4}\right),\ \left(\frac{7}{3},\frac{2}{3}\right)$ **10** 5 **12 a** $y=-2x+4$ **b)** $\left(-\frac{1}{3},4\frac{2}{3}\right)$ **13** 4.8°
14 $2y+3x=12,\ \ y=-4x+9,\ \ y=x+3$ **15 a** (4, 2) **b)** 5 **16 a)** $y-2x-p+7=0$ **b)** $y+3x-3-p=0$ **c)** $p=2,\ q=-1$

Exercise 5F

1 b) $-0.79,\ 3.79$ **2 a)** $-\frac{3}{4},\ -\frac{2}{11},\ \frac{4}{3}$ **c)** $(3,8)$ **3 a)** $-\frac{3}{5},-\frac{3}{5}$ **c)** $(1,2)$ **4** $y=2x$ **5 a)** $-\frac{2}{3}$ **b)** $3x-2y=2$ **6** $5x-6y=2$
7 $3x+5y-4=0;\ \left(\frac{4}{3},0\right)$ **8 a) i)** $-\frac{2}{3}$ **ii)** $3x-2y=29$ **b)** $(5,-7)$ **9 a)** $x+2y=8$ **b)** $x+2y=3;\ (1,1)$
10 a) $x-7y+12=0;\ x+y=12$ **b)** $(9,3)$ **11 a)** $y=\frac{1}{2}x+3$ **b)** $3\sqrt{5}$ **12 a)** $3x-2y=1$ **b)** $y=\frac{1}{2}x$ **c)** $\left(\frac{1}{2},\frac{1}{4}\right)$ **d)** 1:4
13 a) $3x+y=14$ **b)** $(0,14)$ **d)** $(-2,10)$ **e)** $(-6,12)$ **14 a)** $4x-3y=23$ **b)** 10 **c)** 75 **15 i)** $-\frac{1}{2},\ \frac{3}{4},\ -\frac{1}{2},\ \frac{3}{4}$; parallelogram
15 ii) 10 **iii)** $-\frac{4}{3};\ 4x+3y=20$ **iv)** $\left(4\frac{2}{5},\frac{4}{5}\right)$ **v)** 40

Exercise 6A

2 a) $4x^3$ **b)** $6x^5$ **c)** $12x$ **d)** $-15x^2$ **e)** 3 **f)** $12x^5$ **g)** $-14x$ **h)** 0 **i)** $2x^3$ **j)** $4x^5$ **k)** $-\frac{3}{4}x^2$ **l)** $\frac{2}{5}$ **3 a)** $-2x^{-3}$ **b)** $-4x^{-5}$
3 c) $-6x^{-4}$ **d)** $-4x^{-2}$ **e)** $-\dfrac{3}{x^4}$ **f)** $\dfrac{2}{x^3}$ **g)** $-\dfrac{9}{x^4}$ **h)** $\dfrac{2}{x^2}$ **i)** $-\dfrac{3}{x^3}$ **j)** $-\dfrac{27}{2x^4}$ **k)** $\dfrac{3}{x^5}$ **l)** $-\dfrac{2}{5x^2}$ **4 a)** $\frac{1}{2}x^{-\frac{1}{2}}$ **b)** $2x^{-\frac{2}{3}}$

4 c) $-\frac{2}{3}x^{-\frac{5}{3}}$ **d)** $2x^{-\frac{6}{5}}$ **e)** $\frac{7}{2\sqrt{x}}$ **f)** $\frac{1}{3\sqrt[3]{x^2}}$ **g)** $-\frac{2}{5\sqrt{x^3}}$ **h)** $\frac{2}{\sqrt[3]{x^4}}$ **i)** $-\frac{5}{4\sqrt{x^3}}$ **j)** $\frac{3}{\sqrt[5]{x^6}}$ **k)** $\frac{5\sqrt{x^3}}{2}$ **l)** $50x$ **5 a)** $2x+2$

5 b) $6x-5$ **c)** $2x$ **d)** $-12x^2$ **e)** $2x+2$ **f)** $7x^6+12x^3$ **g)** $4x^3-6x$ **h)** $1-\frac{1}{x^2}$ **i)** $10x+\frac{6}{x^4}$ **j)** $5x^4+\frac{9}{x^4}$ **k)** $2x-8x^3$

5 l) $-\frac{3}{x^2}+12x^2$ **6 a)** $3-15x^2$ **b)** $\frac{3}{x^2}$ **c)** $-\frac{4}{x^2}-6x^2$ **d)** $\frac{1}{\sqrt{x}}$ **e)** $\frac{1}{2\sqrt{x}}-\frac{1}{2\sqrt{x^3}}$ **f)** $-8x^{-3}-3$ **g)** $x^{-\frac{2}{3}}+\frac{4}{3}x^{-\frac{4}{3}}$

6 h) $2x^{-\frac{1}{2}}+2$ **i)** $4x^{-\frac{1}{3}}-10x^{\frac{3}{2}}$ **j)** $-\frac{3}{\sqrt[3]{x^4}}+\frac{2}{\sqrt[4]{x^5}}$ **k)** $-\frac{3}{\sqrt{x^3}}-\frac{2}{\sqrt{x}}$ **l)** $\frac{1}{2\sqrt{x}}-\frac{1}{2\sqrt{x^3}}$ **7 a)** 4 **b)** $\frac{1}{\sqrt{x}}-\frac{5}{2x^2}$ **c)** $8x+7$

7 d) $\frac{5}{6\sqrt[6]{x}}$ **e)** $2x^{-\frac{3}{5}}-5x^{\frac{3}{2}}$ **f)** $\frac{3}{\sqrt{x}}+\frac{3}{x^3}$ **g)** $-14x^{-8}+15x^{-4}+1$ **h)** $-\frac{1}{2\sqrt[6]{x^7}}+\frac{1}{2\sqrt[4]{x^5}}$ **i)** $-\frac{10}{x^3}+\frac{2}{x^2}$ **j)** $12x^{\frac{1}{3}}$ **k)** $-8x^{-5}+8x^{-3}$

7 l) $-\frac{3}{2\sqrt{x^3}}-\frac{15}{2\sqrt{x^5}}$ **8 a)** $3x(x+2)$ **b)** $2(1-x)$ **c)** $12x^2-5x^4$ **d)** $\frac{5+3x}{2\sqrt{x}}$ $\left[=\frac{5}{2}x^{-\frac{1}{2}}+\frac{3}{2}x^{\frac{1}{2}}\right]$

8 e) $\frac{3(7x^3-6x+1)}{\sqrt{x}}$ $\left[=21x^{\frac{5}{2}}-18x^{\frac{1}{2}}+3x^{-\frac{1}{2}}\right]$ **f)** $\frac{1}{3}x^{-\frac{2}{3}}(8x-5)$ $\left[=\frac{8}{3}x^{\frac{1}{3}}-\frac{5}{3}x^{-\frac{2}{3}}\right]$ **g)** $\frac{1}{2}x^{-\frac{3}{4}}(9x^2-2)$ $\left[=\frac{9}{2}x^{\frac{5}{4}}-x^{-\frac{3}{4}}\right]$

8 h) $2x-1$ **i)** $2(x+4)$ **j)** $4x+9$ **k)** $4(x-3)$ **l)** $2(x+3)$ **9 a)** $6(x-2)$ **b)** $3x^2(5x^2-1)$ **c)** $\frac{5x^2+3}{2\sqrt{x}}$ $\left[=\frac{5}{2}x^{\frac{3}{2}}+\frac{3}{2}x^{-\frac{1}{2}}\right]$

9 d) $\frac{1}{2}x^{-\frac{3}{2}}(x-1)$ $\left[=\frac{1}{2}x^{\frac{1}{2}}-\frac{1}{2}x^{-\frac{3}{2}}\right]$ **e)** $\frac{x^2-7}{x^2}$ $[=1-7x^{-2}]$ **f)** $-\frac{x+10}{x^3}$ $[=-x^{-2}-10x^{-3}]$ **g)** $\frac{3x^2-2}{x^2}$ $[=3-2x^{-2}]$

9 h) $\frac{2(3x^3+7)}{x^3}$ $[=6+14x^{-3}]$ **i)** $-\frac{3}{5x^2}$ **j)** $\frac{\sqrt{x}(18-35x)}{6}$ $\left[3x^{\frac{1}{2}}-\frac{35}{6}x^{\frac{3}{2}}\right]$ **k)** $4x(x-1)(x-2)$ $[=8x-12x^2+4x^3]$

9 l) $\frac{(x-5)(x+5)}{x^2}$ $[=1-25x^{-2}]$ **10 a)** $3x^2(4x-1)$ **b)** $4x(2x-1)(x-1)$ $[=8x^3-12x^2+4x]$ **c)** $-\frac{1}{2\sqrt{x^3}}$

10 d) $\frac{3x^3-10}{x^3}$ $[=3-10x^{-3}]$ **e)** $\frac{x^2+12}{x^2}$ $[=1+12x^{-2}]$ **f)** $-\frac{2(x^2-13x+30)}{x^4}$ $[=-2x^{-2}+26x^{-3}-60x^{-4}]$

10 g) $\frac{9x^2-1}{2x^2}$ $\left[=\frac{9}{2}-\frac{1}{2}x^{-\frac{1}{2}}\right]$ **h)** $\frac{5x-3}{4\sqrt{x^3}}$ $\left[\frac{5}{4}x^{-\frac{1}{2}}-\frac{3}{4}x^{-\frac{3}{2}}\right]$ **i)** $\frac{(3x-1)(x+1)}{2\sqrt{x^3}}$ $\left[=\frac{3}{2}x^{\frac{1}{2}}+x^{-\frac{1}{2}}\right]$ **j)** $\frac{3x^2-5}{6\sqrt{x^3}}$ $\left[=\frac{1}{2}x^{\frac{1}{2}}-\frac{5}{6}x^{-\frac{3}{2}}\right]$

10 k) $\frac{7}{4\sqrt{x^3}}$ **l)** $\frac{2x^2+5}{9x^2}$ $\left[=\frac{2}{9}+\frac{5}{9}x^{-2}\right]$ **11 a)** $18x$ **b)** $2-120x^4$ **c)** $\frac{2}{x^3}$ **d)** $\frac{12}{x^4}$ **e)** $\frac{3}{4\sqrt{x^5}}+\frac{1}{4\sqrt{x^3}}$ **f)** $-\frac{2}{9\sqrt[3]{x^5}}-6x$

11 g) $10x(2x^2+3)$ **h)** $6(2x+1)$ **i)** $-\frac{2}{x^3}$ **j)** $\frac{15}{4\sqrt{x^5}}$ **k)** $\frac{6(2x-5)}{x^4}$ $[=12x^{-3}-30x^{-4}]$ **l)** $-\frac{12+5x}{12\sqrt{x^5}}$ $\left[=-x^{-\frac{5}{2}}-\frac{5}{12}x^{-\frac{3}{2}}\right]$

12 a) $2ax+b,\ 2a$ **b)** $-\frac{a}{x^2}-\frac{2b}{x^3},\ \frac{2a}{x^3}+\frac{6b}{x^4}$ **c)** $\frac{a}{2\sqrt{x}}-\frac{b}{2\sqrt{x^3}},\ \frac{3b}{4\sqrt{x^5}}-\frac{a}{4\sqrt{x^3}}$ **d)** $2acx+ad+bc,\ 2ac$ **e)** $\frac{2ax+b}{c},\ \frac{2a}{c}$

17 $a=2, b=\pm1$ (or trivially, $a=b=0$) **18 a)** $6x^2-6x-32$ **b)** $-3,\ -\frac{1}{2},\ 5$

Exercise 6B

1 a) 6 **b)** 24 **c)** $\frac{1}{6}$ **d)** $-\frac{1}{9}$ **e)** 4 **f)** $\frac{1}{8}$ **g)** $\frac{2}{3}$ **h)** -1 **i)** -5 **j)** 3 **k)** $\frac{9}{4}$ **l)** 0 **2 a)** $(-2,-8), (2,8)$ **b)** $(-1,3)$ **c)** $(2,17)$

2 d) $\left(-\frac{1}{2},-8\right), \left(\frac{1}{2},8\right)$ **e)** $(-2,4)$ **f)** $(-3,-34), (3,32)$ **g)** $(0,3), \left(\frac{2}{3},\frac{77}{27}\right)$ **h)** $\left(\frac{1}{3},\frac{71}{27}\right), (1,3)$ **i)** $\left(\frac{1}{4},\frac{11}{2}\right)$ **j)** $\left(\frac{1}{9},\frac{11}{9}\right)$

2 k) $(-2,-3), (2,1)$ **l)** $(-1,2)$ **3** $a=\frac{3}{4}, b=-5$ **4** $A=\frac{17}{18}, B=-\frac{13}{3}$ **5** $a=-45, c=24$ **6** $a=12, b=-18$ **7** $A=2, B=4$

8 a) $y=4x-1$ **b)** $y=6x-5$ **c)** $y=-x-6$ **d)** $y=-2x+16$ **e)** $y=-2x-1$ **f)** $2y-3x=12$ **g)** $y=x+1$ **h)** $y=5$

8 i) $y=-x-3$ **j)** $y=9x-16$ **k)** $25y-2x=100$ **l)** $3y+16x=8$ **9 a)** $y=-x$ **b)** $3y+x=8$ **c)** $2y-3x+5=0$

9 d) $y=-3x+33$ **e)** $2y+x=11$ **f)** $4y+x+14=0$ **g)** $54y+x=3$ **h)** $7y-x+27=0$ **i)** $3y+2x+1=0$ **j)** $8y-3x=1$

9 k) $x=-2$ **l)** $4y-4x=13$ **10** $y=-3x+1$ **11** $12y+x=2$ **12 a)** $y=-3x+3, y=3x-12$ **b)** $\left(\frac{5}{2},-\frac{9}{2}\right)$ **13** $(0,-9)$

14 $\left(-\frac{1}{3},-\frac{16}{3}\right)$ **15 a)** $1, 3$ **b)** $y=2, y=6$ **16 a)** $-2, 4$ **b)** $y=15x+38, y=15x-70$ **17** $9y-x+16=0, 9y-x=464$

18 $6y-x+19=0$ **20 a)** $y=3x+1$ **b)** $\left(\frac{1}{5},\frac{8}{5}\right)$ **21** $\left(-\frac{3}{2},\frac{7}{2}\right)$ **22 a)** $y=6x-4$ **b)** $(-2,-16)$ **23** $(-4,-71)$

24 a) $y=x+1$ **b)** $(1,2), (3,4)$ **25** $(-2,1), (2,5)$ **26** $a=3, b=4$ **27** $c=-14, d=-20$ **28** $a=8, b=11$

29 $A=2, B=-3, C=-4, D=7$ **30 c)** $y=x\pm\frac{2}{3\sqrt{3}}$

Exercise 6C

1 a) $(2,-1)$ **b)** $(-3,-4)$ **c)** $(0,6)$ **d)** $\left(\frac{5}{2},-\frac{13}{4}\right)$ **e)** $\left(\frac{3}{4},-\frac{1}{8}\right)$ **f)** $(-1,4), (1,0)$ **g)** $(-2,40), (6,-216)$ **h)** $(1,11), (-3,-21)$

1 i) $(0,5), (2,9)$ **j)** $(-1,2), (0,3), (1,2)$ **k)** $(2,-45)$ **l)** $(-1,8), \left(\frac{1}{2},-\frac{7}{16}\right), (1,0)$ **2 a)** $(1,4)$, min **b)** $(-2,-2)$, min

2 c) $\left(\frac{1}{2},\frac{13}{4}\right)$, max **d)** $(3,-1)$, min **e)** $\left(-\frac{5}{4},-\frac{147}{8}\right)$, min **f)** $(5,0)$, min **g)** $(2,-40)$, min; $(-6,216)$, max

2 h) $\left(\frac{1}{3},\frac{40}{27}\right)$, max; $(3,-8)$, min **i)** $(-5,-97)$, min; $(1,11)$, max **j)** $(-2,-13)$, min; $(0,3)$, max; $(2,-13)$, min

2 k) $(-3, -26)$, min; $(0, 1)$, saddle **l)** $(-3, -127)$, min; $(1, 1)$, max; $(2, -2)$, min **3 a)** $(-1, 5)$, max; $(1, 1)$, min

3 b) $(1, -4)$, min; $(3, 0)$, max **c)** $(3, 8)$, saddle **d)** $(-2, -25)$, min; $(0, -9)$, max; $(2, -25)$, min **e)** $(0, 0)$, saddle; $(6, 432)$, max

3 f) $(-1, 0)$, saddle; $(\frac{5}{4}, -\frac{2187}{256})$, min **4 a)** $(-1, -2)$ max, $(1, 2)$ min **b)** $(2, 12)$ min **c)** $(6, \frac{1}{12})$ max **d)** $(3, -\frac{1}{27})$ min

4 e) $(-\frac{1}{2}, -16)$ min, $(\frac{1}{2}, 16)$ max **f)** $(2, -\frac{3}{8})$ min **8** 20, £10 000 **9** 4 000, £20 800 **10** 55 mph **11** 600 miles

12 1.5 s, 11.25 m **13** 320 m **14 a)** $(10 - x)$ cm **b)** $(x^2 + (10 - x)^2)$ cm^2 **c)** 5 **15 a)** $(9 - 4x)$ cm **b)** $(2x^2 - 12x + 27)$ cm^2

15 c) 3 **16 a)** $(25 - \frac{2}{3}x)$ cm **b)** $(50 - \frac{4}{3}x)$ cm **c)** $(1250 - \frac{200}{3}x + \frac{8}{9}x^2)$ cm^2 **d)** $\frac{300}{17}$ **17 a)** $(500 - x)$ m **b)** 250, 62 500 m^2

18 a) $(40 - 2x)$ m **b)** 10, 200 m^2 **19 a)** $\dfrac{8}{x^2}$ m **b)** $\left(2x^2 + \dfrac{32}{x}\right)$ m^2 **c)** 2 **20 a)** $\dfrac{108}{x^2}$ m **b)** $\left(x^2 + \dfrac{432}{x}\right)$ m^2 **c)** 6 **21 a)** $2x$ cm

21 b) $\dfrac{288}{x^2}$ cm **c)** $\left(4x^2 + \dfrac{1728}{x}\right)$ cm^2 **d)** 6 **22** 6 m, 18 m^3 **23** 12 cm, 172.8 cm^3 **24 a)** $\dfrac{36}{x^2}$ cm **b)** $(40x^2)$ pence

24 c) $\left(\dfrac{1080}{x}\right)$ pence **d)** $\left(\dfrac{4320}{x} + 80x^2\right)$ pence **e)** 3 **f)** £21.60 **25 a)** $\left(\dfrac{5}{6x^2}\right)$ m **b)** £$\left(\dfrac{80}{3x}\right)$ **c)** £$(45x^2)$ **d)** $\frac{2}{3}$ **e)** £60

26 10 cm × 10 cm × 7.5 cm **27** $\frac{5}{3}$, $74\frac{2}{27}$ cm^3 **28** 1

Exercise 6D

1 $y = 7x - 15$ **2** $3x^2 - 8x + 5$ **i)** 4 **ii)** 8 **iii)** $y = 8x - 20$ **iv)** $x + 8y = 35$; 0, $2\frac{2}{3}$ **3 b)** $-1, 1, 2$; $(-1, -3)$, $(2, 0)$

4 iii) $(2, 0)$, $(0, 8)$ **iv)** $(5, 5)$, $(-3, 3)$ **5** $(2, 12)$ **6 a)** $15x^2 - 4x$ **b)** 0, $\frac{4}{15}$ **7 a)** $(1, 0)$, $(2, -1)$ **b)** $1, 2\frac{1}{2}$ **8 i)** $12x^3 + 12x^2$

8 ii) $(-1, -1)$, $(0, 0)$ **iii)** $(-1, -1)$, min; $(0, 0)$, inflexion **9** $(-2, -4)$, max; $(2, 4)$, min; $x < -2$ or $x > 2$ **10 i)** $3x^2 - 12$

10 ii) $(-2, 21)$, $(2, -11)$ **iii)** Investigate sign of $\dfrac{d^2y}{dx^2}$ **v)** $12x + y = 5$ **11 i)** 39; 164 **ii)** $6x^2 - 6x - 36$ **iii)** $(-2, 164)$, $(3, 39)$

11 iv) $(-2, 164)$ max, $(3, 39)$ min **v)** $39 < k < 164$ **12** $\left(-\dfrac{m+2}{m}, 0\right)$, $(0, m+2)$ **13 b)** $40\sqrt{\pi}$ **14 c)** 2 **15 b)** 25 **c)** 2 rad

15 d) 625 cm^2 **16 b)** 10 **c)** 5 cm **d)** $A''(10) = 6 > 0$ **17 b)** 36 **c)** $V''(3) = -12 < 0$ **18 b)** $100x - \dfrac{4x^3}{3}$ **c)** $333\frac{1}{3}$

18 d) $V''(5) = -40 < 0$ **19 b)** 10 **c)** $V''(10) = -30\pi < 0$ **d)** 1000π **20** $\sqrt[3]{\dfrac{1}{\pi}}$ **21 b)** $\dfrac{32\pi}{9}$

Exercise 7A

The constant of integration is omitted in the answers to Questions **1** to **8**.

1 a) $\frac{1}{4}x^4$ **b)** $\frac{1}{5}x^5$ **c)** x^3 **d)** $2x^6$ **e)** $-2x^2$ **f)** $3x^5$ **g)** $\frac{1}{2}x^4$ **h)** $3x$ **i)** $\frac{1}{12}x^6$ **j)** $\frac{1}{6}x^4$ **k)** $-\frac{1}{9}x^3$ **l)** $\frac{2}{3}x$ **2 a)** $-x^{-1}$ **b)** $-\frac{1}{3}x^{-3}$

2 c) $-x^{-2}$ **d)** $2x^{-3}$ **e)** $-\dfrac{1}{2x^4}$ **f)** $\dfrac{1}{4x^4}$ **g)** $-\dfrac{3}{x}$ **h)** $\dfrac{1}{x^2}$ **i)** $-\dfrac{2}{3x^6}$ **j)** $-\dfrac{1}{2x^3}$ **k)** $\dfrac{5}{3x}$ **l)** $-\dfrac{2}{9x^3}$ **3 a)** $\frac{3}{4}x^{\frac{4}{3}}$ **b)** $2x^{\frac{3}{2}}$ **c)** $3x^{\frac{1}{3}}$

3 d) $-5x^{\frac{4}{5}}$ **e)** $-2\sqrt{x^3}$ **f)** $\frac{4}{5}\sqrt[4]{x^5}$ **g)** $6\sqrt[3]{x^2}$ **h)** $-\frac{5}{2}\sqrt[5]{x^4}$ **i)** $\frac{6}{7}\sqrt{x}$ **j)** $\frac{9}{5}\sqrt[3]{x^2}$ **k)** $\frac{2}{5}\sqrt{x^5}$ **l)** $2\sqrt{x^3}$ **4 a)** $\frac{1}{4}x^4 + x^2$ **b)** $x^3 - 2x^2$

4 c) $\frac{1}{4}x^4 - x$ **d)** $6x + \frac{1}{2}x^6$ **e)** $\frac{1}{3}x^3 - \frac{5}{2}x^2 + 3x$ **f)** $\frac{1}{9}x^9 + \frac{1}{3}x^6$ **g)** $\frac{1}{5}x^5 - \frac{3}{2}x^2 + 2x$ **h)** $\dfrac{1}{3}x^3 + \dfrac{1}{x}$ **i)** $x^5 + \dfrac{1}{x^2}$ **j)** $\dfrac{2}{7}x^7 - \dfrac{2}{x^4}$

4 k) $\dfrac{1}{3}x^3 + \dfrac{3}{x}$ **l)** $-\dfrac{5}{x} - 2x - \dfrac{1}{2}x^4$ **5 a)** $2\sqrt{x^3} - 4x$ **b)** $\frac{2}{3}\sqrt{x^3} + 2\sqrt{x}$ **c)** $\frac{9}{4}x^{\frac{4}{3}} - \frac{8}{5}x^{\frac{5}{4}}$ **d)** $10x^{\frac{1}{2}} + 3x^{\frac{2}{3}}$ **e)** $\dfrac{8}{3}\sqrt{x^3} + \dfrac{2}{3x}$

5 f) $\frac{3}{2}\sqrt[3]{x^4} - 12\sqrt{x}$ **g)** $6\sqrt[3]{x^2} - \frac{16}{3}\sqrt{x^3}$ **h)** $\frac{4}{5}\sqrt[4]{x^5} - \frac{4}{3}\sqrt[4]{x^3}$ **i)** $2\sqrt[3]{x^2} - \frac{7}{4}\sqrt[7]{x^8}$ **j)** $\frac{12}{5}\sqrt[6]{x^5} - 10\sqrt[5]{x^4}$ **k)** $\frac{3}{2}x^{\frac{4}{3}} - \frac{9}{7}x^{\frac{7}{3}} - \frac{3}{2}x^{\frac{10}{3}}$

5 l) $-\dfrac{10}{\sqrt{x}} - 6\sqrt[3]{x}$ **6 a)** $\frac{3}{2}x^2 - \frac{1}{3}x^3$ **b)** $\frac{1}{4}x^4 + \frac{5}{3}x^3$ **c)** $\frac{1}{2}x^4 - \frac{1}{6}x^6$ **d)** $\frac{2}{5}\sqrt{x^5} + 2\sqrt{x^3}$ **e)** $\frac{6}{7}\sqrt{x^7} - \frac{6}{5}\sqrt{x^5} + 2\sqrt{x^3}$ **f)** $\frac{6}{7}x^{\frac{7}{3}} + \frac{9}{4}x^{\frac{4}{3}}$

6 g) $\frac{4}{3}x^{\frac{9}{4}} - \frac{24}{5}x^{\frac{5}{4}}$ **h)** $\frac{1}{3}x^3 + 4x^2 + 15x$ **i)** $\frac{1}{3}x^3 - 2x^2 + 4x$ **j)** $\frac{2}{3}x^3 + 10x^2 + 50x$ **k)** $\frac{1}{4}x^4 - \frac{2}{3}x^3 + \frac{1}{2}x^2$ **l)** $\frac{1}{2}x^2 + \frac{4}{3}\sqrt{x^3} - 15x$

7 a) $\frac{5}{3}x^3 - 5x^2$ **b)** $x^6 - \frac{1}{4}x^4$ **c)** $\frac{2}{7}\sqrt{x^7} + \frac{2}{3}\sqrt{x^3}$ **d)** $\frac{6}{5}x^{\frac{5}{3}} + \frac{9}{2}x^{\frac{2}{3}}$ **e)** $x - \dfrac{5}{x}$ **f)** $-\dfrac{1}{x} + \dfrac{2}{x^2}$ **g)** $3x - \dfrac{5}{x}$ **h)** $\frac{2}{3}x^3 - \frac{3}{4}x^2$

7 i) $2\sqrt{x^5} - 8\sqrt{x}$ **j)** $2\sqrt{x^3} - 3\sqrt{x}$ **k)** $\frac{25}{4}x^4 - \frac{20}{9}\sqrt{x^9} + \frac{1}{5}x^5$ **l)** $\frac{2}{5}\sqrt{x^5} - \frac{4}{3}\sqrt{x^3} + 2\sqrt{x}$ **8 a)** $\dfrac{a}{2}x^2 + bx$ **b)** $\frac{2}{3}a\sqrt{x^3} + 2b\sqrt{x}$

8 c) $\frac{1}{3}acx^3 + \frac{1}{2}(ad + bc)x^2 + bdx$ **10 a)** $y = x^3 + x + 2$ **b)** $y = 2x^2 - 3x + 1$ **c)** $y = 2x^3 - 2x^2 - 12$ **d)** $y = 16 + 4x - 3x^2$

10 e) $y = 8 - \dfrac{2}{x} - x$ **f)** $y = \dfrac{5}{x^2} - 7$ **g)** $y = \frac{2}{3}\sqrt{x^3} - 5x + 9$ **h)** $y = \frac{1}{2}x^2 - \frac{1}{3}\sqrt{x^3} - \frac{14}{3}$ **i)** $y = x^3 - \frac{1}{4}x^4 + 4$ **j)** $y = \frac{4}{3}x^3 - 6x^2 + 9x + 4$

10 k) $y = x - 2\sqrt{x} + 6$ **l)** $y = \dfrac{1}{x^2} - \dfrac{1}{x} + \dfrac{11}{4}$ **11** $y = x^3 + 4x + 2$ **12** $y = x^4 - 3x^2 + 4$ **13** $y = 4x^4 + x^2 + x + 2$

14 $y = 9 - \dfrac{5}{x} - 4x$ **15** $f(x) = 2\sqrt{x^3} - 5x + 7$ **16** $y = x^3 - 2x^2 + 5x + 1$ **17** $y = x^4 - 3x^2 + 3$ **18** $y = x^3 - 2x + \dfrac{2}{x}$

19 $a = 3, b = -5, c = 4$

Exercise 7B

1 a) $\frac{8}{3}$ **b)** 81 **c)** 45 **d)** 37 **e)** 21 **f)** -15 **g)** $\frac{3}{8}$ **h)** $\frac{3}{2}$ **i)** $12\frac{2}{3}$ **j)** 7 **k)** $\frac{3}{2}$ **l)** $11\frac{1}{4}$ **2 a)** 16 **b)** 70 **c)** -40 **d)** $-\frac{3}{4}$
2 e) $-\frac{9}{2}$ **f)** $-\frac{15}{2}$ **g)** 0 **h)** $-3\frac{63}{64}$ **i)** $11\frac{1}{5}$ **j)** -9 **k)** $\frac{4}{5}$ **l)** 24 **3 a)** $4\frac{2}{3}$ **b)** $16\frac{1}{2}$ **c)** 20 **d)** $20\frac{5}{6}$ **e)** $\frac{2}{5}$ **f)** $\frac{4}{3}$ **g)** 15
3 h) $4\frac{2}{3}$ **i)** 4 **j)** $53\frac{2}{5}$ **k)** $10\frac{2}{3}$ **l)** $6\frac{1}{4}$ **4 a)** $70\frac{1}{2}$ **b)** $19\frac{1}{2}$ **c)** $18\frac{1}{4}$ **d)** 24 **e)** 52 **f)** 14 **5** Area = 50 **6** $85\frac{1}{3}$ **7** Area = $\frac{1}{6}$
8 Area = $2\frac{2}{3}$ **9** Area = $6\frac{3}{4}$ **10** Area = 9 **11** $30\frac{3}{8}$ **12 b)** $\frac{7}{12}$ **c)** $11\frac{1}{4}$ **13 b)** $\frac{13}{60}$ **c)** $2\frac{14}{15}$ **14 a)** $\frac{16}{3}$ **b)** $\frac{2}{3}$ **c)** $\frac{14}{3}$ **d)** 4
14 e) 2 **f)** $\frac{2}{3}$ **15 a)** P(1, 4), Q(2, 7) **c)** $\frac{1}{6}$ **16 a)** A(−2, 11), B(3, 6) **c)** $20\frac{5}{6}$ **17 a)** C(2, 20), D(4, 32) **c)** $2\frac{2}{3}$ **18 b)** $14\frac{7}{24}$
19 a) (0, 0), (9, 3) **b)** $4\frac{1}{2}$ **20** 32 **21** $21\frac{1}{3}$ **22** 12 **23** 36 **24 a)** P(2, 0), Q(4, 8) **b)** $10\frac{2}{3}$ **25 a)** A(3, 0), B(6, −18) **b)** 36
26 a) P(−1, 0), Q(1, 0), R(2, 3) **b)** $4\frac{1}{2}$ **27 b) i)** 0 **ii)** $25\frac{3}{5}$ **iii)** $5\frac{1}{3}$ **iv)** $43\frac{11}{15}$ **28 b) i)** $-\frac{8}{3}$ **ii)** $\frac{5}{12}$ **iii)** $-2\frac{1}{4}$ **iv)** $3\frac{1}{12}$ **v)** $-5\frac{1}{3}$

Exercise 7C

1 a) 72π **b)** 625π **c)** 8π **d)** $\dfrac{7\pi}{24}$ **e)** 54π **f)** $\dfrac{158\pi}{3}$ **g)** 9π **h)** $\dfrac{348\pi}{5}$ **i)** $\dfrac{352\pi}{3}$ **j)** 42π **k)** $\dfrac{803\pi}{12}$ **l)** $\dfrac{3\pi}{2}$ **2 a)** 288π
2 b) $\dfrac{81\pi}{2}$ **c)** $\dfrac{96\pi}{5}$ **d)** $\dfrac{243\pi}{5}$ **e)** 7π **f)** $\dfrac{14\pi}{3}$ **g)** $\dfrac{\pi}{4}$ **h)** 63π **i)** $\dfrac{104\pi}{3}$ **j)** 8π **k)** $\dfrac{20\pi}{3}$ **l)** $\dfrac{115\pi}{3}$ **3 a)** P(−2, 4), Q(2, 4) **b)** $\dfrac{256\pi}{5}$
4 a) (−1, 2), (1, 2) **b)** $\dfrac{64\pi}{15}$ **5** $\dfrac{28\pi}{3}$ **6** 9π **7** $\dfrac{128\pi}{5}$ **8** $\dfrac{5\pi}{2}$ **9** $\dfrac{25\pi}{2}$ **10 a)** P(3, 9) **b) i)** $\dfrac{162\pi}{5}$ **ii)** $\dfrac{27\pi}{2}$ **11 b)** $\dfrac{\pi}{10}$
12 $\dfrac{1250\pi}{3}$ **13 a)** P(−2, 4), Q(2, 4) **b) i)** $\dfrac{256\pi}{3}$ **ii)** 16π **14 a)** $\dfrac{32\pi}{5}$ **b)** $\dfrac{3\pi}{2}$ **15 a)** A(0, 8), B(6, 8) **b)** $\dfrac{1296\pi}{5}$
16 a) C(0, 6), D(4, 6) **b)** $\dfrac{256\pi}{13}$ **17** $\dfrac{27\pi}{2}$ **18** 9 **19** $\dfrac{8\pi}{3}$ **20** $\dfrac{63\pi}{2}$ **21** $\dfrac{352\pi}{15}$ **22** $\dfrac{56\pi}{5}$ **23** $\dfrac{32\pi}{5}$ **24** ± 6

Exercise 7D

1 a) $9x^2+6+\dfrac{1}{x^2}$ **b)** $3x^3+6x-\dfrac{1}{x}+c$ **c)** $27\frac{1}{2}$ **2 i)** $y=x^3-x^2-x+1$ **ii)** $\left(-\frac{1}{3},\frac{32}{27}\right)$, (1, 0) **3** $y=2x^4-x^2-x+4$
4 a) 6 **b)** $y=2x^3-3x^2-3$ **5 a)** $\frac{3}{2}x^2-\frac{1}{3}x^3+c$ **b)** $4\frac{1}{2}$ **6 a)** (3, −1), min **c)** $1\frac{1}{3}$ **7 a)** (1, 5), max; (2, 4), min **c)** 8
8 $10\frac{2}{3}$ **9 a)** (6, 12) **b)** $13\frac{1}{3}$ **10 a)** $-\frac{1}{4}x^4+\frac{27}{2}x^2-34x+c$ **b)** 34 **c)** 12 **11 i)** $-2x+2$ **ii)** 4 **iii)** $6\,\text{m}^2$
12 b) $\int_0^1 (x-x^2)\,dx$ **c)** $\frac{1}{6}$ **13 i)** (0, 0), (5, 10) **iii)** $20\frac{5}{6}$ **14 a)** A(0, 5), B(1, 0), C(5, 0), D(6, 5) **b)** $x+y=5$ **d)** $14\frac{2}{3}$
16 i) Reflection in $y=1$ or reflection in x-axis followed by translation $\begin{pmatrix}0\\2\end{pmatrix}$ **ii)** $2\frac{2}{3}$ **17 ii)** $\dfrac{16\sqrt{2}}{3}$ **iii)** 4π
18 ii) $125\pi\int_0^{12}(y+4)\,dy$ **iii)** 15 litres

Exercise 8A

1 a) $x^2+y^2-2x-4y-4=0$ **b)** $x^2+y^2-6x-2y-6=0$ **c)** $x^2+y^2+4x-6y+12=0$ **d)** $x^2+y^2-2x+6y-15=0$
1 e) $x^2+y^2+8x=0$ **f)** $x^2+y^2-4x+8y-29=0$ **g)** $x^2+y^2+6x-10y-2=0$ **h)** $x^2+y^2-8x+2y+8=0$
1 i) $x^2+y^2+4y+3=0$ **j)** $x^2+y^2+10x+6y-15=0$ **k)** $x^2+y^2+16x-14y+13=0$ **l)** $x^2+y^2-x-3y-\frac{3}{2}=0$
2 a) (2, 1), 2 **b)** (1, 4), 3 **c)** (−3, 2), 1 **d)** (2, 0), 2 **e)** (0, −3), 5 **f)** (3, −4), 6 **g)** (−7, 5), 9 **h)** (6, 6), 8 **i)** (−8, −6), 10
2 j) (1, −1), 2 **k)** (7, −8), 12 **l)** $\left(0,\frac{5}{2}\right)$, $\frac{3}{2}$ **3** $x^2+y^2-10x-8y+16=0$ **4** $x^2+y^2-2x+14y-119=0$
5 $x^2+y^2-10x-14y+25=0$ **6** $x^2+y^2+4x+6y+9=0$ **7** $x^2+y^2-12x-16y+75=0$ **8** $x^2+y^2-2x-8y+7=0$
9 $x^2+y^2-10y=0$, $x^2+y^2-12x-10y+36=0$ **10 a)** $x=4$ **b)** $y=6$ **c)** (4, 6) **d)** $x^2+y^2-8x-12y+42=0$
11 a) $y=2x-1$ **b)** $x+2y=13$ **c)** (3, 5) **d)** $x^2+y^2-6x-10y-6=0$ **12 a)** $2y-x=8$ **b)** $y=3x+9$ **c)** (−2, 3)
12 d) $x^2+y^2+4x-6y+3=0$ **13** $x^2+y^2-10x-2y+13=0$ **14 a)** (2, 1), 5; (8, 9), 5 **c)** $\left(-4\frac{3}{5},12\frac{1}{5}\right)$, $\left(14\frac{3}{5},-2\frac{1}{5}\right)$

Exercise 8B

1 a) $y=x$ **b)** $3x+5y=13$ **c)** $y=2x+3$ **d)** $x+y=1$ **e)** $x+y+7=0$ **f)** $x=3$ **g)** $y=-2x$ **h)** $y=x+8$ **i)** $x+y=7$
1 j) $3y-2x=23$ **k)** $4y+3x=0$ **l)** $y+3x=9$ **2** $x+2y=3$, $x-2y=7$ **3** $y=3x+5$, $3x+y=3$
4 $2x+5y+28=0$, $5x+2y=21$ **5** 9 **6** 5 **7** $\sqrt{23}$ **8 a)** $\sqrt{17}$ **9 a)** (3, 4), (5, 2) **b)** (−1, −5), (0, 2) **c)** (3, 8) (4, 7)
9 d) (−4, −5), (−1, 10) **10 a)** (−1, −5), (2, 1) **b)** (2, 2), (5, 1) **c)** (1, −1), $\left(1\frac{1}{2},-\frac{1}{2}\right)$ **d)** $\left(-1\frac{3}{5},-2\frac{1}{5}\right)$, (1, 3)

Exercise 8C

1 a) $3, 12$ **b)** 9 **c)** $(-4, 0)$ **2** $x + 6y = 20$ **3** $(-1, 3), 6, 10, 8$ **4** $(-2, 3), 5, 12, 11.2$ **5 a)** $13, (-1, 4)$ **b) i)** 4 **ii)** $(-6, -8)$
6 i) $3\sqrt{5}, (9, 3)$ **ii)** $(3, 6)$ **7 a)** $y = x - 1$ **b)** $1 : 3$ **c)** $(1, -3), 5\sqrt{2}$

Exercise 9A

1 a) $3, 5, 7, 9, 11, 13$; divergent **b)** $1, 4, 7, 10, 13, 16$; divergent **c)** $3, 1, -1, -3, -5, -7$; divergent **d)** $4, 7, 12, 19, 28, 39$; divergent
1 e) $1, \frac{1}{2}, \frac{1}{3}, \frac{1}{4}, \frac{1}{5}, \frac{1}{6}$; convergent $\to 0$ **f)** $\frac{1}{2}, \frac{2}{3}, \frac{3}{4}, \frac{4}{5}, \frac{5}{6}, \frac{6}{7}$; convergent $\to 1$ **g)** $\frac{1}{2}, \frac{1}{5}, \frac{1}{10}, \frac{1}{17}, \frac{1}{26}, \frac{1}{37}$; convergent $\to 0$
1 h) $3\frac{1}{2}, 3\frac{1}{6}, 3\frac{1}{12}, 3\frac{1}{20}, 3\frac{1}{30}, 3\frac{1}{42}$; convergent $\to 3$ **i)** $6, 24, 60, 120, 210, 336$; divergent **j)** $2, 4, 8, 16, 32, 64$; divergent
1 k) $-1, 2, -3, 4, -5, 6$; divergent **l)** $1, -\frac{1}{4}, \frac{1}{9}, -\frac{1}{16}, \frac{1}{25}, -\frac{1}{36}$; convergent$\to 0$ **2 a)** $4n$ **b)** $2n + 3$ **c)** $5n - 1$ **d)** $3n + 5$
2 e) $\frac{1}{n+1}$ **f)** $\frac{1}{3n}$ **g)** $\frac{2}{3n+2}$ **h)** $\frac{n}{n+1}$ **i)** $\frac{n+1}{3n-2}$ **j)** $\frac{2n+1}{6n-1}$ **k)** $\frac{13-n}{5n+2}$ **l)** $\frac{7-3n}{7n-2}$ **3 a)** 2^n **b)** 5×2^n **c)** $5 \times 2^{n-1}$
3 d) $4 \times 3^{n-1}$ **e)** $2 \times (-3)^{n-1}$ **f)** $\left(-\frac{1}{2}\right)^{n-1}$ **g)** n^2 **h)** $\frac{n}{(n+1)^2}$ **i)** $(-1)^n n(n+1)$ **j)** $\frac{n+1}{n(n+2)}$ **k)** $(-1)^n \frac{n-1}{5n-6}$ **l)** n^n
4 a) $5, 7, 9, 11, 13, 15$; divergent **b)** $3, 9, 15, 21, 27, 33$; divergent **c)** $2, 1, 2, 1, 2, 1$; periodic **d)** $3, 7, 15, 31, 63, 127$; divergent
4 e) $3, -1, 11, -25, 83, -241$; divergent **f)** $5, 5, 5, 5, 5, 5$; convergent $\to 5$ **g)** $7, \frac{1}{7}, 7, \frac{1}{7}, 7, \frac{1}{7}$; periodic
4 h) $1, 2, \frac{1}{2}, 8, \frac{1}{32}, 2048$; divergent **i)** $2, 1, -2, 1, -2, 1$; periodic after the first term **j)** $1, -1, 3, 3, 3, 3$; convergent $\to 3$
4 k) $2, \frac{1}{2}, -1, 2, \frac{1}{2}, -1$; periodic **l)** $1, 2, \frac{5}{2}, \frac{29}{10}, \frac{941}{290}, \frac{969\,581}{272\,890}$; divergent **5 a)** $2, 1, 4, 9, 22, 53$ **b)** $3, 4, 7, 11, 18, 29$
5 c) $2, 1, -2, -11, -38, -119$ **d)** $3, 3, 0, 3, -3, 6$ **e)** $-1, 0, -1, 4, -17, 72$ **f)** $5, 4, -1, 16, -67, 316$ **g)** $1, 2, 2, 4, 8, 32$
5 h) $3, 1, 1, -1, 1, -3$ **i)** $2, 3, 3, 9, 27, 405$ **j)** $8, -3, 1, 4, 15, 221$ **k)** $4, 2, \frac{1}{2}, \frac{1}{4}, \frac{1}{2}, 2$ **l)** $7, 3, 2, 1, 1, 0$ **6 a)** $1, 2, 3, 5, 8, 13, 21$
6 c) 610 **7 b)** 171 **8 b)** $233, 610$ **9 b)** $99, 239$ **10 a)** $1, 3, 2, -1, -3, -2, 1, 3, 2, \ldots$ **b)** $1, 2, -3$ **11 a)** $u_{n-1} = 3u_n - u_{n+1}$
11 b) $86, 33, 149$ **12** $u_1 = u_3 = u_5 = 10$ **13** n **14** 2^{n-1} **15** $u_{16} = 65\,539$ **16 a)** 3.45 **c)** -1.45

Exercise 9B

1 a) $1^2 + 2^2 + 3^2 + 4^2 + 5^2$ **b)** $2 + 5 + 8 + 11 + 14 + 17$ **c)** $5 + 11 + 21 + 35$ **d)** $1 \times 2 + 2 \times 3 + 3 \times 4 + 4 \times 5 + 5 \times 6$
1 e) $3^3 + 4^3 + 5^3 + 6^3$ **f)** $5 \times 2 + 6 \times 3 + 7 \times 4 + 8 \times 5 + 9 \times 6 + 10 \times 7$ **g)** $1^2 + 3^2 + 5^2 + 7^2 + 9^2 + 11^2$ **h)** $1 + \frac{1}{2} + \frac{1}{3} + \ldots + \frac{1}{n}$
1 i) $4^4 + 5^5 + 6^6 + 7^7 + 8^8$ **j)** $1 - \frac{1}{2} + \frac{1}{3} - \frac{1}{4} + \frac{1}{5} - \frac{1}{6} + \frac{1}{7}$ **k)** $2 \times 1^2 + 2 \times 3^2 + 2 \times 5^2 + 2 \times 7^2 + 2 \times 9^2$ **l)** $3 + 3 + 3 + 3 + 3$
2 a) $\sum_{r=1}^{5} r$ **b)** $\sum_{r=1}^{7} r^3$ **c)** $\sum_{r=1}^{7} (3r + 4)$ **d)** $\sum_{r=3}^{20} \frac{1}{r}$ **e)** $\sum_{r=5}^{18} r(r+1)$ **f)** $\sum_{r=3}^{n} r^4$ **g)** $\sum_{r=1}^{7} (-1)^{r-1} r$ **h)** $\sum_{r=1}^{9} (-2)^{r+1}$ **i)** $\sum_{r=5}^{n} \frac{r}{r^2 - 1}$
2 j) $\sum_{r=1}^{n} \frac{r}{(r+1)(r+2)}$ **k)** $\sum_{r=1}^{15} (-1)^{r+1}(2r-1)(3r+1)$ **l)** $\sum_{r=1}^{6} \frac{3r-1}{3^r}$

Exercise 9C

1 a) 3 **b)** -11 **c)** Not **d)** Not **e)** 0.1 **f)** $\frac{3}{10}$ **g)** Not **h)** Not **i)** -1 **j)** $1\frac{3}{8}$ **k)** $1\frac{1}{6}$ **l)** a **2 a)** 37 **b)** 65 **c)** -25
2 d) -61 **e)** 3.8 **f)** $85 - 4n$ **g)** -7.1 **h)** 51 **i)** $\frac{n}{6}$ **j)** 24 **k)** $(2n - 1)a$ **l)** $\left(\frac{n+4}{4}\right)x$ **3 a)** 55 **b)** 725 **c)** 837
3 d) 390 **e)** -580 **f)** $\frac{n}{2}(3n + 11)$ **g)** -2775 **h)** $592\frac{1}{2}$ **i)** -2900 **j)** 0 **k)** $n(3n - 2)b$ **l)** $n(19 - 5n)c$ **4 a)** 11 **b)** 21
4 c) 100 **d)** 44 **e)** 8 **f)** 19 **g)** 30 **h)** 28 **i)** 30 **j)** 15 **k)** $n - 2$ **l)** $n + 2$ **5 a)** 5050 **b)** 234 **c)** 225 **d)** 650
5 e) -187 **f)** 35.4 **g)** 120 **h)** 96 **i)** $-71\frac{1}{4}$ **j)** $\frac{n}{2}(n + 1)$ **k)** $n(2n + 1)$ **l)** n^2 **7** $7, 1590$ **8** $-14, 9, 265$ **9** $10, -3, -133$
10 $-9, 210$ **11** $5, 1$ **12** $-12, 4$ **13** $-10, 2, 570$ **14** $37\frac{1}{2}$ **15** 7 **16** $2, 18$ **17** 9 **18** 10 **19** 22nd **20** $3, 5, 7$
21 $4\frac{1}{2}, 5\frac{1}{2}, 13\frac{1}{2}$ **22** 10 **23** $3, 152$ **24** $4, 14$ **25** $4, 216$ **26 a)** 5050 **b)** 1683 **c)** 3367 **27** $16\,000$ **28** 13 **30** 15

Exercise 9D

1 a) 3 **b)** -2 **c)** Not **d)** Not **e)** 1.2 **f)** Not **g)** Not **h)** $\frac{1}{2}$ **i)** -2 **j)** Not **k)** -1 **l)** a **2 a)** 1024 **b)** 729 **c)** 640
2 d) $51\frac{33}{128}$ **e)** $-4\frac{20}{27}$ **f)** $\frac{2}{625}$ **g)** $\frac{1}{2048}$ **h)** $-\frac{1}{243}$ **i)** $3\frac{13}{81}$ **j)** -7 **k)** x^n **l)** $a \times (-1)^{n-1} \times r^{n-1}$ **3 a)** 3069 **b)** -1023
3 c) 2049 **d)** 1275 **e)** 1638 **f)** $1\,111\,111$ **g)** $1\frac{364}{729}$ **h)** $18\frac{17}{36}$ **i)** $\frac{1365}{4096}$ **j)** $7.715\,61$ **k)** $1 - \left(\frac{1}{2}\right)^n$ **l)** $\frac{x(1 - x^n)}{1 - x}, (x \neq 1)$

4 a) 5 **b)** 9 **c)** 7 **d)** 11 **e)** 6 **f)** 6 **g)** 7 **h)** 10 **i)** 9 **j)** 8 **k)** 9 **l)** 8 **5 a)** 765 **b)** 2186 **c)** −728 **d)** 301 **e)** $53\frac{25}{27}$

5 f) $39\frac{11}{16}$ **g)** $\frac{1365}{4096}$ **h)** $\frac{182}{729}$ **i)** 9.487 171 **j)** −78.8125 **k)** $2-\left(\frac{1}{2}\right)^n$ **l)** $5\times\left(\frac{1-(-2)^n}{3}\right)$ **6** 5, 3 **7** −4, 3 **8** $\pm\frac{1}{2}$, ±384

9 ±11, ±77 **10** 5115 **11** 8200 **12** $\frac{1}{7}$ **13** $\frac{1}{11}$ **14** $\frac{1}{29}, -\frac{5}{29}, \frac{25}{29}$ **15** 10 **16** 7 **17** $r=-4$; 8, −32, 128; $r=3$; 8, 24, 72

18 $r=-5$: 5, −25, 125; $r=4$: 5, 20, 80 **19** 5, $\frac{2}{5}$ **20** −3, −2 **21** $5+15+45+135+405$ **22** ±125 **23** −6, −2 **24** 2, 4

25 3, 2; −3, $\frac{13}{7}$ **26** £10 737 418.23 **27** £48 685.18 **28** £6375 **29** 5; $\frac{3}{4}+\frac{3}{2}+3$ **30** 3, $32+16+8$; −2, $-\frac{1}{27}-\frac{1}{9}-\frac{1}{3}$

31 4, $3+6+12$; $-\frac{1}{2}$, $-\frac{3}{2}+\frac{3}{2}-\frac{3}{2}$ **32** 8, $2\pm\frac{3}{\sqrt{2}}$

Exercise 9E

1 a) 2 **b)** $\frac{3}{2}$ **c)** $\frac{1}{4}$ **d)** $\frac{4}{5}$ **e)** $\frac{1}{72}$ **f)** $\frac{1}{8}$ **g)** $\frac{9}{70}$ **h)** $\frac{49}{170}$ **i)** 6 **j)** $-\frac{1}{2}$ **k)** $\frac{1}{1-a}$ **l)** $\frac{9x^2}{1-3x}$ **2 a)** $\frac{5}{9}$ **b)** $\frac{8}{9}$ **c)** $\frac{8}{11}$ **d)** $\frac{34}{333}$

2 e) $\frac{22}{9}$ **f)** $\frac{443}{135}$ **3** 12 **4** $\frac{8}{3}$ **5** $\frac{9}{10}$ **6** $-\frac{1}{2}$ **7** $\frac{1}{2}$ **8** 18, 6, 2; 9, 6, 4 **9** $\frac{2}{3}$ **10** 3 **11** $\frac{1}{3}$ **12** $\frac{1}{2}$ **13** $5+7\frac{1}{2}+10+12\frac{1}{2}$

14 $a=-5, b=25$ or $a=4, b=16$ **15** $p=10, q=50$ or $p=-12, q=72$ **16** $x=-\frac{1}{2}, y=-32$ or $x=16, y=1$

17 $a=4, c=36$ or $a=36, c=4$

Exercise 9F

1 a) Converges to 1 **b)** 2, $\frac{3}{2}$, −1, 2, $\frac{3}{2}$, ... Periodic of period 3

2 i) Periodic of period 2; periodic, after first term, of period 2; divergent **ii)** $\frac{1\pm\sqrt{5}}{2}$ **3** 15, −2 **4 a)** 5 **b)** 26 250 **5 i)** 0, 4

5 ii) 1 500 500 **6 a)** $\frac{7}{3}$ **b)** 28 **7** 2, 3.5, 6.5, 8 **8 b)** £3216 **9 a)** £70 620 **b)** 1062 **10 i)** 2 **ii)** 40 **iii)** $\frac{n(3n+1)}{2}$

10 iv) $\frac{n(9n+1)}{2}$ **11 a)** 4, 210 **b)** 2, 8184 **12 a)** 77 **b)** 5909 **c)** $r>1$ **13 a)** 4950 **b)** $\frac{1}{3}$, 81 **14** $\frac{11}{15}$

15 i) $r=-\frac{2}{3}, a=27$ or $r=2, a=3$ **ii)** 16.2 **16** $p=-2, q=4$ or $p=\frac{5}{2}, q=\frac{25}{4}$ **17 a)** $\frac{16}{3}$ **b) i)** −8, 2 **ii)** $-\frac{1}{2}$, 2 **iii)** 256

17 iv) $170\frac{2}{3}$ **18** −19, 20 **19 a)** $a-2$ **b)** $1<a<3$ **20** $\frac{3}{4}$, 224 cm^3 **21 i)** 1.08 **ii)** £29 985.07 **iv)** £1495.52

22 b) £12 079.98 **23 a)** 0.432 m **b)** 4.76 m **c)** 4.8 m

Exercise 10A

1 a) 10 **b)** 15 **c)** 36 **d)** 6 **e)** 1 **f)** 66 **g)** 35 **h)** 100 **2 a)** $1+4x+6x^2+4x^3+x^4$ **b)** $1+5x+10x^2+10x^3+5x^4+x^5$

2 c) $1+12x+54x^2+108x^3+81x^4$ **d)** $1-3x+3x^2-x^3$ **e)** $1-8x+24x^2-32x^3+16x^4$ **f)** $1-15x+75x^2-125x^3$

2 g) $1+2x+\frac{3}{2}x^2+\frac{1}{2}x^3+\frac{1}{16}x^4$ **h)** $1-\frac{2}{5}x+\frac{1}{25}x^2$ **3 a)** 35 **b)** 36 **c)** 80 **d)** 700 **e)** −540 **f)** −42 **g)** 96 **h)** 80 **i)** $\frac{3}{4}$

4 a) $8+12x+6x^2+x^3$ **b)** $81+108x+54x^2+12x^3+x^4$ **c)** $216-540x+450x^2-125x^3$ **d)** $16+16x+6x^2+x^3+\frac{1}{16}x^4$

4 e) $27x^3+54x^2y+36xy^2+8y^3$ **f)** $32x^5-80x^4y+80x^3y^2-40x^2y^3+10xy^4-y^5$ **g)** $8x^3+60x^2y+150xy^2+125y^3$

4 h) $81x^4-432x^3y+864x^2y^2-768xy^3+256y^4$ **5 a)** 1080 **b)** 44 800 **c)** 6048 **d)** 8960 **e)** −224 **f)** 20 000 **g)** $\frac{5}{2}$ **h)** $-\frac{8}{15}$

6 a) $1-15x+90x^2-270x^3$ **b)** $1+20x+180x^2+960x^3$ **c)** $1-35x+525x^2-4375x^3$ **d)** $32-240x+720x^2-1080x^3$

6 e) $1024-1280x+640x^2-160x^3$ **f)** $64+576x+2160x^2+4320x^3$ **g)** $1+3x+4x^2+\frac{28}{9}x^3$ **h)** $4096+1536x+240x^2+20x^3$

7 a) $2+11x+25x^2$ **b)** $5+29x+69x^2$ **c)** $5-66x+364x^2$ **d)** $486+2997x+7344x^2$ **e)** $5-90x+676x^2$

7 f) $567-756x+216x^2$ **g)** $2+15x+55x^2$ **h)** $32+112x+192x^2$ **8 a)** 20 **b)** −9 **c)** 20 **d)** 54 **e)** 8 **f)** 816

8 g) −191 **h)** 25 **9 a)** $1+4x^3+6x^6+4x^9+x^{12}$ **b)** $1+9x^2+27x^4+27x^6$ **c)** $27-54x^3+36x^6-8x^9$

9 d) $1+2x+3x^2+2x^3+x^4$ **e)** $1+6x+9x^2-4x^3-9x^4+6x^5-x^6$ **f)** $4+12x+5x^2-6x^3+x^4$ **g)** $4+4x-15x^2-8x^3+16x^4$

9 h) $9+12x+10x^2+4x^3+x^4$ **10 a)** −810 **b)** 960 **c)** 240 **d)** 54 **e)** 216 **f)** 490 **g)** 1350 **h)** 2 949 120

11 a) i) $1+4x+6x^2$ **ii)** $1-8x+24x^2$ **b)** $1-4x-2x^2$ **12 a) i)** $1+18x+135x^2$ **ii)** $1-24x+240x^2$ **b)** $1-6x-57x^2$

13 a) i) $243+810x+1080x^2$ **ii)** $1-15x+90x^2$ **b)** $243-2835x+10\,800x^2$ **14 a)** $1-10x+45x^2-120x^3+210x^4$

14 b) 0.904382 **15 a)** $1+28x+364x^2+2912x^3$ **b)** 1.319 **16 a)** $1-24x+252x^2$ **b)** 0.97625 **17 a)** $512+11\,520x+115\,200x^2$

17 b) 523.64 **18** $531\,441-4\,251\,528x+15\,588\,936x^2$, 527 205 **19** $3125-12\,500x+20\,000x^2-16\,000x^3$, 3002

20 a) 1.072 **b)** 0.90 **c)** 299.9448 **d)** 16 676 829 **e)** 1295.14 **f)** 7972.354 **21** $a=5, n=6$ **22** $b=-3, n=5$

23 $c=5, n=4$ **24** $a=-\frac{1}{2}, n=16$ **26 a)** 178 **b)** 82 **c)** 7040 **d)** $60\sqrt{3}$

Exercise 10B

1 a) $1-2x+3x^2-4x^3,\ |x|<1$ **b)** $1+\frac{1}{2}x-\frac{1}{8}x^2+\frac{1}{16}x^3,\ |x|<1$ **c)** $1-6x+24x^2-80x^3,\ |x|<\frac{1}{2}$ **d)** $1+6x+27x^2+108x^3,\ |x|<\frac{1}{3}$

1 e) $1+x+2x^2+\frac{14}{3}x^3,\ |x|<\frac{1}{3}$ **f)** $1-3x^2+6x^4-10x^6,\ |x|<1$ **g)** $1+3x+9x^2+27x^3,\ |x|<\frac{1}{3}$ **h)** $1-3x-\frac{9}{2}x^2-\frac{27}{2}x^2,\ |x|<\frac{1}{6}$

2 a) $\frac{1}{2}-\frac{1}{4}x+\frac{1}{8}x^2-\frac{1}{16}x^3,\ |x|<2$ **b)** $2+\frac{1}{4}x-\frac{1}{64}x^2+\frac{1}{512}x^3,\ |x|<4$ **c)** $\frac{1}{3}+\frac{2}{27}x+\frac{2}{81}x^2+\frac{20}{2187}x^3,\ |x|<\frac{9}{4}$ **d)** $4+x-\frac{1}{16}x^2+\frac{1}{96}x^3,\ |x|<\frac{8}{3}$

2 e) $\frac{1}{8}+\frac{3}{16}x+\frac{3}{16}x^2+\frac{5}{32}x^3,\ |x|<2$ **f)** $2-\frac{1}{4}x-\frac{1}{64}x^2-\frac{1}{512}x^3,\ |x|<4$ **g)** $\frac{1}{6}-\frac{1}{36}x+\frac{1}{216}x^2-\frac{1}{1296}x^3,\ |x|<6$

2 h) $\frac{1}{9}+\frac{4}{27}x+\frac{4}{27}x^2+\frac{32}{243}x^3,\ |x|<\frac{3}{2}$ **3 a)** $1+2x+2x^2$ **b)** $\frac{1}{2}+\frac{5}{4}x-\frac{5}{8}x^2$ **c)** $\frac{1}{4}x+\frac{1}{16}x^2$ **d)** $2+17x+84x^2$ **e)** $8-x-\frac{33}{16}x^2$

3 f) $7+\frac{1}{2}x+\frac{9}{8}x^2$ **g)** $16-72x+9x^2$ **h)** $\frac{1}{16}-\frac{3}{64}x-\frac{5}{128}x^2$ **4 a) i)** $1+8x+24x^2$ **ii)** $1-\frac{1}{2}x-\frac{1}{8}x^2$ **b)** $1+\frac{15}{2}x+\frac{159}{8}x^2$

5 a) i) $1+x-\frac{1}{2}x^2$ **ii)** $1+4x+10x^2$ **b)** $1+5x+\frac{27}{2}x^2,\ |x|<\frac{1}{2}$ **6 a) i)** $1+x+x^2+x^3$ **ii)** $1+2x+4x^2+8x^3$

6 b) $1+3x+7x^2+15x^3$ **7 a) i)** $1-x-x^2-\frac{5}{3}x^3$ **ii)** $1+4x+16x^2+64x^3$ **b)** $1+3x+11x^2+\frac{127}{3}x^3,\ |x|<\frac{1}{4}$

9 $1+2x+2x^2+4x^3$ **10** $c=5, n=-\frac{2}{5}$ **11** $a=\frac{2}{3}, n=\frac{1}{6}$ **12** $1+\frac{1}{2}x-\frac{1}{8}x^2+\frac{1}{16}x^3$, 1.004 987 56 **13** $1-\frac{1}{2}x-\frac{3}{8}x^2$, 0.999 500

14 $\frac{1}{2}+\frac{3}{16}x+\frac{27}{256}x^2$, 0.5019 **16 a)** $1+\frac{1}{200}x-\frac{1}{80\,000}x^2+\frac{1}{16\,000\,000}x^3$ **18 a)** $1+\frac{1}{64}x-\frac{3}{8192}x^2$ **20 c)** $a=\pm\frac{1}{6}, b=\mp\frac{1}{9}$

Exercise 10C

1 -280 **2** $8x+8x^3$ **3 ii)** $-2, -1, 0$ **4 a)** 16 **b)** 270 **c)** -1890 **5** $3-11x+11x^2$ **6 a)** $k=\frac{3}{2}, p=63, q=189$ **b)** 126

7 $1+8ax+28a^2x^2;\ a=1, b=-8$ **8** $x^5-5x^3+10x-\dfrac{10}{x}+\dfrac{5}{x^3}-\dfrac{1}{x^5}$ **9** $x^{18}+3x^{15}+4x^{12};\ \frac{28}{243}$ **10 a)** $a=-12, b=90, c=-540$

10 b) $|x|<\frac{1}{3}$ **11** $1+2t-t^2$ **12 a)** $1+\frac{1}{2}x^2+\frac{3}{8}x^4+\frac{5}{16}x^6$ **b)** 1.414 15 **13 i)** $16-32x+24x^2-8x^3+x^4$

13 ii) $1-6x+24x^2-80x^3;\ |x|<\frac{1}{2}$ **iii)** $a=-128, b=600$ **14 ii)** $1+\frac{1}{8}x+\frac{3}{128}x^2;\ |x|<4$ **iii)** $1+\frac{9}{8}x+\frac{19}{128}x^2$

Exercise 11A

1 a) $\dfrac{5x+7}{(x+3)(x-1)}$ **b)** $\dfrac{x-10}{(x+4)(x-3)}$ **c)** $\dfrac{2(x+13)}{(x+4)(x-5)}$ **d)** $\dfrac{x^2+6}{(x+2)(x-3)}$ **e)** $\dfrac{2x^2-5x-2}{(x+2)(x-2)}$ **f)** $\dfrac{4x+34}{(2x+5)(2x-3)}$

1 g) $\dfrac{5x^2+3x+4}{(x+4)(x-2)}$ **h)** $\dfrac{16x^2-25x+18}{(x^2+3)(2x-5)}$ **2 a)** $\dfrac{3x+8}{(x+4)(x-1)}$ **b)** $-\dfrac{4x+11}{(x+2)(x+4)}$ **c)** $\dfrac{x^3+5x^2-4x-14}{(x+4)(x-2)}$

2 d) $\dfrac{2x^3-17x^2-2x+158}{(x-4)(x-7)}$ **e)** $\dfrac{x^2-x+1}{(x-1)(x-2)}$ **f)** $\dfrac{x^3-5x^2+15x-10}{(x-2)^2}$ **g)** $-\dfrac{3x+4}{(x+2)(x+3)}$ **h)** $\dfrac{x-3}{(2x+5)(x-1)}$

3 $\frac{7}{5}$ **4** $\dfrac{5x-1}{(2x+1)(x-3)}, -\frac{9}{4}$ **5** $\dfrac{x^2+x+3}{(x+1)(x-2)}, -\frac{3}{2}, \frac{17}{4}$ **6** $-\frac{17}{3}, 2$ **7 a)** $(2x-1)(x+2)(x-3)$ **b)** $\dfrac{x+13}{(x+1)^2}$ **c)** $-2, \frac{1}{2}, 3$

8 b) $\dfrac{5x-1}{(x+4)(x-3)}$, **c)** $1, \frac{11}{6}, 2$ **9** $a=2, b=3$ **10** $A=1, B=1$ **11** $P=2, Q=8$ **12** $a=1, b=2$

Exercise 11B

1 a) $\dfrac{2}{x+3}+\dfrac{1}{x-2}$ **b)** $\dfrac{2}{x+4}+\dfrac{3}{x-3}$ **c)** $\dfrac{3}{x+2}-\dfrac{1}{x+1}$ **d)** $\dfrac{2}{2x-1}-\dfrac{3}{3-x}$ **e)** $\dfrac{2}{x+4}+\dfrac{5}{x-2}$ **f)** $\dfrac{3}{2x+5}+\dfrac{1}{x-2}$

1 g) $\dfrac{2}{x+3}+\dfrac{2}{x-3}$ **h)** $\dfrac{3}{x+1}-\dfrac{1}{x-2}$ **2 a)** $\dfrac{1}{x+3}-\dfrac{2}{x+2}+\dfrac{1}{x+1}$ **b)** $\dfrac{2}{x+1}+\dfrac{3}{x-1}-\dfrac{4}{x-2}$ **c)** $\dfrac{1}{3(x+2)}+\dfrac{1}{6(x-1)}-\dfrac{1}{2(x+3)}$

2 d) $\dfrac{1}{2x+1}+\dfrac{2}{3(2x-1)}-\dfrac{1}{3(x-2)}$ **e)** $\dfrac{4}{x+4}-\dfrac{2}{x+3}+\dfrac{3}{2x+5}$ **f)** $\dfrac{1}{x}-\dfrac{1}{2(x+3)}+\dfrac{1}{2(3x+1)}$ **g)** $\dfrac{1}{x+2}+\dfrac{2}{x+1}-\dfrac{1}{x-2}$

2 h) $\dfrac{3}{x-2}-\dfrac{2}{x+3}-\dfrac{1}{x+1}$ **3 a)** $\dfrac{2x+1}{x^2+1}+\dfrac{3}{x-2}$ **b)** $\dfrac{x+3}{x^2+2}-\dfrac{1}{x+3}$ **c)** $\dfrac{2x+3}{x^2+1}+\dfrac{4}{x-5}$ **d)** $\dfrac{2x+5}{2x^2+3}-\dfrac{1}{x+2}$

3 e) $\dfrac{3}{3x^2+1}+\dfrac{2}{4-x}$ **f)** $\dfrac{x}{x^2+3}-\dfrac{2}{2x+1}$ **g)** $\dfrac{3x+2}{x^2+3x+1}-\dfrac{2}{x+3}$ **h)** $\dfrac{x+3}{x^2+x+1}+\dfrac{2}{x-1}$ **4 a)** $\dfrac{2}{x+2}-\dfrac{1}{(x+2)^2}$

4 b) $\dfrac{4}{x-3}+\dfrac{3}{(x-3)^2}$ **c)** $\dfrac{3}{x-4}-\dfrac{2}{(x-4)^2}$ **d)** $\dfrac{3}{x+1}+\dfrac{2}{(x+1)^2}-\dfrac{3}{x+2}$ **e)** $\dfrac{2}{x-2}+\dfrac{3}{(x-2)^2}+\dfrac{4}{x+3}$ **f)** $\dfrac{3}{2x+1}+\dfrac{2}{(2x+1)^2}-\dfrac{3}{x+2}$

4 g) $\dfrac{2}{x-3}-\dfrac{3}{(x-3)^2}-\dfrac{1}{x}$ **h)** $\dfrac{2}{x-2}-\dfrac{3}{(x+3)^2}$ **5 a)** $1+\dfrac{3}{x+5}+\dfrac{2}{x-3}$ **b)** $2-\dfrac{1}{x+2}-\dfrac{4}{x+1}$ **c)** $x+3+\dfrac{1}{x+3}+\dfrac{1}{x-2}$

5 d) $x+\dfrac{1}{2(3x-1)}-\dfrac{1}{2(x-3)}$ **e)** $1+\dfrac{1}{x+4}+\dfrac{2}{x-1}+\dfrac{3}{x-2}$ **f)** $2+\dfrac{3}{x}-\dfrac{4}{x^2}-\dfrac{2}{x+1}$ **g)** $3+\dfrac{2}{x^2+1}+\dfrac{4}{x-3}$

5 h) $x-5+\frac{2}{x+1}+\frac{1}{x+2}-\frac{3}{x+3}$ **6** $\frac{5}{2(x+3)}+\frac{3}{2(x+1)}$ **7** $\frac{1}{2x+1}-\frac{4}{3x+5}+\frac{1}{x-3}$ **8** $\frac{3}{(x+1)^2}-\frac{2}{x+1}$

9 $\frac{2}{(x-3)^2}+\frac{3}{x-3}+\frac{1}{x+2}$ **10** $\frac{3}{x-1}-\frac{2}{x^2+3x+3}$ **11** $1+\frac{3}{x+5}-\frac{2}{x-3}$ **12** $\frac{2}{x-4}-\frac{3}{2x-1}$ **13** $\frac{3}{(x+4)^2}+\frac{2}{x+4}-\frac{1}{x+3}$

14 $x+2+\frac{4}{x+5}+\frac{3}{x-1}$ **15** $\frac{2}{x+5}-\frac{3}{(x+5)^2}$ **16** $2x-3+\frac{5}{(2x+1)^2}$ **17** $\frac{5}{x+3}-\frac{4}{x+2}$ **18** $\frac{3}{x+2}-\frac{4}{x-1}+\frac{2}{x-4}$

19 $\frac{2x+3}{x^2+7}-\frac{1}{x+5}$ **20** $\frac{4}{(x+5)^2}-\frac{3}{x+5}+\frac{2}{2x+3}$ **21** $\frac{3}{x-5}-\frac{1}{x+3}-\frac{2}{x-4}$ **22** $\frac{3}{(x-2)^2}+\frac{1}{x-2}+\frac{3}{x+3}$

23 $\frac{x}{x^2+3}+\frac{2}{x+5}$ **24** $\frac{2}{(x-1)^2}+\frac{1}{x-1}+\frac{4}{x+2}$ **25** $x-\frac{3}{(x-1)^2}+\frac{5}{x+4}$ **26** $\frac{3}{x}+\frac{2}{x-1}+\frac{4}{x-3}$ **27** $\frac{1}{x+3}+\frac{2}{x-5}$

28 $\frac{x}{x^2+x+5}-\frac{2}{2x+5}$ **29** $\frac{3}{x-1}-\frac{4}{x^2+3}$ **30 a)** $\frac{1}{x-5}-\frac{1}{x-4}$ **31 a)** $\frac{1}{x-2}-\frac{1}{x+5}$ **32 a) i)** $\frac{3}{x-2}-\frac{2}{x-4}$

32 a) ii) $\frac{6}{x-4}-\frac{6}{x-2}$ **33 a) i)** $\frac{5}{x+1}-\frac{2}{x+3}$ **ii)** $\frac{1}{x+1}-\frac{1}{x+3}$

Exercise 11C

1 a) $\frac{1}{1+x}+\frac{1}{1-2x}$, $2+x+5x^2+7x^3$, $|x|<\frac{1}{2}$ **b)** $\frac{2}{1-x}-\frac{1}{1+3x}$, $1+5x-7x^2+29x^3$, $|x|<\frac{1}{3}$

1 c) $\frac{2}{1+5x}-\frac{1}{1+3x}$, $1-7x+41x^2-223x^3$, $|x|<\frac{1}{5}$ **d)** $\frac{2}{1+x}+\frac{3}{2-x}$, $\frac{7}{2}-\frac{5}{4}x+\frac{19}{8}x^2-\frac{29}{16}x^3$, $|x|<1$

1 e) $\frac{1}{2+x}-\frac{1}{3+x}$, $\frac{1}{6}-\frac{5}{36}x+\frac{19}{216}x^2-\frac{65}{1296}x^3$, $|x|<2$ **f)** $\frac{1}{1+x^2}+\frac{3}{1-x}$, $4+3x+2x^2+3x^3$, $|x|<1$

2 a) $\frac{1}{1+x+x^2}+\frac{2}{1-2x}$, $3+3x+8x^2+17x^3$ **b)** $\frac{2}{1+3x+x^2}-\frac{1}{1+x}$, $1-5x+15x^2-41x^3$

2 c) $\frac{3}{1-x+x^2}+\frac{2}{1+2x}$, $5-x+8x^2-19x^3$ **3** -65 **4** 2 **5** 244 **6 a)** 2^n+3^n, $|x|<\frac{1}{3}$ **b)** $4^{n+1}+(-5)^{n+1}$, $|x|<\frac{1}{5}$

6 c) $n+3$, $|x|<1$ **7 a)** -4 **b)** $a=-7, b=16$

Exercise 11D

1 $\frac{2}{x-1}-\frac{1}{x}-\frac{1}{x^2}$ **2** $A=2, B=1, C=1, D=-3$ **3 i)** $\frac{3}{x+2}-\frac{1}{2x-3}$ **ii)** $1+\frac{4}{x-2}-\frac{1}{x-1}$ **4** $\frac{1}{x-3}-\frac{1}{x-1}$

5 a) $\frac{1}{1-2x}-\frac{1}{1-x}+\frac{1}{(1-x)^2}$ **b)** $1+3x+6x^2+11x^3$ **c)** $|x|<\frac{1}{2}$ **6** $\frac{3}{1+3x}+\frac{4}{2-x}$, $5-8x+\frac{55}{2}x^2$

7 a) $A=1, B=-3, C=1$ **b)** $1+3x+9x^2+27x^3$, $1-x^2$ **c)** $a=8, b=30$ **8 a)** $\frac{1}{2+x}+\frac{2-x}{1+x^2}$ **b)** $\frac{5}{2}-\frac{5}{4}x-\frac{15}{8}x^2$

Exercise 12A

1 a) $\Leftrightarrow$ **b)** $\Leftarrow$ **c)** $\Rightarrow$ **d)** $\Leftrightarrow$ **e)** $\Leftrightarrow$ **f)** $\Leftrightarrow$ **g)** $\Leftarrow$ **h)** $\Rightarrow$ **i)** $\Leftarrow$ **j)** $\Rightarrow$ **k)** $\Rightarrow$ **l)** $\Leftrightarrow$ **2 a)** If & only if **b)** If & only if

2 c) Only if **d)** If & only if **e)** If **f)** If & only if **g)** If **h)** If **i)** Only if **j)** If & only if **k)** Only if **l)** If & only if

3 a) Nec & suf **b)** Suf **c)** Nec **d)** Suf **e)** Nec & suf **f)** Nec **g)** Nec & suf **h)** Nec **i)** Nec & suf **j)** Suf **k)** Nec & suf

3 l) Nec & suf **4 a)** False **b)** False **c)** True **d)** True **e)** False **f)** True **g)** False **h)** False **i)** True **j)** True **k)** False

3 l) False **m)** True **n)** False **o)** False **p)** True **q)** True **r)** False **6 a)** True **b)** False **c)** True **d)** False

Exercise 12B

2 I is true, II is false **3 b)** False **4 a)** False **b)** True **c)** True **d)** True **e)** False

Exercise 13A

1 a) $6(2x-1)^2$ **b)** $6(3x+4)$ **c)** $20(5x-3)^3$ **d)** $-5(3-x)^4$ **e)** $-18(4-3x)^5$ **f)** $8x(x^2+1)^3$ **g)** $6x^2(x^3-6)$ **h)** $-12x(1-2x^2)^2$

1 i) $-8x^3(4-x^4)$ **j)** $-90x^2(7-5x^3)^5$ **k)** $48x(6x^2-5)^3$ **l)** $-42x(9-7x^2)^2$ **2 a)** $-6(2x-5)^{-4}$ **b)** $-3(3x+2)^{-2}$

2 c) $-4x(x^2+3)^{-3}$ **d)** $6x^2(5-2x^3)^{-2}$ **e)** $-\dfrac{4}{(3+4x)^2}$ **f)** $\dfrac{2x}{(4-x^2)^2}$ **g)** $\dfrac{10}{(3-2x)^2}$ **h)** $-\dfrac{6}{(x+1)^3}$ **i)** $\dfrac{70x}{(2-x^2)^6}$ **j)** $\dfrac{6x}{(3x^2+8)^2}$

2 k) $-60x^2(5x^3-4)^{-5}$ **l)** $\dfrac{12x^3}{(5-3x^4)^3}$ **3 a)** $(2x-1)^{-\frac{1}{2}}$ **b)** $-\frac{1}{3}(6-x)^{-\frac{2}{3}}$ **c)** $2x^2(x^3-2)^{-\frac{1}{3}}$ **d)** $x^4(4-x^5)^{-\frac{6}{5}}$ **e)** $\dfrac{2}{\sqrt{4x-5}}$

3 f) $\dfrac{2x}{3\sqrt[3]{(x^2+3)^2}}$ **g)** $\dfrac{1}{\sqrt{(5-2x)^3}}$ **h)** $-\dfrac{4x}{\sqrt[3]{(x^2+5)^4}}$ **i)** $\dfrac{2}{\sqrt[6]{(4x-7)^7}}$ **j)** $-\dfrac{10(5-4\sqrt{x})^4}{\sqrt{x}}$ **k)** $\dfrac{1}{4\sqrt{x}\sqrt{3+\sqrt{x}}}$ **l)** $\dfrac{1}{3\sqrt[3]{x^2}(4-\sqrt[3]{x})^2}$

4 a) $4(2x+1)(x^2+x-1)^3$ **b)** $\dfrac{3x^2-6}{2\sqrt{x^3-6x}}$ **c)** $\dfrac{3-2x}{(x^2-3x+5)^2}$ **d)** $-\dfrac{2}{\sqrt{x^3}}\left(\dfrac{1}{\sqrt{x}}-1\right)^3$ **e)** $20x(x^2-1)(x^4-2x^2+3)^4$

4 f) $\dfrac{2(1-3x^2)}{\sqrt[3]{(2-x+x^3)^4}}$ **g)** $-\dfrac{6}{x^2}\left(1+\dfrac{3}{x}\right)$ **h)** $4\left(\dfrac{1}{\sqrt{x}}-1\right)(2\sqrt{x}-x)^3$ **i)** $\dfrac{6-3x^2}{4\sqrt[4]{(6x-x^3)^3}}$ **j)** $7(5-12x^3)(2+5x-3x^4)^6$

4 k) $\dfrac{12(1-2x)}{(1+x-x^2)^5}$ **l)** $-\dfrac{1}{2x^2}\sqrt{\dfrac{x}{1+3x}}$

The constant of integration is omitted from the answers to Questions **5** and **6**.

5 a) $\frac{1}{10}(2x-3)^5$ **b)** $\frac{1}{15}(5x+8)^3$ **c)** $\frac{1}{18}(3x-4)^6$ **d)** $(x-7)^3$ **e)** $-\frac{1}{6}(4-x)^6$ **f)** $\frac{1}{28}(6-7x)^4$ **g)** $-\frac{1}{6}(3x-4)^{-2}$ **h)** $\frac{2}{3}(5-9x)^{-1}$

5 i) $-\dfrac{1}{12(2x-1)^6}$ **j)** $\dfrac{3}{1-x}$ **k)** $\frac{1}{3}\sqrt{(2x-3)^3}$ **l)** $18\sqrt[3]{(x-4)^2}$ **6 a)** $\frac{1}{12}(2x-7)^6$ **b)** $\sqrt{2x-1}$ **c)** $\frac{1}{8}(x^2+2)^4$ **d)** $-\frac{1}{18}(4-3x^2)^6$

6 e) $\frac{1}{9}(x^3-4)^3$ **f)** $\dfrac{2}{3-x^2}$ **g)** $\frac{1}{6}\sqrt{(x^4-1)^3}$ **h)** $-\frac{1}{2}\sqrt[3]{(2-3x^2)^4}$ **i)** $\frac{1}{4}(x^{\frac{4}{3}}-2)^3$ **j)** $\dfrac{5}{3-x^5}$ **k)** $\frac{1}{6}(x^2+1)^3$ **l)** $\frac{1}{5}x^5+\frac{2}{3}x^3+x$

8 $-\dfrac{4}{\left(x+\sqrt{4+(x^2-1)^3}\right)^5}\left[1+\dfrac{3(x^2-1)^2}{\sqrt{4+(x^2-1)^3}}\right]$

Exercise 13B

1 a) $2x-1$ **b)** $4x-9$ **c)** $12x+7$ **d)** $-(1+2x)$ **e)** $-(5+12x)$ **f)** $3x^2+8x-2$ **g)** $12x^2-2x-20$ **h)** $3x^2+6x-15$

1 i) $9x^2-22x+31$ **j)** $x(5x^3+3x-10)$ **k)** $25x^4+80x^3-15x^2-6x-12$ **l)** $27x^8-8x^7+24x^5-18x^2+4x$

2 a) $6x(x+1)(x+3)^3$ **b)** $x^2(2+x)(6+5x)$ **c)** $x^3(21x-4)(3x-1)^2$ **d)** $6x(2x+5)(4x+5)$ **e)** $3x^2(12x^2-1)(4x^2-1)^2$

2 f) $20x(2-x^3)(1-2x^3)$ **g)** $6x(25x^2+1)(5x^2+1)^3$ **h)** $x^6(14-95x^3)(2-5x^3)^3$ **i)** $x(8x^2+5x-2)(x^2+x-1)^2$

2 j) $x^2(12-7x+22x^2)(4-x+2x^2)^3$ **k)** $4x^3(15x^2-21x+4)(3x^2-6x+2)^2$ **l)** $5x(11x^3-5x+2)(x^3-x+1)^2$

3 a) $(5x-4)(x+2)(x-5)^2$ **b)** $2(5x+11)(2x-1)^2(x+4)$ **c)** $4(35x-9)(5x+2)^3(4x-3)^2$ **d)** $-2(7+10x)(2-x)^5(5+2x)^3$

3 e) $-(107+315x)(3+5x)(4-7x)^6$ **f)** $4(4x^2-3x+2)(x^2+1)(2x-3)^3$ **g)** $3(15x^2+18x-10)(5x+9)^2(x^2-2)^2$

3 h) $4(32x^2-35x-18)(2x^2-3)^4(4x-7)^5$ **i)** $x(52x^3+45x-16)(x^3-1)^2(4x^2+5)$ **j)** $4x(25x^2-87)(5-x^2)^3(6-5x^2)^5$

3 k) $(26x^2-78x+51)(x^2-3x+1)^4(2x-3)^2$ **l)** $15(7x^4-12x^3+12x^2-12x+12)(5x^2-10x+12)^2(x^3-6)^4$ **4 a)** $\dfrac{3x+2}{2\sqrt{x+1}}$

4 b) $\dfrac{3(2-x)}{\sqrt{3-x}}$ **c)** $\dfrac{3(3x+5)}{\sqrt{5+2x}}$ **d)** $\dfrac{x(5x+12)}{2\sqrt{x+3}}$ **e)** $\dfrac{2x(3-5x)}{\sqrt{3-4x}}$ **f)** $\dfrac{6x+11}{2\sqrt{x+3}}$ **g)** $-\dfrac{9x+14}{\sqrt{2x+5}}$ **h)** $\dfrac{(35x-4)(5x-4)^2}{2\sqrt{x}}$

4 i) $\dfrac{(15x-19)(3x+5)}{2\sqrt{x-2}}$ **j)** $\dfrac{8x-5}{\sqrt{(2x-3)(4x+1)}}$ **k)** $-\dfrac{9+4x}{2\sqrt{(6+x)(3-2x)}}$ **l)** $\dfrac{9x^2-x-6}{\sqrt{(x^2-2)(6x-1)}}$ **5 a)** $-\dfrac{2}{(x-2)^2}$ **b)** $-\dfrac{4}{(x-1)^2}$

5 c) $-\dfrac{7}{(4+x)^2}$ **d)** $\dfrac{11}{(x+2)^2}$ **e)** $\dfrac{13}{(x+4)^2}$ **f)** $\dfrac{10}{(x+2)^2}$ **g)** $\dfrac{11}{(2-5x)^2}$ **h)** $-\dfrac{10}{(2x-1)^2}$ **i)** $\dfrac{x(x+6)}{(x+3)^2}$ **j)** $\dfrac{x(8-x)}{(4-x)^2}$ **k)** $\dfrac{x^2(4x-9)}{(2x-3)^2}$

5 l) $\dfrac{x^4(15-4x)}{(3-x)^2}$ **6 a)** $\dfrac{(9x+2)(3x-2)}{2\sqrt{x^3}}$ **b)** $\dfrac{(25x-1)(5x+1)^2}{2\sqrt{x^3}}$ **c)** $\dfrac{(19x^2+4)(x^2-4)^4}{2\sqrt{x^3}}$ **d)** $-\dfrac{2x+1}{2\sqrt{x}(2x-1)^2}$ **e)** $\dfrac{3x-12\sqrt{x}-2}{2\sqrt{x}(2+x)^3}$

6 f) $\dfrac{5(4x+12\sqrt{x}+1)}{\sqrt{x}(5-4x)^4}$ **g)** $\dfrac{(45x^2-24x-2)(3x^2+2)^3}{\sqrt{(2x-1)^3}}$ **h)** $\dfrac{(3x^2+2x-6)(2-3x)}{\sqrt{(1-x^2)^3}}$ **i)** $\dfrac{3}{2\sqrt{x-2}\sqrt{(x+1)^3}}$ **j)** $\dfrac{11}{2\sqrt{x-3}\sqrt{(2x+5)^3}}$

6 k) $\dfrac{11}{2\sqrt{3+x}\sqrt{(2-3x)^3}}$ **l)** $-\dfrac{x(x^3+3x+6)}{2\sqrt{x^2+1}\sqrt{(x^3-3)^3}}$ **7 a)** $x^2(9-5x)(3-x)$ **b)** $-\dfrac{1}{(2x-1)^2}$ **c)** $\dfrac{(25x-1)(5x-1)}{2\sqrt{x}}$ **d)** $\dfrac{\sqrt{x}+2}{(\sqrt{x}+1)^2}$

7 e) $(25x-24)(5x+3)^2(x-2)$ **f)** $-\dfrac{7}{2\sqrt{3x-2}\sqrt{(x-3)^3}}$ **g)** $\dfrac{7(3-x)x^2}{\sqrt{7-2x}}$ **h)** $\dfrac{x(4-x)}{(2-x)^2}$ **i)** $-\dfrac{1}{\sqrt{x}(\sqrt{x}-1)^2}$

7 j) $(7-9x)(3-x)^3(2+x)^4$ **k)** $\dfrac{3x^2-2x-3}{(3x-1)^2}$ **l)** $\dfrac{4x-1}{2\sqrt{(x-1)(2x+1)}}$ **8 a)** $\dfrac{3(x^2-4x+1)(x-1)^2}{(x-3)^2}$ **b)** $\dfrac{30-12x^2+x^3}{\sqrt{(5-x^2)}\,(6-x)^2}$

8 c) $\dfrac{x^2(7\sqrt{x^5}-40x^2-20\sqrt{x}+120)}{2\sqrt{(4-x^2)}\,(5-\sqrt{x})^2}$

Exercise 13C

1 $y+4x=16, 4y-x=30$ **2** $2y+x=9, y=2x-3$ **3** $\left(-\frac{11}{17}, \frac{45}{17}\right)$ **4** $9\frac{11}{13}$ **5** $\left(-4, \frac{4}{3}\right), \left(2, \frac{2}{3}\right)$ **6** $(2, 2)$ **7** $(0, 0), (4, -8)$

8 $(-1, -1)$ **9** $(-3, 0)$, min; $\left(\frac{1}{3}, \frac{500}{27}\right)$, max **10** $\left(-2, \frac{1}{4}\right)$, max **11** $(10, 2\sqrt{5})$, min

12 $(-2, 0)$ max; $\left(-\frac{4}{5}, -\frac{148176}{3125}\right)$, min; $(1, 27)$, max; $(2, 0)$, min **14** 100, £15 000 **15** 25 cm^2 **16 b)** $a=2, b=-3, y''(3)=\frac{2}{3}>0$

17 b) $\dfrac{32\sqrt{3}}{9}$

Exercise 13D

1 $\dfrac{5}{x-3}-\dfrac{2}{3x+2}, -\frac{157}{32}$ **2** $\dfrac{3x}{x^2+1}-\dfrac{1}{x-2}$ **3** $\dfrac{4}{x-4}-\dfrac{1}{x-1}; \dfrac{1}{(x-1)^2}-\dfrac{4}{(x-4)^2}$ **4 a)** $\frac{3}{2}x^2(1+x^3)^{-\frac{1}{2}}$ **b)** $\frac{4}{3}$

6 a) $\left(-\frac{1}{2}, 0\right), x=1$ **b)** $y=4x+1, \left(1\frac{3}{4}, 8\right)$ **7** $f'(x)=\dfrac{x-3}{2\sqrt{(x-1)^3}}$; $(3, 2\sqrt{2})$, min **8** $-\dfrac{6x}{(x^2-2)^2}$; $\left(0, -\frac{1}{2}\right)$, min **9** $\frac{1}{9}, -1$

Exercise 14A

1 a) $\dfrac{2x}{3y^2}$ **b)** $\dfrac{3y}{2y-3x}$ **c)** $-\dfrac{2xy+y^2}{2xy+x^2}$ **d)** $\dfrac{2-3y}{3(x+y^2)}$ **e)** $\dfrac{4x^3-y^2-6}{2xy}$ **f)** $\dfrac{6x^5-5y^3-9y}{3x(5y^2+3)}$ **g)** $\dfrac{x-2y}{x}$ **h)** $x-1$ **i)** $\dfrac{7x^4(x^2-5y^3)}{1+21x^5y^2}$

2 a) $\frac{1}{10}$ **b)** $\frac{4}{3}$ **c)** $\frac{1}{4}$ **d)** $\frac{4}{27}$ **e)** -6 **f)** 0 **g)** -15 **h)** $-\frac{1}{2}$ **3** $5y-3x+18=0$ **4** $y+2x=5, 2y=x$

5 $18y+x=12, 3y-54x+323=0$ **6** $15x-8y=36, 7y-15x=36$ **7** $(2, 1), (2, 5)$ **8** $x=-4, x=1$

9 a) $(2, -2)$, max; $(2, -4)$, min **b)** $(1, -1)$, max; $(1, -3)$, min **c)** $(0, 2)$, max **d)** $(-4, -8)$, max; $(2, 4)$, min

9 e) $(-1, -1)$, max; $(1, 1)$, min **f)** $\left(\frac{1}{2}, 1\right)$, min; $\left(-\frac{1}{2}, -1\right)$, max **11 b)** $\left(\frac{1}{3}, \frac{7}{3}\right)$, max; $(3, -3)$, min **c)** $\left(\frac{13}{3}, -\frac{5}{3}\right), (-1, 1)$

Exercise 14B

1 a) $y=(x-3)^2$ **b)** $y=3x^2+6x-2$ **c)** $y=\frac{1}{2}x^4+5$ **d)** $y=\dfrac{4}{x}$ **e)** $y=\dfrac{3x^2}{x^2+4}$ **f)** $y=\dfrac{5-x}{3}$ **g)** $y=\dfrac{1-x}{x^2}$ **h)** $y=\dfrac{1-x}{2}$

1 i) $y=\dfrac{x}{5x+1}$ **2 a)** $\dfrac{2}{t}$ **b)** $\dfrac{1}{3t^2}$ **c)** $5\sqrt{t}$ **d)** $(2t-1)^2$ **e)** $\dfrac{3(t-1)}{8}$ **f)** $\dfrac{2\sqrt{t^3}-1}{2-\sqrt{t}}$ **g)** $-2t^2(t+2)$ **h)** $-\dfrac{(3+\sqrt{t})^2}{2}$

2 i) $-\sqrt[3]{\dfrac{(3t-4)^4}{(6t+1)^2}}$ **3 a)** 2 **b)** $-\frac{1}{9}$ **c)** -36 **d)** 1 **e)** $-\frac{3}{50}$ **f)** $\frac{15}{32}$ **g)** -432 **h)** $\frac{3}{22}$ **4** $y=x+19$

5 $y=3x-4, y+3x+10=0$ **6** $12y-x=30, y+12x=75$ **7** $y+x=2, 4y-x+4=0$ **8** $(-1, 0), (15, -4)$

9 $4y-x=19, 4y+x+43=0$ **10** $(-2, 2)$ **11 a)** $2y-x+5=0, y+2x+30=0$ **b)** $\frac{1}{2}\times-2=-1, (-11, -8)$

12 b) $\left(\frac{3}{4}, \frac{9}{4}\right), \left(\frac{5}{4}, -\frac{25}{4}\right)$ **13 b)** $\frac{4}{19}, -1$ **14 a)** $t^2, \dfrac{2t}{3}$ **b)** $\dfrac{t^2+2}{3}, \dfrac{2t}{27}$ **c)** $\dfrac{1-2t}{t}, -\dfrac{1}{2t^3}$ **d)** $\dfrac{t^2-1}{t}, \dfrac{t^2+1}{12t^4}$ **e)** $\sqrt{t}\,(2t+3), \frac{3}{2}(2t+1)$

14 f) $2t^2(2-t), 2t^3(3t-4)$ **g)** $\dfrac{2t}{2t+3}, \dfrac{6}{(2t+3)^3}$ **h)** $2t, \dfrac{1}{3t(t+1)}$ **i)** $2t^2(t+3), 6t^3(t+2)$ **j)** $\left(\dfrac{t+1}{t-1}\right)^2, 4\left(\dfrac{t+1}{t-1}\right)^3$ **k)** $\frac{1}{2}, 0$

14 l) $-\left(\dfrac{1+\sqrt{t}}{1-\sqrt{t}}\right)^2, 4\left(\dfrac{1+\sqrt{t}}{1-\sqrt{t}}\right)^3$ **15 b)** $(27, -13)$, min **16 b)** $(-13, -8)$, min **17 b)** $(0, 25)$, max; $(12, -7)$, min

18 b) $\left(\frac{7}{2}, -4\right)$, max; $\left(\frac{5}{2}, 4\right)$, min

Exercise 14C

1 24 cm^2 s^{-1} **2** 100 cm^2 s^{-1} **3** $\dfrac{10\pi}{3}$ cm^2 s^{-1} **4** $\frac{7}{20}$ cm s^{-1} **5** 4 cm s^{-1} **6** $\frac{6}{25}$ cm s^{-1} **7** $\frac{1}{12}$ cm s^{-1} **8** $\dfrac{1}{20\pi}$ cm s^{-1}

9 $\dfrac{2\pi}{5}$ cm s^{-1} **10** $\dfrac{3}{x}$ cm s^{-1} **11 b)** $\frac{2}{3}$ cm s^{-1} **12** 4 cm^2 s^{-1} **13** $\frac{2}{3}$ cm s^{-1} **14** $\dfrac{3}{4\pi}$ cm min^{-1} **15** $\dfrac{4}{9\pi}$ cm s^{-1} **16** $\dfrac{128}{9\pi}$ cm s^{-1}

17 $\frac{1}{9}$ cm s^{-1} **18** $\frac{1}{2}$ cm s^{-1} **19** $\frac{1}{100}$ cm s^{-1}

Exercise 14D

1 $\frac{1}{7}$ **2 a)** $-\frac{2}{3}$ **b)** $2x+3y=21$ **3 b)** $(-1,0)$ **c)** $-\frac{1}{5}$ **d)** $y=5x-3$ **4 a)** $\dfrac{x-y}{x-4y}$; $(-2,-2)$, $(2,2)$ **b)** $2x-y=3\sqrt{3}$

5 a) $(5,0)$ **c)** $-\frac{9}{4}$ **6 i)** -2 **iii)** $y=2x-6$ **iv)** $(-5,9)$ **7 b)** $2x+y+2=0$ **8 a)** $\left(\frac{1}{4},0\right)$ **c)** $(1,3)$

9 iii) $x+t^2y=4t$ **iv)** Area $=8$ **10 b)** $-\dfrac{1}{t^2}$ **d)** $-\frac{1}{8}$ **11 i)** $\dfrac{5}{18\pi}$ cm s^{-1} **ii)** $33\frac{1}{3}$ cm^2 s^{-1}

Exercise 15A

2 a) $1, 90°$; $-1, 270°$ **b)** $4, 0°$; $2, 180°$ **c)** $8, 270°$; $2, 90°$ **d)** $1, 70°$; $-1, 250°$ **e)** $7, 220°$; $-1, 40°$ **f)** $10, 330°$; $-4, 150°$

2 g) $\frac{1}{2}, 270°$; $\frac{1}{4}, 90°$ **h)** $6, 0°$; $2, 90°$ **3 a)** $-\sin 20°$ **b)** $-\cos 60°$ **c)** $-\tan 20°$ **d)** $\cos 50°$ **e)** $\tan 40°$ **f)** $-\cos 50°$ **g)** $-\sin 20°$

3 h) $\cos 80°$ **4 a)** 17.5°, 162.5° **b)** 45.6°, 314.4° **c)** 63.4°, 243.4° **d)** 120°, 240° **e)** 200.5°, 339.5° **f)** 98.1°, 278.1° **g)** 53.1°, 126.9°

4 h) 270° **5 a)** 30°, 150° **b)** ±70.5° **c)** −116.6°, 63.4° **d)** −18.4°, 161.6° **e)** ±80.4° **f)** 11.5°, 168.5° **g)** ±180° **h)** −5.7°, −174.3°

6 a) 0°, 180°, 360°, 30°, 150° **b)** 90°, 270°, 70.5°, 289.5° **c)** 0°, 180°, 360°, 78.5°, 281.5° **d)** 0°, 180°, 360°, 104.0°, 284.0°

6 e) 30°, 150°, 19.5°, 160.5° **f)** 45°, 225°, 26.6°, 206.6° **g)** 0°, 360°, 101.5°, 258.5° **h)** 26.6°, 206.6°, 33.7°, 213.7°

6 i) 48.2°, 311.8° **j)** 270°, 48.6°, 131.4° **k)** 104.5°, 255.5° **l)** 63.4°, 243.4°, 104.0°, 284.0° **7 a)** 8.7°, 81.3° **b)** 4.7°, 64.7°, 124.7°

7 c) 7.5°, 37.5°, 97.5°, 127.5° **d)** 56.8°, 123.2° **e)** 110.9°, 159.1° **f)** 13.3°, 58.7°, 85.3°, 130.7°, 157.3° **g)** 24.3°, 65.7°, 114.3°, 155.7°

7 h) 18°, 90°, 162° **8 a)** 3.6°, 136.4° **b)** 157.5°, 302.5° **c)** 88.5°, 291.5° **d)** 13.0°, 193.0° **e)** 132.5°, 351.5° **f)** 94.8°, 274.8°

8 g) 29.0°, 209.0° **h)** 164.5°, 315.5° **9 a)** −26.7°, 51.7° **b)** −86°, −6°, 34° **c)** −33.4° **d)** −81.4°, −42.1°, 8.6°, 47.9°

9 e) −75.9°, 14.1° **f)** 54.3° **g)** −75.0°, −39.0°, −3.0°, 33.0°, 69.0° **h)** −6.5°, 53.5°

Exercise 15B

1 a) $\dfrac{1}{\sqrt{2}}$ **b)** $\dfrac{\sqrt{3}}{2}$ **c)** $\dfrac{1}{\sqrt{3}}$ **d)** $-\dfrac{\sqrt{3}}{2}$ **e)** $-\dfrac{2\sqrt{3}}{3}$ **f)** $-\dfrac{1}{\sqrt{3}}$ **g)** -1 **h)** $\sqrt{3}$ **i)** $-\sqrt{2}$ **j)** $\dfrac{\sqrt{3}}{2}$ **k)** $\sqrt{2}$ **l)** $-\frac{1}{2}$

2 a) 60°, 120° **b)** 60°, 300°, 120°, 240° **c)** 135°, 315° **d)** 210°, 330° **e)** 60°, 240°, 120°, 300° **f)** 150°, 210°

2 g) 15°, 75°, 135°, 195°, 255°, 315° **h)** 30°, 120°, 210°, 300° **i)** 30°, 60°, 120°, 150°, 210°, 240°, 300°, 330° **j)** 12°, 192°

2 k) 52°, 232° **l)** 46°, 66°, 166°, 186°, 286°, 306° **3 a)** $\dfrac{\sqrt{5}}{3}$ **b)** $\dfrac{2}{\sqrt{5}}$ **c)** $\dfrac{\sqrt{5}}{2}$ **4 a)** $\dfrac{\sqrt{15}}{4}$ **b)** $\sqrt{15}$ **c)** $\dfrac{4}{\sqrt{15}}$ **5 a)** $\dfrac{3}{\sqrt{10}}$

5 b) $\sqrt{10}$ **c)** $\dfrac{\sqrt{10}}{3}$ **6 a)** $\frac{4}{5}$ **b)** $\frac{4}{3}$ **c)** $\frac{3}{4}$ **7 a)** 71.6°, 251.6° **b)** 59.0°, 239.0° **c)** 135°, 315° **d)** 33.7°, 213.7° **e)** 63.4°, 243.4°

7 f) 149.0°, 329.0° **g)** 0°, 180°, 360°, 78.7°, 258.7° **h)** 90°, 270°, 23.2°, 203.2° **i)** 26.6°, 206.6°, 153.4°, 333.4°

7 j) 18.4°, 198.4°, 161.6°, 341.6° **k)** 51.3°, 231.3°, 128.7°, 308.7° **l)** 63.4°, 243.4° **8 a)** 90°, 30°, 150° **b)** ±70.5°

8 c) ±41.4°, ±60° **d)** 41.8°, 138.2° **e)** 53.1°, 126.9° **f)** ±90°, ±120° **g)** ±90°, 30°, 150° **h)** ±90°, −138.2°, −41.8°, 23.6°, 156.4°

9 a) 71.6°, 251.6°, 116.6°, 296.6° **b)** 63.4°, 243.4°, 166.0°, 346.0° **c)** 48.5°, 98.6°, 228.5°, 278.6° **d)** 63.4°, 243.4°, 143.1°, 323.1°

9 e) 48.6°, 131.4° **f)** 70.5°, 289.5° **g)** 14.0°, 194.0°, 104.0°, 284.0° **h)** 0°, 180°, 360°, 26.6°, 206.6° **i)** 0°, 360°, 109.5°, 250.5°

9 j) 45°, 225°, 153.4°, 333.4° **10 a)** −160.5°, −19.5°, −90°, 30°, 150° **b)** 0°, ±60°, ±120°

10 c) −116.6°, 63.4°, −45°, 135°, −108.4°, 71.6° **d)** 0°, ±75.5°, ±109.5° **e)** 19.5°, 160.5°, 90°

10 f) −108.4°, 71.6°, −63.4°, 116.6°, −26.6°, 153.4° **11 a)** 14.3°, 50.3°, 86.3°, 122.3°, 158.3° **b)** 7.3°, 82.8°, 99.8°, 170.3°

11 c) 4.6°, 49.6°, 94.6°, 139.6°, 27.1°, 72.1°, 117.1°, 162.1° **d)** 16.2°, 43.8°, 136.2°, 163.8°, 30°, 90°, 150° **e)** 30°, 150°, 39.2°, 140.8°

11 f) 4.9°, 40.1°, 22.5°, 112.5°, 94.9°, 130.1° **12** $-\frac{1}{2}, \frac{1}{3}, \frac{1}{2}, 1$ **13** $-2, -\frac{1}{2}, 1, 3$

Exercise 15D

1 i) $\pi-\alpha$ ii) $3\pi+\alpha$, $4\pi-\alpha$ **2 a)** 21.8°, 201.8° **b)** 11.8°, 78.2°, 191.8°, 258.2° **3 a)** $\dfrac{\pi}{18}, \dfrac{5\pi}{18}, \dfrac{13\pi}{18}, \dfrac{17\pi}{18}$ **b)** $\dfrac{3\pi}{4}$

4 a) ±60° **b)** $-22\frac{1}{2}°$, $37\frac{1}{2}°$ **5** 210°, 330° **6 a)** $\frac{1}{3}, -\frac{1}{2}$ **b)** 289.5°, 430.5°, 240°, 480° **7** 3.48 rad, 5.94 rad **8** 210°, 330°

9 ±108.0° **10** $-\frac{1}{2}$, 1, 3; 90°, 210°, 330° **11 a)** x^2-3x-1; $\dfrac{1}{2}, \dfrac{3\pm\sqrt{13}}{2}$ **b)** 60°, 300°, 108°, 252°

Exercise 16A

1 a) $\frac{63}{65}$ **b)** $\frac{56}{65}$ **c)** $-\frac{33}{56}$ **2 a)** $\frac{16}{65}$ **b)** $\frac{56}{65}$ **c)** $\frac{16}{63}$ **3 a)** $\dfrac{31\sqrt{2}}{50}$ **b)** $\dfrac{17\sqrt{2}}{50}$ **c)** $-\frac{17}{31}$ **4** 1 **5** $\frac{3}{11}$ **6** $-\frac{5}{3}$ **7** $\dfrac{1+2\sqrt{3}}{2-\sqrt{3}}(=8+5\sqrt{3})$

8 $\dfrac{1+3\sqrt{3}}{3-\sqrt{3}}\left(=\dfrac{6+5\sqrt{3}}{3}\right)$ **9 a)** $\sqrt{3}$ **b)** $\sqrt{2}-1$ **c)** $\dfrac{1}{2+\sqrt{3}}(=2-\sqrt{3})$ **d)** 1 **e)** $\dfrac{2\sqrt{3}-1}{2-\sqrt{3}}(=4+3\sqrt{3})$ **f)** $\dfrac{\sqrt{3}-\sqrt{2}}{\sqrt{2}-1}$

Exercise 16B

1 a) $\frac{24}{25}$ **b)** $-\frac{7}{25}$ **c)** $-\frac{24}{7}$ **2 a)** $-\frac{119}{169}$ **b)** $\frac{169}{120}$ **c)** $-\frac{119}{120}$ **3 a)** $-\frac{4}{3}$ **b)** $\frac{4}{5}$ **c)** $-\frac{5}{3}$ **4 a)** 0°, 180°, 360°, 80.4°, 279.6°

4 b) 90°, 270°, 41.8°, 138.2° **c)** 90°, 270°, 210°, 330° **d)** 60°, 300°, 109.5°, 250.5° **e)** 48.6°, 131.4° **f)** 90°, 194.5°, 345.5°

4 g) 0°, 180°, 360°, 60°, 300°, 120°, 240° **h)** 0°, 180°, 360°, 60°, 300° **5 a)** 0°, ±120°, ±360° **b)** ±180°, 97.2°, 262.8°

5 c) 0°, ±120°, ±240°, ±360° **d)** ±209.0° **e)** 0°, ±53.1°, 360°, ±306.9°, ±360° **f)** −323.1°, 36.9°, 60°, 300°

22 b) 30°, 150°, 90°, 210°, 330° **23 b)** 0°, 180°, 360°, 75.5°, 284.5° **26** $a = -1, b = 1$

Exercise 16C

1 a) $2\sin 3\theta\cos\theta$ **b)** $2\cos 4\theta\cos\theta$ **c)** $-2\sin 4\theta\sin 2\theta$ **d)** $2\cos 4\theta\sin\theta$ **e)** $-2\sin 6\theta\sin 2\theta$ **f)** $-2\sin\left(\frac{7\theta}{2}\right)\sin\left(\frac{\theta}{2}\right)$

1 g) $2\cos 5\theta\cos 2\theta$ **h)** $-2\cos 5\theta\sin 3\theta$ **2 a)** $\sin 6\theta + \sin 2\theta$ **b)** $\cos 5\theta + \cos\theta$ **c)** $\cos 5\theta + \cos 3\theta$ **d)** $\sin 9\theta - \sin 3\theta$

2 e) $\cos 5\theta - \cos 3\theta$ **f)** $\cos 13\theta + \cos\theta$ **g)** $\cos 8\theta + \cos\theta$ **h)** $\cos 4\theta - \cos\theta$ **3 a)** $\frac{1}{\sqrt{2}}$ **b)** $-\sqrt{\frac{3}{2}}$ **c)** $\frac{1}{\sqrt{2}}$ **d)** $-\sqrt{\frac{3}{2}}$ **e)** $\frac{1}{4}$

3 f) $\frac{2+\sqrt{3}}{4}$ **g)** $\frac{\sqrt{3}+\sqrt{2}}{4}$ **h)** $\frac{\sqrt{2}-1}{4}$ **4 a)** 0°, 40°, 80°, 120°, 160°, 36°, 108°, 180° **b)** 0°, 72°, 144°, 120°

4 c) 0°, 40°, 80°, 120°, 160°, 60°, 180° **d)** 30°, 90°, 150° **e)** 0°, 90°, 180°, 18°, 54°, 126°, 162° **f)** 0°, 72°, 144°, 60°, 180° **g)** 70°

5 a) 90°, 10°, 50°, 130°, 170° **b)** $22\frac{1}{2}$°, $67\frac{1}{2}$°, $112\frac{1}{2}$°, $157\frac{1}{2}$°, 60° **c)** 0°, 90°, 180°, 70°, 110° **d)** 0°, 45°, 90°, 135°, 180°, 20°, 100°, 140°

5 e) 0°, 36°, 72°, 108°, 144°, 180°, 20°, 100°, 140° **f)** $22\frac{1}{2}$°, $67\frac{1}{2}$°, $112\frac{1}{2}$°, $157\frac{1}{2}$°, 35°, 55°, 95°, 115°, 155°, 175°

5 g) 36°, 108°, 180°, 45°, 135°, 0°, 120° **h)** 180°, 10°, 50°, 90°, 130°, 170°

Exercise 16D

1 a) 5, $\frac{4}{3}$ **b)** 13, $\frac{12}{5}$ **c)** $\sqrt{29}$, $\frac{5}{2}$ **d)** $\sqrt{29}$, $\frac{2}{5}$ **e)** $\sqrt{2}$, 1 **f)** 25, $\frac{3}{4}$ **g)** 2, $\frac{1}{\sqrt{3}}$ **h)** $2\sqrt{5}$, 2 **2 a)** 13, 67.4°; −13, 247.4°

2 b) $\sqrt{5}$, 26.6°, $-\sqrt{5}$, 206.6° **c)** 12, 143.1°; 2, 323.1° **d)** $10+\sqrt{5}$, 296.6°; $10-\sqrt{5}$, 116.6°

2 e) $\frac{1}{2-\sqrt{2}}\left(=\frac{2+\sqrt{2}}{2}\right)$, 225°; $\frac{1}{2+\sqrt{2}}\left(=\frac{2-\sqrt{2}}{2}\right)$, 45° **f)** $\frac{1}{4}$, 311.8°; $\frac{1}{10}$, 131.8° **g)** 1, 112.6°; $\frac{3}{29}$, 292.6° **h)** ±∞, near 351.8°

3 a) 90°, 330° **b)** 60.4°, 193.3° **c)** 105°, 345° **d)** 80.0°, 325.2° **e)** 51.6°, 187.9° **f)** 237.7°, 339.2° **g)** 63.8°, 339.8°

3 h) 14.2°, 141.8°, 194.2°, 321.8° **i)** 86.6°, 326.6° **j)** 236.3°, 326.3° **k)** 7.3°, 140.2° **l)** 52.5°, 82.5°, 232.5°, 262.5°

5 a) A = 3, R = 5, $\alpha = \tan^{-1}\frac{4}{3}$ **b)** 6.6°, 120.2°, 186.6°, 300.2° **c)** −83.4°, 30.2°, 96.6°, 210.2°

Exercise 16E

1 a) 1.16, 5.12 **b)** 0.78, 2.37 **c)** 1.33, 4.47 **d)** 2.50, 3.79 **e)** 3.45, 5.98 **f)** 0.54, 3.68 **g)** 0.14, 3.00 **h)** 0.17, 3.31

1 i) 1.98, 4.30 **j)** 3.24, 6.18 **k)** 1.77, 4.51 **l)** 0.34, 3.48 **2 a)** 0.30, 3.04 **b)** −1.13, 0.73 **c)** −1.39, 1.75 **d)** −1.07, 1.67

2 e) −0.75, 3.09 **f)** −0.57, 2.32 **g)** ±0.46, ±2.68 **h)** −2.97, −2.26, −0.88, −0.17, 1.21, 1.93 **i)** ±0.29, ±1.28, ±1.86, ±2.85

2 j) −1.78, −0.21, 1.37, 2.94 **k)** −2.63, −2.18, 0.51, 0.96 **l)** −2.47, −1.42, −0.38, 0.67, 1.72, 2.77

3 a) $\frac{\pi}{12}, \frac{5\pi}{12}, \frac{13\pi}{12}, \frac{17\pi}{12}$ **b)** $\frac{\pi}{4}, \frac{3\pi}{4}, \frac{5\pi}{4}, \frac{7\pi}{4}, \frac{\pi}{2}, \frac{3\pi}{2}$ **c)** $\frac{\pi}{2}, \frac{7\pi}{6}, \frac{11\pi}{6}$ **d)** $\frac{5\pi}{24}, \frac{17\pi}{24}, \frac{29\pi}{24}, \frac{41\pi}{24}$ **e)** $\frac{\pi}{2}, \frac{3\pi}{2}, \pi$

3 f) $0, \pi, 2\pi, \frac{\pi}{3}, \frac{4\pi}{3}, \frac{2\pi}{3}, \frac{5\pi}{3}$ **g)** $\frac{\pi}{18}, \frac{\pi}{2}, \frac{13\pi}{18}, \frac{7\pi}{6}, \frac{25\pi}{18}, \frac{11\pi}{6}$ **h)** $0, \frac{\pi}{4}, \frac{\pi}{2}, \frac{3\pi}{4}, \pi, \frac{5\pi}{4}, \frac{3\pi}{2}, \frac{7\pi}{4}, 2\pi$ **i)** $\frac{7\pi}{12}, \frac{11\pi}{12}$

3 j) $\frac{\pi}{3}, \frac{4\pi}{3}, \frac{2\pi}{3}, \frac{5\pi}{3}$ **k)** $0, \pi, 2\pi, \frac{2\pi}{3}, \frac{4\pi}{3}$ **l)** $\pi, \frac{\pi}{3}, \frac{5\pi}{3}$

Exercise 16F

1 $\frac{29}{11}$ **2** $\frac{84}{85}$ **3 i)** $\frac{20}{21}$ **iii)** 52 cm **4 a)** $\frac{455}{697}$ **b)** $\frac{455}{697}$ **c)** 136 cm **5** −67.5°, 22.5°, 45° **6** 41°, 319° **7** 270°, 56.4°, 123.6°

9 15°, 75°, 195°, 255° **10** 41.4°, 55.8°; $\hat{C} > \hat{A}$ ∴ AB > BC **11 i)** $\sin x$ **ii)** $\sin x$ **iii)** $\sin x$ **iv)** $\sin x$ **v)** $\cos x$ **vi)** $\tan 3x$

12 0°, 180°, 15°, 75°, 105°, 165° **13 ii)** 54.7°, 125.3° **14** $5\cos(\theta + 36.9°)$ **i)** 5, −5 **ii)** 29.6°, 256.7° **15 i)** $37\cos(\theta - 18.9°)$

15 ii) 76.2°, 321.6° **16** 60°, 180° **17** $6\cos\theta - 2\sin\theta$ **a)** 18.7°, 55.1° **b)** 10; 26.5° **18** 57.8°; 92.2°, 332.2° **19 a)** 12

19 b) 13, 1.176 **c)** 3, 1; 0.983 rad **20** $\frac{\pi}{3} < x < \frac{\pi}{2}$ or $\frac{2\pi}{3} < x < \pi$

Exercise 17A

The constant of integration is omitted in the answers to Questions **3** and **6**.

1 a) $3\cos 3x$ **b)** $-2\sin 2x$ **c)** $5\cos 5x$ **d)** $-6\cos 6x$ **e)** $-14\sin 7x$ **f)** $30\sin 5x$ **g)** $4\cos\frac{1}{2}x$ **h)** $-\sin(x+3)$ **i)** $\cos(x-4)$

1 j) $3\cos\left(x+\frac{\pi}{4}\right)$ **k)** $8\sin(4x-7)$ **l)** $12\cos\left(\frac{3x-\pi}{2}\right)$ **2 a)** $2x\cos(x^2)$ **b)** $-3x^2\sin(x^3)$ **c)** $-4x\sin(x^2-1)$

2 d) $18x^2\cos(2x^3+3)$ **e)** $8x\cos(1-x^2)$ **f)** $72x^3\sin(4-3x^4)$ **g)** $2(x-1)\sin(x^2-2x)$ **h)** $3(x^2-2x)\cos(x^3-3x^2)$

2 i) $2(3x-1)\cos(6x^2-4x+1)$ **j)** $14(1-2x^3)\sin(2x-x^4)$ **k)** $\frac{3\cos\sqrt{x}}{\sqrt{x}}$ **l)** $\frac{1}{x^2}\sin\left(\frac{1}{x}\right)$ **3 a)** $\sin 2x$ **b)** $-\frac{1}{4}\cos 4x$

3 c) $-\frac{3}{5}\cos 5x$ **d)** $\frac{1}{2}\sin(2x-1)$ **e)** $2\cos(3x+2)$ **f)** $-\frac{4}{5}\cos\left(\frac{5x-\pi}{4}\right)$ **g)** $\frac{1}{2}\sin(x^2)$ **h)** $-2\cos(x^4)$ **i)** $\frac{3}{2}\sin(x^2-7)$

3 j) $\sin(x^2-4x)$ **k)** $\frac{1}{3}\cos(3x^2-x^3)$ **l)** $-2\cos(\sqrt{x})$ **4 a)** $2\sin x\cos x$ **b)** $-3\sin x\cos^2 x$ **c)** $-\frac{\sin x}{2\sqrt{\cos x}}$ **d)** $\frac{2\sin x}{\cos^3 x}$

4 e) $14\cos x\sin^6 x$ **f)** $18\sin x\cos^5 x$ **g)** $20\cos 5x\sin^3 5x$ **h)** $-3\sin\frac{1}{2}x\cos^5\frac{1}{2}x$ **i)** $-\frac{4\sin 4x}{\sqrt{\cos 4x}}$ **5 a)** $2\cos x(1+\sin x)$

5 b) $4\sin x(3-\cos x)^3$ **c)** $-18\sin x(5+3\cos x)^5$ **d)** $3(\cos x-2\sin 2x)(\sin x+\cos 2x)^2$ **e)** $\frac{\sin x}{(1+\cos x)^2}$ **f)** $-\frac{3\cos x}{\sqrt{1-6\sin x}}$

5 g) $-\frac{9\sin 3x}{(1+\cos 3x)^2}$ **h)** $\frac{12\cos 6x}{\sqrt{(1-\sin 6x)^3}}$ **i)** $3\sin 2x(1+\sin^2 x)^2$ **6 a)** $\sin^4 x$ **b)** $-\frac{1}{3}\cos^3 x$ **c)** $\frac{1}{6}(4-\cos x)^6$ **d)** $\frac{1}{2}(3+\sin x)^4$

6 e) $\frac{1}{1+\cos x}$ **f)** $2\sqrt{4-\sin x}$ **g)** $\frac{1}{3}\sin^6 3x$ **h)** $\frac{1}{2}(5-2\cos 2x)^4$ **i)** $-\frac{1}{3}\sqrt{(6+\cos 4x)^3}$ **j)** $\frac{1}{3}(x-\sin x)^3$ **k)** $\sqrt{x^2+2\cos x}$

6 l) $-\frac{1}{8}\cos^2 2x$ **7 a)** $\sin x+x\cos x$ **b)** $x(2\cos x-x\sin x)$ **c)** $\cos 3x-3x\sin 3x$ **d)** $3x^2(\sin 6x+2x\cos 6x)$

7 e) $\sin^4 x(\sin x+5x\cos x)$ **f)** $6x\cos^3 2x(\cos 2x-4x\sin 2x)$ **g)** $\frac{\sin x-x\cos x}{\sin^2 x}$ **h)** $-\frac{2(x+1)\sin 2x+\cos 2x}{(x+1)^2}$ **i)** $-\frac{\cos x}{(1+\sin x)^2}$

7 j) $\frac{2(1+\sin 2x)}{\cos^2 2x}$ **k)** $\frac{1+2x\sin x\cos x+\cos^2 x}{(1+\cos^2 x)^2}$ **l)** $\frac{1+\sin x+\cos x}{(1+\cos x)^2}$

Exercise 17B

The constant of integration is omitted in the answers to Questions **5** and **8**.

3 a) $2\sec^2 2x$ **b)** $3\sec 3x\tan 3x$ **c)** $-6\operatorname{cosec}^2 6x$ **d)** $-4\sec^2 4x$ **e)** $-5\operatorname{cosec}5x\cot 5x$ **f)** $14\operatorname{cosec}^2 7x$ **g)** $2\sec\frac{1}{3}x\tan\frac{1}{3}x$

3 h) $\sec^2(x+2)$ **i)** $-\operatorname{cosec}(x-1)\cot(x-1)$ **4 a)** $3x^2\sec(x^3)\tan(x^3)$ **b)** $-2x\operatorname{cosec}^2(x^2)$ **c)** $8x^3\sec^2(x^4)$

4 d) $-12x\operatorname{cosec}^2(3x^2)$ **e)** $40x^3\operatorname{cosec}(2x^4)\cot(2x^4)$ **f)** $-\frac{\operatorname{cosec}^2(\sqrt{x})}{\sqrt{x}}$ **g)** $2x\sec^2(x^2+3)$ **h)** $3x^2\operatorname{cosec}(1-x^3)\cot(1-x^3)$

4 i) $36x\sec(6x^2+5)\tan(6x^2+5)$ **j)** $28x^2\operatorname{cosec}(2-x^4)\cot(2-x^4)$ **k)** $6x(x-4)\sec^2(x^3-6x^2)$ **l)** $2(1-x^3)\sec(4x-x^4)\tan(4x-x^4)$

5 a) $\tan 2x$ **b)** $-\frac{1}{3}\operatorname{cosec}3x$ **c)** $-\frac{1}{2}\cot 8x$ **d)** $\frac{1}{2}\sec 6x$ **e)** $-3\tan 3x$ **f)** $-3\operatorname{cosec}4x$ **g)** $\cot(x^2)$ **h)** $4\sec(x^2)$ **i)** $-\cot(x^3)$

5 j) $\frac{1}{2}\operatorname{cosec}(x^2)$ **k)** $\frac{1}{5}\sec(x^5)$ **l)** $-\tan\left(\frac{1}{x}\right)$ **6 a)** $2\sec^2 x\tan x$ **b)** $3\sec^3 x\tan x$ **c)** $-4\operatorname{cosec}^2 x\cot 3x$ **d)** $3\operatorname{cosec}^3 x\cot x$

6 e) $8\sec^4 2x\tan 2x$ **f)** $-18\operatorname{cosec}^2 3x\cot^5 3x$ **g)** $-10\sec^2 5x\tan 5x$ **h)** $-12\operatorname{cosec}^4 3x\cot 3x$ **7 a)** $2\sec^2 x(1+\tan x)$

7 b) $4\operatorname{cosec}x\cot x(2-\operatorname{cosec}x)^3$ **c)** $6\sec x\tan x(1+\sec x)^5$ **d)** $-6\sec^2 2x(3-\tan 2x)^2$ **e)** $12\operatorname{cosec}^2 3x(1+\cot 3x)^3$

7 f) $6\operatorname{cosec}2x\cot 2x(5-\operatorname{cosec}2x)^2$ **g)** $-\frac{\sec^2 x}{2\sqrt{1-\tan x}}$ **h)** $\frac{3\operatorname{cosec}3x\cot 3x}{(1+\operatorname{cosec}3x)^2}$ **i)** $-\frac{6\operatorname{cosec}^2 4x}{\sqrt{(1-\cot 4x)^3}}$ **j)** $\frac{6\operatorname{cosec}^2 x}{(1+\cot x)^4}$

7 k) $-\frac{2\sec 8x\tan 8x}{\sqrt[4]{(1-\sec 8x)^3}}$ **l)** $\frac{2\sec^2 x\tan x}{(1+\sec^2 x)^2}$ **8 a)** $\tan^5 x$ **b)** $-\frac{1}{5}\cot^5 x$ **c)** $\frac{1}{3}(3+\tan x)^3$ **d)** $\frac{2}{5}(1-\cot x)^5$ **e)** $\frac{1}{4}(1+\sec x)^4$

8 f) $\frac{1}{3}(3+\operatorname{cosec}x)^3$ **g)** $-\frac{2}{3}\cot^3 2x$ **h)** $\frac{1}{1+\operatorname{cosec}x}$ **i)** $-\frac{2}{15}\sqrt{(2+\cot 5x)^3}$ **j)** $\frac{1}{5}\sec^5 x$ **9 a)** $\tan x+x\sec^2 x$

9 b) $x^2\operatorname{cosec}x(3-x\cot x)$ **c)** $x\sec x(2+x\tan x)$ **d)** $\cot 6x-6x\operatorname{cosec}^2 6x$ **e)** $x^4(5\tan 3x+3x\sec^2 3x)$ **f)** $6x\sec^4 x(1+2x\tan x)$

9 g) $2\sec 2x(\sec^2 2x+\tan^2 2x)$ **h)** $\sec x(\tan^2 x+\sec^2 x+\sec x)$ **i)** $-\frac{\sec^2 x}{(1+\tan x)^2}$ **j)** $\frac{2x(1-x\tan 2x)}{\sec 2x}$

9 k) $-\frac{x\operatorname{cosec}x\cot x+\operatorname{cosec}x+1}{x^2}$ **l)** $-\frac{\sec x}{\sec x+\tan x}$

Exercise 17C

1 $6y - 9x = 3\sqrt{3} - \pi$ **2** $y = -x; y = x - 2\pi$ **3** $\left(\frac{\pi}{2}, \frac{\pi}{8}\right); \frac{\pi^2}{32}$ **4** $24y + 6\sqrt{3}x = 12 + \sqrt{3}\pi; 6\sqrt{3}y - 24x = 3\sqrt{3} - 4\pi$

5 $\left(-\frac{\pi}{3}, -\sqrt{3}\right), \left(\frac{\pi}{3}, \sqrt{3}\right)$ **6** $\left(\frac{\pi}{3}, \frac{\sqrt{3}}{3}\right), \left(\frac{5\pi}{3}, -\frac{\sqrt{3}}{3}\right)$ **7** $\left(\frac{2\pi}{3}, \frac{5\sqrt{3}}{3}\right), \left(\frac{4\pi}{3}, -\frac{5\sqrt{3}}{3}\right)$ **8** $(0, 0), \left(\frac{2\pi}{3}, \frac{3\sqrt{3}}{4}\right)$

9 $\left(\frac{\pi}{4}, 2\sqrt{2}\right)$ **10** $(0, 0), (2\pi, \pi)$ **11 a)** $\left(\frac{\pi}{6}, \frac{\pi}{6} + \sqrt{3}\right)$, max; $\left(\frac{5\pi}{6}, \frac{5\pi}{6} - \sqrt{3}\right)$, min

11 b) $(0, 1)$, min; $\left(\frac{\pi}{3}, \frac{3}{2}\right)$, max; $(\pi, -3)$, min; $\left(\frac{5\pi}{3}, \frac{3}{2}\right)$, max; $(2\pi, 1)$, min **c)** $\left(\frac{\pi}{2}, 1\right)$, max; $\left(\frac{3\pi}{2}, -\frac{1}{3}\right)$, min

11 d) $\left(\frac{\pi}{6}, \frac{3\sqrt{3}}{16}\right)$, max; $\left(\frac{\pi}{2}, 0\right)$, saddle; $\left(\frac{5\pi}{6}, -\frac{3\sqrt{3}}{16}\right)$, min; $\left(\frac{7\pi}{6}, \frac{3\sqrt{3}}{16}\right)$, max; $\left(\frac{3\pi}{2}, 0\right)$, saddle; $\left(\frac{11\pi}{6}, -\frac{3\sqrt{3}}{16}\right)$, min

12 a) $\frac{\pi}{2} - 1$ **b)** $\frac{1}{3}$ **c)** $\frac{3}{64}$ **d)** -2 **13** 2 **14** $\frac{5(\sqrt{3}-1)}{2}$ **15 a)** $A\left(\frac{\pi}{6}, \frac{1}{2}\right), B\left(\frac{5\pi}{6}, \frac{1}{2}\right)$ **b)** $\sqrt{3} - \frac{\pi}{3}$ **16 a)** $P\left(\frac{\pi}{3}, \frac{\sqrt{3}}{2}\right)$

16 b) $\frac{1}{4}$ **17** $\frac{\pi}{2\sqrt{3}}$ **18** $3y + 6x = 2\pi, 4y - 2x = \pi$ **19** $4y + 5x = \pi + 10, 20y - 16x = 5\pi - 32$ **20** $4y - 3x = 3\pi$

21 b) $\left(0, \frac{\pi}{3}\right)$, max **22 b)** $\left(-1, \frac{3\pi}{4}\right)$, min **23** $4(x + y) = 5$ **24** $4y + 2x = 3; 2y - 4x + 1 = 0$ **25** $\frac{5\sqrt{5}}{2}$

26 b) $\left(\frac{\pi+1}{2}, \frac{\pi}{6} + \sqrt{3}\right)$, max; $\left(\frac{5\pi+1}{2}, \frac{5\pi}{6} - \sqrt{3}\right)$, min **27 b)** $(1 + \sqrt{2}, \sqrt{2})$, max; $(1 - \sqrt{2}, -\sqrt{2})$, min

28 b) $\frac{75\sqrt{3}}{4}\,\text{m}^2$ **29 b)** $100\,\text{cm}^2$ **30 b)** $\frac{2\sqrt{3}\pi a^3}{9}$ **31 b)** $\frac{3\sqrt{3}r^2}{8}$ **32 b)** $\frac{32\pi}{81}\,\text{cm}^3$

Exercise 17D

1 a) $\frac{\pi}{2}, \frac{7\pi}{6}, \frac{11\pi}{6}$ **b)** 0.25, 1.57, 2.89, 4.71 **2 a) ii)** $2\cos 2x$ **b)** $1 - \frac{\sin^2 2x}{2}$ **3 a)** $\sec^2 2x$ **b)** $\frac{1}{2} - \frac{\pi}{8}$ **4 a)** $\cos x \leqslant 1$

4 c) $\theta - \sin\theta$ **5 b)** $\left(\frac{\pi}{2}, 1\right); 0 \leqslant f(x) \leqslant 1$ **6** $1 \leqslant f(x) \leqslant \sqrt{3}; \pi\left(\frac{\sqrt{3}-1}{2} - \frac{\pi}{24}\right)$ **7** 1.32; $A''(1.32) \approx -7.75 < 0$

8 b) $9\pi\cos\theta(2 - 9\sin^2\theta)$ **c)** $\sin\theta = 0, \cos\theta = 1$ or $\sin\theta = \sqrt{\frac{2}{3}}, \cos\theta = \frac{1}{\sqrt{3}}$ **d)** $-12\sqrt{3}\pi, V''(\theta) = -12\sqrt{3}\pi < 0; 2\sqrt{3}\pi$

9 a) $-2\sin t + 2\cos 2t; -\sin t - 4\cos 2t$ **b)** $\frac{1}{2}$ **c)** $4x + 2y = 5\sqrt{2}$ **10 i)** $\frac{\sin t}{1 - \cos t}$

Exercise 18A

1 a) x^9 **b)** p^2 **c)** $9k^6$ **d)** $y^{\frac{5}{6}}$ **e)** c^4 **f)** $\frac{3h^6}{2}$ **g)** $2d$ **h)** $16p^{-4}$ **2 a)** 2 **b)** 3 **c)** 27 **d)** 32 **e)** 25 **f)** 343 **g)** $\frac{1}{5}$ **h)** $\frac{4}{9}$

3 a) $\frac{1}{7}$ **b)** $\frac{1}{9}$ **c)** $\frac{1}{2}$ **d)** $\frac{1}{125}$ **e)** $\frac{3}{2}$ **f)** $\frac{121}{9}$ **g)** $\frac{64}{49}$ **h)** 32 **4 a)** 5 **b)** $\pm\frac{1}{7}$ **c)** -7 **d)** $\frac{9}{5}$ **e)** $\frac{1}{2}$ **f)** ± 4 **g)** $\frac{5}{3}$ **h)** $-\frac{1}{7}$

5 a) 9 **b)** 32 **c)** $\frac{4}{49}$ **d)** $\frac{1}{256}$ **e)** $\frac{1}{2}$ **f)** $\frac{1}{5}$ **g)** $\frac{16}{25}$ **h)** $\frac{1}{3}$ **6 a)** ± 27 **b)** 16 **c)** 16 **d)** $\pm\frac{1}{729}$ **e)** $\frac{1}{625}$ **f)** $\frac{4}{49}$ **g)** $\frac{27}{8}$ **h)** $\frac{49}{4}$

7 a) $-1, 8$ **b)** 16, 81 **c)** 16, 36 **d)** $-\frac{27}{8}, 64$ **e)** $-1, \frac{32}{243}$ **f)** $\frac{1}{16}, 256$ **g)** $-\frac{1}{27}, -\frac{1}{8}$ **h)** 1, 4 **8 a)** $\frac{2}{3}$ **b)** $\frac{2}{5}$ **c)** $\frac{10}{3}$ **d)** $-\frac{13}{7}$

Exercise 18B

1 a) $2\sqrt{3}$ **b)** $5\sqrt{2}$ **c)** $4\sqrt{7}$ **d)** $11\sqrt{3}$ **e)** $16\sqrt{5}$ **f)** $2\sqrt{2}$ **g)** $2\sqrt{2}$ **h)** $9\sqrt{5}$ **2 a)** $\frac{3\sqrt{2}}{2}$ **b)** $\frac{5\sqrt{3}}{3}$ **c)** $\frac{\sqrt{6}}{3}$ **d)** $\frac{\sqrt{14}}{2}$ **e)** $2\sqrt{35}$

2 f) $\frac{\sqrt{30}}{4}$ **g)** $\frac{\sqrt{6}}{3}$ **h)** $\frac{3\sqrt{10}}{5}$ **3 a)** $2 + \sqrt{3}$ **b)** $\frac{3 - \sqrt{5}}{4}$ **c)** $\frac{5 + \sqrt{7}}{9}$ **d)** $\frac{6 - \sqrt{3}}{11}$ **e)** $3 + 2\sqrt{2}$ **f)** $\frac{13 + 2\sqrt{2}}{23}$ **g)** $-(17 + 8\sqrt{5})$

3 h) $\frac{6 - \sqrt{6}}{2}$ **4** $\sqrt{2}$

Exercise 18C

1 a) $\log a + \log b$ **b)** $\log a - \log b$ **c)** $2\log a + \log b$ **d)** $\frac{1}{2}\log a$ **e)** $-2\log a$ **f)** $\log a + \frac{1}{2}\log b$ **g)** $3\log a - \log b$

1 h) $2\log a - 3\log b$ **i)** $\frac{1}{2}\log a - \frac{1}{2}\log b$ **j)** $-\log a - 4\log b$ **k)** $-\frac{1}{2}\log a - \frac{1}{2}\log b$ **l)** $\frac{1}{3}\log a + \frac{1}{6}\log b$ **2 a)** $\log 12$ **b)** $\log 14$

2 c) $\log 5$ **d)** $\log 6$ **e)** $\log 30$ **f)** $\log 2$ **g)** $\log 3$ **h)** $\log 10$ **i)** $\log 4$ **j)** $\log 5$ **k)** $\log\left(\frac{a^2}{bc}\right)$ **l)** $\log\left(\frac{a\sqrt{b}}{c^3}\right)$ **3 a)** 2.32 **b)** 1.77

3 c) 1.52 **d)** 0.65 **e)** −4.42 **f)** 5.78 **g)** 1.87 **h)** 0.28 **4 a)** 0, 2 **b)** 1, 3 **c)** 0.79, 0.5 **d)** −1, 2.32 **e)** −0.25, 0.86 **f)** 1.04

5 a) $(y-1)(y-3)(y-9)$ **b)** 0, 1, 2 **6 a)** $(4a-1)(a-2)(a-5)$ **b)** $-1, \frac{1}{2}, 1.16$ **7** −1, 0, 1 **8** −1.58, 1, 2.32

9 $x = 10\,000, y = 1000$ **11 a)** −1, 2 **b)** 1, 2 **c)** 1, 3 **d)** 0, 1, 2

Exercise 18D

1 b) $y = \frac{1}{2}x + 400$ **2 b)** $y = \frac{1}{2}(x+1)$ **3 b)** $l = 3w + 20$, ignoring the last pair **c)** 20 cm **4 b)** $t = 310 - 2x$

4 c) With more than 155 g of catalyst the reaction would commence in negative time **5 b)** $A = 0.4, B = 20$ **c)** 180

6 a) $h = 3 + 2\sqrt{t}$ **b)** 121 **7 a)** $T = 20 + \frac{1000}{t}$ **b)** 20 °C **8 a)** $f = 20 \times d^{1.5}$ **b)** 7.4 cm **9 a)** $T = 0.05 \times x^{2.4}$ **b)** 13.3 mm

10 a) $A = 100, n = -0.5$ **b)** 7.1 **11** $y = \frac{600\,000}{x^3}$ **12 b)** $A = 4, B = 1.2$ **13 a)** $P = \frac{1}{20} \times 1.5^n$ **b)** 550 000 **14** $N = 2 \times 3^t$

Exercise 18E

1 $t^2 + \frac{1}{t^2}$ **2** −2 **3 a)** $\frac{6}{y}$ **b)** 1, −216 **4 a)** $\frac{1}{2}$ **b)** 1.63 **5 a)** $1\frac{1}{2}$ **b)** 1.431 **6** $20 \times 1.1^{n-1}$; 18 **8** $\ln 3$ **9** (−1, 2), (0, 1)

10 −1, 2; 0.631 **11** −6.1 **12 b)** $u \approx 70, a \approx -18$ **13** 0.8; 13 kg

Exercise 19A

The constant of integration is omitted from the answers to Questions **2**, **4** and **15**.

1 a) $3e^{3x}$ **b)** $10e^{5x}$ **c)** $2xe^{x^2}$ **d)** $\frac{2e\sqrt{x}}{\sqrt{x}}$ **e)** $-\frac{5e^{\frac{5}{x}}}{x^2}$ **f)** $2e^{2x-3}$ **g)** $2xe^{x^2+1}$ **h)** $-12x^3e^{-x^4}$ **2 a)** $\frac{1}{4}e^{4x}$ **b)** $\frac{1}{6}e^{6x}$ **c)** $-\frac{1}{2}e^{-2x}$

2 d) $2e^{3x}$ **e)** $-5e^{-2x}$ **f)** e^{x^2} **g)** $\frac{1}{3}e^{x^3}$ **h)** $\frac{1}{2}e^{x^4}$ **i)** $2e^{\sqrt{x}}$ **j)** $-\frac{1}{3}e^{-x^3}$ **3 a)** $2e^x(1+e^x)$ **b)** $12e^{-3x}(1-e^{-3x})^3$

3 c) $-\frac{e^x}{(e^x+1)^2}$ **d)** $-\frac{4e^{4x}}{\sqrt{1-2e^{4x}}}$ **e)** $12x^2e^{x^3}(2+e^{x^3})^3$ **f)** $\frac{8xe^{4x^2}}{(3-e^{4x^2})^2}$ **g)** $3(1+2e^{2x})(x+e^{2x})^2$ **h)** $12(e^x+e^{3x})(3e^x+e^{3x})^3$

4 a) $\frac{1}{3}(3+e^x)^3$ **b)** $\frac{1}{2}(e^x-4)^4$ **c)** $\frac{2}{(1+e^{-2x})}$ **d)** $-\frac{1}{3}(e^{-x}+7)^3$ **e)** $\frac{2}{3}\sqrt{(4+e^x)^3}$ **f)** $\frac{2}{15}\sqrt{(e^{5x}+2)^3}$ **g)** $-\frac{1}{1-e^{-x}}$ **h)** $\frac{2}{3}\sqrt{e^{3x}-1}$

4 i) $\sqrt{1-e^{-x}}$ **5 a)** $(1+2x)e^{2x}$ **b)** $2x(1+2x)e^{4x}$ **c)** $6x^2(1-x)e^{-3x}$ **d)** $2e^{2x}(1+2e^x)(1+e^x)$ **e)** $3e^{3x}(1-2e^{-x})^2$

5 f) $(9-e^{-x})(1+3e^x)^3$ **g)** $\frac{e^{2x}(2x-1)}{x^2}$ **h)** $\frac{e^{3x}(3x-2)}{x^3}$ **i)** $\frac{2e^{2x}}{(1+e^{2x})^2}$ **j)** $\frac{2e^{4x}(4-3e^x)}{(1-e^x)^2}$ **k)** $\frac{e^{3x}(3+e^{2x})}{(1+e^{2x})^2}$ **l)** $\frac{2e^x}{(1-e^x)^2}$ **15** e^{e^x}

Exercise 19B

The constant of integration is omitted in the answers to Questions **2**, **3** and **11**.

1 a) $\frac{2}{1+2x}$ **b)** $-\frac{4}{1-4x}$ **c)** $\frac{2x}{1+x^2}$ **d)** $\frac{3x^2}{x^3-2}$ **e)** $\frac{3(x^2-1)}{x^3-3x}$ **f)** $\frac{e^x}{e^x+4}$ **g)** $\frac{6e^{6x}}{1+e^{6x}}$ **h)** $\frac{1}{2x}$ **2 a)** $\ln(1+x)$ **b)** $\ln(2+3x)$

2 c) $2\ln x$ **d)** $2\ln(5+x)$ **e)** $2\ln(2x-1)$ **f)** $-\ln(4-x)$ **g)** $\frac{1}{2}\ln(5+6x)$ **h)** $-\frac{5}{3}\ln(2-3x)$ **3 a)** $\ln(x^2+1)$ **b)** $\ln(x^3+4)$

3 c) $-2\ln(2-x^2)$ **d)** $\frac{1}{3}\ln(3+x^3)$ **e)** $\ln(x^2-x)$ **f)** $\frac{1}{2}\ln(x^2-6x+1)$ **g)** $\ln(1+e^x)$ **h)** $\frac{1}{5}\ln(e^{5x}+1)$ **4 a)** $\frac{2(1+\ln x)}{x}$

4 b) $-\frac{6(3-2\ln x)^2}{x}$ **c)** $-\frac{1}{x(1+\ln x)^2}$ **d)** $x(1+2\ln x)$ **e)** $x^2(1+3\ln x)$ **f)** $\frac{x}{1+x}+\ln(1+x)$ **g)** $\ln(2-x)-\frac{x}{2-x}$

4 h) $\frac{2x^2}{3+2x}+2x\ln(3+2x)$ **i)** $\frac{2x^2}{1+x^2}+\ln(1+x^2)$ **j)** $\frac{\ln x-1}{(\ln x)^2}$ **k)** $\frac{\ln x-3}{x^2}$ **l)** $-\frac{1+2\ln x}{x^3}$ **10 a)** $A = 5, B = 3, C = -8$

12 $x + \ln(e^x + e^{-x})\,[= \ln(e^{2x}+1)]$

Exercise 19C

The constant of integration is omitted from the answers to Questions **24** and **28**.

1 $y = 3x + 1$ **2** $3y - x = 3\ln 3 - 2;\ y + 3x = 6 + \ln 3$ **3** $y = e(2x - 1);\ 2ey + x = 2e^2 + 1$ **4** $\frac{\sqrt{5}}{2}e$ **5** $y = e(x - 1);\ ey + x = 1$

6 $(\frac{1}{2}, \frac{1}{4} - \ln 2), (1, 1)$ **7** $\left(\frac{1}{5}, \ln\left(\frac{26}{25}\right)\right), (5, \ln 26)$ **8** $\left(\ln 2, \ln\left(\frac{5}{2}\right)\right)$ **9** $(0, 0), \left(2, \frac{4}{e^2}\right)$ **10** $\left(e, \frac{1}{e}\right)$ **11** $\left(-1, -\frac{1}{2e}\right), \left(3, \frac{e^3}{6}\right)$

12 a) $(1, \ln 4)$, min **b)** $\left(3, \frac{e^3}{27}\right)$, min **c)** (e^2, e^2), max **d)** $\left(-1, \frac{4}{e}\right)$, max; $(1, 0)$, min **13** $\frac{e^6 - 1}{2}$ **14** $2\ln 2$ **15** $2e^4 - 5e^2 + 3e$

16 a) $(2, \frac{1}{3})$ **b)** $\ln 3 - \frac{2}{3}$ **17 a)** $P(1, 2), Q(3, 4)$ **18** $\frac{\pi}{2}(e^4 + 4e^2 - 1)$ **20** $y = -x,\ y = x - 6$ **21** $7y - 4x = 20,\ 4y + 7x = 30$

25 $2y - x = 1,\ y + 2x = 3$ **26** $y = 3x - 5,\ 3y + x = 5$ **27** $(1 + \ln 2)y + x = 2 + \ln 16,\ y = x + 2$ **28** $(0, 0)$, max

29 $(0, 0)$, min; $\left(\frac{1}{e^2}, \frac{1}{e^2}\right)$, max **31 a)** $\ln(1 + \sin x)$ **b)** $-\ln(\sin x + \cos x)$ **c)** $\ln(\sec x)$ **d)** $\ln(1 + \tan x)$ **e)** $\ln(1 + \sec x)$

32 $\left(\frac{\pi}{4}, \frac{e^{\frac{\pi}{4}}}{\sqrt{2}}\right)$, max; $\left(\frac{5\pi}{4}, -\frac{e^{\frac{5\pi}{4}}}{\sqrt{2}}\right)$, min **35** $x + \ln(\sin x + \cos x)$

Exercise 19D

1 a) $= -4xe^{-2x}$ **b)** $(0, 1)$, max **2 a)** $A = -1$ **b)** $\left(1, \frac{1}{e}\right)$ **c)** $1; -2$ **3 a)** $\frac{1}{3}\ln 7$ **4 ii)** $e^{-n} - e^{-(n+1)}$ **7** $\frac{2x - 4}{x^2 - 4x + 5}; \frac{1}{2}\ln\left(\frac{5}{2}\right)$

8 $\frac{\pi}{2}(1 - e^{-4})$ **9** $\frac{3}{4} + \ln 2$ **10 a)** $\frac{\pi}{6}, \frac{5\pi}{6}$ **b)** $-\sqrt{3} \leqslant y \leqslant \sqrt{3}$ **c)** $2x - 6y = \pi$ **d)** $3\ln 2 + \frac{\pi^2}{24}$

Exercise 20A

1 a) $\frac{1}{4}(x + 1)(x - 3)^3$ **b)** $\frac{1}{5}(x - 1)(x + 4)^4$ **c)** $\frac{1}{20}(4x - 19)(x - 1)^4$ **d)** $\frac{1}{16}(2x + 1)(2x - 3)^3$ **e)** $\frac{1}{48}(18x + 23)(2x - 5)^3$

1 f) $x - 3\ln(x + 3)$ **g)** $\frac{1}{x + 1} + \ln(x + 1)$ **h)** $-\frac{4x + 1}{8(2x - 3)^2}$ **2 a)** $\frac{2}{15}(3x - 2)\sqrt{(x + 1)^3}$ **b)** $\frac{2}{15}(3x + 2)\sqrt{(x - 1)^3}$

2 c) $\frac{2}{5}(x - 10)\sqrt{(x + 5)^3}$ **d)** $\frac{1}{15}(7 - 9x)\sqrt{(1 - 2x)^3}$ **e)** $\frac{2}{3}(x - 2)\sqrt{x + 1}$ **f)** $\frac{2}{3}(x + 6)\sqrt{x - 3}$ **g)** $\frac{2}{3}(x + 2)\sqrt{x - 4}$

2 h) $-\frac{2}{3}(x + 19)\sqrt{5 - x}$ **3 a)** $\frac{1}{20}(4x - 3)(x + 3)^4$ **b)** $x + 3\ln(x - 1)$ **c)** $-\frac{2}{15}(3x + 10)\sqrt{(5 - x)^3}$ **d)** $\frac{5}{x + 2} + \ln(x + 2)$

3 e) $\frac{1}{3}(x - 1)\sqrt{2x + 1}$ **f)** $\frac{1}{120}(31 - 10x)(5 - 2x)^5$ **g)** $2x - 15\ln(x + 7)$ **h)** $\frac{2(x + 2)}{\sqrt{x + 1}}$ **i)** $-\frac{2}{3}(x + 14)\sqrt{4 - x}$ **j)** $\frac{6}{3 - x} + \ln(3 - x)$

3 k) $\frac{1}{105}(15x^2 + 5x + 1)(x - 1)^5$ **l)** $-\frac{2}{35}(5x + 2)\sqrt{(1 - x)^5}$ **4 a)** $1 + 2\ln 2$ **b)** 12 **c)** $46\frac{2}{5}$ **d)** $\frac{5}{2} + \frac{1}{8}\ln 5$ **e)** $-\frac{2}{3}$ **f)** $2\frac{4}{5}$

5 a) $\sin^{-1}\left(\frac{x}{2}\right)$ **b)** $\sin^{-1}\left(\frac{x}{5}\right)$ **c)** $\frac{1}{3}\sin^{-1}(3x)$ **d)** $\sin^{-1}(6x)$ **e)** $\frac{1}{3}\sin^{-1}\left(\frac{3x}{2}\right)$ **f)** $\frac{1}{4}\sin^{-1}\left(\frac{4x}{5}\right)$ **g)** $2\tan^{-1}\left(\frac{x}{2}\right)$

5 h) $\frac{1}{10}\tan^{-1}\left(\frac{x}{10}\right)$ **i)** $\frac{1}{3}\tan^{-1}(3x)$ **j)** $\frac{1}{5}\tan^{-1}(5x)$ **k)** $\frac{1}{6}\tan^{-1}\left(\frac{2x}{3}\right)$ **l)** $\frac{1}{2}\tan^{-1}\left(\frac{4x}{7}\right)$ **6 a)** $\frac{\pi}{6}$ **b)** $\frac{\pi}{4}$ **c)** $\frac{\pi}{36}$ **d)** π **e)** $\frac{2\pi}{3}$

6 f) 2π **12 a)** $(3, 3)$ **b)** $\frac{15}{2} - 4\ln 4$ **c)** $4 - 3\ln 3$ **13 a)** $A(2, 2), B(3, 0)$ **b)** $\frac{18}{5}$ **14 a)** $A(3, 9)$ **b)** $30 - 16\ln 4$

15 b) $\frac{11}{30}, \frac{1}{30}$ **16 a)** $A(1, \frac{1}{2}), B(2, \frac{2}{3})$ **c)** $1 - \ln\left(\frac{8}{3}\right)$ **17 a)** $a = 1, b = 2$ **19 c)** π

Exercise 20B

The constant of integration is omitted from these answers.

1 a) $\frac{1}{2}x^2 + 2x - 3\ln(x - 2)$ **b)** $\frac{1}{2}x^2 - 3x + \ln(x + 4)$ **c)** $x^2 + 5x + 6\ln(x - 3)$ **d)** $x^2 + x + 2\ln(2x - 1)$ **e)** $\frac{1}{3}x^3 - x - 2\ln(x - 2)$

1 f) $x^3 - x^2 + \ln(x + 1)$ **g)** $x^3 + 2x^2 - 6x + \ln(x - 3)$ **h)** $\frac{1}{4}x^4 + \frac{1}{3}x^3 + \frac{1}{2}x^2 + x + \ln(x - 1)$ **2 a)** $2\ln(x - 1) - \ln(x - 2)$

2 b) $\ln(x + 3) + \frac{1}{2}\ln(2x - 1)$ **c)** $\frac{1}{4}\ln\left(\frac{x - 2}{x + 2}\right)$ **d)** $\frac{1}{2}\ln(x^2 + 3) - \ln(2x + 1)$ **e)** $\ln\left(\frac{x - 2}{x + 1}\right) - \frac{2}{x - 2}$ **f)** $3\ln\left(\frac{2x + 1}{x + 1}\right) + \frac{1}{x + 1}$

2 g) $\ln\left(\frac{2 - x}{5 - x}\right) + \frac{1}{5 - x}$ **h)** $x + \ln\left(\frac{2x - 3}{x - 1}\right)$ **3 a)** $\ln\left(\frac{3}{2}\right)$ **b)** $\frac{1}{2}\ln 7 - \frac{6}{7}$ **c)** $14 + \ln\left(\frac{4}{3}\right)$ **d)** $\ln\left(\frac{3}{2}\right) - \frac{1}{6}$ **e)** $\frac{1}{2}\ln\left(\frac{4}{3}\right)$

3 f) $6 - \ln\left(\frac{4}{3}\right)$ **g)** $\frac{1}{12} + \frac{1}{4}\ln 2$ **h)** $28 + \ln 10$ **4 a)** $\ln(1+x)$ **b)** $-x - \ln(1-x)$ **c)** $\frac{1}{2}\ln(1+x^2)$ **d)** $\sin^{-1}x$

4 e) $\ln(1-x) + \frac{1}{1-x}$ **f)** $2\sqrt{1+x}$ **g)** $\tan^{-1}x$ **h)** $\frac{1}{1-x}$ **i)** $x - \ln(1+x)$ **j)** $\frac{1}{2}\ln\left(\frac{1+x}{1-x}\right)$ **k)** $-\sqrt{1-x^2}$ **l)** $\frac{2}{3}(x-2)\sqrt{(1+x)}$

6 a) $A(-2, \frac{1}{5})$, $B(2, \frac{1}{5})$ **b)** $\frac{4}{5} - \frac{1}{3}\ln 5$ **7 a)** $P(3,3)$ **b)** $\frac{3}{2} + \ln 4$ **8 b)** $2\ln 3$ **9 a)** $(-\frac{25}{4}, -25)$, $Q(4,16)$

11 $\frac{1}{\sqrt{3}}\tan^{-1}\left(\frac{2x+1}{\sqrt{3}}\right) + \frac{1}{6}\ln(x^2+x+1) - \frac{1}{3}\ln(1-x)$

Exercise 20C

1 a) $\frac{1}{12}(3x+1)(x-1)^3$ **b)** $\frac{1}{20}(4x-1)(x+1)^4$ **c)** $-\frac{1}{5}(1+x)(4-x)^4$ **d)** $\frac{1}{56}(4x-1)(2x+3)^6$ **e)** $\frac{1}{4}(x-2)(x+2)^3$

1 f) $\frac{1}{42}(6x+25)(x-4)^6$ **g)** $\frac{1}{48}(18x-17)(2x+3)^3$ **h)** $\frac{1}{30}(25x+8)(4-x)^5$ **2 a)** $\ln(x-1) - \frac{x}{x-1}$ **b)** $\frac{1}{x+1} + \ln(x+1)$

2 c) $\frac{1}{4}\ln(2x-3) + \frac{1}{4(2x-3)}$ **d)** $\frac{2-9x}{6(x+2)^3}$ **e)** $\frac{1}{3}(x+3)\sqrt{2x-3}$ **f)** $\frac{2}{27}(3x+40)\sqrt{3x-2}$ **g)** $-(2+x)\sqrt{1-2x}$

2 h) $-\frac{2}{15}(3x+8)\sqrt{(4-x)^3}$ **i)** $\frac{1}{3}(1+3x)\sqrt{(3-2x)^3}$ **3 a)** $x\sin x + \cos x$ **b)** $-\frac{x}{2}\cos 2x + \frac{1}{4}\sin 2x$ **c)** $\frac{e^{3x}}{9}(3x-1)$

3 d) $-e^{-x}(x+1)$ **e)** $\frac{(6x-1)}{3}\sin 3x + \frac{2}{3}\cos 3x$ **f)** $\frac{x^2}{4}(2\ln x - 1)$ **g)** $\frac{x^3}{9}(3\ln x - 1)$ **h)** $\frac{2}{9}\sqrt{x^3}(3\ln x - 2)$ **4 a)** 6 **b)** $\frac{\pi}{4}$

4 c) $\frac{e^2+1}{e}$ **d)** $8\frac{2}{3}$ **e)** $124\ln 2 - 15$ **f)** $-\frac{\pi+2}{18}$ **5 a)** $-x^2\cos x + 2x\sin x + 2\cos x$ **b)** $\frac{1}{60}(10x^2 - 12x + 9)(x+3)^4$

5 c) $e^x(x^2 - 2x + 2)$ **d)** $\frac{1}{2}x^2\sin 2x + \frac{1}{2}x\cos 2x - \frac{1}{4}\sin 2x$ **e)** $-\frac{e^{-2x}}{4}(2x^2+2x+1)$ **f)** $(1-2x-x^2)\cos x + 2(x+1)\sin x$

6 a) $\frac{e^2-1}{4}$ **b)** $\pi^2 - 4$ **c)** $\frac{108}{5}$ **d)** $\frac{\pi^2-8}{32}$ **e)** $\frac{586}{15}$ **f)** $\pi^2 - 4$ **7 a)** $A\left(1, \frac{1}{e}\right)$ **b)** $\frac{2e-5}{2e}$ **8 a)** $P(\pi, 0)$, $Q(2\pi, 0)$ **b)** $\pi, 3\pi$

9 $\frac{e^2+1}{4}$ **10 b)** $e - 2$ **11 a)** $x(2 - 2\ln x + (\ln x)^2)$ **b)** $\frac{e^{x^2}}{2}(x^2-1)$ **c)** $\frac{e^x}{2}(\sin x - \cos x)$

Exercise 20D

The constant of integration is omitted from these answers.

1 a) $\frac{1}{2}\ln(\sec 2x + \tan 2x)$ **b)** $\frac{1}{3}\ln(\sec 3x)$ **c)** $\frac{2}{5}\ln(\sin 5x)$ **d)** $-\frac{1}{2}\ln(\operatorname{cosec} 2x + \cot 2x)$ **e)** $\frac{1}{3}\ln(\sin 3x)$ **f)** $-8\ln\left[\operatorname{cosec}\left(\frac{1}{2}x\right) + \cot\left(\frac{1}{2}x\right)\right]$

1 g) $\ln[\sec(x^2)]$ **h)** $-\frac{1}{2}\ln[\operatorname{cosec}(x^2) + \cot(x^2)]$ **i)** $\frac{1}{3}\ln[\sin(x^3)]$ **j)** $\frac{3}{5}\ln[\sec(x^5) + \tan(x^5)]$ **k)** $\frac{1}{3}\ln[\sec(x^3 - 3x)]$

1 l) $-2\ln[\operatorname{cosec}(\sqrt{x}) + \cot(\sqrt{x})]$ **2 a)** $-\frac{1}{6}\cos^6 x$ **b)** $\sin x - \frac{1}{3}\sin^3 x$ **c)** $-\frac{1}{2}\cos 2x + \frac{1}{6}\cos^3 2x$ **d)** $\frac{1}{4}\sin 4x - \frac{1}{12}\sin^3 4x$

2 e) $-\frac{1}{3}\cos^3 x + \frac{1}{5}\cos^5 x$ **f)** $\frac{1}{5}\sin^5 x - \frac{1}{7}\sin^7 x$ **g)** $\frac{1}{6}\sin^3 2x - \frac{1}{10}\sin^5 2x$ **h)** $\frac{1}{2}\sin^4 x$ **3 a)** $\frac{1}{2}x - \frac{1}{4}\sin 2x$ **b)** $\frac{1}{2}x + \frac{1}{8}\sin 4x$

3 c) $\frac{1}{2}x + \frac{1}{12}\sin 6x$ **d)** $\frac{1}{2}x - \frac{1}{16}\sin 8x$ **e)** $\frac{1}{2}x + \frac{1}{24}\sin 12x$ **f)** $\frac{1}{2}x - \frac{1}{2}\sin x$ **g)** $\frac{3}{8}x + \frac{1}{4}\sin 2x + \frac{1}{32}\sin 4x$ **h)** $\frac{3}{8}x - \frac{1}{8}\sin 4x + \frac{1}{64}\sin 8x$

4 a) $-\cot x - x$ **b)** $\frac{1}{2}\tan 2x - x$ **c)** $\frac{1}{4}\tan^4 x$ **d)** $-\frac{1}{10}\cos^5 2x$ **e)** $\frac{1}{2}\tan^2 x + \ln(\cos x)$ **f)** $-\frac{1}{9}\cot^3 3x + \frac{1}{3}\tan 3x + x$

4 g) $\frac{1}{15}\tan^3 5x - \frac{1}{5}\tan 5x + x$ **h)** $\frac{1}{8}\tan^4 2x - \frac{1}{4}\tan^2 2x + \frac{1}{2}\ln(\sec 2x)$ **5 a)** $-\cot x$ **b)** $\frac{1}{3}\tan 3x$ **c)** $-3\cot\left(\frac{1}{3}x\right)$ **d)** $\frac{1}{3}\tan^3 x + \tan x$

5 e) $-\frac{1}{15}\cot^3 5x - \frac{1}{5}\cot 5x$ **f)** $\frac{1}{9}\tan^3 3x + \frac{1}{3}\tan 3x$ **g)** $\frac{1}{5}\tan^5 x + \frac{2}{3}\tan^3 x + \tan x$

5 h) $-\frac{1}{4}\operatorname{cosec}^3 x\cot x - \frac{3}{8}\operatorname{cosec} x\cot x - \frac{3}{8}\ln(\operatorname{cosec} x + \cot x)$ **6 a)** $-\frac{1}{4}\cos 4x + \frac{1}{12}\cos^3 4x$ **b)** $\frac{1}{2}x + \frac{1}{20}\sin 10x$

6 c) $\frac{1}{9}\tan^3 3x - \frac{1}{3}\tan 3x + x$ **d)** $\frac{1}{2}\ln(\sin 2x)$ **e)** $\frac{1}{4}\ln(\sec 4x + \tan 4x)$ **f)** $-\frac{1}{15}\operatorname{cosec}^3 5x$ **g)** $-\frac{1}{3}\cot 3x - x$

6 h) $-\frac{1}{6}\cos 6x + \frac{1}{9}\cos^3 6x - \frac{1}{30}\cos^5 6x$ **i)** $2\ln\left(\sec\left(\frac{1}{2}x\right)\right)$ **j)** $\frac{1}{5}\tan 5x - x$ **k)** $-\frac{1}{4}\ln(\operatorname{cosec} 4x + \cot 4x)$ **l)** $-\frac{1}{15}\cos^5 3x + \frac{1}{21}\cos^7 3x$

7 a) $2\ln(2+\sqrt{3})$ **b)** $\frac{5}{48}$ **c)** $\frac{2\pi+3}{24}$ **d)** $\frac{8}{27}\left(9 - \sqrt{3}\right)$ **e)** $\frac{4}{9}$ **f)** $\frac{4}{3} - \frac{1}{2}\ln 3$ **8 a)** $P\left(\frac{\pi}{3}, \frac{\sqrt{3}}{2}\right)$, $Q(\pi, 0)$ **b)** $\frac{1}{4}, \frac{9}{4}$

10 b) $A(0, 1)$, $B\left(\frac{2\pi}{3}, -\frac{1}{2}\right)$ **c)** $\frac{3\sqrt{3}}{4}$ **12** $\frac{1}{2}\sin^{-1}x + \frac{1}{2}x\sqrt{1-x^2}$

Exercise 20E

The constant of integration is omitted in these answers.

1 a) $\frac{1}{4}(x-1)(x+3)^3$ **b)** $2\ln x - \ln(x+1)$ **c)** $\frac{1}{5}\ln(\sec 5x)$ **d)** $-x\cos x + \sin x$ **e)** $\dfrac{4}{2-x}$ **f)** $\sin^{-1}x - \sqrt{1-x^2}$ **g)** $\frac{1}{2}\ln(x^2-4)$

1 h) $-e^{\cos x}$ **i)** $\frac{1}{4}\ln(\sec 4x + \tan 4x)$ **j)** $\ln x - \dfrac{1}{x^2}$ **k)** $3\ln(x+1) - 2\ln(x-1)$ **l)** $x\ln x - x$ **m)** $-x - 3\ln(3-x)$ **n)** $\frac{1}{3}\sqrt{(x^2+1)^3}$

1 o) $\dfrac{e^{2x}}{4}(2x-1)$ **p)** $\frac{1}{4}(x-5)^4$ **q)** $-\frac{1}{4}\cos^4 x$ **r)** $x + \ln(x-2) - 2\ln(x+3)$ **s)** $\ln(e^x+1)$ **t)** $-\frac{1}{3}\cos 3x + \frac{1}{9}\cos^3 3x$

1 u) $-\frac{2}{15}(4+3x)\sqrt{(2-x)^3}$ **v)** $\frac{2}{3}\sqrt{x^3+2}$ **w)** $\frac{1}{2}x - \frac{1}{8}\sin 4x$ **x)** $\frac{1}{12}\tan^2 6x + \frac{1}{6}\ln(\cos 6x)$ **2 a)** $\frac{1}{2}x + \frac{1}{24}\sin 12x$

2 b) $x - 3\ln(x+2)$ **c)** $-\frac{1}{3}\ln[\operatorname{cosec}(x^3) + \cot(x^3)]$ **d)** $\ln\left(\dfrac{x+3}{x+4}\right)$ **e)** $\frac{1}{20}(x^4-5)^5$ **f)** $-2\ln(5-x)$ **g)** $-\dfrac{e^{-2x}}{4}(2x+1)$

2 h) $\frac{1}{3}x^3 - \frac{1}{2}x^2 - 6x$ **i)** $3\ln(x+2) - 2\ln(x-1) + \dfrac{1}{(x-1)}$ **j)** $x - 4\ln(x+4)$ **k)** $\frac{1}{9}\tan^3 3x$ **l)** $3\sin^{-1}x - \sqrt{1-x^2}$ **m)** $\frac{1}{2}e^{2x}$

2 n) $\dfrac{x}{2}\sin 2x + \dfrac{1}{4}\cos 2x$ **o)** $\frac{2}{15}\sqrt{(5x-1)^3}$ **p)** $\ln x - \tan^{-1}x$ **q)** $\ln(x^3+5x)$ **r)** $-e^{\operatorname{cosec} x}$ **s)** $\ln(1-\cos x)$ **t)** $\frac{2}{15}(3x-4)\sqrt{(x+2)^3}$

2 u) $-\dfrac{1}{4(x^2+2x-5)^2}$ **v)** $\frac{1}{2}\sin 2x - \frac{1}{6}\sin^3 2x$ **w)** $\dfrac{x^6}{36}(6\ln x - 1)$ **x)** $\frac{1}{3}\ln(\operatorname{cosec} 3x) - \frac{1}{6}\cot^2 3x$ **3 a)** $-\frac{1}{2}\cos 2x + \frac{1}{3}\cos^3 2x - \frac{1}{10}\cos^5 2x$

3 b) $x + 5\ln(x-5)$ **c)** $\ln(1+\tan x)$ **d)** $\frac{1}{9}(x^3+3x-2)^3$ **e)** $\dfrac{1}{6}\ln\left(\dfrac{x-3}{x+3}\right)$ **f)** $-\frac{1}{3}\cos^3 x + \frac{1}{5}\cos^5 x$ **g)** $\dfrac{e^{5x}}{25}(5x-1)$

3 h) $-\dfrac{1}{2(x^2+1)}$ **i)** $3\ln(x+1) + \ln x + \dfrac{2}{x}$ **j)** $\frac{1}{5}\ln(\sec 5x + \tan 5x)$ **k)** $\frac{1}{2}e^{x^2}$ **l)** $\frac{1}{80}(8x-3)(2x+3)^4$ **m)** $\frac{2}{3}\sqrt{x}(x-3)$

3 n) $-\dfrac{x}{3}\cos 3x + \dfrac{1}{9}\sin 3x$ **o)** $\frac{1}{2}\ln(x^2+1) + \tan^{-1}x$ **p)** $-\frac{1}{2}\cos(x^2)$ **q)** $\frac{1}{2}\ln(2x+1)$ **r)** $\ln(x+1) + 3\ln(x-3)$ **s)** $3\sqrt{2x+1}$

3 t) $\frac{1}{3}\sec^3 x$ **u)** $\ln(2-x) + \dfrac{2}{2-x}$ **v)** $-\frac{1}{4}e^{-2x^2}$ **w)** $\dfrac{x}{2}(\ln x - 1)$ **x)** $\frac{1}{2}x + \frac{1}{20}\sin 10x$ **4 a)** $2\sqrt{x} - 2\ln(1+\sqrt{x})$

4 b) $2x^{\frac{1}{2}} - 3x^{\frac{1}{3}} + 6x^{\frac{1}{6}} - 6\ln(1+x^{\frac{1}{6}})$ **c)** $\dfrac{x[\cos(\ln x) + \sin(\ln x)]}{2}$ **d)** $\dfrac{2^x}{\ln 2}$ **e)** $-\dfrac{1}{1+\tan x}$

Exercise 20F

1 a) $y = \pm\sqrt{(x+1)^2+c}$ **b)** $y = \pm\sqrt{x^3-2x+c}$ **c)** $y = \sqrt[3]{3x^2+9x+c}$ **d)** $y = \left(\frac{1}{2}x^3+c\right)^2$ **e)** $y = \left(\dfrac{1}{x-1}+c\right)^2$

1 f) $y = \frac{1}{2}[1+(x^2+c)^2]$ **g)** $y = Ae^{x^3}$ **h)** $y = \ln\left(\frac{1}{2}x^2+c\right)$ **i)** $\sin y = A\,e^{\sin x}$ **j)** $y = -\ln(\cos x + c)$ **k)** $\tan y = x^4 + c$

1 l) $\sin y = x + \cos x + c$ **2 a)** $y = 2 + 4x - x^3$ **b)** $y = \sqrt{x^2+2x+9}$ **c)** $y = 3 + \sqrt{x^2+6x+1}$ **d)** $y = \sqrt[3]{20+x-x^2}$

2 e) $y = \dfrac{x^2}{1-x^2}$ **f)** $y = (1+\sqrt{x+1})^2$ **g)** $y = 7e^{x^2}$ **h)** $3\sin y = 5 - 3\cos x + \cos^3 x$ **i)** $\cos y = \dfrac{1}{4} + \dfrac{1}{x}$ **j)** $y = 2 - \dfrac{2}{x}$

2 k) $y = \ln\left(\dfrac{1+2\cos x}{2}\right)$ **l)** $\tan y = 1 + \ln(x^3-7)$ **3** $y = e^{x^2-9} - 1$ **4** $y = \sin^2 x + 2\sin x - 2$

5 $y = \cos x + c$ or $y = \frac{1}{3}\cos^3 x - \cos x + c$ **6 b)** $y = -\dfrac{x}{1+\ln(2+2x)}$ **7 b)** $y = 3e^x - x + 1$

Exercise 20G

1 c) 250 **2 c)** 9 **3 c)** 4 years 11 months **6 c)** 921 000

Exercise 20H

The constant of integration is omitted from these answers.

1 $\frac{5}{18}$ **3 a) i)** $\frac{1}{4}(1+x)^4$ **ii)** $2\frac{2}{5}$ **b)** $\frac{1}{3}(2\sqrt{2}-1)$ **4** $\dfrac{\pi}{12}$ **5** $\dfrac{\sqrt{3}-1}{4}$ **6 a)** $\dfrac{1}{\sqrt{3}}\tan^{-1}\left(\dfrac{\tan x}{\sqrt{3}}\right)$ **b)** $\dfrac{\sqrt{3}\pi}{18}$ **7** $\ln\left(\dfrac{3}{2}\right)$

8 a) i) $\sin^{-1}\left(\dfrac{x}{\sqrt{3}}\right)$ **ii)** $\dfrac{1}{2}\ln(1+x^2)$ **b) ii)** $\dfrac{1}{4}\left(\dfrac{1}{u-2} - \dfrac{1}{u+2}\right)$ **iii)** $\dfrac{1}{4}\ln\left(\dfrac{\sqrt{3-x^2}-2}{\sqrt{3-x^2}+2}\right)$ **9** $\dfrac{1}{x-2} + \dfrac{2}{(x-2)^2}$; $1 + \ln 2$

10 $\dfrac{1}{2-x} + \dfrac{3}{1+3x}$; $3\ln 2$ **11** $\dfrac{1}{x} - \dfrac{2}{x+1} + \dfrac{1}{x+2}$ **12** $\dfrac{1}{x+1} + \dfrac{2}{x-3} + \dfrac{3}{(x-3)^2}$; $\dfrac{3}{2} + \ln\left(\dfrac{3}{8}\right)$ **13 a)** $A = 1, B = -1, C = 2$

14 $A = 2, B = 3, C = -1$; $2x + 5y = 17$ **16 i)** $e^{-x} - xe^{-x}$ **ii)** $\left(1, \dfrac{1}{e}\right)$ **iii)** $1 - \dfrac{3}{e^2}$ **17 a)** $A(2, 0), B(0, 4)$

17 b) $2e^{-x}(x-3)$; $\left(3, -\dfrac{2}{e^3}\right)$ **18 a)** 1 **c)** $\left(e^{\frac{1}{2}}, \dfrac{1}{2e}\right)$, $-\dfrac{2}{e^2}$ **d)** $\frac{1}{2}(1-\ln 2)$ **20 a)** $3x(1+x^2)^{\frac{1}{2}}$ **b)** $y = \dfrac{3}{4-(1+x^2)^{\frac{3}{2}}}$

21 a) $x+3\ln x$ **b)** $y=\left(\frac{x+3\ln x+1}{2}\right)^2$ **22** $y=e^{(1-e^{-x})}$ **23** $\ln y=\ln\left(\frac{2}{(1-x)^2}\right)-x$

24 $x\tan x+\ln(\cos x)$; $y=\frac{1}{1-x\tan x-\ln(\cos x)}$ **25 b)** $t=\ln\left(\frac{x^2+25}{25}\right)+\frac{4}{5}\tan^{-1}\left(\frac{x}{5}\right)$ **27 b)** 6.31

Exercise 21A

1 a) 3.896, 0.006 **b)** 10, 0.947 **c)** 0.4286, 0.007 **2 a) i)** 28.27433388 **ii)** 28.278 **b)** 0.013 **3 a)** 8.345, 8.335
3 b) 61.2345, 61.2335 **c)** 0.0055, 0.0045 **d)** 3.4675, 3.4665 **e)** 6.5, 5.5 **f)** 0.4625, 0.4615 **4 a)** 12.2 cm, 11.8 cm
4 b) 6.4125 cm^2, 5.8125 cm^2 **5** 15.960 025 m^2 **6** 17.2757 m^2 **7** 0.829, 0.828 **8** 2.3077, 2.2858 **9** 3.4435 cm^2, 3.0540 cm^2
10 47.439, 47.090

Exercise 21B

1 a) 1.6 **b)** −2.2 **c)** 0.9 **d)** −0.9 **e)** 0.8 **f)** 1.2 **2 a)** 0.1716 **b)** 0.2915 **c)** 0.1835 **d)** Diverges **e)** Oscillates
3 a) 3, $2.\dot{6}$, 2.875, 2.739 130 435, 2.825 396 825 **b)** $x^2-x-5=0$ **4 a)** 0.2, 0.238 095 2381, 0.235 955 0562, 0.236 074 2706
4 b) $x^2+4x-1=0$ **5 a)** 1.732 050 808, 1.706 072 567, 1.703 532 855, 1.703 284 362, 1.703 260 047 **b)** $3x^2-x-7=0$
6 a) $x^3-2x^2-1=0$ **b)** 2.206 **7 b) i)** 2.646 **ii)** 3.873 **iii)** 6.164 **8** 0.658 **9** 1.334 **10** 0.567 **11** P(1.146, 1.146)
12 1.290

Exercise 21C

1 357 **2 a)**

x	1	2	3	4	5
$\sqrt{1+x^2}$	1.414	2.236	3.162	4.123	5.099

c) 12.8 **3** 2.99 **4** 0.809 **5** 8.91

6 0.682 **7** 80.7 **8** 1.41 **9** 28.0 **10** 15.1 **11** 0.656 **12** 7.21 **13 b)** $\frac{617}{315}$ **c)** 0.659% **14 a) i)** 1.896 **ii)** 2.094 **b)** 2
14 c) 5.2%, 4.7%

Exercise 21D

1 a) 9.775 cm **b)** 7.80625 cm^2 **2 b)** $\ln 2+2-3\approx -0.3<0$; $\ln 3+3-3\approx 1.1>0$
2 c) $x_{n+1}=3-\ln x_n$, $x_0=2.2$; 2.211 542 64, 2.206 309 701, 2.208 678 699, 2.207 605 537, 2.208 091 539; **2.208**

or $x_{n+1}=x_n-\dfrac{x_n+\ln x_n-3}{1+\dfrac{1}{x_n}}$, $x_0=2.2$; 2.207 935 565, 2.207 940 032; **2.208**

2 d) $\ln(2.2075)-3+2.2075\approx -6\times 10^{-4}<0$; $\ln(2.2085)-3+2.2085\approx 8\times 10^{-4}>0$
3 $2^3-4-5=-1<0$; $3^3-6-5=16>0$; (3) gives 2.094 55 **4 a)** $0^3-5\times 0+1=1>0$; $1^3-5\times 1+1=-5<0$
4 b) $x_{n+1}=\frac{1}{5}(x_n^3+1)$ gives 0.225, 0.202 278 125, 0.201 655 3001, 0.201 640 0569, 0.201 639 685; **0.2016**
5 a) $1.5^3-3\times 1.5+1=-0.125<0$; $1.6^3-3\times 1.6+1=0.296>0$ **b)** 1.528 9697, 1.531 654 27; **1.53**
6 $1^5-5\times 1-6=-10<0$; $2^5-5\times 2-6=16>0$; $p=5$, $q=6$, $r=5$; **1.708**
7 $x_1=0.5$; 0.629 960 5279, 0.615 442 0024, 0.618 561 6566; **0.62** **8** 2.218, $xe^x-2e^x-2=0$ **9** 2.14 **10** 1.17 **11 a)** 1.05 **b)** $\frac{4\pi}{3}$
12 318 **13** 35.77 **14** 0.62 **15** 1.744

Exercise 22A

1 a) 5 **b)** $\sqrt{74}$ **c)** 3 **d)** $\sqrt{61}$ **e)** 13 **f)** $2\sqrt{5}$ **g)** $\sqrt{130}$ **h)** $\sqrt{83}$ **2** ±7 **3** ±4 **4** ±3 **5** $\frac{4}{5}\mathbf{i}-\frac{3}{5}\mathbf{j}$ **6** $\frac{5}{\sqrt{89}}\mathbf{i}-\frac{8}{\sqrt{89}}\mathbf{j}$

7 $\begin{pmatrix}-\frac{7}{\sqrt{130}}\\ \frac{9}{\sqrt{130}}\end{pmatrix}$ **8** $\frac{3}{\sqrt{38}}\mathbf{i}-\frac{2}{\sqrt{38}}\mathbf{j}+\frac{5}{\sqrt{38}}\mathbf{k}$ **9** $\frac{1}{\sqrt{14}}\mathbf{i}-\frac{3}{\sqrt{14}}\mathbf{j}+\frac{2}{\sqrt{14}}\mathbf{k}$ **10** $\begin{pmatrix}-\frac{3}{13}\\ \frac{12}{13}\\ -\frac{4}{13}\end{pmatrix}$ **11** $12\mathbf{i}-6\mathbf{j}+4\mathbf{k}$ **12** $\mathbf{i}-2\mathbf{k}$

13 $\begin{pmatrix} \sqrt{5} \\ -\frac{3\sqrt{5}}{5} \\ \frac{\sqrt{5}}{5} \end{pmatrix}$ **14** $5\mathbf{i}+2\mathbf{j}+3\mathbf{k}$ **15** $3\mathbf{i}+3\mathbf{j}-13\mathbf{k}$ **16** $\begin{pmatrix} 8 \\ 1 \\ -8 \end{pmatrix}$ **17** $-3\mathbf{i}+10\mathbf{j}$ **18** $10\mathbf{i}-11\mathbf{j}-\mathbf{k}$ **19** $\begin{pmatrix} -7 \\ 3 \\ 2 \end{pmatrix}$ **20** $-7, -6, 1$

21 a) $\begin{pmatrix} 2 \\ 1 \\ 3 \end{pmatrix}, \begin{pmatrix} 4 \\ 2 \\ 6 \end{pmatrix}$ **b)** $1:2$ **24** $8, 3$

Exercise 22B

1 a) i) $\frac{1}{2}\mathbf{a}$ **ii)** $\mathbf{b}-\mathbf{a}$ **iii)** $\frac{1}{2}(\mathbf{b}-\mathbf{a})$ **iv)** $\frac{1}{2}\mathbf{b}$ **b)** $\overrightarrow{OB}$ is parallel to $\overrightarrow{PQ}$, and OB = 2PQ **2 a)** $\mathbf{b}-\mathbf{a}$ **b)** $\mathbf{b}-\frac{1}{3}\mathbf{a}$ **c)** $\frac{1}{2}\mathbf{a}-\frac{1}{2}\mathbf{b}$
2 d) $\frac{1}{6}\mathbf{a}+\frac{1}{2}\mathbf{b}$ **3 a) i)** $\mathbf{c}-\mathbf{a}$ **ii)** $\frac{1}{4}\mathbf{c}$ **iii)** $-\frac{1}{4}\mathbf{a}$ **iv)** $\frac{1}{4}(\mathbf{c}-\mathbf{a})$
3 b) $\overrightarrow{ST}$ and $\overrightarrow{AC}$ are parallel since both are multiples of $\mathbf{c}-\mathbf{a}$, and ST : AC = 1 : 4 **4 a)** $\frac{1}{2}\mathbf{a}$ **b)** $\mathbf{a}-\mathbf{b}$ **c)** $\frac{3}{2}\mathbf{a}-\mathbf{b}$ **d)** $\frac{1}{2}\mathbf{a}-\mathbf{b}$
5 a) $\frac{2}{3}\mathbf{a}+\frac{1}{3}\mathbf{c}$ **b)** $2:1$ **6 a)** $\frac{4}{5}\mathbf{a}+\frac{4}{15}\mathbf{c}$ **c)** $4:11$ **7 a)** $\frac{1}{6}(\mathbf{a}+\mathbf{b})$ **8 a)** $\frac{2}{3}\mathbf{a}+\frac{1}{3}\mathbf{c}$

Exercise 22C

1 a) -9 **b)** -9 **c)** 26 **d)** -13 **e)** 25 **f)** 13 **2 a)** -3 **b)** -3 **c)** 12 **d)** 21 **e)** 9 **f)** 14 **3 a)** 1 **b)** 3 **c)** 3 **d)** 2
3 e) -13 **f)** 17 **4 a)** -45 **b)** -43 **c)** 24 **d)** 2 **e)** -19 **f)** 91 **5 a)** Perpendicular **b)** Parallel **c)** Parallel **d)** Neither
5 e) Parallel **f)** Perpendicular **g)** Parallel **h)** Neither **6 a)** 75.7° **b)** 45° **c)** 119.2° **d)** 66.4° **e)** 53.1° **f)** 115.3° **g)** 95.3°
6 h) 69.4° **7 a)** 10 **b)** $1\frac{1}{4}$ **8 a)** -12 **b)** 12 **9** $2\frac{4}{7}$ **10** 18 **11** 2 or -5 **12** 1 or 3 **13** $\frac{1}{9\sqrt{5}}(7\mathbf{i}+10\mathbf{j}+16\mathbf{k})$

Exercise 22D

1 $\mathbf{r}=(4+t)\mathbf{i}+(3-2t)\mathbf{j}$ **2** $\mathbf{r}=(5+4t)\mathbf{i}-(2+3t)\mathbf{j}+(3+t)\mathbf{k}$ **3** $\mathbf{r}=(5+t)\mathbf{i}-(1+t)\mathbf{j}$ **4** $\mathbf{r}=(2+3t)\mathbf{i}+(6-8t)\mathbf{j}$
5 $\mathbf{r}=(7t-1)\mathbf{i}+(2+t)\mathbf{j}+3(t-1)\mathbf{k}$ **6** $\mathbf{r}=(4+3t)\mathbf{i}+(1-t)\mathbf{j}$ **7** $\mathbf{r}=(3+2t)\mathbf{i}+(5-t)\mathbf{j}$ **8** $\mathbf{r}=(4+2t)\mathbf{i}-(1+t)\mathbf{j}$
9 $\mathbf{r}=(5+t)\mathbf{i}-2(1+t)\mathbf{j}$ **10 a)** $\mathbf{r}_1=(3+2t)\mathbf{i}+(1+t)\mathbf{j}$ **b)** $\mathbf{r}_2=(5+2s)\mathbf{i}+(1-s)\mathbf{j}$ **c)** $(4, 1\frac{1}{2})$ **11 a)** $\mathbf{r}_1=3(1+t)\mathbf{i}+(t-1)\mathbf{j}$
11 b) $\mathbf{r}_2=(5+2s)\mathbf{i}+(4-3s)\mathbf{j}$ **c)** $(7\frac{4}{11}, \frac{5}{11})$ **12** $10\mathbf{i}+4\mathbf{j}$ **13** $2\mathbf{i}-3\mathbf{j}$ **14 a)** $2\mathbf{i}+5\mathbf{j}+9\mathbf{k}$ **b)** 15.6° **15 a)** $\mathbf{i}+3\mathbf{j}+5\mathbf{k}$
15 b) 58.5° **16 a)** $\mathbf{r}_l=3t\mathbf{i}+(5+t)\mathbf{j}$ **b)** $\mathbf{r}_p=(4-s)\mathbf{i}+3(1+s)\mathbf{j}$ **c)** (3, 6) **17 a)** $\mathbf{r}_l=(t-1)\mathbf{i}+2(t-1)\mathbf{j}$ **b)** $\mathbf{r}_p=2s\mathbf{i}+(5-s)\mathbf{j}$
17 c) (2, 4) **19 a) i)** $\frac{\sqrt{3}}{3}$ **20 a) i)** $\frac{\sqrt{2}}{3}$ **21 a)** $6\mathbf{i}+3\mathbf{j}-9\mathbf{k}$ **b)** $(6\lambda-5)\mathbf{i}+(3\lambda-4)\mathbf{j}-(9\lambda)\mathbf{k}$ **c)** $\mathbf{i}+4\mathbf{j}+2\mathbf{k}$
22 a) $4\mathbf{i}+8\mathbf{j}-4\mathbf{k}$ **b)** $(4\lambda-5)\mathbf{i}+8(\lambda-1)\mathbf{j}-(4\lambda+3)\mathbf{k}$ **c)** $6\mathbf{i}+4\mathbf{j}+2\mathbf{k}$

Exercise 22E

1 b) $\frac{\lambda}{2}\mathbf{a}+\left(1-\frac{3\lambda}{4}\right)\mathbf{b}$; $\frac{2}{3}\mathbf{a}$ **2 i)** $\mathbf{i}+2\mathbf{j}+2\mathbf{k}, -2\mathbf{i}+\mathbf{j}+2\mathbf{k}$ **3 i)** $-4\mathbf{i}+2\mathbf{j}, 4\mathbf{i}+8\mathbf{j}$ **ii)** 26.6° **iii)** $3\mathbf{i}+7\mathbf{j}$ **iv)** 53.2°
4 a) 17; 40.2° **b) i)** $4\mathbf{i}-4\mathbf{j}+3\mathbf{k}$ **ii)** 14.4 **c)** $-5\mathbf{j}+\mathbf{k}$; $\frac{10}{3}\mathbf{i}-\frac{7}{3}\mathbf{j}+\frac{5}{3}\mathbf{k}$ **5 a)** 3, 5 **b)** 10, 48.2° **c)** 5.59
5 d) i) $(1+3\lambda)\mathbf{i}+(2+\lambda)\mathbf{j}+2(1-\lambda)\mathbf{k}$ **iii)** $-2\mathbf{i}+\mathbf{j}+4\mathbf{k}$; $\frac{25}{7}\mathbf{i}+\frac{20}{7}\mathbf{j}+\frac{2}{7}\mathbf{k}$ **6 a)** 5 **b)** $3\mathbf{i}+4\mathbf{j}+2\mathbf{k}$ **c)** $5\mathbf{i}+\mathbf{j}+4\mathbf{k}+t(2\mathbf{i}-3\mathbf{j}+3\mathbf{k})$

7 ii) $\begin{pmatrix} 4 \\ -3 \\ 2 \end{pmatrix}+t\begin{pmatrix} -2 \\ 1 \\ -9 \end{pmatrix}$; (6, −4, 11) **8 ii)** $\mathbf{i}+4\mathbf{j}+2\mathbf{k}$ **9 a)** $5\mathbf{i}+3\mathbf{j}+\lambda(\mathbf{i}+\mathbf{j}-\mathbf{k})$ **c)** $3\mathbf{i}+\mathbf{j}+2\mathbf{k}$ **e)** $4\mathbf{i}+3\mathbf{j}+5\mathbf{k}$

10 a) $\mathbf{i}+2\mathbf{j}+3\mathbf{k}$ **b)** 40.2° **11 b)** $\sqrt{11}$ **c)** 35° **d)** 1.9 **12 a)** $\begin{pmatrix} -3 \\ 1 \\ -7 \end{pmatrix}+t\begin{pmatrix} 4 \\ 1 \\ 6 \end{pmatrix}$ **b)** (1, 2, −1) **c)** $\sqrt{6}$

13 a) $3\mathbf{i}-\mathbf{j}+2\mathbf{k}+s(\mathbf{i}-\mathbf{j}-3\mathbf{k})$ **b)** $6\mathbf{i}-4\mathbf{j}-7\mathbf{k}$ **e)** $3\sqrt{2}$ **14 a)** $\begin{pmatrix} 24 \\ 6 \\ 0 \end{pmatrix}+t\begin{pmatrix} 1 \\ 1 \\ 2 \end{pmatrix}$ **c)** (35, 17, 22) **d)** 181

15 a) $-4\mathbf{i}+2\mathbf{j}+6\mathbf{k}$ **b)** $\frac{3}{7}$ **c)** $\frac{8\sqrt{35}}{7}$ **d)** $8\mathbf{i}+4\mathbf{j}+t(2\mathbf{i}+\mathbf{j}-3\mathbf{k})$ **e)** $\frac{36}{7}\mathbf{i}+\frac{18}{7}\mathbf{j}+\frac{30}{7}\mathbf{k}$

Exercise 23A

1 a) $\frac{1}{2}$ **b)** $\frac{4}{13}$ **c)** $\frac{2}{13}$ **d)** $\frac{17}{26}$ **2 a)** $\frac{1}{2}$ **b)** $\frac{1}{5}$ **c)** $\frac{1}{10}$ **d)** $\frac{3}{5}$ **3 a)** $\frac{11}{36}$ **b)** $\frac{1}{9}$ **c)** $\frac{1}{18}$ **d)** $\frac{13}{36}$ **4** $\frac{7}{12}$ **5** $\frac{23}{30}$ **6** 0.16 **7** $\frac{7}{20}$ **8** $\frac{1}{15}$
9 $\frac{19}{21}$ **11** $\frac{1}{8}$ **12** $\frac{5}{12}$ **14** $\frac{2}{9}$ **15 a)** $\frac{19}{37}$ **b)** $\frac{13}{37}$ **c)** $\frac{7}{37}$ **d)** $\frac{12}{37}$ **16** 11% **17** $\frac{1}{12}$ **18** 0.4 **19** $\frac{3}{32}$ **20** 0.45

Exercise 23B

1 $\frac{2}{5}$ **2** $\frac{3}{25}$ **3** $\frac{1}{4}$ **4** $\frac{4}{5}$ **5** $\frac{3}{5}$ **6** $\frac{15}{23}$ **7** $\frac{5}{11}$ **8** $\frac{5}{9}$ **9** $\frac{1}{10}$ **10** $\frac{31}{36}$ **11** $\frac{3}{7}$ **12** $\frac{7}{9}$ **13** $\frac{2}{7}$ **15** $\frac{5}{18}$
16 $\frac{1}{2}$ **17 a)** $\frac{23}{48}$ **b)** $\frac{8}{23}$ **18 a)** $\frac{37}{60}$ **b)** $\frac{12}{37}$ **19 a)** $\frac{103}{180}$ **b)** $\frac{30}{103}$ **20** $\frac{1}{3}$ **21** 0.105

Exercise 23C

1 i) $\frac{4}{9}$ **ii)** $\frac{2}{9}$ **2** $\frac{1}{105}$ **3** $\frac{97}{105}$ **4 a)** No **b)** 0.1 **c)** 0.5 **5 a)** 0.3 **b)** $\frac{3}{7}$ **6 a)** 0.1 **b)** $P(A|B) \neq P(A)$ **c)** $\frac{2}{7}$ **7 i)** 0.12
7 ii) 0.28 **iii)** 0.7 **iv)** $P(A \cup C) = 0.8 > 0.6[= P(B')]$ **8 c)** $\frac{2}{29}$ **9 ii)** 0.6644

Index